Advances in Fish Tagging and Marking Technology

Advances in Fish Tagging and Marking Technology

Edited by

Jeremy R. McKenzie
National Institute of Water and Atmospheric Research
Viaduct Harbour, Auckland Central 1010
Private Bag 99940, Newmarket, Auckland 1149, New Zealand

Bradford Parsons
Minnesota Department of Natural Resources
1200 Warner Road, St. Paul, MN 55106, USA

Andrew C. Seitz
School of Fisheries and Ocean Sciences, University of Alaska Fairbanks
P.O. Box 757220, Fairbanks, AK 99775, USA

R. Keller Kopf
Institute for Land, Water and Society, Charles Sturt University
Elizabeth Mitchell Drive, Albury NSW 2640, Australia

Matthew G. Mesa
U.S. Geological Survey, Western Fisheries Research Center
Columbia River Research Laboratory
5501 Cook-Underwood Road, Cook, WA 98605, USA

Quinton Phelps
Open Rivers and Wetlands Field Station, Missouri Department of Conservation
3815 East Jackson Boulevard, Jackson, MO 63755, USA

American Fisheries Society Symposium 76
Proceedings of the symposium
Advances in Fish Tagging & Marking Technology
Held in Auckland, New Zealand
24–28 February 2008

American Fisheries Society
Bethesda, Maryland
2012

A suggested citation format for this book follows.

Entire Book

McKenzie, J. R., B. Parsons, A. C. Seitz, R. K. Kopf, M. Mesa, and Q. Phelps, editors. 2012. Advances in fish tagging and marking technology. American Fisheries Society, Symposium 76, Bethesda, Maryland.

Chapter within the Book

Walsh, C. T., I. V. Reinfelds, R. J. West, C. A. Gray, and D. E. van der Meulen. 2012. Distribution and movement of catadromous fish: design and implementation of a freshwater-estuarine acoustic telemetry array. Pages 251–264 *in* J. R. McKenzie, B. Parsons, A. C. Seitz, R. K. Kopf, M. Mesa, and Q. Phelps, editors. Advances in fish tagging and marking technology. American Fisheries Society, Symposium 76, Bethesda, Maryland.

Printed in the United States of America on acid-free paper.

Library of Congress Control Number 2012934224
ISBN 978-1-934874-27-1
ISSN 0892-2284

American Fisheries Society Web site address: *www.fisheries.org*

American Fisheries Society
5410 Grosvenor Lane, Suite 100
Bethesda, Maryland 20814
USA

Table of Contents

Section IV: Radio Telemetry

Section V: Chemical and Biological Markers

Section VI: Archival and Pop-up Satellite Tags

Panel Discussion Summary

Preface

Fish marking and tracking is a fundamental research tool for fisheries management and research. Governments literally spend millions of dollars each year on fish tagging and tagging related research. In June of 1988, over 400 fisheries and aquatic scientists gathered in Seattle, Washington, USA, for the International Symposium and Educational Workshop on Fish-Marking Techniques. That landmark event included presentations on virtually every fish tagging method in use at that time. The ultimate product was the publishing in 1990 of *Fish-Marking Techniques*, American Fisheries Society Symposium 7, arguably one of the most influential fisheries publications in decades.

In the two decades since that Symposium, the world of fisheries science has changed dramatically; the technologies and analytical procedures available for marking and monitoring fisheries have evolved as well. Fish marking technologies on the cutting edge twenty years ago are now commonplace, and new technologies are developed yearly. Clearly, the time had come to once again bring together global expertise on the subject.

Visionary members of the American Fisheries Society, the Australian Society for Fish Biology, and the New Zealand Marine Sciences Society seized this unique opportunity. They realized the need for the symposium, and recognized that a joint venture of the three Societies could accomplish much more. Not only would the range of presentations be greatly expanded, the opportunity to work together on this venture would help bond our Societies together, furthering all of our goals.

New Zealand graciously offered to host the conference, and the University of Auckland was selected as the venue. From there, many people worked diligently on the conference details and program. Over a four day gathering we experienced more than 120 oral presentations, 45 poster presentations, and several panel discussions. The presentations were separated into eight topic areas: Radio Telemetry, chaired by Larry Hildebrand; Satellite Tags, chaired by Malcolm Francis; Archival Tags, chaired by Andrew Seitz; Transponder and Non-electronic Tags, chaired by Jeremy McKenzie; Fisheries Management, co-chaired by Nick Davies and Paul Starr; Acoustic Tags, chaired by Mike Stokesbury; Biological Markers, chaired by Tony Fowler; and Integrated Approaches, chaired by Jennifer Nielsen.

As with any conference, the numerous opportunities to meet, mingle, and network with professionals and students was a real highlight. In this case it was all the more intriguing with our diverse backgrounds yet shared interest in the marking and tagging aspects of fisheries science. Attendance at the Symposium exceeded 230 persons, and 18 countries were represented. We also shared an incredibly unique conference dinner atop Auckland's famous Sky Tower, and took advantage of the opportunities to tour in New Zealand.

Advances in Fish Tagging and Marking Technology contains 31 manuscripts that describe much of the flavor, breadth, and depth of the topics presented. These manuscripts have all been peer reviewed and deemed acceptable for publication. We thank the numerous professionals who served as peer reviewers and thus contributed to the quality of the publications.

None of this would have been possible without the generous support of our host sponsor, the National Institute of Water & Atmospheric Research Ltd., and the wonderful support from our contributing sponsors: New Zealand Ministry of Fisheries, Fisheries Victoria, New South Wales DPI, University of Tasmania and Sonar Data Ltd., Murray Darling Basin Commission, United States Geological Survey, South Australian R&D Institute, Northwest Marine Technology, Hallprint & ENSID Ltd., and the AFS International Fisheries Section. We also learned a great deal from the companies in our trade show, and were able to give them ideas for the future. Finally, we thank the Centre for Continuing Education at the University of Auckland for logistical work and planning.

Brad Parsons
February 2012

Acknowledgments

Sole credit for conceiving the idea of a 2nd International Symposium on Fish-Marking Techniques and championing it to fruition lies with Stan Moberly. The editors would like to both acknowledge and thank him for his outstanding efforts. The Auckland symposium and these proceedings would not have occurred were it not for Stan's vision and commitment to make it happen.

Symbols and Abbreviations

The following symbols and abbreviations may be found in this book without definition. Also undefined are standard mathematical and statistical symbols given in most dictionaries.

A	ampere
AC	alternating current
Bq	becquerel
C	coulomb
°C	degrees Celsius
cal	calorie
cd	candela
cm	centimeter
Co.	Company
Corp.	Corporation
cov	covariance
DC	direct current; District of Columbia
D	dextro (as a prefix)
d	day
d	dextrorotatory
df	degrees of freedom
dL	deciliter
E	east
E	expected value
e	base of natural logarithm (2.71828…)
e.g.	(exempli gratia) for example
eq	equivalent
et al.	(et alii) and others
etc.	et cetera
eV	electron volt
F	filial generation; Farad
°F	degrees Fahrenheit
fc	footcandle (0.0929 lx)
ft	foot (30.5 cm)
ft^3/s	cubic feet per second (0.0283 m^3/s)
g	gram
G	giga (10^9, as a prefix)
gal	gallon (3.79 L)
Gy	gray
h	hour
ha	hectare (2.47 acres)
hp	horsepower (746 W)
Hz	hertz
in	inch (2.54 cm)
Inc.	Incorporated
i.e.	(id est) that is
IU	international unit
J	joule
K	Kelvin (degrees above absolute zero)
k	kilo (10^3, as a prefix)
kg	kilogram
km	kilometer
l	levorotatory
L	levo (as a prefix)
L	liter (0.264 gal, 1.06 qt)
lb	pound (0.454 kg, 454g)
lm	lumen
log	logarithm
Ltd.	Limited
M	mega (10^6, as a prefix); molar (as a suffix or by itself)
m	meter (as a suffix or by itself); milli (10^{-3}, as a prefix)
mi	mile (1.61 km)
min	minute
mol	mole
N	normal (for chemistry); north (for geography); newton
N	sample size

NS	not significant
n	ploidy; nanno (10^{-9}, as a prefix)
o	ortho (as a chemical prefix)
oz	ounce (28.4 g)
P	probability
p	para (as a chemical prefix)
p	pico (10^{-12}, as a prefix)
Pa	pascal
pH	negative log of hydrogen ion activity
ppm	parts per million
qt	quart (0.946 L)
R	multiple correlation or regression coefficient
r	simple correlation or regression coefficient
rad	radian
S	siemens (for electrical conductance); south (for geography)
SD	standard deviation
SE	standard error
s	second
T	tesla
tris	tris(hydroxymethyl)-aminomethane (a buffer)
UK	United Kingdom
U.S.	United States (adjective)
USA	United States of America (noun)
V	volt
V, Var	variance (population)
var	variance (sample)
W	watt (for power); west (for geography)
Wb	weber
yd	yard (0.914 m, 91.4 cm)
α	probability of type I error (false rejection of null hypothesis)
β	probability of type II error (false acceptance of null hypothesis)
Ω	ohm
μ	micro (10^{-6}, as a prefix)
′	minute (angular)
″	second (angular)
°	degree (temperature as a prefix, angular as a suffix)
%	per cent (per hundred)
‰	per mille (per thousand)

Section I

Integrated Approaches

American Fisheries Society Symposium 76:3–14, 2012

How Advances in Tagging Technology Improved Progress in a New Science: Marine Stock Enhancement

KENNETH M. LEBER*
Center for Fisheries Enhancement, Mote Marine Laboratory
1600 Ken Thompson Parkway, Sarasota, Florida 34236, USA

H. LEE BLANKENSHIP
Northwest Marine Technology
P.O. Box 427, Ben Nevis Loop Road, Shaw Island, Washington 98286, USA

Abstract.—In designing research programs, scientists may constrain development of sequential hypotheses because of perceptions about logistical constraints to using new technologies in monitoring or experimental design. Using trusted, familiar methods can supersede asking which hypotheses would have the greatest impact and what method(s) are required to test them. To help maintain a 'problem-oriented' approach, rather than a 'methods oriented' one, we could strive to remain aware of new innovations and applications in research; this is particularly so for tagging technology, when new methods emerge. Research enabled by recent innovations can be incorporated through collaborations with other scientists or by working directly with vendors to implement and refine new tag technologies and applications. Some tagging studies can be improved by using multiple marking methods (e.g. see recent applications of various tag technologies with common snook *Centropomus undecimalis* and red drum *Sciaenops ocellatus* in Florida to evaluate recruitment, mortality, and habitat use of different life stages; Adams et al. 2006; Bennett 2006; Marcinkiewicz, 2007; Brennan et al. 2008; Tringali et al. 2008). Here we consider a few case studies that have implemented a variety of tagging methods to explore poorly understood factors that mediate growth and survival and the effectiveness of hatchery releases to help replenish depleted marine fish stocks.

Introduction

A half century ago, marine hatcheries established in the USA to supplement marine fish stocks were abandoned for lack of evidence of their effectiveness in fisheries management (Richards and Edwards 1986). At that time, success was based on the number of juvenile fish stocked rather than on the number of adults added to the catch or spawner biomass. Worldwide declines in coastal fisheries and advances in marine aquaculture have sparked a resurgence of interest in hatchery-based marine stock enhancement (see symposium proceedings edited by Lockwood 1991; Danielssen et al. 1994; Travis et al. 1998; Howell et al. 1999; Leber et al. 2004; and Bell et al. 2008).

*Corresponding author: KLeber@mote.org

Development of new marine aquaculture technologies coupled with advances in tagging technology, and greater attention to a more cost-effective, conservation-orientated, and scientific approach, have fostered a better understanding of whether and how to use marine stock enhancement in fisheries management (Blankenship and Leber 1995; Howell et al. 1999; Leber et al. 2004; Bell et al. 2008; Lorenzen et al. 2010). Contemporary research in this field is more rigorously evaluating and quantifying ecological and biological impacts of hatchery releases and has begun to address critical uncertainties in stock enhancement theory (e.g., Bell et al. 2008; Lorenzen 2008; Lorenzen et al. 2010). Advances in tagging technology have underpinned many of the scientific advances made in this field in the past two decades.

Over 20 years has now passed since the American Fisheries Society convened the June 1988 "International Symposium and Educational Workshop on Fish-Marking Techniques" at the University of Washington in Seattle, which resulted in the useful publication, Fish Marking Techniques (Parker et al. 1990). One of the most lasting impressions from that symposium was the overview talk presented by Ray Hilborn, Carl Walters and Douglas Jester on the "Value of Fish Marking in Fisheries Management" (Hilborn et al. 1990). In this, they made three important observations:

- that World Fisheries Production had leveled off and that total catch would not rise much above 1988 levels (80 million metric tons per year);
- that aquaculture would play an increasingly important role in fish production; and
- that recreational fishing was beginning to exceed commercial fishing in value in North America.

Of their five challenges to fisheries managers, one in particular—the use of aquaculture production for enhancement and ocean ranching of fishes—helped stimulate new research on the benefits and risks of marine fisheries enhancement. To highlight the challenges facing those trying to use enhancement effectively, we note that in 1988 there were no previous publications in peer-reviewed journals on the effects or effectiveness of stocking fish that spawn in the sea. Over the next two decades, aided largely by advances made in tagging technology, marine stock-enhancement research began to develop an old idea into a new kind of science.

For over a century, two issues had retarded scientific discovery in marine stock enhancement—the small sizes (eggs, yolk sac larvae and postlarvae) of cultured marine organisms being released (e.g., Atlantic cod *Gadus morhua*; haddock *Melanogrammus aeglefinus*; Pollock *Pollachius virens*; flounder *Plueronectes americanus*, and Atlantic mackerel *Scomber scombrus*, Richards and Edwards 1986), and lack of adequate marking methods for evaluating the fate of stocked juvenile fishes (Blankenship and Leber 1995). But Keith Jefferts and Pete Bergman had solved the lack of marking technology for small fishes 2½ decades before this symposium, when they developed the coded wire tag (CWT) system (Jefferts et al. 1963); and both Hager and Noble (1976) and Bilton et al. (1982) had published results of using this system in mark–recapture experiments with salmon. However, lacking a comprehensive reference work on modern tagging methods, most marine aquaculturists and many fishery scientists had no idea how to effectively mark small juvenile fishes. Thus the 1988 AFS symposium in Seattle empowered a generation of field biologists by enhancing awareness of new developments in fish marking methods.

Recognizing this and the need for an update, the conveners of the latest AFS marking symposium, held in Aukland, New Zealand in 2008, are now helping to stimulate new research by compiling examples of studies aided by the variety of marking technologies available today (this volume). The progress in electronics, computer technology, and genetics over the past twenty years has enabled great advances in fish tagging technology. Examples include genetic tags or markers using microsatellite DNA, antennae systems for passive integrated transponder tags, and archival tags that record temperature, depth, longitude and latitude; acoustic tags have been miniaturized and receivers can be spaced across vast areas for large scale recovery; advances in parental genotyping and automated mass marking machines have enabled millions of hatchery fish to be identified so interactions between hatchery and wild counterparts can be monitored. These tools are providing biologists and managers the technology needed to gain further insight into the biology and ecology, survival, and migratory behavior of our fisheries resources.

This paper highlights some of the research advances that have been made by incorporating various tags and marks in studies of hatchery releases into the sea; moreover, we discuss how combining multiple tag innovations is enabling more rapid progress in understanding how to use an undeveloped fishery management tool—marine stock enhancement.

Identifying Optimal Release Strategies

Size-at-release (SAR), release habitat and microhabitat, release season, acclimation, and density dependence are key factors that affect survival of stocked fish and need to be evaluated prior to large scale tests of stock enhancement (Leber 1999). But to do the empirical experiments needed to evaluate these factors, a marking system capable of providing high information content is needed to identify experimental treatment and control groups of hatchery fish.

The CWT had been used to examine size and season effects on hatchery-released smolt survival in salmonids (Hager and Noble 1976; Bilton et al. 1982), but prior to the 1988 AFS Symposium, biologists working with marine spawners seemed largely unaware of this tag or experienced low and declining tag-retention rates, in part because the tag was not suitable for the small size of individuals being stocked (<50 mm) (e.g., Gibbard and Colura 1980; Bumguardner et al. 1990, 1992). However, immediately following the Seattle symposium, biologists in the USA began to evaluate release strategies with juveniles of several marine fishes including red drum *Sciaenops ocellatus* (Willis et al. 1995), striped mullet *Mugil cephalus* (Leber 1994, 1995; Leber et al. 1995), Pacific threadfin *Polydactylus sexfilis* (Leber et al. 1998) and white seabass *Atractoscion nobilis* (Kent et al. 1995).

Those researchers enlisted the assistance offered by Washington Department of Fish and Wildlife (WDFW) biologists Lee Blankenship and Ray Buckley, and with their assistance launched a series of tag-retention experiments that resulted in successful adaptation of coded-wire tags (CWTs) to juvenile life stages of all of those fishes. This empowered these research groups to move rapidly into testing release strategies and evaluating the effectiveness of stocking. A series of CWT retention tests using trial-and-error coupled with scanning EM photography was used to select adipose tissue in the snout as the target site for tagging juvenile striped mullet (Oceanic Institute 1990) and cheek muscle as the target site for red drum and white seabass.

In Hawaii, striped mullet juveniles were graded into five size intervals and in summer 1989, only a year after the Seattle symposium; 10,000 were tagged and released into Maunalua Bay, followed by 85,000 more in 1990 into Maunalua and Kaneohe Bays in Hawaii (Oceanic Institute 1990; Leber 1995). Field sampling followed, using seines and cast nets to monitor the releases, which had been made in open ocean habitats. Recapture results were outstanding. Tag retention averaged 98.6% (97.3–99.6% in five size groups ranging from 45 mm to 130 mm total length [TL]) and biologists recovered 277 tagged hatchery juveniles (4.6% of the mullet collected) in Kaneohe Bay and 456 in Maunalua Bay (15.5% of the mullet collected there) over 46 weeks following the releases in 1990 (Leber 1995). The tag data showed that SAR had a clear and significant effect on recapture rates, with very few of the smaller fish released (45–60 mm TL) recovered after nine weeks at liberty. Critical size for release–recapture experiments in those habitats during summer months was 70 mm TL (Leber 1995).

Thus began a series of factorial, randomized-block, CWT release–recapture experiments designed to gain a better understanding of SAR effects in multiple habitats and the interactive effects of release season and SAR on survival of stocked juvenile striped mullet (Leber et al. 1996, 1997). Based on data recovered from coded wire tags, those experiments showed a clear effect of the seasonal timing of releases on SAR effects, and revealed that pilot-scale releases had actually doubled abundance of striped mullet in nursery habitats. A subsequent evaluation of whether released hatchery mullet were displacing wild mullet at release sites showed that the hatchery mullet were adding to production in nursery habitats, not depressing it (Leber et al. 1995). Subsequent evaluations of the contribution rates of released hatchery fish showed that hatchery fish were providing 15% of the catch in a subsistence mullet fishery in Kaneohe Bay (Leber and Arce 1996). In that study, a sampling program designed to recover cultured mullet from the mullet fishery resulted in the recovery of 214 tagged adult hatchery fish from fishermen. The effects of SAR on recapture frequencies were determined based on CWT information in the recaptured hatchery fish. After teaming with a resource economist, subsequent work with the Leber and Arce (1996) data set revealed that the economically optimal size-at-release, based upon production costs and yields in the fishery of the various SAR groups of hatchery fish released, was 91 mm TL at one release site, 98 mm at another (Leber et al. 2005). The one critical factor we could not evaluate was the effectiveness of releases of large numbers of very small mullet (postlarvae). No marking system with an adequate number of individual codes or batch marks was available at the time for identifying postlarval fishes. Thus, evaluating survival of stocked postlarvae was put on hold for lack of an adequate marking system for distinguishing experimental treatment conditions and replicates in releases of such a small life stage.

Following the development at Oceanic Institute of new aquaculture technology for Pacific threadfin, WDFW biologists helped the Hawaii team adapt CWT technology to this fish and a trial release was conducted with 20,000 tagged juveniles released along the sandy beaches on the Northeast coast of Oahu, Hawaii. In one large experiment, 81,000 Pacific threadfin juveniles were tagged and released to investigate the interactive effects of release season, release habitat, and SAR (Leber et al. 1998). That work provided conclusive evidence that optimal SAR for juvenile Pacific threadfin was dependent upon both release season and release habitat and that habitat alone could cause a complete reversal in size-at-release effects on recapture rate. Subsequent research using CWTs revealed that hatchery-released Pacific threadfin made significant contributions to the fishery on Oahu, HI (Friedlander and Ziemann 2003; Ziemann 2004).

Meanwhile, researchers in Florida and California were also able to successfully apply CWTs in investigations of red drum and white seabass stocking effectiveness and release strategies. Willis et al. (1995) recovered 821 hatchery-released red drum. Results at fixed sampling stations were consistent with the Hawaii results, revealing a large effect of both SAR and release season upon survival of stocked juvenile red drum. The white seabass research in California (Kent et al. 1995; Hervas et al. 2010) revealed similar findings showing a strong relationship between SAR and survival and a clear release-season impact.

The findings in California, Florida, and Hawaii with several species of marine fishes corroborate earlier research results with marine fishes from releases of relatively large juvenile cod marked with external tags in Norway (Svåsand and Kristiansen 1990) and very small juvenile red sea bream *Pagrus major* in Japan, marked with alizarin complexone (Tsukamoto et al. 1989). However, the later studies also revealed significant interactive effects of release strategies (SAR, release season, release habitat) and that significant contributions of marine hatchery fish could be made to fisheries only by optimizing release strategies, which could not have been done with small juveniles without the aid of a high-information content, benign tag.

As more recent advances have been made in CWT technology, additional applications are now possible, for example mass marking to identify every stocked salmon in a river system, which enables better management of mixed stock fisheries (Vander Haegen et al. 2012, this volume).

Addition of a Visible Tag to the Tagging Toolkit

Clearly, much information can be gained using CWT systems alone as a marking tool, and this became one of the "tags of choice" for many stock-enhancement researchers in the 1990s and beyond as research on release strategies expanded to other species of marine organisms, e.g. red snapper *Lutjanus campechanus* (Blaylock et al. 2000), swimming crab *Portunus trituberculatus* (Okamoto 2004), barramundi *Lates calcarifer* (Russell et al. 2004), winter flounder *Pseudopleuronectes americanus* (Fairchild et al. 2005), mandarin fish *Simperca chuatsi* (Zhang and Li 2007), and blue crabs *Callinectes sapidus* (Zohar et al. 2008).

Coded wire tags are usually not externally visible and require lethal removal to recover the code (Oven and Blankenship 1993). In Florida, researchers studying hatchery releases of common snook *Centropomus undecimalis* needed to incorporate an externally visible tag, because the study design required multiple recaptures of small juveniles across seasons and life stages and the ability to distinguish between hatchery fish and wild fish. So laboratory and *in situ* studies of tag retention were initiated in snook, using CWT and color-coded elastomer implants to identify treatment groups of stocked juveniles (Brennan et al. 2005). Using both tag systems, researchers could evaluate juvenile hatchery snook growth, survival, and interactions with wild stocks across a range of release sites and microhabitats in Sarasota Bay. Brennan et al. (2006) also discovered that caging the released hatchery snook for three days in *in situ* enclosures at release sites actually doubled subsequent survival rates over a 12-months period following release from cages (Brennan et al. 2006). This effectively cut the rearing costs of survivors in half, to the delight of resource economists. Subsequent studies revealed dynamic effects of release habitat on growth and survival (Leber and N.P. Brennan, Mote Marine Laboratory, unpublished data).

Elastomers and CWTs were also used in tandem to test a key question in stock enhancement—whether released hatchery fish displace wild fish and thus provide no additive gain in biomass. This analysis found that juvenile hatchery snook were themselves displaced by wild snook, even as surviving hatchery snook contributed to increased production in nursery habitats without affecting densities of wild snook (Brennan et al. 2008), and suggested that a key survival bottleneck had been avoided by size escape from competitive exclusion (the stocked hatchery snook were ~ age 1).

Addition of PIT Tags to the Tagging Tool Kit

To evaluate movement patterns among release sites and microhabitats additional tagging technologies were needed. To monitor movements throughout 24 h, many researchers have implemented transmitting tags and receivers. Juvenile snook are too small at this stage for application of most transmitter tags and acoustic tags. A breakthrough for freshwater researchers working with juvenile fishes came when Zydlewski et al. (2001) documented use of an autonomous antennae system to monitor passive integrated transponder (PIT) tags in Midwestern freshwater systems. After working with the Zydlewski team to learn how to design and build the detector system, Adams et al. (2006) were able to adapt this system for use in brackish water habitats to monitor snook movement patterns in and out of nursery systems in Charlotte Harbor, Florida. Their studies have determined weekly apparent survival rates that will be invaluable in comparing juvenile habitats of different quality (Adams et al. 2006). Snook nursery habitats were small streams and creeks and these systems afforded great locations for implementing the automated PIT tag detection systems.

PIT tags have long been used to evaluate survival of salmonids, and new applications are constantly emerging (e.g., Ostrand et al. 2012, this volume; van den Broek et al. 2012, this volume).

Addition of Acoustic Tags to the Tagging Toolkit

Following the work on snook in their juvenile nursery habitats, Adams et al. (2009) coupled PIT tags with acoustic transmitters and detectors to study movement patterns and site fidelity of sub-adult and adult snook on their spawning grounds in Charlotte Harbor, Florida, and found very high site fidelity and annual return rates. Also using sonic transmitters and detectors, Pine and colleagues evaluated survival, movement and habitat utilization by snook both inside and outside of their nursery habitats in Sarasota Bay, Florida (Pine et al. 2007; Marcinkiewicz 2007). Their work evaluated habitat use and movement patterns of adult snook in seasons with and without large-scale environmental (red tide blooms) and anthropogenic (dredging) disturbance events. The sonic tags afforded a method for revealing habitat utilization patterns and habitat disturbance effects that could not be approached with other methods. Adams et al. (2009), using sonic tags, PIT tags and external dart tags, found snook homing to spawning ground behavior in Charlotte Harbor at a much smaller scale than ever expected for a broadcast spawner; Pine et al. (2007) and Marcinkiewicz (2007) found variable site fidelity in Sarasota Bay, revealing substantial differences between these two bay systems.

Acoustic tags and receiver arrays have expanded the horizon in studies of fish movement and habitat use patterns, as no fewer than eight papers in the present volume can attest.

Utility of Adding Genetic Fingerprinting to the Tagging Toolkit

For all the answers gained researching snook, additional studies are needed to address many challenging and yet unanswered questions. What about the relative effectiveness of stocking large numbers of small postlarvae or very small juveniles in comparison with stocking larger siblings, an issue that had not previously been examined quantitatively? And what effect might stocking very small fish have upon wild snook at release sites? The tag technology needed to resolve these questions has now emerged in the form of genetic fingerprinting. In collaborations with biologists at Florida Fish and Wildlife Conservation Commission's Fish and Wildlife Research Institute (FWRI), we have used new genetic markers to investigate survival and performance of hatchery-reared postlarval red drum stocked into Tampa Bay, Fl. Through our collaboration with FWRI, genetic fingerprinting is now being adapted for use with snook to evaluate the effects and effectiveness of stocking ontogenetic stages too small for other high-information content tags or marks. Although less flexible than CWTs, which we'd used to identify as many as 80 different batch marks in a single experiment (Leber et al. 1998), genetic fingerprinting is a useful method for batch marking all life stages (eggs through adults) for as many experimental treatment and control conditions as can be accommodated by isolating and rearing siblings in separate tanks or ponds. Through the collaboration with FWRI in Florida, we identified 34 discrete spawning groups, which afforded 34 different experimental treatment conditions in the Tringali et al. (2008) studies of red drum. This afforded identities for fish released at 3 different SAR stages, in 4 different release habitats (river miles) and replicated over 5 years in a randomized-block design experiment. Based on microsatellite DNA identities of offspring from red drum broodfish matured and spawned in the FWRI hatchery, FWRI biologists have now documented that stocking red drum postlarvae can be as economically efficient as stocking larger (better surviving) juveniles, but only in certain habitats (only one of the two rivers examined) and at certain times of year (Tringali et al. 2008).

Rapid advances are being made in genetic marking systems and one of the latest is close kin matching, which has enabled even tracking of fish across generations (M. Bravington and P. Grewe, unpublished data).

Discussion

Today's fisheries scientists have tagging tools in their arsenal that now enable research not even feasible in 1988. Tagging systems and innovations have resulted in technologies that are smaller, 'smarter,' more automated, more reliable, and longer-lasting since the 1988 AFS marking symposium. Revolutionary advances in fish tagging and marking technology continue to be made. With each new innovation, seemingly another monitoring breakthrough or logistical constraint to experimentation is solved.

As these innovations occur, it is easy to become attached to the tagging systems that have most enabled research in our own specific arenas. We must not lose sight, though, of the new innovations that are occurring, giving rise to new tag systems. As Platt reminds us in his seminal and still pertinent paper, "Strong Inference" (Platt 1964), we must keep our research focused on 'problem solving;' we must resist becoming 'method oriented' to the point where we miss opportunities to test some of our highest-priority hypotheses. In designing research programs, scientists often and unwittingly constrain development of sequential hypotheses through awareness of logistical constraints inherent in monitoring or experimental design. It is tempting to use familiar methods we trust and ask "which questions can I solve with this method?" rather than asking "which hypotheses would have the greatest impact in my research program and where can I find the method(s) I need to test them?" Scientific fields of research that stay 'problem oriented' make much more rapid progress than those fields that become more 'method oriented' (Platt 1964). To help maintain a 'problem-oriented' approach in our science, we must strive to stay aware of innovations in tagging methods. By the time most of us realize from a journal article or symposium presentation the hot new research directions enabled by a recent innovation, we have long since planned research hypotheses and written grant proposals based on familiar methods. Awareness of tag innovations is enhanced by collaborations.

In these proceedings, there are many tag technologies available, each tailored to different kinds of applications. Stock enhancement studies can clearly be improved by using multiple marking methods. So, too, can many other fields of research. Integration of CWTs, elastomers, genetic fingerprinting, hydro-acoustic tags and PIT tags were needed in the case studies presented here, to explore poorly understood ecological issues and their effects on movement patterns, site fidelity, release strategies, stocking effectiveness, and density dependence in stock enhancement.

Technological bottlenecks have severely hindered development of the science needed for evaluating effectiveness of hatchery releases over most of the past century. Several of the case studies considered here were enhanced by advances in tag technology. Coded wire tag studies helped resolve many size-at-release, release-site and release-season issues. But it took the combination of CWTs, PIT tags, sonic tags, and genetic fingerprinting to resolve the other pressing research questions: survival of stocked postlarvae relative to larger juveniles, density-dependent effects on survival, juvenile habitat effects on movement patterns and site fidelity of spawning populations, and the relative effects of mortality versus dispersal in explaining abundance declines following juvenile recruitment. None of these issues had been examined during nearly a century of stocking fishes, before the development of modern benign tagging technology in the 60s and 70s for evaluating salmonid enhancement. Now, a century after stocking postlarvae began, several critical uncertainties in the field of fisheries enhancement are finally being addressed, including one of the most elusive—survival of postlarval fishes, enabled by advances in tag technologies.

The papers published in these proceedings bring us up to date on the technologies now available for monitoring fish movement, survival and habitat use in their environment. In this paper, we have shown examples of the utility of combining multiple tag systems and stressed the need to keep aware of new technologies as they become available. We note that it was primarily through collaborations and symposia like this one that we gained the insight and experience needed to rapidly incorporate additional tag systems in our research, which enabled much more rapid progress than we would otherwise have made.

Acknowledgments

We thank our many colleagues who have helped us gain insights into new tagging technologies and ways of using these systems to evaluate survival and movement patterns. In particular, we acknowledge the ideas and guidance of Rich Lincoln, Ray Hilborn, Carl Walters and Douglas Jester, who have left indelible impressions of the questions of the day in our field and the need to resolve them by coupling adaptive experimentation and tagging technology.

References

Adams, A. J., R. K. Wolfe, W. E. Pine III, and B. L. Thornton. 2006. Efficacy of PIT tags and an autonomous antenna system to study the juvenile life stage of an estuarine-dependent fish. Estuaries and Coasts 29:311–317.

Adams, A. J., R. K. Wolfe, N. Barkowski, and D. Overcash. 2009. Fidelity to spawning grounds by a catadromous fish, *Centropomus undecimalis*. Marine Ecology Progress Series 389:213–222.

Bell, J. D., K. M. Leber, H. L. Blankenship, N. R. Loneragan, R. Masuda, and G. Vanderhaegen, editors. 2008. A New Era for Restocking, Stock Enhancement and Sea Ranching of Coastal Fisheries Resources. Special Issue, Reviews in Fisheries Science 16(1–3):1–9.

Bennett, J. 2006. Using acoustic technology to estimate natural and fishing mortality of common snook in Sarasota Bay Florida. Master's thesis., Department of Fisheries and Aquatic Sciences, University of Florida, Gainesville.

Bilton, H. T., D. F. Alderdice, and J. T. Schnute. 1982. Influence of Time and Size at Release of Juvenile Coho Salmon (*Oncorhynchus kisutch*) on Returns at Maturity. Canadian Journal of Fisheries and Aquatic Sciences 39:426–447.

Blankenship, H. L., and K. M. Leber. 1995. A responsible approach to marine stock enhancement. Pages 167–175 *in* H. L. Schramm, Jr. and R. G. Piper, editors. Uses and effects of cultured fishes in aquatic ecosystems. American Fisheries Society, Symposium 15, Bethesda, Maryland.

Blaylock, R. B., Leber, K. M., Lotz, J. M., and D. A. Ziemann. 2000. The U.S. Gulf of Mexico Marine Stock Enhancement Program (USGMSEP): the use of aquaculture technology in "responsible" stock enhancement. Bulletin of the Aquaculture Association of Canada 100(3):16–22.

Brennan, N. P., K. M. Leber, H. L. Blankenship, J. Ransier, and R. DeBruler. 2005. An evaluation of coded wire and elastomer tag performance in juvenile common snook under field and laboratory conditions. North American Journal of Fisheries Management 25:437–445.

Brennan, N. P., M. C. Darcy, and K. M. Leber. 2006. Predator-free enclosures improve post-release survival of stocked common snook. Journal of Experimental Marine Biology and Ecology 335(2):302–311.

Brennan, N. P., C. W. Walters, and K. M. Leber. 2008. Manipulations of stocking magnitude: Addressing density dependence in juvenile cohorts of common snook *Centropomus undecimalis*. Reviews in Fisheries Science 16(1–3):215–227.

Bumguardner, B. W., R. L. Colura, A.F. Maciorowski, and G. C. Matlock. 1990. Tag retention, survival, and growth of red drum fingerlings marked with coded wire tags. Pages 286–292 *in* N. C. Parker, A. E. Giorgi, R. C. Heidinger, D. B. Jester, Jr., E. D. Prince, and G. A. Winans, editors. Fish-marking techniques. American Fisheries Society, Symposium 7, Bethesda, Maryland.

Bumguardner, B.W., R.L. Colura and G.C. Matlock. 1992. Long-term coded wire tag retention in juvenile *Sciaenops ocellatus*. Fishery Bulletin 90(2):390–394.

Danielssen D. S., B. R. Howell, and E. Moksness, editors. 1994. Aquaculture and Fisheries Management 24 (Supplement 1).

Fairchild, E. A., J. Fleck, and W. H. Howell. 2005. Determining an optimal release site for juvenile winter flounder *Pseudopleuronectes americanus* (Walbaum) in the Great Bay estuary, NH, USA. Aquaculture Research 36:1374–1383.

Friedlander, A. M. and D. A. Ziemann. 2003. Impact of hatchery releases on the recreational fishery for Pacific threadfin (*Polydactylus sexfilis*) in Hawaii. Fishery Bulletin 101(1):32–43.

Gibbard, G. L. and R. L. Colura. 1980. Retention and movement of magnetic nose tags in juvenile red drum. Annual Proceedings of the Texas Chapter, American Fisheries Society 3:22–29. (Texas Parks and Wildlife Department, Austin.)

Hager, R. C., and R. E. Noble. 1976. Relation of size at release of hatchery-reared coho salmon to age, sex, and size composition of returning adults. Progessive Fish-Culturist 38:144–147.

Hervas, S. K., S. K. Lorenzen, M. A. Shane, and M. A. Drawbridge. 2010. Quantitative assessment of a white seabass *Atractoscion nobilis* stock enhancement program in California: Post-release dispersal, growth and survival. Fisheries Research 105; 237–243.

Hilborn, R.C. J. Walters, and D.B. Jester Jr. 1990. Value of fish marking in fisheries management. Pages 5–7 *in* N. C. Parker, A. E. Giorgi, R. C. Heidinger, D. B. Jester, Jr., E. D. Prince, and G. A. Winans, editors. American Fisheries Society, Symposium 7, Bethesda, Maryland.

Howell, B. R., E. Moksness, and T. Svåsand. 1999. Stock Enhancement and Sea Ranching. Blackwell Scientific Publications Ltd., Oxford. 606 pp.

Jefferts, K. B., P. K. Bergman, and H. F. Fiscus. 1963. A coded-wire identification system for macro-organisms. Nature 198:460–462.

Kent, D. B., M.A. Drawbridge, and R. F. Ford. 1995. Accomplishments and roadblocks of a marine stock enhancement program for white seabass in California. Pages 492–498 *in* H. L. Schramm, Jr. and R. G. Piper, editors. Uses and effects of cultured fishes in aquatic ecosystems. American Fisheries Society, Symposium 15, Bethesda, Maryland.

Leber, K. M. 1994. Prioritization of marine fishes for stock enhancement in Hawaii. The Oceanic Institute. Honolulu, Hawaii.

Leber, K. M. 1995. Significance of fish size-at-release on enhancement of striped mullet fisheries in Hawaii. Journal of the World Aquaculture Society 26:143–153.

Leber, K. M., N. P. Brennan, and S. M. Arce. 1995. Marine Enhancement with striped mullet: Are hatchery releases replenishing or displacing wild stocks? Pages 376–387 *in* H. L. Schramm, Jr. and R. G. Piper, editors. Uses and effects of cultured fishes in aquatic ecosystems. American Fisheries Society, Symposium 15, Bethesda, Maryland.

Leber, K. M., and Arce, S. M. 1996. Stock enhancement in a commercial mullet, *Mugil cephalus* L., fishery in Hawaii. Fisheries Management and Ecology 3:261–278.

Leber K. M., S. M. Arce, D. A. Sterritt, and N. P. Brennan. 1996. Marine stock-enhancement potential in nursery habitats of striped mullet, *Mugil cephalus*, in a Hawaiian estuary. Fishery Bulletin 94:452–471.

Leber K. M., H. L. Blankenship, S. M. Arce, and N. P. Brennan. 1997. Influence of release season on size-dependent survival of cultured striped mullet, *Mugil cephalus*, in a Hawaiian estuary. Fishery Bulletin 95(2):267–279.

Leber K. M., N. P. Brennan, and S. M. Arce. 1998. Recruitment patterns of cultured juvenile Pacific threadfin, *Polydactylus sexfilis* (Polynemidae), released along sandy marine shores in Hawaii. Bulletin of Marine Science 62(2): 389–408.

Leber, K. M. 1999. Rationale for an experimental approach to stock enhancement. Pages 493–508 *in* B. R. Howell, E. Moksness and T. Svåsand, editors. Stock Enhancement and sea ranching. Blackwell Scientific Publications Ltd., Oxford, U.K.

Leber, K. M., S. Kitada, T. Svåsand, and H. L. Blankenship, editors. 2004. Stock enhancement and sea ranching: developments, pitfalls and opportunities. 2nd edition. Blackwell Scientific Publications Publishing, Oxford, U.K., 562pp.

Leber, K. M., R. N. Cantrell and P. S. Leung. 2005. Optimizing cost-effectiveness of size at release in stock enhancement programs. North American Journal of Fisheries Management 25:1596–1608.

Lockwood S. J., editor. 1991. The Ecology and Management Aspects of Extensive Mariculture. International Council for Exploration of the Sea Marine Science Symposium 192.

Lorenzen, K. 2008. Understanding and managing enhancement fisheries systems. Reviews in Fisheries Science 16:10–23.

Lorenzen, K., K. M. Leber, and H. L. Blankenship. 2010. Responsible approach to marine stock enhancement: an update. 2010. Reviews in Fisheries Science 18(2): 189–210.

Marcinkiewicz. L. L., 2007. Examining seasonal movement and habitat occupancy patterns of adult common snook. Master's thesis. Department of Fisheries and Aquatic Sciences, University of Florida, Gainesville, Florida.

Oceanic Institute. 1990. Stock Enhancement of Marine Fish in the State of Hawaii (SEMFISH). Annual Report to NOAA-NMFS,

February 1989 – June 1990. Waimanlo Hawaii, USA.

Okamoto, K. 2004. Juvenile release and market size recapture of the swimming crab Portunus trituberculatus (Miers) marked with coded wire tags. Pages 181–186 in K. M. Leber, S. Kitada, T. Svåsand, and H. L. Blankenship, editors. Stock enhancement and sea ranching: developments, pitfalls and opportunities. 2nd edition. Blackwell Scientific Publications Publishing, Oxford, U.K.

Ostrand, K., G. Zydlewski , W. Gale, and J. Zydlewski. 2012. Long term retention, survival, growth, and physiological indicators of salmonids marked with passive integrated transponder tags. Pages 135–145 *in* J. McKenzie, B. Parsons, A. C. Seitz, R. K. Kopf, M. Mesa, and Q. Phelps, editors. Advances in fish tagging and marking technology. American Fisheries Society, Symposium 76, Bethesda, Maryland.

Oven, J. H., and H. L. Blankenship. 1993. Benign recovery of coded wire tags from rainbow trout. North American Journal of Fisheries Management 13:852–855.

Parker, N. C., A. E. Giorgi, R. C. Heidinger, D. B. Jester, Jr., E. D. Prince and G. A. Winans, editors. 1990. Fish marking techniques. American Fisheries Society, Symposium 7, Bethesda, Maryland.

Pine, W. E., III, L. L. Marcinkiewicz, and J. P. Bennett. 2007. Examining adult snook habitat occupancy, movement and exploitation rates in Sarasota Bay, Florida. Florida Fish and Wildlife Conservation, St. Petersburg.

Platt, J. R. 1964. Strong Inference. Science 146(3642):347–353.

Richards, W. J., and R. E. Edwards. 1986. Stocking to restore or enhance marine fisheries. Pages 75–80 *in* R. H. Stroud, editors. Fish Culture in Fisheries Management. American Fisheries Society, Bethesda, Maryland.

Russell, D. J., M. A. Rimmer, A. J. McDougall and S. E. Kistle. 2004. Stock enhancement of barramundi *Lates calcarifer* (Bloch) in a coastal river system in Australia: stocking strategies, survival and benefit-cost. Pages 490–500 *in* K. M. Leber, S. Kitada, T. Svåsand, and H. L. Blankenship, editors. Stock enhancement and sea ranching: developments, pitfalls and opportunities. 2nd edition. Blackwell Scientific Publications Publishing, Oxford, U.K.

Svåsand, T. and T. S. Kristiansen. 1990. Enhancement studies of coastal cod in western Norway, part 4. Mortality of reared cod after release. Journal du Conseil International pour l'Exploration de la Mer 47:30–39.

Travis, J., F. C. Coleman, C. B. Grimes, D. Conover, T. M. Bert, and M. Tringali. 1998. Critically assessing stock enhancement: an introduction to the Mote symposium. Bulletin of Marine Science 62(2): 305–311.

Tringali, M. D., K. M. Leber, W.G. Halstead, R. McMichael, J. O'Hop, B. Winner, R. Cody, C. Young, C. Neidig, H. Wolfe, A. Forstchen, and L. Barbieri. 2008. Marine Stock Enhancement in Florida: a multi-disciplinary, stakeholder-supported, accountability-based approach. Reviews in Fisheries Science 16(1–3):51–57.

Tsukamoto, K., and six coauthors. 1989. Size-dependent mortality of red sea bream *Pagrus major*, juveniles released with fluorescent otolith-tags in News Bay, Japan. Journal of Fish Biology 35 (Supplement A): 59–69.

Van den Broek, K., J. Smith and G. Wade. 2012. Estimating salmon escapement on the Copper River, Alaska, using mark-recapture with external PIT-tags. Pages 77–94 *in* J. McKenzie, B. Parsons, A. C. Seitz, R. K. Kopf, M. Mesa, and Q. Phelps, editors. Advances in fish tagging and marking technology. American Fisheries Society, Symposium 76, Bethesda, Maryland.

Vander Haegen, G., H. L. Blankenship, and D. Knutzen. 2012. Advances in coded wire tag technology: meeting changing fish management objectives. Pages 43–62 *in* J. McKenzie, B. Parsons, A. C. Seitz, R. K. Kopf, M. Mesa, and Q. Phelps, editors. Advances in fish tagging and marking technology. American Fisheries Society, Symposium 76, Bethesda, Maryland.

Willis, S. A., W. W. Falls, C. W. Dennis, D. E. Roberts, and P. G. Whitchurch. 1995. Assessment of season of release and size at release on recapture rates of hatchery-reared

red drum. Pages 354–365 *in* H. L. Schramm, Jr. and R. G. Piper, editors. Uses and effects of cultured fishes in aquatic ecosystems. American Fisheries Society, Symposium 15, Bethesda, Maryland.

Zhang, B., and Z. Li. 2007. Tagging mandarin fish *Simperca chuatsi* (Basilewsky) with coded wire tags. Journal of Fishery Sciences of China 14:53–58.

Ziemann, D. A. 2004. Enhancement of Pacific threadfin *Polydactylus sexfilis* in Hawaii: interactions between aquaculture and fisheries. Pages 477–489 *in* K. M. Leber, S. Kitada, T. Svåsand, and H. L. Blankenship, editors. Stock enhancement and sea ranching: developments, pitfalls and opportunities. 2nd edition. Blackwell Scientific Publications Publishing, Oxford, U.K.

Zohar, Y., A. H. Hines, O. Zmora, E. G. Johnson, R. N. Lipcius, R. D. Seitz, D. B. Eggleston, A. R. Place, E. J. Schott, J. D. Stubblefield, and J. S. Chung. 2008. The Chesapeake Bay blue crab (*Callinectes sapidus*): a multidisciplinary approach to responsible stock replenishment. Reviews in Fisheries Science 16(1–3):24–34.

Zydlewski, G. B., A. Haro, K. G. Whalen, and S. D. McCormick. 2001. Performance of stationary and portable passive transponder detection systems for monitoring of fish movements. Journal of Fish Biology 58:1471–1475.

American Fisheries Society Symposium 76:15–21, 2012

Viewing Complex Mark and Recapture Data in a Spatial Environment

Russell Bradford*, Scott Cooper, Paavo Jumppanen, and Jason Hartog
CSIRO Marine and Atmospheric Research
GPO Box 1538, Hobart, Tasmania 7001, Australia

Abstract.—Tagging studies have been a fundamental tool in fisheries science for over 100 years. During this time tags have progressed from simple markers through to sophisticated miniature computers capable of storing large amounts of data. Making sense of large amounts of spatial and temporal tagging data lends itself well to graphical interpretation. The Electronic Tag Support Systems (ETSS) team at CSIRO has developed a comprehensive visualization system to suit researcher and stakeholder (funding agencies, fishers, public) needs. Researchers are able to access and visualize the data through a proprietary (MapInfo) GIS package; stakeholders can visualize the data without the need for prior GIS knowledge through a menu-driven graphical web portal.

Introduction

Studying the distribution and stock structure of marine animals through the use of tagging has been a staple of marine science for over 100 years (Everhart et al. 1975). Whereas early passive tags only provided release and recapture location—and if size was recorded at both capture times—growth, today's sophisticated electronic tags can also provide data on habitat choice, physiology, as well as giving intermediate locations between tagging and recapture events. Not surprisingly, modern electronic tags generate large volumes of information. The processing, analysis and storage of this information needs careful consideration (Hartog et al., 2009).

The generation of large volumes of data are not only a feature of tagging programs using modern tagging technologies; large scale conventional mark and recapture studies often produce large volumes of data. For example, the Australian Commonwealth Scientific and Research Organization (CSIRO) has been tagging southern bluefin tuna *Thunnus maccoyii* (SBT) with conventional dart tags since the 1960s. Over this period about 140,000 release locations and 20,000 recapture locations have been recorded. In 1993 CSIRO began deploying electronic data storage tags on southern bluefin tuna, yellowfin tuna *T. albacares* and white sharks *Carcharodon carcharias*. Between 1993 and 2008 the CSIRO released in excess of 1,700 electronic tags. From the approximately 10% of electronic tags recovered more than 90,000 intermediate and recapture positions and 40 million lines of habitat and fish physiology data have been generated.

*Corresponding author: Russ.Bradford@csiro.au

Most of the archival electronic tags in use today require recapture of the tagged individual and subsequent return of the tag in order for the data stored therein to be recovered. As with conventional tagging, there is a high reliance on fishers and the public to return recovered archival tags. Monetary and merchandise rewards are common means of encouraging fishers to return tags. But increasingly fishers also want to know the growth and movement details of "their" fish/animal. Web-based mapping applications, because they offer an effective means of providing fishers feedback and thus maintaining their interest, are now becoming the standard information dissemination tool for voluntary tag recovery programs (or should be).

The positional, environmental and biological data that electronic tags typically provide are particularly suited to graphical interpretation and analysis. We believe the human visual cortex is still one of the best analytical tools available for discerning spatial trends and patterns in tagging data. This belief has strongly shaped our approach for building user interface systems for tagging data interpretation. Modern tagging programs typically have to cater for two end-users: the researcher who has need of a detailed and complex interpretation of the data; the external stakeholder (funding agencies, fishers, public) who, more typically, requires a simplistic or highly synthesized view of the data. Graphical data interpretation software can meet the needs of both class of end-user.

Whereas a researcher will often employ a complex geographic information system (GIS) software to interpret the data, GIS visualizations are generally not interpretable by stakeholders. Instead, we think web-based visualization built around simple and yet flexible user interfaces are a better vehicle for the delivery of relevant and interesting information to the stakeholder

In this paper we provide a brief description of the two parallel systems the CSIRO Electronic Tag Support Systems (ETSS) team have developed to enable researcher and stakeholder end-users to view tagging data within a mapping environment.

Desktop Visualization and Analysis for Stock Assessment Scientists

We chose MapInfo as our desktop data visualization tool. Our choice was largely driven by the ETSS group's existing expertise, however, other similar GIS programs (for example ArcGIS (http://www.esri.com), Manifold (http://www.manifold.net)) would likely have performed equally as well. For simplicity we specifically refer to the components we have developed, but the concept applies equally to any combination of relational database, spatial extension, and GIS.

In order to be spatial interpreted by MapInfo our tagging data required georeferencing. There were two ways this could have been achieved: a) through MapInfo by using the "create points" command on an open data table; or b) by maintaining the tag data in a spatially aware table within the database. Spatially aware tables are simply tables that contain a spatial object (point) which has an associated latitude and longitude along with other geographic attributes (such as projection). Spatial data tables allow data to be coupled with geographic information and are extensions of the traditional relational database tables (Mitchell 2005). In our system the tag data were already stored in an Oracle database, making the selection of the geo-referencing tool "Oracle Spatial" a logical choice, although other databases with spatial components would have worked equally well.

Using Open DataBase Connectivity (ODBC) we are able to establish a live connection to the Oracle database that holds the tag data. The Oracle Spatial extension to the database allowed us to reference point data within the Oracle database to a latitude and longitude. A Perl script was automatically run each night to append any new electronic tag data received during the day to the Oracle Spatial tables and create new point objects. The creation of point objects is a critical step and could not be automated within MapInfo. A semilive connection between MapInfo and Oracle Spatial meant all the new point objects were available to MapInfo; the process requiring no intervention or understanding by the researcher.

A MapInfo workspace was created in which a world map and connections to Oracle Spatial were set. The workspace provided immediate visualization of the tagging data through the use of standard query templates. Basic templates provided simple distance plots, but these could be extended by more capable MapInfo users to provide more complex visualization output. Examples of complex plots might include overlaying environmental data such as sea surface temperature or bathymetry, or may involve producing thematic layers that summarize numeric data.

In addition to MapInfo we also incorporated GeoServer (http://geoserver.org) into our database visualization tool kit. Using GeoServer we were able to transform the spatial data in the tagging database to provide a number of standard representations as web services; one such service being the Keyhole Markup Language (KML). KML data conversion provided the mechanism whereby tagging movements could be rendered on a Google Earth map (http://earth.google.com/). Google Earth provides the capability of connecting to a web service by adding a network link and providing the uniform resource location (url) of the GeoServer instance and the name of the layer and is a simple, yet highly effective data visualization tool most users are familiar with.

Internet Based Visualization of Tagging Data for Fisheries Managers and Stakeholders

Web portal access to tagging data are handled through a custom mapping application using a number of different technologies including javascript client side scripting, AJAX, jQuery, SVG DOM and server side perl scripting.

The core mapping capability is provided by MapServer (http://mapserver.org) through server side perl scripts. Initially developed by the University of Minnesota, MapServer has an unrestricted open source license and comes with an extensive feature set. The open source aspect of MapServer was of pivotal importance as it allowed us to easily customize and extend the performance of the software and fix any errant behavior. MapServer is widely used, well documented, and has an active development program. One of the primary strengths of MapServer is its support for web server-side scripting. A scripting interface is provided for Python, Perl, PHP, Java, and C# among others; in the present system we make use of Perl script. This allows the development of sophisticated and highly customized web applications through combining MapServer functionality with many of the publicly available Perl modules.

MapServer also supports access to network Common Data Form (netCDF) data through the netCDF driver provided in the Geospatial Data Abstraction Library (GDAL—http://www.gdal.org, allowing for the translation of raster geospatial data formats). NetCDF is a common set of software libraries used in climatology, meteorology, and oceanography to provide platform-independent access to array-orientated scientific data. Support for netCDF is important in our system because the CSIRO maintains a large number of physical oceanography related files in netCDF format containing such information as sea surface temperature, bathymetry, and mixed layer depth (Hobday et al. 2006). The netCDF driver currently provided in the GDAL library was unable to correctly interpret and access the data within many of our files. This limitation was resolved through the development of a customized GDAL netCDF driver that extended the capabilities of the existing driver to include access to arbitrarily structured netCDF files, and explicit color coding of data through HSL-based (hue, saturation, lightness) color mapping.

The web portal is implemented as a two page application, the opening page allowing the user to select the tags and data products to map (Figure 1) and the second page implementing a tiled "slippy map" interface (Figure 2) that displays the selected data products and tags. In the mapping page the researcher can zoom and pan the map, change the display of layers, query point track data through mouse-overs and display interactive graphs of archival data through mouse clicking a track point. Standard user interface elements are implemented using the widget set provided by the jQuery javascript framework along with the AJAX request handling used to request tiles and graph data. The mapping and graphing interfaces are implemented using custom client side javascript and SVG DOM and thus requires an SVG enabled web browser. A custom mapping client was chosen over the now ubiquitous OpenLayers out of time constraints, having limited experience with the framework and library documentation being hard to come by. Things have changed considerably since that time and we have plans to re-implement the portal using OpenLayers at some future date, the principal motivation being Internet Explorer compatibility (which does not support SVG DOM), though the graphing interface remains an issue.

The data comes from two primary sources: an oracle database containing tag information; a series of files in various formats resident on a file server. Access to the tagging data are made through the Structured Query Language (SQL) database support within server side perl scripts. Map tiles come in two varieties: pregenerated and live. All tiles that can be pregenerated and stored on the server have been and are stored as SVG files. This includes essentially all the vector layer data with the exception of the tag tracks themselves. The raster data, while cached, is generated live on demand which is out of necessity because our mapping interface allows users to choose how to color the raster layers. This flexibility means there is a near infinite set of possible ways to colorize the data, meaning pregeneration is not practically possible due to storage constraints.

Many of the raster data products available contain a date/time component (e.g., sea surface temperature). In the present system the date/time component of the point object is matched to that of the data view being displayed. For example, the sea surface temperature raster layer displayed corresponds to the date/time of the tag location that has been selected.

The mapping application itself is structured in a manner that easily facilitates the addition or removal of available data views without code alteration. Rendering information for the various layers is specified within an Extensible Markup Language (XML) file. Hypertext mark-up language (HTML) pages are generated from template files using text substitution. This strategy allows the layout of the pages to be changed by the webpage administrator without the need to change any Perl code.

As with many web-based applications there is a need to monitor and control content access. Within the present system provision has been made for both public and private access. Public access is restricted to certain tags and is currently available through a demonstration link (http://www.marine.csiro.au/

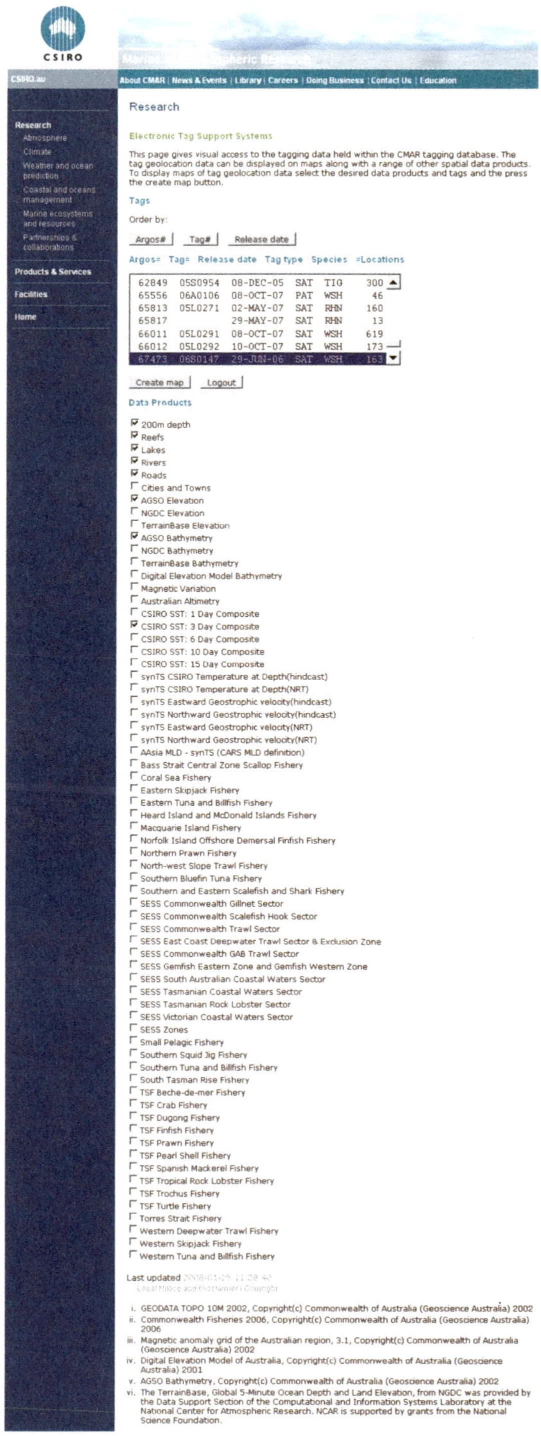

FIGURE 1. The ETSS web access configuration page illustrating the means to select tags to be mapped and data products to be available as layers on the resulting map page. A public access demonstration page is available through http://www.marine.csiro.au/cgi-bin/tags/svgmapping/etss_access_svg_public.pl.

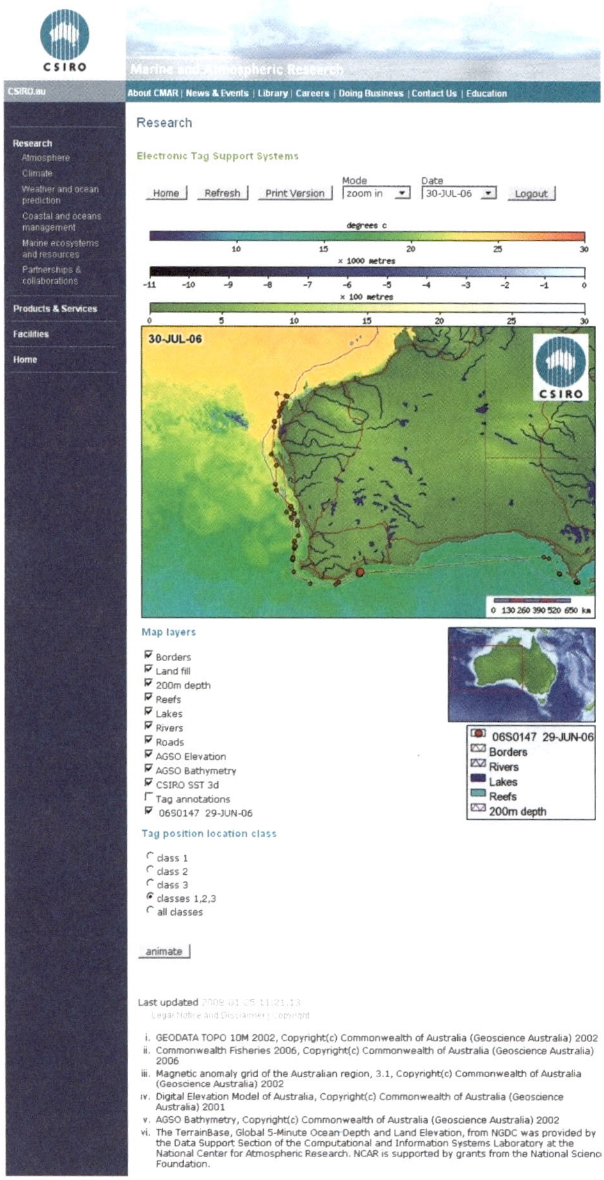

FIGURE 2. Track of a juvenile white shark *Carcharodon carcharias* displayed as a layer in Google Earth; access to an Oracle spatial database via Geoserve convsersion to Keyhole Markup Language.

cgi-bin/tags/svgmapping/etss_access_svg_public.pl). Tag data are added immediately upon release by the principle researcher. Stakeholder access to data are via a login page whereby the user is required to enter a username and password to access an account. This process dictates the level of data access a specific stakeholder will have.

New accounts are initiated by a prospective end-user after which the administrator receives an e-mail notification of a new account in need of activation, which is then vetted. If approved, the administrator assigns a security access level to the account. Once activated by the administrator, the user receives e-mail notification of the activation. This approach was taken to minimize the likelihood of unauthorized entry. Account activation is handled through direct access to an accounts table within the Oracle database. This approach was chosen in preference to using an administrator web page to minimize the possibility of account administration being breached.

Discussion

Advances in electronic tags have resulted in the generation of large volumes of spatially referenced data; this in turn has led to the development of more sophisticated methods of data analysis and interpretation. The human visual cortex is capable of discerning patterns from complex visual data with relative ease and is still in our opinion one of the best tools for interpreting these types of data. The information requirements of fisheries scientists and stakeholders typically differ. Systems for storing, extracting tagging data, and those for visualization and analysis often need to cater for these two classes of end-user.

In our experience one system rarely suits all needs: the researcher is usually interested in delving deep into the data to extract patterns and assign order to the data; the stakeholder is often only interested in the outcomes of research, not the process. We believe it is important to understand the distinction between these two end-users and develop appropriate system to support both. All too often the outcomes of research are only published in reports or scientific papers to which the public has limited access or a limited desire to access. The internet data porting tools developed by the ETSS team we believe have increased the opportunities for the public to see the results of CSIRO tagging research and have helped foster an interest in that research. The combined system of proprietary and Open Source software described in this paper may not suit the needs of other research groups. However, the concept of addressing the needs of both researcher and stakeholder is one we believe all tagging research groups should investigate.

Acknowledgments

Funding for the development of the database and visualization systems was provided by CSIRO. The authors thank Alistair Hobday and Matt Lansdell for their invaluable input to earlier drafts of this manuscript.

References

Everhart, W. H., A. W. Eipper, and W. D. Youngs. 1975. Principles of fisheries science. Cornell University Press, New York.

Hartog, J. R., T. A. Patterson, K. Hartmann, P. Jumppanen, S. Cooper, S., and R. Bradford 2009. Developing integrated database systems for the management of electronic tagging data. Pages 367–380 *in* J. L. Nielsen, H. Arrizabalaga, N. Fragoso, A. Hobday, M. Lutcavage, J. Sibert, editors. Tagging and tracking of marine animals with electronic devices. Reviews: methods and technologies in fish biology and fisheries. Springer.

Hobday, A. J., K. Hartmann, J. Hartog, and S. Bestley. 2006. SDODE: Spatial dynamics ocean data explorer. User Guide, CSIRO Marine and Atmospheric Research, Hobart.

Mitchell, T. 2005. Web mapping illustrated. O'Reilly Media Inc., Sebastopol, California.

Varshney, P. K. 2000. Multisensor Data Fusion. Pages 103–186 *in* R. Loganantharaj, G. Palm, and M. Ali, editors. Intelligent problem solving. Methodologies and Approaches, Springer, Heidelberg, Berlin.

Wald L., A. 1998. European proposal for terms of reference in data fusion. Pages 651–654 *in* International Archives of Photogrammetry and Remote Sensing, volume XXXII, Part 7.

Section II

Conventional Tagging

Foreword to Conventional Tagging Section

Jeremy R. McKenzie

In the 20 years since the last tagging symposium (Parker et al 1990), despite the remarkable development of acoustic and satellite tags, the long-standing or "conventional" tags (including passive integrated transponder (PIT), coded wire (CWT), anchor, and Visible Implant (VI) tags) remain crucial for fisheries management and research and still far outnumber all other types of tags deployed.

Fish tagging studies can be broken down into three broad categories: assessing the effectiveness of enhancement and mitigation programs; determining migration patterns, habitat use, and stock separation; estimating growth, survival, and population size. The use of conventional tags under each of these categories is represented by one or more papers in this section. These papers also exemplify some key advances in conventional tagging technologies, specifically in: the bulk application of tags (Vander Haegen et al.); tag coding and information content (Vander Haegen et al., Marvin, van den Broek et al.); tag detection and real-time processing of recovery data (Marvin, Achord et al., Baker and Smith, Lipsky et al.). These advances now enable the provision of management advice which was difficult, albeit impossible, to attain back in 1988 with conventional tagging technologies.

As remarkable as tagging technological developments have been in the last 20 years, it is important not to lose sight of the fact that technical advancement is born out of management requirement. It is therefore fitting we begin this section with a management paper (Bronte et al.) describing an enhancement program for the Laurentian Great Lakes intended to restore native fish stocks. The necessities of the Pacific Northwest and Great Lakes salmon programs underlay the development of the AutoFish™ CWT system (Vander Haegen et al.), the most technological advanced automated fish tagging system currently available.

The final paper in this section by Weitkamp is a spatial analysis of many years of cumulative Pacific Northwest salmon tagging data. Although interesting in itself, the Weitkamp paper serves to illustrate how long time-series tagging data can provide management insight beyond the original purpose or intent of collection. Weikamp's paper is food for thought in the context that, in each passing year, millions of conventional tags are seeded into world fisheries.

Reference

Parker, N. C., A. E. Giorgi, R. C. Heidinger, D. B. Jester, Jr., E. D. Prince, and G. A. Winans, editors. 1990. Fish-marking techniques. American Fisheries Society, Symposium 7, Bethesda, Maryland.

American Fisheries Society Symposium 76:27–42, 2012

A Coordinated Mass Marking Program for Salmonines Stocked into the Laurentian Great Lakes

Charles R. Bronte*
*U. S. Fish and Wildlife Service, Green Bay Fish and Wildlife Conservation Office
2661 Scott Tower Drive, New Franken, Wisconsin 54229, USA*

Kenneth A. Walch
*U.S. Fish and Wildlife Service, retired
7005 Saddle Back Peak Street, Las Vegas, Nevada 89166, USA*

John M. Dettmers and Marc Gaden
*Great Lakes Fishery Commission
2100 Commonwealth Drive, Suite 100, Ann Arbor, Michigan 48105, USA*

Michael J. Connerton
*New York Department of Environmental Conservation, Cape Vincent Fisheries Station
P.O. Box 292, Cape Vincent, New York 13618, USA*

Marion E. Daniels
*Ontario Ministry of Natural Resources, Lake Ontario Management Unit
Peterborough, Ontario, K9J 8L5, Canada*

Tammy J. Newcomb
*Michigan Department of Natural Resources, Fisheries Division
P. O. Box 30446, Lansing, Michigan 48909, USA*

Abstract.—Fisheries managers in the Laurentian Great Lakes annually stock over 30 million salmonines (salmon and trout) to diversify sport fisheries, restore native fish populations, and control invasive fishes. However, little is known about how well these fish survive, contribute to the fisheries, and reproduce in the wild. As a result, the Council of Lake Committees (CLC), a basin-wide group of fishery managers that operates under the auspices of the Great Lakes Fishery Commission, agreed in 2005 to develop a basin-wide program to mark all stocked salmonines. This marking effort would provide greater insight into survival of stocked fish, the contribution of stocked adults to restoration of native fishes, the ability to manage harvest away from wild fish, and the opportunity to evaluate and improve hatchery operations. After evaluating the available marking and tagging options, the CLC chose to pursue mass-mark-

*Corresponding author: charles_bronte@fws.gov

ing of all stocked salmonines. The program is based on the Pacific Northwest Pacific salmon *Oncorhynchus* sp. marking program and utilizes adipose fin-clips and coded-wire tags. When fully developed, the program aims to tag all hatchery-reared trout and salmon released in to the Great Lakes (about 30 million annually). The full program will require nine automated trailers and four manual trailers (total equipment costs: $13.7 million U.S. in 2007). Annual operations including tags, recovery efforts, and tag extraction services, will cost about $5.9 million. An operational plan has been completed to mark all fish prior to stocking, and to recover tagged fish as part of regular fishery surveys from anglers, commercial fishers, and charter boat operators. A data management system will be developed to cooperatively archive and analyze recovery data to answer questions of lake-wide or basin-wide scope. Efforts are underway to communicate to stakeholders the benefits of mass marking and to secure the necessary funding for equipment and annual operations. Project leaders expect this program to be fully implemented within five years but that schedule is predicated on funding levels and subsequent agency commitment. This project is an excellent example of cooperative, proactive efforts to improve and refine fisheries management across the Great Lakes basin.

Introduction

The Laurentian Great Lakes in central North America (hereafter "Great Lakes") are home to 35 million Canadian and U.S. citizens. The Great Lakes waterways themselves are important for transportation, fishing, municipal and industrial water supplies, and recreational boating. They also form the basis for a strong tourism economy. The fisheries of the region represent an enormous economic, social, and cultural asset to the region. Fisheries date back to preColumbian subsistence use by local aboriginal peoples. During 1820–1960, large scale commercial fisheries, mostly for cisco *Coregonus artedi*, lake trout *Salvelinus namaycush*, and lake whitefish *Coregonus cupleaformis*, developed but collapsed thereafter from overfishing, habitat degradation, and the negative effects of invasive species (i.e., sea lamprey *Petromyzon marinus*, alewife *Alosa pseudoharengus*) on these and other ecologically and economically important native fish (Smith 1968, 1970, 1972). Coordinated management by federal, provincial, and state governments of Canada and the United States rebuilt the fish community by stocking native salmonines for restoration programs (i.e., lake trout) and several nonnative salmonines (salmon and trout) from the Pacific (Chinook salmon *Oncorhynchus tshawytscha*, coho salmon *O. kisutch*, steelhead *O. mykiss*) and Atlantic drainages (*Salmo trutta*), creating a multibillion dollar sport fishery in the process (American Sportfishing Association 2008). While success of fisheries is now measured in dollars and angler satisfaction, information on survival of stocked fish, naturalized fish reproduction and the effects that stocked salmonines have on fisheries and the ecosystem is incomplete throughout the basin.

The development of an integrated stocking and marking strategy for Great Lakes salmonines has been difficult because of a lack of strategic coordination on this issue and variable participation among the eight states, two federal governments, the Province of Ontario, and numerous Native American Tribes. Insufficient funding for marking and recovery efforts in Great Lakes fisheries, combined with past philosophical differences among agencies, have impeded the ability to estimate natural reproduction, survival, movement, and contribution of stocked fish to extant populations.

The size of the Great Lakes also adds to the complexity—it is one of the largest freshwater ecosystems in the world and represents 20% of the world's surface freshwater. The Great Lakes cover more than 244,000 km^2 with 17,600 km of shoreline. These lakes are some of the largest by surface area, and are interconnected by rivers and straits that allow some interchange of fish populations. Water generally moves from the west from lakes Superior and Michigan, east into Lake Huron, and then southeast into lakes Erie and Ontario. The St. Lawrence River eventually carries most of this freshwater to the Atlantic Ocean where it empties into the Gulf of St. Lawrence.

The process for inter-jurisdictional fisheries management in the Great Lakes is described in *A Joint Strategic Plan for Management of Great Lakes Fisheries* (GLFC 2007). The plan seeks to help the jurisdictions identify and work toward achieving shared fish community objectives. Under the plan, each lake is managed cooperatively by a lake committee composed of senior level fishery managers from each of the jurisdictions on that lake. A Council of Lake Committees (CLC), comprised of managers from all jurisdictions in the basin, addresses issues that affect more than one lake or that are of basin-wide concern. At the request of the signatories to the Joint Strategic Plan, these cooperative fishery management efforts are facilitated by the Great Lakes Fishery Commission (GLFC). The GLFC was established in 1955 by the *Convention on Great Lakes Fisheries* between Canada and the United States to control the sea lamprey in the Great Lakes basin and to develop coordinated programs of research. Under the convention, the GLFC is also encouraged to develop and maintain working relationships among agencies.

The CLC addresses basin-wide management issues like the mass marking of stocked salmonines. The need for a coordinated approach to monitoring salmonine populations and their fisheries has been amplified by the continued disruption of food webs from invasive species, stocking levels that did not consider natural recruitment, and the dynamic nature of the Great Lakes (Madenjian et al. 2002; Mills et al. 2003; Dobiesz et al. 2005). Decision-making regarding stocking, tracking progress of lake trout rehabilitation, and assessing the health of aquatic ecosystems require consideration of the efficiencies, accountability, and management options that would result from information obtained from a coordinated mass marking and recovery effort of all stocked salmonines.

In response to the need for a coordinated mass-marking program, the CLC began an initiative in 2003 to develop such a program. During 2003–2008, interagency committees were formed to 1) develop initial management questions that can be answered only with mass-marking, 2) scope out potential marking technologies and applications, 3) develop an implementation plan and budget based on the selection of a marking technology, and 4) develop a practical long term funding strategy. This report provides an overview of the proposed program for mass marking salmonines in the Great Lakes, and the progress made to date.

Great Lakes Salmonine Fish Community

Historically three salmonines (noncoregonine) were native to the Great Lakes: the brook trout *Salvelinus fontinalis* and lake trout, which occurred in all five Great lakes, and the Atlantic salmon *Salmo salar*, which was restricted to Lake Ontario because of Niagara Falls that blocked access to all other lakes. Lotic and adfluvial brook trout populations were reduced during the 1800s, when stream habitat degradation from logging and damming, combined with overfishing reduced or eliminated many Great Lakes populations (MacCrimmon and Gots 1980; Huckins et al. 2008). Lake trout, once the keystone predator in all five lakes (Smith 1972), succumbed to overexploitation and the invasion of the parasitic sea lamprey (Hansen 1999), which bypassed Niagara Falls via the Welland Canal that was built during 1824–1848 and enlarged in 1919. By 1960, lake trout were eliminated from Lakes Ontario, Erie, and Michigan, and from most of Lake Huron (Cornelius et al. 1995; Elrod et al. 1995; Eshenroder et al. 1995; Holey et al. 1995). Although Lake Superior populations were also severely reduced (Lawrie and Rahrer 1973; Bronte and Sitar 2008), the timely arrival of chemical control of sea lamprey (Smith and Tibbles 1980) coupled with controls on exploitation prevented complete population collapse and allowed for subsequent recovery (Hansen et al. 1995; Bronte et al. 2003). Atlantic salmon were lost to Lake Ontario before 1900 from overfishing, drainage modification, and habitat degradation associated with logging and agriculture (Huntsman 1944; Smith 1995).

During the last century, intentional and nonintentional introductions of rainbow smelt *Osmerus mordax* and alewife into the Great Lakes, along with overfishing and habitat degradation led to population collapses of native planktivores that formerly served as prey for native predators (Smith 1972). By the 1960s, these species dominated the planktivore fish community. Chinook salmon, coho salmon, steelhead *Oncorhynchus mykiss*, and brown trout *Salmo trutta*, that were unsuccessfully introduced in the late 1800s and early 1900s, were intensively stocked in the 1960s and thereafter to control alewife through predation (Keller et al. 1990; Tanner and Tody 2002). Successful state and provincial stocking programs first began in Lakes Michigan and Huron (Tanner and Tody 2002), and were then duplicated across all the lakes. The result is a sport fishery worth $7.0 billion dollars (American Sportfishing Association 2008) supported by stocked and naturalized populations of these nonnative salmonines that now make up most of the predator biomass in Lakes Ontario and Michigan (Kocik and Jones 1999). Native predators (brook trout, lake trout, Atlantic salmon) were also stocked for restoration efforts. These fish and splake, a hybrid of brook trout and lake trout, also contribute to the fishery.

Since 1960, 1.36 billion salmon and trout have been stocked (Figure 1) into the Great Lakes. Stocking peaked in 1989 at just over 50 million fish and has declined since, and has averaged around 29 million fish annually during 2000–2007. Reductions in stocking, especially for Chinook salmon and lake trout, were intended to reduce predation on alewives, now the major forage, which declined significantly in Lakes Michigan, Huron and Ontario during 1965–1990 (Madenjian et al. 2008). As both prey and stocked salmonines were declining, natural reproduction of Chinook and coho salmon was documented (Bence and Mohr 2008; Clapp and Horns 2008). Naturally produced Chinook salmon comprised most of the lakewide recreational catch in Lake Huron during 2000–2004 (Bence and Mohr 2008). During 1988–1992, overt mortality and declines in catches of Chinook salmon in Lake Michigan appeared to be related to stress from low standing stocks of alewives that made salmon more susceptible to disease (Clark 1997; Holey et al. 1998). This threat to the centerpiece of the salmonine sport fishery necessitated attempts to balance predator stocking and consumption with prey availability—a task further complicated by unknown levels of natural reproduction by Chinook salmon and other nonnative salmonines (Jones et al. 2008), and variable levels of alewife recruitment (Madenjian et al. 2005). It is for these reasons, coupled with other questions regarding the efficacies of hatchery programs, which led to a collective decision to develop a coordinated mass marking program in the Great Lakes.

Information Needs

The most important questions that have been identified by the CLC can be classified into four broad categories. The first category involves questions related to estimating levels of natural reproduction of both native and nonnative salmonines. Answers will lead to improved estimates of standing stock sizes and prey consumption, and for native species, will gauge progress toward achieving self-sustaining populations. It will also allow for the implementation of selective fisheries, especially for native species under restoration, by restricting harvest to only marked fish and encouraging the release of unmarked, wild fish. The second category seeks to determine among- and within-lake movement of all species to refine estimates of standing stocks and consumption, and track the fate of stocked

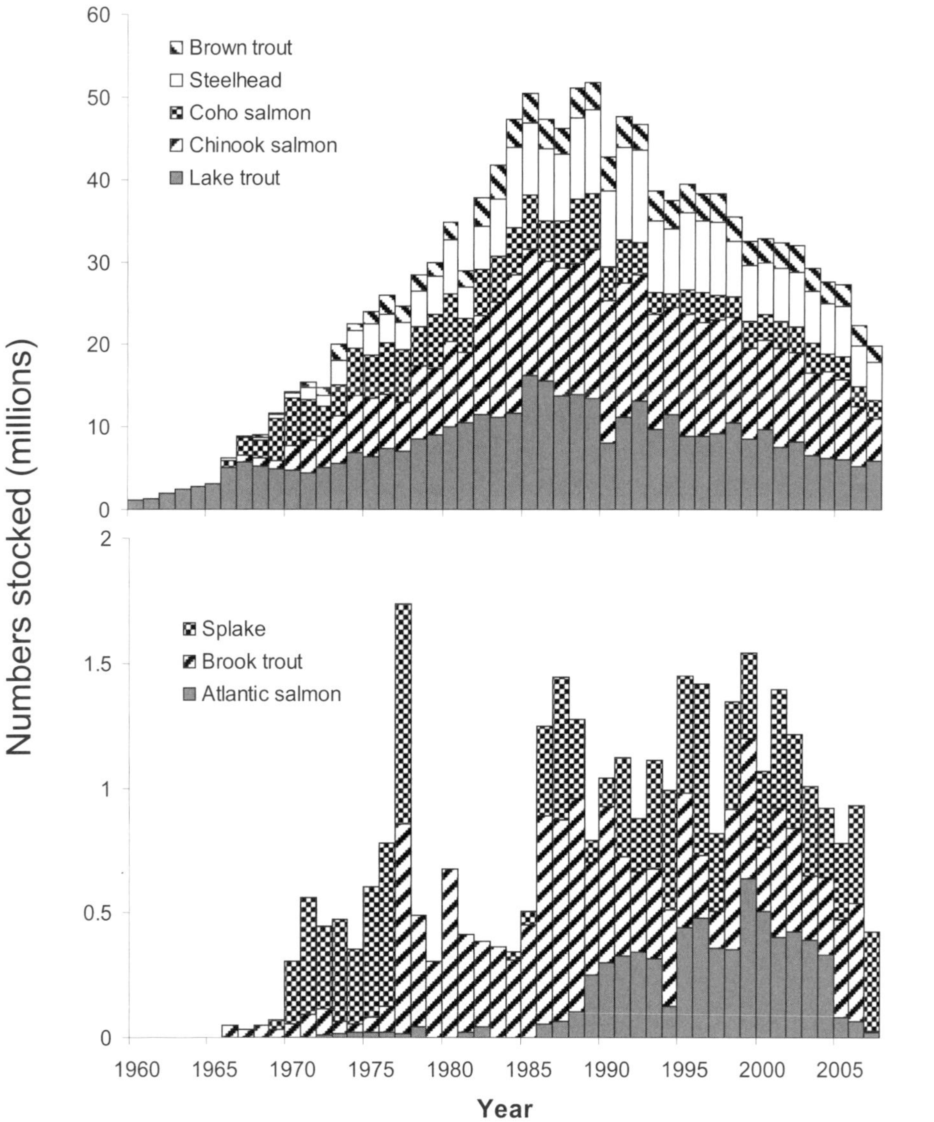

FIGURE 1. Salmonine stocking by species and year in the Laurentian Great Lakes, 1960–2007.

fish across inter-jurisdictional boundaries. These two categories are related to the objective to match stocking and future consumption with available prey. The third category deals with general questions of population dynamics—growth, age at capture, and survival among strains stocked within a species. The last category deals with evaluation of hatchery and stocking practices such as comparisons of rearing densities, stocking locations and methods, and size-at-stocking on return rates, and the differential contribution of groups of fish to specific fisheries. This is important to maximize hatchery operations and product-quality to meet specified fishery outcomes in the most efficient and cost-effective manner.

Marking Method Selection and Evaluation

To meet the management requirements for Great Lakes salmonines and to address all information needs the chosen marking technology must have the following characteristics: 1) a duration commensurate with the longest lived species (lake trout >20 years); 2) the ability to create unique marks to discriminate among many groups; 3) the ability to be applied quickly and uniformly at each facility with minimal handling; 4) a high marking success and retention; 5) be easy to identify or detect; 6) induce negligible mortality at marking and post release; 7) require minimal human handling; 8) allow for unambiguous group assignments; and 9) a proven technology. Commensurate with these requirements six mass marking techniques were considered: 1) antibiotic oxytetracycline (OTC) marks; 2) strontium chloride/isotope marking; 3) passive integrated transponder (PIT) tags; 4) adipose fin clips; 5) thermal marks; and 6) coded-wire tags.

Most of these techniques failed to meet one or more of our selection criteria. OTC marking was eliminated from further consideration because some agencies will not use antibiotics except to treat disease and are concerned about the human health consequences of antibiotics in public waters. Strontium chloride or isotope marking was not considered because of annual environmental variability in isotopic elemental signatures (Rutherford et al. 2002), and the need to characterize isotopic profiles of each group of interest to establish a reference baseline for classification (Schaner et al. 2004). PIT tags are attractive for marking small numbers of fish but not practical for a basin-wide mass-marking program involving millions of fish because tagging rates are slow and unit tag costs are high compared to other tag options (Rutherford et al. 2002). Thermal marking, i.e., depositional changes on sagittae from short-term water temperature changes, was also considered but were rejected due to a limited number of unique marks (S. Campana, Bedford Institute of Oceanography, personal communication), no external visibility, high between-reader error (Bergstedt et al. 1990), high classification error rates, (Hagen et al. 1995; Volk et al. 1999), and variable mark quality and recognition. Adipose fin clips are a simple way to mark fish but were ruled out because they do not provide unique marks to identify separate groups. However, adipose fin clips have been used in conjunction with internal tags as an external identifier of tagged fish.

A coded wire tag (CWT) combined with an adipose fin clip (external mark) was the most attractive marking/tagging method for the objectives of the Great Lakes program. A CWT is a 1.1 mm x 0.20-mm magnetized stainless steel wire marked with a binary code or serial numbers denoting a specific cohort (Jefferts et al. 1963). The tag is injected into the nasal cartilage of a fish prior to stocking, and is extracted at recapture and viewed under a microscope to determine the code. This method has many desirable characteristics: high mark retention rate, a large number of unique marks including batch and individual marks, high marking rate using automated tagging trailers (AutoFish systems, Northwest Marine Technology, Inc., Shaw Island, Washington), no special training or multiple readers needed for mark and tag recognition, minimal mortality at tagging, application can be standardized, and negligible classification error. This tagging and marking method has proven reliable and feasible, and has been used extensively for decades in Pacific Northwest of America for salmon and steelhead management (Johnson 2004; Nandor et al. 2009).

CWT Tagging Equipment

The centerpiece of the Great Lakes program is the automated tagging and marking trailers, known as the AutoFish system, produced by Northwest Marine Technology. The AutoFish System is an alternative to manual clipping and tagging in high-volume stocking applications, however the more traditional manual tagging trailers will still be used in hatcheries with lower production or where requirements (e.g., water supply) for the AutoFish trailer cannot be met. Portable equipment, Mark IV Coded Wire Tag Injectors and Quality Control Devices (QCD) (Northwest Marine Technology, Inc., Shaw Island, Washington), will be used for small marking projects where a manual trailer cannot be accommodated. The QCDs are an accessory to the Mark IVs that automatically sort tagged and untagged fish as they pass through a tag detection tunnel. Both portable Mark IVs and QCDs can serve as replacements for those contained in the manual and AutoFish trailers. Mark IVs already owned by agencies will be reconditioned by the manufacturer and placed back into service.

Size of the Program

The projected number of salmonines to be stocked into the Great Lakes is about 30 million fish annually (Table 1). These fish are reared by tribal, state, federal, and provincial agencies and by nongovernmental organizations (NGOs) (i.e., sport fishing groups that are permitted to rear and stock fish under agency guidelines). Eight species are regularly stocked—Chinook salmon (30% of total), lake trout (29%), and steelhead (18%) make up most of the fish stocked. The states of Michigan (27%), Wisconsin (14%), and New York (11%), the U.S. Fish and Wildlife Service (19%), and the Province of Ontario (11%) stock the most fish. Rearing occurs at 50 separate facilities spread across seven states in the U.S and the province of Ontario in Canada (Figure 2). The agencies propose to CWT and adipose clip all salmonines stocked as opposed to fractional marking. This is important because most marked fish recoveries will be from the sport fishery and agency assessments, where potential sample sizes are smaller compared to collection opportunities in the Pacific Northwest where return data come from large numbers of fish landed by commercial fisheries or from adults returning to hatcheries (i.e., "rack counts") or natal streams. Unlike the Pacific Northwest, most of the Great Lakes salmonine hatcheries are not located on rivers where hatchery returns are possible. Partial tagging of Great Lakes fish would reduce sample sizes and add further uncertainty to any population measures derived from the return data. This is particularly important for studies that have a restricted spatial component or for performance comparisons among hatcheries, strains, or other smaller lots. The program expects to recover over 100,000 coded wire tags annually.

The U.S. Fish and Wildlife Service will execute the program in the U. S. because of

TABLE 1. Total number of each salmonine species proposed to be coded-wire tagged, adipose fin-clipped, and stocked into each Great Lake.

Species	Lake					Species total
	Erie	Huron	Michigan	Ontario	Superior	
Atlantic salmon				751,000		751,000
Brook trout					280,000	280,000
Brown trout		546,250	1,709,250	575,000	72,750	2,903,250
Chinook salmon		1,933,000	4,497,000	2,300,000	600,000	9,330,000
Coho salmon			2,615,000	240,000		2,855,000
Lake trout	200,000	3,250,000	3,610,000	1,440,000	407,000	8,907,000
Steelhead	1,843,600	834,600	1,860,550	830,700	258,750	5,628,200
Splake		44,000	108,000		88,000	240,000
Lake totals	2,043,600	6,607,850	14,399,800	6,136,700	1,706,500	30,894,450

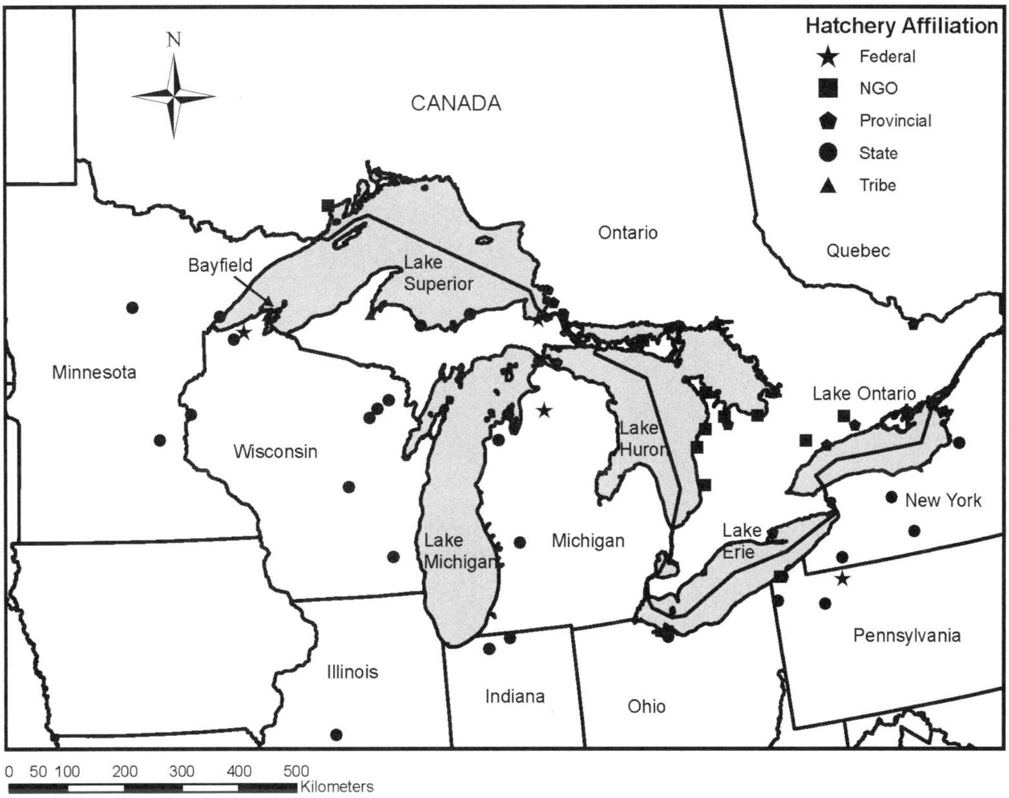

Figure 2. Location of hatcheries by affiliation within each jurisdiction in the U.S and Canada that stock salmonines into the Laurentian Great Lakes. "NGO" refers to non-governmental sportsmen's groups that grow out and stock salmon and trout under permit from the resident management agency.

this agency's history of providing basin-wide in the Great Lakes services for lake trout rehabilitation and sea lamprey assessment and control. In Canada, the program will be run by the Ontario Ministry of Natural Resources because this agency provides all fish stocked by the province and oversees stocking efforts of NGOs.

Survey of CWT Program Requirements

Coded-wire tagging of salmonines has been conducted by federal, state and provincial agencies in the Great Lakes for more than 20 years. In most cases, experimental designs were developed only to answer specific questions for a single agency. Multi-agency or inter-jurisdictional coordination was sporadically exercised, especially for species other than lake trout. Hence, the level of expertise available from Great Lakes fishery agencies was not commensurate with the spatial and numerical scale of the proposed project. An expert from the Pacific Northwest CWT program (see Johnson 1990) was contracted for 1.5 years to assist in planning and program development.

In 2006 the design contractor visited all 50 state, NGO, tribal, federal, and provincial Great Lakes salmonine hatcheries. An assessment of equipment, labor, and infrastructure needed to support marking/tagging operations for the expanded program was made at each facility. Existing marking programs were reviewed and most were found to be compatible with an expanded program. All hatcheries were evaluated to determine: anticipated fish production by species, a tentative marking schedule and marking time period, the number and type of marking/tagging equipment required (manual trailer, automated trailer, or portable units), available electrical supply and costs to upgrade service, available setup area for trailers and electrical service access, and operational plan with hatchery staff input. This information was used to estimate capital equipment, labor, tags and consumables, and a proposed schedule of delivering services and equipment to each facility to mark/tag all fish.

Institutional Arrangements

The inter-jurisdictional and international nature of a basin-wide CWT program necessitates the establishment of multi-agency committees to guide the program and a centralized group to carry out the work of tagging, and data collection and processing. The CLC recommended the formation of two groups: the "Great Lakes Regional Marking Committee" (GLRMC), and the "Data Standards Committee" (DSC). They also recommended that the U.S. Fish and Wildlife Service create the "Great Lakes Fish Marking Laboratory" (GLFML) to provide an organizational framework to meet the marking and information needs of the cooperating agencies, and to implement the program in the U.S. across all states.

The GLRMC will consist of up to two representatives from each state, U.S. tribal management authority, U.S. federal agency, and the Ontario Ministry of Natural Resources, with membership from each jurisdiction consisting of one hatchery expert and one fishery manager or researcher. Once operational, the committee will provide jurisdictional resources to facilitate fish marking/tagging operations and subsequent data collection. Members representing hatchery operations will coordinate logistics between GLFML marking crews and the hatchery personnel from facilities for their respective agency. Members representing management/research interests will serve as liaison between GLFML technical staff and agency personnel for study plan design, tag recovery efforts, and data analysis. The GLRMC will serve as the primary liaison between the CLC, fishery management agencies, and GLFML to ensure coordinated implementation of mass-marking studies. This group will provide official requests to GLFML and OMNR for mass marking studies prioritized by the CLC, and with the GLFML, will develop mass marking study plans, coordinate the logistics among agency hatcheries to ensure timely and efficient tagging operations, coordinate basin-wide efforts for the recovery of CWT fish, and report of progress back to the CLC on mass marking studies.

The DSC will consist of data managers/biologists from all agencies and will be responsible for development of a single relational database that will house all the stocking/marking and recovery data at the GLFML. Other responsibilities include development of protocols, timetables, and requirements for automated data transfer of recovery data from agencies to the basin-wide database, development of standard output reports per agency or study plan requirements, and provide overall guidance on data management, security, and sharing activities and capabilities.

The GLFML will: 1) purchase, operate, and maintain all mass marking equipment used in the U.S.; 2) provide experimental design, data analysis, database management and tag extraction (head shop) services for all marking studies; 3) provide manpower to enhance agency efforts to recover heads from tagged fish captured in sport fisheries or other sources; and 4) provide coordination between the U.S. and Canada. The GLFML laboratory will be centrally located and include support and administration capabilities. A satellite field station and a trailer storage facility will be strategically located elsewhere in the basin. Key support staff will include a biostatistician familiar with mark–recapture experimental design and analytical techniques; a database manager and programmer familiar with the structure, function, output generation, and upkeep of large relational databases; and a program supervisor. The program supervisor will oversee all operations for the U.S. side and be the main interface with the GLRMC. All services provided by the laboratory (database, programming, experimental design, tag extraction) will be available to the Ontario Ministry of Natural Resources program.

Equipment and Operational Costs

The estimated costs to implement a complete and fully-funded Great Lakes mass marking program in the U.S. and Canada is \$13.7 million U.S. (2011 dollars) in capital equipment purchases, \$0.4 million in electrical upgrades to hatcheries, and an annual operating budget of \$5.9 million (details provided in Bronte 2008; Table 2). The program is designed as a "no cost" program to state and tribal agencies in the U.S., with most equipment owned, operated, and maintained by U.S. Fish and Wildlife Service. In Canada, all equipment will be owned, operated, and maintained by the Ontario Ministry of Natural Resources.

The anticipated costs represent a full program that delivers marking and tagging, head recovery and tag extraction, statistical and database support, and data dissemination services to the partner agencies. Costs also include improvements to electrical service at hatcheries with insufficient power based on the requirements of the marking trailers. Budgeted staff positions include full-time biologists and technicians trained in the operation and maintenance of the sophisticated AutoFish trailers. These positions will be supported by seasonal technicians, manual markers, and dedicated field staff assigned to recover tags.

Full-time biologists and technicians, when not marking fish, will operate a tag extraction laboratory, support data management, maintain all equipment, and collect

Table 2. Estimated costs for a basin-wide coded wire tag program in the Great Lakes. Updated from estimates in Bronte (2008).

Item	Estimated cost 2011 (Millions, \$U.S.)
Capital equipment (all equipment including 9 AutoFish and 3 manual trailers)	\$ 13.7
Electrical upgrades at hatcheries	\$ 0.4
Annual operating cost	
Personnel (perm and temp)	\$ 2.2
Travel	\$ 0.8
CWTs	\$ 1.8
Others costs and expendable supplies	\$ 1.1
Total annual operating costs	\$ 5.9
Grand total (capital+electrical+operating)	\$ 20.0

fish heads for future tag extraction. The coordinator will serve as the point person and manager for a spatially and temporally complex program that will span the Great Lakes basin, marking and tagging about 26 million fish annually at 31 state and tribal hatcheries in U.S waters and will synchronize activities with the marking coordinator for the Ontario Ministry of Natural Resources. Statistical and computer support positions will make the development of study designs (done before any fish are marked), collection protocols to meet study objectives, and data analysis (working with key partner biologists) an integral part of the program. These data support positions are considered instrumental to successfully answer management and performance questions on the appropriate spatial and temporal scales.

Mass marking requires significant fiscal resources; annual costs for tags, AutoFish system consumables, equipment maintenance, and operations represent 63% of the annual operating budget. Salaries and benefits make up about 37% of the total budget. Costs of services are comparable to those in the Pacific Northwest, and are based on real historical expenditures for similar programs. Costs will likely range between $150–190 per 1000 marked/tagged fish, which includes head collection and tag extraction services, as well as analytical support. These latter services are not represented in cost estimates from the Pacific Northwest that range between $140–170 per 1000 fish marked/tagged. Higher per unit costs for the Great Lakes are anticipated given the spatial complexity, the higher number of low production facilities compared to the Pacific Northwest, the greater number of species, and the need for many simultaneous marking and tagging events that are widely separated geographically. The intension is to mark all hatchery released fish, significant reductions in cost can only be made by tagging fewer fish in each lake or restricting the program to fewer lakes.

Funding

The cost, size, and complexity of this initiative will require that equipment inventories (capital costs), personnel capabilities, and operational budgets be built over time to establish and maintain the program. Securing the funding has involved building support for the initiative among key Great Lakes stakeholders and communicating the needs to the U.S. Congress. This effort was initiated and organized by the GLFC secretariat. Broad stakeholder support was viewed as essential for several reasons including the need of many agencies to consult citizens before major decisions, to engage citizens in key fishery management initiatives, to explain the connection of the marking program and sound science and fishery management, to achieve ongoing stakeholder participation in recovering the heads of marked fish, and to assure the supporters (e.g., Congress, the U.S. Fish and Wildlife Service) that the program enjoys broad support among agencies and the public.

To build awareness and support among stakeholders, the GLFC facilitated two demonstrations at hatcheries in Wisconsin and Michigan, and involved the use of the AutoFish trailers, which allowed stakeholders to witness actual marking operations. User groups (i.e., Michigan Steelhead and Salmon Fisherman's Association, Michigan Charter boat Association), the media, Congressional staff, and agency personnel were invited and attended. These demonstrations allowed hatchery staff to highlight the technology and to learn for themselves the limits and possibilities of the marking equipment. Hatchery operators and other agency personnel learned more about the feasibility of mass marking. Stakeholders learned about how the equipment would be applied in the Great Lakes basin and what would be expected of them after implementation. The demonstrations also provided the CLC with critical information to be used to secure funding for the project.

Funding for mass marking will require support from members of Congress. Fish chiefs from Wisconsin, Michigan, New York, and Ohio visited more than 30 Congressional offices during 2006–2009 to raise awareness of the initiative and request funds. These efforts proved successful–Congress awarded the U.S. Fish and Wildlife Service $1.72 million (U.S.) to begin equipment acquisition 2008 and an additional $1.5 in 2009. In 2010, the Great Lakes Restoration Initiative—a multi-year program backed up by hundreds of millions of dollars—resulted in nearly $3 million through the competitive *Fish and Wildlife Restoration Act* for the purchase of mass marking equipment. This capital purchase was in addition to $1.0 million appropriated in fiscal year 2010 to begin tagging operations that resulted in tagging 5 million lake trout and 1.1 million Chinook salmon for lakes Michigan and Huron. The Service now has purchased four AutoFish trailers and one manual tagging trailer, and will be able to expand the number and complexity of the tagging studies pending future operational funding. The CLC has requested funds for operations and additional equipment purchases in 2011 and beyond.

Commitments for the mass marking initiative have also been demonstrated by individual state and the provincial fisheries management agencies. In 2008 the New York Department of Environmental Conservation (DEC) purchased the first AutoFish tagging trailer in the Great Lakes Basin. Since then, DEC and the Ontario Ministry of Natural Resources (OMNR) began working cooperatively to mass mark all of the Chinook salmon stocked into Lake Ontario. From 2008–2010, DEC and OMNR have marked 5.84 million Chinook salmon to determine the relative contributions of hatchery and wild Chinook salmon to the sport fishery. A proportion of these fish (2.24 million) also received coded wire tags to evaluate the relative survival of various stocking strategies and to learn more about the movement and degree of straying or homing by returning adults to stocking sites. Funding received by the U.S. Fish and Wildlife Service in 2010 was used to assist New York in purchasing tags and help recover tagged fish from the sport fishery. Future plans for the trailer include similar studies on other species of salmon and trout in Lake Ontario.

Conclusion

Fisheries management on the Great Lakes is biologically and politically complex. Managers must implement science-based programs that balance predators and prey, protect and enhance native fish stocks, and consider many ecosystem factors. They also must balance myriad public interests; work across state, tribal, and provincial boundaries, and navigate the relationship between federal and nonfederal agencies. The Great Lakes mass marking initiative is a major new effort to improve all of these aspects of Great Lakes fishery management, and represents a major paradigm shift for the basin. From the start, the initiative was conceived, developed, and advanced by all agencies. Although improving scientific understanding has been an important driver of the initiative, managers and stakeholders also recognized the significant operational benefits from using automated technology to mark hatchery fish, and the wealth of information that can be obtained on the fish they stock. Recognizing the importance of this initiative, fish chiefs and the CLC launched an effort to educate the public and Congress about mass marking. Efforts have yielded significant awareness and some funding from Congress, but need to continue to fulfill equipment and operational needs for the program. Once fully implemented, the mass marking initiative will truly transform fishery management on the Great Lakes.

Acknowledgments

We thank Jeremy McKenzie and two anonymous reviewers for their comments that improved the paper, and Dale Hanson for preparing Figure 2. This is contribution P-2011-1 of the USFWS Region 3 Fisheries Program. The findings and conclusions in this article are those of the authors and do not necessarily represent the views of the U. S. Fish and Wildlife Service. Mention of trade names does not imply government endorsement.

References

American Sportfishing Association. 2008. Today's angler: a statistical profile of anglers, their targeted species and expenditures. American Sportfishing Association, Alexandria, Virginia.

Bence, J. R., and L. C. Mohr. editors. 2008. The state of Lake Huron in 2004. Great Lakes Fishery Commission Special Publication 08–01.

Bergstedt, R.A., Eshenroder, R.L., Bowen, C., Seelye, J.G., and Locke, J.C. 1990. Mass-marking of otoliths of lake trout sac fry by temperature manipulation. Pages 216–223 *in* N. C. Parker, A. E. Giorgi, R. C. Heidinger, D. B. Jester, Jr., E. D. Prince, and G. A. Winans, editors. Fish-marking techniques. American Fisheries Society, Symposium 7, Bethesda, Maryland.

Bronte, C. R., editor. 2008. Report of the implementation task group on mass marking hatchery-reared salmonines in the Great Lakes. Great Lakes Fishery Commission. Project completion report. Ann Arbor, Michigan. Available: http://www.glfc.org/boardcomm/clc/Mass_marking_report_CLC2008.doc (February 2011).

Bronte, C. R., M. P. Ebener, D. R. Schreiner, D. S. DeVault, M. M. Petzold, D. A. Jensen, C. Richards, and S. J. Lozano. 2003. Fish community change in Lake Superior, 1970–2000. Canadian Journal of Fisheries and Aquatic Sciences 60:1552–1574.

Bronte, C. R., and S. P. Sitar. 2008. Harvest and relative abundance of siscowet lake trout in Michigan waters of Lake Superior, 1929–1961. Transactions of the American Fisheries Society 137:916–926.

Clapp, D. F., and W. Horns. editors 2008. The state of Lake Michigan in 2005. Great Lakes Fishery Commission Special Publication 08–02.

Clark, J. D., Jr., editor. 1997. Status of Chinook salmon in the upper Great Lakes, 1995. Michigan Department of Natural Resources, Fisheries Division, Administrative Report, Lansing.

Cornelius, F. C., K. M. Muth, and R. Kenyon. 1995. Lake trout rehabilitation in Lake Erie: a case history. Journal of Great Lakes Research 21 (Supplement 1):65–82.

Dobiesz, N. E., D. A. McLeish, R. L. Eshenroder, J. R. Bence, L. C. Mohr, M. P. Ebener, T. F. Nalepa, A. P. Woldt, J. E. Johnson, R. L. Argyle, and J. C. Makarewicz. 2005. Ecology of the Lake Huron fish community, 1970–1999. Canadian Journal of Fisheries and Aquatic Sciences 62:1432–1451.

Elrod, J. H., R. O'Gorman, C. P. Schneider, T. H. Echert, T. Schaner, J. N. Bowlby, and L. P. Schleen. 1995. Lake trout rehabilitation in Lake Ontario. Journal of Great Lakes Research 21(Supplement 1):83–107.

Eshenroder, R. L., N. R. Payne, J. E. Johnson, C. Bowen, II, and M. P. Ebener. 1995. Lake trout rehabilitation in Lake Huron. Journal of Great Lakes Research 21(Supplement 1):108–127.

GLFC (Great Lakes Fishery Commission). 2007. A joint strategic plan for management of Great Lakes fisheries (adopted in 1997 and supersedes 1981 original). Great Lakes Fishery Commission Miscellaneous Publication 2007–01.

Hagen, P., K. Munk, B. Van Alen, and B. White. 1995. Thermal mark technology for inseason fisheries management: a case study. Alaska Fishery Research Bulletin 2:143–155.

Hansen, M. J. 1999. Lake trout in the Great Lakes: basinwide stock collapse and binational restoration. Pages 417–453 *in* W. W. Taylor and C. P. Ferreri, editors. Great Lakes fisheries and policy management: a binational perspective. Michigan State University Press, East Lansing.

Hansen, M. J., J. W. Peck, R. G. Schorfhaar, J. H. Selgeby, D. R. Schreiner, S. T. Schram, B. L. Swanson, W. R. MacCallum, M. K. Burnham-Curtis, G. L. Curtis, J. W. Heinrich, and R. J. Young. 1995. Lake trout (*Salvelinus namaycush*) populations in Lake Superior and their restoration in 1959–1993. Journal of Great Lakes Research 21(Supplement 1):152–175.

Holey, M. E., R. E. Elliot, S. V. Marcquenski, J. G. Hnath, and K. D. Smith. 1998. Chinook salmon epizootics in Lake Michigan: possible contributing factors and management implications. Journal of Aquatic Animal Health 10:202–210.

Holey, M. E., R. R. Rybicki, G. W. Eck, E. H. Brown, Jr., J. E. Marsden, D. S. Lavis, M. L. Toneys, T. N. Trudeau, and R. M. Horrall. 1995. Progress toward lake trout restoration in Lake Michigan. Journal of Great Lakes Research 21(Supplement 1):128–151.

Huckins, C. J., E. A. Baker, K. D. Fausch, and J. B. K. Leonard. 2008. Ecology and life history of coaster brook trout and potential bottlenecks in their rehabilitation. North American Journal of Fisheries Management 28:1321–1342.

Huntsman, A. G. 1944. Why did Ontario salmon disappear? Transactions of the Royal Society of Canada 68:82–102.

Jefferts, K. B., P. K. Bergman, and H. F. Fiscus. 1963. A coded wire identification system for micro-organisms. Nature (London) 198:460–462.

Johnson, J. K. 2004. Regional overview of code wire tagging of anadromous salmon and steelhead in Northwest America. Available: http://www.rmpc.org/publications.html (February 2011).

Jones, M. L., J. R. Bence, E. B. Szalai, and W. Dai. 2008. Pages 79–86 *in* D.F. Clapp and W. Horns, editors. The state of Lake Michigan in 2005. Great Lakes Fishery Commission Special Publication 08–02. Ann Arbor, Michigan.

Keller, M., K. D. Smith, and R. W. Rybicki. 1990. Review of salmon and trout management in Lake Michigan. Michigan Department of Natural Resources, Fisheries Special Report 14, Lansing.

Kocik, J. F., and Jones, M. L. 1999. Pacific salmonines in the Great Lakes basin. Pages 455–488 in W.W. Taylor and C.P. Ferreri, editors. Great Lakes fisheries policy and management: a binational perspective. Michigan State University Press, East Lansing.

Lawrie, A. H., and J. F. Rahrer. 1973. Lake Superior: a case history of the lake and its fisheries. Great Lakes Fishery Commission Technical Report 19.

MacCrimmon, H. R., and B. L. Gots. 1980. Fisheries for charrs. Pages 797–839 *in* E. K. Balon, editor. Charrs: salmonid fishes of the genus Salvelinus. Dr. W. Junk, The Hague, The Netherlands.

Madenjian, C. P., G. L. Fahnenstiel, T. H. Johengen, T. F. Nalepa, H. A. Vanderploeg, G. W. Fleischer, P. J. Schneeberger, D. M. Benjamin, E. B. Smith, J. R. Bence, E. S. Rutherford, D. S. Lavis, D. M. Robertson, D. J. Jude, and M. P. Ebener. 2002. Dynamics of the Lake Michigan food web, 1970–2000. Canadian Journal of Fisheries and Aquatic Sciences 59:736–753.

Madenjian, C. P., T. O. Hook, E. S. Rutherford, D. M. Mason, T. E. Croley, I. I., E. B. Szalai, and J. R. Bence. 2005. Recruitment variability of alewives in Lake Michigan. Transactions of the American Fisheries Society 134:218–230.

Madenjian, C. P., R. O'Gorman, D. B. Bunnell, R. L. Argyle, E. F. Roseman, D. M. Warner, and J. D. Stockwell. 2008. Adverse effects of alewives on Laurentian Great Lakes fish communities. North American Journal of Fisheries Management 28:263–282.

Mills, E. L., J. M. Casselman, R. Dermott, J. D. Fitzsimons, G. Gal, K. T. Holeck, J. A. Hoyle, O. E. Johannsson, B. F. Lantry, J. C. Makarewicz, E. S. Millard, I. F. Munawar, M. Munawar, R. O'Gorman, R. W. Owens, L. G. Rudstam, T. Schaner, and T. J. Stewart. 2003. Lake Ontario: food web dynamics in a changing ecosystem (1970–2000). Canadian Journal of Fisheries and Aquatic Sciences 60:471–490.

Nandor, G. F., J. R. Longwill, and D. L. Webb. 2009. Overview of the coded wire tag program in the greater Pacific region of North

America. Pages 5–46 *in* K. S. Wolf, and J. S. O'Neal, editors. Tagging, Telemetry and Marking Measures for Monitoring Fish Populations—a compendium of new and recent science for use in informing technique and decision modalities: Pacific Northwest Aquatic Monitoring Partnership Special Publication 2010–002. Available: http://www.pnamp.org/TTM (February 2011).

Rutherford, E. S., J. D. Iacono, and G. Callaham. 2002. Evaluation of marking procedures to estimate natural reproduction of Chinook salmon in Lake Michigan. Great Lakes Fishery Commission 2002 Project Completion Report, Ann Arbor, Michigan.

Schaner, T., R. O'Gorman, B. F. Lantry, and W. P. Patterson. 2004. Application of stable isotope method to determine origin of lake trout; and description of thermal history of lake trout during first year of life. Great Lakes Fishery Commission 2004 Project Completion Report, Ann Arbor, Michigan.

Smith, S. H. 1968. Species succession and fishery exploitation in the Great Lakes. Journal of the Fisheries Research Board of Canada 25:667–693.

Smith, S. H. 1970. Species interactions of the alewife in the Great Lakes. Transactions of the American Fisheries Society 99:754–765.

Smith, S. H. 1972. Factors of ecologic succession in oligotrophic fish communities of the Laurentian Great Lakes. Journal of the Fisheries Research Board of Canada 29:717–730.

Smith, S. H. 1995. Early changes in the fish community of Lake Ontario. Great Lakes Fishery Commission Technical Report 60.

Smith, B. R., and J. J. Tibbles. 1980. Sea lamprey (*Petromyzon marinus*) in Lakes Huron, Michigan, and Superior: history of invasion and control, 1936–78. Canadian Journal of Fisheries and Aquatic Sciences 37:1780–1801.

Tanner, H. A., and W. H. Tody. 2002. History of the Great Lakes salmon fishery: a Michigan perspective. Pages 139–154 *in* K. D. Lynch, M. L. Jones, and W. W. Taylor, editors. Sustaining North American salmon: perspectives across regions and disciplines. American Fisheries Society, Bethesda, Maryland.

Volk, E. C., S. L. Schroder, and J. J. Grimm. 1999. Otolith thermal marking. Fisheries Research 43:205–219.

American Fisheries Society Symposium 76:43–62, 2012

Advances in Coded Wire Tag Technology: Meeting Changing Fish Management Objectives

GERALDINE VANDER HAEGEN*, LEE BLANKENSHIP,
AND DAVE KNUTZEN
Northwest Marine Technology, Inc.
955 Malin Lane SW, Tumwater, Washington 98501, USA

Abstract.—Coded Wire Tags (CWT) are lengths of magnetized stainless steel wire 1.1 mm long and 0.25 mm in diameter with a code etched along the wire. The tags are injected internally and while their presence can be detected electronically, code recovery requires that the tag be read under a microscope. CWTs have been in widespread use since their invention in the 1960s and their ongoing importance has been partly due to technological advances that meet the needs of fish management and research. We describe three changes in CWT technology that we think have played the most important roles. First was a coding format change from binary to decimal coding that made the tag code directly readable and reduced errors as well as providing a virtually unlimited code capacity. Second was the development of the highly sophisticated AutoFish System for automated injection of CWTs and adipose fin clipping of salmonids without human handling or anesthetic. Finally, we describe the development of new tools for electronic detection of CWT that allows tagged animals of any species to be sorted from untagged animals.

Introduction

Since the first International Symposium and Educational Workshop on Fish Marking Techniques in 1988, the quest for the perfect fish tag has continued. It is unlikely that any one type of tag could ever meet all the demands of fishery researchers and managers, but rapidly developing technology and microelectronics have provided new fish tags with amazing tracking and monitoring capabilities. At the same time, manufacturers of already established tags, including Coded Wire, passive integrated transponder, Visible Implant, and external anchor tags have taken advantage of new technologies and increased the performance and utility of these tags. The number of these tags far exceeds the number of all other types of tags used annually thus they remain crucial to fisheries management and research.

The Coded Wire Tag (CWT; Jefferts et al. 1963) was well established at the time of the first tagging symposium (Johnson 1990). Invented in the 1960s, the CWT is a small piece of stainless steel wire 1.1 mm in length and 0.25 mm in diameter with a code etched on the surface (Figure 1). The tags are magnetized and their presence can therefore be externally detected using electronic magnetometers. Coded Wire Tags are nontransmitting and must be extracted and viewed under a low-powered microscope to retrieve the code. When used appropriately, CWT typically has

*Corresponding author: geraldine.vanderhaegen@nmt.us

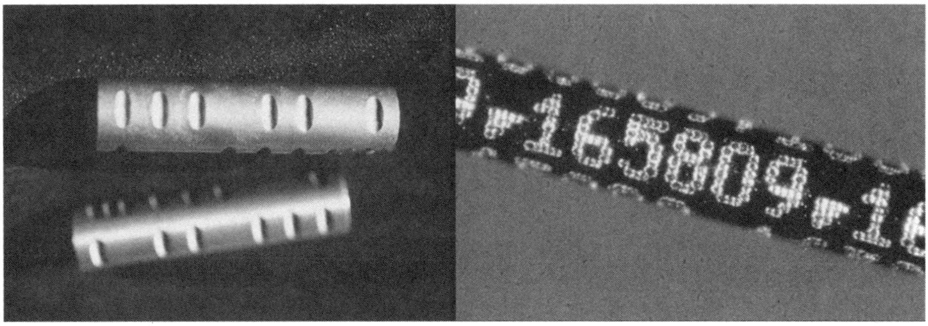

Figure 1. Binary Coded Wire Tag (left); code etched onto the surface of the wire. Laser etched Decimal Coded Wire Tag (Right). Actual size of both types of Coded Wire Tag 1.1 mm long and 0.25 mm in diameter.

no negative impact on the host animal (Blankenship and Thompson 2003; Davis et al. 2004b; Brennan et al. 2005; Vander Haegen et al. 2005). Because of their small size, they can be used in species or life stages that are too small for other tagging methods. The management of Pacific salmon by the United States and Canada relies heavily on the use of CWT (Nandor 2010) and programs around the world have benefited from their development.

In the 1980s, about 40 million CWT were applied annually. Currently, about 60 million CWT are used throughout the world per year. The increasing importance of CWT in fisheries management can be attributed to improvements in application and detection technology. In this paper, we outline three recent technological advances that, in our opinion, have contributed most to the continued importance of CWT to modern fisheries management, these being a change from binary to decimal coding on the tags, the development of the AutoFish System for tag injection and adipose fin clipping, and the development of electronic tag detection for field use.

Development of New Coding Methods

The original CWTs were coded with up to five bands of color that ran along the length of the wire. The strips of color sometimes peeled off, the colors were difficult to distinguish from each other when the tags were recovered, and the colors were inconsistently interpreted by different technicians. While this tag still represented a leap forward in the tagging technology available at the time, the coding was soon changed to a binary code that was etched on the wire using electrical discharge machining. This process provided much more stability to the code and vastly increased the number of codes available. Under the microscope, the binary coding appeared as a series of vertical dashes along the length of the wire. Because CWTs are cut beginning anywhere along the spool of wire, the beginning of the code is not necessarily at the left edge of the tag. However, the length of the tag is longer than the space occupied by the full code, and thus the complete code is always contained within a tag, even if the start of the code is not at the left end of the tag. Binary coding included a "master word" that was used to orient the tag, and to show where the code began on the tag. Below the master word were three lines of coding around the wire, each representing two digits of a six digit code. The dashes and spaces of the coded lines were aligned with the master word to indicate the numeric value of each dash. Decoding the tag required that the tag be correctly oriented and that each component of the binary code be correctly added together. While technicians who interpreted the codes regularly were very adept at this, it was a relatively lengthy process and challenging for people reading the tags infrequently. In many cases, each tag was read at least twice to reduce decoding errors. Most CWT are used for identifying batches of fish so that all of the tags have the same code. Coded Wire Tags are also available for identifying small batches or individual animals, and are known as Sequential CWT (seqCWT). Reading the sequential portion of a binary seqCWT required a special program, or the use of tables to decode the sequence. Finally, because each side of the wire was required to fully interpret the code, if severe scratching occurred during tagging, it could obscure the code.

As laser technology advanced, the format was changed from binary code to decimal digits using lasers to etch the tag code clearly on the wire (Figure 1). The numbers are read directly from left to right, and because the tags can be cut from any spot along the spool of wire, the code is still formatted to be shorter than the length of the tag, and the beginning of the code is marked by a flag. This newest format is easier and quicker to read than the binary code, and reduces reading errors. The most commonly used types of CWT have the entire code on a single data line, so the code is now printed four times around the wire on each tag, reducing the possibility of losing the code if the tag has been damaged. The code on each side is offset two character spaces from the side above it for added redundancy and to facilitate code reading. The coding for decimal seqCWT is also printed clearly on the wire, making them easy to decipher. The change to decimal coding further increased the code capacity so that it is now virtually unlimited.

Development of AutoFish System

Along with technological innovation, the evolving needs of fisheries managers drive advances in tagging and other fisheries related technology. In the Pacific Northwest of the United States and Canada, most state, tribal and federal governments are implementing a fish management initiative, called mass marking, which requires all hatchery fish be identified by removal of the adipose fin. Mass marking serves two important purposes. First, externally marked hatchery fish can be identified by fishermen for retention in a live capture, selective harvest fishery that would enable release of wild fish requiring protection under the U.S. Endangered Species Act. Second, the adipose fin clip is used to distinguish hatchery fish from wild fish which allows the extensive hatchery program in this region to be genetically managed in conjunction with wild fish (Mobrand et al. 2005).

Hatcheries in the Pacific Northwest rear and release about 40 million coho salmon *Oncorhynchus kisutch* and 150 million Chinook salmon *O. tschawytcha* annually that are required to be marked under the mass marking initiative. Further complicating the process, increased requirements for injecting CWT arose as part of the implementation of mass marking. Because there is only a short time between when the fish are large enough to mark and tag, and when they need to be released, most of these hatchery reared Chinook and coho salmon must be processed within a 90 d window. While it is plausible to achieve the yearling coho salmon targets using "traditional" marking and tagging methods (i.e., scissors to excise the adipose fins and manual tagging with a Mark IV Automated Tag Injector), reaching the goals for marking and tagging the Chinook salmon required a new automated approach.

In response to this need, Northwest Marine Technology, the Washington Department of Fish and Wildlife, the Bonneville Power Administration, and the U.S. Fish and Wildlife Service teamed up to develop an automated machine that could be used at hatcheries to help reach the mass marking goals. The AutoFish System is the result of 10 years of research and development. It can do any combination of sorting by size, excising the adipose fin, and injecting CWT in salmonids (57–142 mm TL) at a rate of over 7,500 fish per hour. There is no human handling or anesthetic used in this system and it can be operated by one trained operator and an assistant. AutoFish System is superior to manual marking and tagging because of the consistent handling and sorting methods and the rigorous quality control checks performed on every fish (Hand et al. 2010); throughput is almost twice that of 10 human processors.

AutoFish System is contained within a 44 ft long trailer that can be easily moved between fish rearing facilities (Figure 2). Each AutoFish System has the computer controls at the front of the trailer, with the sorter and six marking and tagging lines in the main work area (Figures 3 and 4). Most trailers also have a work table at the rear where additional personnel can CWT and adipose clip fish manually. This can increase efficiency when there are size categories with few fish, typically on the small and large ends of the population size distribution that would otherwise tie up an entire marking and tagging line with very low throughput.

At the beginning of each tagging project, the trailer is positioned near the ponds holding the fish to be tagged. The trailer is leveled and connected to a power and water supply. Freshwater is continually delivered to the trough and sorter and to each of the marking and tagging lines. Water is not recirculated in the trailer and once it has been used, it is returned back to the hatchery pond with the processed fish. The sorter and tagging lines are calibrated for the size of the fish on hand. The operator will measure about 1000 fish with the sorter to designate the size ranges and install the corresponding components in the marking and tagging lines. This set up takes approximately 4–8 h.

AutoFish System requires precise fish measurement. One of the reasons for this is that to correctly place a CWT in the snout and accurately excise the adipose fin, the size of the fish must be matched to the head mold that correctly positions the fish on the CWT injector and in the adipose fin clipping mechanism. Each head mold and clipping mechanism can accept fish within a particular size range, so the fish must be sorted to those categories. This requirement drove the need for an accurate sorting process before any marking and tagging could begin.

The sorter has four components (Figure 5)—the holding trough, the internal fish delivery pump, the sorter Volitional Entry Device (VED), and the measuring and diverting device. A fish pump or dip net is used to load fish into the aluminum holding trough through a window in the trailer. The internal fish delivery pump lifts the fish up to the sorter VED from the trough. Some water is removed so that the VED does not overflow, but the fish are never completely dewatered.

AutoFish System relies on the VEDs to move fish into the sorting and measuring device and into the marking and tagging lines (Figure 6). The shallow and bright environment with upwelling water in the VED takes advantage of the fish's natural instinct to seek a darker, more protected location, in this case, the entry to the sorting and measuring device. The fish typically swim into the entryway head first. Wire gates and gentle water jets keep the fish moving in single file and slightly separated from each other. The sorting and measuring device (Figure 7) has an internal chute that the fish enter right after leaving the VED. As fish pass down the chute, digital imaging is used to measure the length of each fish. The program compensates for body curves and rejects fish that enter

Figure 2. AutoFish System is completely contained within a mobile trailer that can be moved between tagging sites. The trailer is towed to the hatchery and set up near the rearing ponds holding the fish to be tagged. An external fish pump supplies water to the front of the trailer from the hatchery pond. The water is not reused; it moves through the trailer and flows back to the hatchery pond with the fish.

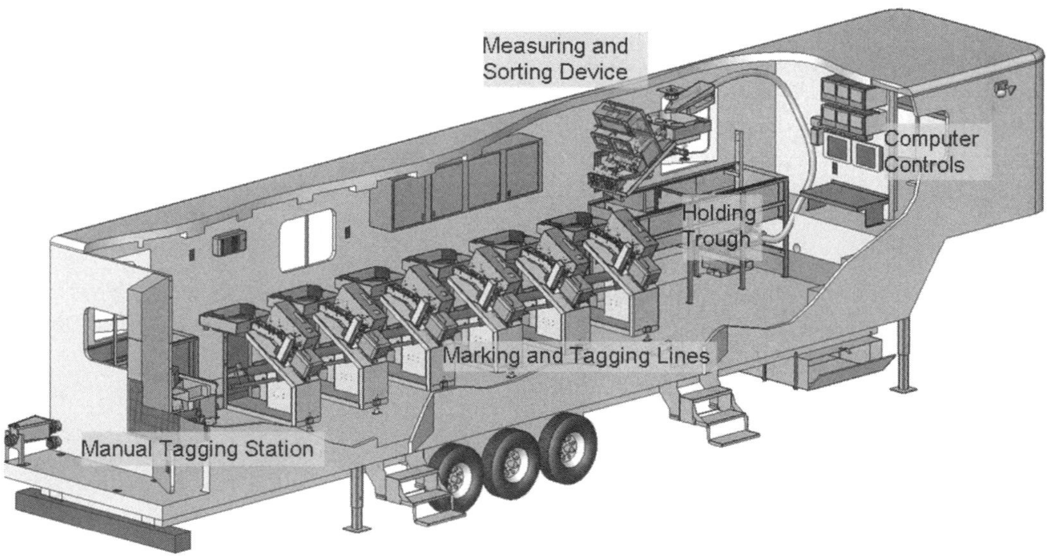

Figure 3. Cut-away view of the interior of an AutoFish System trailer. The main components are the sorter and the marking and tagging lines. The front of the trailer includes some office space and houses the computer controls. The back of the trailer is outfitted with a small work area for manual clipping and tagging. Fish are loaded into the holding trough through a window, then measured and sorted by size. The sorter diverts the fish to the marking and tagging lines for adipose fin clipping and injection of a Coded Wire Tag before they exit the trailer and go back to the rearing ponds.

FIGURE 4. The main working components of AutoFish System are the sorter (left, in background) and the marking and tagging lines (foreground).

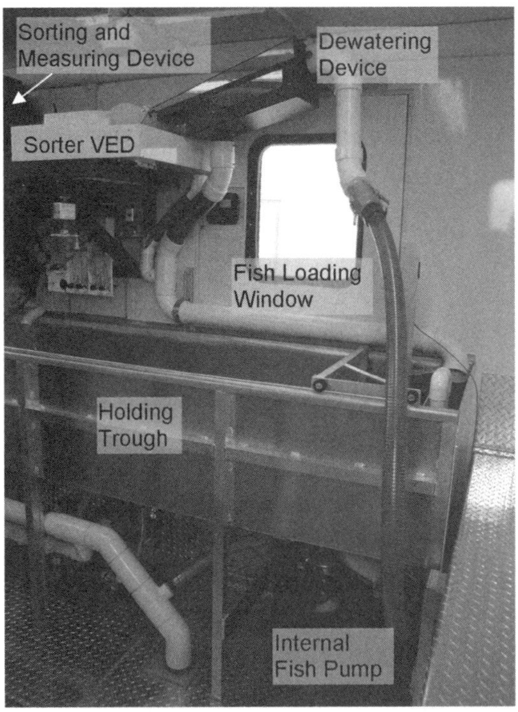

FIGURE 5. The sorter is used to sort fish into precise size categories needed for accurate adipose fin clipping and injection of Coded Wire Tags. The fish are loaded into the holding trough through a window in the side of the trailer, then lifted into the sorter Volitional Entry Device (VED) by the internal fish pump before they move through the measuring and diverting device. As the fish move into the sorter VED, some of the water is removed to keep the VED from overflowing, but the fish are never completely dewatered as they move through the AutoFish System.

Figure 6. Fish are not anesthetized or handled by humans once they are in the trailer. AutoFish System relies on Volitional Entry Devices (VED, the white trays at the top of the sorter and each marking and tagging line) to move the fish through the measuring and diverting device and the marking and tagging lines. The VED is a shallow, bright environment with upwelling water that takes advantage of the fish's natural instinct to seek a darker, more protected location, in this case, the entry to the sorter.

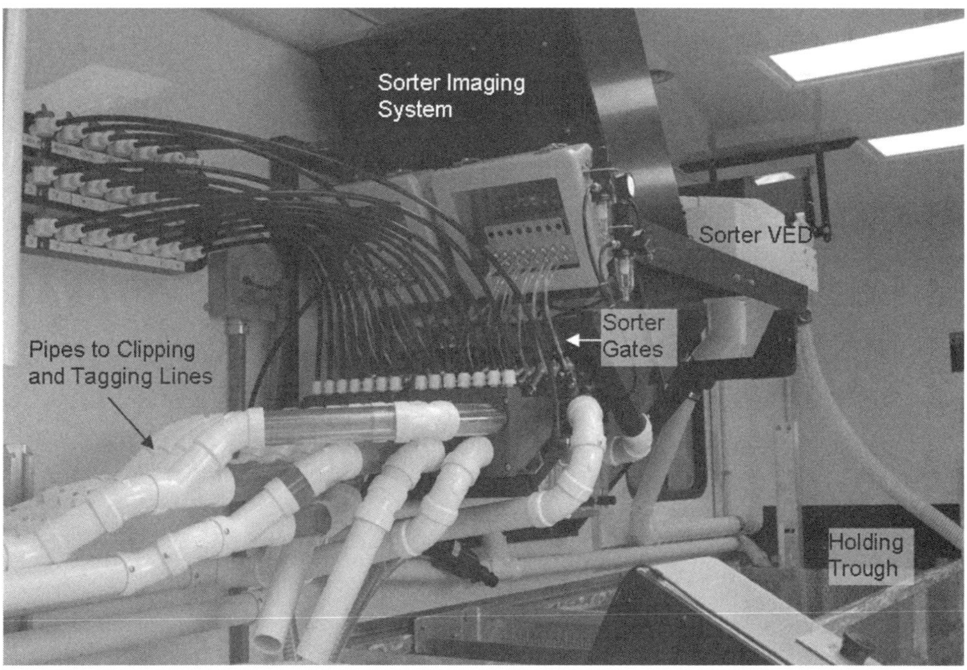

Figure 7. Fish swim into the entry to the sorting and measuring device and immediately go down one of two channels, passing cameras that measure the length of each fish. Based on the length, they are diverted to ports leading to the marking and tagging lines.

tail first or simultaneously with another fish. Based on the length, the fish are diverted to the appropriate marking and tagging line by a series of hydraulically driven gates (Figure 8) and pipes. The sorter can process 10,000 fish per hour, and thus the gates move quickly. Following fish must be separated by at least 80 milliseconds to avoid being struck by the gates. If fish are following closer than 80 milliseconds, the gate movement is momentarily suspended to avoid injuring fish.

AutoFish System is controlled and monitored by computers. Displays show what the cameras are seeing in the sorter VED (Figure 9). An imaging program automatically maintains a constant fish density by continually calculating the surface area of the VED that is covered by fish and activating the internal fish delivery pump to add more fish as needed. Computer displays also show what the sorter cameras are seeing as they calculate fish length. This information can be collected and summarized and can provide hatchery managers with valuable details about their fish. The continual computer tracking also helps the AutoFish System operator monitor and adjust the performance of the system throughout the day.

After sorting, the fish arrive in the second VED that is used to move fish into the marking and tagging lines (Figure 10). Because fish swim volitionally through the AutoFish System, they remain upright and oriented head first, the position needed for marking and tagging. However, the VED cannot ensure that the fish will enter the line one at a time. To achieve this, sensor controlled gates determine when the fish are present and allow the fish forward or hold them back as needed (Figure 11). As a fish enters the water filled gate channel, a sensor triggers the upper gate to open. When the fish is about halfway through, the upper gate begins to close to hold back the next fish in line. There is now

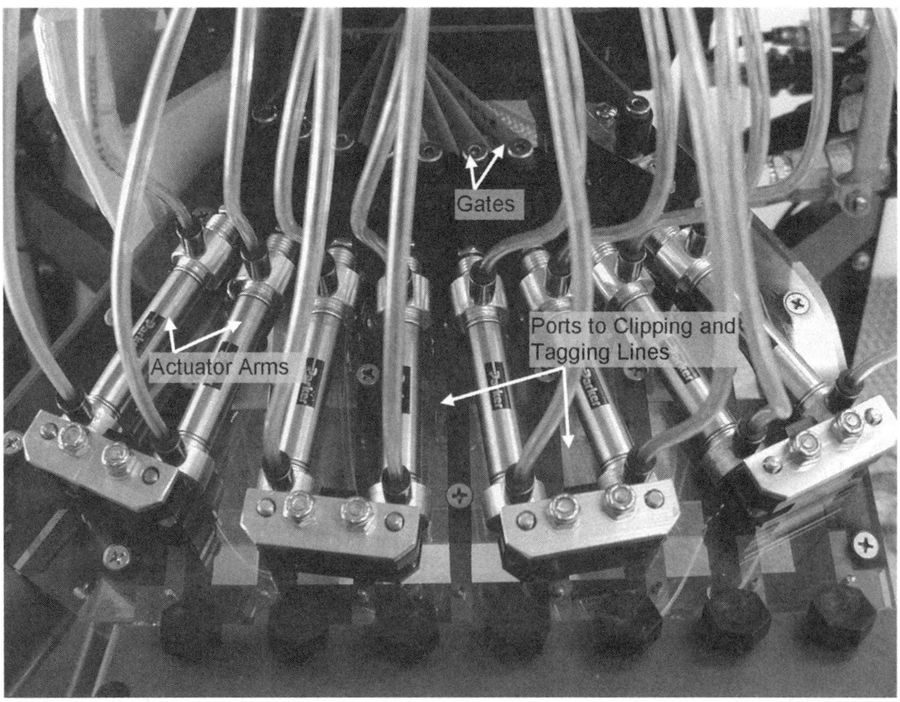

Figure 8. In the sorter, fish are diverted by a series of computer controlled gates. Based on the length of the fish, the pneumatic actuator arms move the gates from side to side to create an opening to the port that leads to the marking and tagging line that is configured for fish of that size range.

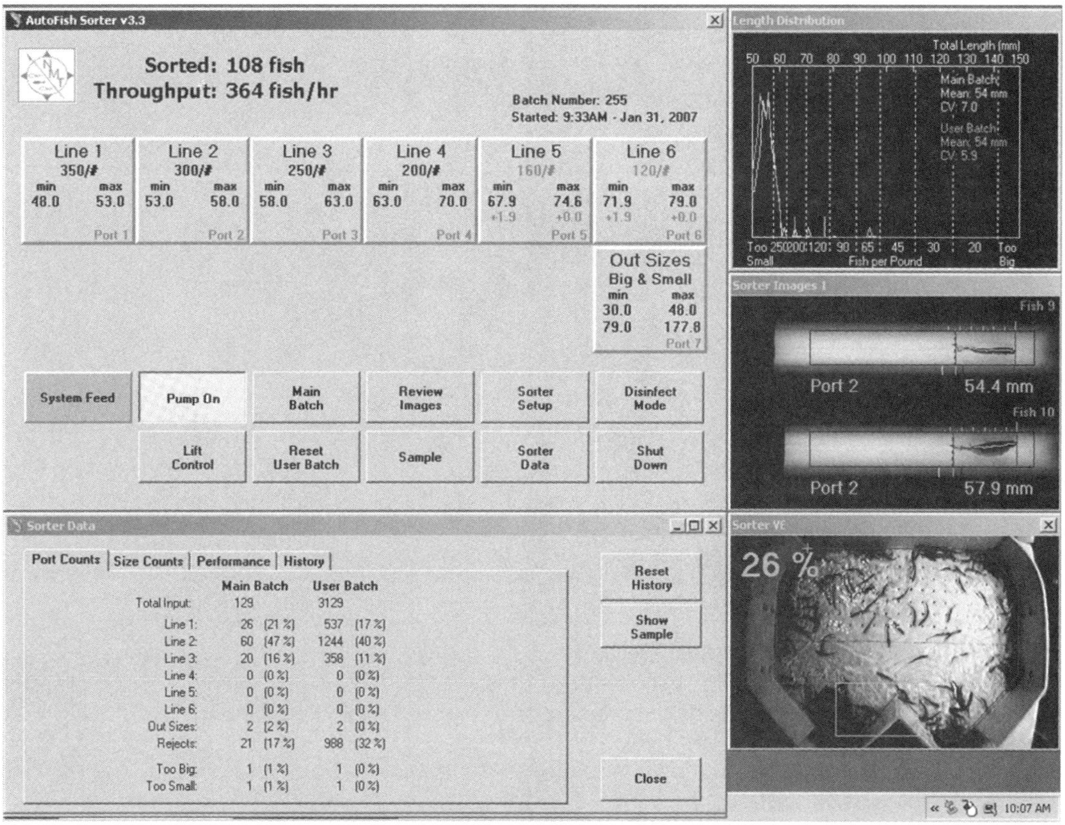

FIGURE 9. The computer programs that control AutoFish System are accessed through touch screens. This example shows the main sorter control screen during set up. The operator uses these screens to designate the size ranges for each marking and tagging line, to calibrate and adjust the measuring and sorting device, to monitor throughput and tagging accuracy, and to troubleshoot certain technical problems. As the fish are being sorted, the computer generates a length distribution curve and displays that data in the "Length Distribution" window, shown on the top right. The "Sorter Images" window (center, right) displays the last fish image for each channel, it's length, and the corresponding port. The screen also displays a live image of the sorter Volitional Entry Device (VED), shown in the "Sorter VE" window (lower right). The sorter software uses this image to estimate the density of fish within the VED and from that estimate, it determines when to open the valve and pump fish from the holding trough into the sorter VED to optimize throughput without overcrowding the VED.

a single fish at the lower gate and several fish lined up behind the upper gate. The single fish is released from the lower gate and slides into the clamping area (Figure 12) where the fish's snout enters the head mold and triggers a sensor indicating that it is properly positioned for fin clipping and tagging. Clamps close to hold the fish. The clamps are lined with foam and fitted for fish in each size-class to ensure that the fish are not injured, but that they are immobilized in water during clipping and tagging. Each fish is restrained for about 1 s.

The clipping arm is operated by a computerized machine imaging system that locates the adipose fin and moves the clipper to that location. The clipper itself is gimbaled so that it will lay flat against the fish's back to ensure that the clip angle is optimized. Computer imaging is used to verify that the adipose fin was completely removed by comparing images before and after fin clipping

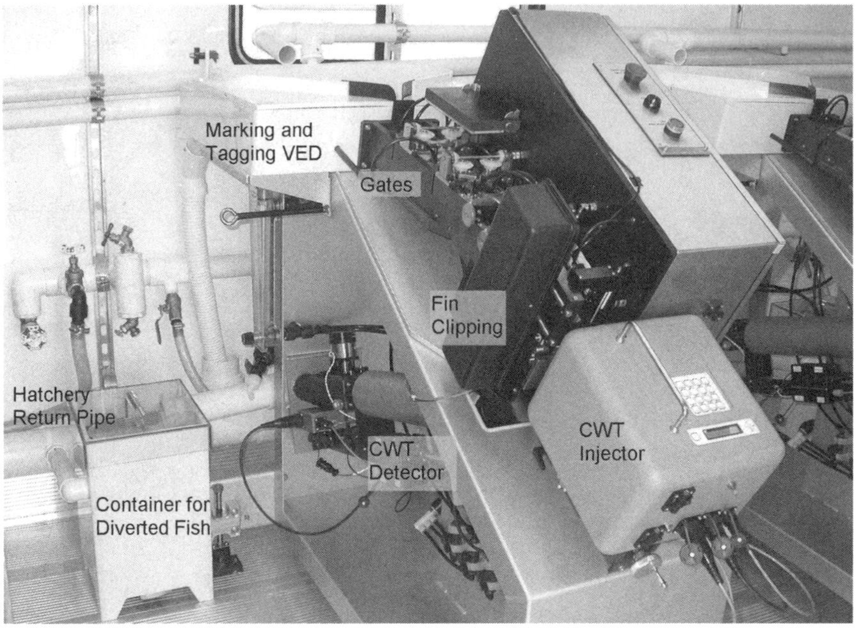

Figure 10. After sorting, the fish are marked and tagged in one of six processing lines in the trailer. The fish arrive at the top of the line in a Volitional Entry Device (VED), and move through a series of gates to a clamping mechanism that holds them during the adipose fin clip and injection of the CWT. Fish are released into the Quality Control Device which verifies the presence of a tag and receives mark verification information from the computer and diverts the fish to either the hatchery receiving pond or to the reject containment system.

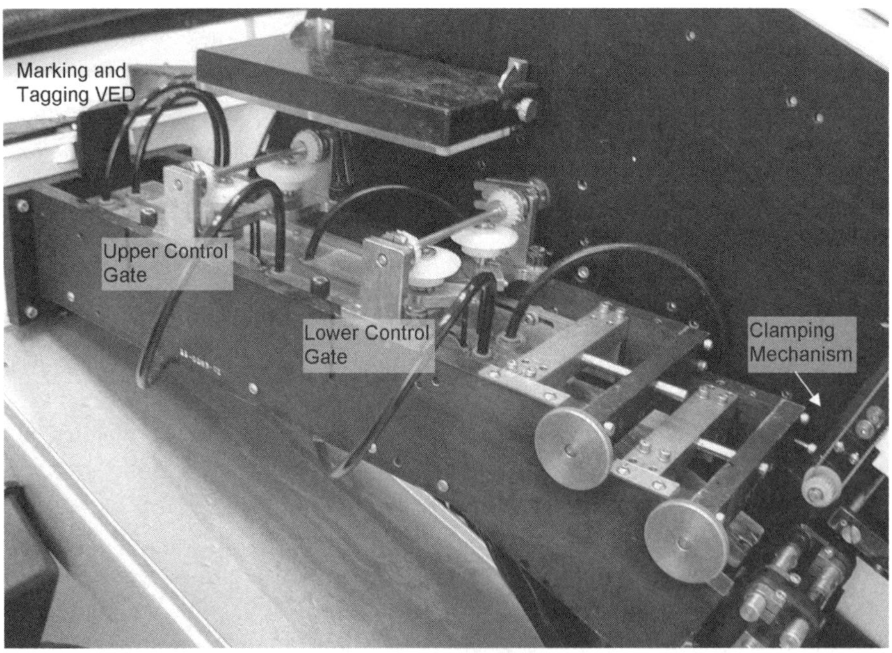

Figure 11. A gate channel mechanism controls the fish between the VE and the processing area. The internal walls and top of the gate channel are adjusted for different sized fish. There are two fish control gates within the channel. The upper gate is designed to ensure the passage of only a single fish down to the lower gate. The lower gate holds the fish until the processing area is clear and ready for the next fish. Ten infrared sensors track the location of fish in the gate channel.

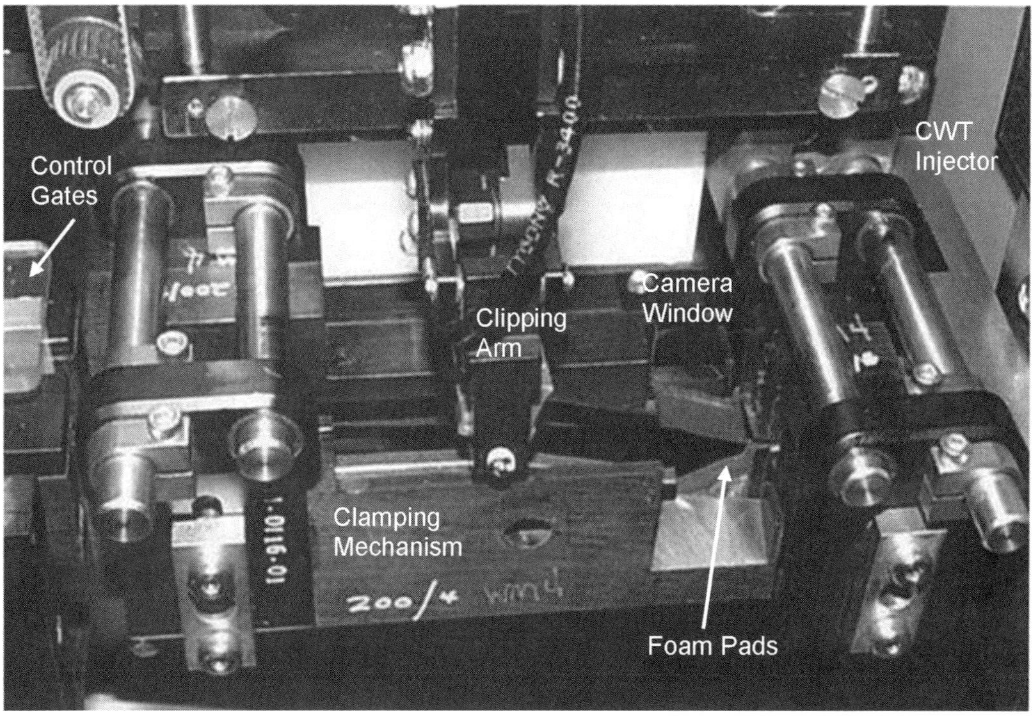

Figure 12. Fin clipping and tagging takes place at the bottom of each line (shown here without its protective cover). This section is composed of several parts that require precise adjustments so that the fish are correctly positioned. The fish are immobilized for about 1 s between foam clamps to ensure they are uninjured while the adipose fin is clipped and the tag injected. Cameras locate the adipose fin and the clipping arm moves to the fin and removes it while a CWT is simultaneously injected. A trapdoor opens and the fish are released down a chute to the Quality Control Device.

(Figure 13). A head mold ensures that the fish is properly positioned for tagging. As the arm moves down to remove the adipose fin, a CWT is simultaneously injected into the snout cartilage. After clipping and tagging, the fish are released from the clamping mechanism and slide downwards into the Quality Control Device (QCD). The QCD is a tunnel magnetometer that is used to verify the presence of a CWT, and it also receives a signal from the imaging system indicating whether or not the adipose fin was fully removed. Fish without a CWT or without an adipose clip are diverted for reprocessing, while the vast majority of the fish are returned directly to the hatchery pond through a pipe leading out the back of the trailer.

AutoFish System has certainly become an important part of salmon management in the Pacific Northwest of the United States and in California, and it now being used in the Great Lakes region of North America (Bronte et al. 2012, this volume) where extensive fin clipping and stock identification with CWT will be used to manage salmonids. The use of AutoFish System is also enabling new research into rearing practices that could increase the efficiency of hatchery programs (Tipping and Seidel 2006).

Development of Electronic Tag Detection

Before the implementation of the mass marking initiative in the Pacific Northwest of the United States, the removal of the adi-

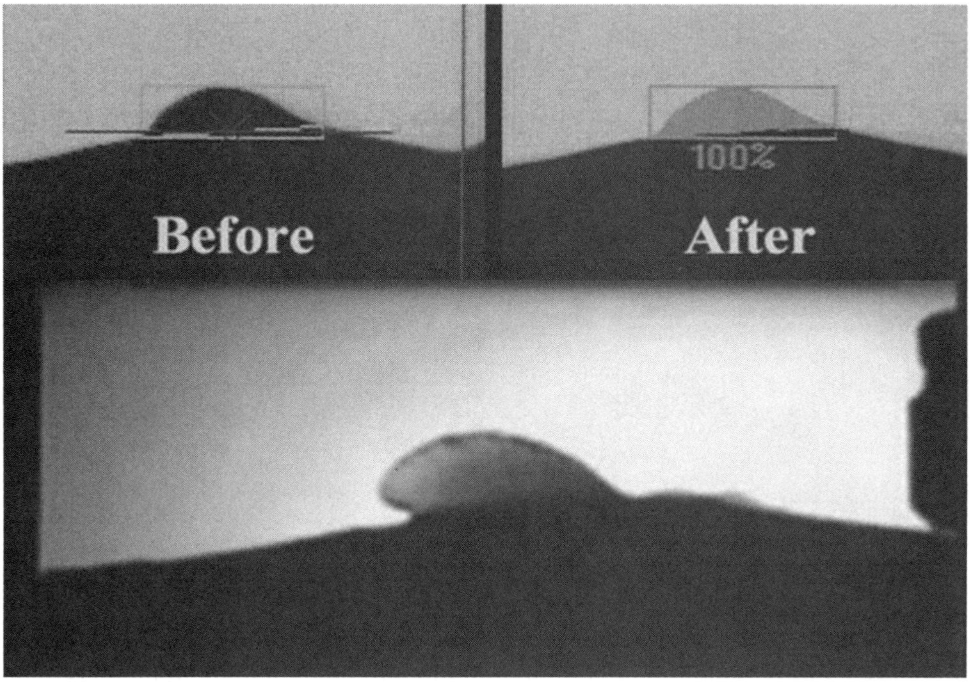

Figure 13. AutoFish System achieves high quality standards through consistent handling and rigorous quality control checks on every fish. AutoFish System uses cameras to find the position of the adipose fin (bottom photo) and to verify that it was correctly removed. Images taken before and after fin clipping (top photos) are compared ensure that the fin was correctly removed. The percentage of missing fin must meet a standard or the fish is rejected for further processing.

pose fin had been sequestered to designate a fish with a CWT. Extensive surveys of other marking and tagging methods found that the most reasonable way to visually identify (mass mark) hatchery fish was by removal of the adipose fin (PSMFC, 1992; PSC 1995). However, given the importance of the data collected by the recovery of CWT, successful implementation of the mass marking program required a new way of finding CWT in adult salmon. Northwest Marine Technology designed some new tools and modified existing ones to provide fishery samplers with a way to detect CWT electronically. The original "V" detector, first introduced in 1971, was followed by the Handheld Wand Detector and tunnel detectors (Figure 14).

The Handheld Wand Detector is a portable field detector that senses the magnetic moment of a CWT when it is moved in close proximity to the tag. Over the last 5 years, the Handheld Wand Detector has benefited from several upgrades to eliminate interference from radio waves, and to improve waterproofing and stability. These improvements resulted in a reliable field sampling tool and an increased detection range, both of which contribute to more tag recoveries.

Tunnel detectors are designed to have the entire fish passed through them and will sound an alarm or, with a gate, divert the fish when a tag is detected. Tunnels can be suitable for live or dead adult fish. Because of their size, they are most useful at sites where large numbers of fish are processed in a short time, such as at hatcheries or fish processing plants.

In addition to providing electronic tag detection for salmonids, the Handheld Wand in

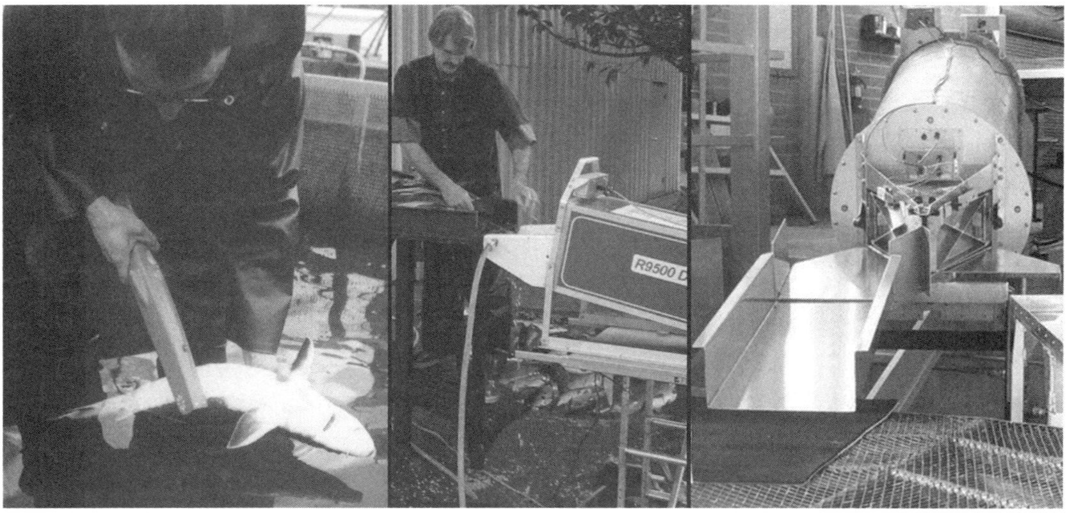

Figure 14. Electronic detection is used to sort fish with tags from those without tags in the absence of an external mark. Handheld Wand Detectors (left) are easily adapted to even remote field sampling locations while tunnel detectors (center and right) are more suitable for sites with large numbers of fish to be sorted. The new T13 Tunnel Detector (right) is the biggest detector available and can accommodate large live adult salmon.

particular has opened the way for researchers working with other species to sort tagged and untagged fish when no other way to designate a fish with a CWT was available. In addition, it allows nonlethal collection of data from CWTs when just detecting the presence or absence of CWT provides the information required. This scheme can be expanded in some cases by putting the tag in different body locations (at the base of different fins, or behind a particular scute in sturgeon, for example) to designate different experimental groups (Bonar et al. 1997). The Handheld Wand can then be used to discriminate the presence of tags at different sites without having to remove the tags.

Expansion of Species Tagged

The CWT remains the smallest tag available and much smaller animals and earlier life stages can thus be identified with CWTs than with other tags. The development of electronic detection, the versatility of the CWT, and its benign nature have resulted in the CWT being adapted for research programs involving hundreds of different species around the world. At the first symposium, peer-reviewed publications described the application of CWT to species from about 13 families. Since then, over 30 more families have been added to the list (Table 1). In general, CWTs have consistently high retention rates for the life of the animal. For example, CWTs have been recovered 24 years after tagging in sturgeon *Scaphirhynchus platorynchus*, 22 years after tagging in lake trout *Salvelinus namaycush* and 10 years after tagging in white seabass *Atractocsion nobilis*.

Coded Wire Tags are used to identify individual animals or batches of animals to accomplish numerous objectives. Because of their benign nature, CWT have been used as standards for evaluating the effects of other marks or tags when double tagging is required (e.g., Van den Avyle and Wallin 2001). Stock enhancement is on the rise as worldwide fish populations decline. However, the approach to enhancement is much more scientifically based than it was 20 years ago, and the suc-

Table 1. Coded Wire Tags have been used in a wide range of species representing more than 40 families. The wide applicability of the tags is due to the nature of the tag (its small size, inert materials, high retention, and lack of biological effect) and to the technological advances in the Coded Wire Tag system, particularly electronic tag detection. Families in which at least one species had been tagged as of the International Symposium and Educational Workshop on Fish-Marking Techniques held in Seattle, Washington, USA in 1988 are indicated with an *.

Family	Representative References (see also www.nmt.us for more references)
Acipenseridae*—sturgeons	Bordner et al. 1990; Isely and Fontenot 2000
Anguillidae—freshwater eels	Thomassen et al. 2000
Astacidae—crayfish	Isely and Eversole 1998
Carangidae—jacks	Ross and Lancaster 2002
Centrarchidae*—sunfishes	Buckmeier 2001; Fletcher et al. 1987; Heidinger and Cook 1988
Centropomidae—snooks	Brennan et al. 2005
Cichlidae*—cichlids	Klar and Parker 1986
Clupeidae*—herrings	Isely and Tomasso 1998; Richardson and Minkkinen 1995
Cyprinidae—carps and minnows	Heidinger and Cook 1988
Engraulidae*—anchovies	Leary and Murphy 1975
Escocidae—pikes	Dorsey 2004
Fundulidae—topminnows	Able et al. 2006
Galaxiidae—galaxiids	Crook and White 1995
Haemulidae—grunts	Verweij and Nagelkerken 2007
Hexagrammidae—greenlings	Buckley et al. 1994
Ictaluridae*—North American catfishes	Heidinger and Cook 1988
Latidae—lates perches	Russell and Hales 1992
Lutjanidae—snappers	Brennan et al. 2007
Majidae*—spider crabs	Bailey and Dufour 1987
Moronidae*—temperate basses	Mattson et al. 1990; Wallin and Avyle 1994
Mugilidae —mullets	Nishimoto et al. 2007
Nephropidae*—clawed lobsters	James-Pirri and Cobb 1999; Uglem and Grimsen 1995; Wickins et al. 1986
Nereidae and Glyceridae—marine worms	Joule 1983
Palinuridae—spiney lobsters	Sharp et al. 2000
Pandalidae*	West and Chew 1968
Paralichthyidae—sand flounders	Bucheister and Latour 2010
Parastacidae	de Graff 2007
Penaeidae—penaeid shrimps	Webb and Kneib 2004
Percichthyidae—temperate perches	Ingram 1993
Percidae*—perches	Heidinger and Cook 1988; Peterson and Key 1992
Petromyzontidae—lampreys	Meeuwig et al. 2007
Pleuronectidae—righteye flounders	Sulikowski et al. 2005
Polynemidae—threadfins	Friedlander and Ziemann 2003
Polyodontidae—paddlefishes	Fries 2001
Pomacentridae—damselfishes	Beukers et al. 1995
Pomatomidae—bluefishes	Able et al. 2003
Portunidae—swimming crabs	Fitz and Wiegert 1991

Table 1. Continued.

Family	Representative References (see also www.nmt.us for more references)
Salmonidae*—trouts and salmons	Blankenship 1990; Elrod and Schneider 1986; Kolari and Hirvonen 2006; Lehtonen et al. 1998
Sciaenidae*—drums and croakers	Miller and Able 2002; Willis et al. 1995
Scorpaenidae—scorpionfishes (rockfishes)	Buckley et al. 1994
Sparidae—porgies	Doupe and Lymbery 2005
Teraponidae—grunters or tigerperches	Ingram 1993
Unionidae—freshwater clams	Layzer and Heinricher 2004
Veneridae—short necked clam	Lim and Sakurai 1999

cess of these programs is often rigorously evaluated. Coded Wire Tags have played an important role in helping researchers to distinguish hatchery from natural origin animals, to estimate the contribution rates of the stocked fish to fisheries or to local populations, and to evaluate the economic feasibility of stock enhancement programs (e.g., Agnalt et al. 2004; Brennan et al. 2007; Davis et al. 2004a; Drawbridge et al. 1995). Coded Wire Tags are suitable because many stocked fish are released when they would be too small for other tags, and because large numbers of fish generally require tagging. The development of electronic detection has been important for this use as it allows researchers to sort tagged from untagged animals, in some cases providing all the information needed without having to kill the animal to retrieve the tag. In some instances, tags can be removed, and even read and reinjected without injuring the animal.

Coded Wire Tags are also widely used to investigate habitat use, movement of fish within their habitat, and growth rates (e.g., Able et al. 2006; Adlerstein et al. 2007; Fadaee et al. 2006; Miller and Able 2002; Verweij and Nagelkerken 2007; Weitkamp and Neely 2002). In these projects, researchers often require individual identification, so they use sequential CWT, and the projects are designed so that characteristics of the fish can be observed between the time of first capture and tagging and the final recapture when the tags are recovered. These projects provide growth and movement data for large numbers of animals, in contrast with projects using electronic tags to get very detailed movement data about a few animals. They also benefit from being able to tag fish that are too small for other tags, and from the ability to sort tagged and untagged fish using electronic detection. In sturgeon and other larger fish, the body location of the tag can be used to designate a batch without having to remove the tag.

Because CWT can be applied to the large numbers of juvenile fish without harm and retained until the adults return to spawn or are captured in fisheries, the tags have been used in many published studies to evaluate whether in-hatchery rearing practices affect post release survival. These include studies comparing the effect different types of rearing containers, methods of feeding the fish, time, size and location of release, etc. on juvenile survival (e.g., Doupe and Lymbery 2005; Garcia et al. 2004; Leber 1995; Tipping et al. 2003; Unwin and Quinn 1993).

Summary

In many ways, the story of the CWT is unique. We know of no other zoological tagging program that has reached the size and produced the quantity of data of the CWT program for Pacific salmon. Few technologies last as long as CWTs before they are replaced. This success is due to a combination of circumstances, of which the technological advances are an important part. Since the tags were invented in the 1960s, virtually every tool associated with using CWT has been upgraded. The invention of new tools, usually in response to the need to collect new types of data, or to adapt the system to the changing needs of ongoing programs, has made the CWT system easier to use and increased its applicability to a wider range of management problems.

Although the inherent design of the CWT has not changed since 1963, improvements in the quality of the wire, particularly in its magnetic properties and the use of decimal coding have made CWTs easier to read, easier to detect, and provided a vast code capacity.

Progressive upgrading of the tag injectors has made them fast, reliable, and accurate. The development of AutoFish System for CWT tagging has kept the CWT system working for management of Pacific salmon when the circumstances of that program changed and is arguably the most sophisticated technology used in fish management anywhere in the world. The AutoFish System was subsequently applied to salmonid management in the Great Lakes region of North America.

Significant developments in electronic tag detection have occurred since the technology was first introduced in 1971. Electronic detection is now an integral part of the management of Pacific salmon, and was a critical development for continuing that CWT program when the tagged and untagged fish could no longer be sorted visually. Electronic detection has also been important for expanding the applicability of CWT to species that cannot easily be marked to indicate that they have a CWT.

While traditional tags may have been used for a relatively long time, this does not indicate that they are outdated and no longer useful. The CWT exemplifies how some longstanding technologies can change and adapt, and will continue to change and adapt, to research and management needs.

References

Able, K. W., S. M. Hagan, and S. A. Brown. 2006. Habitat use, movement, and growth of young-of-the-year *Fundulus* spp. in southern New Jersey salt marshes: Comparisons based on tag/recapture. Journal of Experimental Marine Biology and Ecology 335:177–187.

Able, K. W., P. Rowe, M. Burlas, and D. Byrne. 2003. Use of ocean and estuarine habitats by young-of-year bluefish (*Pomatomus saltatrix*) in the New York Bight. Fishery Bulletin 101(2):201–214.

Adlerstein, S. A., E. S. Rutherford, D. Clapp, J. A. Clevenger, and J. E. Johnson. 2007. Estimating seasonal movements of Chinook salmon in Lake Huron from efficiency analysis of coded wire tag recoveries in recreational fisheries. North American Journal of Fisheries Management 27:792–803.

Agnalt, A.-L., K. E. Jørstad, T. Kristiansen, E. Nøstvold, E. Farestveit, H. Næss, O.I. Paulsen, and T. Svåsand. 2004. Enhancing the European lobster (*Homarus gammarus*) stock at Kvitsoy Islands: perspectives on rebuilding Norwegian stocks. Pages 415–426 *in* K. M. Leber, S. Kitada, H. L. Blankenship, and T. Svåsand, editors. Stock Enhancement and Sea Ranching: Developments, Pitfalls and Opportunities. Blackwell Scientific Publications Publishing, Oxford, U.K.

Bailey, R. F. J., and R. Dufour. 1987. Field use of an injected ferromagnetic tag on the snow crab (*Chionoecetes opilio* O-Fab). Journal du Conseil International de l'Exploration de la Mer 43:237–244.

Beukers, J. S., G. P. Jones, and R. M. Buckley. 1995. Use of implant microtags for studies of populations of small reef fish. Marine Ecology Progress Series 125:61–66.

Blankenship, H. L. 1990. Effects of time and fish size on coded wire tag loss from Chinook and coho salmon. Pages 237–243 *in* N. C. Parker, A. E. Giorgi, R. C. Heidinger, D. B. Jester, Jr., E. D. Prince and G. A. Winans, editors. Fish-marking techniques. American Fisheries Society, Symposium 7, Bethesda, Maryland.

Blankenship, H. L., and D. A. Thompson. 2003. The effect of 1.5-length and double-length coded wire tags on coho salmon survival, growth, homing, and electronic detection. North American Journal of Fisheries Management 23:60–65.

Bonar, S. A., J. Pahutski, B. D. Bolding, D. Fletcher, and M. Divens. 1997. Survival and growth and channel catfish stocked in Washington Lakes. North American Journal of Fisheries Management 17:773–778.

Bordner, C. E., S. I. Doroshov, D. E. Hinton, R. E. Pipkin, R. B. Fridley, and F. Haw. 1990. Evaluation of marking techniques for juvenile and adult white sturgeons reared in captivity. Pages 293–303 *in* N. C. Parker, A. E. Giorgi, R. C. Heidinger, D. B. Jester, Jr., E. D. Prince and G. A. Winans, editors. Fish-marking techniques. American Fisheries Society, Symposium 7, Bethesda, Maryland.

Brennan, N. P., K. M. Leber, and B. R. Blackburn. 2007. Use of coded-wire and visible implant elastomer tags for marine stock enhancement with juvenile red snapper Lutjanus campechanus. Fisheries Research 83:90–97.

Brennan, N. P., K. M. Leber, H. L. Blankenship, J. M. Ransier, and R. DeBruler. 2005. An evaluation of coded wire and elastomer tag performance in juvenile common snook under field and laboratory conditions. North American Journal of Fisheries Management 25(2):437–445.

Bronte, C. R., K. A. Walch, J. M. Dettmers, M. Gaden, M. J. Connerton, M. Daniels, and T. J. Newcomb. 2012. Pages 27–42 *in* J. McKenzie, B. Parsons, A. C. Seitz, R. K. Kopf, M. Mesa, and Q. Phelps, editors. Advances in fish tagging and marking technology. American Fisheries Society, Symposium 76, Bethesda, Maryland.

Buchheister, A., and R. J. Latour. 2010. Turnover and fractionation of carbon and nitrogen stable isotopes in tissues of a migratory coastal predator, summer flounder (*Paralichthys dentatus*). Canadian Journal of Fisheries and Aquatic Sciences 67:445–461.

Buckley, R. M., J. E. West, and D. C. Doty. 1994. Internal micro-tag systems for marking juvenile reef fishes. Bulletin of Marine Science 55:848–857.

Buckmeier, D. L. 2001. Coded wire tag insertion sites for small fingerling black bass. North American Journal of Fisheries Management 21:696–698.

Crook, D. A., and R. W. G. White. 1995. Evaluation of subcutaneously implanted visual implant tags and coded wire tags for marking and benign recovery in a small scaleless fish, *Galaxias truttaceus* (Pisces: Galaxiidae). Marine and Freshwater Research 46:943–946.

Davis, J. L. D., and A. C. Young-Williams, R. Aguilar, B. L. Carswell, M. R. Goodison, A. H. Hines, M. A. Kramer, Y. Zohar, O. Zmora. 2004a. Differences between hatchery-raised and wild blue crabs: Implications for stock enhancement potential. Transactions of the American Fisheries Society 133:1–14.

Davis, J. L. D., A. C. Young-Williams, A. H. Hines, and O. Zmora. 2004b. Comparing two types of internal tags in juvenile blue crabs. Fisheries Research 67:265–274.

de Graff, M. 2007. Tag retention, survival and growth of marron *Cherax tenuimanus* (Crustacea : Decapoda) marked with coded micro wire tags. Marine and Freshwater Research 58:1044–1047.

Dorsey, L. G. 2004. Retention of coded wire tags by age-0 Muskellunge. Journal of Freshwater Ecology 19:333–337.

Doupe, R. G., and A. J. Lymbery. 2005. Genetic covariation in production traits of sub-adult black bream *Acanthopagrus butcheri* after grow-out. Aquaculture Research 36:1128–1132.

Drawbridge, M. A., D. B. Kent, M. A. Shane, and R. F. Ford. 1995. The assessment of marine stock enhancement in southern California: A case study involving the white seabass. Pages 568–569 *in* H. L. Schramm, Jr. and R. G. Piper, editors. Uses and effects of cultured fishes in aquatic ecosystems. American

Fisheries Society, Symposium 15, Bethesda, Maryland.

Elrod, J. H., and C. P. Schneider. 1986. Evaluation of coded wire tags for marking lake trout *Salvelinus namaycush*. North American Journal of Fisheries Management 6:264–271.

Fadaee, B., M. Pourkazemi, M. Tavakoli, H. Joushideh, M. R. B. Khoshghalb, M. R. Hosseini, H. Abdulhay. 2006. Tagging and tracking juvenile sturgeons in shallow waters of the Caspian Sea (less than 10 m depth) using CWT (Coded Wire Tags) and barbel incision. Journal of Applied Ichthyology 22:160–165.

Fitz, H. C., and R. G. Wiegert. 1991. Tagging juvenile blue crabs, Callinectes sapidus, with microwire tags—retention, survival, and growth through multiple molts. Journal of Crustacean Biology 11:229–235.

Fletcher, D. H., F. Haw, and P. K. Bergman. 1987. Retention of coded wire tags implanted into cheek musculature of largemouth bass. North American Journal of Fisheries Management 7:436–439.

Friedlander, A. M., and D. A. Ziemann. 2003. Impact of hatchery releases on the recreational fishery for Pacific threadfin (*Polydactylus sexfilis*) in Hawaii. Fishery Bulletin 101:32–43.

Fries, J. N. 2001. Retention of coded wire tags in four locations in juvenile paddlefish. North American Journal of Fisheries Management 21:962–966.

Garcia, A. P., W. P. Connor, D. J. Milks, S. J. Rocklage, and R. K. Steinhorst. 2004. Movement and spawner distribution of hatchery fall chinook salmon adults acclimated and released as yearlings at three locations in the Snake River basin. North American Journal of Fisheries Management 24:1134–1144.

Hand, D. M., W. R. Brignon, D. E. Olson, and J. Rivera. 2010. Comparing two methods used to mark juvenile Chinook salmon: automated and manual marking. North American Journal of Aquaculture 72:10–17.

Heidinger, R., and S. B. Cook. 1988. Use of coded wire tags for marking fingerling fishes. North American Journal of Fisheries Management 8:268–272.

Ingram, B. A. 1993. Evaluation of coded wire tags for marking fingerling golden perch, *Macquaria ambigua* (Percichthyidae), and silver perch, *Bidyanus bidyanus* (Teraponidae). Australian Journal of Marine and Freshwater Research 44:817–824.

Isely, J. J., and A. G. Eversole. 1998. Tag retention, growth, and survival of red swamp crayfish *Procambarus clarkii* marked with coded wire tags. Transactions of the American Fisheries Society 127:658–660.

Isely, J. J., and Q. C. Fontenot. 2000. Retention of coded wire tags in juvenile shortnose sturgeon. North American Journal of Fisheries Management 20:1040–1043.

Isely, J. J., and J. R. Tomasso. 1998. Estimating fish abundance in a large reservoir by mark-recapture. North American Journal of Fisheries Management 18:269–273.

James-Pirri, M. J., and J. S. Cobb. 1999. Influence of coded micro-wire tags on postlarval of lobsters *Homarus americanus* behavior. Marine and Freshwater Behaviour and Physiology 32:255–259.

Jefferts, K. B., H. F. Fiscus, and P. K. Bergman. 1963. A coded wire identification system for macro-organisms. Nature (London) 198:460–462.

Johnson, J. K. 1990. Regional overview of coded wire tagging of anadromous salmon and steelhead in northwest America. Pages 782–816 *in* N. C. Parker, A. E. Giorgi, R. C. Heidinger, D. B. Jester, Jr., E. D. Prince and G. A. Winans, editors. Fish-marking techniques. American Fisheries Society, Symposium 7, Bethesda, Maryland.

Joule, B. J. 1983. An effective method for tagging marine polychaetes. Canadian Journal of Fisheries and Aquatic Sciences 40:540–541.

Klar, G. T., and N. C. Parker. 1986. Marking fingerling striped bass and blue tilapia with coded wire tags and microtaggants. North American Journal of Fisheries Management 6:439–444.

Kolari, I., and E. Hirvonen. 2006. Long-term retention of coded wire tags in juvenile Arctic charr *Salvelinus alpinus*. Fisheries Management and Ecology 13(3):143–148.

Layzer, J. B., and J. R. Heinricher. 2004. Coded wire tag retention in ebonyshell mussels *Fusconaia ebena*. North American Journal of Fisheries Management 24:228–230.

Leary, D. F., and G. I. Murphy. 1975. Successful method for tagging small, fragile engraulid, *Stolephorus purpureus*. Transactions of the American Fisheries Society 104:53–55.

Leber, K. M. 1995. Significance of fish size-at-release on enhancement of striped mullet fisheries in Hawaii. Journal of the World Aquaculture Society 26:143–153.

Lehtonen, H., H. Peltonen, O. Heikinheimo, E. Saarijarvi, K. Saulamo, M. Vinni, and T. Nurmio. 1998. Application of coded microtags to study growth rates of stocked sympatric whitefish (*Coregonus lavaretus* (L.)) forms. Fisheries Research 39:9–15.

Lim, B. K., and N. Sakurai. 1999. Coded wire tagging of the short necked clam *Ruditapes philippinarum*. Fisheries Science 65:163–164.

Mattson, M. T., B. R. Friedman, D. J. Dunning, and Q. E. Ross. 1990. Magnetic tag detection efficiency for Hudson River striped bass. Pages 267–271 *in* N. C. Parker, A. E. Giorgi, R. C. Heidinger, D. B. Jester, Jr., E. D. Prince and G. A. Winans, editors. Fish-Marking Techniques. American Fisheries Society, Symposium 7, Bethesda, Maryland.

Meeuwig, M. H., A. L. Puls, and J. M. Bayer. 2007. Survival and tag retention of Pacific lamprey larvae and macrophthalmia marked with coded wire tags. North American Journal of Fisheries Management 27:96–102.

Miller, M. J., and K. W. Able. 2002. Movements and growth of tagged young-of-the-year Atlantic croaker (*Micropogonias undulatus* L.) in restored and reference marsh creeks in Delaware Bay, USA. Journal of Experimental Marine Biology and Ecology 267:15–33.

Mobrand, L. E., J. Barr, H. L. Blankenship, D. E. Campton, T. T. P. Evelyn, T. A. Flagg, C. V. W. Mahnken, L. W. Seeb, P. R. Seidel, and W. W. Smoker. 2005. Hatchery reform in Washington State. Fisheries 30(6):11–23.

Nandor, G., J. R. Longwill and D. L. Webb. 2010. Overview of the Coded Wire Tag program in the greater Pacific region of North America. *In* K. S. Wolf, and J. S. O'Neal, editors. PNAMP Special Publication: Tagging, Telemetry and Marking Measures for Monitoring Fish Populations—a compendium of new and recent science for use in informing technique and decision modalities: Pacific Northwest Aquatic Monitoring Partnership Special Publication 2010–002.

Nishimoto, R. T., T. E. Shimoda, and L. K. Nishiura. 2007. Mugilids in the Muliwai: a tale of two mullets. Bishop Museum Bulletin in Cultural and Environmental Studies 3:143–156.

PSC (Pacific Salmon Commission). 1995. Report of the ad-hoc selective fishery evaluation committee to the Pacific Salmon Commission. Pacific Salmon Commission, Vancouver, BC.

PSMFC (Pacific States Marine Fisheries Commission). 1992. Mass marking anadramous salmonids; techniques, options and compatibility with the coded wire tag system. Pacific States Marine Fisheries Commission, Portland, Oregon.

Peterson, M. S., and J. P. Key. 1992. Evaluation of hand-tagging juvenile walleyes with binary-coded wire microtags. North American Journal of Fisheries Management 12:814–818.

Richardson, B. M., and S. P. Minkkinen. 1995. American shad handling, rearing, and marking trials. Pages 557–559 *in* H. L. Schramm, Jr. and R. G. Piper, editors. Uses and effects of cultured fishes in aquatic ecosystems. American Fisheries Society, Symposium 15, Bethesda, Maryland.

Ross, S. W., and J. E. Lancaster. 2002. Movements and site fidelity of two juvenile fish species using surf zone nursery habitats along the southeastern North Carolina coast. Environmental Biology of Fishes 63:161–172.

Russell, D. J., and P. W. Hales. 1992. Evaluation of techniques for marking juvenile barramundi, *Lates calcarifer* (Bloch), for stocking. Aquaculture and Fisheries Management 23(6):691–699.

Sharp, W. C., W. A. Lellis, M. J. Butler, W. F. Herrnkind, J. H. Hunt, M. Pardee-Woodring, and T. R. Matthews. 2000. The use of coded microwire tags in mark-recapture studies of juvenile Caribbean spiny lobster, *Panulirus argus*. Journal of Crustacean Biology 20:510–521.

Sulikowski, J. A., E. A. Fairchild, N. Rennels, W. H. Howell, and P. C. W. Tsang. 2005. The effects of tagging and transport on stress in juvenile winter flounder, *Pseudopleuronectes americanus*: Implications for successful stock enhancement. Journal of the World Aquaculture Society 36:148–156.

Thomassen, S., M. I. Pedersen, and G. Holdensgaard. 2000. Tagging the European eel *Anguilla anguilla* (L.) with coded wire tags. Aquaculture 185:57–61.

Tipping, J. M., A. L. Gannam, T. D. Hillson, and J. B. Poole. 2003. Use of size for early detection of juvenile hatchery steelhead destined to be precocious males. North American Journal of Aquaculture 65:318–323.

Tipping, J. M., and P. R. Seidel. 2006. Reducing length variation in juvenile steelhead and coho salmon through the use of the AutoFish tagging trailer system. North American Journal of Aquaculture 68:81–85.

Uglem, I., and S. Grimsen. 1995. Tag retention and survival of juvenile lobsters (*Homarus gammarus* (L)) marked with coded wire tags. Aquaculture Research 15:837–841.

Unwin, M. J., and T. P. Quinn. 1993. Homing and straying patterns of Chinook salmon (*Oncorhynchus-tshawytscha*) from a New Zealand hatchery—spatial distribution of strays and effects of release date. Canadian Journal of Fisheries and Aquatic Sciences 50:1168–1175.

Van den Avyle, M. J., and J. E. Wallin. 2001. Retention of internal anchor tags by juvenile striped bass. North American Journal of Fisheries Management 21:656–659.

Vander Haegen, G. E., H. L. Blankenship, A. Hoffmann, and D. A. Thompson. 2005. The effects of adipose fin clipping and coded wire tagging on the survival and growth of spring Chinook salmon. North American Journal of Fisheries Management 25:1161–1170.

Verweij, M. C., and I. Nagelkerken. 2007. Short and long-term movement and site fidelity of juvenile Haemulidae in back-reef habitats of a Caribbean embayment. Hydrobiologia 592:257–270.

Wallin, J., and M. J. V. D. Avyle. 1994. Retention of coded wire tags by juvenile striped bass. 1994 Proceedings of the Annual Conference of the Southeastern Association of Fish and Wildlife Agencies 550–554.

Webb, S., and R. T. Kneib. 2004. Individual growth rates and movement of juvenile white shrimp (*Litopenaeus setiferus*) in a tidal marsh nursery. Fishery Bulletin (Seattle) 102:376–388.

Weitkamp, L., and K. Neely. 2002. Coho salmon (*Oncorhynchus kisutch*) ocean migration patterns: insight from marine coded-wire tag recoveries. Canadian Journal of Fisheries and Aquatic Sciences 59:1100–1115.

West, W. Q. B., and K. K. Chew. 1968. Application of the Bergman-Jefferts tag on the spot shrimp, *Pandalus platyceros* Brandt. Proceedings of the National Shellfish Association 58:93–100.

Wickins, J. F., T. W. Beard, and E. Jones. 1986. Microtagging cultured lobsters, *Homarus gammarus* (L.) for stock enhancement trials. Aquaculture and Fisheries Management 1986:259–265.

Willis, S. A., W. W. Falls, C. W. Dennis, D. E. Roberts, and P. G. Whitchurch. 1995. Assessment of season of release and size at release on recapture rates of hatchery-reared red drum. Pages 354–365 *in* H. L. Schramm, Jr. and R. G. Piper, editors. Uses and effects of cultured fishes in aquatic ecosystems. American Fisheries Society, Symposium 15, Bethesda, Maryland.

American Fisheries Society Symposium 76:63–75, 2012

Correcting Visual Implant Elastomer Tag Identification to Better Evaluate Adult Atlantic Salmon Returns from Multiple Stocking Sites in the Penobscot River, Maine, USA

Christine A. Lipsky*
National Marine Fisheries Service, Northeast Fisheries Science Center
17 Godfrey Drive, Suite 1, Orono, Maine 04473, USA

Edward M. Hastings†
Integrated Statistics
16 Summer Street, Woods Hole, Massachusetts 02543, USA

Russell Brown
National Marine Fisheries Service, Northeast Fisheries Science Center
166 Water Street, Woods Hole, Massachusetts 02543, USA

Richard B. Dill
Maine Department of Marine Resources, Bureau of Sea-Run Fisheries and Habitat
650 State Street, Bangor, Maine 04401, USA

Abstract.—Atlantic salmon *Salmo salar* smolts are stocked in the Penobscot River, Maine to supplement declining populations. Since 2000, approximately 550,000 hatchery smolts are annually released into the river, of which 32% of these (175,000) were marked with visual implant elastomer (VIE) tags indicating release time and location. Our study found that tag colors of adult returns were often misread due to confusion between green and yellow marks and between red and pink marks. Using data collected on mark readings in a controlled hatchery environment, we corrected for tag identification errors made in the field, thereby allowing adjustments to be made in the marked returns to accurately assess the number of returns from multiple stocking groups. When adult returns from multiple stocking groups were compared, significant differences were detected between stocking groups, including a three-fold difference between two of the stocking groups. Hence, a simple change in smolt stocking locations and times may be of value in the recovery of the Penobscot River salmon population.

*Corresponding author: Christine.Lipsky@noaa.gov
†Deceased

Introduction

The U.S. Fish and Wildlife Service annually stocks approximately 550,000 one-year-old Atlantic salmon *Salmo salar* smolts into the Penobscot River, Maine, USA. These fish are stocked in several locations throughout the watershed, which vary in both distance to the ocean and number of dams between the stocking site and the ocean. In 2000, NOAA's National Marine Fisheries Service initiated a coordinated smolt marking study to investigate how stocking location and stocking timing affects smolt migration, nearshore marine growth rates, and adult returns to the Penobscot River.

To differentiate fish from the various stocking groups when adult fish return to the river, smolts were batch marked using visual implant elastomer (VIE) tags (Northwest Marine Technology Inc., Shaw Island, Washington). VIE tags are composed of a two-part silicone based material (color elastomer and curing agent) which is mixed as a liquid immediately prior to use, and is then injected into either the adipose tissue or fin rays of a fish. The liquid cures into a pliable bio-compatible solid.

VIE tags have become a very popular way to mark large batches of fish. The tags are relatively inexpensive and negligibly affect fish behavior, growth, and survival (Olsen and Vollestad 2001). VIE tags are available in six fluorescent colors (NMT 2008), and can be applied to various body locations on the fish. This allows researchers to use combinations of color and body location to create multiple distinct marks. VIE tags have been implanted in various body locations in salmonids (Olsen and Vollestad 2001; FitzGerald et al. 2004; Walsh and Winkelman 2004), however, tag retention studies for periods longer than 24 months have only been undertaken for two tagging locations: adipose eye tissue and ventral surface of the mandible (Bailey et al. 1998; Close and Jones 2002; FitzGerald et al. 2004). These studies suggest visibility and retention are better in the adipose eye tissue than in the jaw (Close and Jones 2002; FitzGerald et al. 2004).

A larger number of colors are needed to mark multiple batches of fish when tagging is limited to only one or two body locations. However, a potential problem in using several different colors is that similar colors may be confused (Willis and Babcock 1998; FitzGerald et al. 2004; Curtis 2006). The ability to differentiate colors can vary due to inconsistent lighting conditions (either too bright or too dark), growth of tissue over the VIE mark, observer distance from the tag, and time between tagging and resighting.

The objectives of this study were to determine whether mark identification accuracy differed depending on the environment in which the mark was read, and to develop a cross-classification correction matrix which could be used to adjust numbers of returning marked salmon in order to account for differences in correct color identification between groups. The study was also designed to compare adjusted returns from several different stocking groups to identify stocking strategies that optimize returns.

Methods

Visual implant elastomer marking

We marked age-one Atlantic salmon smolts with a site-specific and release timing-specific VIE mark. Five fluorescent colors (green, red, orange, pink and yellow), and two body locations (right and left adipose eye tissue) were used (Table 1). While these colors are generally visible to the naked eye, their visibility is improved by using a blue LED light and amber filter glasses. An adipose fin clip was applied as a secondary mark to all the VIE-marked smolts to further assist

Table 1. VIE marks by cohort and stocking group. Early releases were stocked over a period of three days in April, and late releases were stocked two to three weeks later in late April and/or May.

Smolt cohort	Stocking Group	VIE tag color	VIE tag location
2000, 2001	Mattawamkeag–early	Green	Right eye
	Mattawamkeag–late	Red	Right eye
	Milo–early	Green	Left eye
	Milo–late	Red	Left eye
	Howland–early	Orange	Left eye
	Howland–late	Yellow	Left eye
	West Enfield (smolt ponds)	Yellow	Right eye
2002, 2003, 2004	Mattawamkeag–early	Green	Right eye
	Mattawamkeag–late	Red	Right eye
	Milo–early	Green	Left eye
	Milo–late	Red	Left eye
	Howland–early	Orange	Left eye
	Howland–late	Orange	Right eye
	West Enfield (smolt ponds)	Pink	Right eye
2005	Milo–early	Green	Left eye
	Milo–late	Red	Left eye
	Howland–early	Green	Right eye
	Howland–late	Red	Right eye
	Additional marked groups	Orange	Right eye
			Left eye
			Both eyes
		Pink	Right eye
			Left eye

in identifying these fish when sampled as returning adults.

To reduce stress prior to marking, fish were anesthetized with a buffered tricaine methanesulfonate (MS-222) solution, and handling of the fish was minimized. The adipose fin was excised, and an air-driven injection system having a 29-gauge needle was used to apply the VIE mark. As the elastomer was injected, the needle was withdrawn, stopping before the needle bevel reached the entry point. Excess elastomer was lightly wiped off to avoid any material trailing from the injection site and curing outside the fish's body. Counts of marked fish were recorded on the injector and were confirmed with hatchery fish counts. We marked all smolts with fork length equal to or greater than 150 mm, while fish smaller than 150 mm were not marked and were returned to the corresponding hatchery pool.

After marking, all fish were released into designated hatchery pools. We sampled at least one percent of the smolts from each pool at the conclusion of marking operations to assess mark quality and mark retention. Short-term retention of VIE marks was categorized as good (mark could easily be identified), poor (mark contained some pigment but was barely visible), or absent (no sign of elastomer). Daily mortality checks were performed on the marked fish until they were released several weeks post marking, and the number of mortalities was subtracted from the marked totals. The number of mortalities did not differ between groups, and the majority of mortalities occurred within a week of marking.

Atlantic salmon smolt stocking

VIE marked smolts were stocked annually in the Penobscot River each spring from 2000 through 2005. Smolts were stocked in groups that differed by release location and release timing. In 2000–2004, six of the seven groups were stocked at three different locations [Mattawamkeag (142 km from the mouth of the Penobscot River [river kilometer 142]), Milo (river kilometer 120), and Howland (river kilometer 99)] during two release periods (early and late) in the spring smolt emigration (Figure 1). Early releases were stocked over a period of three days in April, and late releases were stocked two to three weeks later in late April and/or May. The seventh group of smolts was stocked into riverside holding ponds in West Enfield (101 km from Penobscot Bay) (Figure 1), where smolts were able to migrate volitionally from these "smolt ponds." Smolts remaining in the ponds were flushed out when water temperatures were expected to be above 9.5°C for two to three days. Additional groups of VIE-marked smolts were released into the Penobscot River at various locations during these years, but adult returns from these groups were not analyzed in this study.

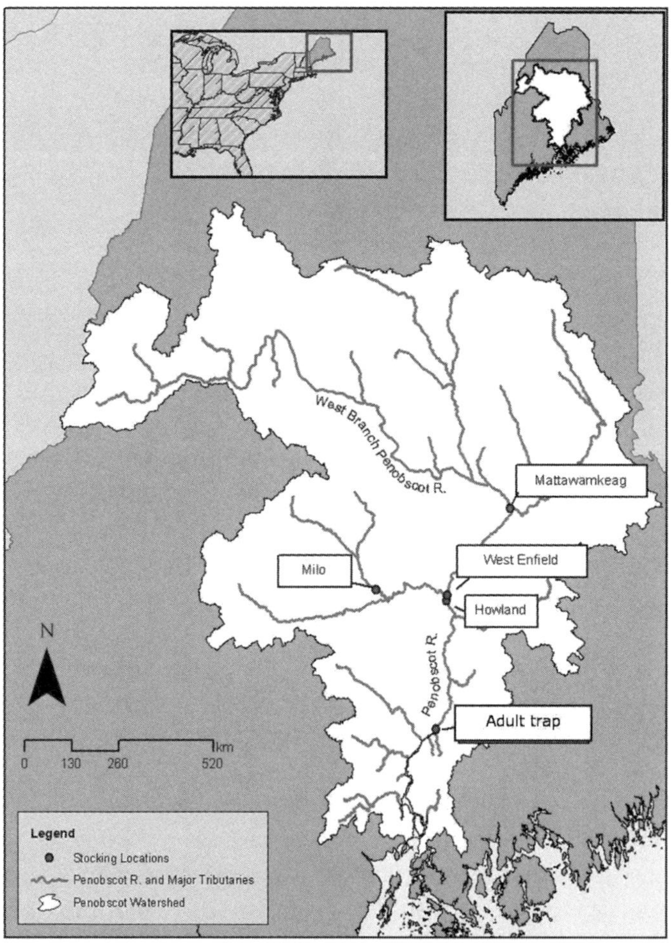

Figure 1. Atlantic salmon *Salmo salar* smolt stocking locations in the Penobscot River, Maine, USA.

In 2005, VIE marked smolts that are being analyzed in this study were released in four identifiable groups: one group during each of the two release periods at the Howland site and one group during each of the two release periods at the Milo site. The Mattawamkeag site and the West Enfield smolt pond site were not used in 2005.

Adult mark identification

Hatchery smolts released into the Penobscot River returned as adults after one year (so-called grilse return fish) or after two years. These returning adults were captured from May to November in a fishway trap located at the Veazie Dam in Veazie, Maine. Fish were retrieved daily from the trap, where they were measured and inspected for marks and tags. VIE-marked fish were initially identified by their adipose fin clip, and then examined for the presence of VIE noting color and location. A portion of the returning adults captured at the Veazie Dam were tagged with passive integrated transponders (PIT tags) for identification purposes and brought to the Craig Brook National Fish Hatchery (CBNFH) for use as broodstock. The remaining marked adults were released upstream of the dam (and will be referred to as nonbroodstock VIE marked fish).

Detection of VIE marks at the Veazie trap site was difficult during inclement weather because of poor lighting and this might have negatively biased recovery results. Other factors compounding the recovery of VIE tags at Veazie are that fish were not anesthetized at the trap and there may be growth of tissue over the VIE mark. We believed, therefore, that detection rates could improve in an environment with better lighting and handling conditions. Because a portion of adult returns to the Veazie trap were brought to the hatchery as broodstock, an opportunity existed to evaluate the accuracy of mark identifications made at Veazie by using the improved viewing conditions at CBNFH.

In 2002 (the first year of returning Penobscot two-sea winter [2SW] VIE marked salmon), VIE mark inspection activities occurred at CBNFH and consisted of inspecting marks in adipose clipped broodstock salmon during hatchery spawning operations in November. Inspecting marks during spawning operations limited the number of times the fish had to be handled and anesthetized. Fish were anesthetized with MS-222 and were examined in covered Swedish pools. For quality control, blind duplicate checks were used for a total of three to four observations per fish during this evaluation.

We observed that some VIE marks had become obscured by pigmented tissue or were too small to see with the naked eye. When a mark was not readily visible, a blue seven-LED flashlight and amber filter glasses were used to aid in detection, along with a mark standard. The marks observed on adipose-clipped fish at CBNFH were cross-referenced (using PIT tag number) with corresponding observations at the Veazie trap.

After we analyzed the 2002 data, efforts were made to enhance mark detection at the Veazie site in several ways. First, a VIE mark summary was provided to crews tending the adult trap so they were aware of all potential marks. Second, a black cover was used to provide a darkened environment for detecting marks and identifying mark color. Third, protocols were applied using the blue 7-LED flashlight, amber filter glasses, and a mark standard. Finally, color-blind crew staffings were avoided.

VIE checks at CBNFH were conducted annually from 2003 to 2007 using the methods implemented in 2002, except that the fish were examined in the hatchery during disease screening in September instead of during the spawning operations in November. In 2005, 2006, and 2007, questionable VIE marks were removed with a scalpel and tweezers when possible and, if color of the removed mark could not be determined with the naked eye, the mark was examined under a microscope.

The color determinations made at CBNFH were assumed to be less biased as the marks were read under more optimal sighting conditions. When the mark color determined at the adult trap matched the mark color identified at the hatchery but the side of the fish differed (left side versus right side), the comparison was still considered a color match, although the difference in fish side was recorded.

We compared all of the 2SW VIE broodstock identifications made at the hatchery with those made at the trap by combining the results from each year into a cross-classification matrix, with trap identifications as columns and hatchery identifications as rows. Each matrix cell was divided by its column total, resulting in the percentage of fish with each classification. Percentages of correct color identifications are located along the diagonal of each matrix. The cross-classification matrices were compared between years using a general linear model (GLM) and a Fisher's least significant difference (LSD) procedure. Similar matrices were combined, and 95% confidence intervals were created for each cell in each matrix (Snedecor and Cochran 1967).

To correct the nonbroodstock 2SW adult VIE returns, the numbers of nonbroodstock adult VIE returns were first arranged in vectors by year, using mark identifications from the Veazie trap. These vectors were then multiplied by the correction matrix, and each row was summed to give adjusted totals for each mark. The adjusted totals were then compared to the original VIE identifications at the trap. A G-test for goodness of fit was used to determine if the adjusted totals were significantly different from the unadjusted totals. A difference between the totals would indicate a need to use the adjusted totals in determining overall returns.

Overall returns from each marked group within a smolt cohort were determined by adding: 1) the adjusted total of nonbroodstock salmon from the field site, 2) the adjusted total of VIE-marked broodstock at the hatchery, and 3) the unadjusted total of one-sea-winter returns (grilse) from the previous year's returns. The number of grilse was not adjusted for several reasons that are discussed in the Results section. All fish (broodstock and nonbroodstock) whose marks could not be detected were prorated equally among all the marked groups and added to the respective group totals, because tag loss was assumed to be equal among all groups.

Adjusted adult returns were analyzed using a general linear model (GLM) using year and stocking group as main effects, and we used a Fisher's least significant difference (LSD) procedure to determine which, if any, stocking groups were different. Adjusted returns from the stocking groups were also compared to the expected number of returns for each group. The number of expected returns was determined by multiplying the proportion of smolts stocked in each group by the total number of adipose-clipped returns from all stocking groups.

Results

During the seven years of the study, we examined 551 2SW Atlantic salmon and 42 grilse at CBNFH. The marked 2SW salmon taken to the hatchery for broodstock accounted for between 35.1% and 79.2% of the adipose-clipped fish captured at the Veazie trap, and the marked grilse taken to the hatchery for broodstock accounted for between 7.6% and 17.5% of the total number of marked grilse captured at the trap. The number of marked broodstock grilse each year ranged from 7 to 16 fish. In several years, the number of different marks exceeded the number of marked broodstock grilse. Therefore, the data were insufficient to create a separate classification matrix for grilse. Blankenship and Tipping (1993) and FitzGerald et al. (2004) reported high visual implant mark detection rates (>92%) at 19 and 17 months posttagging, respectively, using similar tagging and reading conditions. Since the grilse in this study were examined for marks at the Veazie trap at a maximum of 19 months after tagging, all of the VIE mark detections of grilse at the trap were assumed to be correct.

In the accuracy analysis, correct observations refer to those observations that were identical at both the trap and at the hatchery. This includes observations where a mark was recorded as not visible at both sites. A total of five marks were incorrect by the side of the fish on which the mark was detected. These marks were attributed to reader/recorder error. In several cases, a mark was detected on one side of the fish either at the trap or the hatchery, although it was not possible to identify the mark color. These instances were classified as "not visible." On four occasions, both the color and side of the fish were incorrect (for example, trap identification was right eye green, while the hatchery identification was left eye red). In these cases, these data were not included as it was assumed the errors were attributed to PIT tag read errors, which incorrectly linked two different fish.

In 2005, one batch of fish received marks on both sides, and although half of the fish in this group were identified at the trap as having a mark on only one side, it was assumed that both sides were not checked for marks since both marks were easily seen at the hatchery. These were considered correct observations since it was not a case of a poor mark, but rather was a mistake by the viewer at the trap. There were three different orange marks applied that year (left eye, right eye, and both eyes), all which referred to one stocking site and one stocking event.

Red, orange, and pink marks were most often confused, but green and yellow were also often confused, findings similar to Willis and Babcock (1998) and Curtis (2006). Overall identification accuracy of 2SW adults ranged from a low of 55.4% for the 2000 smolt cohort to a high of 82.9% for the 2004 smolt cohort (Table 2). Identification accuracy improved substantially for the 2001

Table 2. Comparison of mark identification at the Veazie trap and Craig Brook National Fish Hatchery. Broodstock fish were PIT tagged at the Veazie trap for later identification at the hatchery.

		Hatchery vs. trap identification		
			Incorrect identification	
Smolt cohort	*N* (Broodstock adults with fin clip and PIT tags	Correct identification (%)	Not visible at trap, visible at hatchery (%)	Mistaken color (%)
2000	56	56.4	25.5	18.2
2001	106	64.4	18.3	17.3
2002	80	83.5	11.4	5.1
2003	90	81.1	13.3	5.6
2004	105	82.9	6.7	10.5
2005	114	78.1	16.7	5.3

cohort and also for the 2002 cohort. Accuracy increased moderately for the 2003 and 2004 cohorts, and then declined for the 2005 cohort. An average of 80.4% of the adults examined at the hatchery had a visible VIE mark, as compared with 85.5% of smolts immediately after marking (Table 3). In addition, we compared marks by the side of the fish on which they were applied. Although there were differences within colors, the differences were not consistent between colors.

Because the 2000 and 2001 matrices were significantly different from the remaining years (GLM, $p < 0.05$), two classification matrices were developed. One matrix was developed for the 2000 and 2001 smolt cohorts (Table 4), and one was developed for the 2002–2005 smolt cohorts (Table 5).

We compared numbers of trap identifications of nonbroodstock VIE marked fish with the adjusted (after applying the classification matrix) nonbroodstock VIE identifications. A G-test for goodness of fit indicated no significant difference between the two groups in the 2000, 2003, 2004, and 2005 cohorts. Because two years showed significant differences between unadjusted and adjusted VIE identifications, we used the adjusted VIE identifications when determining returns from each stocking group each year.

We found that both year and stocking group significantly affected returns from each group (GLM, $P < 0.01$ and $P < 0.05$, respectively). There were annual variations in return rates, although several stocking groups consistently outperformed others. The smolts stocked in the Howland-early groups had greater returns than both Mattawamkeag groups and the Milo-late group. Returns from smolts stocked in the smolt ponds were also consistently greater than those in both Mattawamkeag groups (Table 6).

Discussion

Mark identification rates at the Veazie adult trap were similar to those reported by FitzGerald et al. (2004) and greater than those documented by Close and Jones (2002). However, we were able to improve upon the Veazie trap identification rates by viewing a subsample of marked fish in a controlled environment. By doing this, we successfully identified marks that were not visible in the field, and were able to better visualize and differentiate between tag colors. The use of a secondary mark, an adipose fin clip, facilitated the mark detection process by indicating that a VIE mark should be present on an individual fish.

The increase in mark identification accuracy for the 2001 cohort may be attributable to tag reader experience, although there are conflicting views as to whether experience actually improves detection accuracy (Frederick 1997; Curtis 2006). Identification ac-

Table 3. Percentage of smolts with good marks and 2SW adults with visible marks.

Smolt cohort	Percentage of smolts with good mark	*N*	Percentage of 2SW adults with visible mark	*N*
2000	84.0	6,042	74.5	55
2001	84.7	4,200	79.8	104
2002	90.3	2,401	81.0	79
2003	84.2	2,381	78.9	90
2004	84.3	2,334	88.6	105
2005	85.4	2,100	79.8	114

Table 4. Classification matrix for returning adults from the 2000–2001 smolt cohorts. Table values are proportions of fish with the indicated mark (combination of either left eye (LE) or right eye (RE), and either green (G), red (R), orange (O), yellow (Y), or pink (P)). Tags that were not visible are labeled as NV.

Hatchery Identification	Veazie Trap Identification									
	LEG	REG	LER	RER	LEO	REO	LEY	REY	LEP	NV
LEG	0.64	0	0	0	0	0	0.23	0	0	0.05
REG	0	0.79	0	0	0	0	0	0.06	0	0.03
LER	0	0	0.89	0	0.13	0	0.08	0	1	0.03
RER	0	0	0	0.9	0	0.4	0	0	0	0.06
LEO	0	0	0.11	0	0.75	0	0.31	0	0	0.16
REO	0	0.07	0	0.1	0	0.2	0	0	0	0.1
LEY	0.27	0	0	0	0.06	0	0.38	0	0	0.06
REY	0	0.07	0	0	0	0.2	0	0.88	0	0.03
LEP	0	0	0	0	0	0	0	0	0	0
NV	0.09	0.07	0	0	0.06	0.2	0	0.06	0	0.48

curacy also increased for the 2002 smolt cohort, but this is attributed to eliminating the use of yellow marks after 2001 (Tables 4 and 5), as these were commonly confused with both green and orange marks. This is similar to findings by Willis and Babcock (1998) and Curtis (2006). Detection rates for the green marks remained at 100% in subsequent years. We suggest that using both yellow and green marks in a tagging scheme should be avoided. Red and pink marks were also often confused, which was also found to be the case by Willis and Babcock (1998) and Curtis (2006), and if given the option, pink should be avoided when red marks are used.

Increases in identification accuracy occurred between the 2000 and 2002 cohort, and this improved accuracy is reflected in the classification matrices. Increases in identification accuracy are attributed to several factors. First, an improved mark-reading environment was developed, which included darkening the area, and using a seven-LED blue light, amber glasses, and a mark standard against which mark colors were compared. These adaptations utilized all of the tools suggested by the manufacturer to increase mark reading accuracy (NMT 2008). Second, a list of marks-at-large each year was used by the field crews to limit identifications to the marks applied to that year's cohort, since the marks used changed slightly from year to year.

Identification accuracy decreased slightly in 2005. Early tag loss in this cohort was similar to that of previous years, and the number of color misidentifications was lower than previous years. The decrease is a result of tags that were not visible at the trap but that were visible at the hatchery. The cause of this decrease is not known, however. Studies have shown that skin pigmentation affects the ability to properly identify tags (Close and Jones 2002; Curtis 2006), and it is possible that this cohort had slightly different pigmentation from other cohorts.

Over half of the marks classified as "not visible" on adults are attributable to short-term tag loss, because these marks were classified as either poor or not visible immediately following the smolt marking activities. Bailey et al. (1998) reported that problems they encountered with initial tag retention were due to elastomer that had not properly

Table 5. Classification matrix for returning adults from the 2002–2005 smolt cohorts. Values are proportions of fish with the indicated mark (combination of left eye (LE), right eye (RE), or both eyes (BE) and either red (R), green (G), orange (O), yellow (Y), or pink (P)). Tags that were not visible are labeled as NV.

Hatchery	Veazie Trap Identification										
Identification	LEG	REG	LER	RER	LEO	BEO	REO	LEP	REP	REY	NV
LEG	1	0	0	0	0	0	0	0	0	0	0.08
REG	0	1	0	0	0	0	0	0	0	0	0.03
LER	0	0	0.93	0	0.09	0	0	0.21	0	0	0.03
RER	0	0	0	0.92	0	0	0	0	0.05	0	0.03
LEO	0	0	0	0	0.85	0	0	0.03	0	0	0.05
BEO	0	0	0	0	0	1	0	0	0	0	0
REO	0	0	0	0	0	0	0.93	0	0.05	1	0.03
LEP	0	0	0.03	0	0.06	0	0	0.76	0	0	0.06
REP	0	0	0	0.08	0	0	0.08	0	0.9	0	0.1
REY	0	0	0	0	0	0	0	0	0	0	0
NV	0	0	0.03	0	0	0	0	0	0	0	0.59

TABLE 6. Range of overall adjusted number of returning Atlantic salmon (95% confidence interval) to Veazie, Maine, USA, and mean differences from expected returns. Returns that were less than expected are denoted as a negative percent. Superscripts indicate groups with means that were not significantly different.

	Smolt Cohort					
	2000		2001		2002	
Release Group	Adjusted returns	% Difference from expected	Adjusted returns	% Difference from expected	Adjusted returns	% Difference from expected
Mattawamkeag Early[a]	26.1–27.6	41.2%	40.4–55.2	–6.7%	25.0–28.6	–34.3%
Mattawamkeag Late[a]	11.7–12.9	–35.1%	34.4–45.4	–20.4%	27.9–34.1	–24.5%
Milo Early[abc]	19.0–22.4	7.5%	42.9–68.9	10.0%	41.4–47.6	9.5%
Milo Late[ab]	15.7–17.5	–14.8%	28.1–46.7	–35.5%	33.7–46.4	–2.2%
Howland Early[c]	26.1–29.2	43.0%	69.9–104.4	73.2%	44.1–56.5	23.9%
Howland Late[abc]	12.3–16.2	–25.9%	26.1–57.9	–17.5%	45.8–54.8	23.5%
Smolt Ponds[bc]	8.8–10.8	–24.3%	44.5–58.6	0.5%	35.9–49.6	4.3%

	Smolt Cohort					
	2003		2004		2005	
Release Group	Adjusted returns	% Difference from expected	Adjusted returns	% Difference from expected	Adjusted returns	% Difference from expected
Mattawamkeag Early[a]	13.8–15.9	–55.8%	15.7–16.8	–51.4%		
Mattawamkeag Late[a]	25.3–29.3	–18.9%	25.6–28.7	–18.0%		
Milo Early[abc]	35.7–39.5	12.1%	29.0–31.2	–9.5%	33.3–35.0	–2.7%
Milo Late[ab]	34.3–40.4	11.8%	21.3–27.5	–27.0%	29.3–32.8	–11.4%
Howland Early[c]	25.6–30.1	–17.8%	41.2–46.2	30.3%	38.6–39.9	11.8%
Howland Late[abc]	34.2–39.8	10.3%	36.5–40.0	14.8%	35.7–37.9	2.2%
Smolt Ponds[bc]	48.9–57.3	58.7%	50.9–56.7	60.5%		

cured and by elastomer trailing from the injection site. Taking more time to ensure that all of the elastomer is contained within the target tissue (NMT 2008) may significantly increase initial tag retention, thereby increasing the percentage of visible tags in returning adults. This can be accomplished through improving tagging protocols that emphasize proper tagging technique.

Significant differences in returns between stocking groups were detected. Returns from the smolt ponds were three times those of the Mattawamkeag-early stocking group for the 2003 and 2004 cohorts, and the Mattawamkeag stocking groups provided the fewest adult returns overall. If the management goal is to maximize returns, eliminating the Mattawamkeag site from future stocking activities and focusing future stocking efforts on the Howland site (at both time periods) and the West Enfield smolt ponds may help to achieve this goal.

Although returns between stocking groups were significantly different, it must be noted that the confidence intervals were extremely large due to the small sample sizes in this study. It is also unclear whether the differences between stocking groups are biologically significant. However, given that the returns of marked adults ranged between 129 and 360 for the 2000–2005 smolt cohorts, even a small increase in adult returns generated by simply altering stocking locations may be of value.

Acknowledgments

We acknowledge the staff from U.S. Fish and Wildlife Service for their work rearing the fish, supporting tagging operations, and holding the broodstock, and the individual taggers who administered the tags each year. We thank the staff of the Maine Department of Marine Resources, Bureau of Sea-Run Fisheries and Habitat for their work at the Veazie trap, and the staff of NOAA's National Marine Fisheries Service for assisting with tagging and for conducting VIE checks at CBNFH. In addition, we thank Graham Goulette for his assistance in coordinating the tagging and sampling.

References

Bailey, R. E., J. R. Irvine, F. C. Dalziel, and T. C. Nelson. 1998. Evaluations of visible implant fluorescent tags for marking coho salmon smolts. North American Journal of Fisheries Management 18:191–196.

Blankenship, H. L., and J. M. Tipping. 1993. Evaluation of visible implant and sequentially coded wire tags in sea-run cutthroat trout. North American Journal of Fisheries Management 13:391–394.

Close, T. L., and T. S. Jones. 2002. Detection of visible implant elastomer in fingerling and yearling rainbow trout. North American Journal of Fisheries Management 22:961–964.

Curtis, J. M. R. 2006. Visible implant elastomer color determination, tag visibility, and tag loss: potential sources of error for mark-recapture studies. North American Journal of Fisheries Management 26:327–337.

FitzGerald, J. L., T. F. Sheehan, and J. F. Kocik. 2004. Visibility of visual implant elastomer tags in Atlantic salmon reared for two years in marine net-pens. North American Journal of Fisheries Management 24:222–227.

Frederick, J. 1997. Evaluation of fluorescent elastomer injection as a method for marking small fish. Bulletin of Marine Science 61:399–408.

Northwest Marine Technology, Inc (NMT). 2008. Visible implant elastomer tag project manual: guidelines on planning and conducting projects using VIE and associated equipment. NMT, Shaw Island, Washington. Available: http://www.nmt.us/support/appnotes/ape06.pdf (November 2008).

Olsen, E. M., and L. A. Vollestad. 2001. An evaluation of visible implant elastomer for marking age-0 brown trout. North American Journal of Fisheries Management 21:967–970.

Snedecor, G. W., and W. G. Cochran. 1967. Statistical Methods. 6th edition. The Iowa State University Press, Ames, Iowa.

Walsh, M. G., and D. L. Winkelman. 2004. Anchor and visible implant elastomer tag retention by hatchery rainbow trout stocked into an Ozark stream. North American Journal of Fisheries Management 24:1435–1439.

Willis, T., and R. Babcock. 1998. Retention and *in situ* detectability of visible implant fluorescent elastomer (VIFE) tags in *Pagrus auratus* (Sparidae). New Zealand Journal of Marine and freshwater Research 35:581–590.

American Fisheries Society Symposium 76:77–94, 2012

Estimating Chinook and Sockeye Salmon Escapement on the Copper River, Alaska, Using Mark–Recapture with External PIT-tags

Keith van den Broek*
Native Village of Eyak
P.O. Box 1388, Cordova, Alaska 99574, USA

Jason J. Smith
LGL Alaska Research Associates, Inc.
2000 W. International Airport Road, Suite C1, Anchorage, Alaska 99512, USA

Guy Wade
LGL Alaska Research Associates, Inc.
2000 W. International Airport Road, Suite C1, Anchorage, Alaska 99512, USA

Abstract.—A feasibility study was begun in 2005 to obtain annual escapement information for sockeye salmon *Oncorhynchus nerka* over a minimum 3 years, using fish wheels and mark–recapture techniques already employed for Chinook salmon *Oncorhynchus tshawytscha* since 2001. Failed trials using both traditional spaghetti tags and injected PIT tags led to development of a new type of dorsal tag which which encapsulates a 134.2 KHz PIT tag in the marker of a 70 mm dual-anchor T-Bar tag. This tag was first employed in 2007 to estimate the inriver abundance of Chinook and sockeye salmon returning to the Copper river. For the first sample event, up to three live-capture fish wheels were operated at Baird Canyon for a total of 4,495 h from 18 May to 6 August. During this period, 4,456 adult Chinook salmon and 11,027 adult sockeye salmon were marked. For the second sample event, up to two fish wheels were operated at Canyon Creek near the lower end of Wood Canyon for 3,717 h from 28 May to 19 August. A total of 4,192 Chinook salmon and 56,5511 sockeye salmon were examined for marks. Of these, 459 Chinook salmon and 521 sockeye salmon were recaptures. Using a temporally stratified Darroch estimator, abundance of Chinok salmon measuring 500 mm FL or greater than migrated upstream of Baird Canyon from 18 May to 6 August was 46,349 (SE = 3,283). Using a similar estimator, estimated abundance of sockeye salmon that migrated upstream of Baird Canyon from 18 May to 6 August was 1,290,591 (SE = 92,590). This was the first ever defensible escapement estimate derived for sockeye salmon on the Copper River, and the fifth straight year for Chinook salmon with similar data quality to previous years using traditional spaghetti tags.

*Corresponding author: keith@eyak-nsn.gov

Introduction

The Copper River supports large commercial, subsistence, sport, and personal use fisheries for Chinook *Oncorhynchus tshawytscha* and sockeye salmon *O. nerka* with the majority harvested in an ocean commercial gillnet fishery from mid May through August in the Copper River District at the mouth of the river. The primary management tool utilized by the Alaska Department of Fish and Game (ADF&G) for the Copper River salmon fisheries since 1978 has been a sonar station located at the outlet of Miles Lake (Figure 1). This system has recently been upgraded from Bendix to Dual-Frequency Identification sonar (DIDSON). Counts were validated between these acoustic systems as part of the upgrade, but have never, since inception, been independently validated using other nonacoustic techniques. Neither system is capable of species apportionment between co-mingled salmon species, and as such the sonar only provides an index of overall salmon abundance migrating through the nearshore ensonified area, which may ignore fish migrating further offshore.

Since 2001 the Native Village of Eyak (NVE) has successfully run a two-sample tag-recapture program to monitor the Chinook component of the annual Copper River salmon run. This study makes use of fish wheels to capture salmon for tag and release (Smith et al. 2005).

On the basis of a feasibility analysis in 2005 (van den Broek and Smith 2005), a three year study to estimate the abundance of returning sockeye and Chinook salmon in the Copper River basin was initiated in 2006 (van den Broek et al. 2007). The study was in essence an expansion of the annual NVE tagging program; the goal was to achieve a more statistically robust species separation and validation of the Miles Lake sonar indices. A secondary study goal was to provide supporting data for a concurrent radiotelemetry study.

This paper documents results from the 2007 year of the program in which the primary objective was to estimate the inriver abundance of Chinook and sockeye salmon returning to the Copper River such that the estimates were within 25% of the true escapements 95% of the time. The 2007 study used largely the same Petersen mark–recapture methodologies as the NVE program, but the precision requirements demanded a substantial increase in the numbers of salmon tagged and released. It was therefore necessary to develop a new type of tag that would allow rapid application, minimal fish handling and damage, negligible tag loss between sample events, automatic data entry and easy identification by both automated equipment and human observers.

Methods

The Copper River flows southward through south-central Alaska to the Gulf of Alaska near Cordova, draining an area more than 62,100 km^2. A relatively high proportion of the headwaters are glaciated, resulting in very high unit discharge and sediment loads. The annual mean discharge on the lower Copper River is 1,625 m^3/s with annual peak discharge ranging from 6,681–11,750 m^3/s (1988–1995). High sediment loads cause the water to be unusually turbid and fill the river with numerous ephemeral sandbars and channel braids for most of its length, expanding into a wide and highly braided alluvial flood plain at the delta (Brabets 1997). The site chosen for tag–release was at the upper end of Baird Canyon (Figure 1). The tag recapture location was situated 91 km further upstream at the lower end of Wood Canyon at Canyon Creek (Figure 1). Both locations were chosen to take advantage of the only two major channel constrictions occurring

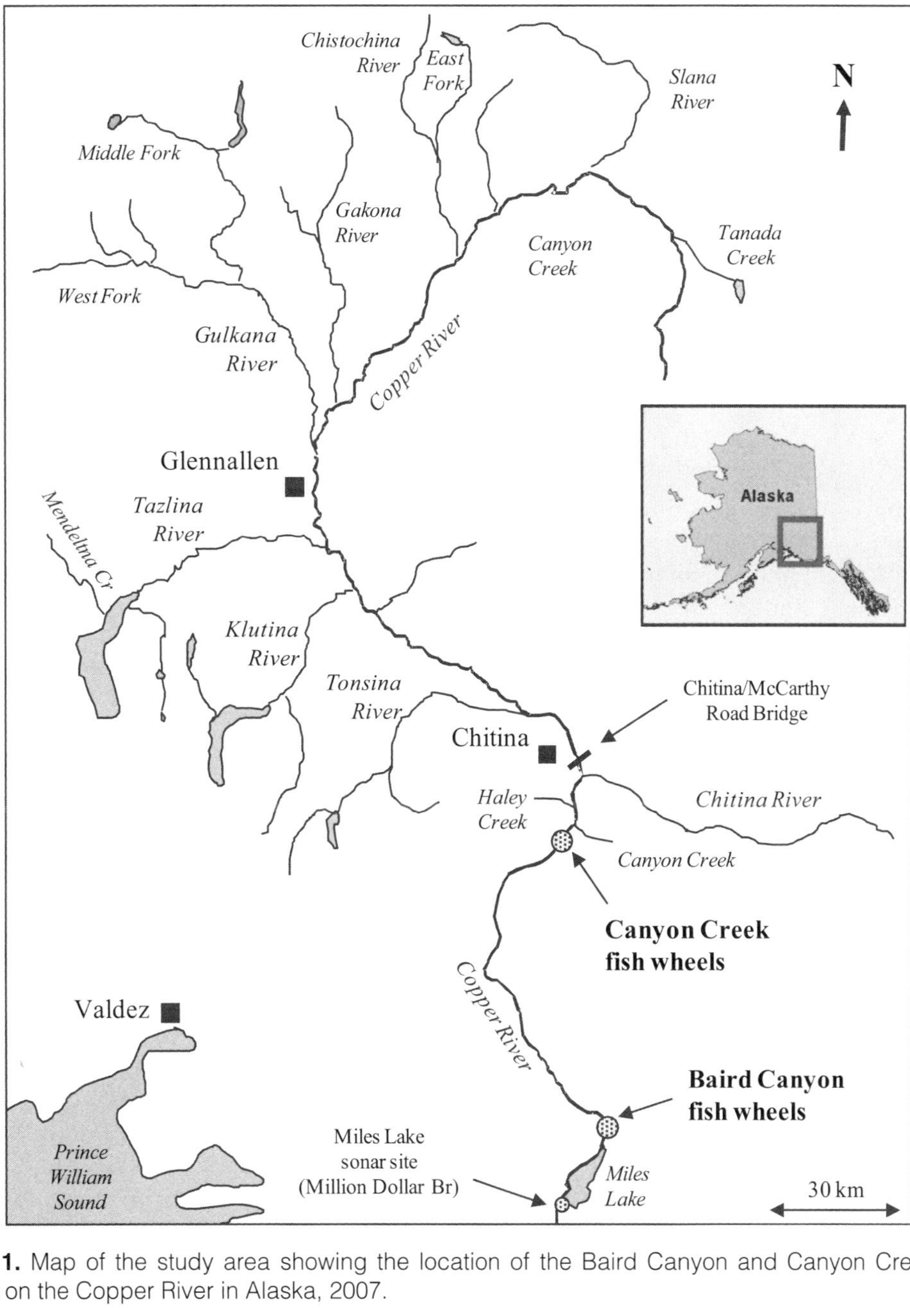

FIGURE 1. Map of the study area showing the location of the Baird Canyon and Canyon Creek fish wheels on the Copper River in Alaska, 2007.

between the sonar site at Miles Lake and the confluence of the Chitina River (Figure 1).

Large, custom built, live-capture fish wheels were utilized for sampling at both locations, with suitable sites selected based on water depth and velocity, accessibility, bankfull width, and protection from floating debris and rockfall. Three fish wheels were operated at the Baird Canyon tag–release location, and two at the Canyon Creek tag-recovery location. Similar methods have been demonstrated on numerous large rivers (Meehan 1961; Donaldson and Cramer 1971; Johnson et al. 1992; Arnason et al. 1996,

Link et al. 1996; Cappiello and Bromaghin 1997; Gordon et al. 1998; Link and Nass 1999; Sturhahn and Nagtegaal 1999), and on the Copper River since 2001 (Smith et al. 2005). The fish wheels were operated 24 h per day except for occasional necessary stoppages. Escape panels at the downstream end of each live tank could be adjusted to regulate catch rates of sockeye salmon and avoid overcrowding, while always retaining 100% of Chinook salmon measuring ≥500 mm FL. Sockeye salmon were only sampled during periods when the escape panels remained closed. All fish were removed from the live tanks and sampled two to four times per day depending on catch rates.

At each sampling session at both sample locations, all adult Chinook and sockeye salmon were counted. All retained Chinook salmon and a predetermined subsample of sockeye salmon were sexed, measured for fork length (FL), and examined for external marks, scars or bleeding. All other captured fish were identified to species, counted, and released.

At the Baird Canyon tag–release location, all retained Chinook salmon greater than 500 mm FL and in good condition were selected for tagging (up to a maximum of 150 per day). Sockeye samples were stratified based on the count obtained at the Miles Lake sonar the previous day; the proportion selected for tagging represented 1–1.2% of this count. Any previously tagged fish recaptured at the Baird Canyon location were noted and released.

Both species were marked with TBA-PIT tags. The tag is composed of a uniquely coded 134.2 kHz FDX-B passive RFID transponder in an 11 mm ruggedized polymer capsule (ENSID Technologies, Inc, Auckland, New Zealand), encapsulated within a 50 mm polyethylene marker, with double T-bar style anchor filaments 25 mm in length (Hallprint Pty Ltd, Victor Harbor, South Australia (Photo 1). The NVE contact details were printed on the side of the tag. The TBA-PIT tags are a new technology specially designed and constructed for these projects. Tags were supplied in magazine clips of 20 tags each, and were applied to fish using a hand held applicator gun with a 16-gauge needle (Avery Dennison). The tip of the needle was sunk into the musculature of the fish 1–2 cm ventral to the insertion of the dorsal fin between the third and fourth pterygiophores, to a depth of 1–2 cm, so that the tag anchors would lodge behind the pterygiophores within the dorsal musculature when ejected from the applicator gun (Photo 2). A 6.35 mm hole was punched in the right operculum on Chinook salmon to act as a secondary mark for quantification of primary tag loss.

At the Canyon Creek tag-recovery location, all retained Chinook and sockeye salmon were examined for a primary mark, and all Chinook salmon were examined for a secondary mark, in addition to the biometrics described previously. A 6.35 mm hole was punched in the left operculum on all inspected Chinook salmon in order to quantify multiple observations of both tagged and untagged fish. All recaptured Chinook and sockeye salmon retained the lower anchors of the primary tag, while the upper marker of the tag was clipped immediately below the transponder in order to retain this component for reuse, while still identifying multiple observations of tagged recovered fish.

The recovery samples were stratified by time and a series of abundance estimates for both species in each time step were derived using the method described by Darroch (1961). The computer program SPAS (Arnason et al. 1996) was used to derive abundance estimates and standard errors. Initially, data were pooled into strata based on where major breaks occurred in the daily mark and recapture rates over the season. Chi-square tests were then used to compare mark and recapture rates of adjacent strata; and homogeneous strata were pooled. If SPAS failed to generate an estimate using the strata produced by

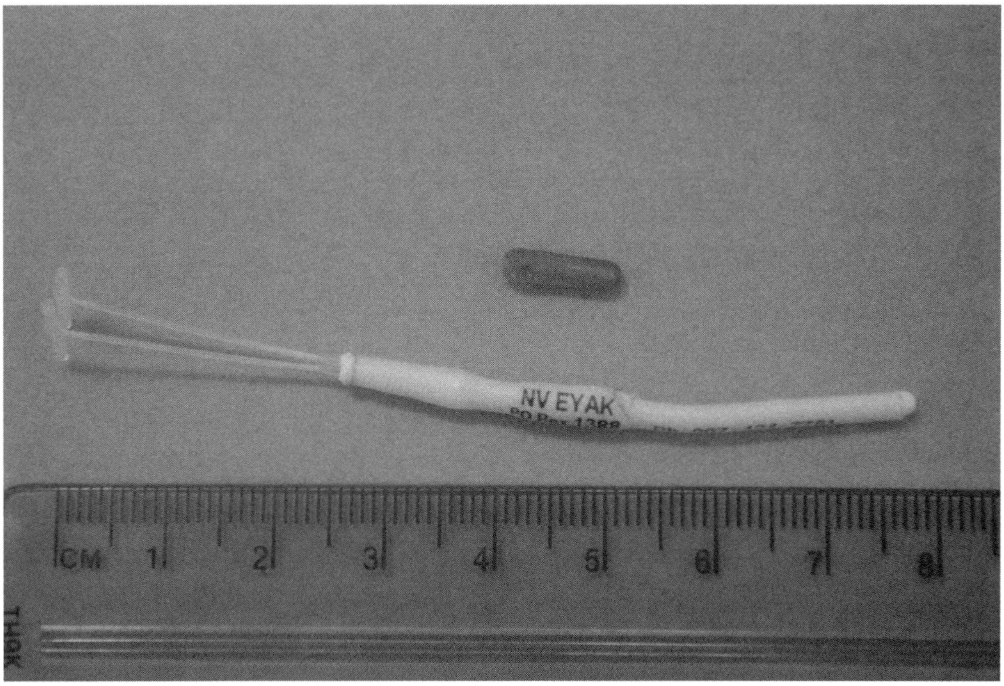

Photo 1. Dual anchor TBA-PIT tag, showing PIT component that is encapsulated within the polyethylene marker.

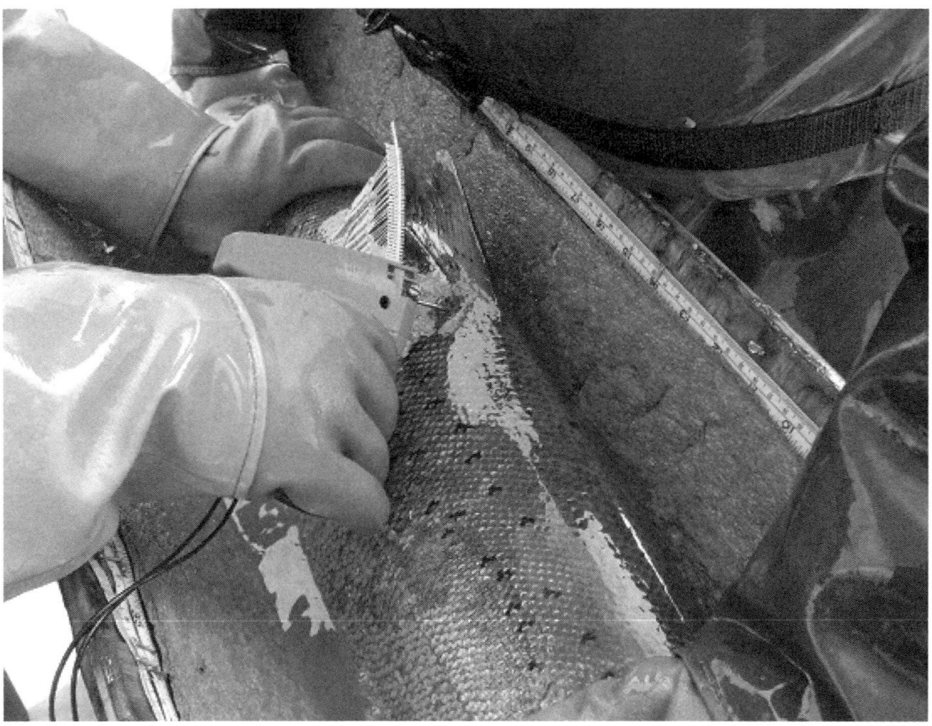

Photo 2. Application of TBA-PIT tag to an adult Chinook salmon.

this process, then one or more heterogeneous strata would be pooled. If necessary, different combinations would be compared in SPAS to determine how sensitive the abundance estimate was to strata poolings.

The use of mark–recapture to derive estimates of Copper River salmon abundance is contingent upon validity of the following assumptions (after Ricker 1975; Seber 1982):

(a) mortality due to handling and tagging was negligible;

(b) tagged fish were not more or less vulnerable to recapture than untagged fish;

(c) tagged fish did not lose their tags;

(d) the population was closed, i.e., immigration or emigration of tagged fish between the tagging and recovery sites was negligible;

(e) tagged fish mixed completely with untagged fish between the sampling events; and

(f) the proportion of tags released into the two salmon populations was homogeneous with respect to fish-length and time.

There was no explicit test for assumption (a); however, efforts were taken to reduce sampling induced stress or injury through frequent sampling sessions and use of escape panels, and visibly stressed or injured fish were not tagged. We believe the distance between the tag and recapture sites (91 km) was sufficient to reduce the potential for handling-induced catchability changes in tagged fish (i.e., "trap happiness" or "trap shyness") such that assumption (b) was met. In the case of Chinook salmon we were able to ascertain if assumption (c) was met via the secondary mark (operculum punch). It was not practical to operculum mark the PIT-tagged sockeye due to the large amount of additional handling required in order to individually inspect large numbers of this species at the recovery site. We therefore make the assumption that sockeye tag-loss was the same as that observed for double tagged Chinook. Buzby and Deegan (1999) showed T-Bar tag retention in Artic Grayling Thymallus arcticus to be 93% when re-sampled a year after initial tagging. With the short migration time for Chinook salmon (range 5–30 d, median 9–13 d; Smith et al. 2005) of our sample group between sites, and the use of double anchor T-Bar tags, it is assumed that tag retention between events in our experiment would be much higher. Operculum punches should not begin healing in this short time, but technicians were trained to inspect for any sign of a previous operculum punch (ODFW 2005). We think it reasonable that assumption (d) was met as the life history of salmon isolates those fish returning to the Copper River as a "closed" population. The second event occurred downstream of any major spawning tributaries or inriver fisheries, and it is assumed that any incidental mortality (predation) or out-migration to minor spawning tributaries occurred equally for tagged and untagged fish; thus the abundance estimates were germane to the tagging location at Baird Canyon.

The Copper River is highly braided in some sections between the release and recovery locations. We believe the braded structure of the river would promote adequate mixing of tagged and untagged salmon (assumption e). However, we were able to test for violations of the mixing assumption by comparing mark and recapture rates by bank of mark versus recapture using contingency tables and chi-square tests Furthermore, studies from 1999 to 2001 showed equal mixing of tagged and untagged Chinook salmon between the lower end of Wood Canyon and the CSS fishery (Evenson and Wuttig 2000; Wuttig and Evenson 2001; Savereide and Evenson 2002), a much shorter and more constricted portion of the river.

We were able to test for violations of the homogeneity assumption (f) in relation to length using a Kolmogorov-Smirnov (K-S) two-sample tests (Zar 1984) to compare the cumulative length-frequency distributions of marked and recaptured fish. It was assumed if the K-S test failed to reject the null hypotheses that the length distributions of the marked and recaptured fish were the same and size-stratification in abundance analyses would not be required (Bernard and Hansen 1992).

Fishing effort for sockeye salmon was based on periods when the escape panels were closed and was distributed between fish wheels and riverbanks (spatially) at consistent intervals (temporally) throughout each day to reduce the potential of bias being introduced into the experiment. The mark-rates as observed at the Canyon Creek tag-recovery location for specific time intervals were compared, again using contingency table analysis, to determine if the distribution of marked salmon was homogenous in the two populations in respect to time (i.e., that assumption f was met). Any significant violation of the temporal homogeneity assumption would be addressed by temporal stratification in the analysis.

Results

Chinook Salmon

Apart from minor fish wheel stoppages for repairs and moves, fishing effort at the Baird Canyon fish wheels was continuous throughout the study period for Chinook salmon (Table 1), and a total of 5,294 Chinook salmon were captured (Figure 2). Of these, 4,456 (84.2%) in healthy, unstressed and uninjured condition and measuring ≥500 mm FL were tagged and released (Figure 2). The number of tags applied on a single day peaked at 159 fish on 31 May.

A total of 4,192 Chinook salmon were examined for primary and secondary marks at the Canyon Creek fish wheels (Figure 3). Of those examined, 459 (10.9%) were recaptures, or fish that had been tagged at Baird Canyon (Figure 3). The number of Chinook salmon examined for marks at Canyon Creek peaked at 179 fish on 4 June and the number of recaptures peaked at 27 fish on 27 and 28 June (Figure 3). The median travel time of Chinook salmon tagged at Baird Canyon and recaptured at Canyon Creek was 12.2 d (range: 3.6–56.7 d).

Of the 286 Chinook salmon captured twice or more at the Baird Canyon fish wheels, 155 (54.2%) were recaptured within 1 d of being tagged with 90% not exceeding 9 d (Figure 4). The longest delay between captures was 32.8 d (Figure 4). These results suggest the majority of tagged Chinook did not remain long at the Baird Canyon release site after tagging; we assume from this that tagging did not significantly affect upstream migration behavior.

No Chinook salmon were captured at Canyon Creek with an operculum punch and

Table 1. Operational period and fishing effort for the Copper River fish wheels, 2007.

Location	Fish Wheel #	River Bank	Start Date	End Date	Fishing Effort (h)	Percent Operational
Baird Canyon	1	West	18-May	16-Jul	1,148	81.5
	2	East	21-May	24-Jul	1,522	98.6
	5	West	21-May	06-Aug	1,825	99.6
Canyon Creek	3	East	28-May	10-Aug	1,776	99.7
	4	West	29-May	19-Aug	1,941	98.6

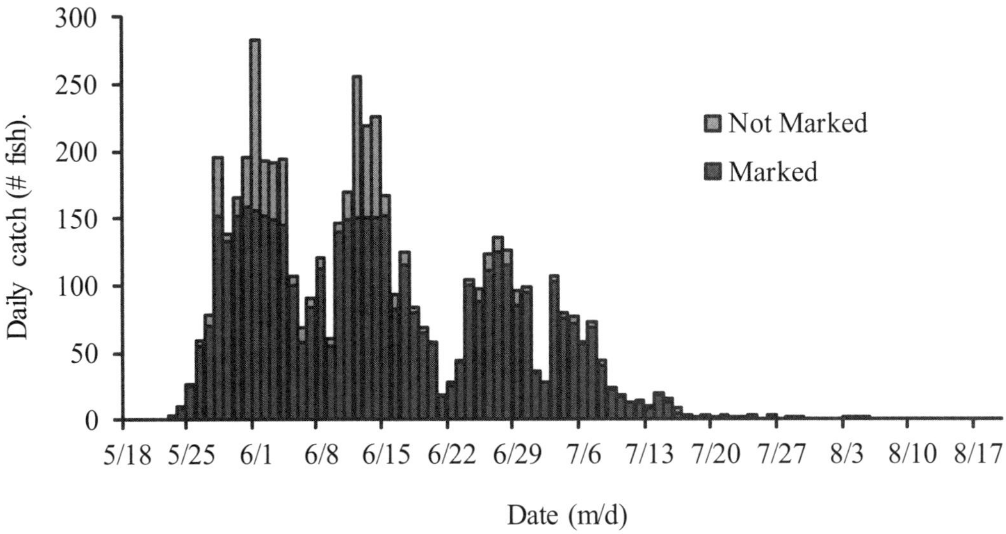

FIGURE 2. Daily breakdown of the total Chinook salmon caught (5,294) and marked (4,456) at the Baird Canyon fish wheels on the Copper River, 2007.

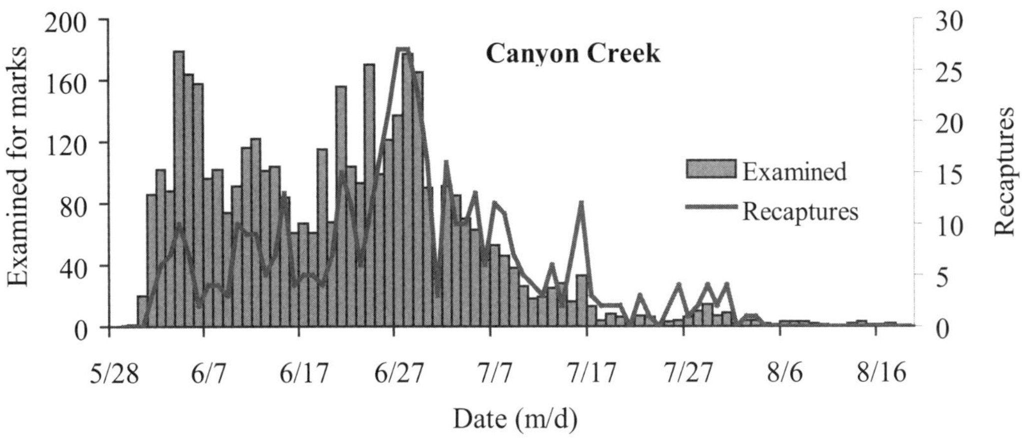

FIGURE 3. Daily breakdown of the total Chinook salmon examined (4,192) and total tagged Chinook recaptured (459) at the Canyon Creek fish wheels on the Copper River, 2007.

no TBA-PIT tag. On the basis of this observation we have assumed tag-loss in Chinook was zero between sampling events.

Tagged Chinook salmon appeared to move equally between banks. Recapture rates of Chinook salmon that were tagged on the west bank (9.6%) of the river at Baird Canyon were not significantly different than recapture rates of fish tagged on the east bank at the 5% level (11.2%; $\chi^2 = 2.96$, df = 1, $P = 0.085$; Table 2). A second test to assess the proportionality of movement across the river using Chinook salmon recaptures by bank of release and bank of recovery showed no significant difference ($\chi^2 = 0.01$, df = 1, $P = 0.928$; Table 3). We believe the failure to reject these tests supports the adequate spatial mixing hypothesis (e) for tagged Chinook and there was no need for spatial stratification in the analysis.

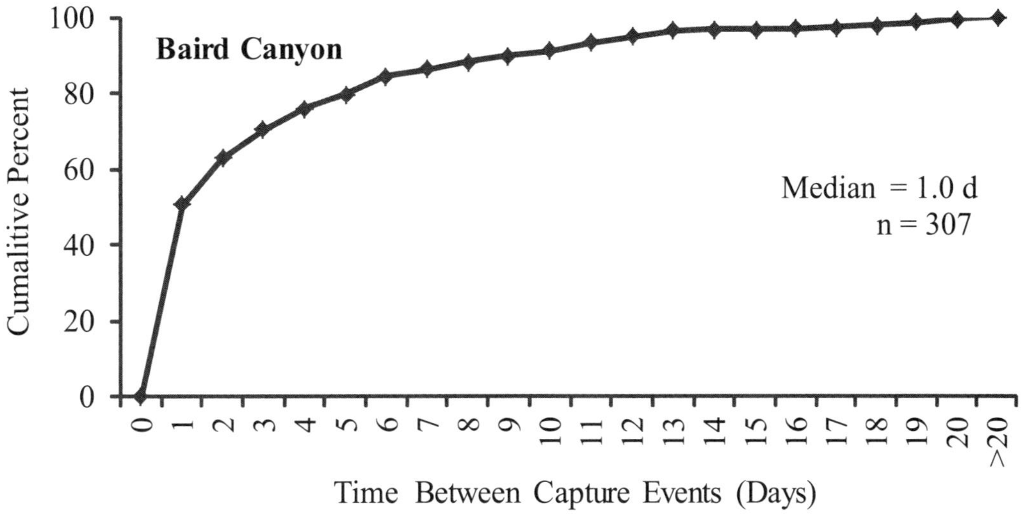

FIGURE 4. Cumulative percent of the number of tagged Chinook salmon recovered more than once at the Baird Canyon release location fish wheels by days at liberty.

TABLE 2. Number of Chinook salmon recaptured, by bank of release, and the results of a test to compare recapture rates of fish marked on the east and west banks of the Copper River, 2007.

Bank of Release	Recaptured [a]	Not Recaptured	Tagged	Recapture Rate
West (FW1&5)	274	2,585	2,859	0.096
East (FW2)	179	1,418	1,597	0.112
Total	453	4,003	4,456	0.102

Chi-square = 2.96 df = 1 p-value = 0.085

[a] Excludes 6 recaptures where the bank of tagging was unknown.

TABLE 3. Number of Chinook salmon recaptured, by bank of release and bank of recovery, and the results of a test to compare for equal movement across the river, 2007.

Bank of Release	Bank of Recovery: West	Bank of Recovery: East	Recaptured [a]
West	73	201	274
East	47	132	179
Total	120	333	453

Chi-square = 0.01 df = 1 p-value = 0.928

[a] Excludes 6 recaptures where the bank of tagging was unknown.

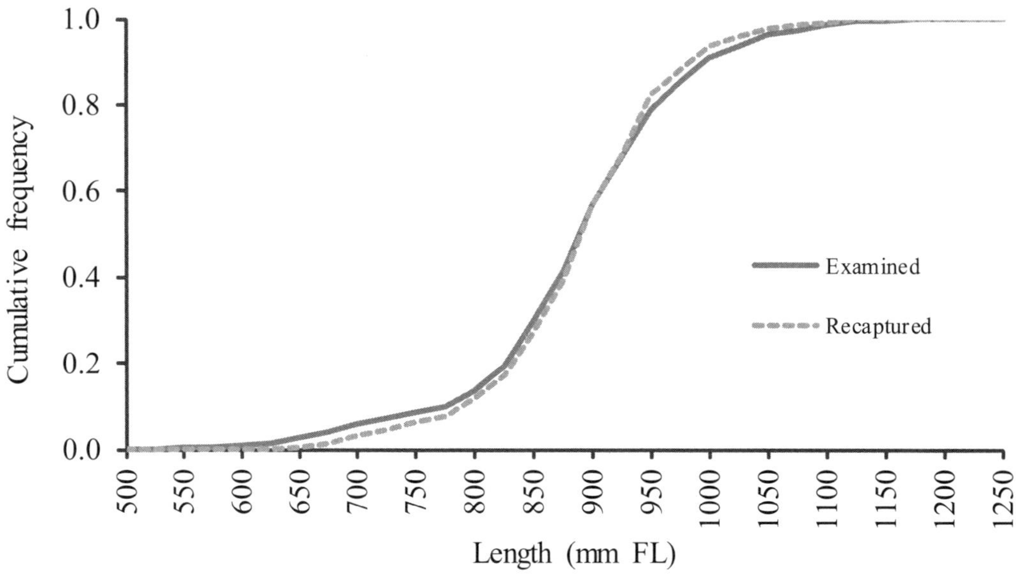

Figure 5. Cumulative length-frequency distributions of recaptured and examined Chinook salmon at Canyon Creek, 2007.

Table 4. Capture history of Chinook salmon that were tagged, examined, and recaptured at the Copper River fish wheels, 2007. Bold text indicates the data used for the chi-square tests.

Period of marking	Period of recapture						
	5/30-6/9	6/10-6/24	6/25-7/4	7/5-7/14	7/15-8/17		
Recaps	46	122	169	69	53	459	
Unmarked	1,023	1,391	897	290	131	3,732	
Examined	1,069	1,513	1,066	359	184	4,191	
Mark rate	0.043	0.081	0.159	0.192	0.288		χ^2 = 174.7, df = 4, P = 0.000

No significant difference (Dmax = 0.040, P = 0.527) was found between cumulative length-frequency distributions of Chinook examined and tagged Chinook recaptured in the second event (Figure 5). We concluded from this that Chinook mark-rates were homogenous by length (assumption e) and no stratification by size was necessary to estimate abundance.

Capture statistics were summarized by period of marking and recapture over the study period (Table 4). The probability of a fish being marked at Baird Canyon was not independent of time of capture. A progressive increase in the population mark-rate is clearly evident in the results (Figure 6). To account for the temporal mark-rate heterogeneity the abundance analysis was stratified into five time intervals; a chi-square test confirmed the difference in the mark-rates between time periods was significant (χ^2 = 174.7, df = 4, P = 0.000; Table 4). Separate abundance estimates were derived for each recovery period using a maximum likelihood Darroch estimator and these were then combined. The total number of Chinook salmon,

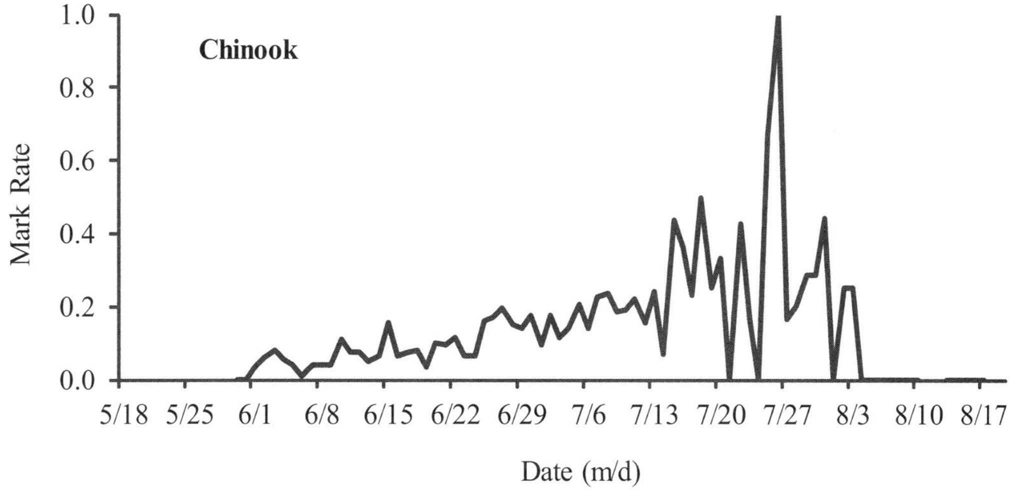

Figure 6. Daily proportion of Chinook salmon examined at the Canyon Creek fish wheels that were marked (mark rate), 2007.

measuring 500 mm FL or greater, migrating upstream of Baird Canyon between 18 May and 6 August was 46,349 (SE = 3,283; 95% CI = 39,914–52,784).

Sockeye Salmon

A total of 19,888 sockeye salmon were captured at the Baird Canyon fish wheels (Figure 7). Of these, 11,027 (55.4%) in healthy, unstressed and uninjured condition were tagged and released. Apart from a few days early and late in the season when catches were low, the number of sockeye salmon tagged daily represented 1.0–1.2% of the previous day's counts at Miles Lake DIDSON sonar. The number of tags applied on a single day peaked at 438 on 3 June (Figures 7).

A total of 56,551 sockeye salmon were examined for primary and secondary marks at the Canyon Creek fish wheels (Figure 8). Of those examined 521 (0.9%) were recaptures (Figure 8). The number of sockeye examined for marks at Canyon Creek peaked at 1,942 fish on 7 June and the number of recaptures peaked at 19 fish on 4 June (Figure 8). The median travel time of sockeye salmon tagged at Baird Canyon and recaptured at Canyon Creek was 9.5 d (range: 3.6–30.6 d).

Of the 161 sockeye salmon captured twice at the Baird Canyon fish wheels, 136 fish (84.5%) were recaptured within 1 d of being tagged with 90% not exceeding 9 d at liberty before recapture (Figure 9). The longest delay between capture events was 5.1 d. These results suggest the majority of tagged sockeye did not remain long at the Baird Canyon release site after tagging; we assume from this that tagging did not significantly affect upstream migration behavior.

Similar to Chinook salmon, sockeye salmon appeared to move equally between banks. Recapture rates of sockeye salmon that were tagged on the west bank (4.5%) of the river at Baird Canyon were not significantly different than recapture rates of fish tagged on the east bank (4.3%; $\chi^2 = 0.34$, df = 1, $P = 0.561$; Table 5). A second test to assess equal movement across the river using sockeye salmon recaptures by bank of release and bank of recovery showed no significant difference ($\chi^2 = 0.43$, df = 1, $P = 0.511$; Table 6). We believe the failure to reject these tests supports the adequate spatial mixing hypothesis (e) for tagged sockeye

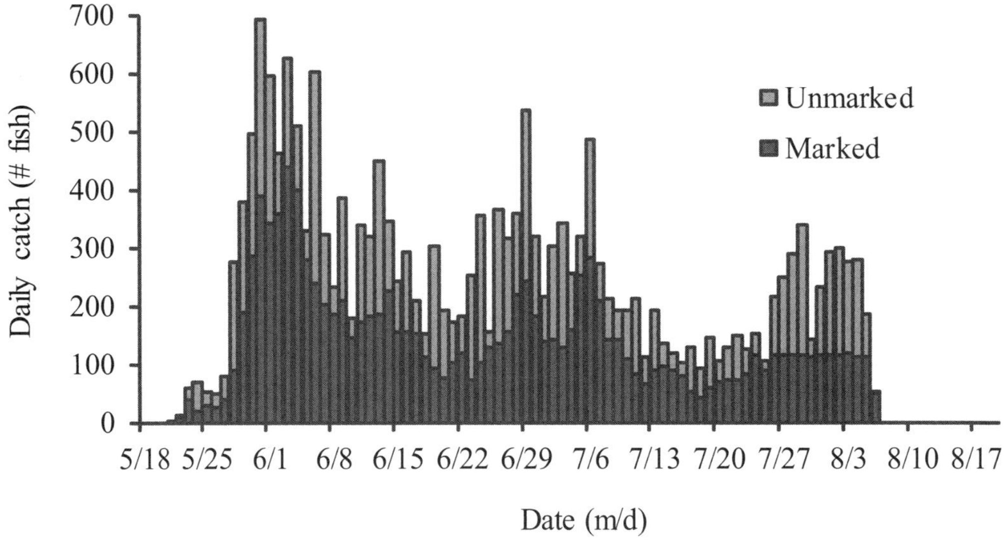

Figure 7. Daily breakdown of the total sockeye salmon caught (19,888) and marked (11,027) at the Baird Canyon fish wheels on the Copper River, 2007.

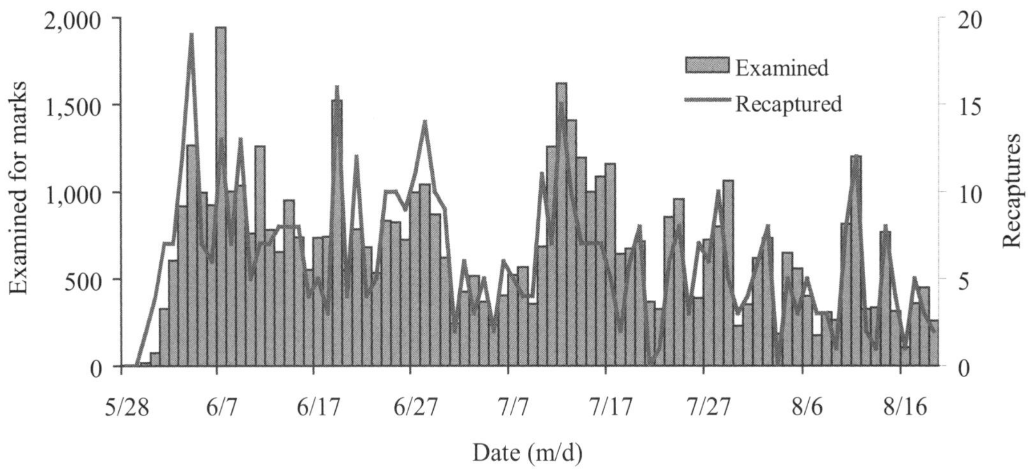

Figure 8. Daily breakdown of the total sockeye examined (56,551) and total tagged sockeye recaptured (521) at the Canyon Creek fish wheels on the Copper River, 2007.

and there was no need for spatial stratification in the analysis.

No significant difference (Dmax = 0.047, P = 0.298) was found between cumulative length-frequency distributions of sockeye examined and tagged sockeye recaptured in the second event (Figure 10). We concluded from this that sockeye mark-rates were homogenous by length (assumption e) and no stratification by size was necessary to estimate abundance.

With the exception of the initial high peak, likely resulting from low in-river abundance at the beginning of the run (2 out of 18 fish examined on 30 May were marked) there is little evidence of a progressive trend in the

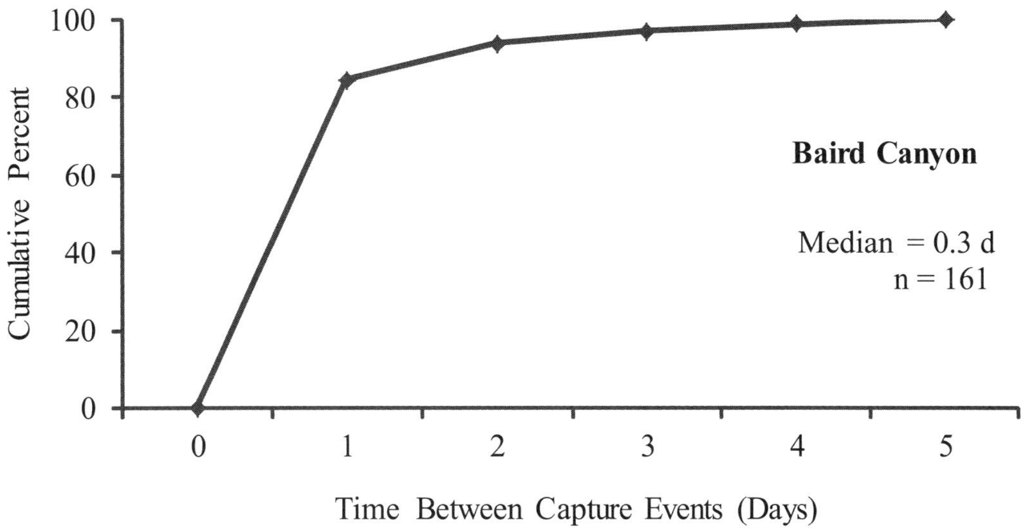

Figure 9. Cumulative percent of the number of tagged sockeye salmon recovered more than once at the Baird Canyon release location fish wheels by days at liberty.

Table 5. Number of sockeye salmon recaptured, by bank of release, and the results of a test to compare recapture rates of fish marked on the east and west banks of the Copper River, 2007.

Release location	Recaptured[a]	Not recaptured	Total released	Recapture Rate
West (FW1&5)	316	6,688	7,004	0.045
East (FW2)	172	3,851	4,023	0.043
Total	488	10,539	11,027	0.044

Chi-square = 0.34, df = 1, P = 0.561

[a]Excludes 33 recaptures with unknown release locations.

Table 6. Number of sockeye salmon recaptured, by bank of release and bank of recovery, and the results of a test to compare for equal movement across the river, 2007.

	Bank of Recovery		
Bank of Release	West	East	Recaptured[a]
West	150	166	316
East	87	85	172
Total	237	251	488

Chi-square = 0.43 df = 1 p-value = 0.511

[a]Excludes 33 recaptures with unknown release locations.

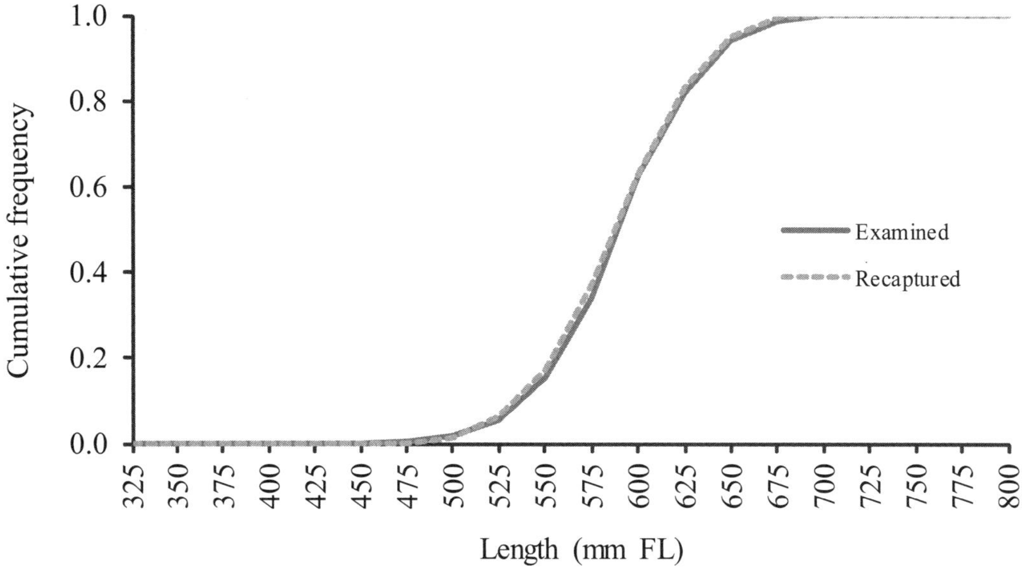

Figure 10. Cumulative length-frequency distributions of recaptured and examined sockeye salmon at Canyon Creek, 2007.

sockeye mark-rates through time (Figure 11). As with Chinook analysis, we stratified the recovery period into five time intervals and compared the mark-rates using a chi-square test (Table 7). The probability of a fish being marked at Baird Canyon was not independent of time of capture period (χ^2 = 35.0, df = 5, P = 0.000) and varied from 0.007 to 0.016 (Table 7). To account for the temporal mark-rate heterogeneity the abundance analysis was stratified into five time intervals (Table 7) and separate abundance estimates were derived for each recovery period using a maximum likelihood Darroch estimator. The total number of sockeye salmon migrating upstream of Baird Canyon between 18 May and 6 August was 1,290,591 (SE = 92,590; 95% CI = 1,109,115–1,472,067).

Discussion

Abundance estimates for Chinook (46,349; CV = 7.1%) and sockeye salmon (1,290,591; CV = 7.2%) in 2007 exceeded the precision levels specified in the study objectives. It was necessary to stratify the abundance analyses into 5 time intervals, to account for temporal heterogeneity in the mark-rates over the 4 month release/recovery period. Tag-loss and tag mortality effects were minimal; there was no evidence of disproportionate tagging in relation to length or lack of spatial mixing of tagged fish. We believe all the major assumption criteria for estimating abundance from tagging were met and therefore our estimates of the number of Chinook and sockeye migrating between Baird Canyon and Canyon Creek over May to August 2007 had minimal bias.

Given the high catch rates observed for both species during both sampling events, it was relatively easy to capture a sufficient number of fish to satisfy the project requirements. The sockeye salmon mark–recapture estimate was 39% (364,153 fish) greater than the 2007 Miles Lake sonar count of 926,438 salmon. If the estimated number of Chinook salmon that migrated above Baird Canyon (46,349) was subtracted from the Miles Lake sonar count, then the sockeye salmon mark–

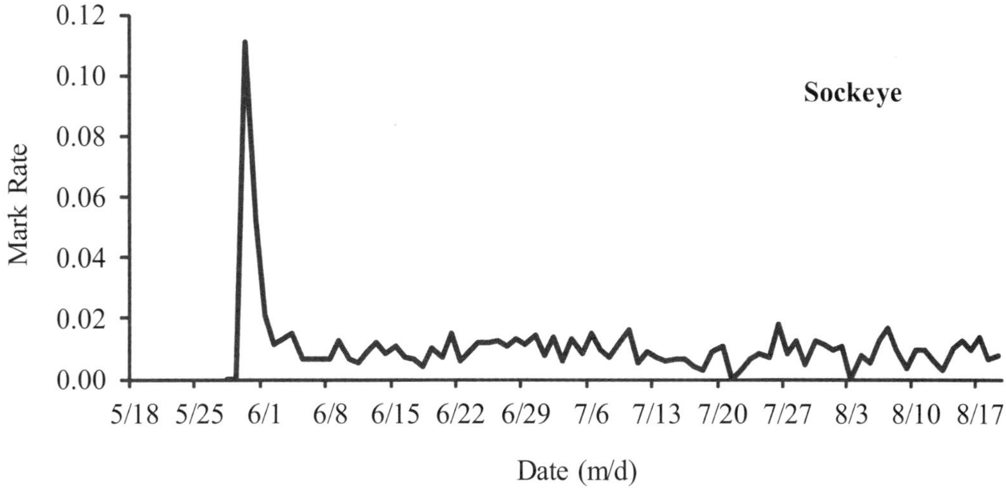

FIGURE 11. Daily proportion of sockeye salmon examined at the Canyon Creek fish wheels that were marked (mark rate), 2007.

TABLE 7. Capture history of sockeye salmon that were tagged, examined, and recaptured at the Copper River fish wheels, 2007. Bold text indicates data used for chi-square tests.

	Period of recapture						
Period of marking	5/28-6/4	6/5-6/18	6/19-7/10	7/11-7/25	7/26-8/6	8/7-8/19	
Unknown	5	1	5	11	5	6	33
Recaps	51	101	162	92	62	53	521
Unmarked	3,160	12,962	14,119	13,553	6,625	5,611	56,030
Examined	3,211	13,063	14,281	13,645	6,687	5,664	56,551
Mark rate	0.016	0.008	0.011	0.007	0.009	0.009	0.009

$\chi^2 = 35.0$, df = 5, P = 0.000

recapture estimate was 47% greater than the sonar count. A large proportion of the difference between the mark–recapture estimate and Miles Lake count was attributed to the first week of June when high abundances of sockeye salmon were passing through the study area. Recapture rates were relatively low for fish tagged at Baird Canyon from 2 to 5 June (0.009–0.028) and mark rates were also relatively low for fish examined at Canyon Creek from 5 to 18 June (0.004–0.013). As a result, 27% (349,084 fish) of the mark–recapture estimate was derived from fish tagged in the 2–5 June stratum. In comparison, only 14% (127,888 fish) of the total Miles Lake sonar count occurred from 1 to 4 June (which assumes a 1-d travel time between the sonar site and Baird Canyon).

The Chinook salmon monitoring program historically employed traditional spaghetti tags sewn through the dorsal musculature of fish with a length of monofilament. On average, 2–4,000 fish were marked each season, with up to 150 per day at the peak of the run. All data were recorded by hand and subsequently transferred to a computer database. Average tagging time per individual was approximately 5 min, plus data postprocessing which was considerable and often fraught with data entry errors primarily due to the harsh sampling environment. This was an adequate methodology given the relatively small sample sizes, but was not feasible for use with sockeye salmon, which would require a sample size exceeding 10,000 individuals per season, and upwards of 300 per day for a large proportion of the run. Use of a more rapidly applied T-bar tag would reduce tagging time and fish handling, but still involve high risk of data entry errors which could jeopardize experiment results with such large sample sizes. The only way to minimize data entry errors would be to reduce the amount of information entered (e.g., color coded weekly tag groups versus uniquely numbered tags), or develop a means of automatic electronic data entry. Reducing information was not desirable because of the valuable migration rate and other data provided by uniquely identified individuals.

Injected PIT tags seemed to be a viable option, with the ability for rapid tag application to adult salmon and automatic electronic data entry to a PDA with integrated tag scanner. The method was tested over a 10-d feasibility study in 2005, and a full-scale mark–recapture experiment for the duration of the 2006 sockeye run. No defensible estimate could be derived from this experiment (van den Broek et al. 2007). The main shortcoming to this approach was in technical difficulties related to the recovery system. With no means of visual identification of marked fish, and insurmountable burden associated with physically scanning every fish at the recovery site, the only option was to use a fully automated recovery system. This allowed no manual failsafe, no means of validating the efficacy of the system (read rates of test tags passed through the system were often as low as 60%) and no means of quantifying tag loss. Furthermore, the tagging effort at the first sample event proved to be more tedious and less efficient than initially believed. Tags were often lost during insertion into the body cavity, or immediately migrated back out of the insertion point. Tag retention between events was therefore assumed to be poor, but with no means of quantification. Severe bleeding was often observed after tagging, but with no way to quantify mortality after release. Finally, the public relations aspect of the experiment was dismal, with no means for participation in harvest reports by inriver fishery users, and more importantly, several postseason complaints and expressions of concern from fishery users finding PIT tags in food fish.

Development of the TBA-PIT tag was the single greatest factor in determining the success of this program. This novel tag design offers an overwhelming suite of advantages over other tags: (a) rapid tag application with minimal fish handling stress/injury and visual confirmation of tagging success; (b) automatic and accurate data entry to a handheld PDA with integrated tag scanner at both sample events; (c) visible external mark allows recaptures to be easily isolated from examined fish at the recovery site without handling every individual; (d) dual anchor bars improve tag retention over traditional T-bar tags (which are already highly effective); (e) external markers can be observed by fishery participants and printed with instructions for harvest reporting; (f) uniquely coded tags provide additional information on migration rates and allow detailed temporal stratification for data analysis; and (g) transponder technology offers potential for fully automated recovery systems as technology improves, and in-stream spawning distribution monitoring systems for a relatively low incremental cost as compared to radiotelemetry.

Acknowledgments

This project was approved by the Federal Subsistence Board, managed by the U.S. Fish and Wildlife Service, Office of Subsistence Management, and funded by the U.S. Forest Service as FIS 07–503 at $372,456 in 2007 (3 Year Total $1,112,275)

Additional matching funds of $50,000 were provided by ADF&G. John H. Clark and State House Representative Bill Thomas were instrumental in securing these funds each year for the duration of the study.

David Hall of Hallprint Pty Ltd and EN-SID Technologies Inc provided invaluable technical support in the development and implementation of the new TBA-PIT tagging technology. This project would not have been successful without his effort and dedication.

We appreciate the general public support of our research. Over 1,200 individual fishery users returned tags captured in the inriver fisheries. Employees of the Prince William Sound Aquaculture Corporation assisted by collecting and returning tags from broodstock collected for the Gulkana sockeye salmon hatchery. Employees of the National Park Service assisted by collecting and returning tags on fish passing the Long Lake counting weir.

We appreciate the more than 20 NVE employees who worked 7 d weeks for more than 3 months straight, often in harsh weather, conducting the sampling and solving the problems that led to the overwhelming success of this study.

References

Arnason, A. N., C. W. Kirby, C. J. Schwarz, and J. R. Irvine. 1996. Computer analysis of data from stratified mark-recovery experiments for estimation of salmon escapements and other populations. Canadian Technical Report of Fisheries and Aquatic Sciences 2106.

Bernard, D., and P. A. Hansen. 1992. Mark-recapture experiments to estimate the abundance of fish. Alaska Department of Fish and Game, Division of Sport Fish, Special Publication No. 92–4, Anchorage.

Brabets, T. P. 1997. Geomorphology of the lower Copper River, Alaska. United States Geological Survey, Denver, Colorado.

Buzby, K., and L. Deegan. 1999. Retention of Anchor and Passive Integrated Transponder Tags by Arctic Grayling. North American Journal of Fisheries Management 19:4:1147–1150.

Cappiello, T. A. and J. F. Bromaghin. 1997. Mark-recapture abundance estimates of fall-run chum salmon in the Upper Tanana River, Alaska, 1995. Alaska Fishery Research Bulletin 4(1):12–35.

Darroch, J. N. 1961. The two-sample capture-recapture census when tagging and sampling are stratified. Biometrika 48:241–260

Donaldson, I. J. and F. K. Cramer. 1971. Fishwheels of the Columbia. Binfords and Mort, Portland, Oregon.

Evenson, M. J., and K. G. Wuttig. 2000. Inriver abundance, spawning distribution, and migratory timing of Copper River Chinook salmon in 1999. Alaska Department of Fish and Game, Division of Sport Fish, Fishery Data Series No. 00–32, Anchorage.

Gordon, J. A., S. P. Klosiewski, T. J. Underwood, and R. J. Brown. 1998. Estimated abundance of adult fall chum salmon in the Upper Yukon River, Alaska, 1996. U.S. Fish and Wildlife Service, Fairbanks, Alaska.

Johnson, R. E., R. P. Marshall, and S. T. Elliott. (1992). Chilkat River chinook salmon studies, 1991. Alaska Department of Fish and Game, Division of Sport Fish, Anchorage.

Link, M. R., K. K. English, and R. C. Bocking. 1996. The 1992 fishwheel project on the Nass River and an evaluation of fishwheels as an inseason management and stock assessment tool for the Nass River. Canadian Manuscript Report of Fisheries and Aquatic Sciences 2372.

Link, M. R., and B. L. Nass. 1999. Estimated abundance of Chinook salmon returning to the Nass River, B.C., 1997. Canadian Manuscript Report of Fisheries and Aquatic Sciences 2475.

Meehan, W. R. 1961. The use of a fishwheel in salmon research and management. Transactions of the American Fisheries Society 90:490–494.

ODFW (Oregon Department of Fish and Wildlife). 2005. Coastal salmon spawning survey procedures manual 2005. Oregon Department of Fish and Wildlife, Corvallis.

Ricker, W. E. 1975. Computation and interpretation of biological statistics of fish populations. Fisheries Research Board of Canada Bulletin 191.

Savereide, J. W., and M. J. Evenson. 2002. Inriver abundance, spawning distribution, and migratory timing of Copper River Chinook salmon in 2001. Alaska Department of Fish and Game, Fishery Data Series No. 02–28, Anchorage.

Seber, G. A. F. 1982. The estimation of animal abundance and related parameters, 2nd edition. Charles Griffin, London.

Smith, J. J., M. R. Link, and B. D. Cain. 2005. Development of a long-term monitoring project to estimate abundance of Chinook salmon in the Copper River, Alaska, 2001–2004. Alaska Fishery Research Bulletin 11(2):118–134.

Sturhahn, J. C., and D. A. Nagtegaal. 1999. Results of the chinook assessment study conducted on the Klinaklini River during 1998. Canadian Manuscript Report of Fisheries and Aquatic Sciences 2497.

van den Broek, K. M. and J. J. Smith. 2005. Estimating the inriver abundance of Copper River sockeye salmon (feasibility study). 2005 annual report. Alaska Department of Fish and Game, Juneau, Alaska.

van den Broek, K. M., J. J. Smith and G. Wade. 2007. Estimating the inriver abundance of Copper River sockeye salmon, 2006 annual report. U.S. Fish and Wildlife Service, Office of Subsistence Management, Fisheries Resource Monitoring Program (Study No. 06–502), Anchorage, Alaska.

Wuttig, K. G., and M. J. Evenson. 2001. Inriver abundance, spawning distribution, and migratory timing of Copper River Chinook salmon in 2000. Alaska Department of Fish and Game, Fishery Data Series No. 01–22, Anchorage.

Zar, J. H. 1984. Biostatistical analysis, 2nd edition. Prentice-Hall, Saddle River, New Jersey.

American Fisheries Society Symposium 76:95–134, 2012

The Success of the Columbia Basin Passive Integrated Transponder (PIT) Tag Information System

David P. Marvin*
Pacific States Marine Fisheries Commission
205 SE Spokane Street, Portland, Oregon 97202, USA

Abstract.—Passive integrated transponder (PIT) tags have been used since 1987 to evaluate the survival and behavior of juvenile anadromous Pacific salmon *Oncorhynchus* spp. and steelhead *O. mykiss* as they migrate to the Pacific Ocean from the Columbia Basin. Through 2008, PIT tags have been implanted into 22 million parr and smolts originating from hatcheries and natal streams throughout the Basin. During their downstream migration, more than 8.5 million PIT-tagged fish have been passively detected as they are bypassed at one or more large hydroelectric dams in the main-stem Columbia and Snake rivers. As PIT tag technology has improved, larger detection systems have been installed in the adult fish passage facilities at many of these dams and elsewhere throughout the Basin. Tagged salmon and steelhead are commonly detected at multiple locations in the Columbia and Snake rivers and their tributaries during their downstream and subsequent upstream migrations. In addition to these passive detection events, more than 620,000 tagged fish have been physically recaptured, providing an opportunity to observe and record changes to the morphology and physiology of an individual tagged fish before re-releasing it. Almost 850,000 PIT tags have been recovered from dead fish or detected under circumstances that confirm the mortality of the tagged hosts. All of these separate PIT tag mark and recovery events are reported to the Columbia Basin PIT Tag Information System (PTAGIS), where the individual events for each tag are catalogued in the PTAGIS database. Standardized methods have been developed to systematically record these mark and recovery tag events, and to ensure that the data are accurately and consistently reported to PTAGIS by contributors from dozens of agencies and organizations in the Columbia Basin. There is essentially no delay between the acquisition of these validated field data and their incorporation into the PTAGIS database. Researchers and resource managers have immediate access, through the PTAGIS Web Portal, to the comprehensive sets of individual and correlated PIT tag events reported in the Columbia Basin.

Introduction

The Columbia Basin, located in the Pacific Northwest region of North America, encompasses portions of seven states in the United States and part of the Canadian province of British Columbia (Figure 1). Drained by the Columbia and Snake rivers and their tributaries, the Basin historically supported large populations of anadromous Pacific salmon *Oncorhynchus* spp. and steelhead *O. mykiss* (Chapman 1986). The construction of numerous dams throughout the "most hydroelectrically developed river system in the world" (Wagner 2007) has re-

*Corresponding author: Dave.Marvin.USA@gmail.com

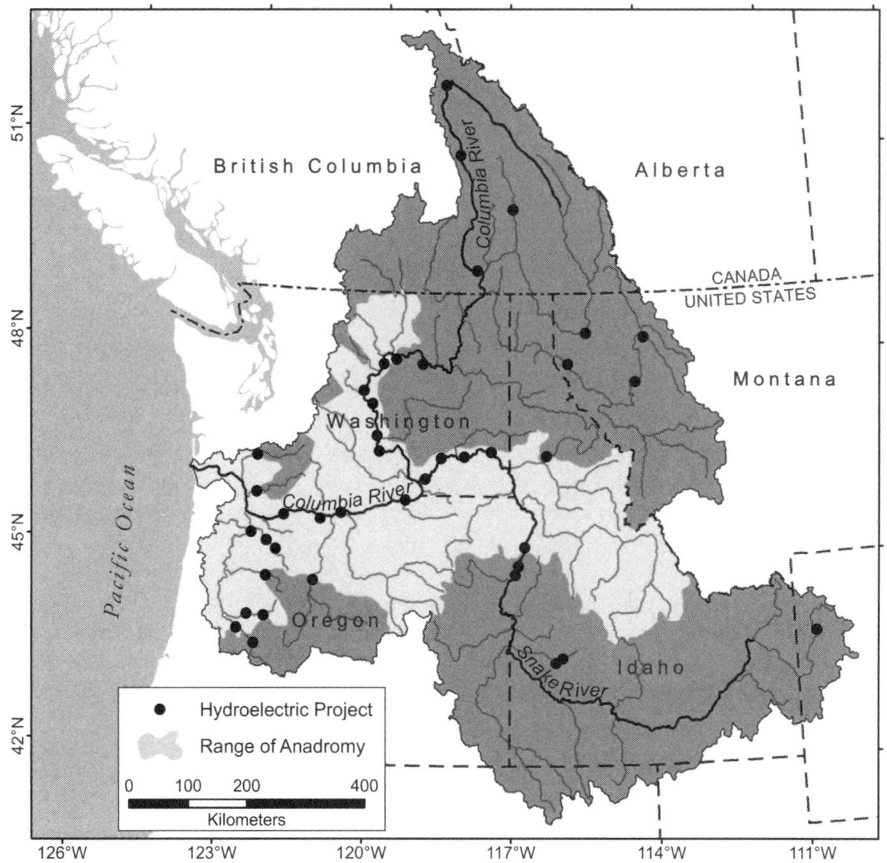

FIGURE 1. Major hydroelectric dams in the Columbia Basin, and the current extent of spawning and rearing habitat available to anadromous salmonids.

sulted in the decline in many of these populations, due both to the permanent loss of spawning and rearing habitat above impassible dams, and to the cumulative mortality incurred as salmon and steelhead smolts pass through as many as nine main-stem dams and reservoirs during their migration downstream to the ocean (Raymond 1988). Since the completion of Bonneville Dam on the lower Columbia River in 1939, researchers have continually evaluated the impacts of hydroelectric dam operations on the passage behavior and survival of anadromous salmonids in the Columbia Basin (Williams et al. 2005). For more than 20 years, the use of passive integrated transponder (PIT) tags has been crucial to the design and successful implementation of many of these studies.

Within their current range of anadromy, salmon and steelhead in the Columbia Basin may migrate more than 1,500 km to and from the Pacific Ocean. Researchers from more than 30 research and management entities have marked millions of steelhead and Chinook salmon *O. tshawytscha*, and hundreds of thousands of coho salmon *O. kisutch* and sockeye salmon *O. nerka*, in order to monitor the movements and disposition of individual tagged fish as they migrate through the Basin. Most migrating fish encounter between one and nine of the 13 hydroelectric dams located on the lower main-stem Snake and/or Columbia rivers. Seven of the 13 dams have a juvenile fish bypass system (JBS) at the powerhouse that diverts a large portion of downstream migrants away from the turbines and around the dam. Each JBS is instrumented with a sophisticated PIT tag detection system that passively and benignly detects essentially 100% of the tagged fish as they are bypassed around the dam.

The highly migratory nature of salmon and steelhead within the Columbia Basin necessitated the development of a cooperative exchange of the data acquired from PIT tag mark and release activities, and from any subsequent active recovery or passive detection events. During their freshwater migrations, these fish encounter "a web of jurisdictions and competing sovereign interests" that includes the federal governments of the United States and Canada, multiple tribal nations, and dozens of resource management agencies (Johnson et al. 2007). In 1987, the first year PIT tags were used to mark fish in the Columbia Basin, researchers from three fisheries management agencies tagged a combined total of 2,500 Chinook and steelhead smolts, and released those fish from six distinct locations in the Basin. Twenty one years later, in 2008, researchers from dozens of agencies collectively marked 2.5 million juvenile and adult salmon and steelhead, releasing those fish from hundreds of locations within the Basin. The increasing scale and scope of studies utilizing PIT tags, and the need by researchers and resource managers throughout the Basin to reliably and consistently exchange these enormous data sets, led to the creation of the Columbia Basin PIT Tag Information System (PTAGIS).

This paper presents a review of the development of PIT tag technology and its application to fisheries research and monitoring studies in the Columbia Basin; a detailed description of PIT tag mark and release, interrogation, recapture, and mortality events; an explanation of how these event data are systematically acquired by, correlated within, and distributed from the PTAGIS database; and a discussion of how these data are used to monitor and manage fisheries resources in the Basin.

The Use of PIT Tag Technology in the Columbia Basin

In 1984, the National Marine Fisheries Service (now NOAA Fisheries) began investigating the use of a 400 kHz full-duplex (FDX) PIT tag to identify and monitor individual salmon and steelhead smolts during their migrations in the Columbia Basin. Prentice et al. (1990a) described many of the attributes that the 400 kHz FDX PIT tag provided to fisheries researchers.

• The tag contained no internal power source; it was activated only as it passed through an electro-magnetic field generated by an external transceiver. Because of the passive nature of the tag, its operational life was expected to exceed that of even the most long-lived salmon species.

• Without the need for an internal power source, the tag components were contained within a glass capsule 12.0 mm long by 2.1 mm in diameter. Tags were inserted into the body cavity of juvenile fall Chinook and sockeye salmon as small as 55 mm fork length (FL) with no observed jeopardy to growth or survival.

• The rapid decoding of the FDX PIT tag's 10-character datagram (tag code) allowed effective detection of tagged fish passing through a detector antenna at velocities up to 3.7 m/s. An automated detection system was developed to issue a time stamp and record the tag code of a PIT-tagged fish as it swam through each antenna within a hydroelectric dam's JBS (Prentice et al. 1990b). Automating these "interrogation" events negated the need to re-handle a tagged fish subsequent to the initial marking event.

• Each tag was preprogrammed by the manufacturer with one of approximately 34 billion (10^9) unique hexadecimal codes. Manual transcription errors of the codes could be avoided by using a hand-held or table-top reader to decode and automatically record the tag code at the time of marking (Prentice et al. 1990c).

One limitation of the 400 kHz PIT tag was its short (18 cm) detection range (Prentice et al. 1990b). While this distance was sufficient to efficiently detect tagged salmonid smolts as they passed though the 30 cm (inside diameter) antennas commonly deployed in a main-stem dam's JBS, the 400 kHz transceivers could not effectively detect tagged fish passing through the larger orifices and other passageways located in the adult fishways at those dams. Additionally, the parity check imbedded in the 400 kHz tag code (Prentice et al. 1990a) could detect and self-correct some, but not all, decoding errors. During the 1990s, the radio frequency identification (RFID) industry developed new PIT tag technologies that improved the tag's read range and code fidelity. In 1999, following extensive coordination, the fisheries research community in the Columbia Basin converted from the use of the 400 kHz PIT tag model to the new RFID industry standard 134.2 kHz ISO-B FDX tag (hereafter denoted as the 134.2 ISO tag). This conversion required the replacement of all of the fixed interrogation systems installed in the juvenile fish bypass systems at six hydroelectric dams, and the staged replacement of the 400 kHz detectors installed in adult fish traps at two dams. The conversion to the 134.2 ISO PIT tag and detection system provided researchers with many additional or improved capabilities.

• The 134.2 ISO tag can be detected up to 1.0 m from the center of an antenna in a fixed interrogation system. This is a five-fold increase from the 18-cm range of the 400 kHz system, and has resulted in the deployment of larger antennas at some locations within the juvenile fish bypass systems. More significantly, large-geometry antennas have been installed in the adult fishways at many dams in the Basin, allowing researchers to monitor the up-stream passage of tagged adult salmon and steelhead detected at these sites. Under special operating conditions, 134.2 ISO tags can be detected as they pass through a 4.6 m square antenna at the Bonneville Dam Second Powerhouse Corner Collector flume.

• The 134.2 ISO tag can be detected up to 32 times per second (Zydlewski et al. 2006). This ability to resolve the tag code within 32 ms makes it possible to reliably decode multiple tags passing through antennas nearly simultaneously at velocities up to 5.0 m/s.

• Improving on the capabilities of the 400 kHz tag, the 134.2 ISO tag can be programmed with any of 274 billion unique identification codes. Laboratory tests of the 134.2 ISO tag conducted by NOAA Fisheries and the Pacific States Marine Fisheries Commission (PSMFC) documented an electronic mis-read rate of less than 0.000001 mis-identified detection events (S. Downing, NOAA Fisheries, personal communication). This capacity for a very large supply of unique tag identifiers is critical to the integrity of PIT tag mark and recapture activities in the Columbia Basin, where at least 2 million fish have been tagged each year since 2003.

• During the conversion to the 134.2 ISO technology in 1999, all of the original 400 kHz antennas in the juvenile fish bypass systems were replaced by PSMFC field engineering staff. Each new antenna was custom-wrapped to either concentrate or dissipate the antenna field as necessary to optimize its detection capability within a given flume or pipe. This optimization of the antenna field during installation has proven to have significantly reduced the time and effort required to tune and maintain the 134.2 ISO detection systems, compared to the 400 kHz systems.

• Regular improvements to both the 134.2 ISO tag and the transceiver equipment have resulted in significant increases in the read range of the tag, with commensurate increases in the sizes of interrogation system antennas. Future technological advances are expected to continue to provide additional improvements to detection performance.

The 134.2 ISO PIT tag is slightly larger than the 400 kHz tag it replaced. The current model of the 134.2 ISO tag commonly used in the Columbia Basin is the same diameter (2.1 mm) as the 400 kHz tag, but 0.5 mm longer. These 12.5 mm PIT tags can be used to mark salmonids as small as 60 mm FL (Zydlewski et al. 2006), although the Columbia Basin PIT Tag Steering Committee (PTSC) conservatively recommends a minimum tagging size of 70 mm FL for Pacific salmon and steelhead (PTAGIS 2004).

PIT Tag Mark and Recovery Events

Since 1987, researchers in the Columbia Basin have consistently used only 400 kHz or 134.2 kHz ISO FDX PIT tags. This adherence to a common technology has helped ensure that a tag inserted into a host salmon or steelhead can be reliably detected and reported during all subsequent recovery events anywhere in the Basin. A simple PIT tag event model is used to correlate the initial mark and release (tagging) event with any subsequent passive detection (interrogation) and physical recovery (recapture or mortality) events for every tagged salmon and steelhead in the Basin (Figure 2). The components of this event model are described below.

Mark and release (tagging) events

Juvenile salmon and steelhead are typically tagged using the techniques outlined by Prentice et al. (1990c). During the 1980s and early 1990s, PIT tags were inserted into the body cavities of parr or smolts using a 12-gauge hypodermic needle attached either to a modified syringe for delivery of a single tag, or to an electro-mechanical or pneumatic "auto-tagger" injector fitted with a gravity-fed magazine containing up to 150 tags. However, the marking speed afforded by the auto-tagger was offset by concerns about the use of a single needle to inject tags into multiple fish. The needles became dull with repeated use, creating a puncture wound rather than a clean incision. Also, with no disinfection between injections, the auto-tagger increased the risk for the horizontal transmission of disease at hatcheries and when mixed stocks and species of migrating fish were tagged at traps and dams. By 1993, researchers had discontinued the use of auto-tagger devices when marking mixed stocks of fish, and an auto-tagger was last used at a hatchery in 1998. In the Columbia Basin, fish are generally tagged using individual needles disinfected between each use (Achord et al. 2007).

The reported rate of direct mortality due to insertion of the PIT tag is generally low, as is shedding of the tag after marking. Dare (2003) reported a cumulative mortality rate, due to either handling or tagging, of less than 1.0% during a 28-d study of 145,000 Chinook tagged at a hatchery. During that study, the tag retention rate for fish marked during the first day was 98.1%, and increased to 99.996% overall as the inexperienced tagging personnel improved their skills. Achord et al. (2007) reported an average direct mortality rate of 0.1% for wild Chinook parr tagged at

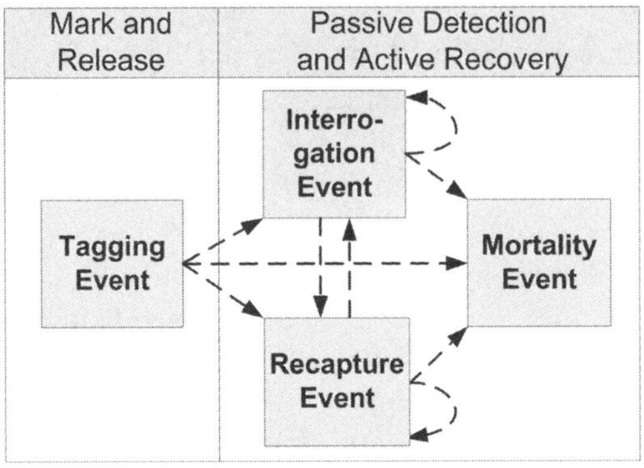

FIGURE 2. PIT tag event model. After a tagged salmon or steelhead is released, it may be passively detected (interrogated) or physically recaptured one or more times during its migration to and from the ocean. The retrieval of a dead tagged fish, or the ex corporeal detection of its tag, is reported as a mortality event. The occurrence of a detection or recovery event is independent of any previous event.

multiple study locations in the Salmon River Basin between 1991 and 2003.

Juvenile salmon and steelhead are marked with PIT tags and released at over 500 locations distributed throughout most of their range in the Columbia Basin (Figure 3). Fish tagged at hatcheries in the Basin may be released directly from the hatchery or from a satellite acclimation pond, with the release event occurring weeks or months after the fish are tagged. Conversely, naturally-reared parr and smolts are usually collected, tagged, and then immediately released back into their natal streams. Mixed stocks of fish are collected, tagged, and released at numerous instream traps and weirs. Smolts are regularly intercepted and tagged at many of the dams on the Columbia and Snake rivers and their tributaries. The number of tagged fish released in each of the Columbia Basin's major watersheds reflects the research and monitoring efforts in those regions (Figure 4). However, that marking effort is not necessarily proportional to the size of the various populations of salmon and steelhead in the Basin, nor representative of their distributions (Table 1).

Researchers began using PIT tags to mark returning adult salmon and steelhead in 2000, when 134.2 ISO detection systems were first installed in fish ladders at dams in the Columbia Basin. Most often these adult fish are tagged in order to augment the population of upstream migrants that were previously tagged during their juvenile life stages. Far fewer salmonids are marked as adults than are tagged during the parr or smolt life stages. Approximately 35,000 PIT tags have been inserted into adult salmon and steelhead—less than 0.02% of all fish tagged in the Basin.

Incidental numbers of other fish species in the Columbia Basin have been marked with 400 kHz or 134.2 ISO FDX PIT tags. Approximately 13,000 bull trout *Salvelinus confluentus* have been tagged, as have 11,500 anadromous coastal cutthroat trout *O. clarkii clarkii* and adfluvial westslope cutthroat trout *O. clarkii lewisii*, along with smaller numbers of brook trout *Salvelinus fontinalis*, mountain whitefish *Prosopium williamsoni*, northern pikeminnow *Ptychocheilus oregonensis*, adult American shad *Alosa sapidissima*, and Pacific lamprey *Lampetra tridentata*. In total, 42,000 of these other species have been marked with FDX PIT tags, accounting for 0.02% of all fish tagged.

More than 99.96% of the 22 million PIT tags used in the Columbia Basin since 1987 have been inserted into juvenile salmon and

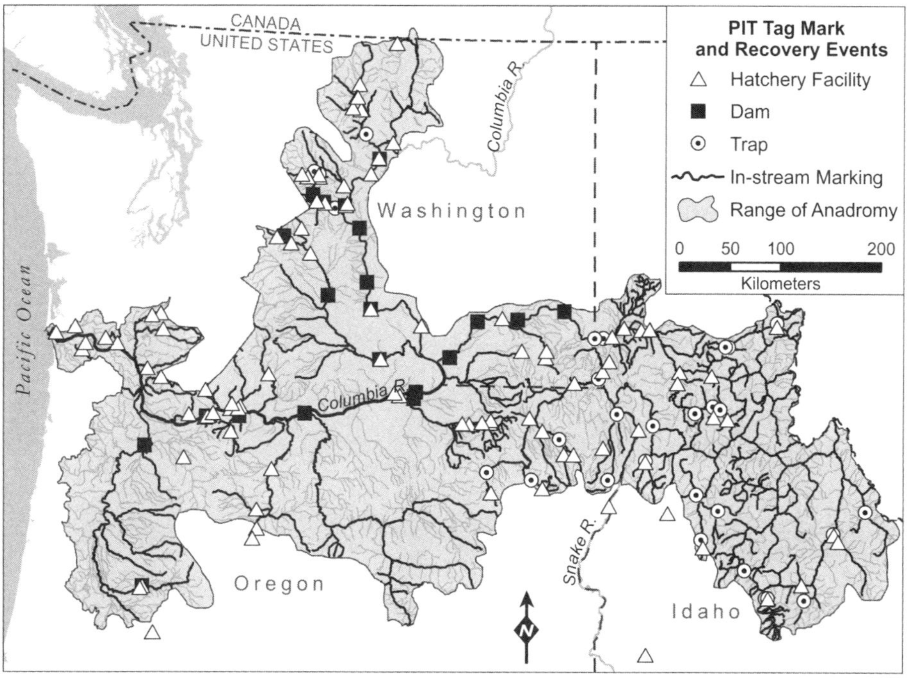

FIGURE 3. Location of PIT tag marking and recovery activities in the Columbia Basin. These activities occur on or at more than 550 rivers and streams, dams, traps, and hatchery rearing and release facilities throughout the Basin within the range of anadromous salmon and steelhead.

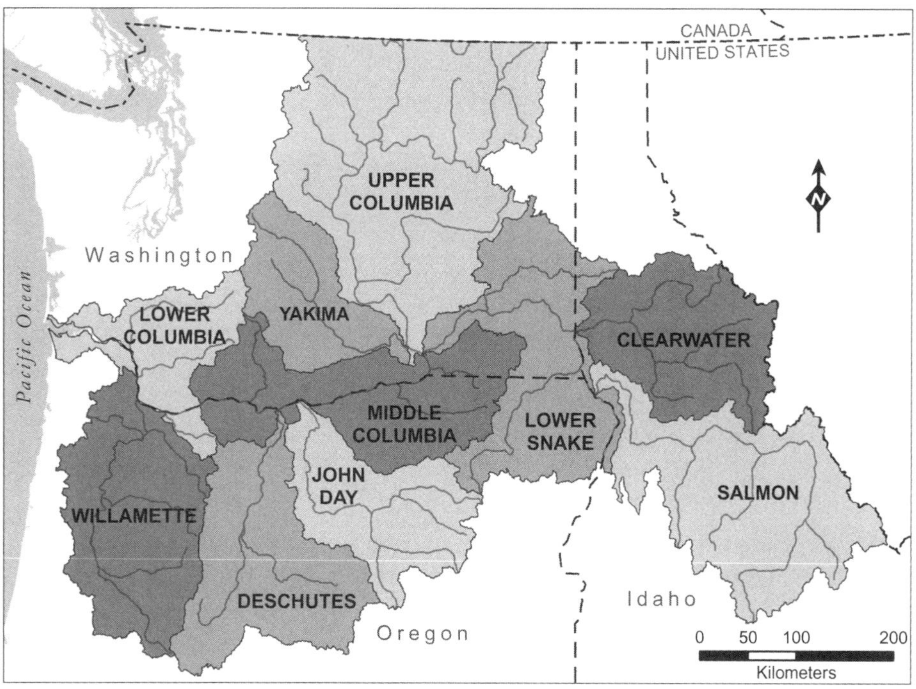

FIGURE 4. Sub-basins within the lower portion of the Columbia Basin accessible to anadromous salmonids.

TABLE 1. Numbers of PIT-tagged fish released in the Columbia Basin, by sub-basin, 1987–2008.

Year[a]	Sub-basins						
	Lower Snake	Upper Columbia	Salmon	Clearwater	Middle Columbia	Yakima	Other sub-basins
1987	20,179		7,673				
1988	6,689		6,259	9,919	25,089		
1989	29,622	4,997	40,615	24,618	20,281	3,979	
1990	10,889	7,857	33,519	28,730		23,241	
1991	14,985	7,960	30,437	31,428		32,727	
1992	18,988	11,677	29,623	27,279	28	30,645	
1993	74,221	24,530	48,790	23,229		29,693	
1994	206,050	35,102	60,098	59,548	28,240	1,812	1,775
1995	379,021	37,584	70,013	68,092			
1996	220,569	51,000	77,732	77,066			6,024
1997	211,914	40,420	146,879	101,718	103,731	3,692	10,704
1998	332,637	147,133	143,940	113,927	186,112	14,803	12,568
1999	447,826	222,761	183,137	153,748	412,942	80,875	8,730
2000	319,361	176,542	177,425	117,484	394,683	87,976	16,044
2001	280,718	300,845	169,292	129,372	120,005	78,636	28,680
2002	429,327	613,287	323,825	127,807	154,520	79,956	58,849
2003	432,895	1,193,003	389,027	136,390	45,411	93,662	44,311
2004	293,841	1,148,706	246,117	143,482	65,711	79,342	39,922
2005	415,986	775,654	252,736	198,454	91,186	115,884	53,171
2006	909,538	155,864	258,004	411,485	100,354	128,368	50,879
2007	470,312	229,287	273,734	238,472	111,177	129,372	70,071
2008	976,622	235,033	360,629	408,077	232,876	169,001	91,510
Total	6,502,190	5,426,915	3,323,385	2,630,325	2,092,346	1,183,664	493,238

[a] Releases are reported by the year during which the tagged fish were expected to migrate to the ocean.

steelhead (Table 2). These tagged fish then migrated downstream in the main-stem Snake and Columbia rivers to the Pacific Ocean. Researchers tagged those fish with the expectation that a significant portion of the migrants would be routed into a juvenile fish bypass system at one or more of the main-stem hydroelectric dams, and that the tagged fish would be detected by the interrogation systems installed in those bypass systems.

Automated passive detection (interrogation) events

As salmonid smolts migrate downstream in the Snake and/or Columbia rivers, they encounter up to nine hydroelectric dams (Figure 5). When these downstream migrants encounter a main-stem dam, they either pass over the spillway or enter the powerhouse. Seven of the eight dams in the lower Columbia and Snake rivers (Bonneville, John Day, McNary, Ice Harbor, Lower Monumental, Little Goose, and Lower Granite dams) are equipped with fish diversion screens in the powerhouse turbine intakes. The diversion screens typically guide 70% or more of the smolts away from the turbines (Ferguson et al. 2005) into a collection channel within the powerhouse, and from there through a primary dewatering system into a low-flow (<1.0 cms) bypass pipe. This assemblage of diversion screens, collection channel, and bypass pipe is collectively referred to as a juvenile fish bypass system (JBS). These juvenile fish bypass systems benignly concentrate and divert a large portion of the smolts that pass these dams (Figure 6). The bypassed fish are directed back to the river in the tailrace below the dam, or are routed to a juvenile fish facility (JFF) that further concentrates the bypassed fish in order to sample a portion of them (Figure 7). The juvenile fish facilities at Lower Granite, Little Goose, Lower Monumental, and McNary dams are used to collect salmon and steelhead smolts, which are then transported downstream around the remaining dams to a release location below Bonneville Dam (Matthews et al. 1977). According to data obtained from the Fish Passage Center (FPC) on-line database (available at www.fpc.org), more than one million smolts per day may be collected at a single transportation facility during the peak of the spring migration.

Prentice et al. (1990b) evaluated automated 400 kHz PIT tag interrogation systems in the juvenile fish facilities at McNary and Lower Granite dams during 1985 and 1986, and reported tag detection efficiencies between 95% and 100%, with detection accuracy exceeding 99%. These favorable results led to the deployment, in 1987, of permanent interrogation systems at McNary and Lower Granite dams and a new installation at Little Goose Dam. Since 1993, new juvenile fish monitoring and transportation facilities at Lower Monumental, John Day, and Bonneville dams have been designed and constructed with PIT tag interrogation systems integrated directly into each passage route. These permanent installations employ a larger number of PIT tag detectors than the systems evaluated in 1985 and 1986, providing more redundancy and a higher probability of tag detection. Using the empirical methods described by Prentice et al. (1993), the cumulative detection efficiency consistently exceeds 99.99% for tagged fish that pass through four or more of the PIT tag detectors currently deployed within a JFF. With the use of 134.2 ISO tags and detectors, the detection accuracy rate is essentially 100%.

Prentice et al. (1990b) compared the recovery rates of PIT-tagged fish and fish marked with freeze brands, the standard marking technique used during the 1980s to monitor the arrival timing and passage distributions of juvenile salmon and steelhead at key dams in the Columbia and Snake rivers. Fish marked with PIT tags were passively interrogated at a higher rate than the freeze-

TABLE 2. Numbers of PIT-tagged fish released in the Columbia Basin, by species, 1987–2008.

	Species						
Year[a]	Chinook	Steelhead	Coho	Sockeye	Bull trout	Cutthroat trout[b]	Other species
1987	20,752	5,166		1,934			
1988	29,836	17,975		145			
1989	88,552	31,581		3,979			
1990	73,448	23,344		7,441			
1991	79,639	23,280	6,040	8,578			
1992	72,046	29,338	4,455	12,401			
1993	132,383	34,633		33,440			
1994	242,387	143,186		3,884			
1995	478,646	80,489		8,134			
1996	333,242	80,396	5,338	16,348			
1997	440,251	127,083	47,359	4,267			98
1998	768,033	85,141	76,009	21,108	569		260
1999	1,081,523	353,891	60,960	13,243	326		76
2000	939,838	242,462	95,272	8,707	295		2,940
2001	917,538	132,731	49,500	5,893	126	338	1,422
2002	1,550,757	168,465	56,683	8,205	464	1,040	1,957
2003	1,628,977	631,901	58,143	11,305	1,126	2,101	1,146
2004	1,284,365	653,460	67,378	7,930	1,014	1,936	1,038
2005	1,030,062	772,866	83,889	8,939	2,394	1,959	2,962
2006	1,563,673	355,259	78,579	10,548	1,726	1,893	2,814
2007	1,072,350	325,573	79,820	39,502	2,567	979	1,634
2008	1,868,057	483,682	88,884	28,056	2,520	1,219	1,330
Total	15,675,565	4,801,902	858,309	263,987	13,127	11,465	17,677

[a] The year during which the tagged fish were expected to migrate to the ocean.
[b] Includes both coastal cutthroat trout *Oncorhynchus clarkii clarkii* and westslope cutthroat trout *O. clarkii lewisii*.

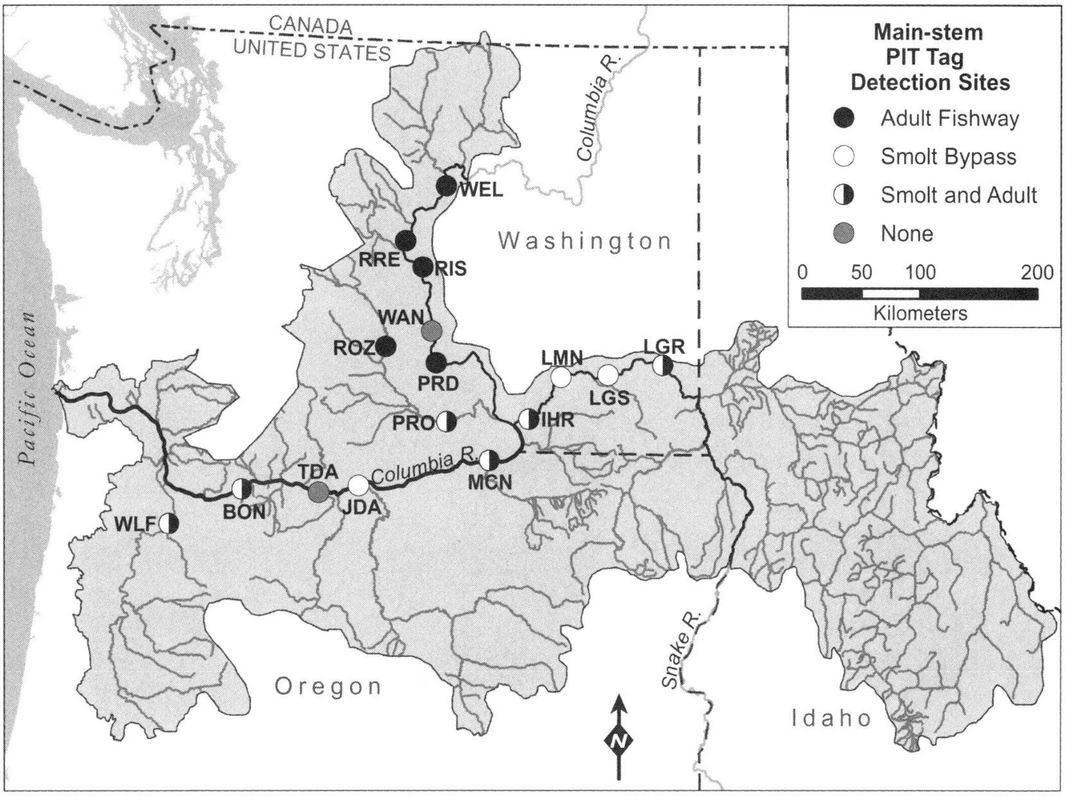

FIGURE 5. Locations of permanent PIT tag interrogation systems at dams in the Columbia Basin. BON = Bonneville Dam; IHR = Ice Harbor Dam; JDA = John Day Dam; LGR = Lower Granite Dam; LGS = Little Goose Dam; LMN = Lower Monumental Dam; MCN = McNary Dam; PRD = Priest Rapids Dam; PRO = Prosser Dam; RIS = Rock Island Dam; ROZ = Roza Dam; RRE = Rocky Reach Dam; TDA = The Dalles Dam; WAN = Wannapum Dam; WEL = Wells Dam; WLF = Willamette Falls fishway and Sullivan Dam bypass. On the Columbia River, only The Dalles and Wannapum dams have no PIT tag detection capabilities.

branded fish were recovered, with less statistical variability between replicates, and without the need to handle numerous unmarked fish in order to collect a sufficient number of the small proportion of freeze-branded fish in the Columbia Basin's out-migrant salmonid population. By 1992, PIT tags had replaced freeze brands as the primary mark used to monitor the downstream migrations of anadromous salmonids in the Basin (Downing et al. 2001).

The conversion to the 134.2 ISO tag in late 1999, and the increased read range afforded by that tag, made it feasible to install PIT tag detectors on each of the large (91.4–121.9 cm inside diameter) primary fish bypass pipes at Little Goose, Lower Monumental, McNary, John Day, and Bonneville dams. By locating four detectors in series on the primary bypass pipe, more than 99.5% of the tagged fish passing a dam are detected as they pass through its JBS (Axel et al. 2005). These detectors provide the only interrogation capabilities during those intervals when fish are bypassed but the JFF is not in operation. There is no dedicated fish sampling facility in the JBS at Ice Harbor Dam, so detectors installed on its primary bypass pipe in

FIGURE 6. McNary Dam is located below the confluence of the Snake and Columbia rivers. Some salmon and steelhead smolts migrating seaward pass over the spillway, while others enter the powerhouse. Some of the latter are diverted into the juvenile fish bypass system, visible at the lower right of **(a).** Fish diverted around the powerhouse through the primary bypass pipe pass through four PIT tag antennas (circled in **(b)**), and can then be discharged directly back into the tailrace (center of **(b)**) or routed into the collection facility and transported downstream by truck or barge.

FIGURE 7. The juvenile fish collection facility at Lower Monumental Dam. Fish are conveyed in pipes and flumes through this and similar facilities. PIT tag detectors are integrated into all of the possible routes through the facility to confirm the path taken by each tagged fish.

2005 provide the only PIT tag interrogation opportunity for bypassed fish at that site.

The current deployment of PIT tag detectors at the seven juvenile fish bypass and monitoring facilities provides comprehensive interrogation capabilities for all fish routed through those facilities (Table 3). At a dam with a typical smolt bypass system and monitoring facility (Figure 8), tagged fish are first interrogated as they pass through the four detectors on the primary bypass pipe. If the fish are routed into the JFF, they pass through detectors located in each segment of the separate routes within the collection and transportation facility. Depending on facility operations and research needs, PIT-tagged fish may be detained for specific research projects; discharged from the JFF into the tailrace; held in a raceway with untagged fish until they are transported downstream; or routed with untagged fish to a sampling facility where they are anesthetized, examined, and then released back into the general population (Ferguson et al. 2005).

In addition to simply auditing the passage of PIT-tagged fish through a fish facility, the detectors can also affect the passage route and ultimate disposition of fish within the facility. Since 1993, tagged fish have been automatically segregated from the nontagged population and returned to the tailrace at the smolt transportation sites in the Columbia Basin. At these sites, the detection of a PIT tag automatically activates a mechanical gate to selectively divert the tagged fish away from the untagged, transported population (Marsh et al. 1999). For survival studies, this passive "re-release" of the tagged fish back to the river is integral to the "Complete Capture History Protocol" experimental design identified by Burnham et al. (1987). The ability to mechanically segregate tagged fish from other fish was further refined in 1996, when NOAA Fisheries developed a separation-by-code (SxC) computer application (MULTIMON) that selectively triggers one or more diversion gates based on the recognition of individual PIT tag codes. The MULTIMON software can quickly (<1.5 ms) query a lookup database containing several million tag codes, identify an individual PIT tag candidate, and immediately activate a gate to route that tagged fish to a specific pathway through the facility (Downing et al. 2001).

While there are multiple routes of dam passage available to juvenile fish migrating downstream, all adult migrants at a dam must ascend a fish ladder to pass upstream. The limited read distance of the 400 kHz interrogation systems precluded their deployment in the adult fishways at main-stem dams except for installations in adult fish traps located in side-channels at Lower Granite Dam (Prentice et al. 1990b) and Bonneville Dam (McCutcheon et al. 1994). The increased read range available with the first production model of the 134.2 ISO tag led to the deployment of interrogation systems in the main fish ladder channels at McNary and Bonneville dams. Antennas installed in the submerged orifices of the ladder weirs reliably detected the tagged fish that passed through those orifices, but tagged migrants that passed over those weirs were not interrogated (Figure 9). Subsequent improvements in 134.2 ISO detector and tag technology made it feasible to install larger antennas at an adult fish facility's counting window or in vertical slot weirs located near the ladder exit (Figure 10). Serial arrays of PIT tag detectors are now installed in every adult fishway at Bonneville, McNary, Priest Rapids, Rock Island, Rocky Reach, and Wells dams on the Columbia River, as well as at Ice Harbor and Lower Granite dams on the lower Snake River. These interrogation systems are capable of detecting all tagged fish ascending the adult fishways at each of these dams (Figure 11).

The MULTIMON SxC application is used at Bonneville and Lower Granite dams to identify and collect target fish from the adult fish ladders. As at the juvenile fish fa-

Table 3. Numbers of PIT-tagged fish released in the Columbia Basin, numbers and percentages of tags detected, and the average detections per tag, 1987–2008.

Year[a]	Tagged Fish Released	Detection Events[b]	Unique Tags Detected	Percentage of Tags Detected[c]	Average Detections per Tag[d]
1987	27,852	1,522[e]	1,522	5%	1.00
1988	47,956	24,478	24,471	51%	1.00
1989	124,112	58,941	55,266	45%	1.07
1990	104,236	36,031	33,583	32%	1.07
1991	117,537	53,354	47,889	41%	1.11
1992	118,240	41,454	39,066	33%	1.06
1993	200,463	107,547	77,285	39%	1.39
1994	389,457	249,044	182,614	47%	1.36
1995	567,282	393,607	236,090	42%	1.67
1996	435,329	275,074	164,983	38%	1.67
1997	619,058	339,325	217,845	35%	1.56
1998	951,120	681,097	405,577	43%	1.68
1999	1,510,019	1,012,063	621,685	41%	1.63
2000	1,289,515	718,000	546,993	42%	1.31
2001	1,107,548	918,619	556,839	50%	1.65
2002	1,787,571	1,450,893	804,737	45%	1.80
2003	2,334,699	1,350,629	987,725	42%	1.37
2004	2,017,121	966,295	659,786	33%	1.46
2005	1,903,071	1,035,876	662,383	35%	1.56
2006	2,014,492	1,310,567	824,822	41%	1.59
2007	1,522,425	964,914	626,167	41%	1.54
2008	2,473,748	1,403,682	963,972	39%	1.46
Total	21,662,851	13,393,012	8,741,300		

[a] Releases are reported by the year during which the tagged fish were expected to migrate to the ocean; detection events are reported in the year they occurred.

[b] A detection event is defined as a unique PIT tag detected at a distinct interrogation site. If the same tag is detected at two (or more) separate sites, then two (or more) detection events are reported for that tag.

[c] Percentage of tagged fish that were subsequently detected at one or more interrogation sites. Since 1993, most PIT-tagged fish have been diverted back to the river after being detected at one of the transportation sites in the Snake and Columbia rivers, increasing the probability they will be detected again at another site downstream.

[d] Average number of detection events per tagged fish detected at least once.

[e] The 1987 detection data collected by NOAA Fisheries from LGR and MCN were not recorded in the PTAGIS database; only the detection data from LGS are available from PTAGIS for 1987.

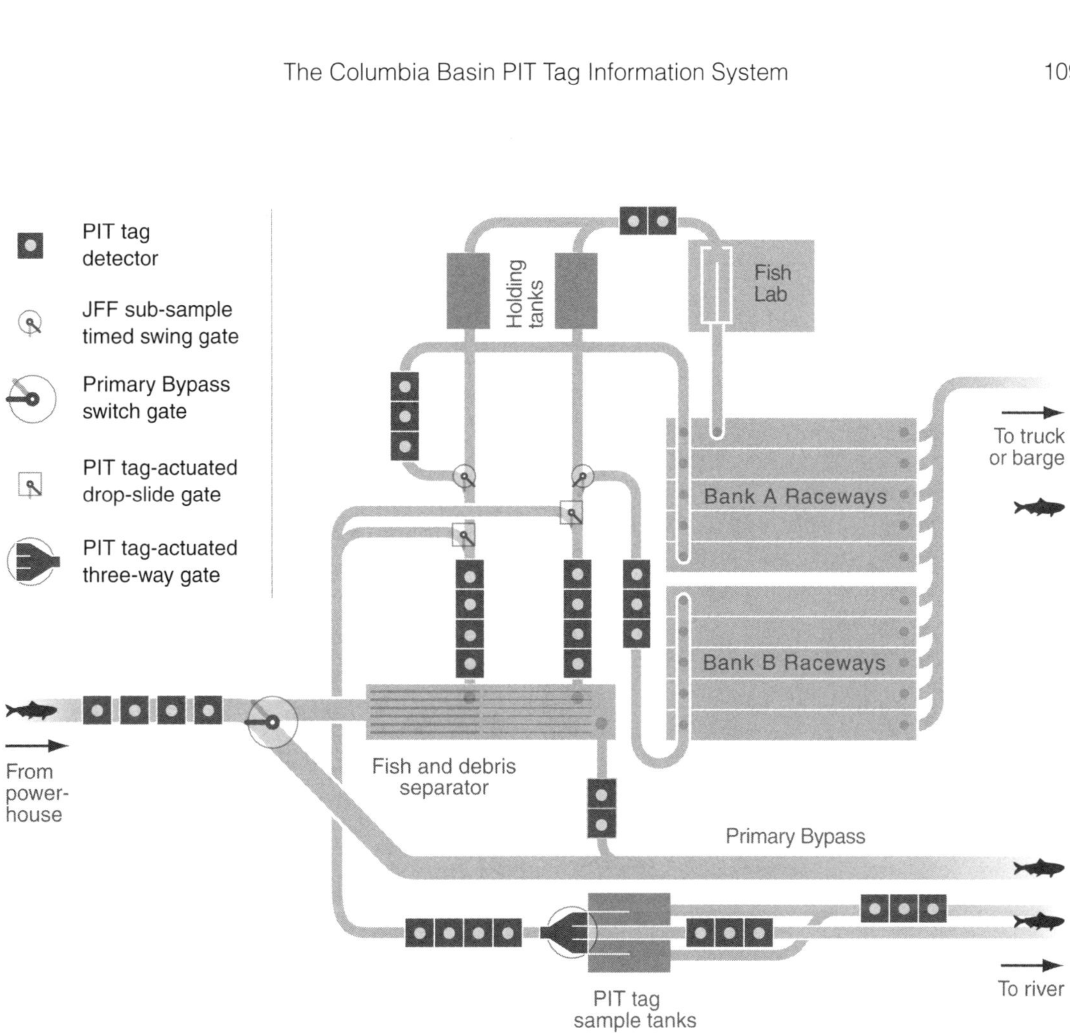

FIGURE 8. Schematic of a typical juvenile fish collection and transportation facility. As they exit the powerhouse, fish in the large diameter (91–122 cm) primary bypass pipe pass through four PIT tag antennas. Fish directed over the fish and debris separator pass through at least two, and as many as 11, antennas as they are routed to their various destinations within the facility. The separation-by-code system allows the selective diversion of tagged fish at PIT tag-actuated gates. The diverted fish are typically routed directly back to the river, or collected and held for examination.

cilities, preselected PIT-tagged fish are automatically shunted into the adult fish trap at Lower Granite Dam when their detection triggers the diversion gate. At Bonneville Dam, a member of the trapping crew manually activates the gate in response to one of two distinct audible and visible signals generated from MULTIMON. One signal alerts the crew to the presence of a target tagged fish, which can then be manually routed into a collection tank. The other signal warns the crew of a tagged fish to be avoided so it can instead return to the main ladder to continue its upstream migration without interruption. The SxC look-up lists used at the Bonneville Dam adult fish trap routinely contain more than 2 million tag codes, predominantly from studies that require that their fish pass unhindered through the trap and back into the ladder (S. Lee, University of Idaho, personal communication).

The continued improvements in detection performance have made it possible to deploy progressively larger antennas in juvenile fish

FIGURE 9. Partial overflow weirs in one of the adult fish ladders at Bonneville Dam. Fish can pass over the notches in the top of the weir wall, or through the orifices at the base of the weir **(a)**. PIT tag antennas were retro-fitted into the orifices at several of the weirs; the antenna inserts were designed to match the hydraulic characteristics of the original orifices **(b)**. The ladder may remain in continuous operation for 12 months or more **(c)**, during which time the PIT tag antennas are also expected to function continuously, as they cannot be accessed for repair or replacement.

bypass systems and adult fishways. To date, the largest 134.2 ISO PIT tag interrogation system in the Columbia Basin is installed in the Second Powerhouse Corner Collector at Bonneville Dam (B2CC). The corner collector flume operates with a flow of 146 cms at a velocity of 10.7 m/s. A single antenna spans the full 4.6 m width of the bypass flume (Figure 12). This antenna, fabricated specifically for this location and driven by a custom-designed transceiver, detects approximately 70% of the tagged fish passing through the B2CC flume (S. Downing, NOAA Fisheries, personal communication).

The juvenile fish bypass systems and adult fishways at the hydroelectric dams in the main-stem Columbia and Snake rivers provide the world's most advanced fish passage technology (Wagner 2007). These passage facilities benignly pass anadromous salmonids and other fish around the dams with minimal delay. The PIT tag interrogation systems installed at these dams are fully integrated into the passage facilities. These systems efficiently detect tagged fish without impacting operations at juvenile fish facilities or impeding passage in adult fishways. The separation-by-code subsystems in these passage facilities are highly effective at diverting and collecting target PIT-tagged salmon and steelhead while minimizing any impacts to untagged and nontarget tagged fish.

At some hatcheries and acclimation sites in the Columbia Basin, juvenile salmon and steelhead volitionally exit the rearing facility to begin their downstream migration (Figure 13). These releases may span many weeks, and are often unattended. The release dates

Figure 10. One of four slotted weirs reconstructed with PIT tag detectors in an adult fishway at Bonneville Dam. In order to provide an electrically noise-free environment, the original steel-reinforced weir **(a)** was demolished and replaced with new concrete poured around fiberglass reinforcement bars secured with plastic ties **(b)**. Custom antennas (71 cm wide and up to 3.5 m high, inside dimensions) mimic the dimensions of the original weir slots **(c)**. All fish must pass through each of the four instrumented slotted weirs **(d)**; this redundancy provides 100% detection capability for tagged fish ascending the ladder.

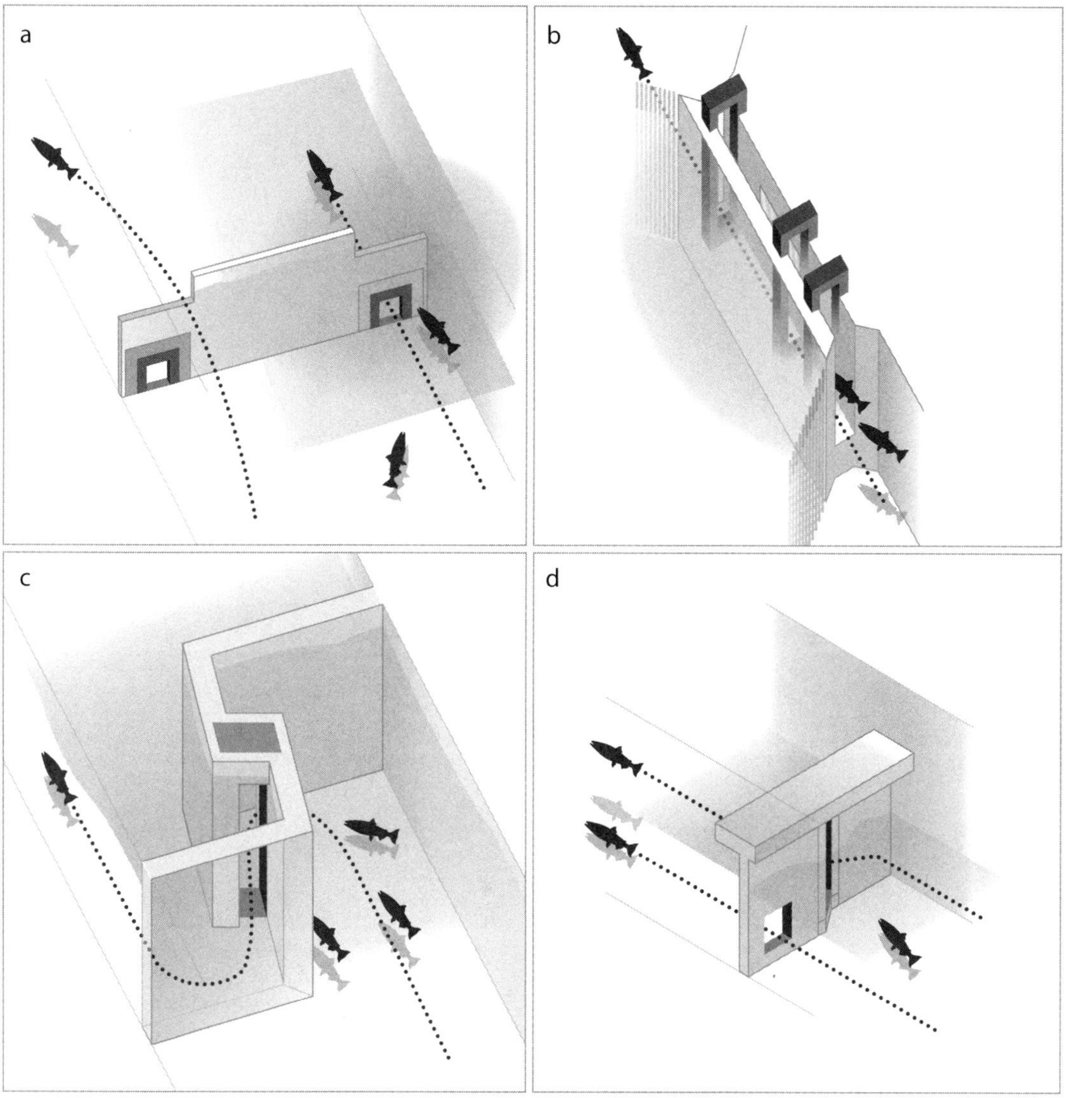

Figure 11. PIT tag antenna locations in adult fishways. At partial or complete overflow weirs **(a)**, tagged fish avoid detection if they pass over the top of the weir. All fish must pass through a series of antennas at a fish counting window channel **(b)**. Similarly, all fish must pass through the full-slot antennas in one or more serpentine weirs **(c)**. Combination weirs provide a submerged orifice and vertical slot for passage **(d)**, and both are instrumented with antennas.

Figure 12. The Bonneville Dam Second Powerhouse Corner Collector bypasses salmon and steelhead smolts from the forebay of the Second Powerhouse through a concrete flume downstream 850 m into the center of the river **(a)**. The bypass flume is 4.6 m wide **(b)**. A section of the concrete was demolished, electrically isolated, and reconstructed with nonferrous reinforcement, prior to the installation of a single highly-customized PIT tag antenna **(c)**. This antenna, shown before the internal structural bracing was removed **(d)**, spans the entire width of the flume and has a measured detection efficiency of 70%.

are generally reported beginning with the date fish are first able to exit the facility, and concluding with the date the facility is dewatered or cleared of any remaining fish. At 12 of these facilities, PIT tag detectors are deployed at the outfall of the hatchery pond or raceway to monitor the egress of the tagged portion of the population (Table 4). An accurate release date can be obtained for each tagged fish using the date and time of its interrogation event at the facility exit. This makes it possible to compute a precise travel time for each tagged host, from its last detection at the hatchery to its next downstream interrogation event. If the tagged fish are representative of the untagged population at that hatchery or pond, then these detections can also be used to describe the release distribution of the untagged population.

Automated PIT tag detection systems are installed at many other locations in the Columbia Basin. Interrogation systems at Prosser and Roza dams on the Yakima River, and at the Sullivan Dam-Willamette Falls complex on the Willamette River, monitor the passage of tagged fish through the juvenile and/or adult fish passage facilities at those locations. Similar detection systems are installed at other water diversion dams located on smaller streams. Various pass-through antenna designs (Zydlewski et al. 2006) and pass-over and hybrid designs (Connolly et al. 2008) are used at numerous in-river PIT tag detection systems to monitor the upstream

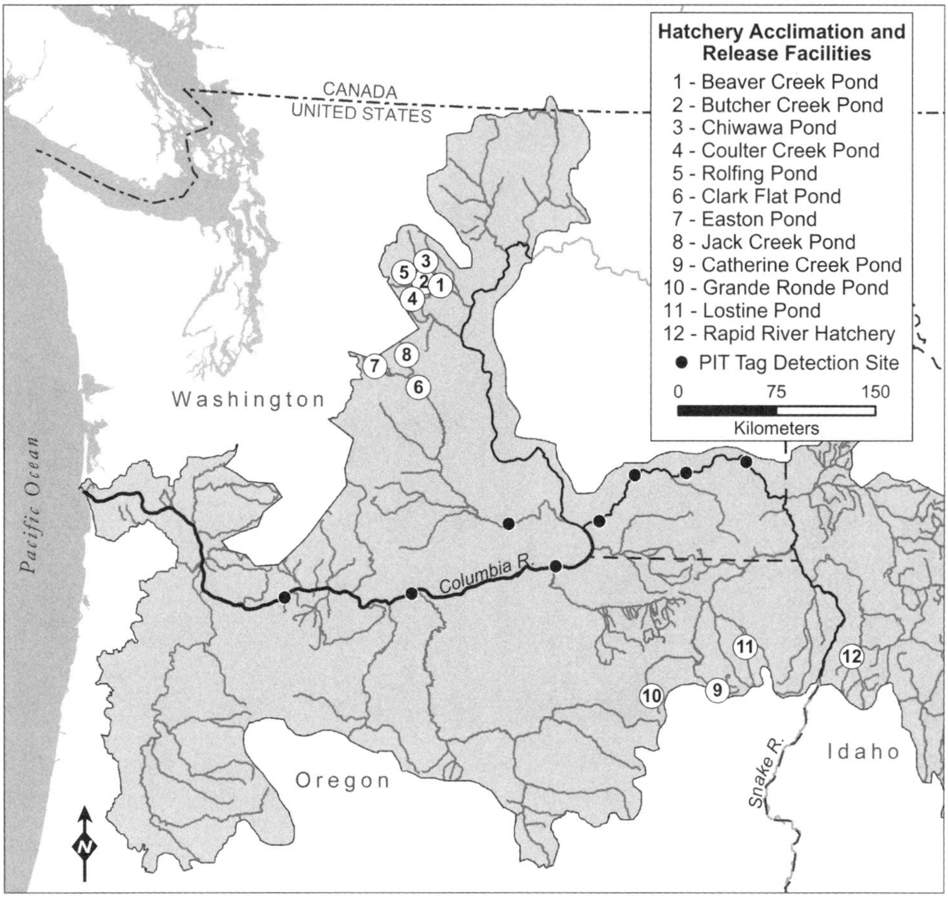

FIGURE 13. Locations of hatchery and satellite acclimation facilities in the Columbia Basin with volitional releases monitored for PIT tag detections.

and downstream movements of anadromous salmon and steelhead, adfluvial bull trout, and other species. NOAA Fisheries operates a surface trawl equipped with a PIT tag interrogation system in its cod end to detect tagged smolts migrating downstream into the Columbia River estuary (Ledgerwood et al. 2004). The detection data from these various locations augments the interrogations reported from the fish passage facilities at the mainstem hydroelectric dams.

Physical recovery (recapture and mortality) events

During the migrations subsequent to their tagging and release events, PIT-tagged fish may be physically recovered and visually inspected. These recovery events provide an opportunity to observe changes in the morphology and physiology of individual tagged fish. Recovery events can also be used in conjunction with, or in lieu of, interrogation events to audit the movements of individual fish. Previously-tagged fish that are intercepted and then allowed to continue their migration are reported to PTAGIS as "recapture" events. Recoveries from dead or moribund fish, and recoveries (or in situ detections) of bare tags separated from their host fish, are reported as "mortality" events.

Since 1987, 622,600 PIT-tagged fish in the Columbia Basin have been recaptured,

Table 4. Numbers of PIT-tagged fish detected during volitional releases from hatchery rearing and acclimation facilities in the Columbia Basin, 1999–2008.

	Release Year										
Release Site	1999	2000	2001	2002	2003	2004	2005	2006	2007	2008	Site Total
Yakima River sites											
Clark Flat Pond	3,053	14,408	12,974	12,224	6,934	12,996	12,918	12,980	13,071	15,192	116,750
Easton Pond	5,000	14,598	12,938	12,450	6,436	12,963	12,982	12,766	12,936	11,657	114,726
Jack Creek Pond		11,425	12,891	12,981	23,166	13,050	13,056	10,691	13,122	11,776	122,158
Grande Ronde River sites											
Catherine Creek Pond		3,435	7,155	12,908	12,073	12,754	12,486	15,003	10,627	13,603	100,044
Grande Ronde Pond				1,210	1,290	836		439	1,571	1,467	6,813
Lostine Pond				8,574	8,098	12,503	12,321	12,080	5,487	5,239	64,302
Wenatchee River sites											
Beaver Creek Pond								6,231	5,918		12,149
Butcher Creek Pond						5,282	5,293			4,957	15,532
Chiwawa Pond									8,653	6,855	15,508
Coulter Creek Pond								5,123	4,794		9,917
Rolfing Pond						3,678	6,067			4,420	14,165
Rapid River Hatchery	957	47,883	53,140	172,142	177,750	51,101	50,362	95,515	102,314	105,022	856,186
Annual Total	9,010	91,749	99,098	232,489	235,747	125,163	125,485	170,828	178,493	180,188	1,448,250

handled and observed, and then re-released to continue their migration. Of these, 30,100 fish (4.8%) were recaptured two or more times. The recapture of a previous-tagged fish is of significant interest to the researcher who initially release those animals, particularly in those cases where the recapture record reports the size and condition of the tagged specimen. A recapture event also augments the detection history for the tagged fish, especially if a smolt is recaptured at an intermediate location between the point of release and one of the interrogation sites in the lower Snake or Columbia rivers.

Previously-tagged salmon and steelhead are regularly recaptured in the juvenile fish collection facilities and adult fish traps at the main-stem dams on the Columbia and Snake rivers (Table 5). In the juvenile fish collection facilities, tagged fish are regularly entrained along with the facility's general sub-sample of all migrant salmonids. Tagged fish are also recaptured in the adult trapping facilities at Bonneville, Priest Rapids, and Lower Granite dams. The separation-by-code systems at both the juvenile and adult fish sampling facilities allow researchers to target and recapture specific previously-tagged fish. In all cases, the condition and morphological attributes of the recaptured fish are recorded, and the fish are then returned to the river to continue their migration.

To date, 844,500 mortality events have been reported to PTAGIS. These mortality events include tags recovered from adult salmon and steelhead during spawning activities at hatcheries; tags retrieved from adult fish harvested in recreational and commercial (including ocean troll) fisheries; tags recovered from salmon and steelhead that were selected and sacrificed for research purposes; and accumulations of bare tags retrieved from sumps, drains, and at other collection points in the juvenile and adult fish passage facilities. However, these various recoveries account for only 31,800 (3.8%) of the mortality events reported to PTAGIS.

Most mortality events (812,700, 96.2%) are the result of in situ detections of PIT tags deposited on islands in the Columbia and Snake rivers, and elsewhere, populated by colonies of avian piscivores (predominantly Caspian terns *Sterna caspia*, double-crested cormorants *Phalacrocorax auritis*, American white pelicans *Pelecanus erythrorhynchos*, and gulls *Larus* spp.). In 1996, researchers began surveying these islands to detect or physically recover the PIT tags deposited there by breeding or roosting birds (Ryan et al. 2001, 2003; Antollos et al. 2005). Annual surveys to collect or detect tags are now conducted on 18 of these islands (Figure 14). The results of these surveys may include tags that were deposited, but not reported, in previous years, in addition to new tags that had been deposited on the islands since the last survey (Table 6). Researchers generally self-censor their detection data to remove PIT tag codes recorded during previous surveys, and thus avoid reporting redundant mortality events to the PTAGIS database.

Of the 22 million PIT-tagged steelhead and salmon released into the Columbia Basin between 1987 and 2008, 8.7 million were passively detected at one or more interrogation sites, generating 96.9 million records detailing the passage route and ultimate disposition of each detected fish. After their release, 622,600 tagged fish were physically recaptured and then allowed to continue their migration. The deaths of 844,500 tagged fish have been directly confirmed from the retrieval of the tagged fish or can be inferred from the detection of the bare tags. All of these disparate PIT tag–release and recovery event data are stored to, and can be accessed and correlated through, the Columbia Basin PIT Tag Information System.

Table 5. Numbers of PIT-tagged fish recaptured at main-stem dams in the Columbia Basin, 1990-2008. BON = Bonneville Dam; JDA = John Day Dam; MCN = McNary Dam; IHR = Ice Harbor Dam; LMN = Lower Monumental Dam; LGS = Little Goose Dam; LGR = Lower Granite Dam; RIS = Rock Island Dam; RRE = Rocky Reach Dam.

	Middle Columbia River dams			Lower Snake River dams				Upper Columbia River dams	
Year	BON	JDA	MCN	IHR	LMN	LGS	LGR	RIS	RRE
1990							1,238		
1991							2,709		
1992						53	513		
1993					42	3,579	526		
1994		343			2,557	177	1,187		
1995	303	978	7,546	22		21	1,781	15	
1996	164	660	115		191	3,456	3,062		
1997	391	390	83		357	2,039	959	28	
1998	102	133	484		303	2,811	6,300	104	
1999		1,224	132		215	453	5,542	1,805	11
2000	97	2,976	121	70	204	56	2,605	2,241	3,884
2001	1,340		2,279	4,978	89	699	1,679	469	
2002	5,585	922	4,611	78	79	523	690	2,526	
2003	38	13	1,707	332	57	523	929	4,820	1,539
2004		18	6,884	65		1,144	519	3,625	1,856
2005	77	24	11,465	34		736	931	5,168	4,871
2006	863		1,027			658	1,305	1,464	411
2007	90		22	20	50	301	766	888	502
2008	26,512		438	319	219	608	5,891	1,410	475
Total	35,562	7,681	36,914	5,918	4,363	17,837	39,132	24,563	13,549

The Columbia Basin PIT Tag Information System

Since 1987, millions of Columbia Basin PIT tag data events have been reported by individuals working for dozens of organizations from hundreds of locations across the Basin. Researchers and resource managers expect complete access to these comprehensive historic data sets, and immediate access to all new data events, through the Columbia Basin PIT Tag Information System. These demands have been addressed through the deliberate and coordinated development of standard protocols and procedures that ensure the consistent acquisition, integration, and dissemination of accurate and reliable PIT tag event data.

Within the Columbia Basin, researchers use a single PIT tag technology to mark juvenile anadromous salmonids, and adhere to a common set of tagging and handling protocols. They have cooperatively developed a common data dictionary in order to systematically describe their PIT tag release and recovery events. This data dictionary is incorporated directly into the robust data collection software application used to record and report these event data. This same data dictionary is also referenced by the automated procedures

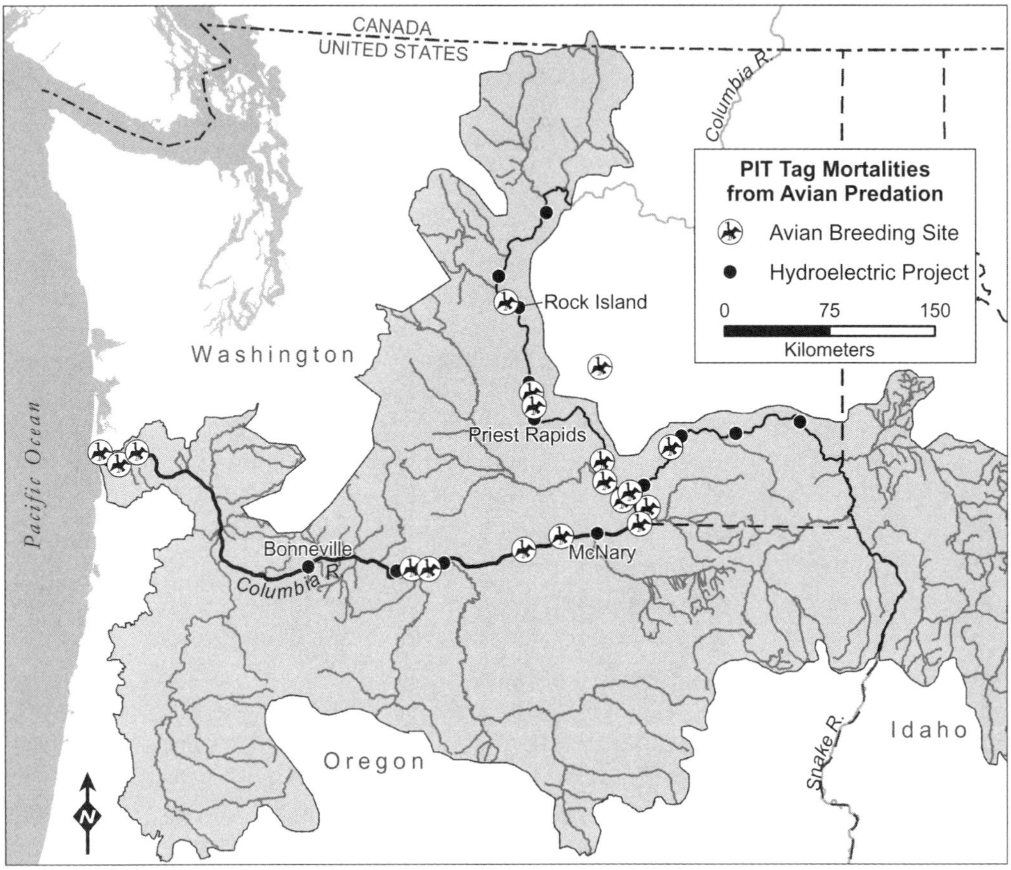

Figure 14. Locations of islands in the Columbia and Snake rivers, with breeding colonies of piscivorous avian predators, where PIT tags have been recovered or detected.

that corroborate the raw tag event data as they are presented to the PTAGIS database. These automated validation and corroboration processes negate the need to manually inspect or manipulate the event data before they are incorporated into the database. Other software applications automatically record the passive detection events and report those events directly to the PTAGIS database. The individual and correlated data events for every tag are immediately available, from preformatted reports or by direct query of the database, from the PTAGIS Web Portal. Each of these elements, described in detail below, is crucial to the ability of researchers to systematically collect and exchange PIT tag event data within the Columbia Basin.

Agency and project coordination

Since 1987, the Columbia Basin's federal, state, and tribal fisheries resource management organizations have collaborated to share their PIT tag release and recovery data. A PIT Tag Work Group (PTWG), comprised of representatives of the regional fisheries management entities, was established in 1987 to address issues involving the exchange of PIT tag release and detection data. The PTWG directed the development of a prototype centralized database to receive and maintain all of the Columbia Basin's PIT tag event data. In 1991, PSMFC was selected by the Basin's fisheries management agencies to develop this prototype into the permanent PIT Tag

Table 6. Numbers of PIT tags recovered or detected in situ at avian breeding locations in the Columbia Basin, 1996–2008.

	PIT Tag Recovery Location			
Year[a]	Lower Columbia[b]	Middle Columbia[c]	Snake-Columbia[d]	Upper Columbia[e]
1996	358			
1997	505			
1998	49,760	632	9,953	
1999	43,712	14,120	13,862	
2000	38,846	5,703	7,845	1,645
2001	23,271	3,041	30,573	1,997
2002	24,485	5,211	19,324	2,697
2003	58,822		17,838	3,276
2004	49,499	7,344	31,569	4,470
2005	46,848		33,356	9,379
2006	46,611	1,911	16,134	2,112
2007	36,993	7,078	33,715	3,743
2008	77,150	2,821	23,429	1,023
Total	496,860	47,861	237,598	30,342

[a] The calendar year in which tags were detected or recovered. The tags may have been deposited in a previous year.

[b] Three islands in the Columbia River estuary.

[c] Four islands in the middle Columbia River between Bonneville and McNary dams.

[d] Seven islands near the confluence of the Snake and Columbia rivers.

[e] Four islands in or near the upper Columbia River between the forebay of Rock Island Dam downstream to Priest Rapids Dam.

Information System database. The regional agencies chose PSMFC to administer the PTAGIS database due to PSMFC's expertise maintaining other large-scale, multi-agency databases (Johnson 1990; Daspit et al. 1997), and because PSMFC had no resource management responsibilities that might conflict with the management of the event data. The PIT Tag Operations Center (PTOC) was established at PSMFC's headquarters in Portland, Oregon to house the PTAGIS database system and management staff.

Prior to 1992, NOAA Fisheries was responsible for collecting PIT tag detection data and maintaining the interrogation systems at Lower Granite, Little Goose, and McNary dams. Those data collection and equipment maintenance responsibilities were transferred from NOAA Fisheries to PTOC during 1991–1992. Since 1993, those responsibilities have been administered from a separate PTOC field office located in Kennewick, Washington. The PTOC field staff are currently responsible for the operation of the interrogation systems at all seven of the juvenile fish bypass systems in the main-stem Columbia and Snake rivers; at the adult fish passage facilities at Bonneville, McNary, Ice Harbor, and Lower Granite dams; and in the fish passage facilities at two irrigation diversion dams and at three hatchery acclimation sites in the Yakima River basin. The interrogation systems maintained by PTOC provide essentially all of the PIT tag detection data for salmon and steelhead smolts during their migration to the Pacific Ocean, and most of the detection data for those tagged fish returning from the ocean to the lower Columbia and Snake rivers.

The PIT Tag Operations Center also administers the PIT tag separation-by-code (SxC) systems at the six dams on the lower Columbia and Snake rivers with juvenile fish sampling and collection facilities, and at the adult fish traps at Bonneville and Lower Granite dams. To ensure that SxC activities do not interfere with regional fish passage policies or with operations at the fish bypass facilities, researchers must coordinate their SxC requests with the Basin's fisheries resource managers as well as with the project biologists and facility operators at each dam. The PTOC staff develop and deploy the site-specific tag databases and other configuration files used to automatically execute the various diversion actions as individual target fish are detected. The PTOC staff also audit the SxC diversion activities to confirm that the appropriate PIT-tagged fish are diverted as requested, that the rate of bycatch is minimized, and that the diversion events do not interfere with the normal sampling and collection activities at the facilities.

Large-scale release and recovery programs such as PTAGIS "should be overseen by a representative board of the cooperating agencies, and it should act as coordinator for their common interests" (Nielsen 1992). In 1993, the Columbia Basin PIT Tag Steering Committee (PTSC) was established by charter between PSMFC and the Columbia Basin Fish and Wildlife Authority (CBFWA), whose membership includes the Basin's federal, state, and tribal fish and wildlife managers. The PTSC charter formalized the roles and responsibilities of PTOC and the Committee members from CBFWA's constituent agencies, who in turn provide representation for all of the PTAGIS data contributors in the Basin (Table 7). The goals of the PTSC are to coordinate the Columbia Basin's PIT tag marking activities and interrogation system operations and, through PTOC, to provide a comprehensive database of PIT tag events accessible to all interested parties (PTAGIS 1993). The PTSC was responsible for coordinating the successful transition from the old 400 kHz PIT tag technology to the 134.2 kHz ISO technology now used throughout the Basin. The current activities of the PTSC include: providing technical guidance for PTOC activities; reviewing new and revised models of PIT tags and approving their use in the Basin; and promoting consistent, high-quality tagging techniques. The PTSC representatives define the standard attributes maintained in the various domains of descriptors used to systematically describe the PIT tag event data contributed to the PTAGIS database. The coordinated use of these sets of "valid codes" helps ensure that the individual attributes of PIT tag event data records are consistently recognized by all entities in the Columbia Basin.

Study or project managers intending to submit their data to PTAGIS must first register as "tag data coordinators." Each coordinator is responsible for the veracity of the PIT tag events he or she contributes to the PTAGIS database, and for resolving data conflicts with other coordinators. Coordinators are frequently called on to address inquiries posed by other researchers regarding the objectives, methods and constraints of their marking studies, as PIT tag event data are often subject to review and re-use by other analysts.

Data Collection and Corroboration

The standards defined by the PTSC ensure that all PIT tag event data are systematically recorded and reported to PTAGIS. Since

Table 7. Federal, state, tribal, academic, and other entities that contribute PIT tag mark & release and recovery data to the Columbia Basin PIT Tag Information System. Native American tribal interests are formally coordinated through the Columbia River Inter-Tribal Fish Commission. The interests of the non-government entities are coordinated through either their respective state agency representative or one of the federal representatives on the PIT Tag Steering Committee.

Tribal resource management entities	**State resource management agencies**
Colville Confederated Tribes	[a]Idaho Dept. of Fish and Game
[a]Columbia River Inter-Tribal Fish Comm.	[a]Oregon Dept. of Fish and Wildlife
Confed. Tribes of the Umatilla Indian Reser.	[a]Washington Dept. of Fish and Wildlife
Confed. Tribes of the Warm Springs Reser.	
Nez Perce Tribe	**Federal management & research agencies**
Shoshone-Bannock Tribes	[a]NOAA Fisheries
Yakama Nation	U.S. Army Corps of Engineers
	U.S. Bureau of Reclamation
Electrical power utilities	U.S. Dept. of Agriculture
Chelan County Public Utility District	[a]U.S. Fish and Wildlife Service
Grant County Public Utility District	U.S. Geological Survey
Portland General Electric	
	Other research entities
Academic institutions	Biomark, Inc.
Idaho Cooperative F&W Research Unit	Kintama Research Corporation
Utah Cooperative F&W Research Unit	Pacific Northwest Laboratories (Battelle)
University of Idaho	Real Time Research
Non-profit resource advocacy groups	**Other government entities**
Ducks Unlimited	Columbia River Estuary Study Taskforce
Sea Resources	[a]Fish Passage Center
Wild Fish Conservancy	[a]Pacific States Marine Fisheries Commission

[a]PIT Tag Steering Committee member agency.

each tagged fish is uniquely and permanently identified, the host's physical attributes at the time of marking can be directly correlated with its activities during all subsequent recovery or interrogation events. For these correlations to be of practical use, researchers must be confident that the tagging and recovery attributes assigned to a PIT-tagged fish are accurate, and that the tag code is correctly identified and reported. The software applications used in the Columbia Basin to record PIT tag events were developed to address these concerns.

Streamlined and ergonomic data entry procedures have been developed for the PIT tag marking process that minimize handling time in order to reduce stress to the host fish. Tagging, recapture, and mortality event data are collected quickly and efficiently with the PITTAG3 (P3) software, a Microsoft Windows© application developed by PTOC. Similar to the PITTAG program that it replaced (Prentice et al. 1990c), the P3 application can receive input from peripheral data input devices connected to the computer, including one or more PIT tag detectors, an electronic balance, and a digitizer tablet. The digitizer tablet can be used simultaneously both as a measuring board to record the fish's fork length, and as a menu interface from which to select predefined attributes and commands. Direct data input from these peripheral devices accelerates data input, compared to manual entry from the keyboard, and eliminates transcription errors. Customizable configuration profiles allow the P3 application to be easily scaled for use in a variety of data collection situations, such as mass-marking thousands of fish at a hatchery, marking hundreds of fish of mixed stocks or species from a screw trap, or recording the occasional recapture of a tagged fish observed at an adult fish weir.

The P3 application includes a comprehensive suite of functions that can immediately detect and quickly resolve errors or anomalies with each data entry. The application incorporates the dictionary of valid codes specified by the PTSC, and uses the definitions from each code domain (species, rearing status, etc.) to populate a constrained drop-down list of values for the corresponding field in the P3 data entry form. The P3 application does not permit an ad hoc entry into a data field governed by one of these domains; this constraint ensures that only the standardized attributes are recorded and reported to the PTAGIS database. (Attributes specific to a single research project can be recorded to a separate, unregulated field in the data entry form.) The length and, frequently, the weight of the host fish is recorded during the tagging process. Through the P3 application, the tag data coordinator can specify, by individual species or stock, a range of acceptable length and weight values. An entry outside this range triggers an immediate visual and audible alert to the data entry personnel so the value can be verified or corrected while the tagged fish is still in hand. Since each tagged fish in the Columbia Basin is identified by its unique tag code, the P3 application detects incomplete or outstanding tag code assignments, and includes routines to remedy duplicate tag code assignments. These various verification routines enable the data entry personnel to quickly identify and resolve data entry issues that might otherwise impact the veracity or integrity of a tag event record. These routines do not impede the data collection process unless an error is detected.

The P3 application also validates the set of global attributes assigned to a set of tag event data records. The application will generate a finalized data file only after it has validated the entire data set. Using a proprietary PTAGIS File Transfer Protocol (PTTP), the validated file is transmitted from the P3 client computer directly to the PTAGIS database server. The PTTP transfer ensures that the file's integrity is not compromised between the client computer and the server. The data content and format constraints enforced by the P3 application guarantee that the data will be automatically incorporated into the PTAGIS database without modification or transformation. The database server returns a status record to the P3 client computer, and a separate e-mail message to the tag data coordinator, confirming that the various PIT tag events contained in the data file were successfully integrated into the PTAGIS database tables.

As tag data events are presented to the database, the PIT tag code from each new tagging event is evaluated to ensure it is unique from all other tag codes previously reported to PTAGIS. If not, the entire record is identified as a duplicate tagging event and quarantined in a separate database table. A duplicate tag code assignment generally results from failing to identify a previously-tagged fish during data entry; the tag data coordinator can resolve this issue by recasting the tag event record as a recapture event and resubmitting the data file containing that event. If the duplicate tag entry was the result of the re-use or mis-identification of another valid tag, then the coordinators responsible for the conflicting tag records must work together to establish the true ownership of the disputed tag. The original and duplicate tag event records are maintained separately within the PTAGIS database until the issue is resolved.

Occasionally, a PIT tag code is detected at an interrogation site for which PTAGIS has no mark-release information. This most often occurs when a detection is reported prior to the receipt of the tagging event record for that tag code. If PTAGIS receives an interrogation event for an unknown tag code, an "orphan" record is created in the tagging event table for that tag code. The orphan record is automatically replaced with the actual tagging event record when those data are received by PTAGIS. This process maintains referential integrity between the tag event and any subsequent interrogation events, regardless of the order in which those events are reported to PTAGIS.

Researchers in the Columbia Basin confidently assume that their PIT-tagged fish will be detected if they pass through an interrogated juvenile or adult fish passage facility at a dam on the main-stem Columbia or Snake rivers. At the interrogation sites administered by PTOC, researchers expect that tagged fish will pass serially through multiple antennas; that, cumulatively, these antennas will detect every tagged fish; and that those detection (and detector diagnostic) records will be automatically relayed to the PTAGIS database regularly throughout the day (Figure 15). These interrogation systems operate consistently and continuously throughout the year. The PTOC field engineers continually evaluate the PIT tag detection operations at these interrogation sites to ensure that all compo-

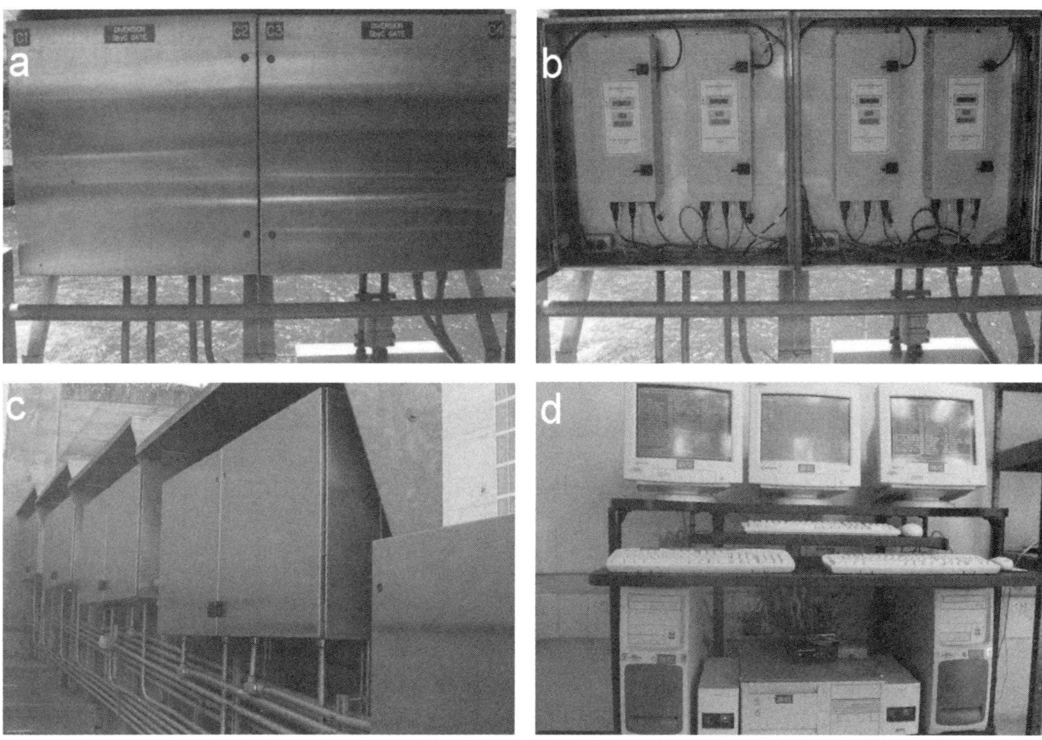

Figure 15. Typical installation of detection equipment at interrogation sites maintained by the PIT Tag Operations Center. Permanent enclosures **(a)** protect the transceivers from extreme environmental conditions. Pairs of identical transceivers are installed in each enclosure **(b)**. All power and signal cables are permanently connected, and protected by rigid conduit **(c)**. The data collection platform consists of designated primary and secondary computers (pictured in **(d)** with a separate dedicated utility computer), both of which record identical versions of the detection data. At separation-by-code sites, the secondary computer automatically assumes control of the PIT tag diversion gates if the primary computer fails.

nents are functioning optimally. By remotely monitoring the diagnostic records, and physically inspecting the components at each site on a regular basis, the PTOC staff are able to identify and correct most deficiencies before an issue can escalate and impact the ability to detect a tagged fish. Any facility activities or equipment failures that might impact tag detection, including those rare events that result in a decrease in detection efficiency or loss of detection data, are reported in a site-specific event log.

Two software programs are used to record interrogation events. The MULTIMON application, developed by NOAA Fisheries (Downing et al. 2001), is used at the eight PTOC-administered sites where separation-by-code (SxC) diversion activities occur. The majority of interrogation sites in the Columbia Basin are not engaged in SxC diversion activities but simply report PIT tag detections; data at these sites are collected with MINIMON, a Microsoft Windows© application developed by PTOC. The interrogation event information recorded by the two applications is essentially identical, consisting of the tag code, the transceiver and antenna identifiers, and a time stamp. Both applications are fully automated and designed to operate indefinitely without interruption or manual intervention.

At most sites, including those at all of the main-stem dams, the PIT tag antennas are arrayed serially to provide redundant interrogation opportunities, reducing the risk of missing a tagged fish as it passes through them. At these sites, the risk of data loss is further mitigated by equipment redundancy in the data collection process. Each PIT tag detection record is transmitted to a pair of computers, one of which is designated the "primary" data repository. The "secondary" computer stores a complete copy of the detection data, which can be retrieved in the event the primary computer fails. (In the MULTIMON application, the secondary computer also provides automated fail-over control of the SxC operations if the primary computer fails.) Both the MULTIMON and MINIMON applications write a time stamp into the data file at regular intervals, to confirm that the application is active and responsive during long intervals with no reported detection activity. In addition, a physical or virtual reference tag is activated on each detector at a regular interval. The periodic detection of these "timer" tags confirms that the respective transceiver and antenna are operating properly and continuously; this confirmation is especially crucial if tagged fish are only occasionally present at the detector. All reported detection data, including timer tag records, are written to permanent ASCII data files on the primary and secondary computers. Each transceiver also generates status and diagnostic messages on a regular schedule, and these messages are also written to the data files.

At the interrogation sites maintained by PTOC, the data files are automatically uploaded to PTAGIS every three hours. Many of the other interrogation sites in the Columbia Basin also automatically submit their data on a regular schedule, ranging from every three hours to once per day. The transceiver diagnostic messages in each interrogation data file are procedurally parsed and processed as PTAGIS receives the file. These transceiver alerts and other feedback mechanisms identify potential sources of detector error or failure so the issue can be resolved before a deleterious event can occur. The diagnostic error messages are automatically broadcast to a list of designated contacts.

All validated PIT tag event records are automatically integrated into the PTAGIS database without any modification of the original tag event data. The records in each P3 tag data file are processed as soon as the file contents are validated. The contents of tag data files are automatically catalogued and audited as they are loaded into the PTAGIS database, and the data contributors are sent an e-mail confirming the receipt and proper disposition of their PIT tag event records. Interrogation data are queued as they are received and validated, and then batch-loaded into the database on a regular schedule, generally within no more than three hours of the detection event. All of the original data files are permanently archived on the PTAGIS file server, where they are publicly-accessible for review and retrieval. By direct inspection of these data files, researchers and others can independently compare, and confirm the veracity of, the event data reported to and retrieved from the PTAGIS database.

Retrieving PIT tag event data and other information from PTAGIS

Within the PTAGIS database, PIT tag interrogation, recapture, and mortality events are automatically correlated with the initial tagging and release event for each host fish. The database is designed and automatically maintained to provide rapid and comprehensive access to the millions of tagging and interrogation events, and to the hundreds of thousands of recapture and mortality events. Access to these event data are provided through the PTAGIS Web Portal (www.ptagis.org).

PTAGIS publishes preformatted reports that contain incremental and cumulative year-to-date summaries of tagging and interrogation events. These reports are automatically generated on regular daily and weekly schedules. Some of these reports were specifically developed to be retrieved by other Columbia Basin data services that mirror selected portions of the PTAGIS data events, but all of the reports are available to the general public.

The Web Portal also provides public access to light-weight interactive queries that selectively retrieve specified PTAGIS event records directly from the database. A "Complete Tag History" query, for example, returns the comprehensive set of data events for a single PIT tag. Other queries provide access to the tagging and interrogation data for the relatively small number of tagged fish that are detected at individual adult fishways or instream arrays. The results from these general-use queries are displayed directly on the user's computer screen or downloaded to their computer, after which the query results are discarded.

Researchers and other visitors to the Web Portal wanting more complex or specialized data sets can create a personal account that provides access to the complete, correlated archive of tagging, interrogation, recapture, and mortality events. Account holders can define their own customized queries of individual and correlated data events. Each query definition is saved to the user's account and preserved between sessions. Saved queries can be run, revised, and re-run. The query results are preserved in comma-separated-value files that can be downloaded and manipulated on a client computer. The tag codes returned in the output from one query can be used as input to other queries. Account holders can also upload lists of tag codes for use as a query parameter. New tag code lists can be created by combining lists, subtracting one existing list from another, or intersecting two or more lists to return just the codes common to all of the lists. These customizable tag code lists provide account holders with a powerful method to refine their queries directly from their PTAGIS account.

The PTAGIS Web Portal and the PTOC (www.ptoccentral.org) Web pages provide on-line access to PTAGIS procedural instructions, contact information for data contributors, detailed histories of detection activities at interrogation sites, and other documents and support services. From the Web Portal, users can download the MINIMON and P3 applications, and the current code sets used to validate the data entries to those applications. The Web Portal and PTOC pages provide direct access to the current and historic event logs for the automated PIT tag interrogation sites. These pages also provide diagrams of the major interrogation and separation-by-code sites; the diagrams help the viewer visualize the complicated paths of passage and PIT tag detection at these sites. Web Portal users can view the protocols and tag groups identified in current and historic separation-by-code requests, review the study description for an SxC request, and retrieve the contact information for that study's principal investigator. From these pages, researchers and data analysts can locate the contact information for any of the Columbia Basin's tag data coordinators. This comprehensive access to the information about, and individuals responsible for, the events reported to PTAGIS is intended to encourage and facilitate direct communication between the Columbia Basin's PIT tag data contributors and other constituents.

The Use of PIT Tags in Columbia Basin Research and Monitoring Activities

The systematic and collaborative use of PIT tags has expanded the opportunities to monitor and manage fisheries resources in the Columbia Basin. Marking fish with PIT tags provides researchers the ability to monitor the migration timing, passage distributions, and survival of individual stocks of juvenile salmon and steelhead to key dams on the main-stem Columbia and Snake rivers (Muir et al. 2001; Paulsen and Fisher 2001; Paulsen and Fisher 2003; Smith et al. 2003; Connor et al. 2004; Zabel and Achord 2004; Achord et al. 2007). The use of PIT tags provides reliable and precise estimates of downstream travel time (Giorgi et al. 1997; Williams et al. 2005) and survival (Muir et al. 2001; Smith et al. 2002; Williams et al. 2005) through individual river reaches within the Columbia and Snake rivers. Evaluations of fish passage routes and structures at dams in the Columbia Basin frequently require or include PIT tag marks. Between 1989 and 2004, PIT tags were used in 23 evaluations of juvenile salmon and steelhead survival through spillways at these main-stem dams, and in 11 turbine survival studies (Ferguson et al. 2005). These assessments of populations can be extended to investigate characteristics of individual marked hosts; Williams et al. (2005) described an inverse relationship between the lengths of fish tagged at Lower Granite Dam and the efficiency with which those fish were then mechanically guided away from the turbines and into the juvenile bypass systems at Little Goose, Lower Monumental, and McNary dams.

With the introduction of the 134.2 ISO tag and the resultant ability to reliably detect PIT tags in salmonids returning from the ocean to the Columbia Basin, researchers have been able to correlate the tag events of juvenile salmonids with their subsequent detection events as adults (Williams et al. 2005). A tag event history can be compiled for an individual salmon or steelhead as the juvenile host is detected in the juvenile bypass system at one or more dams during its seaward migration, and then as the host fish returns from the ocean, migrates upstream, and is detected in the adult fishway of one or more dams. These individual detection histories provide opportunities for robust analyses not possible with other marking techniques. Sandford and Smith (2002) observed significantly higher smolt-to-adult return (SAR) rates for tagged Snake River Chinook and steelhead that were never detected during their downstream migration when compared to SAR rates for similar groups of fish that were detected in multiple juvenile fish facilities as they were bypassed around main-stem dams in the Basin.

Most tagged salmon and steelhead detected in adult fish ladders were marked as parr or smolts. The detection of these adult fish during their upstream migration provides the basis for the computation of SAR ratios used to evaluate the efficacy of various management strategies, such as transporting Chinook and steelhead smolts downstream around the dams in the Columbia and Snake rivers (Buchanan et al. 2006; Muir et al. 2006). Tag detections in the adult fishways also provides metrics for the rate and timing of upstream migrations. Salinger and Anderson (2006) evaluated the upstream migration times, between Bonneville and Lower Granite dams, of individual adult Chinook that had been tagged as juveniles. Keefer et al. (2008a) used data from 12 populations of Chinook tagged as parr or smolts to develop separate adult migration timing distributions for each of those populations upon their return to the Columbia River and detection at Bonneville Dam.

Since 2000, researchers have used PIT tags to mark adult salmon and steelhead during their upstream migration. Returning adults are tagged in order to augment the population of upstream migrants that were previously tagged as parr or smolts. Additional fish are tagged and then tracked as they ascend the Columbia and Snake rivers in order to determine the age composition and escapement rates (Fryer 2008a) and migration timing and survival rates (Fryer 2008b) of those tagged populations. Researchers use PIT tags to monitor the behavior of reconditioned steelhead kelts in the Yakima River Basin (W. Bosch, Yakama Nation, personal communication). Steelhead kelts have been tagged to evaluate their re-ascension rates at dams in the lower Columbia and Snake rivers (Evans et al. 2008; Keefer et al. 2008b).

A PIT tag is often paired with a radio or acoustic tag in telemetry studies to evaluate the movement and survival of smolts though fish passage routes and structures at the mainstem hydroelectric dams. The radio or acoustic tag provides fine-scale spatial resolution of the tagged host's location at the evaluated facility; the PIT tag provides the possibility for a "fail-over" detection at one or more downstream dams if the radio tag or acoustic tag is not detected at the telemetry study exit point (Ferguson et al. 2006). A fish tagged with both a PIT tag and acoustic tag can potentially be monitored using the acoustic tag in the Columbia River Estuary (McComas et al. 2005, 2008) or along the continental shelf of the Pacific Ocean (Welch 2007; Welch et al. 2008), and monitored using the PIT tag during its freshwater migration downstream (and possibly again upstream) through one or more interrogation systems. Researchers inserted a PIT tag and an archival data storage tag into adult Chinook and steelhead they captured at Bonneville Dam, and then used the PIT tag separation-by-code system at Lower Granite Dam to automatically recover the double-tagged fish from the adult fish trap and retrieve the archival tags (Johnson et al. 2005). Burke and Jepson (2006) used PIT tags, alone and with radio tags, to compare the escapement estimates and fallback rates for the single- and double-tagged adult Chinook and steelhead released at Bonneville Dam.

The MULTIMON SxC application is used at six juvenile fish facilities on the Columbia and Snake rivers to identify and collect individual PIT-tagged salmon and steelhead from the population of untagged smolts passing through those facilities (Connor et al. 2001; Muir et al. 2006). This selection capability allows researchers to efficiently identify and recapture individual tagged fish in order to measure the growth and/or evaluate changes in the condition and physiology of the downstream migrants (Achord et al. 1996; Connor et al. 2004; Achord et al. 2007). At Lower Granite, Little Goose, Lower Monumental and McNary dams, the MULTIMON application is also used to facilitate transportation studies (Williams et al. 2005; Buchanan et al. 2006; Muir et al. 2006) by systematically routing a portion of previously identified tagged fish to transportation vessels while diverting the remaining tagged fish back to the river to continue their downstream migration. Keefer et al. (2005) used the MULTIMON SxC system in the adult fish trap at Bonneville Dam to select tagged fish known to have originated in the Snake River, and inserted radio transmitters into those known-origin fish to measure escapement, harvest, and tag loss rates between Bonneville and Lower Granite dams. Using the SxC system at Bonneville Dam to identify and then radio-tag adult salmon and steelhead that had been either transported as smolts from Lower Granite Dam or allowed to migrate downstream in-river, Keefer et al. (2008c) found a significant difference in the homing success rates between those two treatment groups during their upstream migration. Keefer et al. (2008d) used known-origin tagged fish, collected with the SxC system into the Bonneville Dam adult fish trap, to evaluate the overwintering behavior and survival of adult summer steelhead returning to the interior Columbia and Snake river basins.

Northern pikeminnow are marked with PIT tags in order to assess the efficacy of a controlled harvest designed to reduce the impact of these piscivores on juvenile salmonids. During the 2007 Pikeminnow Sport Reward Program field season, PIT tags were detected in 382 northern pikeminnow recovered in the fishery. Of those 382 tags, 102 were later determined to have been inserted into salmon or steelhead smolts that were then ingested by the northern pikeminnow (Porter 2008). The opportunity to document such predation events is limited; a laboratory study by Petersen and Barfoot (2003) reported that PIT tags in salmonids consumed by northern pikeminnow were evacuated within 60 hours of ingestion, with the majority of tags evacuated within 36 hours.

There has been one exceptional mortality event reported to PTAGIS. A PIT tag from a steelhead released in May, 2005 from a hatchery on the Columbia River 567 km upstream from the Pacific Ocean was recovered 23 months later from a sooty shearwater *Puffinus griseus* chick harvested on one of the Titi islands in New Zealand (PTAGIS 2007). An adult shearwater likely intercepted the tagged steelhead in the North Pacific, carried the tag in its crop to its breeding location near Stewart Island, New Zealand, and ultimately regurgitated the tag while feeding its chick.

Summary

The success of the Columbia Basin PIT Tag Information System is the result of the deliberate development and coordinated integration of its core components. PIT tags are an effective and versatile tool to monitor the migrations of individual salmon and steelhead passing the numerous hydroelectric dams located on the Basin's main-stem rivers. The interrogation systems integrated into the fish bypass facilities at these dams are highly reliable and efficient, resulting in the detection of essentially all tagged fish. Collaborative efforts by the Basin's fisheries management agencies to monitor the migrations of tagged fish have resulted in a high level of coordination of those monitoring activities, including the development and maintenance of strong standards to ensure that tag event data are consistently recorded throughout the Basin. The event data are recorded using software applications that streamline the data collection process while providing safeguards to ensure the veracity and integrity of those data. The uniform format of the reported tag event data makes it possible to immediately and automatically incorporate those verified records into the PTAGIS database. All of these data are accessible, through structured reports and custom queries, from the PTAGIS Web Portal. Ultimately, the success of PTAGIS can be measured as its ability to provide complete access to the comprehensive set of all PIT tag events recorded in the Columbia Basin since 1987, from marking and release events reported more than 20 years ago to interrogation events that may have occurred just 20 minutes previously.

The use and utility of PIT tags as a research and management tool in the Columbia Basin continues to expand. NOAA Fisheries is developing a project to intercept and tag adult salmonids as they return to the mouth of the Columbia River. An 8.5 mm 134.2 ISO tag is now available, providing researchers with a new tool to mark smaller fish than can be tagged with the 12.5 mm model. The ability to detect tags in the high volume, high velocity, B2CC discharge flume has spurred development of detection systems that could be deployed in spillbays, turbine intakes, and other potential passage routes at hydroelectric dams in the Columbia and Snake rivers. Simple interrogation systems, relatively inexpensive to build and maintain, are being installed directly into streams and small rivers to obtain detection information for both juvenile and adult anadromous salmonids as well as other species. Each of these developments expands the opportunities to mark and detect a larger portion of salmon and steelhead during their juvenile and adult life stages within the Basin. The live recovery and re-release, in the central North Pacific Ocean, of a steelhead tagged at Lower Granite Dam two years earlier (Davis et al. 2008) demonstrates the potential to expand the scope of PIT tag mark and recovery events into the marine environment outside of the Columbia Basin.

In recent years, researchers have used 134.2 kHz half-duplex PIT tags, as well as 134.2 ISO FDX tags, to monitor populations of adfluvial and resident fish in various watersheds within the Columbia Basin. While some of these researchers report their PIT tag event data to PTAGIS, most of this information has been separately compiled and maintained by the individuals or agencies that conduct those studies. If researchers and fisheries managers desire to integrate and exchange the PIT tag event data for these studies, the success of PTAGIS will require the continued ability to satisfy the needs of its current constituents while accommodating new sets of biological attributes obtained from, and different PIT tag technologies employed by, other research and monitoring activities in the Columbia Basin.

Acknowledgments

I gratefully acknowledge the assistance provided to me by the staff of the PIT Tag Operations Center, past and present, and especially the contributions of Carter Stein, Don Warf, Scott Livingston, and John Tenney. Brett Holycross (PSMFC) displayed great patience and diligence in his preparation of the maps of PIT tag event locations in the Columbia Basin. Quinton Phelps and two anonymous reviewers provided extensive comments that significantly improved an earlier draft of this manuscript. The Columbia Basin PIT Tag Information System is a project (#1990–080–00) of the Columbia Basin Fish and Wildlife Program, and is funded by the Bonneville Power Administration.

References

Achord, S., G. M. Mathews, O. W. Johnson, and D. M. Marsh. 1996. Use of passive integrated transponder (PIT) tags to monitor migration timing of Snake River Chinook salmon smolts. North American Journal of Fisheries Management 16:302–313.

Achord, S., R. W. Zabel, and B. P. Sandford. 2007. Migration timing, growth, and estimated parr-to-smolt survival rates of wild Snake River spring-summer Chinook salmon from the Salmon River Basin, Idaho, to the lower Snake River. Transactions of the American Fisheries Society 136:142–154.

Antollos, M., D. D. Roby, D. E. Lyons, K. Collis, A. F. Evans, M. Hawbecker, and B. A. Ryan. 2005. Caspian tern predation on juvenile salmonids in the mid-Columbia River. Transactions of the American Fisheries Society 134:466–480.

Axel, G. A., E. F. Prentice, and B. P. Sandford. 2005. PIT-tag detection system for large-diameter juvenile fish bypass pipes at Columbia River Basin hydroelectric dams. North American Journal of Fisheries Management 25:646–651.

Buchanan, R. A., J. R. Skalski, and S. G. Smith. 2006. Estimating the effects of smolt transportation from different vantage points and management perspectives. North American Journal of Fisheries Management 26:460–472.

Burke, B. J., and M. A. Jepson. 2006. Performance of passive integrated transponder tags and radio tags in determining dam passage behavior of adult Chinook salmon and steelhead. North American Journal of Fisheries Management 26:742–752.

Burnham, K. P., D. R. Anderson, G. C. White, C. Brownie, and K. H. Pollock. 1987. Design and analysis methods for fish survival experiments based on release-recapture. American Fisheries Society, Monograph 5, Bethesda, Maryland.

Chapman, D. W. 1986. Salmon and steelhead abundance in the Columbia River in the nineteenth century. Transactions of the American Fisheries Society 115:662–670.

Connolly, P. J., I. G. Jezorek, K. D. Martens, and E. F. Prentice. 2008. Measuring the performance of two stationary interrogation systems for detecting downstream and upstream movement of PIT-tagged salmonids. North American Journal of Fisheries Management 28:402–417.

Connor, W. P., T. C. Bjornn, H. L. Burge, A. R. Marshall, H. L. Blankenship, R. K. Steinhorst, and K. F. Tiffan. 2001. Early life history attributes and run composition of PIT-tagged wild subyearling Chinook salmon recaptured after migrating downstream past Lower Granite Dam. Northwest Science 75(3):254–261.

Connor, W. P., S. G. Smith, T. Andersen, S. M. Bradbury, D. C. Burum, E. E. Hockersmith, M. L. Schuck, G. W. Mendel, and R. M. Bugert. 2004. Postrelease performance of hatchery yearling and subyearling fall Chinook salmon released into the Snake River. North American Journal of Fisheries Management 24:545–560.

Dare, M. R. 2003. Mortality and long-term retention of passive integrated transponder tags by spring Chinook salmon. North American Journal of Fisheries Management 23:1015–1019.

Daspit, W., P. Crone, and D. Sampson. 1997. Pacific Fisheries Information Network. Chapter 6 *in* D. Sampson and P. Crone, editors. Commercial Fisheries Data Collection Procedures for U.S. Pacific Coast Groundfish. NOAA Technical Memorandum NMFS-NWFSC-31. National Oceanic and Atmospheric Administration, Silver Spring, Maryland. Available: www.nwfsc.noaa.gov/publications/techmemos/tm31/tm31.html (February 2009).

Davis, N. D., M. Atcheson, D. Marvin, and M. Fukuwaka. 2008. Interception of a PIT-tagged Columbia River Basin steelhead in the central North Pacific Ocean. North Pacific Anadromous Fish Commission, Doc. 1106, Vancouver, British Columbia.

Downing, S. L., E. L. Prentice, R. W. Frazier, J. E. Simonson, and E. P. Nunnalee. 2001. Technology developed for diverting passive integrated transponder (PIT) tagged fish at hydroelectric dams in the Columbia River Basin. Aquacultural Engineering 25:149–164.

Evans, A. F., R. H. Wertheimer, M. L. Keefer, C. T. Boggs, C. A. Peery, and K. Collis. 2008. Transportation of steelhead kelts to increase iteroparity in the Columbia and Snake rivers. North American Journal of Fisheries Management 28:1818–1827.

Ferguson, J. W., G. M. Matthews, R. L. McComas, R. F. Absolon, D. A. Brege, M. H. Gessel, and L. G. Gilbreath. 2005. Passage of adult and juvenile salmonids through federal Columbia river power system dams. U.S. Department of Commerce. NOAA Technical Memorandum NMFS-NWFSC-64.

Ferguson, J. W., R. F. Absolon, T. J. Carlson, and B. P. Sandford. 2006. Evidence of delayed mortality on juvenile Pacific salmon passing through turbines at Columbia River dams. Transactions of the American Fisheries Society 135:139–150.

Fryer, J. K. 2008a. Estimation of mid-Columbia summer Chinook salmon escapement and age composition using PIT tags in 2007. Technical report 08–01. Columbia River Inter-Tribal Fish Commission. Portland, Oregon. Available: www.critfc.org/tech/08-01report.pdf (February 2009).

Fryer, J. K. 2008b. Use of PIT tags to determine upstream migratory timing and survival of Columbia Basin sockeye salmon in 2007. Technical report 08–02. Columbia River Inter-Tribal Fish Commission. Portland, Oregon. Available: www.critfc.org/tech/08-02-report.pdf (February 2009).

Giorgi, A. E., T. W. Hillman, J. R. Stevenson, S. G. Hays, and C. M. Peven. 1997. Factors that influence the downstream migration rates of juvenile salmon and steelhead through the hydroelectric system in the mid-Columbia River Basin. North American Journal of Fisheries Management 17:268–282.

Johnson, E. L., T. S. Clabough, D. H. Bennett, T. C. Bjornn, C. A. Peery, and C. C. Caudill. 2005. Migration depths of adult spring and summer Chinook salmon in the lower Columbia and Snake rivers in relation to dissolved gas supersaturation. Transactions of the American Fisheries Society 134:1213–1227.

Johnson, J. K. 1990. Regional overview of coded wire tagging of anadromous salmon and steelhead in Northwest America. Pages 782–816 *in* N. C. Parker, A. E. Giorgi, R. C. Heidinger, D. B. Jester, Jr., E. D. Prince, and G. A. Winans, editors. 1990. Fish-marking techniques. American Fisheries Society; Symposium 7, Bethesda, Maryland.

Johnson, D. H., B. M. Shrier, J. S. O'Neal, J. A. Knutsen, X. Augerot, T. A. O'Neil, and T. N. Pearsons. 2007. Salmonid field protocols handbook: techniques for assessing status and trends in salmon and trout populations. American Fisheries Society, Bethesda, Maryland.

Keefer, M. L., C. A. Peery, W. R. Daigle, M. A. Jepson, S. R. Lee, C. T. Boggs, K. R. Tolotti, and B. J. Burke. 2005. Escapement, harvest, and unknown loss of radio-tagged adult salmonids in the Columbia River-Snake River hydrosystem. Canadian Journal of Fisheries and Aquatic Sciences 62:930–949.

Keefer, M. L., C. A. Peery, and C. C. Caudill. 2008a. Migration timing of Columbia River spring Chinook salmon: effects of temperature, river discharge, and ocean environment. Transactions of the American Fisheries Society 137:1120–1133.

Keefer, M. L., R. H. Werthheimer, A. F. Evans, C. T. Boggs, and C. A. Peery. 2008b. Iteroparity in Columbia River summer–run steelhead (*Oncorhynchus mykiss*): implications for conservation. Canadian Journal of Fisheries and Aquatic Sciences 65:2592–2605.

Keefer, M. L., C. C. Caudill, C. A. Peery, and S. R. Lee. 2008c. Transporting juvenile salmonids around dams impairs adult migration. Ecological Applications 18(8):1888–1900.

Keefer, M. L., C. T. Boggs, C. A. Peery, and C. C. Caudill. 2008d. Overwintering distribution, behavior, and survival of adult summer steelhead: variability among Columbia River populations. North American Journal of Fisheries Management 28:81–96.

Ledgerwood, R. D., B. A. Ryan, E. M. Dawley, E. P. Nunnallee, and J. W. Ferguson. 2004. A surface trawl to detect migrating juvenile salmonids tagged with passive integrated transponder tags. North American Journal of Fisheries Management 24:440–451.

Marsh, D. M., G. M. Matthews, S. Achord, T. E. Ruehle, and B. P. Sandford. 1999. Diversion of salmonid smolts tagged with passive integrated transponders from an untagged population passing through a juvenile collection system. North American Journal of Fisheries Management 19:1142–1146.

Matthews, G. M., G. A. Swan, and J. R. Smith. 1977. Improved bypass and collection system for protection of juvenile salmon and steelhead trout at Lower Granite Dam. Marine Fisheries Review 39(7):10–14.

McComas, R. L., D. A. Frost, S. G. Smith, J. W. Ferguson, T. J. Carlson, and T. Aboellail. 2005. A study to estimate salmonid survival through the Columbia River estuary using acoustic tags. Report by Fish Ecology Division, Northwest Fisheries Science Center, National Marine Fisheries Service, National Oceanic and Atmospheric Administration to the U.S. Army Corps of Engineers, Portland, Oregon.

McComas, R. L., G. A. McMichael, J. A. Vucelick, L. G. Gilbreath, J. P. Everett, S. G. Smith, T. J. Carlson, and J. W. Ferguson. 2008. A study to estimate salmonid survival through the Columbia River estuary using acoustic tags, 2006. Report by Fish Ecology Division, Northwest Fisheries Science Center, National Marine Fisheries Service, National Oceanic and Atmospheric Administration to the U.S. Army Corps of Engineers, Portland, Oregon.

McCutcheon, C. S., E. F. Prentice, and D. L. Park. 1994. Passive monitoring of migrating adult steelhead with PIT tags. North American Journal of Fisheries Management 14:220–223.

Muir, W. D., S. G. Smith, J. G. Williams, E. E. Hockersmith, and J. R. Skalski. 2001. Survival estimates for migrant yearling Chinook salmon and steelhead tagged with passive integrated transponders in the Lower Snake and Lower Columbia Rivers, 1993–1998. North American Journal of Fisheries Management 21:269–282.

Muir, W. D., D. M. Marsh, B. P. Sandford, S. G. Smith, and J. G. Williams. 2006. Post-hydropower system delayed mortality of transported Snake River stream-type Chinook salmon: unraveling the mystery. Transactions of the American Fisheries Society 135:1523–1534.

Nielsen, L. A. 1992. Methods of marking fish and shellfish. American Fisheries Society, Special Publication 23, Bethesda, Maryland.

Paulsen, C. M., and T. R. Fisher. 2001. Statistical relationship between parr-to-smolt survival of Snake River spring-summer Chinook salmon and indices of land use. Transactions of the American Fisheries Society 130:347–358.

Paulsen, C. M., and T. R. Fisher. 2003. Detecting juvenile survival effects of habitat actions: power analysis applied to endangered Snake River spring–summer Chinook (*Oncorhynchus tshawytscha*). Canadian Journal of Fisheries and Aquatic Sciences 60:1122–1132.

Petersen, J. H., and C. A. Barfoot. 2003. Evacuation of passive integrated transponder (PIT) tags from northern pikeminnow consuming tagged juvenile Chinook salmon. North American Journal of Fisheries Management 23:1265–1270.

Porter, R. G. 2008. Report on the predation index, predator control fisheries, and program evaluation for the Columbia River Basin experimental northern pikeminnow management program. 2007 Annual Report. Pacific States Marine Fisheries Commission. Available: www.pikeminnow.org/reports/2007-Pikeminnow-AR.pdf (February 2009).

Prentice, E. P., T. A. Flagg, and C. S. McCutcheon. 1990a. Feasibility of using implantable passive integrated transponder (PIT) tags in salmonids. Pages 317–322 *in* N. C. Parker, A. E. Giorgi, R. C. Heidinger, D. B. Jester, Jr., E. D. Prince, and G. A. Winans, editors. 1990. Fish-marking techniques. American Fisheries Society; Symposium 7, Bethesda, Maryland.

Prentice, E. P., T. A. Flagg, C. S. McCutcheon, and D. F. Brastow. 1990b. PIT-tag monitoring systems for hydroelectric dams and fish hatcheries. Pages 323–334 *in* N. C. Parker, A. E. Giorgi, R. C. Heidinger, D. B. Jester, Jr., E. D. Prince, and G. A. Winans, editors. 1990. Fish-marking techniques. American Fisheries Society; Symposium 7, Bethesda, Maryland.

Prentice, E. P., T. A. Flagg, C. S. McCutcheon, D. F. Brastow, and D. C. Cross. 1990c. Equipment, methods, and an automated data-entry station for PIT tagging. Pages 335–340 *in* N. C. Parker, A. E. Giorgi, R. C. Heidinger, D. B. Jester, Jr., E. D. Prince, and G. A. Winans, editors. 1990. Fish-marking techniques. American Fisheries Society; Symposium 7, Bethesda, Maryland.

Prentice, E. P., D. J. Maynard, P. Sparks-McConkey, C. S. McCutcheon, D. Neff, W. Steffens, F. W. Waknitz, A. L. Jensen, L. C. Stuehrenberg, S. L. Downing, B. P. Sandford, and T. W. Newcomb. 1993. Study to determine the feasibility of a new fish tagging system. Annual Report 1988–1989. Report (Contract DE-AI79–84BP11982, Project

83–319) to Bonneville Power Administration, Portland, Oregon. Available: pisces.bpa.gov/release/documents/documentviewer.aspx?doc=11982-4 (February 2009).

PTAGIS (Columbia Basin PIT Tag Information System). 1993. Charter for the Columbia Basin PIT Tag Steering Committee. Pacific States Marine Fisheries Commission. Portland, Oregon. Available: ftp.ptagis.org/Documents/PTSC/PTSC_charter.pdf (February 2009).

PTAGIS (Columbia Basin PIT Tag Information System). 2004. Columbia Basin PIT Tag Specification Document. Pacific States Marine Fisheries Commission. Portland, Oregon. Available: ftp.ptagis.org/Documents/PTSD/2004_Spec_Doc_Master.pdf (February 2009).

PTAGIS (Columbia Basin PIT Tag Information System). 2007. Titi Recovery. Pacific States Marine Fisheries Commission. Portland, Oregon. Available: php.ptagis.org/wiki/index.php/Titi_Recovery (February 2009).

Raymond, H. L. 1988. Effects of hydroelectric development and fisheries enhancement on spring and summer Chinook salmon and steelhead in the Columbia River Basin. North American Journal of Fisheries Management 8:1–24.

Ryan, B. A., J. W. Ferguson, R. D. Ledgerwood, and E. P. Nunnallee. 2001. Detection of passive integrated transponder tags from juvenile salmonids on piscivorous bird colonies in the Columbia River Basin. North American Journal of Fisheries Management 21:417–421.

Ryan, B. A., S. G. Smith, J. M. Butzerin, and J. W. Ferguson. 2003. Relative vulnerability to avian predation of juvenile salmonids tagged with passive integrated transponders in the Columbia River estuary, 1998–2000. Transactions of the American Fisheries Society 132:275–288.

Salinger, D. H., and J. J. Anderson. 2006. Effects of water temperature and flow on adult salmon migration swim speed and delay. Transactions of the American Fisheries Society 135:188–199.

Sandford, B. P., and S. G. Smith. 2002. Estimation of smolt-to-adult return percentages for Snake River Basin anadromous salmonids, 1990–1997. Journal of Agricultural, Biological, and Environmental Statistics 7(2):243–263.

Smith, S. G., W. D. Muir, J. G. Williams, and J. R. Skalski. 2002. Factors associated with travel time and survival of migrant yearling Chinook salmon and steelhead in the Lower Snake River. North American Journal of Fisheries Management 22:385–405.

Smith, S. G., W. D. Muir, E. E. Hockersmith, R. W. Zabel, R. J. Graves, C. V. Ross, W. P. Connor, and B. D. Arnsberg. 2003. Influence of river conditions on survival and travel time of Snake River subyearling fall Chinook salmon. North American Journal of Fisheries Management 23:939–961.

Wagner, P. G. 2007. Fish counting at large hydroelectric projects. Pages 173–195 *in* D. H. Johnson, B. M. Shrier, J. S. O'Neal, J. A. Knutsen, X. Augerot, T. A. O'Neil, and T. N. Pearsons, editors. Salmonid field protocols handbook: techniques for assessing status and trends in salmon and trout populations. American Fisheries Society, Bethesda, Maryland.

Welch, D. W. 2007. Pacific Ocean Shelf Tracking (POST) Project: results from the acoustic tracking study on survival of Columbia River salmon, 2005–2006. Bonneville Power Administration. Available: pisces.bpa.gov/release/documents/documentviewer.aspx?doc=00021107-1 (February 2009).

Welch D. W., E. L. Rechisky, M. C. Melnychuk, A. D. Porter, C. J. Walters, S. Clements, B. J. Clemens, R. S. McKinley, and C. Schreck. 2008. Survival of migrating salmon smolts in large rivers with and without dams. Public Library of Science: Biology. PLoS Biol 6(10):e265 doi:10.1371/journal.pbio.0060265. Available: www.plosbiology.org/article/info:doi/10.1371/journal.pbio.0060265 (February 2009).

Williams, J. G., S. G. Smith, R. W. Zabel, W. D. Muir, M. D. Scheuerell, B. P. Sandford, D. M. Marsh, R. A. McNatt, and S. Achord. 2005. Effects of the federal Columbia River power system on salmonid populations. U. S.

Department of Commerce. NOAA Technical Memorandum NMFS-NWFSC-63.

Zabel, R. W., and S. Achord. 2004. Relating size of juveniles to survival within and among populations of Chinook salmon. Ecology 85(3):795–806.

Zydlewski, G. B., G. Horton, T. Dubreuil, B. Letcher, S. Casey, J. Zydlewski. 2006. Remote monitoring of fish in small streams: a unified approach using PIT tags. Fisheries 31(10):492–02.

American Fisheries Society Symposium 76:135–145, 2012

Long Term Retention, Survival, Growth, and Physiological Indicators of Juvenile Salmonids Marked with Passive Integrated Transponder Tags

Kenneth G. Ostrand*, Gayle B. Zydlewski[1], and William L. Gale[2]

Abernathy Fish Techology Center, U.S. Fish and Wildlife Service
1440 Abernathy Creek Road, Longview, Washington, 98632, USA

Joseph D. Zydlewski[3]

Columbia River Fisheries Program Office, U.S. Fish and Wildlife Service
1211 SE Cardinal Court, Suite 100, Vancouver, Washington 98683, USA

Abstract.—To track individuals in situ, over 12 million salmon and trout have been marked with passive integrated transponder (PIT) tags in the Columbia River Basin, USA. However, few studies have examined long term tag retention as well as tag effects on juvenile salmon and trout. We marked juvenile coho salmon *Oncorhynchus kisutch* (*N* = 207), steelhead (anadromous rainbow trout) *O. mykiss* (*N* = 221), cutthroat trout *O. clarkii* (*N* = 202) and bull trout *Salvelinus confluentus* (*N* = 180) with 12, 19, or 23 mm PIT tags and examined tag retention, survival, growth, and physiological performance over a six month period in a laboratory environment. PIT tag retention rates were high for coho salmon (100%), steelhead (95%), cutthroat trout (97%), and bull trout (99%), regardless of tag size. Survival was also high for coho (99%), steelhead (99%), cutthroat trout (97%), and bull trout (88%) and did not vary among tag sizes. Short term individual growth rates for coho salmon marked with 12 mm tags were significantly higher than those marked with 19 mm and 23 mm PIT tags. Likewise, steelhead trout individual growth rates were lower for fish marked with 23 mm PIT tags followed by 19 and 12 mm tags. Conversely, long-term growth rates were positive and not affected by tag size. There were no significant effects of tag size or marking on coho gill Na+, K+, -ATPase activity (μmol ADP x mg protein^{-1} h^{-1}) and plasma osmolality (μmol kg^{-1}) or bull trout hepatosomatic indices. Our study suggests that marking juvenile salmonids with PIT tags results in high retention with little effect upon their survival, growth, and important physiological indicators regardless of tag size in a laboratory environment.

*Corresponding author: Kenneth_Ostrand@fws.gov
[1]Current address: School of Marine Sciences, The University of Maine, 5741 Libby Hall, Orono, Maine 04469, USA
[2]Current address: Mid-Columbia River Fishery Resource Office, 7501 Icicle Road, Leavenworth, Washington 98826, USA
[3]Current address: U.S. Geological Service Cooperative Fish and Wildlife Research Unit, The University of Maine, Orono, Maine 04469, USA

Introduction

Since the 1980s, passive integrated transponder (PIT) tags have been used to support the collection of various biological and population demographic data in a variety of animal models (Gibbons and Andrews 2004). For example, this technology has been used extensively in the Pacific Northwest for monitoring the behavior and survival of juvenile and adult salmonids in the Columbia River Basin (Zabel et al. 2005). Over 15 million salmon and trout have been PIT tagged in the Columbia River Basin since 1987 (Pacific States Marine Fisheries Commission 2005). Though 12-mm PIT tags are most commonly used, to maximize detection distance many researchers employ a variety of tag sizes including 19-mm and 23-mm PIT tags (Roussel et al. 2000; Zydlewski et al. 2001; Hill et al. 2006). The distance at which a PIT tag can be detected is particularly important in applications where fish are not recaptured but are remotely detected via a fixed or mobile antenna. Since data collected from PIT tag detections is often used to determine stock identity (Jenkins and Smith 1990; Achord et al. 1996), movements (Ombredane et al. 1998; Zydlewski et al. 2001), migration rates and routes (Achord et al. 1996; Kennedy et al. 2007a), abundance (Achord et al. 1996), growth (Peterson et al. 1994), mortality (Kennedy et al. 2007a), and stocking success (Wills 2006), it is important that the PIT tags themselves do not directly or indirectly alter these data.

Although PIT tag retention rates (Buzby and Deegan 1999; Gries and Letcher 2002; Peterson et al. 1994) and short term post tagging survival (Prentice et al. 1990; Peterson et al. 1994; Ombredane et al. 1998; Gries and Letcher 2002; Bateman and Gresswell 2006) and growth (Prentice et al. 1990; Peterson et al. 1994; Ombredane et al. 1998; Bateman and Gresswell 2006) have been examined and commonly reported in the literature, the effects of tags on long term growth and important physiological processes for salmonids have only been described only preliminarily. Given the importance of these processes on salmonid growth and survival it is important to understand if PIT tag marking imparts negative effects upon smoltification, growth, and energy storage. In addition, the results from these earlier studies were often confounded by differences in environment, tag size, surgical and implantation technique, study duration, and species and fish life history stage (Bateman and Gresswell 2006) making it difficult to independently evaluate the effect of marking salmonids with PIT tags. As a result we examined the long-term retention and effect of different sized PIT tags (12, 19, 23 mm) on the survival, growth, and physiological performance of juvenile steelhead (anadromous rainbow trout) *Oncorhynchus mykiss*, coho salmon *O. kisutch*, cutthroat trout *O. clarkii*, and bull trout *Salvelinus confluentus* in a laboratory environment.

Methods

Coho salmon and steelhead trout eggs were obtained from Washington Department of Fish and Wildlife (WDFW) Big Creek Hatchery, cutthroat trout eggs were obtained from WDFW Cowlitz Complex, and adfluvial bull trout eggs were obtained from the U.S. Fish and Wildlife Service's Creston National Fish Hatchery (NFH). Eggs were transported to Abernathy Fish Technology Center (AFTC) incubated and hatched in flow through trays maintained on well water (12.5°C) and a natural photoperiod. Juvenile coho salmon (N = 207), steelhead (N = 221), cutthroat (N = 202) and bull trout (N = 180) were reared on artificial salmonid feeds until they reached a desired size (range = 100–150 mm, Fork Length, Florida). Individual fish were anesthetized, measured, weighed, and randomly implanted with a 12, 19, or 23 mm

PIT tag (134.2 khz ISO; Destron Fearing Inc.) or left untagged (controls). The PIT tags were inserted by cutting a small (≤5 mm) ventral opening into the body cavity just behind the pectoral fin insertion (Gries and Letcher 2002). A PIT tag was placed into the body cavity in a lateral ventral position behind the pectoral fin insertion but in front of the pelvic fin insertion. Coho and steelhead were marked with all three differently sized tags whereas bull trout were marked with 12 and 23 mm PIT tags and cutthroat trout with 23 mm tags. Equal numbers of fish were placed within three replicate circular tanks (1.2 m diameter) that were supplied with approximately 1 l/min well water at ambient temperature (10 ± 2°C). Fish were fed Bio-Oregon dry pellets daily at 1% body weight. Feeding levels were adjusted according to total biomass and water temperature monthly. Tanks were examined daily for expelled tags and fish mortality. During each month, for up to 8 months, fish were anesthetized, measured, weighed, and identified via PIT tag code to determine growth rates. Growth rates were determined between all sampling dates by calculating specific growth (Busacker et al. 1990).

During the spring (March, April, and May) coho salmon gill Na^+, K^+-ATPase (ATPase) activity and plasma osmolality (mOsm × Kg H_2O^{-1}) were determined as relative indicators of seawater readiness and smoltification. ATPase activity was determined via a small gill biopsy using the method of McCormick (1993) and immediately frozen on dry ice. Gill samples were stored at –80°C and gill ATPase activity measured spectrophotometrically using the method of McCormick (1993). To measure the relative seawater tolerance, plasma osmolality was determined after 24 h salt water exposures (32 ppt). Fish were killed by an overdose of anesthetic, and blood collected using heparinized syringes. Plasma was obtained by centrifugation at 10,000 rpm, 4°C for 5 min. Osmolality was determined using an Osmette II osmometer.

We determined the hepatosomatic index (HSI) as an indicator of glycogen stores of adfluvial bulltrout. At the termination of the experiment fish were measured (FL), weighed, and livers were dissected, removed from the viscera, weighed, and HIS was calculated as: HIS = (fish liver weight (g)/fish body weight (g)) × 100.

Analysis of Variance (analysis of variance (ANOVA)) and *t*-tests were used to test for differences among the treatments for the response variables (FL, weight, growth rate, ATPase, plasma osmolality, and LR). Chi-square tests were used to compare differences ($P < 0.05$) in tag retention and mortality among treatments for each species. Statistical comparisons of growth rates included only fish marked with PIT tags because untagged control fish could not be individually identified. Data were log transformed to meet assumptions of normality and homogeneity of variances. Significant ANOVAs ($P < 0.05$) were followed by Fisher LSD mean separation tests for pairwise comparisons.

Results

All fish increased in FL and gained weight during the experiment regardless of species or PIT tag size (Figure 1). There were no significant differences in FL or weight among treatment groups for any species, excluding cutthroat trout. Although cutthroat trout marked with 23 mm PIT tags were similar in FL and weight at the onset and completion of the experiment they differed during September (FL, $P = 0.047$; weight, $P = 0.016$) and October (FL, $P = 0.014$; weight, $P = 0.007$). Percent weight gain did not significantly differ between the control fish and those marked with different sized tags, in spite of species (Table 1).

Species specific growth rates of fish marked with PIT tags may be negatively af-

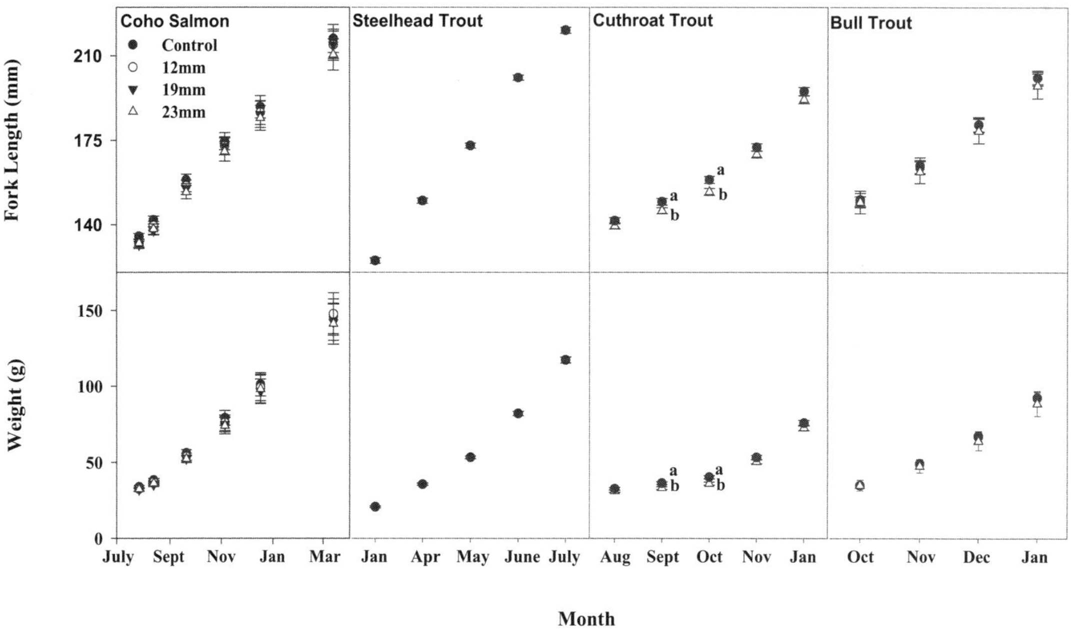

FIGURE 1. Fork length (mean ± 1 SE) and weight (mean ± 1 SE) of coho salmon, steelhead, cutthroat, and bull trout unmarked or marked with a 12, 19, or 23 mm passive integrated transponder tag. Letters denote significant ($P < 0.05$) differences between fish marked with different sized PIT tags on a respective date.

fected following a marking event (Figure 2). The individual growth rates for coho salmon during the first growth interval were significantly lower for the fish marked with 19 mm and 23 mm PIT tags as compared to those marked with 12 mm PIT tags. Individual growth rates did not differ between coho salmon marked with 19 mm and 23 mm PIT tags during the first growth interval. Likewise, steelhead trout individual growth rates were slowest for fish marked with 23 mm PIT tags followed by 19 and 12 mm tags. During the third growth interval (May to June) steelhead marked with 23 mm PIT tags grew significantly faster than those marked with 19 or 12 mm tags that did not significantly differ from each other. Bull trout individual growth rates did not differ among fish marked with different sized PIT tags.

All PIT tagged coho salmon subjected to 24 h seawater challenges survived. Although ATPase activity was significantly higher ($P = 0.003$) in April when compared to March and May, it did not differ among the groups marked with different sized tags (Figure 3A). Plasma osmolality was significantly higher in later months (Figure 3B). Nevertheless, there were no significant differences ($P = 0.16$) in plasma osmolality among the coho salmon marked with PIT tags of various sizes.

Bull trout HSI did not differ among controls and the two marked groups (Figure 4). Fish marked with 23 mm PIT tags had a larger variance of relative liver weight than fish marked with either 12 mm PIT tags or controls.

We observed low mortality and high PIT tag retention for all the species, regardless of tag size (Table 1). Mortality did not significantly differ within each species marked with different sized PIT tags. Mortality was higher for bull trout (range = 10–15%) than for the other species (range = 0–5%) we examined. Likewise PIT tag retention was high and did

TABLE 1. Mean (± 1 SE) initial fork length (FL), initial weight, final weight gain, mortality, and passive integrated transponder (PIT) tag retention for coho salmon and steelhead, cutthroat, and bull trout marked with different sized PIT tags. There were no statistical differences among fish marked with different sized PIT tags regardless of species.

Species	Treatment	*N*	FL (mm)	Weight (g)	Weight Gain (%)	Mortality (%)	PIT tag Retention (%)
Coho salmon	Control	69	135 (1.24)	34 (0.82)	329.7	5.8	*
	12 mm	69	132 (1.58)	32 (1.04)	355.5	0.0	100
	19 mm	69	131 (1.47)	31 (1.00)	345.7	1.5	100
	23 mm	69	132 (1.72)	32 (1.09)	337.2	2.9	100
Steelhead trout	Control	75	125 (1.15)	20 (0.61)	468.9	1.3	*
	12 mm	75	124 (1.17)	20 (0.59)	492.5	1.3	98.7
	19 mm	75	124 (1.11)	20 (0.60)	477.7	0.0	97.3
	23 mm	75	123 (1.20)	20 (0.64)	477.6	1.3	89.0
Cutthroat trout	Control	101	136 (1.22)	31 (0.82)	131.9	5.0	*
	12 mm	*	*	*	*	*	*
	19 mm	*	*	*	*	*	*
	23 mm	101	141 (1.29)	32 (0.96)	134.5	3.0	100
Bull trout	Control	60	150 (2.56)	34 (1.80)	165.6	12.0	*
	12 mm	60	149 (2.20)	34 (1.60)	164.3	15.0	99.5
	19 mm	*	*	*	*	*	*
	23 mm	60	149 (2.56)	34 (1.83)	153.9	10.0	100

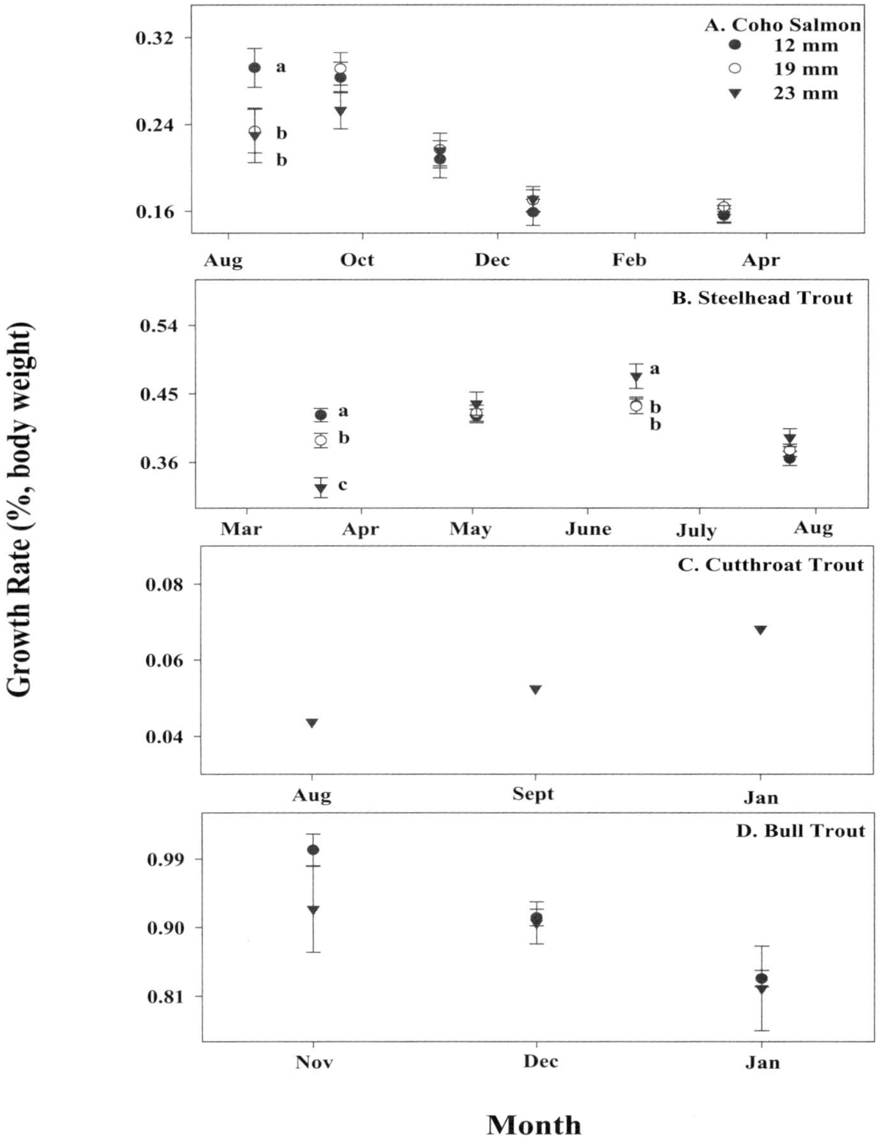

Figure 2. Individual growth rate (mean ± 1 SE) of coho salmon (A), steelhead (B), cutthroat (C), and bull trout (D) marked with a 12, 19, or 23 mm passive integrated transponder tag (PIT). Letters denote significant ($P < 0.05$) differences between fish marked with different sized PIT tags.

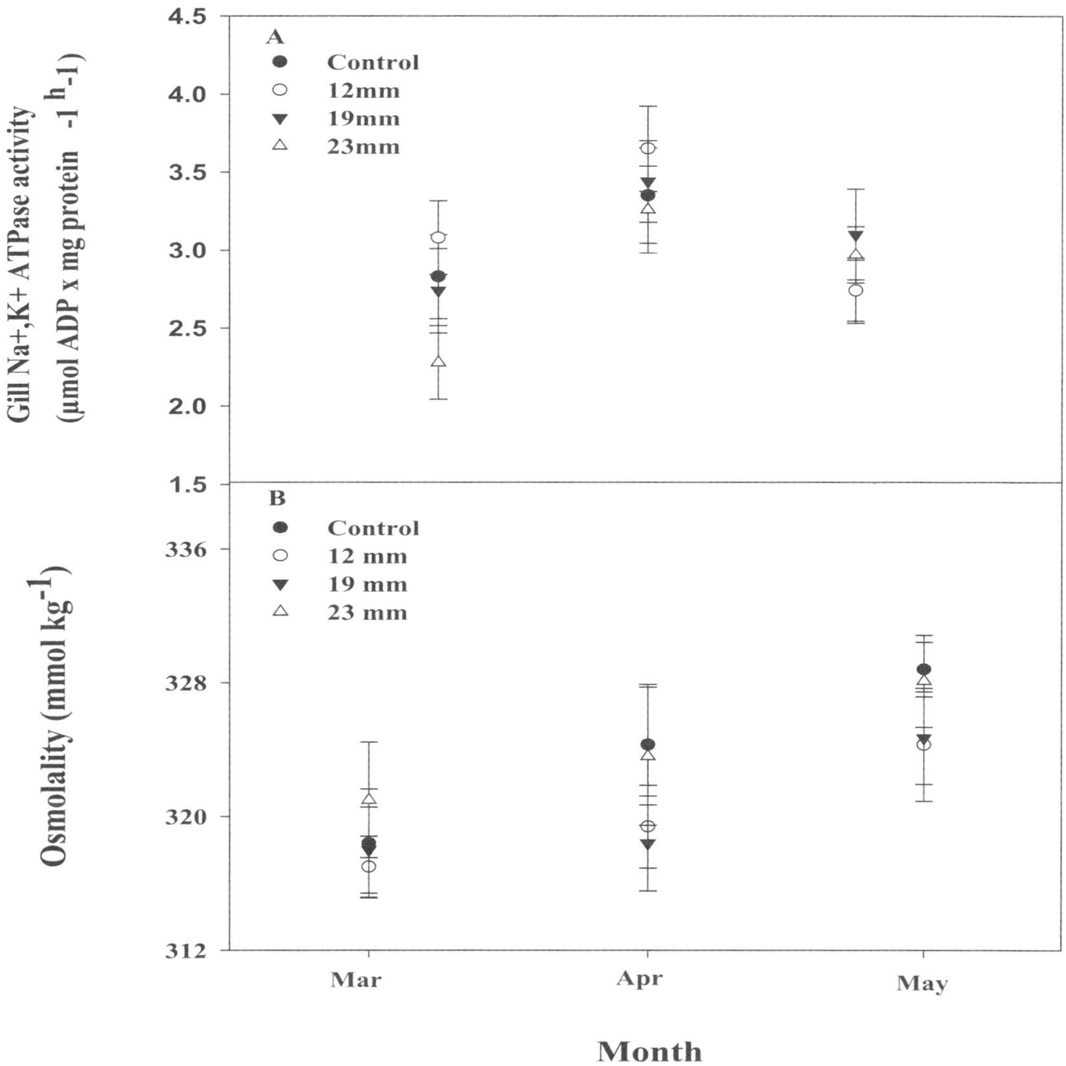

Figure 3. Mean (± 1 SE) Gill Na^+, K^+ -ATPase activity (A) and plasma osmolality (B) of coho salmon marked with a 12, 19, or 23 mm passive integrated transponder tags (PIT).

not differ among the treatments, regardless of species. PIT tag retention was the lowest for steelhead marked with 23 mm PIT tags (89%) as compared to all other treatments (range = 97–100%).

Discussion

Our results suggest that there appears to be little effect of tagging on long term juvenile fish growth, physiological indicators, mortality, and tag retention in a laboratory environment. Individual growth rates appear to be somewhat influenced by tag size shortly after marking. Therefore, researchers examining growth rates soon after tagging need to be cognizant of this potentially negative effect. As technology continues to improve and these tags become smaller, even short term effects on individual growth will likely become inconsequential. Our results suggest that the use of PIT tagged fish as indicators of stock identity (Jenkins and

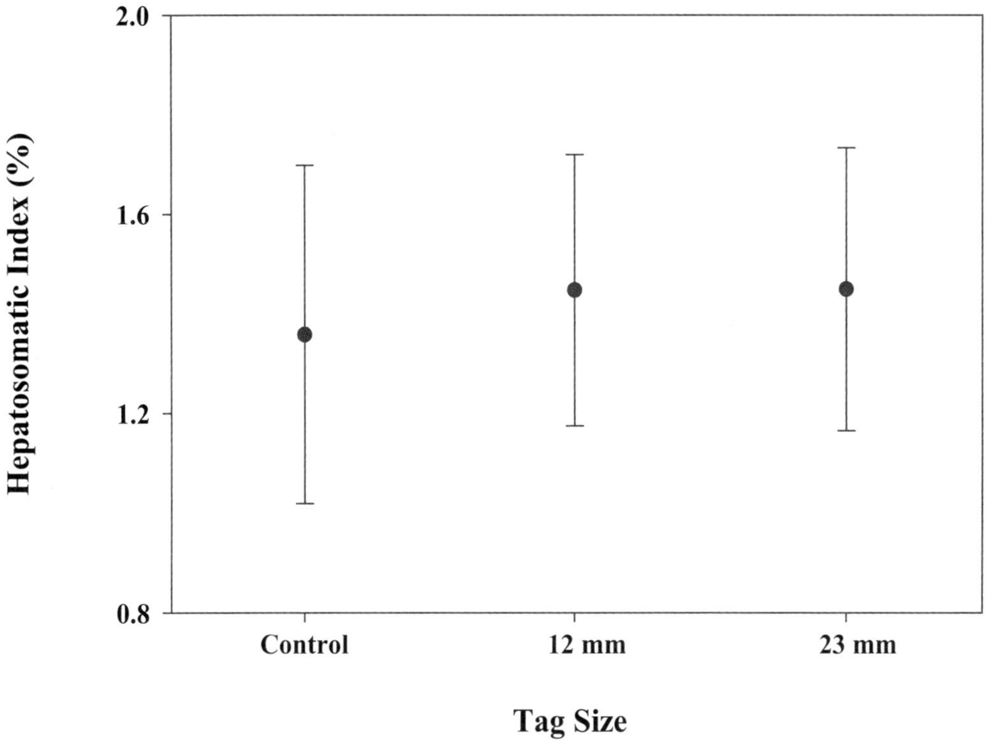

Figure 4. Relative liver weight (mean ± 1 SE) for bull trout marked with either 12, or 23 mm passive integrated transponder tags (PIT). The length specific standard weight (Ws) developed by Hyatt and Hubert in (2000) was used to determine bull trout relative liver weight.

Smith1990; Achord et al. 1996), movements (Ombredane et al. 1998; Zydlewski et al. 2001), migration rates and routes (Achord et al. 1996; Kennedy et al. 2007a), abundance (Achord et al. 1996), growth (Peterson et al. 1994), mortality (Kennedy et al. 2007a), and stocking success (Wills 2006) do not directly or indirectly influence or bias these data during juvenile life history stages. Nevertheless, it is important to note that PIT tag loss may be much higher during later life history stages (e.g., saltwater phase). PIT tagged fish appear to have higher mortality and lower smolt-to-adult recruit survival (SARS) than nonPIT tagged fish (see Knudsen et al. 2009; Williams et al. 2005). Thus our study results should be restricted to or cautiously extrapolated beyond juvenile life history stages.

Growth rates of fish, particularly coho salmon and steelhead, were negatively affected by PIT tagging shortly after tagging. Our results are similar to those observed for Chinook salmon *Oncorhynchus tshawytscha* (Prentice et al. 1990), brown trout *Salmo trutta* (Ombredane et al. 1998), Eurasian perch *Perca fluviatilis* (Baras et al. 2000), mottled sculpin *Cottus bairdii* (Ruetz et al. 2006), and steelhead trout (Bateman and Gresswell 2006). Furthermore, our results suggest that the negative influence on short term growth rates appears to be exacerbated as PIT tag size to body weight ratios increase. Growth rates have been reported to be negatively affected when PIT tag to body weight ratios exceed 4% (Baras et al. 2000; Ruetz et al. 2006). In our study, PIT tag to body weight ratios were less (range = 2% to 0.5%) than those previously reported, yet individual growth rates were still negatively effected in observations shortly after tagging. Fish that were implanted with heavier and larger sized tags had significantly slower growth shortly after tagging than controls or those implanted with smaller tags. This suggests that tag size, as well as the tagging event (Kennedy et al. 2007b) contributes to the observed growth patterns we observed following marking. Interestingly, it appears that once fish have recovered they are able to compensate for short term reductions in growth and resume growth patterns similar to that of control fish. Nevertheless, our results taken together with those of past studies suggest that a short period of reduced growth after tagging may be expected and should be considered when response variables are collected and measured shortly after tagging.

Our results suggest that PIT tagging does not impart significantly negative effects on coho smoltification or juvenile adfluvial bull trout glycogen stores. There were no significant differences in ATPase activity or plasma osmolality among coho salmon controls and those marked with different sized tags. In our study coho salmon were tagged prior to the initiation of smoltification. We suspect that the stress of the tagging event and the tag itself were minimal and therefore did not significantly affect the parr–smolt transformation. Our results for ATPase activity and plasma osmolality were similar to those reported for other salmonids reared under hatchery conditions (Hoar 1988; Hill et al. 2006; Kennedy et al. 2007a). Although PIT tag marking fish did not appear to affect smoltification, fish that are marked at the onset or during the smoltification process may be negatively influenced possibly slowing emigration, encouraging residualism, or increasing mortality (Kennedy et al. 2007b); however, additional research will be needed to examine this hypothesis. Likewise, bull trout HSI did not differ among controls and the two treatment groups. Nevertheless, the increased variance associated with fish PIT tagged with 23 mm tags suggests that a proportion of our treatment population may have faired worse than the rest. Given bull trout's aggressive nature, the effect of social status combined with tag size may explain the observed variance. Thus PIT tagging conducted during critical periods of bull trout energy storage or utilization such as migratory or overwintering periods (Berry 1994; Ratcliff et al. 1996) may affect survival rates following marking events. We encourage researchers to consider the timing of PIT tag marking events as to avoid potentially affecting critical physiological processes.

Juvenile fish survival and PIT tag retention rates in this laboratory study were high and comparable to those reported in the literature. Although most studies on juvenile fish have been much shorter in duration than our study, our mortality rates were similar to those reported for Atlantic salmon *Samo salar* (5.7%, 12 mm PIT tags, Gries and Letcher 2002; 21.2%, 23 mm PIT tags, Roussel et al. 2000), Chinook salmon (1%, 12 mm PIT tags, Achord et al. 1996; 12 mm PIT tags, 3.6%, Prentice et al. 1990), sockeye salmon *Oncorhynchus nerka* (3.6%, 12 mm PIT tags, Prentice et al. 1990), and steelhead (2%, 23 mm PIT tags, Hill et al. 2006; 14%, 23 mm PIT tags, Bataman and Gresswell 2006). Like our results, previously reported tag retention rates (range = 30–100%) were relatively high for juveniles (Prentice et al. 1990; Peterson et al. 1994; Ombredane et al. 1998; Bataman and Gresswell 2006; Knudsen et al. 2009). Collectively, these results and our study indicate that low mortality and high retention rates will occur if good fish handling practices are employed and fish are allowed to fully recover prior to release.

Acknowledgments

We thank J. Holmes, J. Poole, and P. Larson of the U.S. Fish and Wildlife Service.

This information is distributed solely for the purpose of predissemination peer review under applicable information quality guidelines. It has not been disseminated by the Service. It does not represent and should not be construed to represent any Service determination or policy.

References

Achord, S., G. M. Matthews, O. W. Johnson, and D. M. Marsh. 1996. Use of passive integrated transponder (PIT) tags to monitor migration timing of Snake River Chinook salmon smolts. North American Journal of Fisheries Management 16:302–313.

Baras, E., C. Malbrouck, M. Houbart, P. Kestemont, and C. Mélard. 2000. The effect of PIT tags on growth and physiology of age-0 cultured Eurasian perch *Perca fluviatilis* of variable size. Aquaculture 185:159–173.

Bateman, D. S., and R. E. Gresswell. 2006. Survival and growth of age-0 steelhead after surgical implantation of 23-mm passive integrated transponders. North American Journal of Fisheries Management 26:545–550.

Berry, D. K. 1994. Alberta's bull trout management and recovery plan. Alberta Environmental Protection Fish and Wildlife Service Management Division, Edmonton, Alberta 22p.

Busacker G., I. Adelman, and E. Goolish. 1990. Growth. Pages 363–387 *in* C. B. Schreck and P. B. Moyle, editors. Methods for fish biology. American Fisheries Society, Bethesda, Maryland.

Buzby, K., and L. Deegan. 1999. Retention of anchor and passive integrated transponder tags by Arctic grayling. North American Journal of Fisheries Management 19:1147–1150.

Gries, G., and B. H. Letcher. 2002. Tag retention and survival of age-0 Atlantic salmon following surgical implantation with passive integrated transponder tags. North American Journal of Fisheries Management 22:219–222.

Gibbons, J. W., and K. M. Andrews. 2004. PIT tagging: simple technology at its best. BioScience 54:447–454.

Hill, M. S., Zydlewski, G. B., Zydlewski, J. D., and J. M Gasvoda. 2006. Development and evaluation of portable PIT tag detection units: PITpacks. Fisheries Research 77:102–109.

Hoar, W. S. 1998. The physiology of smolting salmonids. Pages 275–343 *in* W. S. Hoar and D. Randall, editors. Fish Physiology. Volume XIB. Academic Press, New York.

Jenkins, W. E. and T. I. J. Smith. 1990. Use of PIT tags to individually identify striped bass and red drum brood stocks. Pages 341–345 *in* N. C. Parker, A. E. Giorgi, R. C. Heidinger, D. B. Jester, Jr., E. D. Prince, and G. A. Winans, editors. Fish-marking techniques. American Fisheries Society, Symposium 7, Bethesda, Maryland Maryland.

Kennedy, B. M., W. L. Gale, and K. G. Ostrand. 2007a. Evaluation of clove oil concentrations for use as an anesthetic during field processing and passive integrated transponder implantation of juvenile rainbow steelhead. Northwest Science 81:147–154.

Kennedy, B. M., W. L. Gale, and K. G. Ostrand. 2007b. The relationship between smolt development and migration timing to avian predation risk of steelhead (*Oncorhynchus mykiss*) in a large estuary. Canadian Journal of Fisheries and Aquatic Sciences 64:1506–1516.

Knudsen, C. M., M. V. Johnston, S. L. Schroder, W. J. Bosch, and D. E. Fast. 2009. Effects of passive integrated transponder tags on smolt-to-adult recruit survival, growth, and behavior of hatchery spring Chinook salmon. North American Journal of Fisheries Management 29:658–669.

McCormick, S. D. 1993. Methods for non-lethal gill biopsy and measurement of Na +, K+ -ATPase activity Canadian Journal of Fisheries and Aquatic Sciences 50:656–658.

Ombredane, D., J. Bagliniere, and F. Marchand. 1998. The effects of passive integrated transponder tags on survival and growth of juvenile brown trout (*Salmo trutta* L.) and their use for studying movement in a small river. Hydrobiologia. 371– 372:99–106.

Peterson, P., E. Prentice, and T. Quinn. 1994. Comparison of sequential coded wire and passive integrated transponder tags for assessing overwinter growth and survival of juvenile coho salmon. North American Journal of Fisheries Management 14:870–873.

Prentice E., T. Flagg, and C. McCutcheon. 1990. Feasibility of using implantable passive integrated transponder (PIT) tags in salmonids. Pages 317–322 *in* N. C. Parker, A. E. Giorgi, R. C. Heidinger, D. B. Jester, Jr., E. D. Prince, and G. A. Winans, editors. Fish marking techniques. American Fisheries Society, Symposium 7, Bethesda, Maryland.

Pacific States Marine Fisheries Commission (PSMFC). 2005. PIT tag information system (PTAGIS). Pacific States Marine Fishereis Commission, Gladstone, Oregon.

Ratcliff, D. E., S. L. Thiesfeld, W. G. Weber, A. M. Stuart, M. D. Riehle, and D. V. Buchanan. 1996. Distribution, life history, abundance, harvest, habitat, and limiting factors of bull trout in the Metolius River and Lake Billy Chinook, Oregon, 1983–1994. Oregon Department of Fish and Wildlife Fisheries Division, Information Report No. 96–97.

Roussel, J. M., A. Haro, and R. A. Cunjak. 2000. Field test of a new method for tracking small fishes in shallow rivers using passive integrated transponder (PIT) technology. Canadian Journal of Fisheries and Aquatic Sciences 57:1326–1329.

Ruetz, C. R., B. M. Earl, and S. L. Kohler. 2006. Evaluating passive integrated transponder tags for marking mottled sculpins: effects on growth and mortality. Transactions of the American Fisheries Society 135:1456–1461.

Williams, J. G., S. G. Smith, R. W. Zabel, W. D. Muir, M. D. Scheuerell, B. P. Sandford, D. M. Marsh, R. A. McNatt, and S. Achord. 2005. Effects of the federal Columbia River power system on salmonid populations. U.S. Department of Commerce, NOAA Tech. Memo. NMFS-NWFSC-63.

Wills, T. C. 2006. Comparative abundance, survival, and growth of one wild and two domestic brown trout strains stocked in Michigan rivers. North American Journal of Fisheries Management 26:535–544.

Zabel, R. W., Wagner, T., Congleton, J. L., Smith, S. G., and J. G. Williams. 2005. Survival and selection of migrating salmon from capture-recapture models with individual traits. Ecological Applications 15:1427–1439.

Zydlewski, G. B., Haro, A., Whalen, K. G., and S. D. McCormick. 2001. Performance of stationary and portable passive transponder detection systems for monitoring of fish movements. Journal of Fish Biology 58:1471–14.

American Fisheries Society Symposium 76:147–161, 2012

Stock Identification Applications of Conventional Tagging Data for Atlantic Cod in the Gulf of Maine

Shelly M. L. Tallack*
Gulf of Maine Research Institute
350 Commercial Street, Portland, Maine 04101, USA

Abstract.—In the Gulf of Maine region, Atlantic cod *Gadus morhua*, are managed as three separate stocks: in U.S. waters, the Gulf of Maine (5Y) and Georges Bank (5Z) stocks and in Canadian waters, the Bay of Fundy stock (4X). The Northeast Regional Cod Tagging Program (NRCTP), a large-scale, international collaborative tagging program, was initiated in 2003 and provides the first region-wide, international snapshot of cod movements, mixing, and growth across all three management areas. As the 2008 stock assessment of Atlantic cod was approaching, these mark–recapture data (114,473 tag releases and >6,500 recaptures) were being analyzed for stock identification purposes. Analysis of raw and weighted data indicate exchanges between different areas which are likely related to spawning behavior, maturation, and environmental conditions. Two core assumptions when defining a stock are that (1) the stock is self-sustaining and that (2) neighboring stocks exist in isolation; the migration patterns observed in the current study may substantially violate both assumptions. With additional ongoing assessments into spatial variation in key life history characteristics of growth and natural mortality, the NRCTP exemplifies the role of conventional mark–recapture data in the complex process of stock identification. The geographic scale and quality of data collected during the NRCTP also confirm the value of international, industry-science collaborative research initiatives; involving this data in both stock assessments and future management initiatives is rewarding to the ~250 commercial and recreational fishermen who tagged cod for this program.

Introduction

In recent decades, the combination of depressed Atlantic cod *Gadus morhua* populations and intensive fishing practices has caused great concern regarding the health and resilience of North Atlantic cod stocks; the New England and Canadian Maritime cod populations are no exception. In the Gulf of Maine region, cod are managed as three separate stocks: the Gulf of Maine (5Y) and Georges Bank (5Z) stocks in U.S. waters and the Bay of Fundy/Western Scotian Shelf stock (4X) in Canadian waters (Figure 1). Total stock biomass estimates for all three cod stocks have shown a steady decline since the 1960s (Mayo et al. 2002), resulting in dramatic management restrictions on the region's groundfish industry (Figure 2). Despite an indication that cod abundance in some New England fishing grounds was increasing around 2001–2002, declines have been seen in years since, and regional stock-rebuilding programs are still underway (Mayo and O'Brien 2006).

*Corresponding author: stallack@gmri.org

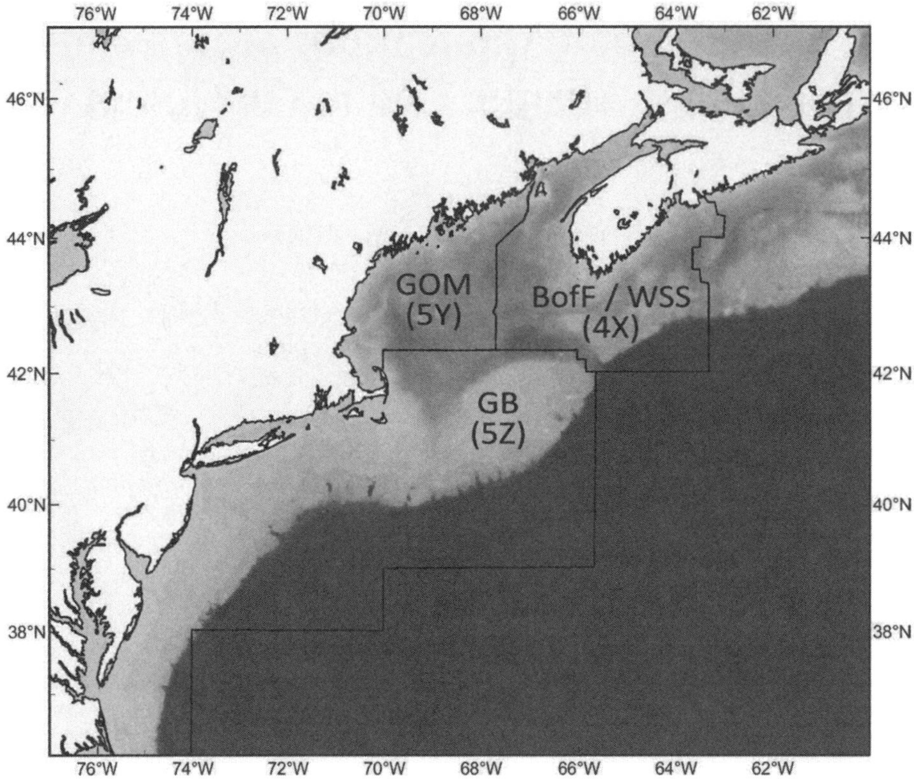

Figure 1. Atlantic cod stock management areas in the Gulf of Maine transboundary region between the U.S. and Canada: 5Y and 5Z are in U.S. waters, 4X is in Canadian waters.

The slow recovery of the region's cod stocks has highlighted a need for additional contemporary information to supplement the traditional data associated with fishery-independent surveys (e.g., biomass estimates) and harvesting (e.g., landings, discards and exploitation rates). Regional scoping meetings led to the recommendation for an intense, region-wide tagging initiative to provide a snapshot on cod movements, migration and growth (NEAQ 2001) under the current fishing climate. Accordingly, the Northeast Regional Cod Tagging Program (NRCTP) was initiated.

Coordinated by the Gulf of Maine Research Institute (GMRI), the NRCTP represents a partnership of eight different research organizations (~40 scientists) and over 250 fishermen along the Northeast seaboard of the USA and Canada. By June 2006 the NRCTP had achieved its original key objectives of: (1) developing a collaborative cod tagging program between scientists and fishermen in the Gulf of Maine region, including Canada; (2) obtaining and analyzing data from T-bar tag recaptures to improve the region's understanding of cod distribution, movement, and growth; (3) making the tagging data available to the public via an online GIS mapping interface; and (4) recommending further testable hypothesis for continuing tagging studies (Tallack 2006). GMRI's focus since June 2006 has been to prepare, analyze, and deliver the NRCTP data to the Northeast Fisheries Science Center (NEFSC) for consideration ahead of the 2008 Groundfish Assessment Review Meeting (GARM), when the stock assessments for 19 groundfish (including Atlantic cod) were to be finalized.

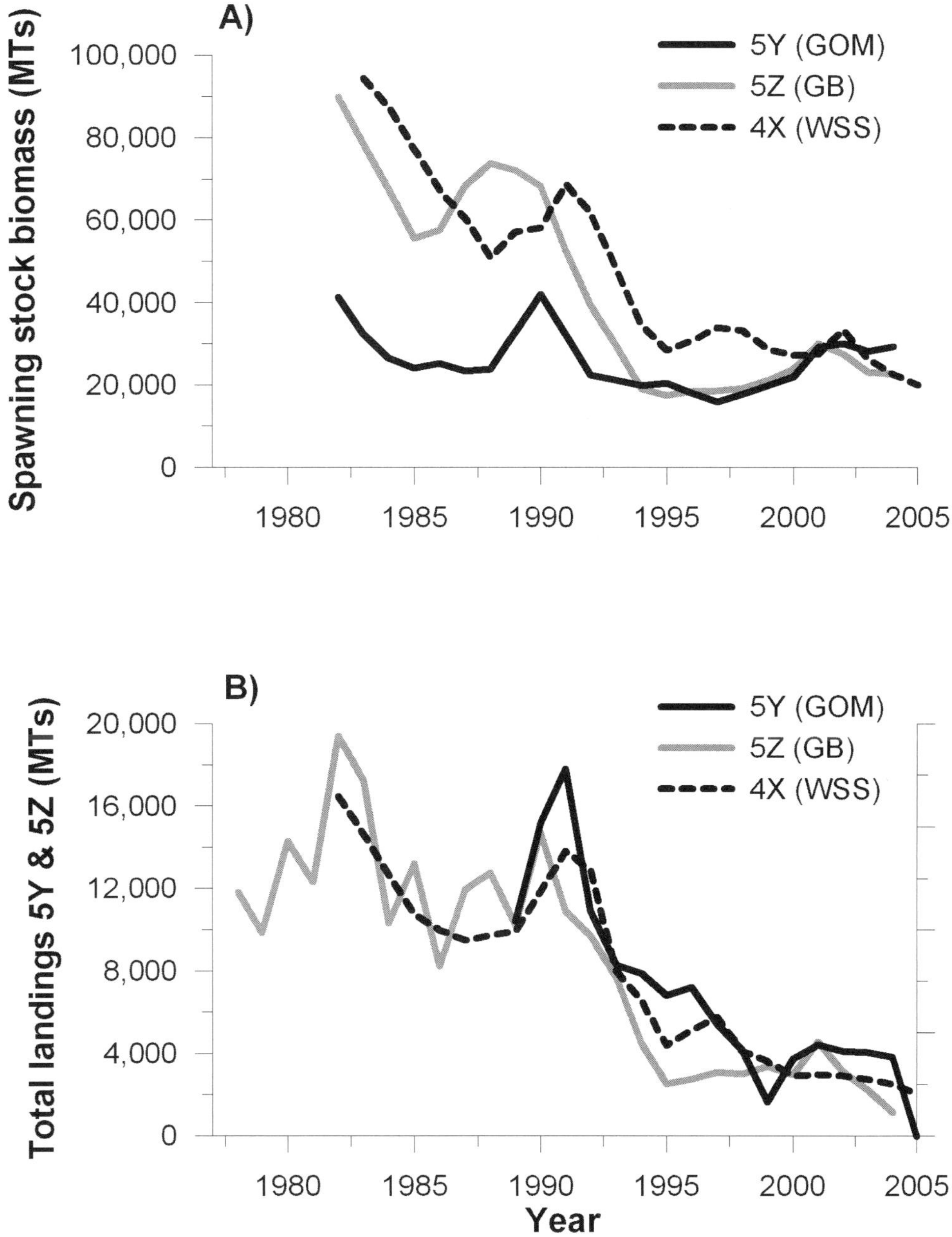

Figure 2. Atlantic cod spawning stock biomass estimates in metric tons (MT) (A) and landings data in metric tons (MT) (B) show signs of depression in the Gulf of Maine region for all three stocks (5Y, 5Z, and 4X) in recent decades.

Cod tagging studies are not new to the Gulf of Maine region, but most historical studies have been temporally discontinuous and have tended to have a localized focus, e.g. Nantucket Shoals (Rich 1925; Schroeder 1930; Smith 1902), Georges Bank (Gröger et al. 2007; Higgins 1927, 1928, 1931), Massachusetts Bay (Howell et al. 2008), Nova Scotia (Cadigan and Brattey 2001, 2003; MacKenzie 2003), Bay of Fundy (Hunt and Nielson 1993; Hunt et al. 1999). Findings from these various tagging studies have not all been published and the raw datasets are neither uniform, collated or necessarily electronically available. Most cod tagging studies in the Gulf of Maine region were not specifically designed to evaluate the relative movement between areas and therefore, when published, most findings have been presented in a qualitative, descriptive manner. Descriptive analyses of tagging data are helpful for understanding the movement patterns and identifying the key migration routes for cod. However, it is difficult to apply such qualitative results in the context of stock assessments which seek to quantify the combined effects of estimated 'inputs' (e.g., reproduction and growth) and 'removals' (e.g., natural mortality and fishing mortality) on the overall size and structure of a fish population. Mark–recapture data can provide supplementary estimates of some of these parameters including: fish growth (Tallack 2009a; Taylor and Tallack, in review), natural mortality (*M*) and fishing mortality (*F*) (Miller and Tallack 2007). All these parameters are key for the accurate performance of most stock assessment models.

Depending on the study design, tagging data can also provide additional information on further stock inputs and removals associated with fish movements. While cod are thought to show considerable site fidelity during spawning seasons (Howell et al. 2008; Robichaud and Rose 2004), during nonspawning seasons cod have been recaptured at much greater distances from their assumed spawning grounds (Robichaud and Rose 2004). This potential for long-distance travel can result in the exchange of cod between stock management areas, particularly in the international transboundary region of the Gulf of Maine. It is therefore important for both scientists and managers to know the extent of cod movements between management areas so as to better understand the connectivity between stocks and thus, the implication that exploiting cod in one management area will have on cod in neighboring management areas. In recent years, efforts have been made to summarize the findings of cod movements for the region generally (Hunt et al. 1999; Loehrke and Cadrin 2007). Of particular value was the first regional attempt at quantifying movements between areas after standardizing recapture rates in different areas by fishing effort (Hunt et al. 1999).

This paper provides an overview of how conventional tag data from the NRCTP has been considered for application to the 2008 GARM, with particular emphasis on stock identification. Movement information is presented as evidenced by data collected for the period of March 2003 through October 2007. Findings will be described as both (1) qualitative descriptions of cod movements within the Gulf of Maine region, and (2) quantitative estimates of mixing and exchange between stock management areas using a weighting method adapted after Hunt et al. (1999). Both forms of analyses were relevant to the 2008 GARM and were presented in detail at the preGARM workshops (Tallack 2007).

Study Site

The NRCTP deployed tags throughout the Gulf of Maine (inshore and offshore) and also on fishing grounds in Southern New England waters (e.g., Cape Cod waters, Great South Channel and Georges Bank) and the Canadian Maritimes, (e.g., Bay of Fundy and

Browns Bank). Each tagging organization targeted fishing grounds which have been traditionally considered key spawning and/or feeding grounds, or were for other reasons believed to be of interest with regard to the movement of Atlantic cod.

Methods

This paper will not provide precise detail on all analytical methods applied, since the detail for certain methods and findings are already documented elsewhere (Miller and Tallack 2007; Tallack 2006, 2007, 2009a, 2009b; Taylor and Tallack, in review). Instead, this paper highlights the variety of approaches taken over the months in order to deliver relevant findings to the stock assessment working groups.

Field and data collection methods

Cod tagging trips took place year round from March 2003 through July 2005, with peak tagging seasons being spring and autumn of each year. All participating fishermen ($N > 250$) and scientists ($N = 42$) were trained in and applied a standardized tagging technique developed by participating scientists during the early months of the Program.

Upon initial capture, the cod were immersed in a 'live well'—tank with fresh-flowing seawater—to minimize the effects of aeration. All cod were tagged providing that they were >38 cm and were in good condition (i.e., lively, minimal to zero injury, no obvious barotrauma and able to swim down in the live well). These fish were individually removed from the live well and placed on a wet measuring board where they were tagged with Hallprint T-bar anchor tags (type TBA-2) using Avery Dennison tag fastening guns (model # 08958).

Three tag colors were used: yellow for regular or 'low' reward (LR) tags (95.7%), blue for 'high' reward (HR) tags (1.9%) and orange tags (2.3%) for monitoring reporting differences between the USA and Canada. All three colors were released in both U.S. and Canadian waters, so that fishermen could not assume which country their fish had been released in; this approach was to minimize the potential for any reporting bias since fishermen from either country may not want to show that 'their' fish also migrate to the other country's waters. A total of 18,305 (15.9%) cod were also double-tagged (DT) to enable the estimation of tag loss, or tag shedding; the remaining 84.0% were single-tagged (ST).

The data recorded for each fish at the time of tagging included: capture location, date, depth and water temperature, total fish length to the nearest cm (industry data collectors were often more familiar with the imperial system and thus, to minimize measurement error, their measurements were often recorded to the nearest half-inch, but were converted by an internal conversion script in the database), spawning condition (presence of milt or eggs as observed externally) and any additional comments regarding factors such as injuries (new or old) or parasites.

The first cod recapture data were reported in April 2003; individuals returning tags were sent a report on the cod they had reported in addition to either a T-shirt, hat or mug for LR tags (with entry into a monthly lottery worth $200 for each of five monthly winners), or with $100 for HR tags. The data requested for recaptures included: (1) tag #, (2) recapture date, (3) recapture location, (4) fish length, (5) spawning condition, (6) water depth, (7) and (8) additional comments; not all recaptures had all eight pieces of data, but items 1–4 were considered the most crucial.

Qualitative movement analysis

The analysis of raw, unweighted mark–recapture data were used to yield descriptive visualizations of cod movements. Cod mark–

recapture data were analyzed for trends in size effects, displacement and distance traveled, direction of travel, seasonal effects, differences between release locations, and evidence of homing movements or site fidelity. These factors were reviewed in combination for all major fishing grounds (Figure 3) in order to identify the primary "migration corridors" for Atlantic cod in the Gulf of Maine region.

Quantitative movement analysis

Providing estimates of exchange rates of cod between management areas was not an original goal for the NRCTP, and thus, tags were not released in proportion to the biomass present. Consequently, it has been necessary to develop a weighting method which could correct for potential biases on the quantification and interpretation of the movements observed (Tallack 2007). For tag releases these biases include: biomass present in a release area, tag loss/shedding and tag-induced mortality estimates and how these impact the expected number of recaptures. For tag recaptures, potential biases include: fishing effort and landings by tagging area, in addition to reporting rates by gear type and by tagging area.

This tailor-made weighting method was based on methods described by Hunt et al. (1999), but has evolved into a more complex weighting procedure which firstly corrects

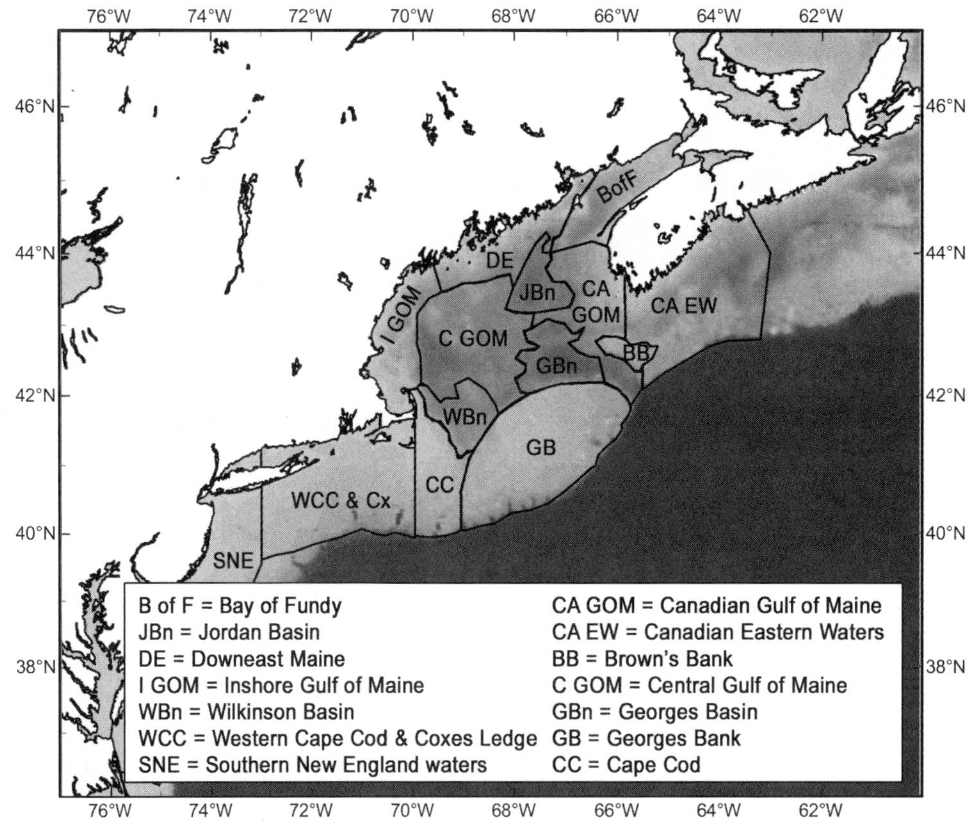

FIGURE 3. Fine-scale areas that have been defined using a combination of information on traditional fishing grounds, bathymetric characteristics and bottom type; these fine-scale areas are not standardized management areas.

for tag shedding and tag-induced mortality and then standardizes tag releases within each management area relative to three separate indices of population size: (1) biomass (derived from NOAA and DFO surveys), (2) catch (incorporating landings and exploitation rates) and (3) biomass estimated through Virtual Population Analysis (VPA) (Figure 4). The decision to run three biomass scenarios was made because the fishery-independent survey biomass estimates for one particular region was very low and yet the number of cod tagged and released in this same region were very high ($N > 50,000$ over two years); it was thus feared that the survey-derived biomass estimate would excessively downweight these ~50,000 tags. Consequently, the additional biomass estimates applied were based on fishery-dependent data (i.e., cod landings and effort which are likely associated with fishermen selecting areas of high fish availability), and the combined approach of VPA-derived biomass estimates. Tag recaptures are then standardized using exploitation rates (VPA F) and reporting rates for each management area.

The weighting model can be run on a variety of subsets of data which are filtered by: fish size, tagging year and different geographical scales. The stock management areas (Figure 1) are appropriate for stock assessment questions, but finer scale resolutions (e.g., fishery statistical squares or fishing grounds) are sometimes relevant and can also be applied with this model.

The model generates cod movement matrices for five weighting scenarios (Table 1): (1) unweighted observed recaptures, (2) unweighted releases corrected for reporting rate and exploitation rate only (i.e., unweighted expected recaptures), and then the three release weighting scenarios also corrected for reporting rate and exploitation rate. Each movement matrix collates estimates of: residency (i.e., released and recaptured in Area X), emigration (i.e., released in Area X, but recaptured elsewhere) and immigration (i.e., released elsewhere but recaptured in Area X).

Results

By October 2007, tag returns totaled 6,540 (~5.6%). Considerable inter-annual

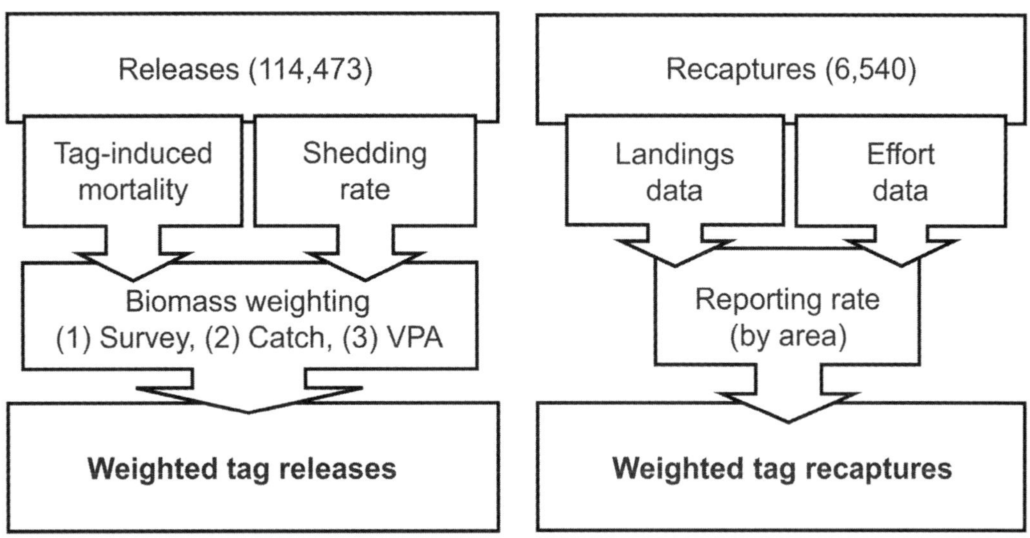

Figure 4. A depiction of the weighting model developed for the Northeast Regional Cod Tagging Program.

Table 1. The five different scenarios applied within the weighting model; weighting on the recapture end is identical in all scenarios after the first which is raw, unweighted observations only.

Weighting Scenario	Releases	Recaptures
RAW (Observed data)	-	-
RAW (Corrected recaptures)	-	F* and RR
Weighted - Survey (Biomass*)	Swept area biomass	F* and RR
Weighted - Fishing (Catch)	Catch/VPA F	F* and RR
Weighted - VPA (Biomass)	VPA biomass	F* and RR

* = VPA derived F (instantaneous fishing mortality)

variation is seen in the reporting of recaptures (Table 2), but on average, ~2% of cod are returned per year for up to two years for each year's releases; thereafter, a considerable decay in return rates is seen. The total proportion of high-reward tags returned was more than double (11.7%) that of low-reward (i.e., regular reward) tags (5.0%) (Table 2).

For the analysis described in this paper, data were downloaded for the period of March 2003 through October 2007. Only recaptures reported with the tag number, recapture date and recapture location were used; any cod reported with approximate information were also omitted. To ensure that time at large was sufficient to enable posttagging mixing (i.e., allowing fish to re-acclimate to their surroundings and resume normal movements after the interruption of being captured/tagged), data were filtered for a minimum of 30 d at liberty (DAL).

Qualitative movement analysis

The regional snapshot of the movements recorded from March 2003 through October 2007 (Figure 5) indicates considerable variation in both distance and direction traveled, but in this unfiltered state, it tells us little about the seasonality or trends in the movements observed. A variety of analyses were used to summarize cod movements in terms of: displacement and distance traveled, direction of travel, size effects, differences between release locations, seasonal effects, and identification of homing movements. After reviewing these findings by region, fish size, season and year, it was possible to identify trends in movements between the different areas. These trends were then compiled to generate a qualitative depiction of the "migration corridors" for cod in the Gulf of Maine region (Figure 6). The timing of movements through the migration corridors appear to be related to spawning behavior, maturation, and environmental conditions. These migrations are summarized as follows:

Georges Bank.—Small fish recruit to Georges Bank from the Cape Cod waters (5Z), but most adult fish then stay offshore on Georges Bank with seasonal northward movements into the Bay of Fundy (4X) during spring and summer, and a southward return to Georges Bank for the autumn and winter.

Bay of Fundy.—For the spring and summer, most movements are confined to the Bay of Fundy (4X), but come autumn, these cod head south to Georges Basin (4X) and Georges Bank (5Z/4X) waters. Little exchange was seen between Bay of Fundy and Gulf of Maine (5Y) cod, except in Passamaquoddy Bay (part of Downeast (DE) in Figure 3).

TABLE 2. Returns of Low-reward tags (LR) and High-reward tags (HR) for each year's releases. A decay in tag returns is seen after two years at large and inter-annual variation is evident. Note that recaptures for 2007 represent just 9 months of the year.

LR tags		Recapture years					
Release years	*N* Releases	2003	2004	2005	2006	2007	Total
2003	47492	1.8%	2.3%	0.9%	0.2%	0.1%	5.3%
2004	58983	0.0%	1.7%	1.9%	0.7%	0.2%	4.5%
2005	5758	0.0%	0.0%	4.2%	3.2%	0.5%	7.9%
Total	112233	0.8%	1.8%	1.6%	0.6%	0.2%	5.0%
HR tags		**Recapture years**					
Release years	*N* Releases	2003	2004	2005	2006	2007	Total
2003	0	-	-	-	-	-	-
2004	1729	0.0%	4.5%	4.1%	1.7%	0.1%	10.4%
2005	511	0.0%	0.0%	10.0%	4.7%	1.2%	15.9%
Total	2240	0.0%	3.4%	5.4%	2.4%	0.4%	11.7%

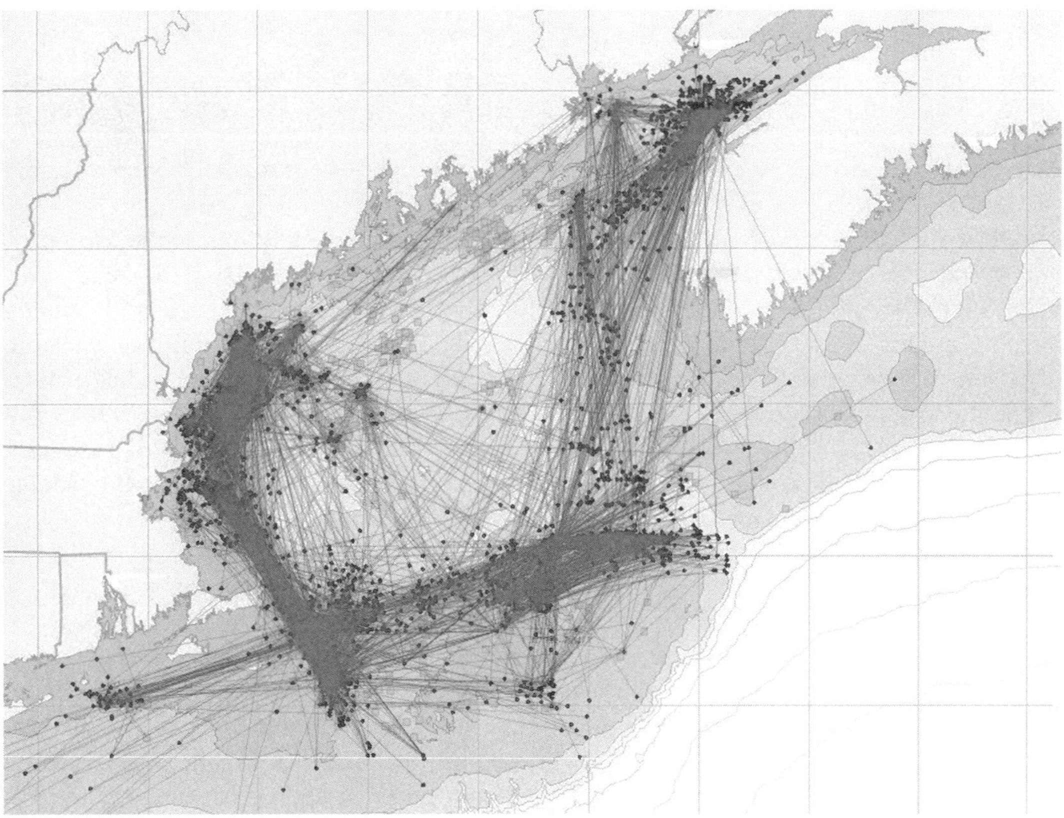

FIGURE 5. Raw, unfiltered mark–recapture data representing 114,467 tagged cod releases and 6,540 recaptures from March 2003 through October 2007.

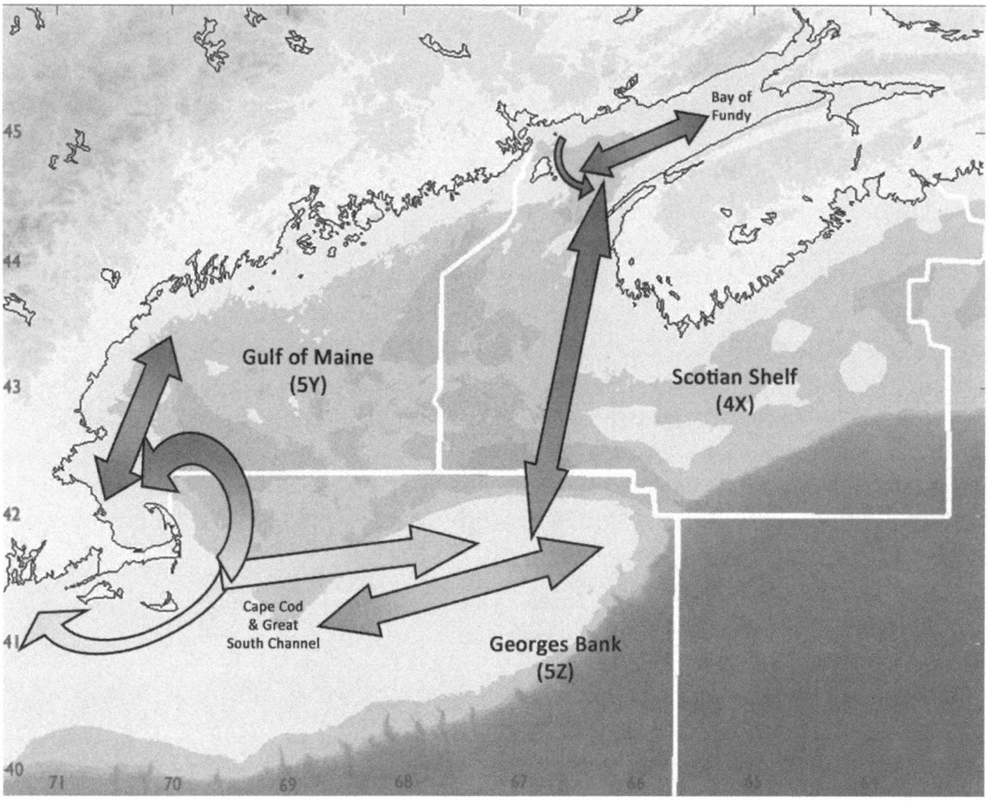

Figure 6. The primary 'migration corridors' proposed for cod movements in the Gulf of Maine region.

Inshore Gulf of Maine.—The Program's largest cod were tagged in these inshore Gulf of Maine (5Y) waters and of these cod, few were recaptured outside of the inshore Gulf of Maine waters. Instead, a shuffle northwards and southwards along the New Hampshire and Maine coastlines was observed. The NRCTP's data provides considerable evidence that the inshore Gulf of Maine (5Y) population recruits small fish from the Cape Cod region (5Z).

Cape Cod and Nantucket Shoals.—The majority of fish tagged here were sub-legal in size and movements for the first year at large were typically fairly localized. During the winter of 2003–2004, a number of Cape Cod releases were reported from inshore waters off Rhode Island and Connecticut—this migration is thought to represent young fish seeking warmer waters during the particularly cold season of the 2003/2004 winter. This migration was less pronounced the following winter. These smaller, preadult fish appear to diverge as they mature and join adult populations; Cape Cod (5Z) fish were seen to recruit eastwards to the Georges Bank (5Z) area, but also northwards to the inshore Gulf of Maine area (5Y).

Quantitative movement analysis

Since considerable inter-annual variation in recaptures was observed throughout the study period (Table 2) it was decided that pooling the data across years (2003–2007) would be most appropriate for providing a regional, contemporary estimate of cod mixing throughout the Gulf of Maine region. The weighting model was then applied to cod of all sizes, cod released at a sub-legal size (<53 cm) and cod released of legal size (>53 cm). Tag shedding was estimated at ~15% and while tagging-induced mortality was not assessed in this study, values documented in the literature (Brattey and Cadigan 1998, 2001, 2004; Cadigan and Brattey 2000; Lear and Rice 1987; Righton et al. 2006) were averaged to yield an assumed rate of ~9%. The discrepancy in return rates of high-reward tags versus low-reward tags was used to calculate reporting rates by area for utilization in the weighting model. A range of residency, emigration and immigration estimates were obtained using each weighting model scenario, but since this paper intends to impart an overview, only the means of these estimates (Table 3) are discussed following.

All sizes.—Cod from management area 5Z appear to migrate equally frequently to 5Z and 4X (~7%), while cod from 5Y are about twice as likely to migrate to 5Z (~9%) than to 4X (~4%). Cod from 4X were recaptured in 5Z about four times more (~8%) than they were recaptured in 5Y (~2%). The highest residency was indicated for Area 4X.

Legal cod.—The average residency of legal cod was highest for 4X (~91%), while residencies for 5Y and 5Z were lower at ~86% and ~85% respectively. When cod did leave 4X, they contributed most to the 5Z management area (~7%) and in exchange ~10% of cod from 5Z were recaptured in 4X, compared with ~6% in 5Y. The lowest exchange was seen between 5Y and 4X and

Table 3. Mean estimates of residency (on the diagonal), emigration and imigration for cod in three size categories: all cod, sub-legal cod (released <53 cm) and legal cod (released >53 cm).

All fish	Release areas		
Recapture areas	5Y	5Z	4X
5Y	87%	7%	2%
5Z	9%	86%	8%
4X	4%	7%	90%
Sub-legal (<53 cm)	Release areas		
Recapture areas	5Y	5Z	4X
5Y	94%	11%	2%
5Z	5%	87%	9%
4X	1%	2%	89%
Legal (>53 cm)	Release areas		
Recapture areas	5Y	5Z	4X
5Y	86%	6%	2%
5Z	9%	85%	7%
4X	5%	10%	91%

when 5Y cod emigrated outside of the management area, they were found in 5Z (~9%) almost twice as often as in 4X (~5%).

Sub-legal cod.—Residency was highest for 5Y sub-legal cod (~94%) and lowest for 5Z sub-legal cod (~87%).There was little movement of younger fish from 5Y to 5Z (~5%) while ~11% of sub-legal cod from 5Z emigrated to 5Y. There was very little interaction between 4X and 5Y, but ~9% of small cod recaptured in 5Z came from 4X.

Discussion

The NRCTP represents a solid example of how international, industry-science collaborative research initiatives have the potential to play a significant role in the collection, analysis and delivery of data which is valuable to both stock assessments and stock management.

Qualitative movement analysis

The regional snapshot of cod movements in its raw, un-weighted and unfiltered state (Figure 5) is highly vulnerable to misinterpretation, since tag returns are dependent on: 1) fishermen's awareness of the program, 2) tag detection, 3) the geographic distribution of tagging effort relative to biomass and relative to seasonal fishing effort, and 4) compliance to report tags with precise recapture information (this varies greatly from individual to individual). However, as the NRCTP recapture data have grown, inter-annual movement trends were detected which enabled the description of "migration corridors" for Atlantic cod in the Gulf of Maine region (Figure 6). The timing of movements indicates that exchanges between the different areas are likely related to spawning behavior, maturation and environmental conditions.

Of the movements observed, some were anticipated: for example, the exchanges of fully recruited cod between the Bay of Fundy (Canada) and Georges Bank (U.S.), and in particular, the north- south shuffle of cod along the New Hampshire and Maine coastlines which is in line with the localized movement patterns documented in another recent cod tagging study for this area (Howell et al. 2008). However, the smaller cod in the Cape Cod area demonstrated a surprising split in migration patterns: 1) northwards into the inshore Gulf of Maine waters, and 2) eastwards out onto Georges Bank.

Overall, the migration patterns evidenced appear to violate two core stock definition assumptions: (1) that the stock is self-sustaining and (2) that neighboring stocks exist in isolation. The movements recorded between all three management areas are strong evidence that at certain times of year, cod interact and mix across both management and national boundaries. However, the extent to which these two stock definitions are violated can only be verified through quantitative analysis.

Quantified movement analysis

An in depth discussion of the weighting method and detailed results has not been presented here, but greater detail on the techniques applied is described in (Tallack 2007). The key message is that it is possible to provide a quantified estimate of movements between management areas which can be verified along with more descriptive findings. The value of using both analytical approaches is that with the descriptive, qualitative approach it is possible to identify areas of potential interest or concern, while subsequent quantification of movements between these areas can help determine the magnitude of the events being seen.

In the case of the Gulf of Maine, the most surprising finding was the migration trend of small cod from Cape Cod heading northwards to the inshore Gulf of Maine area; this was not a widely anticipated movement theory. The value of tagging across a large size range, and then sub-setting data by fish size has been shown since movement estimates for legal sized fish or all fish combined masked this movement. This divergence in migration patterns is not accounted for in the current stock assessment models and will require close attention in the future, since the Gulf of Maine stock (5Y) is apparently dependent on both internal spawning events and the recruitment of young fish from the Georges Bank stock (5Z). Thus, exploitation on the 5Z stock, particularly in the Cape Cod region, has the potential to impact the Gulf of Maine stock. What remains to be determined is whether the levels of migration seen are sufficiently high to warrant a change in stock management measures (e.g., changed boundaries or effort controls); i.e. what level of mixing or exchange between stocks is significant?

A detailed presentation of these movement and mixing findings were presented during the preparatory preGARM workshops (August and October 2007) which investigated how tagging data for cod and other groundfish species might be incorporated into each species' final assessment; this paper has summarized the key findings and potential data applications. In addition to the weighting method, the NRCTP data were investigated using a finite-state continuous-time approach to migration and mortality (Miller and Tallack 2007); this analysis generated outputs in the form or probabilities for movement, fishing mortality, natural mortality, reporting rate and tag shedding.

In addition to movement trends, spatial variation in key life history characteristics (e.g., growth, maturation, natural mortality) is also highly relevant to attempts at identifying stock structure (Begg 1998). The NRCTP provides growth data from ~4,500 individual fish throughout the Gulf of Maine region and analysis using the maximum likelihood GROTAG model (Francis 1988) has shown variation in size structure for Atlantic cod categorized by management area (Tallack 2009a). Ongoing analysis is also applying a more recent and complex model which accounts for fishing selectivity associated with both gear used and historical catch at age data (after Taylor et al. 2005); the result should be an improved and minimally biased estimate of growth by statistical square and stock area (Taylor and Tallack, in review). Natural mortality and fishing mortality have also been investigated to provide supplementary parameter estimates for the upcoming stock assessments (Hart and Miller 2008; Miller and Tallack 2007).

The role of conventional tags

The NRCTP has demonstrated that while many high-tech electronic tags are now available to researchers, simple T-bar anchor tags still represent a valuable tool in the field of mark–recapture studies; their low price enables them to be deployed in a large number of fish and their simplicity is appropriate for large-scale collaborative projects which involve high numbers of nonscientists. Extensive collaboration with industry and widespread dispersal of tags were two of the NRCTP's core goals, and these would not have been attainable had the Program opted for more expensive electronic tags, which also require more stringent control over tag insertion procedures. However, with the regional snapshot of cod movements now available from the T-bar tags, future tagging using electronic tags (e.g., acoustic, data storage tags or GPS tags) will be beneficial, particularly for fine-tuning our knowledge of cod movements and behaviors in areas for which results were surprising, e.g. the Cape Cod waters.

Acknowledgments

The Northeast Regional Cod Tagging Program is funded by NOAA Fisheries, Northeast Region. Owing to the collaborative nature of this Program a complete list of Program Partners and collaborators is available at http://www.codresearch.org. In addition, the following individuals are thanked for their assistance with data management and the analysis presented within this paper: Pat Foote and Sarah Whitford (GMRI), Steve Cadrin, Mark Terceiro, Loretta O'Brien and Ralph Mayo (Northeast Fisheries Science Center, Woods Hole), Don Clark (Department of Fisheries and Oceans, Canada) and Alistair Hobday (CSIRO, Tasmania).

References

Begg, G. A.. 1998. A Review of Stock Identification of Haddock, *Melanogrammus aeglefinus*, in the Northwest Atlantic Ocean Marine Fisheries Review 60:1–15.

Brattey, J., and N. G. Cadigan. 1998. Exploitation rates and population size of cod in Placentia Bay, Subdivision 3Ps: estimates from a new mark-recapture study. CSAS Research Document. Canada Department of Fisheries and Oceans, Ottawa, 98/20.

Brattey, J., and N. G. Cadigan. 2001. Estimation of short-term tagging mortality of adult Atlantic cod (*Gadus morhua*). ICES CM, 2001/O: 03:1–14.

Brattey, J., and N. G. Cadigan. 2004. Estimation of short-term tagging mortality of adult Atlantic cod (*Gadus morhua*). Fisheries Research 66:223–233.

Cadigan, N. G., and J. Brattey. 2000. Lower bounds on the exploitation of cod (*Gadus morhua*) in NAFO Divs. 3KL and Subdiv. 3Ps during 1997–1999 from tagging experiments. CSAS Research Document. Canada Department of Fisheries and Oceans, Ottawa.

Cadigan, N. G., and J. Brattey. 2001. Estimation of the exploitation rates and migration rates for cod (*Gadus morhua*) in NAFO Divisions 3KL and Sub-division 3Ps during 1997–2000 from tagging experiments. ICES CM, 2001/O:04:1–14.

Cadigan, N. G., and J. Brattey. 2003. Semiparametric estimation of tag loss and reporting rates for tag-tecovery experiments using exact time-at-liberty data. Biometrics 59:869.

Francis, R. I. C. C. 1988. Maximum likelihood estimation of growth and growth variability from tagging data. New Zealand Journal of Marine and Freshwater Research 22:42–51.

Gröger, J. P., R. A. Rountree, U. H. Thygesen, D. Jones, D. Martins, Q. Xu, and B. J. Rothschild. 2007. Geolocation of Atlantic cod (*Gadus morhua*) movements in the Gulf of Maine using tidal information. Fisheries Oceanography 16:317–335.

Hart, D., and T. J. Miller. 2008. Analyses of tagging data for evidence of decreased fishing mortality for large Gulf of Maine Cod, *Gadus morhua*. Supplementary Paper. Northeast Fisheries Science Center, Woods Hole, Massachusetts, GARM-III BRP TOR 4.7.

Higgins, E. 1927. Progress in biological inquiries, 1926. Including Proceedings of the Divisional Conference, January 4 to 7, 1927. Report of the U.S. Commissioner of Fisheries for the fiscal year 1927. Woods Hole, Massachussetts, App 7:515–681.

Higgins, E. 1928. Progress in biological inquiries, 1927. Report of the U.S. Commissioner of Fisheries for the fiscal year 1928. Woods Hole, Massachussetts, App 6:199–247.

Higgins, E. 1931. Progress in biological inquiries, 1930. Report of the U.S. Commissioner of Fisheries for the fiscal year 1931. Woods Hole, Massachussetts, App 3:553–626.

Howell, W. H., M. Morin, N. Rennels, and D. Goethel. 2008. Residency of adult Atlantic cod (Gadus morhua) in the western Gulf of Maine. Fisheries Research 91:123–132.

Hunt, J. J., and J. D. Nielson. 1993. Is there a separate stock of Atlantic Cod in the western side of the Bay of Fundy? North American Journal of Fisheries Management 13:421–436.

Hunt, J. J., W. T. Stobo, and F. Almeida, 1999. Movement of Atlantic cod, *Gadus morhua*, tagged in the Gulf of Maine area. Fishery Bulletin 97:842–860.

Lear, W. H., and J. C. Rice. 1987. Proportions of Atlantic cod (*Gadus morhua*) of the 2J3KL stock components taken by the inshore fishery as estimated by offshore tagging. CAFSAC Research Document 3:1–26.

Loehrke, J., and Cadrin, S. X., 2007. A review of tagging information for stock identification of cod off New England. DRAFT REPORT: Submitted to National Marine Fisheries Service, Pre-GARM working group. Northeast Fisheries Science Center, Woods Hole, Massachusetts, WPC2.

MacKenzie, D. 2003. Northern cod discovered "belly up" in Newfoundland..were they frozen, starved or suffocated? Available: www.fisherycrisis.com/DFO/frozencod.htm (October 2011).

Mayo, R. K., and L. O'Brien, 2006. Status of Fishery Resources off the Northeastern US: Atlantic cod. Northeast Fisheries Science Center Webpage: Status of Fishery Resources off the Northeastern US. Northeast Fisheries Science Center: NEFSC—Resource Evaluation and Assessment Division, Woods Hole, Massachusetts.

Mayo, R. K., S. E. Thunberg, S. E. Wigley, and S. X. Cadrin. 2002. The 2001 assessment of the Gulf of Maine Atlantic cod stock. Woods Hole Laboratory Ref. Doc. Woods Hole, Massachusetts, 02-02.

Miller, T. J., and S. M. L. Tallack, 2007. Estimating instantaneous rates of regional migration and mortality from conventional tagging data. DRAFT REPORT: Submitted to National Marine Fisheries Service, Pre-GARM working group. Northeast Fisheries Science Center, Woods Hole, Massachusetts, WP3D.

NEAQ. 2001. The New England Aquarium Final Report: Recommendations for Conducting a Collaborative Cod-Tagging Program for New England and Maritime Canada. Available: www.codresearch.org/NEAQ_Final_Report.htm (October 2011).

Rich, W. H. 1925. Progress in biological inquiries, July 1 to December 31, 1924. Report of the U. S. Commissioner of Fisheries for the fiscal year 1925. Woods Hole, Massachussetts, App 3:37–64.

Righton, D., O. S. Kjesbu, and J. Metcalfe. 2006. A field and experimental evaluation of the effect of data storage tags on the growth of cod. Journal of Fish Biology 68:385–400.

Robichaud, D., and G. A. Rose. 2004. Migratory behaviour and range in Atlantic cod: inference from a century of tagging. Fish and Fisheries 5:185–214.

Schroeder, W. C. 1930. Migrations and other phases in the life history of the cod off southern New England. Bulletin of the U.S. Bureau of Fisheries 46:1–136.

Smith, H. M. 1902. Notes on the tagging of four thousand adult cod at Woods hole, Massachusetts. Report of the U. S. Commissioner of Fisheries Woods Hole, Massachussetts 27:193–208.

Tallack, S. M. L. 2006. The Northeast regional cod tagging program. Final report: submitted to National Marine Fisheries Service, Cooperative Research Partners Program, Contract No. EA133F-03-CN-0016.

Tallack, S. M. L. 2007. A description of tagging data from the Northeast Regional Cod Tagging Program (WP3A) and preliminary applications of weighting and mixing analysis (WP3C). DRAFT REPORT: Submitted to National Marine Fisheries Service, Pre-GARM working group. Northeast Fisheries Science Center, Woods Hole, Massachusetts, WP3A & WP3C.

Tallack, S. M. L. 2009a. Regional growth estimates of Atlantic cod, *Gadus morhua*: applications of the maximum likelihood GROTAG model to tagging data in the Gulf of Maine (USA/Canada) region. Fisheries Research 99:137–150.

Tallack, S. M. L. 2009b. Proceedings from a workshop to identify future research priorities for cod tagging in the Gulf of Maine. Northeast Fisheries Science Center Reference Document. US Deptartment of Commerce 09–09.

Taylor, N., and S. Tallack. In review. Integrating bioenergetics parameter estimates from tagging data into the von Bertalanffy, gear selectivity and fishing mortality parameters of Gulf of Maine, Georges Bank and Southern Scotian Shelf/Bay of Fundy cod stocks. Canadian Journal of Fisheries and Aquatic Sciences.

Taylor, N. G., C. J. Walters, and S. J. D. Martell. 2005. A new likelihood for simultaneously estimating von Bertalanffy growth parameters, gear selectivity, and natural and fishing mortality. Canadian Journal of Fisheries and Aquatic Sciences 62:215–223.

American Fisheries Society Symposium 76:163–176, 2012

In-Stream Monitoring of PIT-tagged Wild Spring/Summer Chinook Salmon Juveniles in Valley Creek, Idaho

STEPHEN ACHORD*, BENJAMIN P. SANDFORD, STEVEN G. SMITH, WILLIAM R. WASSARD, AND EARL F. PRENTICE

Northwest Fisheries Science Center
National Oceanic and Atmospheric Administration
National Marine Fisheries Service
2725 Montlake Boulevard E., Seattle, Washington 98112, USA

Abstract.—Recent advances in passive integrated transponder (PIT) tag technology have allowed the development of in-stream fish-monitoring systems. We installed two such systems in Valley Creek near its confluence with the Salmon River in summer 2002. In the summers of 2003–2005 we collected and PIT tagged wild spring/summer Chinook salmon parr *Oncorhynchus tshawytscha* in natal rearing areas upstream from the monitors. Although subsequent detection numbers between fall 2003 and spring 2006 were low and variable, they were sufficient to determine timing and estimate survival. We defined migrational groups by period of detection: late summer and fall (August–October), winter (November–February), and the following spring (March–June). Combining 3 years of data, the mean proportions of fish detected during these three respective detection periods were 60.6, 27.7, and 11.7%. Mean probability estimates of survival from Valley Creek to Lower Granite or Little Goose Dams were 9.2, 23.4, and 40.8% for the respective late summer and fall, winter, and spring periods. Estimated overall mean probabilities of survival were 46.6% from tagging as parr to movement into the mouth of Valley Creek and 17.3% from Valley Creek to Lower Granite Dam. The overall mean parr-to-smolt survival estimate from tagging to arrival at Lower Granite Dam was 9.0%. The unexpectedly high proportion of fish migrating in winter has important implications for fish monitoring studies that use rotary screw or scoop traps: these traps are generally inoperable during winter near most natal rearing areas and thus may result in biased estimates of fish population status and migration timing. Advancements in technologies and methodologies to instream PIT-tag monitoring systems will improve data quality to assist recovery planning for threatened and endangered fish species.

Introduction

Declines of several stocks of salmon and steelhead in the Columbia River basin over the last three decades have led to listings under the Endangered Species Act (ESA) and have prompted regional fish and wildlife programs to develop strategies to reverse these trends (NPPC 1980, 1994). Additionally, the National Marine Fisheries Service (NMFS) has issued biological opinions on actions

*Corresponding author: steve.achord@noaa.gov

taken by federal agencies to alleviate pressures on these stocks (NMFS 1995, 2001, 2004). These agencies have identified the importance of monitoring ESA-listed juvenile salmon to provide information needed for recovery of these stocks. Development and application of passive integrated transponder (PIT) tag technology (Prentice et al. 1990a, 1990b, 1990c), which permits identification of individual marked fish, has provided a tool to acquire some of this needed information. Many studies of juvenile fish migration timing, survival, biology, and behavior rely on PIT data, particularly in the Columbia River basin (Achord et al. 1996; Skalski et al. 1998; Muir et al. 2001a, 2001b; Achord et al. 2003; Zabel and Achord 2004; Achord et al. 2007; and Monzyk et al. 2009).

For example, in the Snake River portion of the Columbia River basin (Figure 1), many wild juvenile stream-type Chinook salmon *Oncorhynchus tshawytscha* are captured, PIT tagged, and released annually in their natal streams for long-term monitoring (Achord et al. 1996, 2007; Jonasson et al. 2006; Monzyk et al. 2009). Although in principle the PIT tag can be read at any time, previous monitoring opportunities for PIT-tagged fish have been restricted by the short read-range of early PIT tags. Until recently, detections were lim-

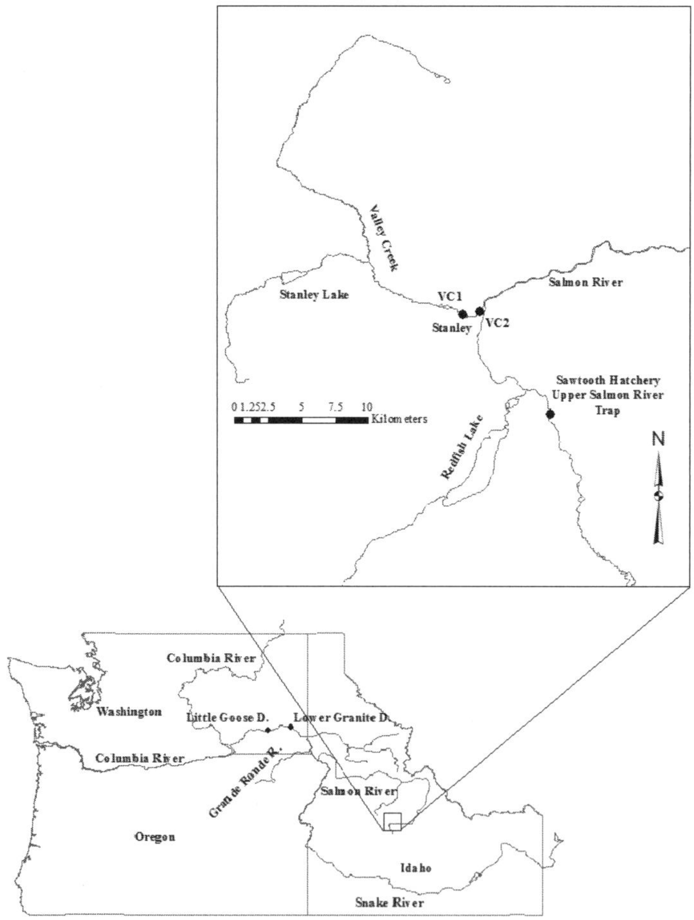

Figure 1. A study area map of the Pacific Northwest, USA with inset showing the Valley Creek in-stream PIT-tag monitoring sites study area.

ited to the juvenile fish bypass systems of some dams (Prentice et al. 1990a) and to a submerged PIT-tag detection trawl operating near the mouth of the Columbia River (Ledgerwood et al. 2004). These detection locations are often hundreds of kilometers from juvenile rearing areas. Thus, although these detections provide valuable information, large voids remain in our understanding of fish biology, behavior, and ecology between juvenile rearing areas and mainstem dams.

To obtain such information, NMFS began developing an in-stream monitor in 2000, when the full-duplex, 134.2-kHz PIT tag was introduced as the standard for basin-wide research. The PIT-tag is a type of RFID (radio frequency identification) tag developed and tested by NMFS during the mid-1980s (Prentice et al. 1990b). Using the standard PIT tag in 2000, a prototype in-stream system was developed at NMFS Manchester Research Station (Downing et al. 2004). In 2001, the first in-stream monitoring system was field tested in Rattlesnake Creek, a tributary of the Little White Salmon River in Washington State. Subsequent studies and development at this location occurred concomitantly with our studies in Valley Creek (Connolly et al. 2008).

In 2002, we continued work to improve this system and began development of a stand-alone system for use in remote areas lacking access to the power grid. Valley Creek near Stanley, Idaho, was selected for evaluation of the stand-alone system because of its year-round accessibility, moderate discharge, and climate extremes. Moreover, the stream constitutes natal rearing habitat for wild spring/summer Chinook salmon. The Chinook salmon population in Valley Creek has a long-term monitoring history (Achord et al. 1996, 2007) and is part of the evolutionarily significant unit (ESU) of Snake River spring/summer Chinook salmon, which was listed as threatened under the U.S. Endangered Species Act in 1992. In an interagency effort to monitor and recover these fish, NMFS, Oregon Department of Fish and Wildlife (ODFW), and Idaho Department of Fish and Game (IDFG) have PIT tagged fish from this ESU at the parr stage since 1988 (Achord et al. 1996, 2007; Jonasson et al. 2006; Kiefer and Lockhart 1997).

The goals of our in-stream PIT-tag monitoring work were to develop, install, and evaluate in-stream monitoring systems that could operate for weeks without service, and to investigate migration timing, travel time, and survival probabilities for juvenile Chinook salmon migrants that display various life history strategies in Valley Creek. In this paper, we describe in-stream PIT-tag monitoring systems and present methods for analyses of these systems. We report results from work during 2003–2006 and give a brief project update for work during 2007–2010.

Methods

Study sites and equipment

In the first week of August each year from 2003 to 2005, wild Chinook salmon parr were collected, PIT tagged, and released in Valley Creek in areas from stream km 4 to stream km 18 (Figure 1). For the duration of our study, we used the Digital Angel[1] "BE" PIT tag. Achord et al. (1996, 2007) provide details on collection, handling, and PIT tagging procedures.

In-stream PIT-tag monitoring systems were deployed in July 2002 at two sites in Valley Creek near the town of Stanley in central Idaho. Valley Creek is a medium-sized stream lying within the larger Salmon River drainage (36,000 km^2). Its drainage area is 381 km^2 with a mean annual discharge of 5.46 m^3/s, based on mean daily flows from 1993 to 2006 (U.S. Geological Survey, sta-

[1]Reference to trade names does not imply endorsement by the National Marine Fisheries Service, NOAA.

tion 13295000). Elevations of the two monitoring sites were 1,901 and 1,904 m, with the lower site located 739.4 km upstream from Lower Granite Dam and 1,434 km from the Pacific Ocean.

The two monitoring sites were located approximately 1.2 km apart. The upstream or "on-grid" site (VC1) was located approximately 1.6-km upstream from the confluence of Valley Creek with the Salmon River. The site had access to electric power and a landline phone (Figure 2).

Following this initial deployment, the instream PIT-tag monitoring system was modified extensively from mid-2002 to mid-2003. Various antenna designs were constructed and evaluated, and details of this work are reported by Downing et al. (2004). Concurrent systems were reported by Zydlewski et al. (2006), Bond et al. (2007), Horton et al. (2007), and Connolly et al. 2008. Ultimately, we developed a "hybrid" system, which uses pivots to combined features of the earlier pass-through and pass-by models. After development and testing, we deployed rectangular antennas of 7.6-cm PVC pipe (3.0-m long × 0.6-m wide). At each site, one long side of the antenna was anchored to the substrate in the main thalweg (Figures 2 and 3) using a frame secured with four 0.9-m steel stakes. Pivoting devices on the antennas allowed the unanchored side of the rectangle to move up and down with the water level. Wetted width of the stream at both sites averaged about 17–21 m during low to medium flows, thus the antennas covered about 14–18% of the stream width.

The antennas were connected to transceivers via specially designed electrical cable, which was 15-m long at the upstream site and

Figure 2. The upper PIT-tag monitoring site (VC1) on Valley Creek, approximately 1.6 km from the confluence with the Salmon River.

Figure 3. The lower PIT-tag monitoring site (VC2) on Valley Creek, approximately 0.4 km from the confluence with the Salmon River. The eight solar panels obscure the view of the utility trailer.

35-m long at the downstream site. All electronics at both locations were housed in 2.4-m long enclosed trailers (Figure 2). Both systems were set up to automatically interrogate, store, and transmit data to the PIT Tag Information System (PTAGIS), a regional database operated by the Pacific States Marine Fisheries Commission (PSMFC 1996-). In addition, both systems could be monitored remotely via "timer" tags built into the systems, which constantly tested antenna function and sent electronic alerts if problems were detected.

The "off-grid" stand-alone system (VC2) was more complicated than the VC1 "on-grid" system because it was located where no power or communication utilities were available. After considerable development, its components included a model FS1000A 24-V DC transceiver, notebook computer, wireless Ethernet link to the VC1 computer, power inverter, DC power disconnect and breaker panel, generator relay, and data control board.

The stand-alone system relied on three areas of developing technology: the antenna system, power supply, and transceiver. For a power supply, we mounted eight solar panels on an outside frame on the trailer (Figure 3), along with two 284-L propane tanks and a generator. Inside the trailer were four 12-V DC 110-amp h^{-1} glass-mat batteries, which were connected in series and shielded from other electronics. Complete details of the system and issues related to the tag, transceiver, and antenna are reported by Downing et al. (2004), Zydlewski et al. (2006), Horton et al. (2007), and Connolly et al. (2008).

During 2003–2004, 2004–2005, and 2005–2006, both systems were operational year-round, although work to improve their performance continued through 2006.

Data Analysis

To evaluate detection efficiencies of the monitoring systems, we used an indirect method similar to that described by Connolly et al. (2008). Detection efficiencies of the two in-stream monitors were inferred from detection probabilities of fish released above the monitors. Fish detection probabilities were estimated from detection or nondetection at downstream monitors. Thus, for all tagged fish we compiled a detection history, which included the two in-stream monitors in Valley Creek as well as monitors in the juvenile collection facilities at downstream dams.

We estimated survival probabilities for the PIT-tagged population from release as parr in Valley Creek to arrival at Lower Granite Dam as smolts. This migration corridor was divided into two smaller segments: (1) stream segment, from the point of release to the lower in-stream PIT-tag monitor, and (2) river segment, from the lower stream monitor to Lower Granite or Little Goose Dam.

Stream segment

To estimate survival in the stream segment, we collapsed the detection histories for each fish to include detection at VC1 and/or detection at VC2 and/or detection at a site downstream. This allowed eight possible detection histories (combinations of "detected" and/or "not detected" on three sites after release). Counts of fish with each detection history were fitted to a multinomial model, with cell probabilities parameterized as functions of detection and survival probability. The model was similar to the Cormack–Jolly–Seber (CJS) model (Cormack 1964; Jolly 1965; Seber 1965) used extensively for PIT-tagged fish survival estimation in the Columbia River basin, except that we assumed 100% survival over the 1.2-km stream segment between VC1 and VC2.

Evidence from detection data showed that detection at VC1 was not independent of detection at VC2; however, an independent probability of detection for each fish is a critical assumption required by the multiple recapture model. The assumption that survival was 100% between VC1 and VC2 allowed us to model the dependency between these detection probabilities.

To evaluate this dependency, two different models were possible, each distinguishable by the three estimable parameters for in-stream detection. The first model included an overall (mean) detection probability at VC1 and two separate detection probabilities at VC2, depending on whether or not the fish was detected at VC1. The second model was the reverse, with an overall detection probability at VC2 and two separate detection probabilities at VC1, depending on whether or not the fish was detected at VC2. The survival probability estimate from point of release to VC2 was the same for both models, as was the probability of detection somewhere downstream from VC2. We used both models and report detection probability estimates from both. Parameter estimates and associated standard errors for all models were calculated using program USER 2.1 (Lady et al. 2003).

River segment

For the river segment, through the use of auxiliary data, we estimated separate probabilities of survival for each of the three periods when fish were detected by in-stream monitors: late summer and fall (August–October), winter (November–February), and spring (March–June). To obtain these estimates, we first grouped detected fish by seasonal period of in-stream detection in Valley Creek. Then, for each group, we compiled a temporal distribution of detections at Lower Granite Dam (i.e., the number of fish from each seasonal period detected at the dam on each day). Each daily count at the dam was then divided by the estimated probability of detection at Lower Granite Dam that day (see below) to derive an estimate of the total number of fish from each seasonal group that passed Lower Granite Dam on that day. Daily passage estimates were then summed to give an estimate of the total number of fish from each seasonal group that survived to Lower Granite Dam. This number was divided by the number in the group (i.e., total number detected at VC1 and VC2 during the seasonal period) to derive the estimate of survival from Valley Creek to Lower Granite Dam.

We summed the estimated numbers passing Lower Granite Dam from each seasonal detection period at Valley Creek to get an overall estimate of the number of fish detected at the stream that passed the dam. We divided this number by the total number of detections to estimate the overall probability of survival for PIT-tagged fish released in a given year.

Daily detection probabilities at Lower Granite Dam were estimated with auxiliary data using the method of Schaefer (1951) as modified by Sandford and Smith (2002). The auxiliary data were for all wild Chinook salmon tagged and released in the Snake River Basin upstream from the dam. For each day of the migration season, we estimated numbers of all wild Chinook salmon PIT-tagged and released upstream from the dam that passed the dam detected or undetected. Thus a series of daily probabilities of detection was developed as follows:

1) Fish detected on day *i* at Little Goose Dam that had previously been detected at Lower Granite Dam were tabulated according to day of passage at Lower Granite Dam.

2) Fish detected on day *i* at Little Goose Dam that had *not* previously been detected at Lower Granite Dam were assigned to an estimated day of passage at Lower Granite Dam, assuming that their passage distribution at Lower Granite Dam was proportionate to that of detected fish.

3) This process was repeated for all days with detections at Little Goose Dam.

4) Detected and nondetected fish passing Lower Granite Dam on day *i* were summed.

5) Detection probability on day *i* was estimated by dividing the number of fish detected on day *i* by the sum of detected and (estimated) nondetected fish passing that day.

We modified the method slightly (see Sandford and Smith 2002) for estimates in the tails of the passage distribution where the above process was not applicable (e.g., for days when no detections occurred at Little Goose Dam).

Bootstrap methods were used to derive standard errors for the estimated probability of survival from Valley Creek to Lower Granite Dam (Achord et al. 2007). Auxiliary data were used to derive bootstrap distributions of daily detection probability estimates. Lower Granite Dam detection data for each Valley Creek group were used for bootstrap distributions of passage at Lower Granite Dam.

Results

Released numbers of PIT-tagged Chinook salmon parr from tagging years 2003 to 2005 in Valley Creek ranged from 2,218–2,511 (Table 1). Mean fork lengths of these fish at tagging ranged from 62.8 to 64.9 mm. Overall, 20% of the tagged fish were held for 24 h in live cages to measure post tagging mortality and tag loss. Mortality averaged 0.3% and tag loss was zero.

Table 1. Tagging and release numbers of wild Chinook salmon parr and estimated survival probabilities from release to the mouth of Valley Creek and the detection probabilities at both in stream PIT-tag monitoring sites for tagging years 2003–2005.

Tagging Year	Total tagged and released	Estimated survival probability (%) from release to detection at Valley Cr (SE)	Estimated detection probability	
			VC1 (SE)	VC2 (SE)
2003	2,498	44.9 (7.1)		
Overall			0.060 (0.012)	0.182 (0.031)
Fish detected on other			0.098 (0.021)	0.299 (0.056)
Fish not detected on other			0.051 (0.012)	0.175 (0.031)
2004	2,511	41.3 (4.9)		
Overall			0.105 (0.016)	0.267 (0.034)
Fish detected on other			0.105 (0.018)	0.266 (0.042)
Fish not detected on other			0.105 (0.020)	0.267 (0.038)
2005	2,218	53.7 (7.6)		
Overall			0.045 (0.009)	0.119 (0.019)
Fish detected on other			0.099 (0.025)	0.259 (0.060)
Fish not detected on other			0.038 (0.008)	0.113 (0.019)
Average 2003–2005		46.6 (3.7)		
Overall			0.070 (0.018)	0.189 (0.043)
Fish detected on other			0.101 (0.002)	0.275 (0.012)
Fish not detected on other			0.065 (0.021)	0.185 (0.045)

The total number of individual tagged fish detected at one or both in-stream monitoring sites between August of the tagging year and June of the following year ranged from 182 to 357 (Table 2). Based on detections at downstream dams, overall efficiency of the two monitoring systems was 22.6% in 2003, 34.4% in 2004, and 14.9% in 2005. Migration from the release site to the in-stream monitors was protracted; significant numbers of fish were detected in late summer and fall the year of release and during the following winter and spring. The CJS model and the similar model used in our study typically rely on the assumption that probabilities of survival and detection are equal for all tagged animals. This assumption is not likely valid for the protracted migrations we observed. However, the data indicated that detection probabilities at the in-stream sites were relatively stable among the three migration periods. Therefore, our overall estimated survival probabilities were reasonable estimates of *average* survival for each annual group comprised of all three seasonal migration timing groups.

The estimated overall probability of detection was lower at VC1 (the on-grid site) than at VC2 (the stand-alone site) in all three years (Table 1). The average estimated detection probability was 0.070 at VC1 and 0.189 at VC2. Estimated detection probability at VC2 averaged 0.275 for fish detected at VC1 and 0.185 for fish not detected at VC1 (Table 1). Estimated detection probability at VC1 averaged 0.101 for fish detected at VC2 and 0.065 for fish not detected at VC2. For annual groups of tagged wild Chinook salmon parr released upstream, estimated survival

Table 2. Numbers and proportions (by season) of wild Chinook salmon juveniles detected at the Valley Creek monitors and the estimated survival probabilities of Valley Creek (VC) detected fish to Lower Granite (LGR) or Little Goose (LGO) Dams and the overall survival probabilities from release to the dam(s).

Tagging and detection years	Detected in Valley Creek N	Detected in Valley Creek (%)	Estimated number passing LGR (spring after tagging)	Estimated survival probability (%): Detection at VC to LGR or LGO (SE)	Estimated survival probability (%): Release to LGR or LGO (SE)
2003–2004					
Total	251		33.1	13.2 (2.7)	5.7 (1.5)
Late summer/fall (Aug–Oct)	148	59.0	9.5	6.4 (3.0)	
Winter (Nov–Feb)	81	32.3	14.1	17.4 (6.1)	
Spring (Mar–Jun)	22	8.8	9.5	43.1 (12.5)	
2004–2005					
Total	357		53.2	14.9 (2.4)	8.6 (1.6)
Late summer/fall (Aug–Oct)	276	77.3	32.0	11.6 (2.4)	
Winter (Nov–Feb)	50	14.0	12.2	24.4 (8.8)	
Spring (Mar–Jun)	31	8.7	9.0	29.0 (12.6)	
2005–2006					
Total	182		43.1	23.7 (4.2)	12.7 (2.9)
Late summer/fall (Aug–Oct)	83	45.6	7.9	9.6 (4.1)	
Winter (Nov–Feb)	67	36.8	19.1	28.4 (7.6)	
Spring (Mar–Jun)	32	17.6	16.1	50.3 (13.6)	
Average 2003–2006				17.3 (3.3)	9.0 (2.0)
Late summer/fall (Aug–Oct)		60.6		9.2 (1.5)	
Winter (Nov–Feb)		27.7		23.4 (3.2)	
Spring (Mar–Jun)		11.7		40.8 (6.3)	

from the point of release to the mouth of Valley Creek (as migrants) ranged from 41.3 to 53.7% and averaged 46.6% (Table 1).

The highest proportion of in-stream detections of tagged fish occurred in late summer or fall (August through October) in all years and averaged 60.6% of the total detections (Table 2). The average proportion of detections that occurred in winter (November through February) was 27.7%, and the average in spring (March through June) was 11.7%. Although lower detection efficiencies were expected during higher flows, almost all fish were detected from August to early May under low to moderate flow conditions, prior to peak flows from mid-May to June.

By seasonal detection period, estimates of average annual survival in the river segment were 9.2% for late summer or fall, 23.4% for winter, and 40.8% for spring (Table 2). In 2006 (tagging year 2005), we based survival estimates on detections at Little Goose Dam because of the low number of detections at Lower Granite Dam. Since survival estimates between the two dams were high in 2006 for all PIT-tagged wild Chinook salmon (95.6%; Faulkner et al. 2007), we did not adjust these estimates. Overall annual estimated survival from detection at the mouth of Valley Creek to arrival at either lower Snake River dam ranged from 13.2 to 23.7% and averaged 17.3% (Table 2).

To estimate overall annual survival from point of release in August to arrival at either lower Snake River dam the following spring, we multiplied the stream segment survival estimate by the river segment estimate (Table 1). Overall estimated annual survival ranged from 5.7 to 12.7% and averaged 9.0% (Table 2).

Juvenile Chinook salmon passed the stream monitors predominately during hours of darkness in all years (2003–2006). The overall mean annual proportion of fish passage from 1800 to 0600 was 93.5% (range 91.7–94.7%).

Juvenile Chinook salmon traveled predominately in a downstream direction. Total numbers detected at both monitoring sites during 2003–2004, 2004–2005, and 2005–2006, were 20, 29, and 14, respectively. The overall mean yearly travel time for these fish between upstream and downstream monitors was 8 h and 21 min (range 19 min-110.4 d). Based on travel times, these fish appeared to actively migrate from Valley Creek. Only one fish likely overwintered between the monitors.

Discussion

This project had both technical and biological goals. While we accomplished the technical goal of producing a stand-alone monitoring system that could operate for several weeks without service, further years of study will be required to meet the biological objectives. The project has provided initial data on migration timing and duration, travel time, and survival probability for juvenile Chinook salmon migrants from Valley Creek. However, only 8–14% of the tagged juvenile Chinook salmon was detected at these monitors during these 3 study years. By increasing coverage or efficiency of the antennas, fish should be detected at higher rates.

To produce adequate precision for estimates of biological parameters will require either more antennas, larger antennas, or greater numbers of fish tagged. The low precision of our estimates to date precludes meaningful analyses relating finer scale fish movement and survival to water quality parameters such as temperature, pH, conductivity, dissolved oxygen, turbidity, and flow. Nevertheless, these data are collected near the monitors and will be available for such analyses in future years.

We observed extremely low 24-h post-tagging mortality and no tag loss in all three tagging years. Actual parr-to-smolt tag loss in our study could have been similar to the 2.0% observed by Knudsen et al. (2009) in hatchery spring Chinook salmon presmolts held 70–125 d after tagging. However, even if this were the case, it would not have changed the accuracy of our survival estimates substantially. Dare (2003) found less tag loss (less than 1.0%) than Knudsen et al. in similar fish held for a shorter time after tagging (28 d). We are not aware of any study of long-term tag retention and delayed mortality during the parr-to-smolt life stage for wild Chinook salmon in natural conditions.

Investigators from ODFW using screw traps near natal rearing areas in streams of the Grande Ronde River drainage (Figure 1) found that most wild juvenile Chinook salmon migrations occur in fall and spring, with few observed in winter (Jonasson et al. 2006). However, winter trapping operations have been intermittent for various reasons, including harsh environmental conditions. Studies by the IDFG using traps in the upper Salmon River (8 km upstream from Valley Creek; Figure 1) have also found that parr, presmolts, and smolts migrate in summer, fall, and spring (Walters et al. 1999; Venditti et al. 2005, 2006); however, trapping generally did not occur in winter (~November–

February) in the Salmon River drainage. We believe our study provides the first evidence of significant migration of presmolts from this ESU during winter (proportional average 28%). This downstream movement was probably not an effort to reach the ocean, but rather to search out warmer water temperatures and better feeding conditions prior to initiation of the smolt stage in spring.

Regardless of the motivation, this behavior was unexpected, and has important implications for fish monitoring studies throughout Idaho and the Pacific Northwest. Because rotary screw and scoop traps are inoperable during winter in most areas, considerable proportions of fish that move downstream during winter may be passing undetected. At some locations, agencies may consider using a combination of traps and in-stream monitors to improve fish monitoring studies. The varied life history strategies displayed by this species probably act as a survival mechanism by spreading the population spatially and temporally throughout downstream areas of the Salmon River.

In the upper Salmon River, significant numbers of parr/presmolts migrate in late summer and fall (Walters et al. 1999; Venditti et al. 2005, 2006). Migration behavior of Valley Creek fish is similar to that of fish in the upper Salmon River. We observed migratory behavior during early August and November in a substantial proportion of these fish. Migration in early August, as well as in November prior to trap removal, has also been observed in upper Salmon River fish (Kiefer and Lockhart 1995, 1997). We believe the migration patterns observed for these fish in the upper Salmon River indicate that a considerable number probably migrate during winter in this stream. However, as in Valley Creek, migration behavior during a given season probably varies from year to year.

Because of the low sample sizes obtained with instream monitoring systems to date, we are continuing work to improve performance of tags, antennas, transceivers, and power supply systems. Future research needs include development of standardized procedures to measure tag-reading efficiencies and to evaluate the effectiveness of antennas placed in streams under variable conditions. These protocols are in turn tied to behavioral responses of various fish species to these structures, as well as differing stream habitats and flow conditions. An example of standardized protocol for instream monitoring of detection efficiencies and effectiveness in small streams is presented by Zydlewski et al. (2006). This might be translated for use in moderate-sized streams. There is also a need to develop additional statistical methodologies for PIT-tag monitoring systems in streams and rivers.

Project Update

Monitoring results from 2007 to 2010 have not substantially changed the biological results presented in this paper. What has changed is the explosion of in-stream PIT tag monitoring systems usage throughout the Pacific Northwest. Most in-stream monitoring systems, including the Valley Creek systems, now use multiplex transceivers, which can power and decode data from multiple antennas. Technological advancements in both tags and monitoring systems have increased detection range, bringing a new set of challenges for successful in-stream PIT tag monitoring. Most of these issues were related to mechanical and electro-magnetic interference, and most have been rectified. In 2007–2008, the first year of using multiple antennas, in-stream monitors operated intermittently. Presently, we have two full years of analyzed data from the Valley Creek monitors using the new system with multiple antennas. In 2008–2009 and 2009–2010, overall detection efficiencies of the two monitoring systems were 45.6% and 37.7%, respectively, compared to a yearly average of 24% from 2003 to 2006. Migration patterns of juvenile fish (groups) moving out of Valley Creek and their estimated survival rates to the dams in 2008–2009 and 2009–2010 were similar to earlier years. Development and improvements to the Valley Creek monitoring systems will continue.

Acknowledgments

Thanks to Scott Davidson, Richard Nelson, Eric Hockersmith, Jeffrey Moser, and Galen Wolf of the National Marine Fisheries Service, and to Jerry Lockhart and Toby Tabor of the Nez Perce Tribe for their help in installing monitoring gear in Valley Creek. We thank Don Warf, Dave Marvin, Darren Chase, and Scott Livingston of the Pacific States Marine Fisheries Commission for installing communications equipment. Special thanks to Conservation Officer Gary Gadwa for his dedicated voluntary efforts to keep the sites operational. We thank Nathan Dumdei for help in preparing the map for Figure 1.

We also appreciate the cooperation of several agencies to make this project possible, including the U.S. Forest Service, Sawtooth National Recreation Area, U.S. Army Corps of Engineers, National Marine Fisheries Service, U.S. Fish and Wildlife Service, Idaho Department of Fish and Game, Idaho Department of Water Resources, Idaho Department of Environmental Quality, Shoshone-Bannock Tribes, and the City of Stanley, Idaho. We thank landowners George McKovich and Jack See for their continued cooperation during research and development of in-stream PIT-tag monitors. Finally, we thank the Bonneville Power Administration for funding this research.

References

Achord, S., P. S. Levin, and R. W. Zabel. 2003. Density-dependent mortality in Pacific salmon: the ghost of impacts past? Ecology Letters 6:335–342.

Achord, S., G. M. Matthews, O. W. Johnson, and D. M. Marsh. 1996. Use of passive integrated transponder (PIT) tags to monitor migration timing of Snake River Chinook salmon smolts. North American Journal of Fisheries Management 16:302–313.

Achord, S., R. W. Zabel, and B. P. Sandford. 2007. Migration timing, growth, and estimated parr-to-smolt survival rates of wild Snake River spring-summer Chinook salmon from the Salmon River basin, Idaho, to the lower Snake River. Transactions of the American Fisheries Society 136:142–154.

Bond, M. H., C. V. Hanson, R. Baertsch, S. A. Hayes, and R. B. MacFarlane. 2007. A new low-cost instream antenna system for tracking passive integrated transponder (PIT)-tagged fish in small streams. Transactions of the American Fisheries Society 136:562–566.

Connolly, P. J., I. G. Jezorek, and K. D. Martens. 2008. Measuring the performance of two stationary interrogation systems for detecting downstream and upstream movement of PIT-tagged salmonids. North American Journal of Fisheries Management 28:402–417.

Cormack, R. M. 1964. Estimates of survival from the sightings of marked animals. Bometrika 51:429–438.

Dare, M. R. 2003. Mortality and long-term retention of passive integrated transponder tags by spring Chinook salmon. North American Journal of Fisheries Management 23:1015–1019.

Downing, S. L., E. F. Prentice, E. P. Nunnallee, and B. F. Jonasson. 2004. Development and evaluation of passive integrated transponder tag technology, 2000–2002. Report of the National Marine Fisheries Service to the Bonneville Power Administration. Available: www.efw.bpa.gov/searchpublications (January 2008).

Faulkner, J. R., S. G. Smith, W. D. Muir, D. M. Marsh, and J. G. Williams. 2007. Survival estimates for the passage of spring-migrating juvenile salmonids through Snake and Columbia river dams and reservoirs, 2006. Report of the National Marine Fisheries Service to the Bonneville Power Administration. Available: www.efw.bpa.gov/searchpublications (January 2008).

Horton, G. E., T. L. Dubreuil, and B. H. Letcher. 2007. A model for estimating passive integrated transponder (PIT) tag antenna efficiencies for interval-specific emigration rates. Transactions of the American Fisheries Society 136:1165–1176.

Jolly, G. M. 1965. Explicit estimates from capture-recapture data with both death and immigration—stochastic model. Biometrika 52:225–247.

Jonasson, B. C., A. G. Reischauer, F. R. Monzyk, E. S. Van Dyke, and R. W. Carmichael. 2006. Investigations into the early life history of naturally produced spring Chinook salmon and summer steelhead in the Grande Ronde River subbasin. Report of the Oregon Department of Fish and Wildlife to the Bonneville Power Administration. Available: www.efw.bpa.gov/searchpublications (January 2008).

Kiefer, R. B. and J. N. Lockhart. 1995. Intensive evaluation and monitoring of Chinook salmon and steelhead production, Crooked River and upper Salmon River sites: annual report 1993. Report of the Idaho Department of Fish and Game to the Bonneville Power Administration. Available: www.efw.bpa.gov/searchpublications (January 2008).

Kiefer, R. B. and J. N. Lockhart. 1997. Intensive evaluation and monitoring of Chinook salmon and steelhead production, Crooked River and upper Salmon River sites: annual report 1994. Report of the Idaho Department of Fish and Game to the Bonneville Power Administration. Available: www.efw.bpa.gov/searchpublications (January 2008).

Knudsen, C. M., M. V. Johnston, S. L. Schroder, W. J. Bosch, D. E. Fast, and C. R. Strom. 2009. Effects of passive integrated transponder tags on smolt-to-adult recruit survival, growth, and behavior of hatchery spring Chinook salmon. North American Journal of Fisheries Management 29:658–669.

Lady, J. M., P. Westhagen, and J. R. Skalski. 2003. USER 2.1 User Specified Estimation Routine. Columbia Basin Research, University of Washington, Seattle, Washington. Available: www.cbr.washington.edu/paramest/docs/user/UserManual/UserManual.pdf (January 2008).

Ledgerwood, R. D., B. A. Ryan, E. M. Dawley, E. P. Nunnallee, and J. W. Ferguson. 2004. A surface trawl to detect migrating juvenile salmonids tagged with passive integrated transponder tags. North American Journal of Fisheries Management 24:440–451.

Monzyk, F. R., B. C. Jonasson, T. L. Hoffnagle, P. J. Keniry, R. W. Carmichael, and P. J. Cleary. 2009. Migration characteristics of hatchery and natural spring Chinook salmon smolts from the Grande Ronde River basin, Oregon, to Lower Granite Dam on the Snake River. Transactions of the American Fisheries Society 138:1093–1108.

Muir, W. D., S. G. Smith, J. G. Williams, and E. E. Hockersmith. 2001a. Survival estimates for migrant yearling spring Chinook salmon and steelhead tagged with passive integrated transponders in the lower Snake and lower Columbia Rivers, 1993–1998. North American Journal of Fisheries Management 21:269–282.

Muir, W. D., S. G. Smith, J. G. Williams, and B. P. Sandford. 2001b. Survival of juvenile salmonids passing through bypass systems, turbines, and spillways with and without flow deflectors at Snake River dams. North American Journal of Fisheries Management 21:135–146.

NMFS (National Marine Fisheries Service). 1995. Proposed recovery plan for Snake River salmon. Available from NMFS Northwest Regional Office, 525 N. E. Oregon Street, Suite 500, Portland, Oregon 97232.

NMFS (National Marine Fisheries Service). 2001. Operation of the federal Columbia River power system including the juvenile fish transportation program and the Bureau of Reclamation's 31 projects, including the entire Columbia Basin Project. Endangered Species Act–Section 7 Consultation. Biological opinion. (Available from NMFS Northwest Regional Office, Hydro power Program, 525 N. E. Oregon Street, Suite 500, Portland, Oregon 97232.)

NMFS (National Marine Fisheries Service). 2004. Consultation on Remand for Operation of the Columbia River Power System and 19 Bureau of Reclamation Projects in the Columbia Basin (Revised and reissued pursuant to court order, NWF v. NMFS, Civ. No. CV 01–640-RE (D. Oregon)). Endangered Species Act–Section 7(a) (2) Consultation. Biological Opinion. (Available from NMFS Northwest Regional Office, Hydro power Program, 525 N. E. Oregon Street, Suite 500, Portland, Oregon 97232.)

NPPC (Northwest Power Planning Council). 1980. Pacific Northwest Electric Power Planning Council and Conservation Act. 1980. Columbia River Basin Fish and Wildlife Program. (Available from Northwest Power Planning Council, 851 SW 6th Avenue, Suite 1100, Portland, Oregon 97204)

NPPC (Northwest Power Planning Council). 1994. Columbia River Basin Fish and Wildlife Program. (Available from Northwest Power Planning Council, 851 SW 6th Avenue, Suite 1100, Portland, Oregon 97204)

Prentice, E. F., T. A. Flagg, and C. S. McCutcheon. 1990a. PIT-tag monitoring systems for hydroelectric dams and fish hatcheries. Pages 323–334 *in* N. C. Parker, A. E. Giorgi, R. C. Heidinger, D. B. Jester, Jr., E. D. Prince, and G. A. Winans, editors. Fish-marking techniques. American Fisheries Society, Symposium 7, Bethesda, Maryland.

Prentice, E. F., T. A. Flagg, C. S. McCutcheon, and D. F. Brastow. 1990b. Feasibility of using implantable passive integrated transponder (PIT) tags in salmonids. Pages 317–322 *in* N. C. Parker, A. E. Giorgi, R. C. Heidinger, D. B. Jester, Jr., E. D. Prince, and G. A. Winans, editors. Fish-marking techniques. American Fisheries Society, Symposium 7, Bethesda, Maryland.

Prentice, E. F., T. A. Flagg, C. S. McCutcheon, D. F. Brastow, and D. C. Cross. 1990c. Equipment, methods, and an automated data-entry station for PIT tagging. Pages 335–340 *in* N. C. Parker, A. E. Giorgi, R. C. Heidinger, D. B. Jester, Jr., E. D. Prince, and G. A. Winans, editors. Fish-marking techniques. American Fisheries Society, Symposium 7, Bethesda, Maryland.

PSMFC (Pacific States Marine Fisheries Commission). 1996-. Columbia Basin PIT Tag Information System (PTAGIS). Interactive database. Available: www.ptagis.org (April 2010).

Sandford, B. P., and S. G. Smith. 2002. Estimation of smolt-to-adult return percentages for Snake River Basin anadromous salmonids, 1990–1997. Journal of Agricultural Biological and Environmental Statistics 7(2):243–263.

Schaefer, M. B. 1951. Estimation of the size of animal populations by marking experiments. U.S. Fish and Wildlife Service Fishery Bulletin 52:191–203.

Seber, G. A. F. 1965. A note on the multiple recapture census. Biometrika 52:249–259.

Skalski, J. R., S. G. Smith, R. N. Iwamoto, J. G. Williams, and A. Hoffman. 1998. Use of passive integrated transponder tags to estimate survival of migrant juvenile salmonids in the Snake and Columbia rivers. Canadian Journal of Fisheries and Aquatic Sciences 55:1484–1493.

Venditti, D. A., K. A. Apperson, A. Brimmer, N. Brindza, C. Gass, A. Kohler, and J. Lockhart. 2005. Idaho supplementation studies; brood year 2002 cooperative report. Technical report 2002–2004 to Bonneville Power Administration, Project No. 198909800, Contract No. 6630–4. (Available: www.efw.bpa.gov/searchpublications/).

Venditti, D. A., A. Kohler, C. Bretz, N. Brindza, J. Lockhart, A. Brimmer, and K. A. Apperson. 2006. Idaho supplementation studies; brood year 2003 cooperative report. Technical report 2003–2005 to Bonneville Power Administration, Project No. 198909800, Contract No. 6630–6. (Available: www.efw.bpa.gov/searchpublications/).

Walters, J., J. Hansen, J. Lockhart, C. Reighn, R. Keith, and J. Olson. 1999. Idaho supplementation studies. Five year report 1992–1996 to Bonneville Power Administration, Project No. 198909800, Contract No. 6630 (Available: www.efw.bpa.gov/searchpublications/).

Zabel, R. W., and S. Achord. 2004. Relating size of juveniles to survival within and among populations of Chinook salmon. Ecology 85:795–806.

Zydlewski, G. B., G. Horton, T. Dubreuil, B. Letcher, S. Casey, and J. Zydlewski. 2006. Remote monitoring of fish in small streams: a unified approach using PIT tags. Fisheries 31:492–502.

American Fisheries Society Symposium 76:177–190, 2012

Inter-Stream Movements and Seasonal Activity of Two Large Galaxiids within a Small Stream System

Cindy F. Baker* and **Joshua P. Smith**
National Institute of Water and Atmospheric Research Limited (NIWA)
10 Silverdale Road, Hillcrest Hamilton, Waikato 3216, New Zealand

Abstract.—The extent of intra- and inter-stream movements of two large galaxiids, banded kokopu *Galaxias fasciatus* and giant kokopu *Galaxias argenteus*, were monitored using a passive integrated transponder (PIT) tag system. In total, 103 banded kokopu and 51 giant kokopu were tagged in three small lowland streams. Within one stream, stationary recording antennae were installed at six locations, and fish movements were monitored for two years. Site fidelity was seen in less than 20% of both kokopu species, with around 40% of both banded and giant kokopu found to frequently undertake intra-stream movements to new cover locations. Movements ranged up to 1000 m for both species with seasonal patterns exhibited. In addition to local movements, inter-stream movements were also recorded in approximately 25% of giant kokopu and 5% of banded kokopu monitored. The extent of intra- and inter-stream movements exhibited by both species highlights the need to consider adult fish migrations in both stream restoration and fish passage design, to ensure upstream and downstream movements of adult galaxiids are unhindered.

Introduction

The banded kokopu *Galaxias fasciatus* and giant kokopu *Galaxias argenteus* are two large galaxiid species that are endemic to New Zealand (McDowall 1990). Both species are normally amphidromous, with larvae hatching in freshwater and then migrating to the sea where they spend the winter rearing into the juvenile form (McDowall 1990). During spring each year, the juveniles return to freshwater in search of habitat for growth to adulthood (McDowall 1990). During their freshwater journey, many juveniles are harvested as part of New Zealand's commercial and recreational whitebait fishery. The distribution and abundance of both species is considered to be in decline from loss of habitat, mainly through land-use changes (McDowall 1990), barriers to upstream migration (Boubée et al. 2000), and the introduction of exotic species (Townsend and Crowl 1991; Crowl et al. 1992). Although banded kokopu are still common throughout New Zealand, giant kokopu are now classified as a threatened species (Hitchmough 2002).

Knowledge of native fish movements is essential for river management, especially protection and restoration of biodiversity. Traditionally, large galaxiid populations have been thought to show high site fidelity once resident in the freshwater system, with many fish spending their entire adult life (up to 20 years) within single pools or short stream reaches. However, recent studies have shown movement patterns of stream resident giant kokopu can be highly variable within a population, and while many fish do show high site

*Corresponding author: c.baker@niwa.co.nz

fidelity, others frequently relocate to different pools within the stream (Hansen and Closs 2005; David and Closs 2001, 2002, 2003). Presently, little comparative data exists for other large galaxiid species, such as banded kokopu, and all studies to date are based on sampling only a limited number of fish and localities.

Recent advances in telemetry have allowed movement patterns of many fish species within lotic systems to be discerned (Paragamian 1989; Nykänen et al. 2004; Popoff and Neumann 2005; Sakaris et al. 2005; Cucherousset et al. 2005). The results of most tagging studies suggest that for many species, the geographic range utilized can show considerable variation between different populations (Gowan et al. 1994; Daugherty and Sutton 2005; Popoff and Neumann 2005; Sakaris et al. 2005). For example, landlocked salmonid populations were traditionally thought to comprise a totally stream resident or a resident and migratory fraction (Gerking 1959). Stream resident populations were thought to have limited movement and often complete their life cycle in short stream reaches (Miller 1957; Bachman 1984). However, recent studies have shown that many types of intra- and inter-stream migrations occur both inside and outside of the spawning season (Gowan et al. 1994; Young 1994).

For many fish species, movement patterns appear plastic and adaptive to a wide variety of environmental situations (Gowan et al. 1994). Mechanisms that cause such movements are complex, but given that substantial heterogeneity exists within stream environments, these movements most likely play a role in optimizing foraging, growth, mate selection and subsequent survival of fish populations. As such, these different migratory behaviors may well be a feature characterizing all stream resident large galaxiid species. Therefore, understanding the timing and extent of local or large scale movements is necessary in order to enhance the diversity and abundance of galaxiids species within small tributary streams, especially with respect to designing in-stream passage facilities. The aim of this study was to use a Passive Integrated Transponder (PIT) tag system to determine the extent of intra- and inter-stream movements of banded kokopu and giant kokopu within a small lowland stream system.

Methods

Study site

Banded kokopu and giant kokopu were tagged in three unnamed tributaries (93–94 km inland) of the Waikato River, which drains the Hakarimata Range in the North Island of New Zealand (175°15'E, 37°64'S; 175°15'E, 37°63'S; 175°16'E, 37°63'S; Figure 1). The largest and most southern tributary was the primary study site, with the upper section classed as a first order stream and the lower reaches a second order stream (Figure 1). The two neighboring tributaries were small, first order streams. Each stream was, on average, between 1 and 2 m in width (see Baker and Smith 2007 for detailed habitat characteristics of the study streams). Stream flow was monitored within the primary study stream using an Odyssey capacitive water level recorder (Dataflow Systems Pty Ltd, Christchurch, New Zealand). Stream gaugings were undertaken at regular intervals to obtain a ratings curve from which stream flow could be calculated using the hourly water level readings.

Tagging of fish

Fish were captured using a battery-powered electric fishing machine (EFM 300, NIWA Instrument Systems, Christchurch, New Zealand). All three streams were fished from the first point of access (between 100 and 200 m upstream of the confluence with

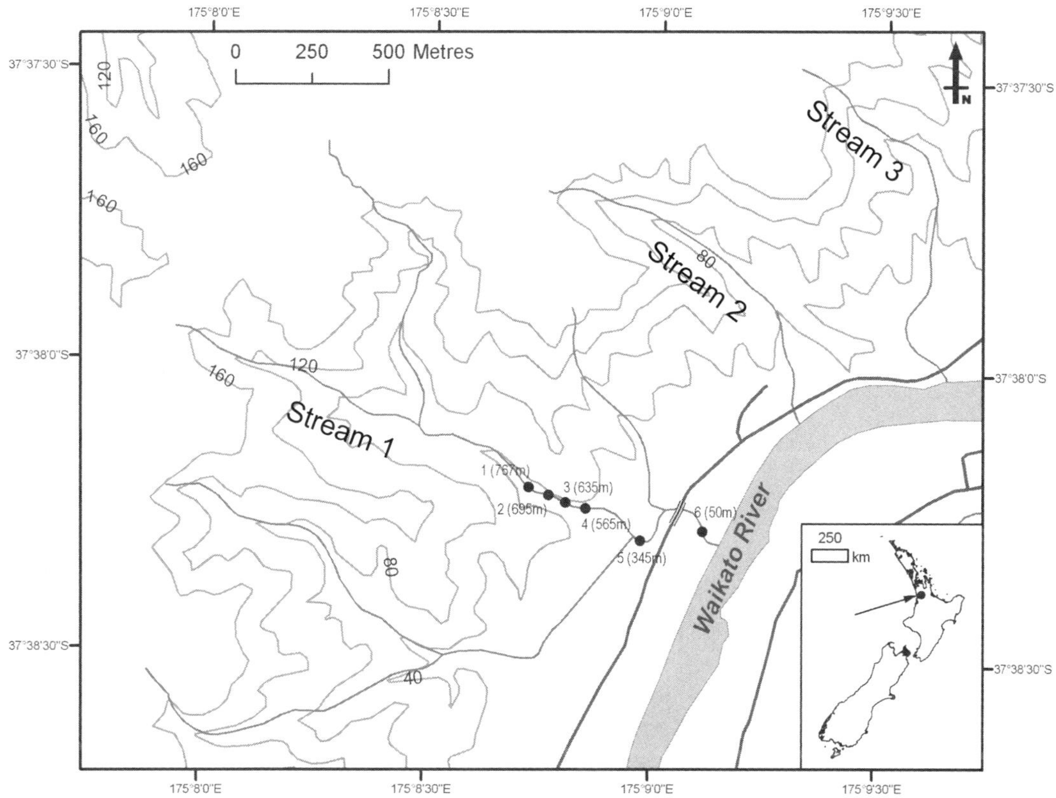

FIGURE 1. Study streams draining the Hakarimata Range, North Island, New Zealand. The primary study stream (stream 1) shows the positions of the six fixed antennae used to monitor movements of tagged fish. The distance upstream (m) of the confluence with the Waikato River that each aerial is located is given in parentheses.

the Waikato River) up to the headwaters where the stream became too narrow to allow fishing (<0.5 m wide). Each stream was fished twice (February 2004 and February 2005) and only adult, sexually mature (>150 mm) banded kokopu and giant kokopu were tagged. Implantation was conducted in situ. Each fish was anesthetized using 0.2% Phenoxyethanol mixed in stream water. Small, 23-mm PIT tags (Texas Instruments, Austin, Texas, USA) were manually inserted into the body cavity of sedated fish through a 3-mm ventral incision. Following implantation, the incision was treated with a topical antiseptic. Fish were then left in a bucket with clean stream water to recover before being released at the same point of capture. The distance upstream of the confluence with the Waikato River was recorded using a Hip-chain (Topometric Products Ltd, Vancouver, Canada).

In total, 103 banded kokopu and 51 giant kokopu were tagged between the three streams (Table 1). Within the primary study stream, 65 banded kokopu and 33 giant kokopu were tagged (Table 1), with 60% of the banded kokopu, and 48% of the giant kokopu tagged in the first year during February 2004. Fish within streams 2 and 3 were all tagged in February 2004.

PIT tag detection and logging system

Fish movement was monitored within the primary study stream only, fish tagged within

Table 1. The number of giant kokopu *Galaxias argenteus* and banded kokopu *Galaxias fasciatus* tagged within each stream. The distance (m) fished for each stream is shown in parentheses.

Species	Stream 1 (1950 m)	Stream 2 (780 m)	Stream 3 (1040 m)
Banded kokopu	65	26	12
Giant kokopu	33	13	5

streams 2 and 3 (Figure 1) were used to determine if fish from neighboring tributaries would move into the study stream. Within Stream 1, a stationary system, as described in Castro-Santos et al. (1996) and Boubée and Williams (2006), was assembled to detect tagged fish moving up or downstream. Six antennae were positioned in the lower third of the stream (Figure 1). To determine if inter-stream movements were occurring, antenna 6 was located approximately 50 m from the confluence with the Waikato River. Little habitat for either kokopu species is found below this antenna site. The antennae loops, which were 1 m high and between 2 and 2.5 m wide, were constructed from high-current power cable (630/0.12 mm) strung through 20-mm PVC pipe for mechanical protection and ease of anchoring. As both banded and giant kokopu are drift feeders that remain under cover while inactive (Baker and Smith 2007; West et al. 2005; David and Closs 2003), the antennae were positioned in habitat that was deemed unsuitable as a cover location for both species in an attempt to prevent tagged fish from 'sitting' within the antennae field while inactive. Each antenna was attached to a TIRFID/RDR, Series-2000 tuner and reader, connected to either a computer running TIRIS Multireader Program version 2.0 (Conte Anadromous Fish Research Centre–USGS-BRD, Turners Falls, Massachusetts, USA) or a Palm M130 (Hewlett-Packard, Palo Alto, California, USA) running TILogger (Oregon RFID, Portland, Oregon, USA). Each PIT tag carries a unique code that is recorded by the software, along with a time and date stamp, each time a tagged fish passes through an antenna. Movements of tagged fish past the antennae were monitored continuously for two years from February 2005 to February 2007.

In addition to a stationary recording system, three to four times a year the length of the primary study stream was scanned with a portable TIRIS PIT tag reader and antennae unit (Texas Instruments, Austin, Texas, USA). The position of each tagged fish was recorded, noting whether they had moved from their original capture location or previously recorded position. Fish that were regularly moving through the stationary antennae system were monitored with the portable unit on a fortnightly basis to keep track of their chosen cover location.

At the conclusion of the study, between March and May 2007, relocating all tagged fish within the primary study stream was attempted. The length of the stream was electric fished on two occasions and swept with the portable PIT unit on two further occasions.

Results

Within the primary study stream, 63% of banded kokopu, and 76% of giant kokopu were detected in 13 trials using the portable PIT reader system. These results were corroborated by the stationary antenna system which detected 14% of banded kokopu, and 48% of giant kokopu that either moved in and out of the array, through the array, or were housed within the array for the entire study period. All fish that were recorded by the portable PIT system as moving into or through the stationary array were also detected by the stationary PIT logging system. There was no difference in the behavior of either kokopu species evident between fish tagged in 2004 compared to those tagged in 2005.

Over the two year study period, approximately 20% of banded kokopu and 12% of giant kokopu within the primary study stream showed strong site fidelity and remained within their pool of capture (Figure 2). However, approximately 40% of both banded and giant kokopu were found to frequently change cover locations (Figure 2). Movements between cover locations ranged up to 1000 m for both species. Based on fish moving past antenna 6, close to 25% of giant kokopu were assumed to have left the study stream (Figure 2). In comparison, only 5% of banded kokopu were assumed to have left the stream (Figure 2). At the conclusion of the study, 37% of banded kokopu and 24% of giant kokopu were unable to be relocated and these fish had not been recorded by either the stationary or portable reader systems during the two year period.

Fish regularly passing the stationary antennae exhibited seasonal movements. During summer months local movements of one banded kokopu ranged up to 132 m, as the fish regularly moved past antenna 2 but did not pass antennae 1 or 3 (Figure 3). Larger movements were exhibited during the late autumn and winter months with this fish extending its downstream distance past antenna 4 on five occasions and therefore covering a distance of at least 130 m but possibly up to 422 m based on subsequent antennae locations (Figure 3). These movements were all undertaken at discharges above base flow.

Fish that were assumed to have left the study stream did so between March and October each year, with the bulk of movement between May and July (Table 2; Figure 4). The time taken for each fish to move past all six antennae varied from less than one day to five weeks (Table 2). Very little downstream movement occurred at base flow (Figure 4), and most movements occurred between dusk and dawn (Table 2). Once fish had initiated

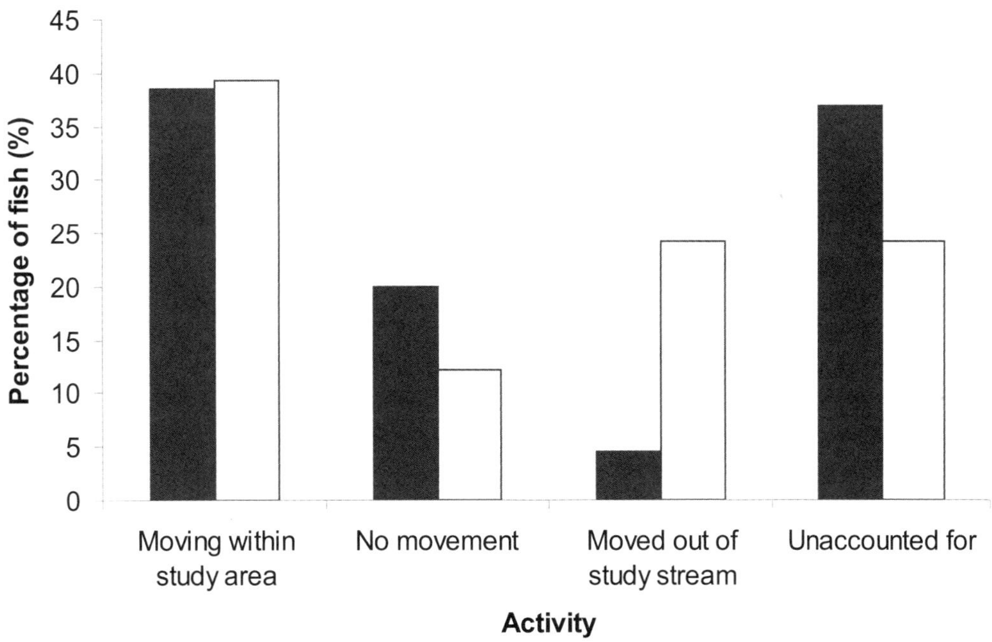

FIGURE 2. Percentage of tagged banded kokopu *Galaxias fasciatus*, ■ and giant kokopu *Galaxias argenteus*, □ exhibiting no movement, small scale movements, large scale movements as well as fish unaccounted for during the study period.

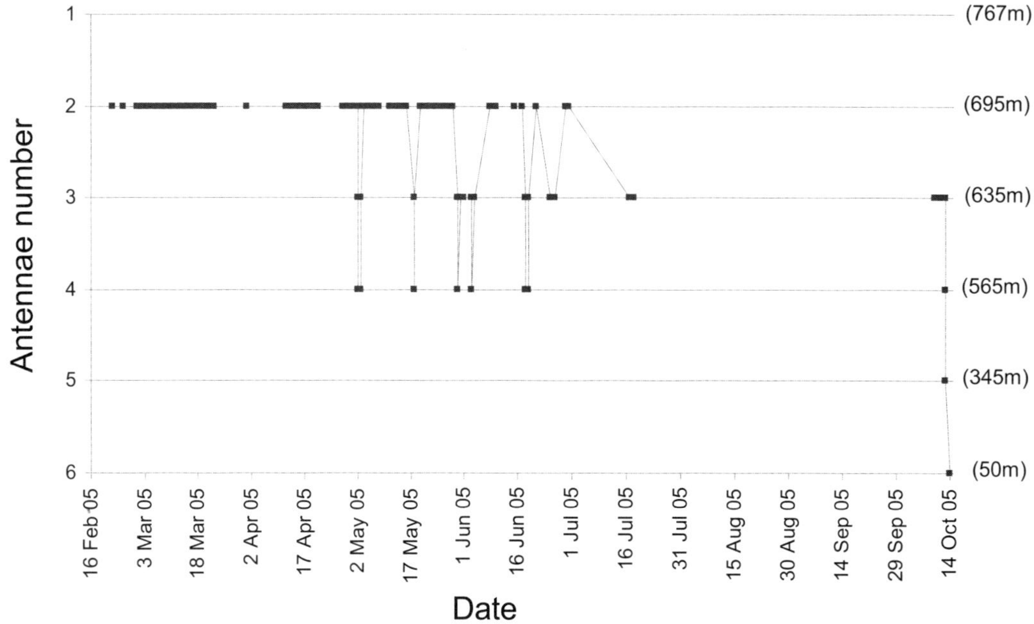

FIGURE 3. Movements of an individual banded kokopu *Galaxias fasciatus* past the stationary antennae from February 2005 to October 2005. The distance (m) each antenna is located upstream of the confluence with the Waikato River is given in brackets. Solid squares represent when each antenna was activated.

TABLE 2. The dates and times (NZ standard time) giant kokopu *Galaxias argenteus* and banded kokopu *Galaxias fasciatus* (that were assumed to leave the study stream) were first detected by a stationary antenna on their downstream journey, and when they subsequently moved past antenna 6. The tagging location represents the distance (m) upstream of the confluence with the Waikato River where each fish was tagged.

Kokopu species	Tagging location (m)	First detected by an antenna Date	Time	Moved past antenna 6 Date	Time
Giant	668	2 May 2005	17:41	3 May 2005	07:00
Giant	1332	18 May 2005	05:18	18 May 2005	18:50
Giant	824	18 May 2005	04:16	18 May 2005	05:22
Giant	1262	22 June 2005	08:35	24 June 2005	20:21
Giant	1097	20 July 2005	19:50	20 August 2005	03:55
Giant	1065	10 October 2005	20:11	10 October 2005	21:01
Giant	1084	27 May 2006	08:14	2 July 2006	02:48
Giant	556	12 June 2006	12:06	12 June 2006	13:13
Banded	621	2 May 2005	17:41	3 May 2005	07:00
Banded	891	18 July 2005	06:04	18 July 2005	19:22
Banded	602	13 October 2005	10:55	20 October 2005	20:14

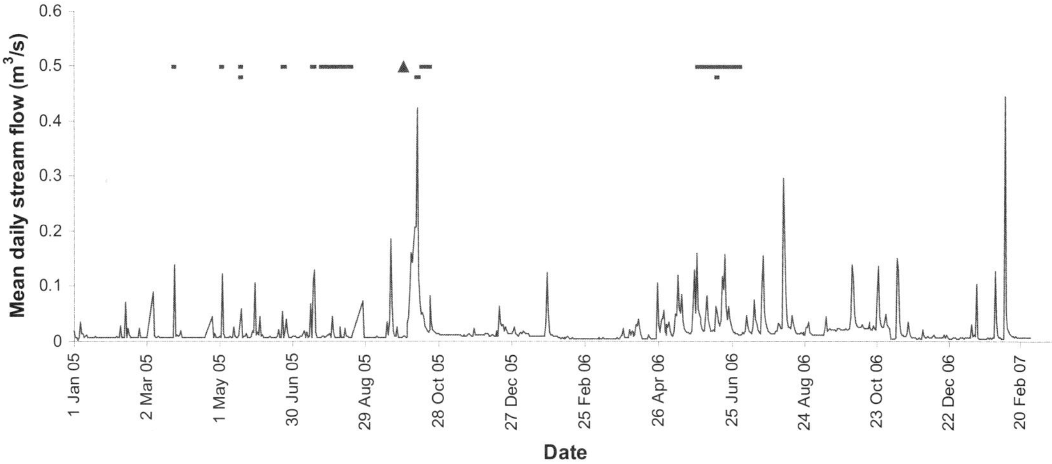

FIGURE 4. Mean daily flow within the primary study stream over the monitoring period. Solid ovals and bars indicate tagged fish assumed to have left the study stream. The length of each oval or bar covers the start and end date of each fish's movement to the Waikato River. Ovals/bars on top of each other represent two fish with overlapping movement dates. The solid triangle represents the entry into the study stream of one tagged giant kokopu from stream 2.

their downstream migration, close to 20% of fish were found to fluctuate between upstream and downstream movements before passing antenna 6. Once fish had passed antenna 6, they were not detected again.

Of the fish tagged in the two neighboring tributaries, one giant kokopu from stream 2 entered the study stream and moved past antenna 6 on the 28 September 2005 at 20:00 h This fish remained within the study stream for just over an hour, moving back past antenna 6 again at 21:20 h (Figure 4).

Seasonal and diel activity patterns over an annual cycle were collected from two giant kokopu and three banded kokopu. Similar patterns were exhibited by all fish of each species. Representative activity patterns are shown for an individual of both species in Figure 5. Giant kokopu exhibited a clearly defined cycle of nocturnal activity during autumn and winter (Figure 5), with around 90% of all records collected during these seasons (Table 3). Although giant kokopu were also nocturnally active during the warmer spring and summer months, few records were collected (Table 3), with most centered on the early hours of the morning and the first few hours after dark (Figure 5). Similarly, banded kokopu also exhibited a clearly defined cycle of nocturnal activity during autumn and winter (Figure 5), with over 80% of all records collected during these seasons (Table 3). During spring and summer, a less defined cycle of activity was exhibited by banded kokopu, with many diurnal movements recorded during spring and only movements between 05:00 and 08:00 h, and 18:00 and 20:00 h, were recorded during summer (Figure 5).

Discussion

A range of movement patterns were exhibited by both banded and giant kokopu. The majority of fish from each species changed cover locations frequently, undertaking regular local movements of up to 1000 m, while a smaller percentage of each species showed high site fidelity and remained within the pool of capture throughout the two year study. Both species also undertook larger

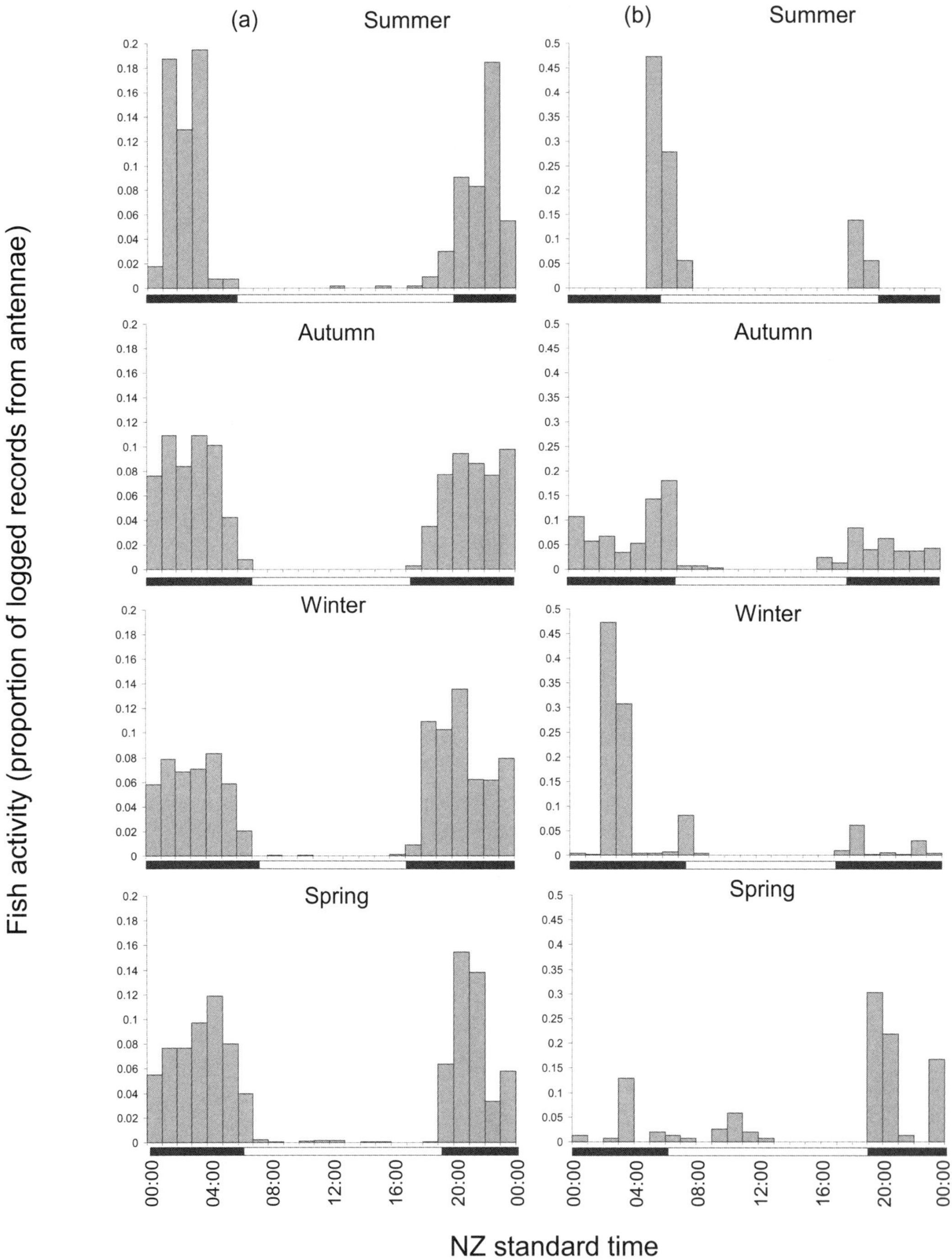

FIGURE 5. Seasonal and diel activity of an individual giant kokopu *Galaxias argenteus* (a) and an individual banded kokopu *Galaxias fasciatus* (b) over the period of a year. The bar at the bottom of each graph indicates light and dark periods. Summer represents December to February, Autumn represents March to May, Winter represents June to August, and Spring represents September to November.

Table 3. Percentage of records (movements past an antenna) obtained for each season by an individual banded kokopu *Galaxias fasciatus* and giant kokopu *Galaxias argenteus*.

Season	Species	
	Banded kokopu (%)	Giant kokopu (%)
Summer (Dec – Feb)	3.2	2.0
Autumn (Mar – May)	35.8	46.4
Winter (Jun – Aug)	47.1	44.4
Spring (Sep – Nov)	13.9	7.2

scale downstream movements past the lowest antenna in the study stream.

It is unlikely that the emigration seen by both species has resulted from mortality, with dead fish being washed out of the system during higher discharge events. At the beginning of the study, dead catfish implanted with PIT tags (fish were trimmed of all fins and barbels and were a similar size and weight to the study galaxiids) were released within the study stream between 800 and 1,400 m upstream of the confluence with the Waikato River (NIWA, unpublished data). No carcass was recorded further downstream than antenna 2 (695 m upstream of the confluence with the Waikato River) throughout the study period. Detection of catfish using the portable PIT tag reader system found carcasses became lodged among debris dams and within pockets in the substrate (NIWA, unpublished data).

As the banded and giant kokopu exhibiting downstream migrations were not detected again after moving past antenna 6, and as little habitat exists for either species between antenna 6 and the Waikato River confluence, it is likely that both kokopu species actively left the stream. Given that one giant kokopu entered the study stream from an adjacent tributary, it is possible that fish leaving the study stream are exhibiting inter-stream migrations. David and Closs (2002) also found a giant kokopu relocated to a neighboring stream during flood flows. Although David and Closs (2002) only recorded one fish leaving the two study streams, their study may underestimate the amount of emigration from the streams as less than 10 fish were tagged within each stream. The present study suggests these inter-stream movements are more prevalent in giant kokopu, as they were seen in close to a quarter of the tagged fish, whereas only 5% of tagged banded kokopu left the study stream.

The inter-stream migrations were seasonal, with most occurring during autumn and winter, and as such, these movements may have been spawning migrations. The spawning biology of both galaxiid species is not well documented, but large galaxiids are known to spawn annually and banded kokopu nests have been observed during May and June (Mitchell and Penlington 1982; Charteris et al. 2003). No observations of spawning have been documented for giant kokopu. However, there have been several observations of downstream movements of ripe fish during late autumn and early winter (McDowall 1990). Additionally, McDowall and Kelly (1999) have suggested from otolith analyses that spawning occurs between early June and early August. Presently, the factors driving inter-stream movements of galaxiids are unknown. Additional research would be required to better understand the primary factors driving movement of banded and giant kokopu from different tributary streams.

Site fidelity of both galaxiid species was less than previous studies have documented (David and Closs 2002; Hansen 2005; West et al. 2005). In the present study, the higher mobility of both galaxiid species may relate to the dynamic nature of cover within the study stream. David and Closs (2002) found giant kokopu exhibited deliberate movements during high discharge events, suggesting fish may chose to vacate pools that have become unsuitable. Because New Zealand has a highly variable climate, many first and second order streams experience regular floods. As shown by the flow record (Figure 4), over the monitoring period, the study stream experienced regular flash floods. That is, the rise and fall of water from the base flow was rapid. Although the size and frequency of floods will vary between years, the flow patterns depicted over the study period are fairly representative of annual flood occurrence within streams draining the Hakarimata Range. Within the study stream, both banded and giant kokopu have been shown to actively select debris dams as their preferred cover (Baker and Smith 2007). These debris dams were highly dynamic throughout the study period (Baker, personal observations) and the loss of cover during high discharge events may cause fish to relocate.

In streams where the metabolic requirements of fish are pressured, higher levels of mobility may result. Hansen (2005) found intra-specific competition for food resources was a major factor influencing movement of subordinate giant kokopu between pools. Within the present study stream, generally only one banded or giant kokopu resided in each pool (Baker and Smith 2007) and both intra- and inter-specific competition for food resources may play a role in the higher mobility displayed by both kokopu species.

Spawning behavior and mate selection may also account for some local movements by banded and giant kokopu. Larger intra-stream movements, which were correlated to high stream flows, were exhibited by fish during May to July, and this is when both galaxiid species are thought to spawn (McDowall and Kelly 1999; David and Closs 2002; Charteris et al. 2003). Additionally, the timing of spawning events for banded kokopu and shortjaw kokopu has been linked with high rainfall and flood events. Charteris et al. (2003) found the nests of both species were laid above the base flow water level, among terrestrial vegetation.

The diel movements of gaint kokopu during the present study were mostly nocturnal, with most activity recorded during autumn and winter. In contrast, David and Closs (2003) found giant kokopu were less active but primarily nocturnal during winter, with increasing diurnal activity exhibited during summer. David and Closs (2003) also recorded a change in microhabitat use by giant kokopu between winter and summer, from low velocity deep water, to shallower higher velocity areas. Differences in activity patterns between this study and that of David and Closs (2003) may reflect the type of telemetry utilized. The capacity of the PIT system to continuously monitor fish activity is limited by the need for fish to pass through the stationary antennae in order to be logged (Greenberg and Giller 2000). In comparison, the radiotelemetry monitoring conducted by David and Closs (2003) allowed continuous data on individual fish locations to be collected. In the present study, the reduced activity recorded during summer and spring may be a result of fish changing or expanding their feeding locations away from the stationary antennae. Alternatively, differences in behavior may be attributable to individual variation between fish. David and Closs (2003) found summer diel activity of giant kokopu was highly variable both between and within individuals. High levels of individual variation in diel activity have also been observed in temperate drift feeding salmonids (Metcalfe et al. 1999; Greenberg and Giller 2001). With such a small sample size and the limitations of the PIT system in recording continuous behavior, interpreting differences between studies is speculative at best. This is also true for interpreting the seasonal and diel activity of banded kokopu. This is the first study to record diel activity of banded kokopu and PIT tag records suggest their activity patterns may be similar to those found in giant kokopu by David and Closs (2003), with increased diurnal activity shown in spring and summer, and only nocturnal activity exhibited during winter. This is not surprising, as banded kokopu and giant kokopu are both drift feeding stream fish (Hansen et al. 2004; West et al. 2005) that occupy similar microhabitats within the study stream (Baker and Smith 2007).

High proportions of both banded and giant kokopu were not detected throughout the study, and could not be relocated at the conclusion of the study by either electro-fishing or by the portable PIT tag reader system. The survival of fish species implanted with PIT tags, and subsequent tag retention, has been extensively studied, with around 90% survival reported in a wide range of fish species, and with between 88 and 99% of fish retaining their tags (Wagner et al. 2007; Bateman and Gresswell 2006; Feldheim et al. 2002; Gries and Letcher 2002; Ombredane et al. 1998). West et al. (2005) reported a recapture rate of 90% of marked banded kokopu three months after tagging, suggesting that mortality was low and tag retention was high. Factoring in dropped tags and fish mortality, around 10% of banded kokopu were still unaccounted for. This discrepancy could result from the inability of electric fishing to capture all fish within the stream, and the limitations of the portable PIT system. Capture efficiency of salmonids using electroshocking has been estimated at 30% for the first pass, with each successive pass decreasing considerably (Peterson et al. 2004). In addition, the portable PIT system is limited by its detection range and several studies have shown that stream fish display pronounced escape behavior towards the antenna (Cucherousset et al. 2005; Roussel et al. 2000). Furthermore, Roussel et al. (2000) revealed that the detection efficiency was reduced by the presence of dense riparian vegetation and overhanging branches that hampered the operator's ability to maneuver the antenna. In this regard, the structural complexity of the cover location may prevent or reduce effective coverage with the antenna.

Within New Zealand, one of the most significant causes of the decline in freshwater fish populations is anthropogenic barriers such as dams, culverts and weirs that prevent fish from accessing otherwise suitable habitats. The results of the present study highlight the potential for anthropogenic barriers such as culverts and weirs to prevent not only inter-stream migrations but also shorter range local movements of adult fish. Within New Zealand, fish passes are designed to facilitate the upstream movement of juvenile fish (Baker and Boubée 2006). Migratory juveniles of native fish species either possess the ability to 'climb' the wetted margins of in-stream obstacles, or can swim through shallow water depths (Boubée et al. 1999; Boubée et al. 2000). Based on these behaviors, ramp style fish passes have been designed to facilitate juvenile fish passage (Boubée et al. 2000; Baker and Boubée 2006). These ramp style passes, however, may be ineffective at passing large adult galaxiids as adult banded and giant kokopu are thought to be unable to climb obstacles, and the swimming ability of these fish may be limited in shallow water. Therefore, in order to conserve these large galaxiids, stream management should ensure the upstream and downstream passage of adult galaxiids is unhindered.

In summary, both banded and giant kokopu exhibited extensive intra- and possibly inter-stream movements that may play an important role in mate selection and the location of suitable habitat. Both inter-stream migrations and local movements of adult fish should be considered in stream restoration and fish passage design, to facilitate the migrations of adult galaxiids.

Acknowledgments

We thank landowners Bob and Rebecca Copsey, Wayne Andrews, John Gumley, and Tainui Group Holdings Ltd. for stream access, and Jacques Boubée (NIWA) for assistance with experimental setup. This study was funded by the New Zealand Foundation for Research, Science and Technology (Contract No. C01X0305).

References

Bachman, R. A. 1984. Foraging behavior of free-ranging wild and hatchery brown trout in a stream. Transactions of the American Fisheries Society 113:1–32.

Baker, C. F., and J. A. T. Boubée. 2006. Upstream passage of inanga *Galaxias maculatus* and redfin bullies *Gobiomorphus huttoni* over artificial ramps. Journal of Fish Biology 69:668–681.

Baker, C. F., and J. P. Smith. 2007. Habitat use by banded kokopu (*Galaxias fasciatus*) and giant kokopu (*Galaxias argenteus*) co-occurring in streams draining the Hakarimata Range, New Zealand. New Zealand Journal of Marine and Freshwater Research 41:25–33.

Bateman, D. S., and R. E. Gresswell. 2006. Survival and growth of age-0 steelhead after surgical implantation of 23-mm passive integrated transponders. North American Journal of Fisheries Management 26:545–550.

Boubée, J. A. T., and E. K. Williams. 2006. Downstream passage of silver eels at a small hydroelectric facility. Fisheries Management and Ecology 13:165–176.

Boubée, J. A. T., E. K. Williams, and J. Richardson. 2000. Fish passage guidelines for the Auckland region. Auckland Regional Council, Auckland, New Zealand.

Boubée, J. A. T., I. G. Jowett, S. Nichols, and E. K. Williams. 1999. Fish passage at culverts: a review with possible solutions for New Zealand indigenous species. Department of Conservation, Wellington, New Zealand.

Castro-Santos, T., A. Haro, and S. Walk. 1996. A passive integrated transponder (PIT) tag system for monitoring fishways. Fisheries Research 28:253–261.

Charteris, S. C., R. Allibone, and R. G. Death. 2003. Spawning site selection, egg development, and larval drift of *Galaxias postvectis* and *G. fasciatus* in a New Zealand stream. New Zealand Journal of Marine and Freshwater Research 37:493–505.

Crowl, T. A., C. R. Townsend, and A. R. McIntosh. 1992. The impact of introduced brown and rainbow trout on native fish: the case of Australasia. Reviews in Fish Biology and Fisheries 2:217–241.

Cucherousset, J., J.-M. Rousell, R. Keeler, R. A. Cunjak, and R. Stump. 2005. The use of two new portable 12-mm PIT tag detectors to track small fish in shallow streams. North American Journal of Fisheries Management 25:270–274.

Daugherty, D. J., and T. M. Sutton. 2005. Seasonal movement patterns, habitat use, and home range of flathead catfish in the lower St. Joseph River, Michigan. North American Journal of Fisheries Management 25:256–269.

David, B. O., and G. P. Closs. 2001. Continuous remote monitoring of fish activity with restricted home ranges using radiotelemetry. Journal of Fish Biology 59:705–715.

David, B. O., and G. P. Closs. 2002. Behavior of a stream-dwelling fish before, during, and after high-discharge events. Transactions of the American Fisheries Society 131:762–771.

David, B. O., and G. P. Closs. 2003. Seasonal variation in diel activity and microhabitat use of an endemic New Zealand stream-dwelling galaxiid fish. Freshwater Biology 48:1765–1781.

Feldheim, K. A., S. H. Gruber, J. R. C. de Marignac, and M. V. Ashley. 2002. Genetic tagging to determine passive integrated transponder tag loss in lemon sharks. Journal of Fish Biology 61:1309–1313.

Gerking, S. D. 1959. The restricted movement of fish populations. Biological Reviews 34:221–242.

Greenberg, L. A., and P. S. Giller. 2000. The potential of flat-bed passive integrated transponder antennae for studying habitat use by stream fishes. Ecology of Freshwater Fish 9:74–80.

Greenberg, L. A., and P. S. Giller. 2001. Individual variation in habitat use and growth of male and female brown trout. Ecography 24:212–224.

Gowan, C., M. K. Young, K. D. Fausch, and S. C. Riley. 1994. Restricted movement in resident stream salmonids: a paradigm lost? Canadian Journal of Fisheries and Aquatic Sciences 51:2626–2637.

Gries, G., and B. H. Letcher. 2002. Tag retention and survival of age-0 Atlantic salmon following surgical implantation with passive integrated transponder tags. North American Journal of Fisheries Management 22:219–222.

Hansen, E. A. 2005. Distribution, movement, growth and individual behaviours of a drift feeding stream fish in relation to food supply. Doctoral dissertation. University of Otago, Dunedin, New Zealand.

Hansen, E. A., and G. P. Closs. 2005. Diel activity and home range size in relation to food supply in a drift-feeding stream fish. Behavioral Ecology 16:640–648.

Hansen, E. A., B. D. David, and G. P. Closs. 2004. Diel patterns of feeding and prey selection in giant kokopu (*Galaxias argenteus*). New Zealand Journal of Marine and Freshwater Research 38:341–345.

Hitchmough, R. 2002. New Zealand Threat Classification System Lists 2002. Threatened Species Occasional Publication 23. Department of Conservation, Wellington, New Zealand.

McDowall, R. M. 1990. New Zealand Freshwater Fishes. A natural history and guide. Heinemann Reed, Auckland, New Zealand.

McDowall, R. M., and G. R. Kelly. 1999. Date and age at migration in juvenile giant kokopu, *Galaxias argenteus* (Gmelin) (Teleostei: Galaxiidae) and estimation of spawning season. New Zealand Journal of Marine and Freshwater Research 33:263–270.

Metcalfe, N. B., N. H. C. Fraser, and M. D. Burns. 1999. Food availability and the nocturnal vs. diurnal foraging trade-off in juvenile salmon. Journal of Animal Ecology 68:371–381.

Miller, R. B. 1957. Permanence and size of home territory in stream-dwelling cutthroat trout. Journal of the Fisheries Research Board of Canada 14:687–691.

Mitchell, C. P., and B. P. Penlington. 1982. Spawning of *Galaxias fasciatus* Gray (Salmoniformes: Galaxiidae). New Zealand Journal of Marine and Freshwater Research 16:131–133.

Nykänen, M., A. Huusko, and M. Lahti. 2004. Changes in movement, range and habitat preferences of adult grayling from late summer at early winter. Journal of Fish Biology 64:1386–1398.

Ombredane, D., J. L. Bagliniere, and F. Marchand. 1998. The effects of passive integrated transponder tags on survival and growth of juvenile brown trout (*Salmo trutta* L.) and their use for studying movement in a small river. Hydrobiologia 371/ 372:99–106.

Paragamian, V. L. 1989. Seasonal habitat use by walleye in a warmwater river system, as determined by radiotelemetry. North American Journal of Fisheries Management 9:392–401.

Peterson, J. T., R. F. Thurow, and J. W. Guzevich. 2004. An evaluation of multipass electrofishing for estimating the abundance of stream-dwelling salmonids. Transactions of the American Fisheries Society 133:462–475.

Popoff, N. D., and R. M. Neumann. 2005. Range and movement of resident holdover and hatchery brown trout tagged with radio transmitters in the Farmington River, Connecticut. North American Journal of Fisheries Management 25:413–422.

Roussel, J.-M., A. Haro, and R. A. Cunjak. 2000. Field test of a new method for tracking small fishes in shallow rivers using passive integrated transponder (PIT) technology. Canadian Journal of Fisheries and Aquatic Sciences 57:1326–1329.

Sakaris, P. C., R. V. Jesien, and A. E. Pinkney. 2005. Brown bullhead as an indicator species: seasonal movement patterns and home ranges within the Anacostia River, Washington, D.C. Transactions of the American Fisheries Society 134: 1262–1270.

Townsend, C. R., and T. A. Crowl. 1991. Fragmented population structure in a native New Zealand fish: an effect of introduced brown trout? Oikos 61:347–354.

Wagner, C. P., M. J. Jennings, J. M. Kampa, and D. H. Wahl. 2007. Survival, growth and tag retention in age-0 muskellunge implanted with passive integrated transponders. North American Journal of Fisheries Management 27:873–877.

West, D. W., I. G. Jowett, and J. Richardson. 2005. Growth, diet, movement and abundance of adult banded kokopu (*Galaxias fasciatus*) in five Coromandel, New Zealand streams. New Zealand Journal of Marine and Freshwater Research 39:915–929.

Young, M. K. 1994. Brown trout mobility in south-central Wyoming streams. Canadian Journal of Zoology 72:2078–2083.

American Fisheries Society Symposium 76:191–214, 2012

Marine Distributions of Coho and Chinook Salmon Inferred from Coded Wire Tag Recoveries

Laurie Weitkamp*
Northwest Fisheries Science Center, U.S. National Marine Fisheries Service
2032 SE OSU Drive, Newport, Oregon 97365, USA

Abstract.—The coded wire tag (CWT) database contains detailed information on millions of Pacific salmon *Oncorhynchus* spp. released from hatcheries or smolt traps and recovered in the Pacific Ocean and its tributaries. I used this information to compare marine distribution patterns of hatchery coho *O. kisutch* and Chinook *O. tshawytscha* salmon, based on recoveries of an estimated 1.99 million tagged salmon in coastal areas from southern California to the Bering Sea. Both species show distinct region-specific distribution patterns. Within these release regions, coho and Chinook salmon marine distributions were often similar, with fish distributed largely in local waters. In other regions, Chinook salmon were distributed father north than coho salmon originating from the same region. Only in two regions did the two species have fundamentally different marine distributions, with coho south of, and Chinook salmon north of, the natal stream. The analysis also revealed several "hot spots" of salmon diversity, identified by numerically few recoveries that represented many of the hatcheries used in the analysis. These hotspots may serve as important reservoirs for the continued existence of populations that are particularly vulnerable to climate change due to their restricted marine distributions. Although CWT technology is primitive by modern standards, the enormous amount of data collected in a consistent fashion over decades and contained in an online database provides a unique and underutilized opportunity to address many elusive questions about Pacific salmon.

Introduction

The primary objective of most fish tagging programs is to determine the movements or migratory patterns of fish (Guy et al. 1996). While this information is typically collected for management purposes (to determine who's catching whose fish), knowing where an individual is located during each stage of its life is crucial to understanding the factors affecting its survival and reproductive success, and therefore the long term survival of entire populations or species (Leggett 1985). This is particularly true for anadromous Pacific salmon *Oncorhynchus* spp. because of their use of both freshwater and marine habitats (Groot and Margolis 1991).

As many of the papers in this book illustrate, increasingly complex electronic data storage tags can determine precise movements and conditions (e.g., temperature, depth) experienced by tagged fish, often without having to retrieve the tag. No doubt, continued rapid evolution in tag technology will result in further increases in tag sophis-

*Corresponding author: laurie.weitkamp@noaa.gov

tication and decreases in tag size. However, widespread use of these electronic tags is limited by the relatively high cost of the tags and the large size of fish necessary to minimize the effects of tag deployment, leading to comparatively small sample sizes from which to base the behavior of whole populations or species (e.g., Neuenfeldt et al. 2007; Block et al. 2008).

For many applications, electronic tags provide greater information than can be obtained from simpler tagging technologies. However, tagging programs employing simple tags that have been ongoing for decades provide a unique opportunity to understand fish migration behavior because of the large number of tags deployed and retrieved over multiple years. This wealth of data allows rigorous examination of fish distributions and the factors affecting it, such as the influence of fish age, stock, release location, or environmental conditions. Tag choice represents an important tradeoff between the amount of information gained from a few technologically advanced tags versus that gained from many primitive tags.

One such old and simple tag technology is the coded wire tag (CWT); 1 mm-long pieces of metal wire etched with a code (Jefferts et al. 1963) are inserted in the nasal cartilage of juvenile Pacific salmon prior to release from the hatchery or in wild fish captured during out-migration. Each tag code is associated with both fish (e.g., species, stock, size, age) and release (e.g., location, date, number tagged) information. Fish bearing the same code (termed a "release group") typically number in the thousands of tagged individuals.

Coded wire tags are recovered from salmon caught in commercial, sport, or tribal fisheries in both marine and freshwater environments, at weirs or hatcheries, and during spawning ground surveys. Tags are extracted, the tag code visually read, and appropriate recovery information (e.g., tag code, date, location, fishery type, fish size, and sex) is reported to the Regional Mark Processing Center, who maintains the Regional Mark Information System (RMIS) database (RMIS 1977).

Coded wire tags have been used continually for Pacific salmon management since the late 1960s, resulting in perhaps the world's largest fish tagging program (Guy et al. 1996). In recent years (late 1990s-early 2000s), 50 million Pacific salmon (39 million Chinook salmon *O. tshawytscha*, 10 million coho salmon *O. kisutch*) tagged with CWTs have been released and 275,000 recovered annually (Johnson 2004). The online RMIS database contains information on over one billion tagged salmon released from over three hundred hatcheries along the west coast of North America, of which tens if not hundreds of millions were subsequently recovered (RMIS 1977).

The CWT program is designed specifically for, and actively used by, salmon managers to determine species-, stock-, and run-specific salmon distributions, and harvest and survival rates. Managers accordingly structure fisheries based on CWT-derived spatial or temporal differences in marine distributions or migratory behavior between stocks (e.g., PSC 2009). The CWT database has also been used to address issues outside the management arena, such as salmon movements (Norris et al. 2000; Morris et al. 2007), homing fidelity (Hard and Heard 1999; Candy and Beacham 2000), and marine survival (e.g., Coronado and Hilborn 1998; Ryding and Skalski 1999; Hobday and Boehlert 2001; Magnusson and Hilborn 2003).

Weitkamp and Neely (2002) and Weitkamp (2010) used the CWT database to document the marine distributions of an estimated 2.4 million coho and Chinook salmon released from hatcheries or tagged during out-migration (wild populations) along the west coast of North America and recovered in coastal waters. Both studies used nearly identical methodologies (e.g., hatchery and release group selection criteria, recovery areas, years of interest). The fundamental finding from both studies was the same: salmon—either hatchery or wild—originating from a particular freshwater region shared a common marine recovery pattern, which was distinct from that of conspecifics from adjacent regions, with often abrupt transitions between regions. Many of the release regions with common recovery patterns identified for coho salmon were the same as those for Chinook salmon, such as Cook Inlet (Alaska), Southeast Alaska, northern British Columbia, the west coast of Vancouver Island, Upper Fraser River, Puget Sound, Lower Columbia River, Oregon Coast, and southern Oregon/northern California.

Here, I use these two datasets to explicitly compare the marine distributions of coho and Chinook salmon from the west coast of North America, using hatchery populations from Cook Inlet, Alaska, to the Klamath River, California. Like the earlier studies, I effectively use thousands of coastal fishers from southern California to the Bering Sea as samplers of the marine environment to detect the presence and abundance of millions of tagged coho and Chinook salmon. Similarities and differences in marine distributions provides unique insight into salmon marine biology and prospects for these species in the future. Although CWTs are an old and passive tagging technology, the consistent release and recovery of millions of tagged salmon over the last four decades provides an unprecedented—and underutilized–opportunity to address enduring ecological questions about Pacific salmon.

Methods

The objective of this study was to directly compare marine distribution patterns of coho and Chinook salmon determined from CWT recoveries used in our earlier analyses (Weitkamp and Neely 2002; Weitkamp 2010). The current analysis was limited to hatchery salmon and to regions where both coho and Chinook salmon hatcheries meeting the criteria were available (Cook Inlet to the Klamath River, excluding the Upper Columbia, Snake, and Willamette rivers).

Like the earlier single species analyses, this analysis is based on two key concepts: 1) where the hatchery is located (freshwater release region), and 2) where the tagged fish are recovered (marine recovery area). The freshwater release regions were based on the general geographic location or conservation unit of the hatchery (Weitkamp et al. 1995; Myers et al. 1998; Holtby and Ciruna 2007). By contrast, the recovery location of each tagged salmon was assigned to a marine recovery area, which has very specific boundaries corresponding to fishery management areas (see methods below). Accordingly, hatcheries were located in freshwater release regions and their tagged fish were recovered in marine recovery areas. Hatcheries used in this analysis and their release regions are provided in Appendix Table 1.

Like the earlier analyses, using the CWT database assumes that all appropriate fisheries were consistently sampled for tagged fish, expansions for sampling effort were correct, all recovered tags were read and reported, and there was no bias in these factors between regions, years of interest, or species. Because the database is maintained for management purposes, which presumably requires the most complete and accurate data possible, I believe that the above assumptions are sufficiently valid and did not unduly influence the results.

The methods used to select hatcheries and run timing groups, release groups, and recoveries were nearly identical for both coho and Chinook salmon and are briefly described here.

Release and recovery criteria

Hatcheries or hatchery run-timing[1] groups (HRGs, hereafter both are referred to as HRGs) were selected to represent coho and Chinook salmon populations along the west coast of North America and that released and recovered large numbers of tagged salmon. These hatcheries recovered a minimum of 1,000 tagged salmon (adjusted for sampling effort—see below) over at least 3 years. To minimize the potentially confounding effects due to unusual rearing practices, use of exotic stocks, or transportation of fish prior to release, release groups were excluded from the analysis if they were experimental (type *E* or *B* in the database), used stocks with names other than the hatchery, stream, or local river basin name, and occurred anywhere other than the hatchery or hatchery stream. By employing these rules, each release group effectively served as a replicate for that HRG. In several cases, hatcheries from the same geographic location were combined if no hatcheries or run timings meeting the criteria were available (e.g., spring Chinook salmon from the Upper Fraser River and Cook Inlet; Appendix Table 1).

Recoveries of CWT release groups were selected to determine typical distributions of salmon in coastal waters. Marine (as defined in the database) recoveries during years 1979–1993 (coho) or 1979–1994 (Chinook) were used because both sampling for CWTs and fishing effort were more or less constant during this time (PSC 2009; PFMC 2007b). Prior to 1979, sampling for CWTs was incomplete, while in 1995 several major coho and Chinook salmon fisheries were greatly curtailed (e.g., West Coast Vancouver Island troll fishery). To simplify the analysis, each of the approximately 7,770 marine recovery location codes listed in the database was assigned to 21 recovery areas (Figure 1). These recovery areas were selected to be approximately equal size, represent geographically distinct areas where possible, and have the same boundaries as fishery management areas to minimize the number of recovery location codes that overlapped the recovery areas. All recoveries were adjusted to account for sampling effort but not unmarked fish (both fields are provided in the database); accordingly, this analysis is restricted to tagged fish, not the larger [untagged] salmon population. Recoveries were also restricted to the dominant ocean age of salmon (ocean age = recovery year – release year) to minimize age-based variation in recovery patterns. Dominant ocean ages were 1 year for coho and either 1–3 or 2–4 years for Chinook salmon.

Once selected recoveries were assigned to the 21 recovery areas, marine distributions were estimated as the proportion of recoveries (R_{ij}) by HRG *j* in recovery area *i* over all years as:

$$R_{ij} = \frac{\sum_k r_{ijk}}{\sum_i \sum_k r_{ijk}}$$

where r_{ijk} is the estimated number of recoveries from HRG *j* in recovery area *i* in year *k*. This formulation gives equal weight to all recoveries, regardless of the year in which fish were recovered and provides the primary data set for the analysis: the proportion (or percent by multiplying portions by 100) of recoveries in each of the 21 recovery areas for each coho or Chinook salmon HRG.

Although mark–recapture studies require inclusion of fishing or recovery effort (Quinn and Deriso 1999), this analysis does not include it because to do so would have made it prohibitively complicated. For example, fishing effort for all the fisheries that recover CWTs is not readily available, although scaled catch is (PFMC 2007a; PSC

[1]Run timing refers to the time (season) when Pacific salmon return to freshwater as adults. Most coho salmon return in the fall (fall run timing), but Chinook salmon stocks used in this analysis returned in the spring, summer, or fall, or had spring, summer, and fall run timing, respectively.

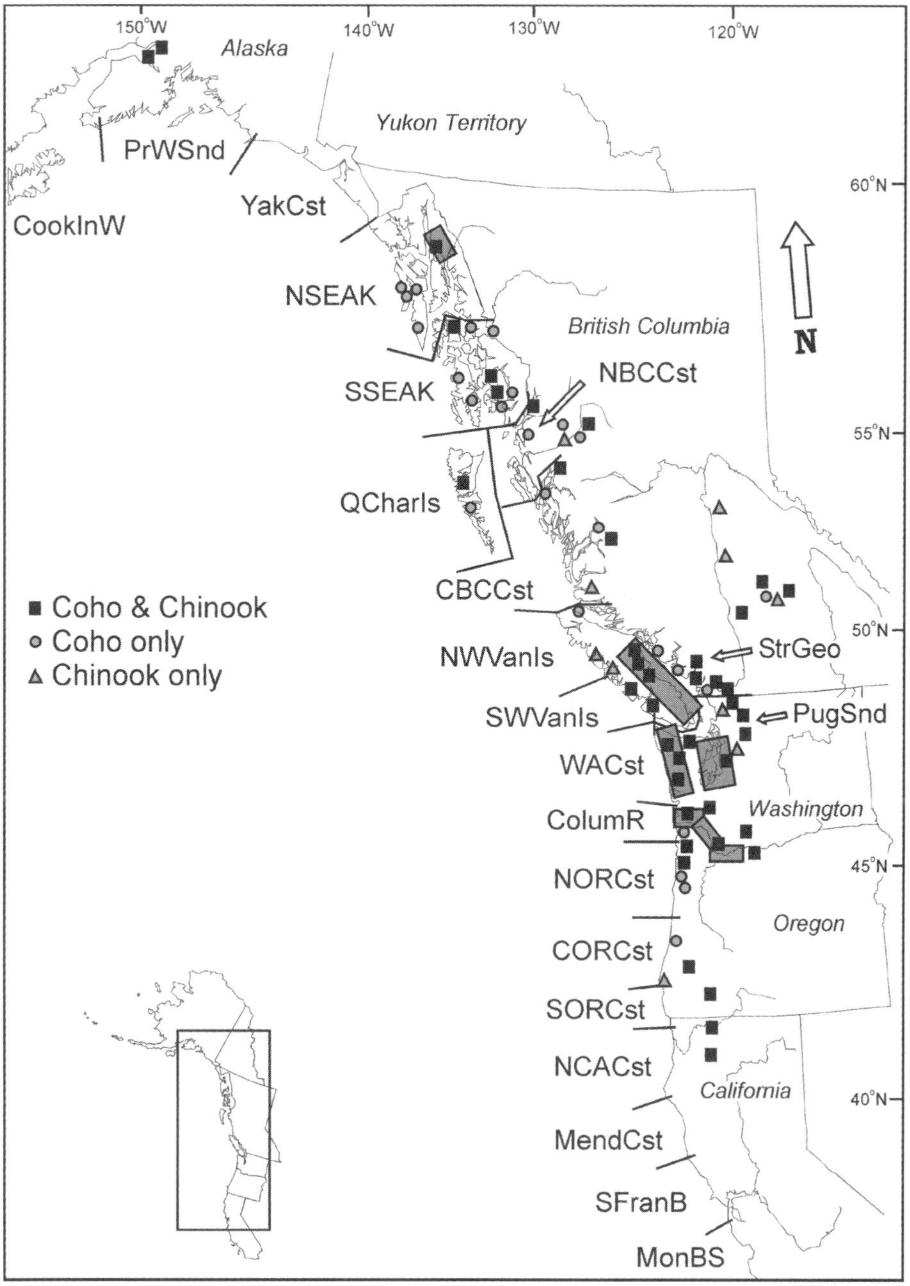

FIGURE 1. Locations of the 21 recovery areas (indicated by black lines) and hatcheries (symbols) used in the analysis. These hatcheries are coded to indicate those that released both coho and Chinook, coho only, or Chinook salmon only. Areas with too many hatcheries to show individual locations are indicated by shading. Marine recovery area abbreviations are CookInW: Cook Inlet west; PrWSnd: Prince William Sound; YakCst: Yakutat Coast; NSEAK: northern Southeast Alaska, SSEAK: southern Southeast Alaska; QCharls: Queen Charlotte Islands; NBCCst: northern British Columbia Coast; CBCCst: central British Columbia Coast; NWVanIs: Northwest Vancouver Island; SWVanIs: Southwest Vancouver Island; StrGeo: Strait of Georgia; PugSnd: Puget Sound; WACst: Washington Coast; ColumR: Columbia River; NORCst: North Oregon Coast; CORCst: Central Oregon Coast; SORCst: Southern Oregon Coast; NCACst: Northern California Coast; MendCst: Mendocino Coast; SFranB: San Francisco Bay; MonBS: Monterey Bay south.

2009). However, using this scaled catch still presents a computational challenge because the fisheries that capture the CWTs used in this analysis are numerous and diverse, each with their own temporal seasons, fish size or mesh limits, gear types (nets, hook-and-line), and catch limits. For example, the coho and Chinook salmon Fisheries Regulation Assessment models (FRAM) currently used to manage fisheries from Southeast Alaska to California, include 198 coho and 78 Chinook salmon fisheries, respectively, at 4–5 time periods per year (PFMC 2007a). In addition, these FRAM include most—but not all—of the fisheries used in this analysis. Clearly, trying to calculate effort or employ available scaling for this extensive suite of fisheries is not a trivial task. The implications of this omission are considered in the discussion.

From these recovery data, I calculated weighted mean distance ("distance" [km]) for each HRG as the distance between the geographic center of the "home [natal] recovery area" (in which their home stream enters the ocean) and the geographic center of each recovery area, multiplied by the proportion of recoveries that occurred in that recovery area, summed across all recovery areas. All distances between recovery areas were estimated as travel by water only using great circle (spherical geometry) distances. Because recovery area was the smallest geographic unit used, recoveries in the home recovery area had a distance of zero.

Statistical analyses

Pairwise Bray-Curtis similarities were estimated between all coho and Chinook salmon HRGs, based on the percent of recoveries in each of the 21 marine recovery areas, described above. This multivariate similarity measure varies from 0% (no recoveries in common) to 100% (identical recoveries). Comparison of these pairwise Bray-Curtis similarities calculated between species, release regions, or HRGs were evaluated using either a Mann-Whitney (MW) test for difference in medians followed by Bonferroni test when statistically significant differences were detected, or Kruskal Wallace (KW) One-Way analysis of variance (ANOVA) on Ranks followed by a KW Multiple Comparison test (Zar 1984).

I also used nonmetric multiple-dimensional scaling (MDS) to graphically explore the relationship between coho and Chinook salmon marine distribution patterns. The MDS ordination technique is based on pairwise Bray-Curtis similarities and places all points in MDS space in relation to their similarity (i.e., points further apart in MDS space are less similar than those closer together). In all MDS analyses, random starting locations were used for each of 25 iterations to find the best solution; minimum stress was attained in multiple iterations suggesting a true minimum solution. Both multivariate techniques (Bray-Curtis similarities and MDS) were run using PRIMER-E software (Clarke and Gorley 2006).

Finally, I used Spearman correlation coefficients (r) to evaluate the relationships between variables in two comparisons (Zar 1984). First, between the number of tagged fish and the number of HRGs recovered in each recovery area, and secondly between the total number tagged fish recovered and the number of recovery areas in which they were recovered by HRG. Data sets used in both comparisons were not transformed.

Results

Recovery statistics

Marine distribution patterns were estimated from approximately 1.99 million coded wire tagged coho and Chinook salmon released from hatcheries from Cook Inlet, Alaska, to the Klamath River, California, and recovered in coastal waters along the west coast of North America. These fish include an estimated 1.47 million coho salmon released from 89 hatcheries and 0.52 million Chinook salmon representing 64 HRGs, with 45 hatcheries in common between the two species (Table 1; Figure 1; Appendix Table 1). Compared to Chinook salmon HRGS, coho salmon had nearly twice as many recoveries per hatchery on average (16,519 versus 8,196 fish) but were recovered in slightly fewer recovery areas (11.4 versus 11.9) and closer to the natal stream (103 versus 273 km) (Table 1).

By marine recovery area, more coho salmon were recovered in each area on average (70,010) than Chinook salmon (24,978), however the percent of possible HRGs represented by these recoveries was similar between coho (54% or 49 of 89 HRGs) and Chinook salmon HRGs (57% or 36 of 64 HRGs; Table 1).

Marine recovery patterns

The average marine distribution patterns (based on the proportion of recoveries in the 21 recovery areas) of coho and Chinook salmon by freshwater release region are illustrated in Figure 2. (Marine distributions of coho and Chinook salmon originating from Cook Inlet are not illustrated; both species were primarily recovered (>90%) in the three northern-most recovery areas, Cook Inlet West [CookInW], Prince William Sound [PrWSnd], and Yakutat Coast [YakCst].) Examination of these maps suggest three broad categories based strictly on geographic location of the recoveries, and three categories based on comparative distribution patterns between coho and Chinook salmon from each region.

Examined on a strictly geographic basic, coho and Chinook salmon possessed one of three distributions: 1) restricted northern, 2) broad central, or 3) restricted southern (Figure 2), although both species from the same region did not necessarily follow the same pattern (see below). For example, coho and Chinook salmon in the restricted northern group had marine distributions restricted to northern recovery areas and were rarely caught south of the Central British Columbia Coast (CBCCst) recovery area; fish with this recovery pattern originated from Southeast Alaska and northern British Columbia release regions.

The broad central group consisted of salmon that were broadly recovered across central recovery areas (Northwest Vancouver Island [NWVanIs], Southwest Vancouver Island [SWVanIs], Strait of Georgia [StrGeo],

Table 1. Summary statistics for coho and Chinook salmon hatchery run-timing groups (HRGs) used in the analysis, and summarized by recovery area (RA). Included are the number (no.) of HRGs used, estimated number of recoveries and average distance of recoveries, and estimated number of fish and HRGs recovered by recovery area. Recovery areas and location of HRGs are provided in Figure 1.

Characteristic	Coho	Chinook
No. HRGs	89	64
No. fish recovered/HRG	16,519	8,196
No. recovery areas/HRG	11.4	11.9
Avg. distance (km)	103	273
No. fish/RA	70,010	24,978
No. HRGs/RA (%)	49 (54)	36 (57)
Maximum no. HRGs/RA (%)	74 (83)	58 (91)

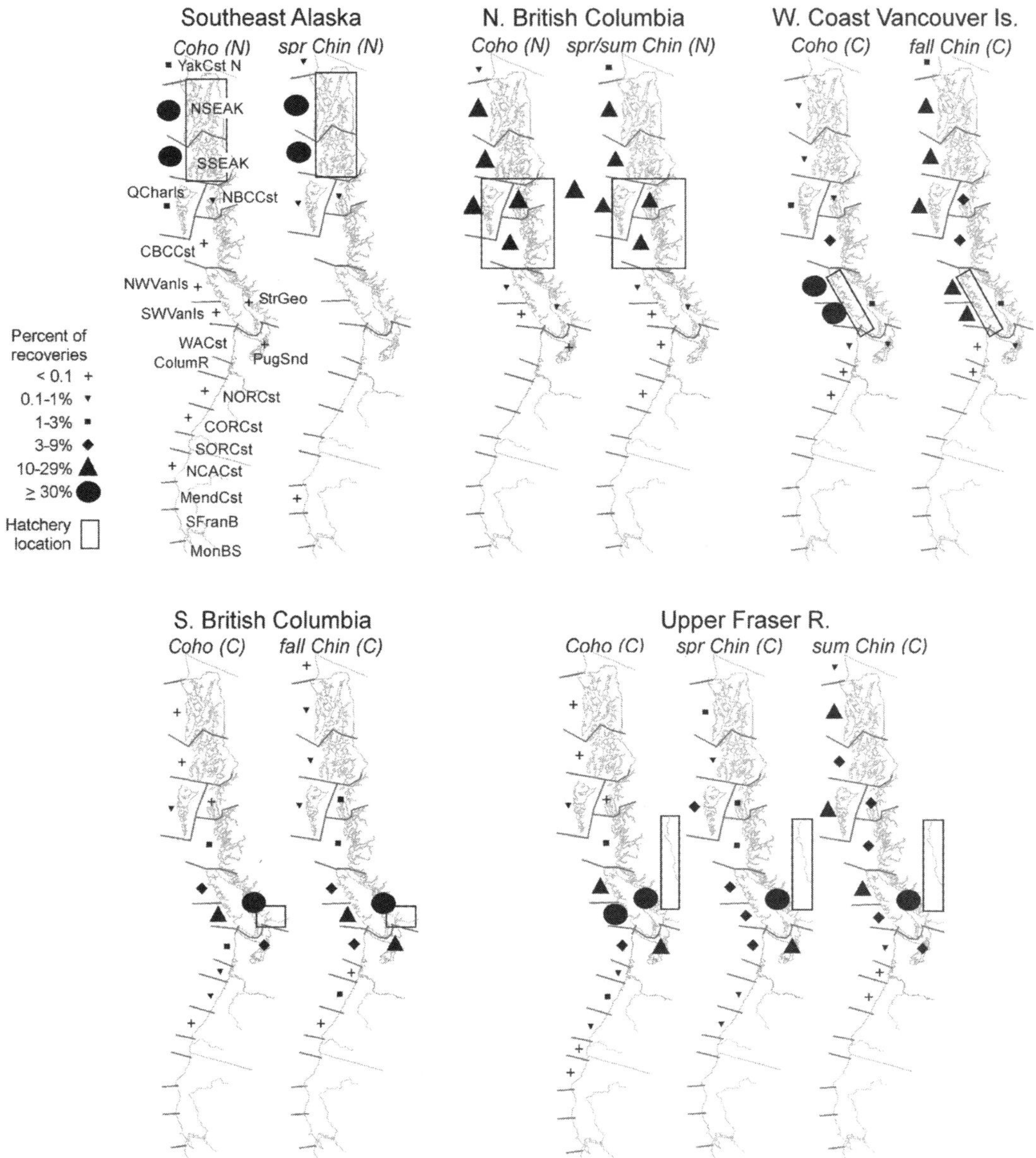

FIGURE 2. Maps illustrating the average percent of recoveries in each marine recovery area from Yakutat Coast (YakCst) to Monterey Bay south (MonBS) for coho and Chinook salmon hatcheries grouped by freshwater release region and run timing; rectangles indicate the approximate location of the hatcheries. Chinook salmon run timing is indicated as spring (spr), summer (sum), and fall. Letters in parentheses following the species name indicate the three general recovery patterns described in the text: restricted northern (N), broad central (C), and restricted southern (S). Recovery areas with less than 0.001% of total recoveries are not shown; recoveries in recovery areas Cook Inlet West and Prince William Sound are included in Yakutat Coast (YakCst) recovery area. Recovery area abbreviations are provided in Figure 1.

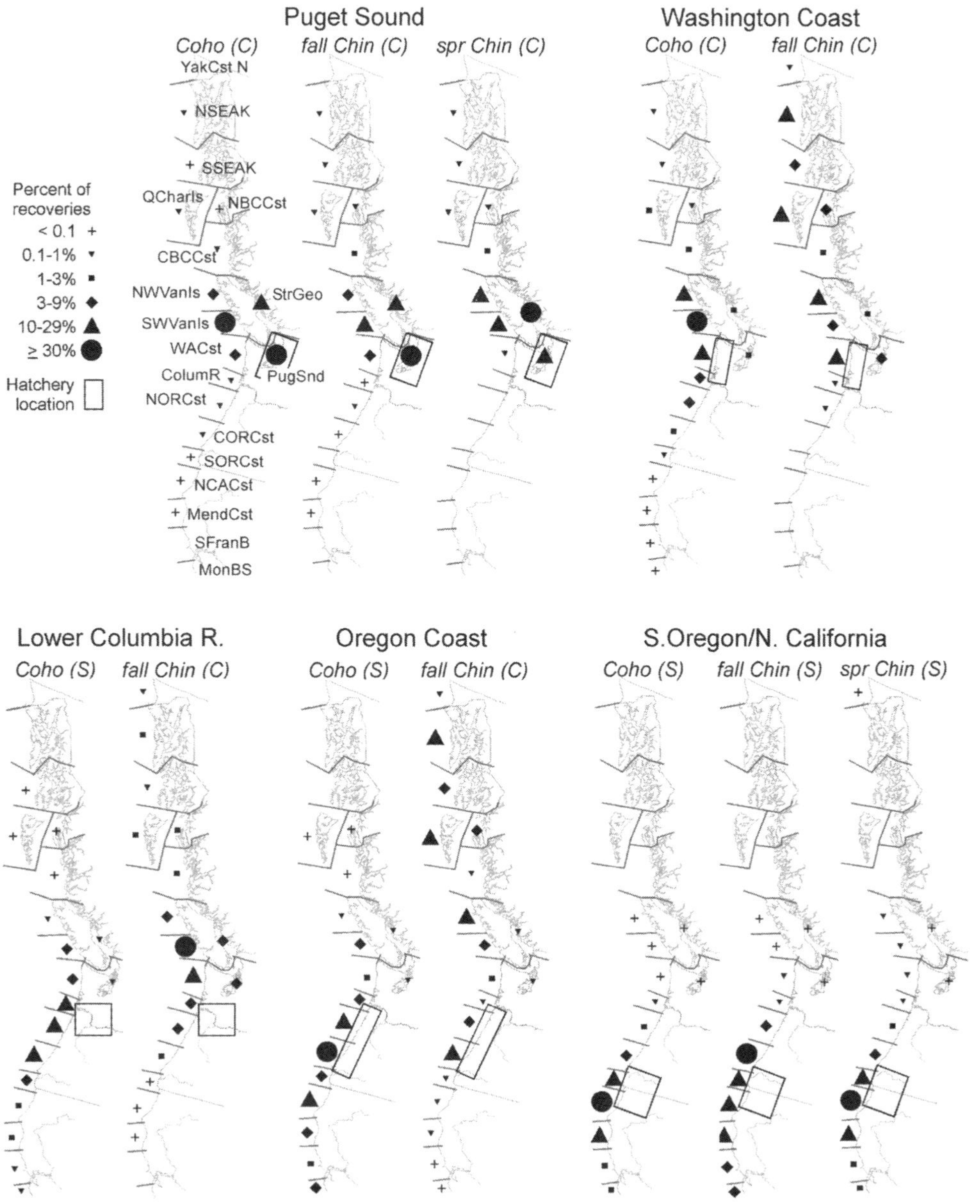

FIGURE 2. Continued.

Puget Sound [PugSnd], Washington Coast [WACst]), with few recoveries south of the northern Oregon Coast (NORCst). Some Chinook salmon HRGs in this group were also recovered in Southeast Alaska (northern Southeast Alaska [NSEAK] and southern Southeast Alaska [SSEAK]), and northern British Columbia Coast (NBCCst) recovery areas. Coho and Chinook salmon in this group originated from HRGs located from the west coast of Vancouver Island to either the Washington Coast (coho) or Oregon Coast (Chinook).

The southern group had distributions largely restricted to recovery areas south of the WACst recovery area; this group consisted of coho salmon originating from the Lower Columbia River and the Oregon Coast, and both coho and Chinook salmon from southern Oregon/northern California (Figure 2).

Direct comparison of coho and Chinook salmon distributions within each release region also suggests three patterns: 1) similar distribution patterns for both coho and Chinook salmon; 2) Chinook salmon distributions that are more widespread northwards than coho salmon distributions; and 3) coho and Chinook salmon distribution patterns that are fundamentally different (Figure 2). Coho and Chinook salmon released from many regions displayed this first general pattern (similar recovery patterns between species), including those from Southeast Alaska, northern British Columbia, southern British Columbia mainland (including the Lower Fraser River), Upper Fraser River (Chinook spring run), Puget Sound, and southern Oregon/northern California release regions. For example, both species originating from Southeast Alaska HRGs were largely recovered in Southeast Alaska (NSEAK and SSEAK), those from both southern British Columbia mainland and Puget Sound were largely caught in StrGeo, PugSnd, NWVanIs, and SWVanIs, and those from southern Oregon/northern California were primarily recovered south of the Columbia River (ColumR) recovery area (Figure 2).

The second pattern (wider distribution of Chinook than coho salmon) applies to salmon originating from the west coast of Vancouver Island, Upper Fraser River (Chinook summer run), and the Washington Coast. In this case, coho salmon from these areas were largely caught near or slightly north of the natal recovery area, while Chinook salmon recoveries occurred in the same areas as coho salmon, but also included significant recoveries in NSEAK, SSEAK, Queen Charlotte Islands (QCharIs), and NBCCst. Two freshwater release areas (Lower Columbia River and Oregon Coast) representing the third pattern (fundamentally different coho and Chinook patterns) both had coho salmon that were largely caught south of the natal stream while Chinook salmon were primarily caught north of the natal stream (Figure 2).

Recoveries by recovery area

Examining the data by marine recovery area rather than individual HRG or freshwater release region indicates that recovery areas recovering large numbers of fish were not necessarily represented by large numbers of HRGs (Figure 3), even though the number of both fish and HRGs recovered in each recovery area were correlated (coho $r = 0.77$, $p < 0.01$; Chinook $r = 0.63$, $p < 0.01$). For example, while the StrGeo recovery area accounted numerically for the most coho salmon (25.1%, 368,844 tagged individuals) and the most Chinook salmon (19.4%, 101,570), it failed to provide the most HRGs for either species (79.8% and 89.1% for coho and Chinook, respectively), despite the high number of recoveries (Figure 3). Instead, the NWVanIs recovery area accounted the most coho salmon HRGs (83.1%, 74 of 89 HRGs), while SSEAK recovery area accounted for the greatest number of Chinook salmon HRGs (90.6%, 58 of 64). In both cases, these recovery areas with diverse representation actually

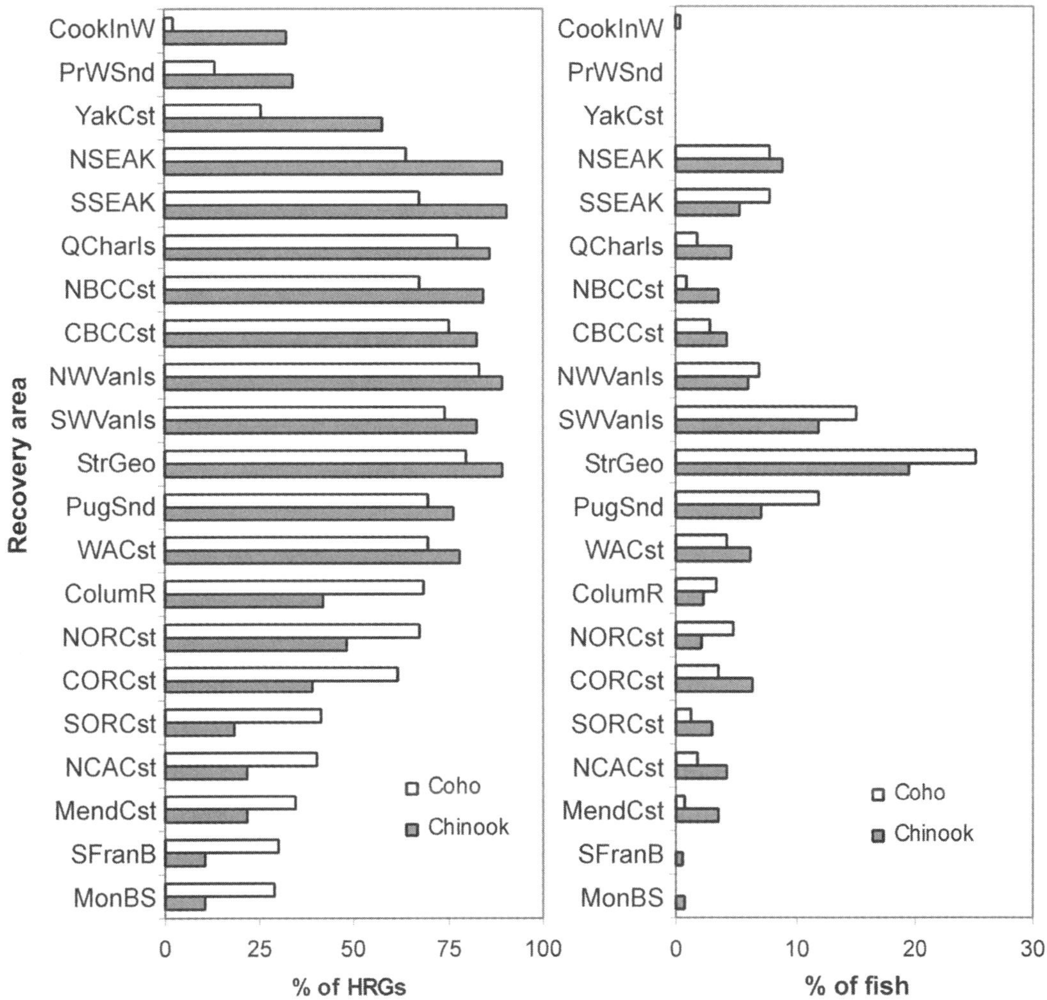

FIGURE 3. Percent of hatchery run-timing groups (HRGs; left) and percent of the number of tagged fish (right) recovered by each recovery area (approximately arranged from north to south) for coho and Chinook salmon HRGs. Recovery area abbreviations are provided in Figure 1.

recovered fairly few fish numerically (6.8% of coho, 5.2% of Chinook).

The recovery area data also indicate several "hot spots" where relatively few tagged salmon were recovered but these recoveries came from a large number of HRGs. For example, the three northern most recovery areas (CookInW, PrWSnd, YakCst) recovered only 0.6% of all coho and 0.5% of all Chinook salmon by number, but these recoveries represented 26% (23) and 63% (40) of coho and Chinook salmon HRGs, respectively. Similarly, the two southern most recovery areas (San Francisco Bay [SFranB], Monterey Bay south [MonBS]) recovered 0.4% of coho and 1.2% of Chinook salmon by number, but represented 30% (27) and 11% (7) of coho and Chinook salmon HRGs, respectively. Other recovery areas that captured relatively few salmon numerically (<2.7% of coho, <4.6% of Chinook) but were represented by large numbers of HRGs (>68% of coho salmon HRGs, >83% of Chinook salmon HRGs) included NBCCst, QCharls, and CBCCst (Figure 3).

Statistical analysis

Multivariate statistical analysis of recovery patterns indicated many of the same patterns suggested by visual examination of the maps (Figure 2) described above. Mean pair-wise Bray-Curtis similarity calculated between all coho and Chinook salmon HRGs ($n = 5{,}704$) indicated that marine distribution patterns of coho and Chinook salmon from the same region were much higher (mean = 53.4%; $n = 500$) than those from different regions (mean = 18.4%; $n = 5{,}204$; Two-sample t-test, $t = 30.7$, $p < 0.01$). Furthermore, mean similarity between coho and Chinook salmon released from the same hatchery was higher (61.1%; $n = 45$) than similarity between the two species released from the same region but different hatcheries (52.7%; $n = 455$; MW $Z = 3.1$, $p < 0.01$). However, mean similarity between coho and Chinook salmon from the same region overall (53.4%) was less than that of either only coho salmon from the same region (73.8%, $n = 379$) or only Chinook salmon from the same region (68.8%, $n = 162$; KW $H = 205.2$, $p < 0.01$). This suggests that although coho and Chinook salmon originating from the same region often had similar marine distributions, they were less alike than conspecifics.

Mean Bray-Curtis similarities between coho and Chinook salmon released from the same freshwater release regions also indicated regional differences (Table 2), consistent with the similarities or differences between coho and Chinook salmon distributions noted above. Mean similarity between species within freshwater release regions showed significant differences by region (KW one-way ANOVA on ranks, $H = 214$, $p < 0.01$), with mean similarity in many release regions being statistically different from each other ($p <$

TABLE 2. The number of hatchery run-timing groups (HRGs) and mean Bray-Curtis similarity calculated by release region (Figure 2) between coho only, Chinook only, and coho and Chinook salmon. Bray-Curtis similarity is based on the percent of recoveries in the 21 recovery areas (Fig. 1); it varies from 0% (no recoveries in common) to 100% (all recoveries in common).

Release Region	No. HRGs		Bray-Curtis similarity (%)		
	Coho	Chinook	Coho	Chinook	Coho v. Chin.
Cook Inlet	2	1	73.8		79.9
Southeast Alaska	15	4	59.8	47.7	55.1
Northern British Columbia	10	7	70.0	54.7	49.6
West Coast Vancouver Island	3	5	79.0	73.0	43.9
East Coast Vancouver Island	4	7	62.8	67.9	61.2
Southern British Columbia mainland	7	4	85.1	60.9	73.3
Upper Fraser River	4	6	93.3	68.2	54.1
Puget Sound	11	9	77.5	77.2	69.3
Washington Coast	10	5	79.6	65.6	40.1
Lower Columbia River	14	7	80.3	75.8	31.2
Oregon Coast	6	4	85.4	42.8	39.0
Southern Oregon/ northern California	3	5	91.4	88.6	64.8

0.05). For example, mean similarity between coho and Chinook salmon from the release region with the lowest similarity, Lower Columbia River (31.2%, distributions were fundamentally different) was statistically lower ($p < 0.05$) than most release regions with relatively high values, while mean similarity for the region with one of the highest values, southern British Columbia mainland (73.3%, similar coho and Chinook salmon distributions), was statistically higher than that from regions with low values.

The MDS analysis also indicated clear separation of groups based on region (Figure 4), consistent with region-specific recovery patterns described above (Figure 2); the "stress" of this analysis (an indication of fit) was reasonably low (0.13), suggesting that spatial patterns portrayed in two dimensional MDS space represented relationships in the original similarity matrix. In many instances, both coho and Chinook salmon HRGs from the same freshwater release region were fairly close together in MDS space, indicating high similarity (e.g., those originating from Cook Inlet, Southeast Alaska, northern British Columbia, and southern Oregon/northern California). In other cases, HRGs from a particular release region formed discrete groups, but the coho and Chinook salmon were wide-

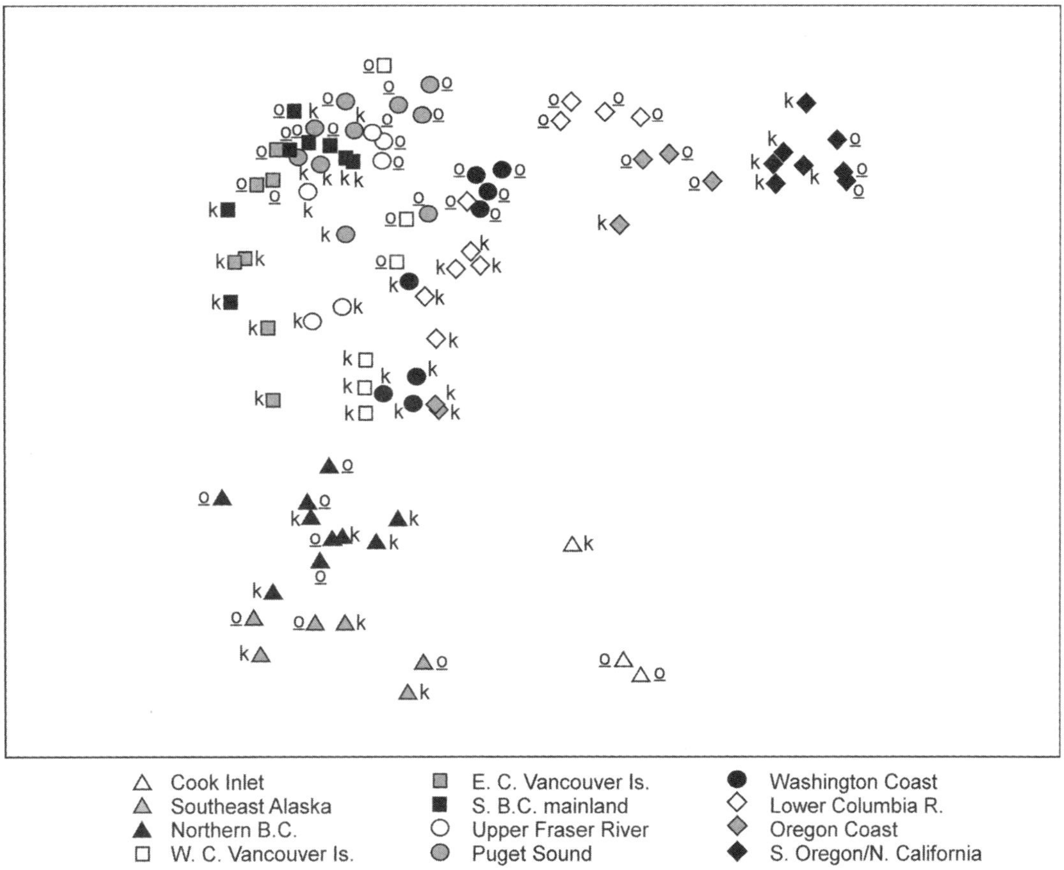

Figure 4. Nonmetric multidimensional scaling (MDS) for the 89 coho and 64 Chinook salmon hatchery run timing groups (HRGs) based on recoveries in the 21 recovery areas. This analysis arranges all HRGs in two dimensional space so that points closer together are more similar than those further apart. Each HRG is coded by release region (Fig. 2); symbols labeled with a "k" are Chinook salmon HRGs, those with an "o" are coho salmon HRGs.

ly separated from each other. For example, Chinook salmon originating from the Washington Coast, Oregon Coast, west coast of Vancouver Island, southern British Columbia mainland and the Upper Fraser River tend to be nearer each other than they were to coho salmon HRGs from the same regions, while the coho salmon from each region formed fairly coherent groups (Figure 4). Both coho and Chinook salmon from southern British Columbia mainland and Puget Sound and coho salmon from the Upper Fraser River produced one large overlapping group indicating high resemblance in marine distribution patterns, consistent with single species analyses (Weitkamp and Neely 2002; Weitkamp 2010).

Discussion

Previous single-species analyses of coho (Weitkamp and Neely 2002) and Chinook salmon (Weitkamp 2010) marine distributions showed that tagged individuals originating from a particular freshwater release region shared a common marine distribution pattern that was distinct from those of adjacent release regions, with often abrupt changes in distributions between regions. In most cases, these release regions are similar to those based strictly on genetics (Beacham et al. 2006; Van Doornik et al. 2007).

Here, I expand on these earlier studies by explicitly comparing the marine distributions of coho and Chinook salmon from the same freshwater release regions. This comparison indicates that marine distributions of salmon originating from the same release region were often similar, with both species largely recovered in or near the natal recovery area. In other regions, Chinook salmon were more widely distributed northwards than coho salmon. Only in two release regions (Lower Columbia River, Oregon Coast) were the distributions fundamentally different, with coho salmon distributed south of, and Chinook salmon north of, the home recovery area. The geographic spread of coho and Chinook salmon recoveries across the full North Pacific latitudinal range suggests three distributional patterns exist: those with restricted northern distributions, those with broad central distributions, and those with restricted southern distributions. The analysis also indicated several "hotspots" of salmon diversity, in which relatively few tagged salmon were recovered but these recoveries represented large numbers of HRGs.

In many respects, it is surprising that the marine distributions of coho and Chinook salmon originating from the same region were as similar as they were, given the older average age and more complex life history (e.g., multiple ages at maturity, multiple run timings) of Chinook salmon compared to coho salmon. This life history variation was shown to influence Chinook salmon marine distributions (Weitkamp 2010); older Chinook salmon were recovered significantly farther from the natal stream than younger fish and spring or summer run HRGs were often recovered further from the home stream than were fall run HRGs. Chinook salmon used in this analysis were typically recovered at older ocean ages (either 1–3 or 2–4 years) than coho salmon (1 year), yet in many regions distribution patterns were similar, contrary to the expectation that Chinook salmon, because of their older age, would be farther from the home stream than coho salmon. Similarly, in cases where Chinook salmon were recovered farther north than coho salmon from the same region, it was fall- or summer-run HRGs that had the more northern distribution, and most spring-run HRGs had similar distributions as coho salmon.

These observations suggest that distributional differences between coho and Chinook salmon were not due entirely to age at recovery or run timing. Instead, these inconsistencies emphasize the dominant role that freshwater region of origin plays in marine distributions. In addition, Chinook salmon HRGs that were used in the earlier analysis but eliminated from this analysis because of absence of comparable coho salmon HRGs (Upper Columbia, Snake, and Willamette rivers; Central Valley [Sacramento and San Joaquin rivers], California)(Weitkamp 2010) were recovered far from the home stream and were often represented by multiple run types, perhaps leading to a slightly different view of distribution diversity than can be drawn from the subset used here.

It is also interesting that the greatest change in recovery patterns for coho salmon occurred between fish originating from the Washington Coast and the Lower Columbia River (at approximately Cape Disappointment, 46° 16.6'N), while the parallel break for Chinook salmon was in southern Oregon at approximately Cape Blanco (42° 50.2'N; Nicholas and Hankin 1988). Consequently, coho and Chinook salmon originating from the Lower Columbia River and Oregon Coast have fundamentally different marine distributions (Figure 2) and wide separation in MDS space (Figure 4). These locations represent important life history transition points for both salmon species (e.g., Weitkamp et al. 1995; Myers et al. 1998); however, neither stands out as exceptional within the range of genetic diversity observed in either species (e.g., Waples et al. 2004; Van Doornik et al. 2007). Accordingly, although these locations represent major transition points in phenotypic diversity in the form of marine distributions, they may be too recent to reflect genetic diversity (Waples et al. 2004; Garcia de Leaniz et al. 2007).

Comparison of coho and Chinook salmon distribution patterns also provides evidence for a strong genetic basis to the distributions. As discussed above, both species have similar release regions in which each expresses a common marine distribution pattern, and the patterns are often similar between species. However, if marine distributions are determined largely by an individual's response to the particular environmental conditions it experiences (e.g., Kallio-Nyberg et al. 1999; Hodgson et al. 2006), then this similarity across species would not be as consistent as it was. Further evidence for a genetic rather than environmental basis to distribution patterns is provided by high interannual stability in marine distributions across years for both species (Weitkamp 2010; L. Weitkamp unpublished data) despite large variation in both ocean conditions and salmon survival between years (Johnson 1988; Beamish et al. 2000; Magnusson and Hilborn 2003).

Instead, similarity in marine distributions between species suggests a common evolutionary response to the challenges of shared marine environments, which have evolved over thousands of generations (Thorpe 1999). In some release regions (e.g., Southeast Alaska, Puget Sound, southern Oregon/northern California) the response by both species to marine environments was nearly identical (as indicated by similar marine distributions), while in other regions Chinook salmon were distributed farther north than coho salmon (e.g., west coast of Vancouver Island, Washington Coast), and in two regions (Lower Columbia River, Oregon Coast) the response was fundamentally different between species. This does not suggest that coho and Chinook salmon use of marine habitats is similar—Chinook salmon are typically deeper in the water column (Orsi and Wertheimer 1995) and more piscivorous than coho salmon (Welch and Parsons 1993; Daly et al. 2009). However, the results do suggest that some marine environments favor both species, while others favor one but not the other, although both patterns depend on salmon origin. Clearly, the reasons why salmonids distribute themselves in marine environments as they do are challenging questions to answer.

Analysis assumptions

By explicitly excluding effort or scaling when calculating marine distributions, this analysis makes the critical assumption that the number of CWTs recovered in a particular recovery area reflects the abundance of tagged fish in that area relative to other recovery areas. Clearly, this assumption is faulty because catch is a function of both salmon abundance *and* fishing effort; Weitkamp (2010) provides several examples indicating effort influenced interannual variability in Chinook salmon marine distributions. However, several measures were taken to minimize the influence of violations of this assumption. In particular, the years of interest were deliberately selected to cover as long a period as possible when fishing effort coastwide was more or less constant. In addition, recoveries were calculated across all years, rather than by year. Analysis of interannual variation in marine distributions of Chinook salmon indicate that they were surprisingly stable between years (Weitkamp 2010), a result which would be unlikely if observed patterns were largely the result of variation in effort.

One section of the analysis where variation in fishing effort likely has greater influence on the results occurs in comparisons of the abundance of tagged salmon or HRGs between recovery areas (Figure 3), because effort is not constant between recovery areas. To further complicate matters, the total number of recovered tagged fish varies by HRG (Appendix Table 1). Although one might expect HRGs represented by relatively few fish (e.g., < 2,000) would be recovered from fewer recovery areas than those represented by large numbers of fish, these two variables were only weakly correlated in either coho ($r = 0.23$, $p < 0.05$), or Chinook salmon ($r = 0.35$, $p < 0.05$). This suggests that the distribution of tagged salmon across recovery areas is not strictly a function of the number of tagged fish used. Furthermore, I deliberately avoided comparisons of the abundances of HRGs within recovery areas, which would have required HRG-specific information about both tagging rates and effort.

Although the relative number of fish recovered in each recovery area likely reflects effort to some degree, the general pattern of declining recoveries at the extreme ends of the geographic range (Figure 3) is unlikely to be strictly an artifact of fishing effort, given that fisheries typically target salmon where they are abundant, and these extreme recovery areas are not consistently low for both coho and Chinook salmon. In addition, recovery areas in which large numbers of tagged coho and Chinook salmon were recovered (Figure 3) are also areas where the catch of coho and Chinook salmon is high (PFMC 2007b; PSC 2009).

It is also unlikely that "hotspots"—recovery areas in which few fish representing many HRGs were recovered—are largely the result of either fishing effort or HRG-specific abundances. The HRGs recovered in these hotspots were not restricted to those with many total recoveries (Appendix Table 1) but also included those with relatively few total recoveries. These hotspots generally included at least a few individuals from most HRGs originating from particular release regions, while other release regions were poorly represented. For example, the coho salmon recovered in the QCharls recovery area included at least one individual from 95% of the HRGs from Southeast Alaska through the Washington Coast (61 of 64 possible HRGs), but were represented by only 30% of the HRGs from the Lower Columbia River, Oregon Coast, and California (7 of 23). Similarly, Chinook salmon HRGs recovered in QCharls recovery area were represented by 96% of the HRGs from northern British Columbia to the Oregon Coast (52 of 54), only half of Southeast Alaska HRGs (2 of 4), and none from southern Oregon/northern California (0 of 6).

Finally, several lines of evidence suggest the marine distribution patterns presented here largely reflect coho and Chinook salmon abundances along the west coast of North America, with relatively little influence due to fishing effort or other factors. For example, studies using the same CWT database but which deliberately incorporated fishing effort in estimations of Chinook salmon ocean distributions produced similar distributions to those produced without fishing effort (Weitkamp 2010), and demonstrated that fishing effort has only minor influence compared to other factors such as region, fish age, year, or environmental effects (Norris et al. 2000; Hyun et al. 2007). Similarly, results from the Chinook salmon FRAM, which includes estimates of stock-specific abundances coastwide, produced similar ocean distributions patterns to those provided here (PSC 2009). Furthermore, if fishing effort had been included in the analysis, differences in marine distributions would be expected to be more distinct between release regions, not less so. While it obviously would have been preferable to have included effort in this analysis, the results provided here without it are still informative and present considerable insight into how coho and Chinook salmon respond to marine environments.

Analysis implications

The results provided here have implications for Pacific salmon in general, how salmon may respond to climate change, and emphasize the often unappreciated value of simple but continuous tagging programs. First, the strong similarity between coho and Chinook salmon marine distributions for fish originating from multiple regions along the west coast of North America suggests a pattern that likely applies to both species around the Pacific Rim, but also to other species of Pacific salmon (e.g., chum *O. keta*, sockeye *O. nerka,* or pink salmon *O. gorbuscha*), and perhaps other fishes. Although coho and Chinook salmon have more coastal distributions than other Pacific salmon (Groot and Margolis 1991), available data indicate that these other salmon species have similar region-specific marine distributions, at least at gross spatial scales (Myers et al. 1996). Data provided here suggest that these similarities may exist at finer geographic scales as well, but it may only be feasible to detect these patterns using recovery location information from the release of millions of tagged fish.

These results also provide insight into the evolutionary response of coho and Chinook salmon to climate change, which is one of the least understood aspects of their biology (Naish and Hard 2008). Climate change is expected to increase ocean water temperatures and alter coastal productivity, potentially forcing the southern limit of salmon marine distributions northward (Welch et al. 1995). However, for each population there will be some critical rate of environmental change that exceeds the population's ability to adapt (Naish and Hard 2008).

First, limited interannual variation in Chinook salmon marine distribution patterns despite poor ocean conditions suggests that salmon distributions may be slow to respond to potentially deteriorating conditions, leading to poor survival rather than altered distributions (Weitkamp 2010).

Second, coho salmon populations from the Columbia River, Oregon, and California, and Chinook salmon populations from southern Oregon and California may be especially vulnerable to climate change because of their restricted southern distributions (Figure 2), should productivity for salmon in the California Current decline in the future (e.g., Schwing et al. 2006).

Third, although current marine distributions exhibited by these fish reflect a successful evolutionary response to coastal ocean conditions (Thorpe 1999), it is unclear whether these populations will be able to successfully adapt to future conditions. The existence of salmon "hotspots" noted earlier suggests that a few individuals do not follow the "typical" marine distribution pattern for their release region but migrate far from the home stream. If these fish represent the ones able to survive in future conditions, then perhaps promoting their survival by limiting fisheries in these areas is warranted.

Although CWTs are extremely primitive compared to the latest electronic tags, it is only with extensive and sustained tagging programs that patterns such as those described here can be detected. In this sense, the CWT database may serve as a critical "reality check" for other tagging programs that rely on hundreds or thousands–rather than millions–of tag deployments and recoveries. Although advances in tagging technology have revolutionized our understanding of how migratory fishes use marine habitats, the CWT database provides the necessary information to address many issues in salmon biology, but this valuable resource has largely been untapped.

Acknowledgments

The author thanks all the people who tag, collect, read, and report coded wire tags and maintain the database in an easily accessible format. Much of the tedious work conducted by Kathleen Neely in our coho salmon analysis was directly used in this analysis. This manuscript was greatly improved by constructive comments from C. Jordan, M. Ford, J. McKenzie, and one anonymous reviewer. Funding was provided by NOAA Fisheries' Northwest Fisheries Science Center.

References

Beacham, T. D., J. R. Candy, K. L. Jonsen, J. Supernault, M. Wetklo, L. Deng, K. M. Miller, and R. E. Withler. 2006. Estimation of stock composition and individual identification of Chinook salmon across the Pacific Rim by use of microsatellite variation. Transactions of the American Fisheries Society 135:861–888.

Beamish, R. J., G. A. McFarlane, and J. R. King. 2000. Fisheries climatology: understanding decadal scale processes that naturally regulate British Columbia fish populations. Pages 94–139 *in* P. J. Harrison and T. R. Parsons, editors. Fisheries oceanography: an integrative approach to fisheries ecology and management. Blackwell Scientific Publications, Oxford, U.K.

Block, B. A., H. Dewar, S. B. Blackwell Scientific Publications, T. D. Williams, E. D. Prince, C. J. Farwell, A. Boustany, S. L. Teo, A. Seitz, A. Walli, and D. Fudge. 2008. Migratory movements, depth preferences, and thermal biology of Atlantic bluefin tuna. Science 293:1310–1314.

Candy, J. R., and T. D. Beacham. 2000. Patterns of homing and straying in southern British Columbia coded-wire tagged chinook (*Oncorhynchus tshawytscha*) populations. Fisheries Research 47:41–56.

Clarke, K. R., and R. N. Gorley. 2006. PRIMER v6: User Manual/Tutorial. PRIMER-E, Plymouth, U.K.

Coronado, C., and R. Hilborn. 1998. Spatial and temporal factors affecting survival in coho and fall chinook salmon in the Pacific Northwest. Bulletin of Marine Science 62:409–425.

Daly, E., R. Brodeur, and L. Weitkamp. 2009. Ontogenetic changes in diets of juvenile and sub-adult coho and Chinook salmon in the Northern California Current. Transactions of the American Fisheries Society 138:1420–1438.

Garcia de Leaniz, C., I. A. Fleming, S. Einum, E. Verspoor, W. C. Jordan, S. Consuegra, N. Aubin-Horth, D. Lajus, B. H. Letcher, A. F. Youngson, J. H. Webb, L. A. Vøllestad, B. Villanueva, A. Ferguson, and T. P. Quinn. 2007. A critical review of adaptive genetic variation in Atlantic salmon: implications for conservation. Biological Review 82:173–211.

Groot, C., and L. Margolis, editors. 1991. Pacific salmon life histories. University British Columbia Press, Vancouver, B.C..

Guy, C. S., H. L. Blankenship, and L. A. Nielsen. 1996. Tagging and marking. Pages 353–384 *in* B. R. Murphy and D. W. Willis, editors. Fisheries techniques, 2nd Edition. American Fisheries Society, Bethesda, Maryland.

Hard, J. J., and W. R. Heard. 1999. Analysis of straying variation in Alaskan hatchery chinook salmon (*Oncorhynchus tshawytscha*) following transplantation. Canadian Journal of Fisheries and Aquatic Sciences 56:578–589.

Hobday, A. J., and G. W. Boehlert. 2001. The role of coastal ocean variation in spatial and temporal patterns in survival and size of coho salmon (*Oncorhynchus kisutch*). Canadian Journal of Fisheries and Aquatic Sciences 58:2021–2036.

Hodgson, S., T. P. Quinn, R. Hilborn, R. C. Francis, and D. E. Rogers. 2006. Marine and freshwater climatic factors affecting interannual variation in the timing of return migration to fresh water of sockeye salmon (*Oncorhynchus nerka*). Fisheries Oceanography 15:1–24.

Holtby, L. B., and K. A. Ciruna. 2007. Conservation units for Pacific salmon under the wild salmon policy. Canadian Science Advisory Secretariat, Research Document 2007/070.

Hyun, S. Y., K. W. Myers, and A. J. Talbot. 2007. Year-to-year variability in ocean recovery rate of Columbia River upriver bright fall Chinook salmon (*Oncorhynchus tshawytscha*). Fisheries Oceanography 16:350–362.

Jefferts, K. B., P. K. Bergman, and H. F. Fiscus. 1963. A coded wire identification system for macro-organisms. Nature 198:460–462.

Johnson, J. K. 2004. Regional overview of coded wire tagging of anadromous salmon and steelhead in Northwest America. Regional Mark Processing Center, Portland, OR, U.S.A., unpublished report. Available: www.rmis.org (February 2011).

Johnson, S. L. 1988. The effects of the 1983 El Niño on Oregon's coho (*Oncorhynchus kisutch*) and chinook (*O. tshawytscha*) salmon. Fisheries Research 6:105–123.

Kallio-Nyberg, I., H. Peltonen, and H. Rita. 1999. Effects of stock-specific and environmental factors on the feeding migration of Atlantic salmon (*Salmo salar*) in the Baltic Sea. Canadian Journal of Fisheries and Aquatic Sciences 56:853–861.

Leggett, W. C. 1985. The role of migration in the life history evolution of fish. Contributions in Marine Science, Supplement 27:277–295.

Magnusson, A., and R. Hilborn. 2003. Estuarine influence on survival rates of coho (*Oncorhynchus kisutch*) and chinook salmon (*Oncorhynchus tshawytscha*) released from hatcheries on the U.S. Pacific coast. Estuaries 26:1094–1103.

Morris, J. F., M. Trudel, M. E. Thiess, R. M. Sweeting, J. Fisher, S. A. Hinton, E. A. Fergusson, J. A. Orsi, E. V. Farley, and D. W. Welch. 2007. Stock-specific migrations of juvenile coho salmon derived from coded-wire tag recoveries on the Continental Shelf of Western North America. Pages 81–104 *in* C. B. Grimes, R. D. Brodeur, L. J. Haldorson, S. M. McKinnell, editors. The ecology of juvenile salmon in the Northeast Pacific Ocean. American Fisheries Society, Symposium 57, Bethesda, Maryland.

Myers, J. M., R. G. Kope, G. J. Bryant, D. Teel, L. J. Lierheimer, T. C. Wainwright, W. S. Grant, F. W. Waknitz, K. Neely, S. T. Lindley, and R. S. Waples. 1998. Status review of Chinook salmon from Washington, Idaho, Oregon, and California. NOAA Technical Memorandum NMFS-NWFSC-035, U.S. Department of Commerce, Washington, D.C.

Myers, K. W., K. Y. Aydin, R. V. Walker, S. Fowler, and M. Dahlberg. 1996. Known ocean ranges of stocks of Pacific salmon and steelhead as shown by tagging experiments, 1956–1995. Fisheries Research Institute Report FRI-UW-9614. University of Washington, Seattle, Washington.

Naish, K. A., and J. J. Hard. 2008. Bridging the gap between the genotype and the phenotype: linking genetic variation, selection and adaptation in fishes. Fish and Fisheries 9:396–422.

Neuenfeldt, S., H. H. Hinrichsen, A. Nielsen, and K. H. Andersen. 2007. Reconstructing migrations of individual cod (*Gadus morhua* L.) in the Baltic Sea by using electronic data storage tags. Fisheries Oceanography 16:526–535.

Nicholas, J. W., and D. G. Hankin. 1988. Chinook salmon populations in Oregon coastal river basins: Descriptions of life histories and assessment of recent trends in run strengths. Oregon Department of Fish and Wildlife, Fish Division Information Report 88–1, Corvallis, Oregon.

Norris, J. G., S. Y. Hyun, and J. J. Anderson. 2000. The ocean distribution of the Columbia River upriver bright fall chinook salmon stocks. North Pacific Anadromous Fish Commission Bulletin 2:221–232.

Orsi, J. A., and A. C. Wertheimer. 1995. Marine vertical distribution of juvenile chinook and coho salmon in southeastern Alaska. Transactions of the American Fisheries Society 124:159–169.

Pacific Fishery Management Council (PFMC). 2007a. Fisheries Regulation Assessment Model (FRAM) technical documentation for coho and Chinook. (Document prepared for the Council and its advisory entities.) Pacific Fishery Management Council, Portland, Oregon.

Pacific Fishery Management Council (PFMC). 2007b. Review of 2006 ocean salmon fisheries. Pacific Fishery Management Council, Portland, Oregon.

Pacific Salmon Commission (PSC). 2009. 2009 Annual report of the exploitation rate analysis and model calibration. PSC Report TC-Chinook (09)-3, Vancouver, B.C.

Quinn, T. J., and R. B. Deriso. 1999. Quantitative fish dynamics. Oxford University Press, Oxford, U.K.

Regional Mark Information System Database (RMIS). 1977. Portland (OR, U.S.A.): Regional Mark Processing Center, Pacific States Marine Fisheries Commission. Available: http://www.rmpc.org (February 2011).

Ryding, K., and J. Skalski. 1999. Multivariate regression relationships between ocean conditions and early marine survival of coho salmon (*Oncorhynchus kisutch*). Canadian Journal of Fisheries and Aquatic Sciences 56:2374–2384.

Schwing, F. B., N. A. Bond, S. J. Bograd, T. Mitchell, M. A. Alexander, and N. Mantua. 2006. Delayed coastal upwelling along the U.S. West Coast in 2005: A historical perspective. Geophysical Research Letters 33(L 22S01):doi:10.1029/2006GL026911.

Thorpe, J. E. 1999. Flexible life-history strategies: a context for understanding migration in salmonids. Bulletin of the Tohoku National Fisheries Research Institute 62:151–164.

Van Doornik, D. M., D. J. Teel, D. R. Kuligowski, C. A. Morgan, and E. Casillas. 2007. Genetic analyses provide insight into the early ocean stock distribution and survival of juvenile coho salmon off the coasts of Washington and Oregon. North American Journal of Fisheries Management 27:220–237.

Waples, R. S., D. J. Teel, J. M. Myers, and A. R. Marshall. 2004. Life-history divergence in Chinook salmon: historic contingency and parallel evolution. Evolution 58:386–403.

Weitkamp, L. A. 2010. Marine distributions of Chinook salmon from the West Coast of North America determined by coded wire tag recoveries. Transactions of the American Fisheries Society 139:147–170.

Weitkamp, L., and K. Neely. 2002. Coho salmon (*Oncorhynchus kisutch*) ocean migration patterns: insight from marine coded-wire tag recoveries. Canadian Journal of Fisheries and Aquatic Sciences 59:1100–1115.

Weitkamp, L. A., T. C. Wainwright, G. J. Bryant, G. B. Milner, D. J. Teel, R. G. Kope, and R. S. Waples. 1995. Status review of coho salmon from Washington, Oregon, and California. NOAA Technical Memorandum NMFS-NWFFSC-24, U.S. Department of Commerce, Washington, D.C.

Welch, D. W., A. Chigirinsky, and Y. Ishida. 1995. Upper thermal limits on the oceanic distribution of Pacific salmon (*Oncorhynchus* spp.) in the spring. Canadian Journal of Fisheries and Aquatic Sciences 52:489–503.

Welch, D. W., and T. R. Parsons. 1993. $d^{13}C$-$d^{14}N$ values as indicators of trophic position and competitive overlap for Pacific salmon (*Oncorhynchus* spp.). Fisheries Oceanography 2:11–23.

Zar, J. H. 1984. Biostatistical analysis, 2nd Edititon. Prentice-Hall, Englewood Cliffs, New Jersey.

Appendix

Appendix Table 1. Coho and Chinook salmon hatcheries and run times used in the analysis, grouped by freshwater release region. Provided are the total estimated number of recovered fish for each hatchery/run time and the number of marine recovery areas from which fish were recovered. Run times (when adults return to freshwater) are: spring (Spr), summer (Sum), and fall. The southern British Columbia mainland release region also includes the Lower Fraser River, Puget Sound also includes Hood Canal, and eastern Strait of Juan de Fuca, while the Oregon Coast does not include areas south of Cape Blanco. National Fish Hatchery is abbreviated as NFH.

Release region	Hatchery name	Species	Run Time	Total recoveries	No. recovery areas
Cook Inlet	Cook Inlet[1]	Chinook	Spr	976	9
Cook Inlet	Elmendorf	coho	Fall	3,206	5
Cook Inlet	Fort Richardson	coho	Fall	2,531	7
Southeast Alaska	Burnett Inlet	coho	Fall	3,487	6
Southeast Alaska	Crystal Lake	Chinook	Spr	9,841	6
Southeast Alaska	Crystal Lake	coho	Fall	3,835	5
Southeast Alaska	Deer Mountain	Chinook	Spr	3,301	7
Southeast Alaska	Deer Mountain	coho	Fall	12,401	5
Southeast Alaska	Earl West Cove	coho	Fall	12,792	7
Southeast Alaska	Gastineau	coho	Fall	28,698	5
Southeast Alaska	Hidden Falls	coho	Fall	4,165	7
Southeast Alaska	Klawock	coho	Fall	19,611	7
Southeast Alaska	Medvejie	coho	Fall	5,612	5
Southeast Alaska	Nakat Inlet	coho	Fall	17,377	7
Southeast Alaska	Neets Bay	Chinook	Spr	3,154	6
Southeast Alaska	Neets Bay	coho	Fall	27,253	7
Southeast Alaska	Port Armstrong	coho	Fall	16,760	5
Southeast Alaska	Sheldon Jackson	coho	Fall	2,563	4
Southeast Alaska	Snettisham	Chinook	Spr	2,020	3
Southeast Alaska	Snettisham	coho	Fall	8,706	5
Southeast Alaska	Tamgas Cr.	coho	Fall	40,594	14
Southeast Alaska	Whitman Lake	coho	Fall	13,971	8
Northern British Columbia	Bella Bella	coho	Fall	5,064	8
Northern British Columbia	Fort Babine	Chinook	Sum	3,130	9
Northern British Columbia	Fort Babine	coho	Fall	3,895	9
Northern British Columbia	Hartley Bay	coho	Fall	4,215	8
Northern British Columbia	Kincolith	Chinook	Spr	1,737	6
Northern British Columbia	Kincolith	coho	Fall	4,533	9
Northern British Columbia	Kispiox R.	coho	Fall	444	6
Northern British Columbia	Kitimat R.	Chinook	Spr	7,331	11
Northern British Columbia	Kitimat R.	coho	Fall	12,209	11
Northern British Columbia	Masset	Chinook	Sum	1,214	10
Northern British Columbia	Masset	coho	Fall	1,646	7
Northern British Columbia	Oweekeno	Chinook	Sum	1,380	9
Northern British Columbia	Pallant Cr.	coho	Fall	7,305	8
Northern British Columbia	Snootli Cr.	Chinook	Sum	5,787	13

Release region	Hatchery name	Species	Run Time	Total recoveries	No. recovery areas
Northern British Columbia	Snootli Cr.	coho	Fall	9,515	10
Northern British Columbia	Terrace	Chinook	Sum	3,207	11
Northern British Columbia	Toboggan Cr.	coho	Fall	2,637	6
West Coast Vancouver Island	Clayoquot	Chinook	Fall	2,452	11
West Coast Vancouver Island	Conuma R.	Chinook	Fall	15,738	12
West Coast Vancouver Island	Conuma R.	coho	Fall	16,561	10
West Coast Vancouver Island	Gold R	Chinook	Fall	2,061	11
West Coast Vancouver Island	Nitinat R.	Chinook	Fall	5,781	13
West Coast Vancouver Island	Nitinat R.	coho	Fall	5,467	11
West Coast Vancouver Island	Robertson Cr	Chinook	Fall	50,365	15
West Coast Vancouver Island	Robertson Cr	coho	Fall	29,176	12
East Coast Vancouver Island	Big Qualicum R.	Chinook	Fall	14,055	11
East Coast Vancouver Island	Big Qualicum R.	coho	Fall	43,953	12
East Coast Vancouver Island	Chemainus R.	Chinook	Fall	17,777	11
East Coast Vancouver Island	Cowichan R.	Chinook	Fall	17,477	11
East Coast Vancouver Island	Nanaimo R.	Chinook	Fall	6,906	11
East Coast Vancouver Island	Port Hardy	coho	Fall	5,473	10
East Coast Vancouver Island	Puntledge R.	Chinook	Fall	5,275	9
East Coast Vancouver Island	Puntledge R.	Chinook	Sum	4,417	8
East Coast Vancouver Island	Puntledge R.	coho	Fall	47,467	13
East Coast Vancouver Island	Quinsam R.	Chinook	Fall	11,248	13
East Coast Vancouver Island	Quinsam R.	coho	Fall	99,229	16
Southern British Columbia mainland	Capilano R.	Chinook	Fall	17,204	11
Southern British Columbia mainland	Capilano R.	coho	Fall	107,498	12
Southern British Columbia mainland	Chehalis R.	Chinook	Fall	10,250	14
Southern British Columbia mainland	Chehalis R.	coho	Fall	23,442	13
Southern British Columbia mainland	Chilliwack R.	Chinook	Fall	9,390	14
Southern British Columbia mainland	Chilliwack R.	coho	Fall	65,335	13
Southern British Columbia mainland	Inch Cr.	coho	Fall	25,350	12
Southern British Columbia mainland	Sechelt	coho	Fall	10,355	12
Southern British Columbia mainland	Sliammon R.	coho	Fall	6,326	11
Southern British Columbia mainland	Tenderfoot Cr.	Chinook	Sum	1,065	8
Southern British Columbia mainland	Tenderfoot Cr.	coho	Fall	30,943	13
Upper Fraser River	Clearwater R.	Chinook	Sum	2,509	11

Release region	Hatchery name	Species	Run Time	Total recov-eries	No. recovery areas
Upper Fraser River	Clearwater R.	coho	Fall	2,695	11
Upper Fraser River	Eagle R.	Chinook	Sum	6,812	10
Upper Fraser River	Eagle R.	coho	Fall	16,856	15
Upper Fraser River	Fraser above Hope[2]	Chinook	Spr	984	12
Upper Fraser River	Quesnel R.	Chinook	Sum	1,379	11
Upper Fraser River	Shuswap R.	Chinook	Sum	4,647	13
Upper Fraser River	Spius Cr.	Chinook	Spr	1,804	10
Upper Fraser River	Spius Cr.	coho	Fall	11,332	12
Upper Fraser River	Thompson R.	coho	Fall	13,927	13
Puget Sound	Dungeness	coho	Fall	7,794	13
Puget Sound	Elwha	Chinook	Fall	2,063	11
Puget Sound	Elwha	coho	Fall	3,227	12
Puget Sound	George Adams	coho	Fall	18,522	10
Puget Sound	Grovers Creek	Chinook	Fall	5,187	13
Puget Sound	Hoodsport	Chinook	Fall	2,313	15
Puget Sound	Issaquah	Chinook	Fall	7,474	12
Puget Sound	Kendall Cr.	Chinook	Spr	2,400	9
Puget Sound	Kendall Cr.	coho	Fall	18,603	12
Puget Sound	Marblemount	Chinook	Spr	2,518	10
Puget Sound	Marblemount	coho	Fall	11,966	12
Puget Sound	Minter Cr.	coho	Fall	20,881	12
Puget Sound	Quilcene NFH	coho	Fall	16,166	11
Puget Sound	Samish	Chinook	Fall	18,345	13
Puget Sound	Soos Cr.	Chinook	Fall	5,661	12
Puget Sound	Soos Cr.	coho	Fall	47,325	14
Puget Sound	Tulalip	coho	Fall	73,454	12
Puget Sound	Voights Cr.	coho	Fall	45,692	12
Puget Sound	Wallace R.	Chinook	Sum	4,360	10
Puget Sound	Wallace R.	coho	Fall	42,080	14
Washington Coast	Bingham Cr.	coho	Fall	5,748	14
Washington Coast	Chalaat Cr.	coho	Fall	2,264	14
Washington Coast	Educket Cr.	coho	Fall	5,742	15
Washington Coast	Forks Cr.	Chinook	Fall	4,267	14
Washington Coast	Forks Cr.	coho	Fall	13,865	17
Washington Coast	Hoko Falls	Chinook	Fall	1,760	12
Washington Coast	Humptulips	Chinook	Fall	5,778	14
Washington Coast	Humptulips	coho	Fall	13,065	14
Washington Coast	Makah NFH	coho	Fall	10,454	16
Washington Coast	Naselle	coho	Fall	10,636	16
Washington Coast	Nemah	coho	Fall	4,706	16
Washington Coast	Quinault Lk & NFH	Chinook	Fall	6,575	14
Washington Coast	Quinault Lk & NFH	coho	Fall	14,969	17

Release region	Hatchery name	Species	Run Time	Total recov-eries	No. recovery areas
Washington Coast	Sol Duc	Chinook	Spr	2,205	15
Washington Coast	Sol Duc	coho	Fall	11,512	16
Lower Columbia River	Abernathy	Chinook	Fall	4,172	14
Lower Columbia River	Big Cr.	coho	Fall	15,418	15
Lower Columbia River	Bonneville	Chinook	Fall	1,867	13
Lower Columbia River	Bonneville	coho	Fall	5,949	14
Lower Columbia River	Cowlitz Salmon	Chinook	Fall	4,791	17
Lower Columbia River	Cowlitz Salmon	Chinook	Spr	46,188	18
Lower Columbia River	Cowlitz Salmon	coho	Fall	26,264	17
Lower Columbia River	Eagle Cr NFH	coho	Fall	16,917	14
Lower Columbia River	Elochoman	coho	Fall	4,542	14
Lower Columbia River	Grays R.	coho	Fall	11,524	13
Lower Columbia River	Klaskanine	coho	Fall	9,436	15
Lower Columbia River	Klickitat	Chinook	Fall	1,352	14
Lower Columbia River	Klickitat	coho	Fall	5,172	14
Lower Columbia River	Lewis R.	coho	Fall	10,219	14
Lower Columbia River	Oxbow	coho	Fall	3,004	14
Lower Columbia River	Sandy	coho	Fall	35,425	16
Lower Columbia River	Spring Cr NFH	Chinook	Fall	24,674	14
Lower Columbia River	Toutle R.	coho	Fall	4,695	13
Lower Columbia River	Washougal	Chinook	Fall	2,311	14
Lower Columbia River	Washougal	coho	Fall	10,226	13
Lower Columbia River	Willard NFH	coho	Fall	4,202	13
Oregon Coast	Elk R	Chinook	Fall	10,671	19
Oregon Coast	Fall Cr.	coho	Fall	8,319	14
Oregon Coast	Nehalem	coho	Fall	6,006	15
Oregon Coast	Rock Cr.	Chinook	Spr	4,848	19
Oregon Coast	Rock Cr.	coho	Fall	6,937	15
Oregon Coast	Salmon R.	Chinook	Fall	5,829	16
Oregon Coast	Salmon R.	coho	Fall	2,818	14
Oregon Coast	Siletz	coho	Fall	4,254	13
Oregon Coast	Trask R.	Chinook	Fall	1,983	15
Oregon Coast	Trask R.	coho	Fall	8,057	15
Southern Oregon/northern CA	Cole Rivers	Chinook	Fall	2,646	10
Southern Oregon/northern CA	Cole Rivers	Chinook	Spr	24,590	14
Southern Oregon/northern CA	Cole Rivers	coho	Fall	6,703	10
Southern Oregon/northern CA	Iron Gate	Chinook	Fall	17,185	13
Southern Oregon/northern CA	Iron Gate	coho	Fall	9,015	11
Southern Oregon/northern CA	Trinity R.	Chinook	Fall	34,637	13
Southern Oregon/northern CA	Trinity R.	Chinook	Spr	13,219	13
Southern Oregon/northern CA	Trinity R.	coho	Fall	20,019	11

[1]Includes fish released from both Elmendorf and Ft. Richardson hatcheries.

[2]Includes fish released from Deadman River, Eagle River, Penny, and Quesnel River hatcheries.

Section III

Acoustic Tags and Arrays

Foreword to Acoustic Tags and Arrays Section

Matthew G. Mesa

Acoustic tags are small, sound-emitting devices that allow remote monitoring of fish in three dimensions. The tags are useful for collecting simple presence–absence data or can be used for detailed studies of fish behavior at a relatively small spatial scale. The tags, which have no trailing antenna, are effective in many environments, including lakes, streams, estuaries, and oceans. Unlike radio-tags, acoustic tags have a large detection range in water and can be used effectively in deep or marine environments. Acoustic tags, when coupled with sophisticated arrays of hydrophones and receivers or mobile tracking with only a single hydrophone, allow the simultaneous tracking of hundreds of individuals. Modern acoustic telemetry systems offer a long tag battery life (some greater than 3 years), many sizes of tags (down to 0.5 g), and the ability to collect data on fish temperature, depth, and other metrics.

Because of the many advantages acoustic tags may have over radio tags (for example, see http://www.htisonar.com/spec_sheets/ATvsRadioTags_2007.pdf), the use of acoustic telemetry in fisheries studies seems to be gaining in popularity and has perhaps surpassed radio telemetry (see Cooke et al., however, for relevant insight). This has certainly been true in the Columbia River basin, where studies of juvenile salmonid behavior are now dominated by use of the Juvenile Salmon Acoustic Telemetry System (JSATS; see McMichael et al. 2010) and radio telemetry seems to be a thing of the past. The call for papers for the acoustic tags and arrays session requested presentations on technology development, diversification of use, and innovative or unique approaches to using this technique for the study of fish behavior. Included in these proceedings are eight papers from a diverse, international set of authors that clearly achieved the goals of the session. The first paper by Michelle Heupel and Dale Webber provides an overview of trends in acoustic telemetry, including case studies showcasing the diversity of this technology and comments on its use in the future. Following this are two papers on the design and use of acoustic telemetry arrays, one describing a new method for estimating positions of tagged animals using the Synthetic Aperture Positioning System (SYNAPS) vessel-based active tracking method (Nielsen et al.) and another on the effectiveness of a large freshwater-estuarine array of acoustic receivers to describe the movements and behavior of two Australian catadromous fishes (Walsh et al.). The papers by Haugen et al. and Wood et al. present two case studies using acoustic telemetry describing the habitat use and migratory behavior of various fishes and the influence of environmental variables. The section concludes with two "methods" papers—both dealing with aspects of detection probabilities for acoustic telemetry studies (Melnychuk and Hobday and Pincock). We trust that this section will provide the reader with new insights and ideas on the use of acoustic telemetry for behavioral and movement studies of fish in a variety of aquatic environments.

Reference

McMichael, G. A., M. B. Eppard, T. J. Carlson, J. A. Carter, B. D. Ebberts, R. S. Brown, M. Weiland, G. R. Ploskey, R. A. Harnish, and Z. D. Deng. 2011. The juvenile salmon acoustic telemetry system: a new tool. Fisheries 35(1):9–22

American Fisheries Society Symposium 76:219–231, 2012

Trends in Acoustic Tracking: Where are the Fish Going and How Will We Follow Them?

Michelle R. Heupel*
School of Earth and Environmental Science, James Cook University Townsville, Qld, 4811, Australia

Dale M. Webber
VEMCO Ltd.
77 Chain Lake Drive, Halifax, Nova Scotia, B3S 1E1, Canada

Abstract.—Acoustic tracking of fishes has been conducted for several decades, but within the last 20 years technology has advanced greatly and many researchers have shifted to long-term (multi-year) passive monitoring studies as opposed to active tracking. The number of publications produced using this approach has varied over the last 20 years with dramatic increases in publications using passive telemetry data since the late 1990s. Passive monitoring has been used to address a wide array of topics including behaviorally focused research into home range analysis and presence patterns, physiological studies and technical uses of equipment or methodology. Studies have included over 80 different species including invertebrates, teleost and chondrichthyan fishes, marine reptiles and marine mammals. Research has been conducted in marine, estuarine, and freshwater systems with varying habitats and substrate. This diverse set of studies represents a wide array of research applications and uses of this technology. This text examines the use of acoustic technology and passive acoustic arrays to define where the technology began, describe past advances and comments on the future use of this technology in marine and aquatic ecology.

Introduction

Numerous methods have been applied to define the movements, presence and behavior patterns of fish in nature. These include tagging individuals with unique identification tags, such as passive integrating transponders (PIT), radio and acoustic tracking, and more recently, satellite and archival tagging. Each of these methods has costs and benefits. For example, mark–recapture studies can reveal general fish movement patterns over time based on release and recapture locations. However, these data do not provide information on what the individual did between these periods, no information is provided about individuals that are not recaptured, and recapture rates are usually low. Although this method is limited in the information it provides, it has the benefit of using high numbers of low cost tags and can be less labor intensive than other methods if conducted in conjunction with a directed fishery. Without the assistance

*Corresponding author: michelle.heupel@jcu.edu.au

of a fishing industry, mark–recapture studies can require substantial effort to release and recapture individuals. In contrast, acoustic transmitters are more expensive than traditional tags, however, electronic tagging approaches provide greater amounts of data and tracking resolution and have variable labor requirements. Satellite and archival tagging programs are generally less labor intensive, but the transmitters are expensive. Archival or data storage tags can store historical data, however these tags are relatively expensive and recapture rates are usually low. This text considers the application of acoustic monitoring and its evolution.

The advent of acoustic telemetry in the 1950s changed the landscape of fish ecology by allowing scientists to directly follow individual's fitted with acoustic transmitters (Trefethen et al. 1957). This allowed researchers to monitor where an individual was within a study region for as long as the receiver was in contact with the transmitter. This approach provides more detailed information on individual movement, but is typically only used in short-term studies (i.e., several hours to several days) as actively tracking individuals is labor intensive and requires near constant contact with the subject. In an effort to increase the duration of manual tracking and reduce labor some authors choose to relocate individuals over time rather than follow them continuously (e.g., Hightower et al. 2001; Sackett et al. 2007). This use of the technology can be equated to "acoustic recaptures" and mimics traditional mark–recapture data, although sampling is typically at a much greater rate and under greater control (i.e., researcher dictates sampling rate) than conventional mark–recapture studies.

In the 1980s, with the development of low-power microprocessor circuitry, telemetry suppliers developed smaller and more powerful transmitters and more sophisticated receivers. The trend continued in the 1990s with the onset of transmitters that could transmit acoustic signals with unique digital signatures and complex sensor data (i.e., depth, temperature, swim speed). These developments enabled a new approach to acoustic tracking generally referred to as passive acoustic monitoring. Passive acoustic monitoring involved the use of moored remote data-logging receivers to continuously record the presence of acoustic transmitters that transmit identification codes. Data-logging receivers have the added ability to monitor more than one individual simultaneously, an impossibility in active tracking where one focal individual is followed. Passive monitoring systems provide long-term data on multiple individuals, are less labor intensive than active tracking studies, and typically require less cooperation with outside groups (e.g., tag reporting by fishers). In the last decade the dramatic rise of the development and use of relatively low-cost receiver technology employed has expanded coverage in freshwater to river and ocean environments. These features made passive acoustic monitoring attractive to scientific users who applied this approach to a wide range of topics and species.

Two types of passive acoustic monitoring have dominated the field: 1) recording presence/absence of individuals, and 2) positioning systems. Presence/absence monitoring involves the use of independent data-logging receivers that record detection of an individual. These data can then be compiled from many receivers to provide a representation of presence and movement patterns (e.g., Simpfendorfer et al. 2002, 2008, Hedger et al. 2008). Positioning systems (e.g., Vemco Ltd: VPS and VRAP, Lotek: MAP and ALPS, and HTI systems) provide detailed, fine-scale resolution tracks of an individual's movements based on differences in signal detection time by a series of linked receivers (Sauer et al. 1997; Steig 2000; Klimley et al. 2001). Triangulation requires synchronized high resolution time data (fractions of seconds) and as such often necessitates short-term deployments, regular maintenance and/or sentinel transmitters to help ground-truth the system and account for electronic clock drift. These two monitoring approaches provide different data and are typically applied at different scales. For example, positioning systems often have strict limits on how far apart linked receivers can be deployed and thus usually necessitate a small study site, although manufacturers are working on improved systems that can cover broader areas. Independent units can be used to cover large areas but cannot produce a movement path as produced by synchronized systems.

In addition to the many advances and benefits of passive monitoring there are also limitations to what can be achieved. Receiver detection range can be affected by environmental conditions and use of this technology is limited by the ability to deploy and retrieve the equipment (Heupel et al. 2006a). This can lead to limitations in the ability to conduct research in both shallow (<1 m) and deep (>50 m) water. Additionally, research into highly mobile species can be difficult and is often limited to a restricted portion of the range of an individual. This limits the amount of data collected on an individual when compared to other technologies such as satellite tagging that can reflect the entire movement path of an individual. Thus data collected by passive monitoring may be biased due to deployment in specific locations that may or may not represent the full extent of movement range for the species in question. Interpretation of data are also dependent on understanding what the study is actually examining. For example, if few individuals remain in a specific monitored region and others leave the area which behavior most appropriately describes the species? Are those that stay atypical? Since only a portion of the population is monitored it is difficult to define which observation is more accurate. Additionally, choice of study site could bias results if it includes habitats that attract or repel individuals. These limitations should be considered in the use and analysis of passive monitoring systems.

Here we examine the application of passive acoustic monitoring over the last 20 years. This review is not intended to be exhaustive, but is representative of the type of research conducted using passive monitoring and how it has evolved through time. Examples of studies and key species will be presented along with potential future directions for this technology and field of research.

Methods

A survey of relevant literature was conducted using the Web of Science and Google Scholar. Papers were found via keyword searches with the terms: fish, telemetry, acoustic, tracking, receiver and hydrophone. Publications in refereed scientific journals or special editions and compilations were used to define the representative research areas. Articles were chosen for inclusion based on reading the abstract or full manuscript to determine if passive monitoring techniques were used. Selected publications were sorted by year of publication, publication journal, geographic region, study species and study topic. Each of these areas was used to provide a summary of passive monitoring from 1987 to the end of 2007.

Results and Discussion

A total of 154 research papers were found in 48 scientific volumes. Five of these were special publications or conference proceedings, and the remaining 43 were scientific journals. Passive monitoring results were disseminated through a variety of scientific publications reflecting the broad applicability of this approach. Six scientific journals contained 10 or more acoustic monitoring publications: Hydrobiologia (21), Journal of Fish Biology (18), Marine Biology (17), Marine Ecology Progress Series (16), Marine Technology Society Journal (10) and Environmental Biology of Fishes (10).

Due to the recent availability of this technology, publications were limited in the early years of this analysis (Figure 1a). From 1987 to 1997, 11 publications used passive monitoring (average: 1.6 papers/year) while the following period (1998–2007) saw 143 publications (average: 14.3 papers/year). The increase in publications after 1997 is likely due to reductions in receiver cost as newer models became available and competition increased. Early units were expensive (i.e., several thousand dollars) and had limited battery life. The advent of cheaper, easy to use equipment with increased battery life—and therefore deployment times—provided more options for use. Also, new transmitter models with digital coding algorithms, smaller size, more acoustic power and longer life enabled population rather than individual-focused studies. Over time, demand for acoustic monitoring products increased and the implementation of this technique expanded. Increased utilization of the technology led to greater research productivity and scientific output within the field as this approach was employed in a wide range of studies.

Geographic information was available for 139 articles. Articles that lacked geographic data were typically methods based projects that did not report a location or the studies were conducted in the open ocean and could not be attributed to a region or nation. Study regions represented 15 countries and were largely based in the USA and Canada (Figure 1b). Higher numbers of studies in the USA and Canada may be the result of greater access to the technology and the larger number of scientists compared to smaller nations. However, economic circumstances are the most likely explanation for this distribution with no publications originating from developing nations. Although several countries have greater numbers of publications than others, the use of this technology is widespread and will continue to be employed throughout the world as research efforts expand.

Passive monitoring publications included approximately 94 different study species. The most commonly studied groups were elasmobranchs, salmon, tuna and cod. The majority of passive monitoring research has focused on marine fishes (71 species). In addition to fishes, studies examined 13 species of invertebrate, one sea turtle, and one marine mammal. Freshwater studies included approximately 23 species, the majority of which were salmon. Variability in species examined in these publications is likely based on several factors. First, general research on high-value fishery species is well represented in the literature. Second is size of the individual. Although small transmitters are now available, individuals would historically need to be of sufficient size to carry an acoustic transmitter. This partially explains the higher number of studies on large species such as elasmobranchs and tuna. A third consideration is that other technologies are favored in some environments, particularly in relation to freshwater research where radio and passive integrated transponder (PIT) tags have commonly been employed (e.g., Zurstadt and Stephan 2004; Zydlewski et al. 2006; Jones and Stuart 2007; Teixeira and Cortes 2007). The availability of additional technologies means that studies on similar topics and species are conducted using other methods. Finally, publications will reflect the research programs of various individuals and priorities of funding bodies.

Over 20 different topics of study were identified with the breadth of topics addressed exemplifying the wide utility of this technique and ability to address numerous scientific questions with data gathered from passive monitoring. Synthesis of the topics of study using passive monitoring was difficult since many studies addressed more than one

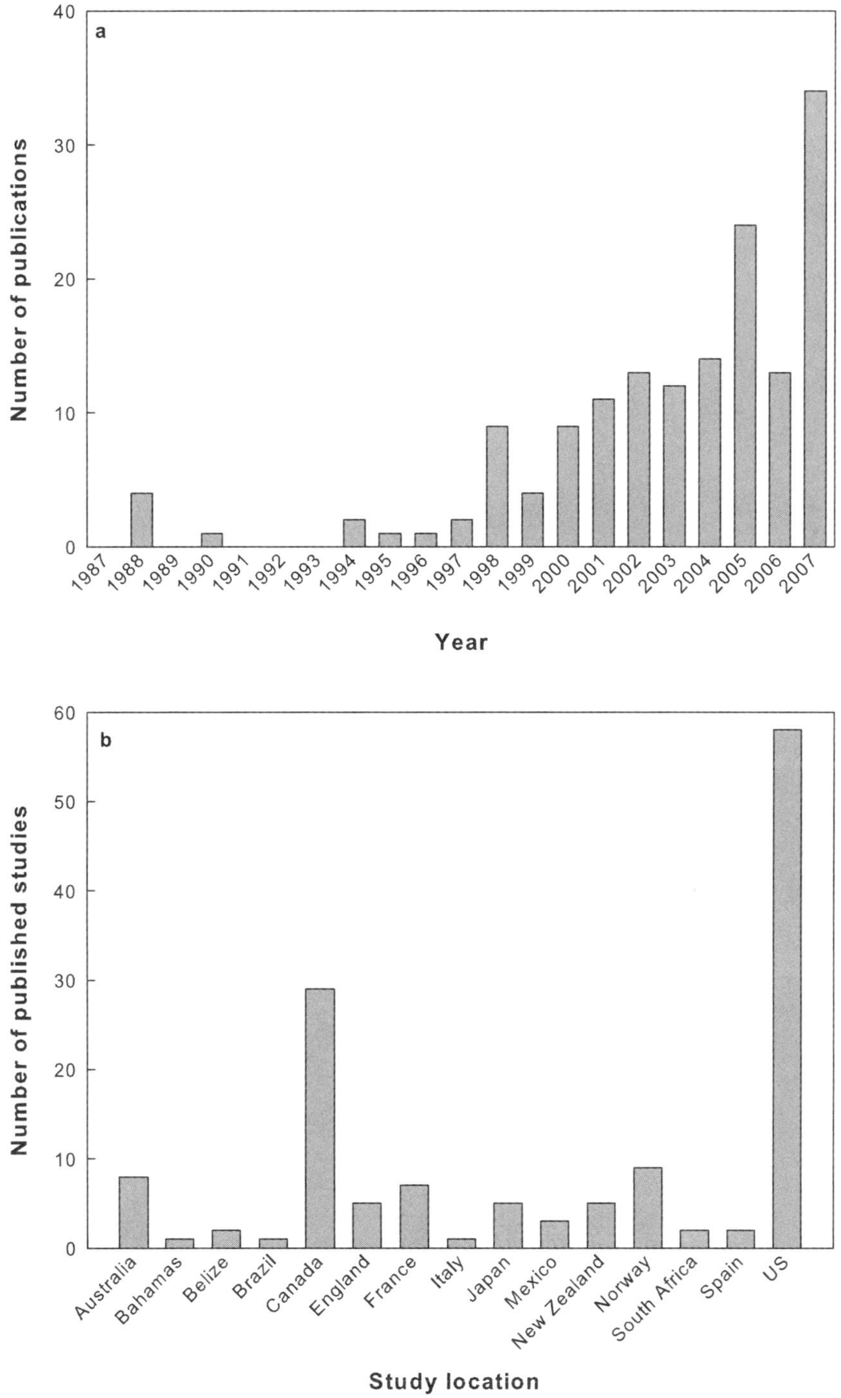

Figure 1. a) Number of passive monitoring publications in peer-reviewed scientific journals from 1987 to 2007 and; b) number of passive monitoring publications by country of origin for the period from 1987 to 2007.

aspect of animal behavior or movement. For example, a single study could examine site fidelity, habitat use and home range dynamics. The most commonly used keyword was "movement." This was a general category but typically included examination of direction of movement, rate of movement and duration of movement. Other studies encompassed movement as part of the analysis, but the type of movement examined was given a more specific term. For example, at least 22 publications related to migration. This was followed by habitat use (18), marine reserve areas (17) and home range analysis (15). Ten or more publications related to behavior, presence, movement in relation to environmental factors and mortality/survival. Other topics included: site fidelity, homing, diel and diurnal activity patterns, movement around fish aggregating devices, distribution, and seasonality of movement. Over 20 papers described methodology of acoustic monitoring systems, study design, equipment design, equipment testing, transmitter attachment, data analysis, etc.

Examples of Applications of Acoustic Monitoring Technology

The following examples highlight some applications of acoustic telemetry on commonly studied groups.

Migration

The high economic and cultural value of salmon, in conjunction with their impacted habitats and uncertain population status, make them a priority study species especially in regard to migration. Passive monitoring research of salmon populations has focused on migration within a river including movement of smolts to the sea (e.g., Russel et al. 1998; Thorstad et al. 2003; Lacroix et al. 2004, 2005) or their return as adults (e.g., Welch et al. 2003; Crossin et al. 2007). Passive monitoring has been used to examine migration movements within a river and navigation around habitats or habitat alterations such as power stations and dams (e.g., Russel et al. 1998; Steig 2000; Thorstad et al. 2003). Other studies have focused on whether hatchery-reared smolts have different migratory patterns to wild individuals to measure the utility of hatchery reared stock enhancement (Finstad et al. 2005; Thorstad et al. 2007). Several studies have been conducted, and research continues into, the survival and mortality rate of salmon smolts during their downriver migration and subsequent time at sea (Lacroix et al. 2005; Crossin et al. 2007; Thorstad et al. 2007).

Migration studies have also examined environmental conditions in relation to movement. For example, Lacroix et al. (2005) determined that salmon smolts remain aggregated as they leave a river and tend to leave on ebb tides. Individuals faced with a strong flood tide showed interruption in migration and remained present within coastal habitats. These results suggest environmental conditions may play a large role in whether salmon make it to open ocean habitats. Examination of whether freshwater releases from a power station affected individuals during their return migration, however, revealed that freshwater flow did not cause large changes in behavior, but subtle changes could not be ruled out (Thorstad et al. 2000). More recently Crossin et al. (2007) examined the biology of returning salmon to define physiological factors in migration. This project combined physiological results with acoustic monitoring data to help explain movement and better define processes driving individual movement patterns.

Thus passive monitoring has wide applicability for examining the migration patterns of species. This approach is now being used on a broader spatial scale to monitor the movements of highly migratory species such as tuna and sharks.

Home range/activity space

Analysis of home range and the size of activity space is one of the most commonly studied topics in passive monitoring research. These data are used to examine the amount of space used, duration of time present and habitat preferences while present in the study area. For example, several studies have examined the use of shallow coastal habitats by small inshore shark species (Heupel and Hueter 2002; Heupel et al. 2004, 2006b). These studies examined the movement, habitat use and home range of small coastal sharks to define key habitats and habitat use patterns. This approach has been widely used and is based on standardized analysis methods (i.e., minimum convex polygon calculation of home range). The ease of collecting and analyzing data for this use is increasing the use of passive monitoring systems to collect these data.

Use of environmental features

As open-ocean, truly-pelagic species, tuna movement patterns are difficult to define outside of traditional mark–recapture or satellite tag data. The desire to apply acoustic monitoring techniques to such mobile animals has driven researchers to focus research on specific locations to define tuna movement and behavior. For example, Klimley et al. (2003) examined the movements and presence of tuna at a seamount and found that individuals remained near the seamount for periods of several days to over a year. The authors suggest this site may act as a migratory landmark or feeding destination. In another attempt to define tuna movement patterns, Willis and Hobday (2007) examined the location of tuna in relation to upwelling events along the southern coast of Australia. Movement of individuals away from the monitored area was correlated with upwelling and concluded to be based on hunting or feeding strategies. Similarly, Chapman et al. (2005) and Wetherbee et al. (2007) used passive monitoring to examine sharks on coral reefs to define presence and movements in relation to this habitat. These studies took advantage of the fact that these mobile species were site-attached to a reef, allowing long-term acoustic monitoring studies to be conducted.

Several studies of tuna movements have examined their presence and movements in relation to fish aggregating devices (FADs) (Ohta and Kakuma 2005; Schaefer and Fuller 2005; Dagorn et al. 2007). FADs are floating or fixed platforms that attract a variety of species and are a key component of commercial fisheries around the world. Attaching an acoustic receiver to a FAD is an ideal method of monitoring the residence time of tagged fish near the FAD and associations between species. Acoustic monitoring of yellowfin and bigeye tuna in Japan revealed that individuals remained at a group of FADs for up to 55 d with presence at individual FADs lasting about seven days. Smaller individuals were present longer than larger individuals and excursions from the area were recorded throughout the monitoring period (Ohta and Kakuma 2005). This suggests smaller individuals may be more reliant on FADs. In addition, Schaefer and Fuller (2005) found that tuna were more aggregated during the day at monitored FADs and appeared to disperse more after dark when they were presumed to be feeding. Differences in depth distribution were also observed with individuals occupying greater depths at night than during the day. These data suggest that tuna are actively selecting when and how to use FADs as a food source or refuge area.

Results of this research show that acoustic monitoring can be employed to examine fast swimming pelagic species if employed at key locations. Although these data only cover a portion of an individual's movement area, they demonstrate how acoustic monitoring can provide insight into how specific regions are used and motivations for using them. These studies have extended beyond mark–recapture style data and are beginning to correlate movement patterns with behavior and environmental features. These findings will be key to understanding the ecology and life history of pelagic populations.

Based on these applications of passive monitoring it is clear these data can be used to help answer applied management and conservation questions. Data regarding use of an area may be key to defining fishery practices (i.e., time and space closures) and thus work to preserve target populations.

Habitat preference/MPAs

Research on cod populations using acoustic monitoring has examined fine-scale movements and habitat use in an attempt to define habitat preferences and determine if marine protected areas would be effective for this species (e.g., Cote et al. 1998, 2003, 2004). Studies have revealed that prior to winter migrations cod maintained small home ranges and showed preference for boulder or kelp habitats over open (sand or gravel) substrates. Passive monitoring has shown that juvenile cod probably use complex habitat as a means of predator avoidance and become more flexible in their use of habitat as they increase in age and size (Cote et al. 1998, 2003, 2004) suggesting MPAs may work for juvenile cod. Lindholm et al. (2007) also examined movement and presence of Atlantic cod revealing that up to 35% of cod remained at a deep boulder reef for up to 95 d. Based on consistent use of the habitat and site fidelity it was suggested that marine protected areas may be effective if a reef and its surrounding halo region are included.

Pecl et al. (2006) examined the efficacy of a closed area to protect spawning southern calamari *Sepioteuthis australis* from fishing pressure and permit egg laying. Passive monitoring revealed that individuals spawned over a series of months traveling 100s of kilometers in spawning regions during that period. However, although squid moved extensively during the spawning period, movements tended to be confined to a set region suggesting that closed areas may be effective at protecting this species at a vulnerable time in its life history.

Ecological role

Studies of ray species have been successfully employed to define movement patterns, presence and site fidelity (Vaudo and Lowe 2006; Collins et al. 2007). These projects examined the movements of benthic and semi-pelagic ray species to define their ecological role within specific habitats. For example, Collins et al. (2007) examined cownose ray movement from the perspective of prey exploitation to address claims that this species was responsible for the decimation of oyster reefs. This research combined with diet analysis revealed cownose rays are opportunistic feeders and are unlikely to cause large scale damage to oyster populations along the Gulf coast of Florida. Vaudo and Lowe (2006) examined the presence of round stingrays in relation to outfall from a power plant and discussed implications of coastal development on stingray presence and movement patterns. These projects related ray movement and ecology to human aspects (e.g., fishery concerns based on ray feeding and habitat alteration) as well as defining the role of these species within habitats. In this instance, passive monitoring was employed to define the role of populations within a broader ecosystem framework. This approach is often reliant on good data regarding other components of the system, but through collaboration this is often achievable.

Diel and circadian patterns

Several studies of commercially important invertebrates have examined diel patterns to help define spawning and mating activities. Examination of arrow squid *Nototodarus gouldi* movement and behavior patterns by Stark et al. (2005) revealed that individuals move in and out of coastal bays regularly suggesting a mobile and dynamic population. There was also a relationship between movement and photoperiod with individuals more active during the night than in the day, and although squid are often known to school, no schooling or group movement was observed. Similarly, Aitken et al. (2005) found cuttlefish to be more active and travel greater distances at night. Individual cuttlefish, however, were found to forage a greater proportion of the day than the night and spent more than 95% of the day resting. This project used acoustic monitoring in a complex environment and included analysis of jet pressure of individuals to provide additional data to interpret behavioral patterns. Other patterns of movement can also be defined using acoustic monitoring. Lynch and Rochette (2007) examined movement of green crabs *Carcinus maenas* to reveal circadian activity patterns. Individual activity increased during rising tides and decreased with falling tides. Without the use of passive monitoring these detailed data would not be available to define the biological patterns of these species.

Future Directions

In retrospect it is clear that acoustic telemetry technology has changed since its advent in the 1950s and has become an increasingly popular research tool in the 21st Century. Advances in the technology have seen smaller transmitters become available (<0.5 g), more sophisticated tag programming options, more advanced receivers, and complex 3D positioning systems. Systems can remain remote and send data via radio, telephone, cable or satellite links to decrease download time and travel distances for users. These technological advances have already revolutionized aquatic and marine research and many more advances are likely on the horizon.

Receiver technology is already changing with longer lasting, more robust units available from most manufacturers. Wireless, acoustic modem and satellite downloading options are becoming available making long-term, remote deployment a more realistic option. A recent survey of acoustic monitoring users revealed that researchers are interested in the ability to use higher resolution 3D telemetry systems over a larger spatial scale, want better deployment options and improved data handling packages (Heupel et al. 2005). Some of these requests are likely already under development by manufacturers and as technology improves, so will receiver performance and options. Recent advances in receiver miniaturization have led to the development of mobile receivers that can be attached to large fish and mammals. The first such development and use of the business card tag (BCT) (Klimley et al. 2009) and the recent ARX acoustic receiver (Guttridge et al. 2010) will lead to new insights into inter- and intraspecific behavior of marine and freshwater species by recording close spatial associations.

Although receiver technology is likely to continue to evolve, the greatest technological changes are likely to come in the form of transmitter options. As technology decreases in size and novel sensors are developed, they will be applied to acoustic transmitters. As transmission speed increases it will be possible to transfer larger amounts of data from a transmitter to a receiver. Transmitter options that are already being discussed include the ability to archive data, integrating pH sensors for defining feeding events, integrating 3D acceleration, and developing transmitters with the ability to communicate with other transmitters and report when they heard another transmitter. These advances, plus many more, will likely become available to the wider research community. Users have asked for advances such as the ability to monitor environmental conditions (e.g., salinity) and the ability to detect transmitters from multiple manufacturers (see Heupel et al. 2005). As transmitter technology evolves and users provide greater input to define research needs, improved transmitters will emerge. The combination of improved receiver technology and advanced transmitter technology ensures that projects considered impossible in the 20th Century are likely to become feasible options in the 21st.

The nontechnical areas of acoustic monitoring that require the greatest growth and effort in the future fall into two categories: data management and user networking. First, data management and analysis tools and techniques continue to be an issue for users. Solutions are becoming available for these areas, but the future of this aspect of acoustic monitoring is wide open with new techniques and data analysis options being explored and developed. Researchers should explore ways to fuse data from different sources (environmental and oceanographic-physiology-movement in spatial/temporal scales) to analyze population and individual behavior. Advancement in this area of acoustic monitoring research will broaden the utility and applicability of data for research and management needs in the future.

The final area of future growth in acoustic monitoring research is in user networking. This is an aspect of acoustic monitoring that must be fostered and encouraged. As large monitoring systems are deployed around the world, networks of collaborating scientists are being developed. These collaborations are resulting in exciting discoveries such as the detection of a 30–60 year old white sturgeon in the Fraser River, British Columbia, Canada more than 1000 kilometers from its tagging location in the Klamath River, California, USA (Welch et al. 2006). On a smaller scale, this type of approach is more difficult to implement, but no less critical. Movements of individual fish from one array region to another do occur, and without data sharing this highly valuable information would be lost. For example, juvenile bull sharks *Carcharhinus leucas* fitted with transmitters in the Caloosahatchee River, central Florida, were detected on receiver arrays over 100 km from the site of tagging (Heupel, unpublished data). Similarly, Hight and Lowe (2007) reported movement of a female leopard shark from Catalina Island, California, to a receiver array at Carlsbad, California over 100 km away. Without communication between the scientists who maintain these arrays, large scale movement events would have gone undetected. These examples highlight the need for users to be aware of the research being conducted around them, be open to sharing data and helping to broaden the results of the user community. This is a difficult task that will require regional or national networks of collaborating scientists and a means of effectively sharing data. Collaboration of users has many potential benefits beyond increased data gathering. Collaborative approaches may provide better leverage for funding and/or more power to leverage coastal observing systems into existing array areas. This approach has been applied by physical oceanographers for decades and is now an area that independent biologists and ecologists must consider. Collaboration may also decrease redundant effort and project costs when users are working in close proximity. Our ability to work together and share results that will benefit other users, and eventually the acoustic monitoring community as a whole, is a key area of growth for this discipline and one that needs to be explored.

The future of acoustic monitoring is certain to change and evolve beyond current approaches and technology. As technology improves it will be possible to study more remote regions and as information is gained more questions will follow. This field has proven to provide powerful data to answer scientific questions, management questions and contribute to conservation strategies. The ability of scientists to work together in the future will remove boundaries and allow larger, more integrated studies to be conducted and more detailed data to be collected. As we continue to explore our oceans, acoustic monitoring is likely to play an increasingly important role.

Summary

Acoustic monitoring technology can be applied to a diverse array of questions concerning marine and aquatic species biology and ecology. This text examined the available literature in this field of study to define the use and applicability of acoustic monitoring techniques. The broad range of species studied and environments in which this approach has been employed suggest that it will continue to have great utility in the coming years. Analysis of this topic identifies the increased use of this approach over time, likely foreshadowing an even greater amount of literature to come in the future. As technology improves and studies expand so will the scientific basis of this research technique.

References

Aitken, J. P., R. K. O'Dor and G. D. Jackson. 2005. The secret life of the giant Australian cuttlefish *Sepia apama* (Cephalopoda): Behavior and energetics in nature revealed through radio acoustic positioning and telemetry (RAPT). Journal of Experimental Marine Biology and Ecology 320:77–91.

Chapman, D., E. K. Pikitch, E. A. Babcock, and M. S. Shivji. 2005. Marine reserve design and evaluation using automated acoustic telemetry a case study involving coral reef-associated sharks in the Mesoamerican Caribbean. Marine Technology Society Journal 39:42–53.

Collins, A. B., M. R. Heupel, and P. J. Motta. 2007. Residence and movement patterns of cownose rays *Rhinoptera bonasus* within a southwest Florida estuary. Journal of Fish Biology 71:1159–1178.

Cote, D., D. A. Scruton, G. H. Niezgoda, R. S. McKinley, D. F. Rowsell, R. T. Lindstrom, L. M. N. Ollerhead, and C. J. Whitt. 1998. A coded acoustic telemetry system for high precision monitoring of fish location and movement: Application to the study of nearshore nursery habitat of juvenile Atlantic cod (*Gadus morhua*). Marine Technology Society Journal 32:54–62.

Cote, D., L. M. N. Ollerhead, D. A. Scruton, and R. S. McKinley. 2003. Microhabitat use of juvenile Atlantic cod in a coastal area of Newfoundland determined by 2D telemetry. Marine Ecology Progress Series 265:227–234.

Cote, D., S. Moulton, P. C. B. Frampton, D. A. Scruton, and S. McKinley. 2004. Habitat use and early winter movements by juvenile Atlantic cod in a coastal area of Newfoundland. Journal of Fish Biology 64:665–679.

Crossin, G. T., S. G. Hinch, S. J. Cooke, D. W. Welch, S. D. Batten, D. A. Patterson, G. Van Der Kraak, J. M. Shrimpton, and A. P. Farrell. 2007. Behavior and physiology of sockeye salmon homing through coastal waters to a natal river. Marine Biology 152:905–918.

Dagorn, L., K. N. Holland, and D. G. Itano. 2007. Behavior of yellowfin (*Thunnus albacares*) (*T. obesus*) tuna in a network of fish aggregating devices (FADs). Marine Biology 151:595–606.

Finstad, B., F. Økland, E. B. Thorstad, P. A. Bjorn, and R. S. McKinley. 2005. Migration of hatchery-reared Atlantic salmon and wild anadromous brown trout post-smolts in a Norwegian fjord system. Journal of Fish Biology 66:86–96.

Guttridge, T., L., S. H. Gruber, J. Krause and D. W. Sims. 2010. Novel acoustic technology for studying free-ranging shark social behavior by recording individuals' interactions. PLoS ONE 5(2).

Hedger, R. D., F. Martin, J. J. Dodson, D. Hatin, F. Caron, and F. G. Whoriskey. 2008. The optimized interpolation of fish positions and speeds in an array of fixed acoustic receivers. ICES Journal of Marine Science 65:1248–1259.

Heupel, M. R., and R. E. Hueter. 2002. The importance of prey density in relation to the movement patterns of juvenile sharks within a coastal nursery area. Marine and Freshwater Research 53:543–550.

Heupel, M. R., C. A. Simpfendorfer, and R. E. Hueter. 2004. Estimation of shark home ranges using passive monitoring techniques. Environmental Biology of Fishes 71:135–142.

Heupel, M. R., C. A. Simpfendorfer and C. G. Lowe. 2005. Passive acoustic telemetry technology: Current applications and future directions. Mote Marine Laboratory Technical Report No. 1066. Available: http://www.vemco.com/pdf/vr2_ws_proceedings.pdf (October 2010)

Heupel, M. R., J. M. Semmens, and A. J. Hobday. 2006a. Automated acoustic tracking of aquatic animals: scales, design and deployment of listening station arrays. Marine and Freshwater Research 57:1–13.

Heupel, M. R., C. A. Simpfendorfer, A. B. Collins, and J. P. Tyminski. 2006b. Residency and movement patterns of bonnethead sharks, Sphyrna tiburo, in a large Florida estuary Environmental Biology of Fishes 76:47–67.

Hight, B. V., and C. G. Lowe. 2007. Elevated body temperatures of adult female leopard sharks, *Triakis semifasciata*, while aggregating in shallow nearshore embayments: evidence for behavioral thermoregulation? Journal of Experimental Marine Biology and Ecology 352:114–128.

Hightower, J. E., J. R. Jackson, and K. H. Pollock. 2001. Use of telemetry methods to estimate natural and fishing mortality of striped bass in Lake Gaston, North Carolina. Transactions of the American Fisheries Society 130:557–567.

Jones, M. J., and I. G. Stuart. 2007. Movements and habitat use of common carp (*Cyprinus carpio*) and Murray cod (*Maccullochella peelii peelii*) juveniles in a large lowland Australian river. Ecology of Freshwater Fish 16:210–220.

Klimley, A. P., B. J. LeBoeuf, K. M. Cantara, J. E. Richert, S. F. Davis, and S. Van Sommeran. 2001. Radio-acoustic positioning as a tool for studying site-specific behavior of the white shark and other large marine species. Marine Biology 138:429–446.

Klimley, A. P., S. J. Jorgensen, A. Muhlia-Melo and S. C. Beavers. 2003. The occurrence of yellowfin tuna (*Thunnus albacares*) at Espiritu Santo seamount in the Gulf of California. Fishery Bulletin 101:684–692.

Klimley, A. P., C. G. Meyer, and L. C. Dagorn. 2009. Inter-animal telemetry: results from first deployment of acoustic 'business card' tags. Endangered Species Research 10:287–293.

Lacroix, G. L., P. McCurdy, and D. Knox. 2004. Migration of Atlantic salmon post-smolts in relation to habitat use in a coastal system. Transactions of the American Fisheries Society 133:1455–1471.

Lacroix, G. L., D. Knox, and M. J. W. Stokesbury. 2005. Survival and behavior of post-smolt Atlantic salmon in coastal habitat with extreme tides. Journal of Fish Biology 66:485–498.

Lindholm, J., P. J. Auster, and A. Knight. 2007. Site fidelity and movement of adult Atlantic cod *Gadus morhua* at deep boulder reefs in the western Gulf of Maine, USA. Marine Ecology Progress Series 342:239–247.

Lynch, B. R., and R. Rochette. 2007. Circatidal rhythm of free-roaming sub-tidal green crabs, Carcinus maenas, revealed by radio-acoustic positional telemetry. Crustaceana 80:345–355.

Ohta, I., and S. Kakuma. 2005. Periodic behavior and residence time of yellowfin and bigeye tuna associated with fish aggregating devices around Okinawa Islands, as identified with automated listening stations. Marine Biology 146:581–594.

Pecl, G. T., S. R. Tracey, J. M. Semmens, and G. D. Jackson. 2006. Use of acoustic telemetry for spatial management of southern calamary *Sepioteuthis australis*, a highly mobile inshore squid species. Marine Ecology Progress Series 328:1–15.

Russel, I. C., A. Moore, S. Ives, L. T. Kell, M. J. Ives, and R. O. Stonehewer. 1998. The migratory behavior of juvenile and adult salmonids in relation to an estuarine barrage. Hydrobiologia 371/ 372:321–333.

Sackett, D. K., K. W. Able, and T. M. Grothues. 2007. Dynamics of summer flounder, Paralichthys dentatus, seasonal migrations based on ultrasonic telemetry. Estuarine, Coastal and Shelf Science 74:119–130.

Sauer, W. H. H., M. J. Roberts, M. R. Lipinski, M. J. Smale, R. T. Hanlon, D. M. Webber and R. K. O'Dor. 1997. Choreography of the squid's "nuptial dance." Biological Bulletin 192:203–207.

Schaefer, K. M., and D. W. Fuller. 2005. Behavior of bigeye (*Thunnus obesus*) tunas within aggregations associated in the equatorial eastern Pacific. Marine Biology 146:781–792.

Simpfendorfer, C. A., M. R. Heupel, and R. E. Hueter. 2002. Estimation of short-term centers of activity from an array of omnidirectional hydrophones and its use in studying animal movements. Canadian Journal of Fisheries and Aquatic Sciences 59:23–32.

Simpfendorfer, C. A., M. R. Heupel, and A. B. Collins. 2008. Variation in the performance of acoustic receivers and its implication for positioning algorithms in a riverine setting. Canadian Journal of Fisheries and Aquatic Sciences 65:482–492.

Stark, K. E., G. D. Jackson, and J. M. Lyle. 2005. Tracking arrow squid movements with an automated acoustic telemetry system. Marine Ecology Progress Series 299:167–177.

Steig, T. W. 2000. The use of acoustic tags to monitor movements of juvenile salmonids approaching a dam on the Columbia River. Biotelemetry 15:296–304.

Teixeira, A., and R. M. V. Cortes. 2007. PIT telemetry as a method to study the habitat requirements of fish populations: application to native and stocked trout movements. Hydrobiologia 582:171–185.

Thorstad, E. B., F. Økland, D. Rowsell, and R. S. McKinley. 2000. A system for automatic recording of fish tagged with coded acoustic transmitters. Fisheries Management and Ecology 7:281–294.

Thorstad, E. B., F. Økland, B. O. Johnsen and T. F. Næsje. 2003. Return migration of adult Atlantic salmon, *Salmo salar*, in relation to water diverted through a power station. Fisheries Management and Ecology 10:13–22.

Thorstad, E. B., F. Økland, B. Finstad, R. Sivertsgård, N. Plantalech, P. A. Bjørn and S. McKinley. 2007. Fjord migration and survival of wild and hatchery-reared Atlantic salmon and wild brown trout post-smolts. Hydrobiologia 582:99–107.

Trefethen, P. S., J. W. Dudley, and M. R. Smith. 1957. Ultrasonic tracer follows tagged fish. Electronics 30:156–160.

Vaudo, J. J., and C. G. Lowe. 2006. Movement patterns of the round stingray *Urobatis halleri* (Cooper) near a thermal outfall. Journal of Fish Biology 68:1756–1766.

Welch, D. W., G. W. Boehlert, and B. R. Ward. 2003. POST—the Pacific Ocean salmon tracking project. Oceanologica Acta 25:243–253.

Welch, D. W., S. Turo, and S. D. Batten. 2006. Large-scale marine and freshwater movements of white sturgeon (*Acipenser transmontanus*). Transactions of the American Fisheries Society 135:140–143.

Wetherbee, B. M., S. H. Gruber, and R. S. Rosa. 2007. Movement patterns of juvenile lemon sharks *Negaprion brevirostris* within Atol das Rocas, Brazil: a nursery characterized by tidal extremes. Marine Ecology Progress Series 343:283–293.

Willis, J., and A. J. Hobday. 2007. Influence of upwelling on movement of southern bluefin tuna (*Thunnus maccoyii*) in the Great Australian Bight. Marine and Freshwater Research 58:699–708.

Zurstadt, C. F., and K. Stephan. 2004. Seasonal migration of westslope cutthroat trout in the middle fork salmon river drainage, Idaho. Northwest Science 78:278–285.

Zydlewski, G. B., G. Horton, T. Dubreuil, B. Letcher, S. Casey, and J. Zydlewski. 2006. Remote monitoring of fish in small streams: a unified approach using PIT tags. Fisheries 31:492–502.

American Fisheries Society Symposium 76:233–250, 2012

Mobile Positioning of Tagged Aquatic Animals Using Acoustic Telemetry with a Synthetic Hydrophone Array (SYNAPS: Synthetic Aperture Positioning System)

Julie K. Nielsen*
University of Alaska Fairbanks, Juneau Center for Fisheries and Ocean Sciences
17101 Point Lena Loop Road, Juneau, Alaska 99801 USA
and
U.S.Geological Survey Alaska Science Center
3100 National Park Drive, Juneau, Alaska 99801 USA

George H. Niezgoda
Lotek Wireless, Inc., 115 Pony Drive, Newmarket, Ontario L3Y 7B5 Canada

S. James Taggart
U.S.Geological Survey Alaska Science Center
3100 National Park Drive, Juneau, Alaska 99801 USA

Steven J. Cooke
Fish Ecology and Conservation Physiology Laboratory, Department of Biology, Carleton University, 1125 Colonel By Drive, Ottawa, Ontario K1S 5B6 Canada

Peter Anson
Lotek Wireless, Inc.
115 Pony Drive, Newmarket, Ontario L3Y 7B5 Canada

Caleb T. Hasler and Kyle C. Hanson
Fish Ecology and Conservation Physiology Laboratory, Department of Biology, Carleton University, 1125 Colonel By Drive, Ottawa, Ontario K1S 5B6 Canada

Gord Carl
Lotek Wireless, Inc.
115 Pony Drive, Newmarket, Ontario L3Y 7B5 Canada

*Corresponding author: jknielsen@alaska.edu

Abstract.—We present a new acoustic telemetry method for efficiently estimating positions of tagged marine and freshwater animals using vessel-based active tracking. Vessel-based tracking can require considerable time and effort, which limits both study area size and the number of tagged animals possible for telemetry studies. However, the recently-developed SYNAPS (Synthetic Aperture Positioning System) tracking method allows efficient collection of fine-scale movement information for many tagged animals within large study areas, and therefore enables large surveys to be conducted in a methodical and cost-effective manner. SYNAPS computes position estimates of tagged animals by means of hyperbolic positioning using the geographic location of a mobile hydrophone synchronized with signal detections to create a synthetic hydrophone array. Here we describe the process of tracking with SYNAPS, quantify accuracy and precision of position estimations, and provide guidelines for tracking procedures. SYNAPS position estimates were compared with known positions of fixed tags in both marine (Alaska, USA) and freshwater lake (Ontario, Canada) systems. Accuracy of position estimates ranged from 1.2 m using hull-mounted hydrophones and survey-quality GPS equipment to 23.4 m using towed hydrophones and a navigation-grade GPS receiver. This new tool will facilitate spatially explicit management applications such as aquatic protected area design and essential fish habitat designation by increasing the ability of acoustic telemetry to characterize movement of marine animals at different scales.

Introduction

Understanding the geographic distribution and movement of animals is important for effective management and conservation of ecological systems. However, understanding the spatial ecology of aquatic animals is difficult because tools used for their direct observation (e.g., submarines, remotely operated vehicles, camera sleds, animal-borne imagery, or scuba) are often expensive, labor intensive, and constrained to a very limited field of view. Movement of aquatic animals such as fish and invertebrates can be inferred through surveys that determine population distribution for multiple size classes in different seasons, but to understand the mechanisms that underlie their movement, it is usually necessary to determine the movement of individual animals. Traditional methods of studying individual movement generally rely on recovering tagged animals (Lucas and Baras 2000). However, that method provides location information for only two points in time, release and re-capture, and can only be used on species that are readily re-captured (e.g., through scientific sampling or fisheries).

Technological advances in recent decades have provided researchers with the ability to use electronic tags to study movement of individually tagged animals (Sibert and Nielsen 2001). Acoustic telemetry, which employs tags that actively transmit an acoustic signal that can be received by moored or mobile hydrophones, is frequently used to study the spatial ecology of fish and other aquatic organisms in freshwater and marine systems. Receiver arrays can be deployed as checkpoints or curtains and document the movement of animals as they pass these reception zones at scales ranging from hundreds of meters to kilometers (Thorstad et al. 2000; Welch et al. 2002; Domeier 2005). However, some types of hydrophone arrays with overlapping detection ranges can provide continuous, fine-scale position estimates of tagged animals within a small study area, typically less than a square kilometer (Niezgoda et al. 2002; Cooke et al. 2005; Rigby et al. 2005; Jorgensen et al. 2006; Hanson et al. 2007; Jørgensen et al. 2007). As long as the tagged animal moves inside the confines bounded by the fixed array footprint, continuous two or three-dimensional estimates of movements can be made based on the time it takes the signal from the tagged animal to reach each hydrophone within the reception range. Fixed arrays with the capacity of positioning fish within the scale of several meters work well for animals that are relatively stationary, have home ranges that are contained within the reception range of multiple array hydrophones, or are physically constrained within a small area (e.g., lakes).

Mobile tracking methods can be used in large, open systems where tagged animals can easily move outside the boundaries of a fixed array's reception range. However, the significant effort necessary to obtain position estimates prohibits large study areas and/or large numbers of tagged animals from being monitored. During mobile tracking, hydrophones are towed, affixed to a tracking vessel, or periodically dropped overboard to monitor signals from a single tagged animal. The position of the tagged animal can be estimated using directional bearings to maneuver the tracking vessel to the location that produces the greatest signal power detections (Stone et al. 1992; Holland et al. 1996; Clabough et al. 2007; Ng et al. 2007), by means of triangulating two or more bearing estimates (Collazo and Epperly 1995), or by simply inferring position from vessel location at the time of signal detection (Hoolihan 2005; Taggart et al. 2008). Regardless of the method employed, the time necessary to obtain a position estimate increases in proportion to the desired estimate precision. Moreover, since mobile tracking typically relies heavily on the operator, subjectivity and human fatigue can impact data quality. The need to focus on one animal at a time, the time necessary to achieve accurate position estimates, and the endurance capacity of human operators create constraints that often limit the size and scope of a study.

Here we present a new telemetry method, SYNAPS (Synthetic Aperture Positioning System), which offers an efficient and automated approach to obtaining accurate position estimates for many tagged animals in large study areas using mobile tracking. We describe the SYNAPS telemetry method and present results from controlled field experiments in marine (Alaska, USA) and freshwater lake (Ontario, Canada) systems that demonstrate its performance. In addition, we provide advice for customizing equipment configurations and vessel maneuvers to achieve different research objectives.

Background: SYNAPS Theory and Application

SYNAPS is a postprocessing method for calculating position estimates for an underwater tag using a signal detection record and the GPS track of the vessel moving in the vicinity of the tag. SYNAPS position estimation occurs in two steps (detailed below). First, a synthetic hydrophone array, or aperture, is created using the detection record and GPS locations of the boat. This synthetic array is conceptually analogous to a fixed hydrophone array that can produce very precise position estimates for a tagged animal within its boundaries. Second, hyperbolic positioning, a common method for calculating geographic positions based on signal time difference of arrival at different locations, is utilized to construct an error surface for estimating the position of the tag. Hyperbolic positioning is also used to obtain position solutions for fixed hydrophone arrays (Niezgoda et al. 2002). By incorporating depth information from pressure-sensing tags, a complete three-dimensional position solution for tagged animals can be formed. Metrics for assessing the quality of position estimates are based on the spatial arrangement of the hydrophones in the synthetic array as well as characteristics of the resulting error surface.

The method is based on the MAP telemetry system (Lotek Wireless Inc., Newmarket, Ontario), which uses CDMA signal technology (code division multiple access, also commonly used for cell phone signal transmission). The MAP telemetry system allows increased tracking efficiency through 1) automatic tag detection and decoding by a receiver, 2) reception of transmissions from all tags on the same frequency, and 3) the ability to receive simultaneous transmissions from multiple tags without signal collision detection losses (Niezgoda et al. 2002; Cooke et al. 2005). The MAP telemetry system also provides a constant, precise tag burst rate and a high resolution tag detection timestamp on the receiver, which are both necessary for estimating tag positions using SYNAPS.

Creating the synthetic array

The first step in SYNAPS processing consists of creating a synthetic spatial array of hydrophones from a temporal sequence of hydrophone detections. Individual signal detections are linked to the geographical location of the hydrophone when the signal was detected using the detection timestamp and the vessel GPS track (Figure 1). The time corresponding to the precise burst interval is subtracted from consecutive timestamps in the detection record. After the portion of the time difference that is caused by the tag burst interval is removed from each time stamp, only time-of-arrival (TOA) information, which is a function of the distance between the tag and hydrophone, remains. In

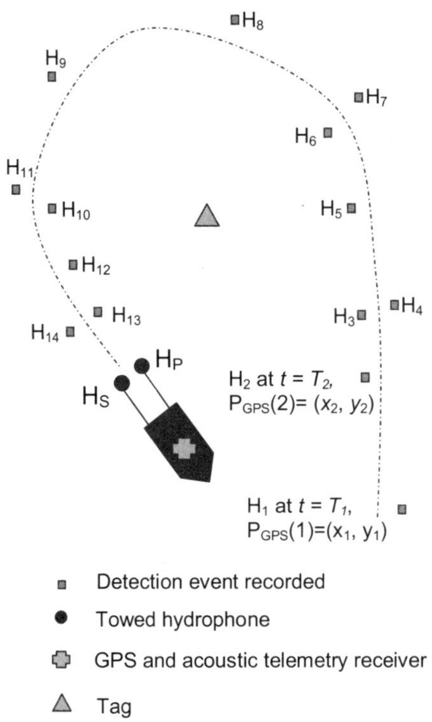

FIGURE 1. Synthetic array consisting of 14 hydrophone detections created by a vessel towing port and starboard hydrophones (HP and HS) in the vicinity of a stationary test tag.

this manner, the temporal collection of detection events by one or two hydrophones is converted into a spatial array of detection events at the same point in time (analogous to many hydrophones), where each location on the synthetic array contains signal TOA information.

Hydrophone positions used to create the synthetic array are derived from the GPS location of the vessel. The SYNAPS program interface provides options for calculating hydrophone positions for both vessel-mounted and towed hydrophone configurations. Vessel-mounted hydrophone positions are calculated using X and Y offsets from the GPS antenna. To estimate the geographical location of towed hydrophones, SYNAPS employs an inertial model that uses the vessel GPS position, the depth of the hydrophone, the horizontal GPS offset, and the distance at which the hydrophone is towed behind the vessel.

Calculating tag position estimates

The second step in SYNAPS processing consists of generating an error surface based on the TOA measurements at each hydrophone on the synthetic array. TOA measurements cannot be directly used to compute fish positions because the tag and receiver do not share a common clock. Instead, we employ the time-difference-of-arrival (TDOA) between pairs of hydrophones in the array in combination with the principle of hyperbolic positioning to compute tagged animal locations (Oppermann et al. 2004). TDOA values, or the net difference in time required for the signal to reach each hydrophone in the pair, are converted to distance values (range-distance-of-arrival, RDOA) using the speed of sound in water. For each pair of hydrophones in the array, a hyperbolic curve describes the possible locations of the tag that would produce observed RDOA values (Figure 2). The common intersection point of hyperbolic curves from all hydrophone pairs in the array corresponds to the estimated position of the tag.

Although there are a number of approaches that can be taken in deriving a solution to the hyperbolic positioning (Oppermann et al. 2004), SYNAPS utilizes a set of equi-spaced grid points representing hypothetical tagged animal locations. For each hypothesized tag location, expected RDOA values are calculated based on distances between the hypothesized location and synthetic array hydrophones. The mean-squared-error (MSE) between observed and expected RDOA values at each SYNAPS grid point is then used to transform the grid into a two-dimensional error surface. The grid point with the minimum MSE value is selected as the estimated tag position. The distance between grid points can be changed during processing in SYNAPS to support specific spatial resolution requirements for different applications.

Quantification of position estimation accuracy and precision

SYNAPS processing also provides two major metrics that describe the quality of the resulting position estimates. First, the dilution-of-precision metric (DOP) that is used to describe the precision of hyperbolic positioning estimates in fixed arrays (Niezgoda et al. 2002) can be applied to position estimates obtained using synthetic arrays from SYNAPS. The DOP is a function of hydrophone array geometry and fundamental time measurement resolution of the receiver (Figure 2). The second metric, the "residual," is based on characteristics of the error surface (e.g., shape and size of the minimum error zone) that reflect the stability of the position estimate.

SYNAPS application

The SYNAPS position estimating process is based on the assumption that an ini-

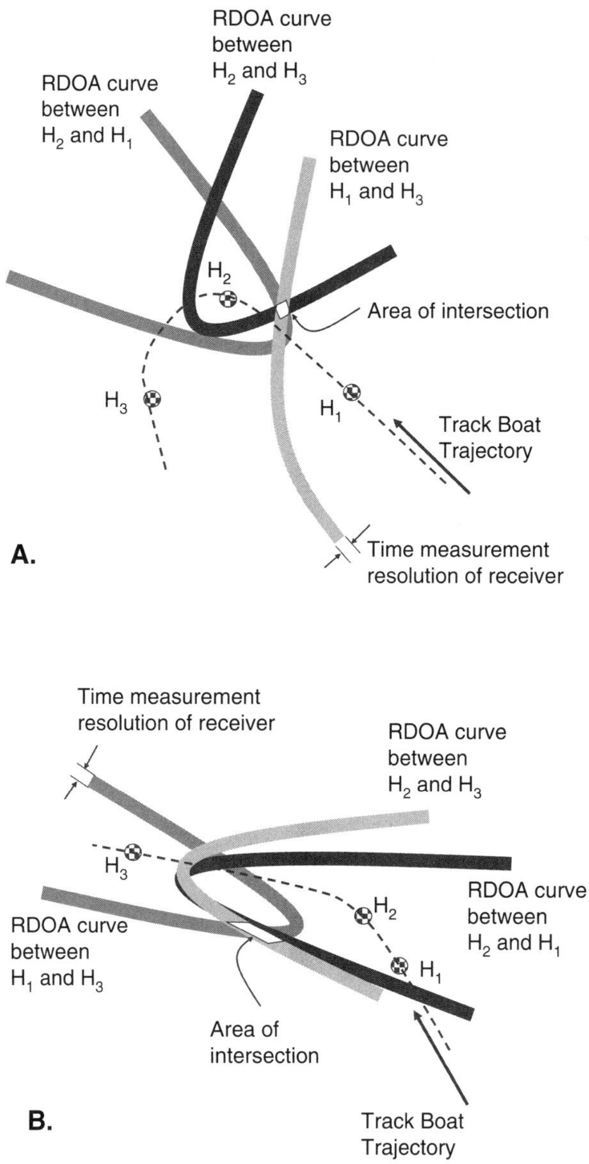

Figure 2. Hyperbolic positioning for a synthetic array with three hydrophones. Intersection of RDOA (range difference of arrival) curves indicates the position of the tag. The width of the RDOA lines is determined by the time resolution capability of the receiver. DOP (dilution of precision) is the size of the area where RDOA curves intersect; Figure **A** has higher precision (smaller intersection area) than Figure **B**.

tial error surface (derived from a minimum number of hydrophones) can evolve to provide a unique and numerically stable position estimate by adding a sufficient number of hydrophones to the synthetic array (Figure 3). SYNAPS uses threshold values for the DOP and the residual to determine whether a stable position estimate has been obtained from a given synthetic array.

In practice, the SYNAPS program lets

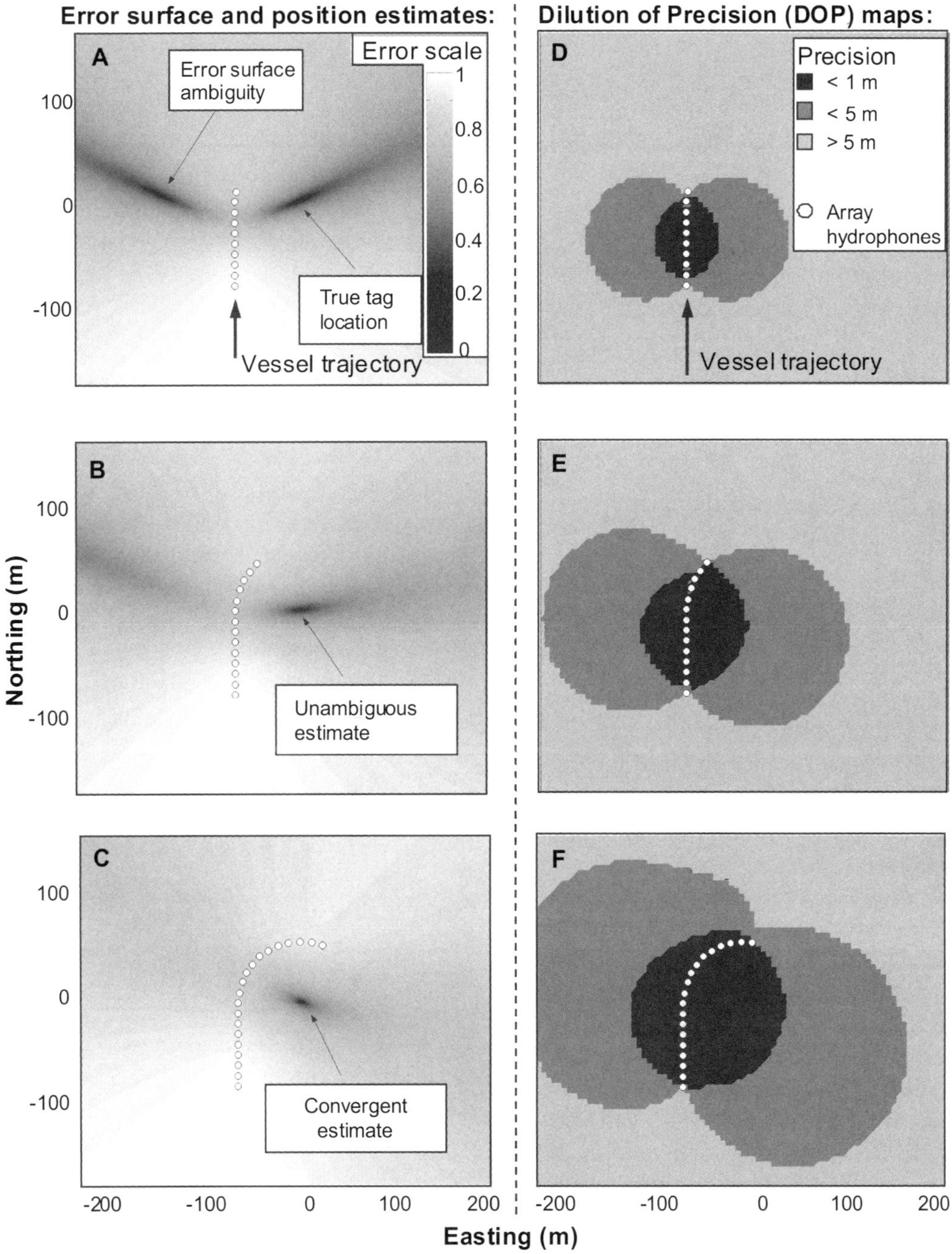

FIGURE 3. Interpolated error surfaces (**A–C**) and DOP (dilution of precision) maps (**D–F**) for a theoretical vessel maneuver around a tag located at (0,0) demonstrate a stable position solution following addition of subsequent hydrophones to a synthetic array. A straight-line trajectory for the first 10 hydrophones in the detection record produces an ambiguous error surface with two mirrored minimum error locations (**A**), despite a precision level of less than 5 m in the vicinity of the tag, as shown by the DOP map (**D**). Changing the vessel trajectory and increasing the number of hydrophones in the synthetic array to 20 eliminates positional ambiguity (**B**) and improves precision performance in the area of the tag (**E**). Optimal array geometry yields a stable, "convergent" solution (**C**) and sub-meter precision in the area around the tag (**F**).

the user specify values for processing parameters, merges a telemetry detection record with a GPS file, calculates positions, and generates a text file containing position estimates. The program begins processing a tag detection record by selecting a user-specified minimum number of hydrophones in an array and using those hydrophones to calculate a position estimate and estimate precision. If neither threshold for position quality is met, the program adds the next hydrophone in the detection record to the synthetic array and re-calculates the position. Processing continues in this manner until either a threshold value (for either metric) is achieved or the user-specified maximum number of hydrophones in an array is reached. The program then selects the next minimum number of hydrophones from the detection record (no hydrophones from the first synthetic array are included in the second) and begins calculating using hydrophones from the second synthetic array in the detection record. This process continues until the end of the detection record is reached.

Repeating SYNAPS processing of a detection record using different input values and averaging the resulting position estimates can result in a more robust position estimate, as altering certain processing parameter values can create synthetic arrays with different numbers and spatial arrangement of hydrophones. The key input parameters that can be varied during processing to provide different combinations of hydrophones in synthetic arrays for a given detection record are: 1) minimum and maximum number of hydrophones in a synthetic array, 2) specification of which hydrophones are to be used in the calculation (port only, starboard only, or both), and 3) maximum allowable time gap in GPS or telemetry record set.

Other input parameters that typically are not changed during processing for the same detection record include SYNAPS grid size, sound speed in water, depth of the tag (obtained from pressure-sensing tags), tag burst interval, hydrophone position offsets from GPS locations, and the inertial coefficient used to estimate the location of towed hydrophones. SYNAPS grid spacing can be adjusted to accommodate different levels of precision in hydrophone positions, estimate precision level requirements versus processing time, and estimated movement rate of the tagged animal.

Methods: Performance Assessment

In order to assess SYNAPS performance, we conducted field tests with stationary test tags using different equipment configurations (i.e., hydrophones and GPS) and study area locations. Fieldwork was conducted in a freshwater lake in Ontario, Canada, a glacial fjord in southeastern Alaska, and a sheltered marine harbor in southeastern Alaska.

Field testing methods

Warner Lake, Ontario.—In August 2007 we conducted a controlled experiment in Warner Lake, a freshwater lake in eastern Ontario contained within the boundaries of the Queen's University Biological Station (QUBS; 44°31'N, 76°22'W). The primary objective of the Warner Lake test was to assess the upper limits of SYNAPS accuracy and precision performance using differential GPS and hull mounted hydrophones. The experiment was conducted using a 4 m aluminum hulled boat equipped with a DGPS unit (Trimble TSC; Trimble, Sunnyvale, California) with 0.5 m precision, a two-port telemetry receiver (MAP600 RT, Lotek Wireless Inc., Newmarket, Ontario) and 2 hydrophones mounted in the bow. Rigid PVC piping was used to offset the hydrophones at a fixed distance from the gunwales. Vertical sections of PVC piping were also used to maintain the hydrophones at a constant depth of 1 m below the keel of the boat. A coded acoustic tag (Lotek MAP11_4, 5 s burst interval) was moored in 2 m of water and a survey of the location made using the DGPS with its antenna placed directly over the moored location (UTM 390264 m ± 0.5 m, 4931530 m ± 0.5 m). GPS records were collected with the tracking boat moving at an average speed of 1 m per second (m/s) along a circular trajectory around the moored tag. Overall, approximately 180 detection records were collected per hydrophone over a 15-min detection period. Telemetry and GPS data were postprocessed to produce a set of convergent position solutions using a 1 m SYNAPS grid setting. Accuracy was determined by calculating differences between the actual and estimated locations on the x-axis, the y-axis, and total (net) distance.

Glacier Bay, Alaska.—From December 2005 through May 2006, test tags were deployed as part of a telemetry study in Glacier Bay, Alaska (58°50'N, 136°06'W) to characterize seasonal movement of red king crabs and adult female Tanner crabs (S. James Taggart, Julie K. Nielsen, and Thomas C. Shirley, Texas A&M University Harte Research Institute, unpublished data). This study required eight tracking trips throughout the winter and spring of 2006. Prior to conducting tracking during each trip, a test tag was deployed in the study area to ensure that the telemetry equipment was functioning properly. We compared test tag positions estimated using SYNAPS to actual test tag locations to measure the accuracy and precision of SYNAPS using towed hydrophones and a 3 m resolution GPS.

Test tags (Lotek MAP16_50, 5 s burst interval) were attached to buoy line 2 m above an anchor and deployed from the vessel. The vessel position when the anchor was dropped was recorded with a WAAS-enabled Garmin GPS 76 (Garmin International Inc., Olathe, Kansas). Slight error in the true locations of test tags occurred as a function of the 3 m precision of the GPS used to record vessel location at test tag deployment combined with a possible slight drift of the test tag anchor in tidal currents before it reached the bottom (test tag depths ranged from 78 to 147 m).

Tracking was conducted using a 17 m research vessel, the *R/V Alaskan Gyre*. Two omni-directional hydrophones (LHP-1, Lotek Wireless, Inc., Newmarket, Ontario), one on each side of the vessel, were towed at an average depth of 20 m using depressor v-fins; hydrophones were located approximately 40 m aft of the vessel while under tow at 2 m/s. A WAAS-enabled Garmin 76 GPS with 3 m resolution was used to record vessel position every second. Telemetry data were recorded using a two-port telemetry receiver (Lotek MAP600 RT).

To account for different vessel trajectories, and thus large variation in the spatial arrangements of hydrophones in synthetic arrays, between test tag maneuvers on different tracking trips, two different synthetic array size ranges were used in SYNAPS processing. For each tag, first a minimum of 15 and a maximum of 20 hydrophones were used, followed by a minimum of 20 and a maximum of 30 hydrophones. Data were processed in SYNAPS using a grid size of 10 m, which was chosen as a conservative estimate of the error that results from using vessel GPS positions with 3 m resolution to estimate towed hydrophone positions using the inertial model. Solutions were calculated using arrays comprised of 1) both hydrophones, 2) port only, and 3) starboard only. Accuracy was determined by calculating differences between the actual and estimated locations on the *x*-axis, the *y*-axis, and total (net) distance for position estimates resulting from all different processing parameter values

Taku Harbor, Alaska.—Because obtaining a precise independently surveyed location for the test tags deployed in Glacier Bay was difficult, additional research was necessary to provide a controlled assessment of SYNAPS accuracy and precision in a marine environment. Therefore, in June 2007 we conducted additional field research to determine the upper limits of SYNAPS performance using a WAAS-enabled Garmin 76 GPS with 3 m resolution. Research was conducted in Taku Harbor (58°04'N, 134°01'W), located 35 km southeast of Juneau, Alaska.

During these field experiments we strove to reduce possible error sources that could hinder unbiased measurement of accuracy and precision. We chose a sheltered study site that would provide the ability to moor test tags at a precisely known position and constant depth. We reduced error in hydrophone positioning by deploying a hydrophone over the side of the vessel at a shallow depth instead of towing it behind the vessel at greater depths, as described above for the Glacier Bay test tag experiments. Finally, we eliminated adverse effects of poor spatial arrangement of hydrophones in the virtual arrays by maneuvering the vessel in a circular trajectory around the test tags.

Three test tags (MAP16_332, 5 s burst interval) were deployed at a depth of 8 m from the center of a 45-m floating dock that was secured to pilings. The dock was not connected to shore, so it could be circled by the tracking vessel. Positions of the test tags were obtained by averaging position estimates from a WAAS-enabled Garmin 76 for 10 min.

Tracking was conducted using the 13.4 m *S/V Ijsselmeer*. An omni-directional hydrophone was deployed on the port side of the tracking vessel at a depth of 2 m using a 13.6 kg hydrodynamic sounding weight. A WAAS-enabled GPS with 3 m resolution was used to record vessel position every second. The GPS antenna was attached to the cleat directly above the hydrophone suspension point. At tracking speed (1.3 m/s), the hydrophone was 0.5–0.75 m aft of the GPS antenna and was very stable in the water column. Telemetry data were recorded using a two-port telemetry receiver (MAP600 RT, Lotek Wireless Inc. Newmarket, Ontario). The vessel circled the tags in a 100 m radius. The average depth in the study area was 30 m.

Detection records for the three test tags contained 1,363, 1,212, and 1,534 signal detections. Telemetry and GPS data were post-processed to produce a set of convergent position solutions using a minimum of 20 and a maximum of 30 hydrophones for 1 m, 5 m, and 10 m SYNAPS grids.

Results: Performance Assessment

Field testing

Warner Lake, Ontario.—Results from field verification experiments in Warner Lake indicate that SYNAPS is capable of providing sub-meter position estimates, equivalent to those obtained using a differential GPS, if hydrophone positions are known precisely. The mean error in SYNAPS position estimates was 1.0 m (Table 1; Figure 4). The mean precision of the estimates (DOP ± 0.8 m) was similar to the precision of the DGPS used to obtain the control location of the test tag (Table 1).

The DOP values that qualify position solution precision for these estimates approach the maximum time resolution of the receiver (i.e., 1/4800 s, or approximately 0.6 m when scaled by a sound speed of 1,480 m/s). This level of precision is a direct result of using a circular boat trajectory to create an optimal arrangement of hydrophones in synthetic array geometries; it demonstrates the importance of boat maneuvers for SYNAPS performance.

Glacier Bay, Alaska.—Using towed hydrophones and a 3 m precision GPS, we were

TABLE 1. Accuracy and precision of SYNAPS test tag comparisons from Warner Lake, Ontario for A) hydrophone 1 only, B) hydrophone 2 only, and C) both hydrophones (see Figure 4).

Synthetic Array	Time (s)	X error (m)	Y error (m)	Total error (m)	DOP* (±m)	Number Hydrophones
A) Hydrophone 1 only						
1	135	2	–1	2.2	1.5	20
2	165	0	0	0.0	0.6	20
3	146	1	1	1.4	0.7	20
4	152	1	1	1.4	0.6	20
5	106	1	1	1.4	1.0	20
B) Hydrophone 2 only						
1	150	0	–1	1.0	0.6	20
2	175	0	0	0.0	0.6	20
3	126	2	0	2.0	0.6	20
4	186	1	0	1.0	0.6	20
5	166	1	–1	1.4	1.0	20
C) Hydrophones 1 and 2						
1	125	1	–2	2.2	1.3	32
2	90	1	0	1.0	0.6	20
3	80	0	0	0.0	0.7	21
4	80	1	1	1.4	0.7	26
5	80	–1	–1	1.4	0.9	29
Mean	130.8	0.7	–0.1	1.2		
(± SE)	(9.4)	(0.2)	(0.2)	(0.2)	0.8	21.9

*Dilution of precision

able to achieve better than 20 m accuracy for moored test tag position estimations. Test tags deployed in different locations throughout the study area and at different times throughout the research project had similar error in position estimation. Mean distance between estimated and actual positions for each test tag ranged from 10.4 to 23.4 m and averaged 16.7 m (Table 2).

Vessel maneuvers around the test tag deployed on March 3, 2006 (Figure 5), at a depth of 85 m, provided the lowest total error of all test tags (Table 2). This is an example of precision produced with optimal arrangement of hydrophones resulting from a circular tracking vessel trajectory. Position estimates differed from known locations by an average of –6.2 m on the x-axis and –2.4 m on the y-axis (Table 3).

Taku Harbor, Alaska.—Results indicate that the upper limits of SYNAPS accuracy and precision are roughly equivalent to the resolution of the GPS used to obtain positions of hydrophones. Using a GPS with 3 m precision to determine hydrophone locations, mean distance between estimated and actual positions for each test tag ranged from 3.9 to 4.4 m using a 1 m SYNAPS grid (Table 4).

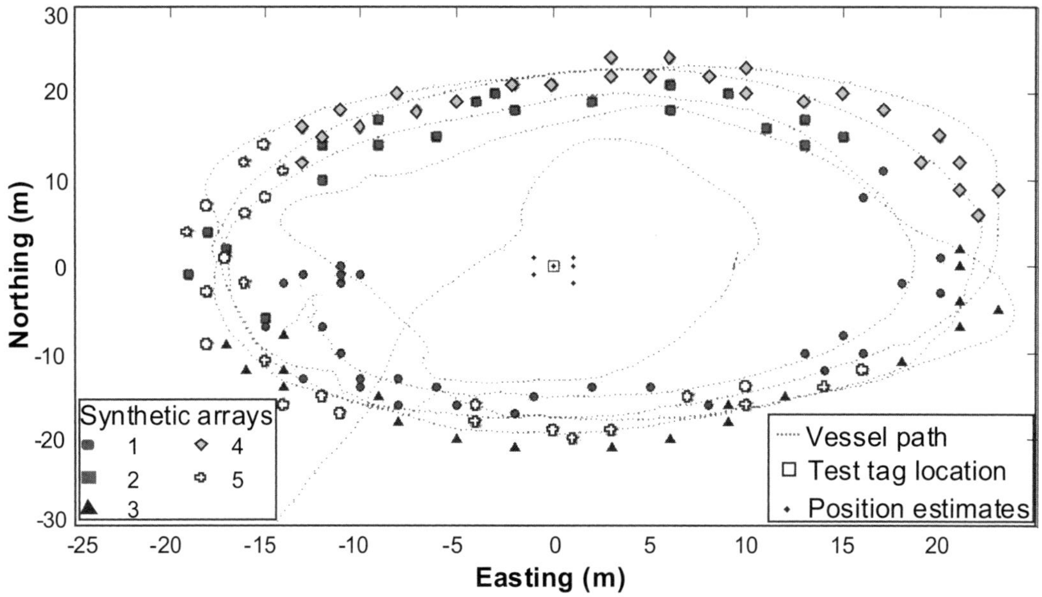

FIGURE 4. Synthetic arrays and estimated test tag positions from field testing at Warner Lake, Ontario. The five synthetic arrays were created using both hydrophones 1 and 2 (see Table 1C for accuracy and precision of estimates resulting from these arrays).

TABLE 2. Accuracy and precision of SYNAPS test tag comparisons in Glacier Bay, Alaska, which were deployed during six tracking trips in 2006.

Test tag Date	Depth (m)	Avg. X error (m) ± (SE)	Avg. Y error (m) ± (SE)	Avg. total error (m) ± (SE)	Avg. DOP* (±m)	Avg. no. hydro-phones
9 February	115	–8.0 (0)	2.0 (10.0)	16.2 (1.8)	11.53	21.7
3 March	85	–6.2 (2.5)	–2.4 (1.8)	10.4 (1.8)	11.58	22.1
27 March	135	9.7 (2.2)	–10.5 (1.8)	18.0 (1.6)	11.28	17.6
11 April	138	12.3 (4.7)	–4.6 (3.8)	23.4 (3.2)	11.84	20.7
27 April	147	5.6 (6.6)	3.9 (3.5)	19.7 (2.7)	11.14	21.6
11 May	78	–0.4 (2.4)	5.7 (3.0)	12.4 (2.4)	10.10	20.8

* Dilution of precision

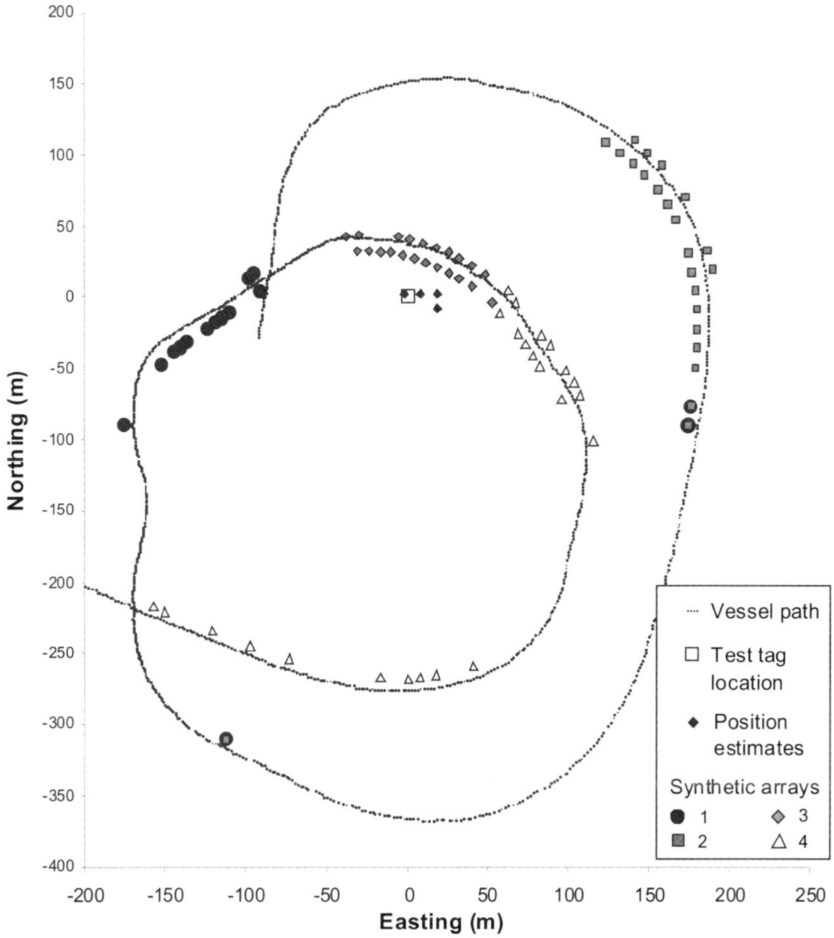

FIGURE 5. Synthetic arrays and estimated positions for a Glacier Bay test tag set at 85 m depth on 3 March 2006, using two towed hydrophones (see Table 3C for accuracy and precision of estimates resulting from these arrays).

Discussion

SYNAPS is a flexible new telemetry technique that can be used to study animal movement at different scales. It fills a gap in methods available for studying movement at scales that fall between the fine-scale, continuous positioning provided by fixed arrays (Cooke et al. 2005) and the large-scale, presence/absence positioning provided by acoustic curtains (Welch et al. 2002). It is particularly valuable for studying the movement of mobile animals in unconstrained areas, such as large lakes or marine systems, where tagged animals could quickly move beyond the range of fixed arrays.

SYNAPS can be performed with a variety of equipment configurations that allow individual researchers to address trade-offs between scale and precision and respond to challenges from specific study area conditions. Using SYNAPS, locations of tagged animals can be determined using towed hydrophones that can search large study areas efficiently with an accuracy of 20 m or better, whereas sub-meter accuracy can be obtained using slower-moving vessels with hull-mounted hydrophones and a differential

Table 3. Comparison of SYNAPS estimates to actual tag location (actual–estimated) for a test tag deployed 3 March 2006 at a depth of 85 m in Glacier Bay (see Figure 5 for estimates in part C). A 10 m SYNAPS grid was used in the estimate calculations.

Synthetic Array	Time (s)	X error (m)	Y error (m)	Total error (m)	DOP* (±m)	Number hydrophones
A) Hydrophone 1 only						
1	200	−3	−10	10.4	11.23	17
2	165	−3	0	3.0	13.19	25
3	160	7	−10	12.2	12.89	25
4	295	−13	−10	16.4	9.27	20
B) Hydrophone 2 only						
5	215	1	2	2.2	11.23	20
6	155	−9	2	9.2	11.08	30
C) Hydrophones 1 and 2						
7	556	−8	8	11.3	10.94	15
8	305	−18	−2	18.1	11.97	23
9	65	2	−2	2.8	12.67	22
10	275	−18	−2	18.1	11.36	24
Mean (±SE)	239.1	−6.2 (2.5)	−2.4 (1.8)	10.4 (1.8)	11.58	22.1

*Dilution of precision

GPS. For individual research plans, choice of hydrophone configuration, GPS resolution, number of hydrophones (one or two), tracking protocols, and SYNAPS processing parameters can be decided based on 1) precision required to achieve research goals, 2) study area characteristics (size, depth, wave height, water column stratification), 3) tagged animal movement rate, and 4) tracking vessel size.

Equipment configuration options

GPS.—Because hydrophone positions are always derived from the vessel GPS, the quality of the GPS instrument used can greatly affect estimate precision. For projects requiring the highest precision (e.g., underwater surveying or fine-scale behavioral studies), a differential GPS with sub-meter precision should be used. Otherwise, a handheld WAAS-enabled GPS can provide 5–20 m precision, depending on hydrophone configuration, for stationary targets or slow-moving animals.

Hydrophone configuration.—Hydrophone configuration options consist of 1) hull-mounted, 2) deployed over the gunwale of the vessel, and 3) towed behind the vessel on v-fins. Although rigid hull attachment removes a large portion of the hydrophone position uncertainty compared to towed hydrophones, there are significant drawbacks for using this configuration. Unless the hydrophone is mounted through the hull, tow speeds can be severely limited as a result of structural vibration in the mounting apparatus (authors, personal observation). Although outfitting a vessel with through-hull hydro-

Table 4. Accuracy and precision of SYNAPS estimates from test tags at in Taku Harbor, Alaska, calculated using SYNAPS grid sizes of A) 10 m, B) 5 m, and C) 1 m.

Tag number	Avg. X error (m) ±(SE)	Avg. Y error (m) ±(SE)	Avg. total error (m) ±(SE)	Avg. DOP* (±m)	Avg. no. hydrophones
A) 10 m grid					
1	1.6 (0.7)	–2.4 (0.6)	7.1 (0.4)	7.3	20.7
2	0.4 (0.7)	–1.6 (0.9)	6.1 (0.7)	7.9	20.5
3	1.2 (0.8)	–0.9 (0.5)	5.8 (0.7)	7.8	20.3
B) 5 m grid					
1	0.3 (0.8)	–1.5 (0.6)	5.3 (0.8)	3.8	20.2
2	0.7 (0.5)	–1.4 (0.5)	5.0 (0.4)	4.8	21.2
3	0.8 (0.4)	–2.1 (0.4)	4.5 (0.2)	3.8	20.7
C) 1 m grid					
1	0.2 (0.4)	–1.7 (0.4)	3.9 (0.2)	1.4	22.5
2	0.5 (0.5)	–2.1 (0.5)	4.4 (0.3)	2.9	22.9
3	1.7 (0.3)	–3.0 (0.3)	4.3 (0.2)	1.3	22.5

*Dilution of precision

phones is viable, high costs and lack of equipment portability make this option restrictive. Simply suspending the hydrophones over the side of the vessel on a weighted line will work for small vessels that have only 1–2 m between the water and the gunwale, however hydrophones become vulnerable to contact with the vessel in swells or stormy weather. Towed hydrophones allow higher tracking vessel speeds than rigid, hull-mounted configurations. Towed hydrophones may provide better signal reception in deep waters (>100 m) or where the water column is stratified by positioning the hydrophones at a greater depth in the water column. The flexible SYNAPS program interface allows researchers to deploy hydrophone configurations that will meet the needs of specific applications.

Vessel maneuvers.—The spatial arrangement of hydrophones in the synthetic array greatly affects the accuracy and precision of position estimates, and tracking protocols that balance required precision and scale should be designed accordingly. If greater precision is required, the tracking vessel should attempt to achieve a semicircle trajectory around the tagged animal. If precision is not required, straight-line transects that cover large areas in a short amount of time may be conducted. Setting a zigzag transect course through the study area is an example of a tracking protocol that provides greater precision with transect searches of large study areas.

Number of hydrophones.—Although SYNAPS can be employed using only one hydrophone, using two hydrophones can increase its efficiency. Using the MapRT receiver (Lotek Wireless Inc) it is possible to obtain bearings to tagged animals in real-time if the vessel receives a signal on both hydrophones simultaneously (Clabough et al. 2007). Real-time navigation in the vicinity of a tagged animal allows the operator to ensure that an optimal (or acceptable) spatial arrangement of hydrophones, (e.g., a circular trajectory around the tagged animal) is achieved.

Potential applications of the SYNAPS method

SYNAPS is useful for studying the movement of sedentary or relatively slow-moving animals in large study areas. For example, SYNAPS was used to obtain position estimates for 30 female and male red king crabs and 50 adult female Tanner crabs in a 100 km^2 study area in Glacier Bay, Alaska over a six month telemetry study (S. James Taggart, Julie K. Nielsen, and Thomas C. Shirley, Texas A&M University Harte Research Institute, unpublished data).

SYNAPS would also be an appropriate tool for animals that are capable of faster swim speeds, but have sedentary periods (e.g., Pacific halibut *Hippoglossus stenolepis*), or for animals that exhibit defined home ranges (e.g., demersal rockfish such as certain *Sebastes* sp. or lingcod *Ophiodon elongatus*). Although SYNAPS has yet to be field-validated for faster-moving animals such as whales or salmon, altering the SYNAPS grid size, tag burst interval, vessel speed, or hydrophone separation may allow reasonable solution precision for more mobile animals.

Summary

SYNAPS constitutes an advance in telemetry technology that will allow researchers to answer a wider range of ecological questions for aquatic animals. Although mobile tracking methods have existed for 30 years (Arnold and Dewar 2001), SYNAPS is capable of yielding more precise position estimates with much less effort than is expended for other methods. In addition, if tagged animals are aggregated, one vessel trajectory can provide precise positions for many tagged animals. This positioning efficiency, when combined with the ability of the MAP telemetry system to detect many tagged animals at the same frequency and instant in time, provides researchers with a telemetry tool that allows fine-scale positioning of many animals over much larger study areas.

The availability of movement information for aquatic animals is becoming increasingly important for management and conservation. For example, designing and assessing effectiveness of aquatic protected areas requires quantifying movement rates (e.g., spillover rates or ontogenetic movement from nursery to adult areas). In addition, it is helpful to be able to understand mechanisms that underlie such movement (Cooke et al. 2004). Because the ability to obtain precise position estimates in an efficient manner is an important part of achieving a mechanistic understanding of animal distribution and movement (Cooke 2008), SYNAPS can be expected to provide significant contributions to the management and conservation of aquatic species.

Acknowledgments

We dedicate this paper to George Niezgoda, who inspired us with his innovative ideas, energy, intellect, and humor. We acknowledge his significant contributions to acoustic telemetry, aquatic research, and education. His recent passing leaves a void that will be hard to fill.

We thank the West Coast & Polar Regions Undersea Research Center, the USGS Alaska Science Center, the Alaska Department of Fish and Game, and the NOAA Alaska Fisheries Science Center for funding research that assisted the development of SYNAPS. We also thank Queen's University Biological Station, the Canada Foundation for Innovation, Carleton University and Mitch Sisak of Lotek Wireless, Inc. for assistance and support of field validation procedures in Ontario. Captain Jim de La Bruere, Jennifer Mondragon, Alexander Andrews, Thomas Shirley, and many others helped with the collection of data in Glacier Bay, Alaska. We are grateful to J.J. Vollenweider and Dean Courtney (NOAA Ted Stevens Marine Research Institute) for loaning their telemetry equipment during this project. Any use of trade names is for descriptive purposes only and does not imply endorsement by the U.S. Government.

References

Arnold, G., and H. Dewar. 2001. Electronic tags in marine fisheries research: a 30 year perspective. Pages 65–88 *in* J. R. Sibert, and J. L. Nielsen, editors. Electronic tagging and tracking in marine fisheries. Kluwer Academic Publishers, Dordrect, The Netherlands.

Clabough, T. S., C. C. Caudill, C. A. Peery, T. C. Bjornn, and B. J. Burke. 2007. Associations between adult salmon and steelhead body temperature during upstream migration and estimated water temperatures in lower Granite Reservoir during cold water releases from Dwarshak Reservoir, 2004. Idaho Cooperative Fish and Wildlife Research Unit for U.S. Army Corps of Engineers, Technical Report 2007–3.

Collazo, J. A., and S. P. Epperly. 1995. Accuracy tests for sonic telemetry studies in an estuarine environment. Journal of Wildlife Management 59:181–188.

Cooke, S. J. 2008. Biotelemetry and biologging in endangered species research and animal conservation: relevance to regional, national, and IUCN Red List threat assessments. Endangered Species Research 4:165–185.

Cooke, S. J., S. G. Hinch, M. Wikelski, R. D. Andrews, L. J. Kuchel, T. G. Wolcott, and P. J. Butler. 2004. Biotelemetry: a mechanistic approach to ecology. Trends in Ecology and Evolution 19:334–343.

Cooke, S. J., G. H. Niezgoda, K. Hanson, C. D. Suski, F. J. S. Phelan, R. Tinline, and D. P. Philipp. 2005. Use of CDMA acoustic telemetry to document 3-D positions of fish: relevance to the design and monitoring of aquatic protected areas. Marine Technology Society Journal 39:17–27.

Domeier, M. L. 2005. Methods for the deployment and maintenance of an acoustic tag tracking array: an example from California's Channel Islands. Marine Technology Society Journal 39:74–80.

Hanson, K. C., S. J. Cooke, C. D. Suski, G. H. Niezgoda, F. J. S. Phelan, R. Tinline, and D. P. Philipp. 2007. Assessment of largemouth bass (*Micropterus salmoides*) behaviour and activity at multiple spatial and temporal scales utilizing a whole-lake telemetry array. Hydrobiologia 582:243–256.

Holland, K. N., C. G. Lowe, and B. M. Wetherbee. 1996. Movements and dispersal patterns of blue trevally (*Caranx melampygus*) in a fisheries conservation zone. Fisheries Research 25:279–292.

Hoolihan, J. P. 2005. Horizontal and vertical movements of sailfish (*Istiophorus platypterus*) in the Arabian Gulf, determined by ultrasonic and pop-up satellite tagging. Marine Biology 146:1015–1030.

Jorgensen, S. J., D. M. Kaplan, A. P. Klimley, S. G. Morgan, M. R. O'Farrell, and L. W. Botsford. 2006. Limited movement in blue rockfish *Sebastes mystinus*: internal structure of home range. Marine Ecology Progress Series 327:157–170.

Jørgensen, T., S. Løkkeborg, A. Fernö, and M. Hufthammer. 2007. Walking speed and area utilization of red king crab (*Paralithodes camtschaticus*) introduced to the Barents Sea coastal ecosystem. Hydrobiologia 582:17–24.

Lucas, M. C., and E. Baras. 2000. Methods for studying the spatial behaviour of freshwater fishes in the natural environment. Fish and Fisheries 1:283–316.

Ng, C. L., K. W. Able, and T. M. Grothues. 2007. Habitat use, site fidelity, and movement of adult striped bass in a southern New Jersey estuary based on mobile acoustic telemetry. Transactions of the American Fisheries Society 136:1344–1355.

Niezgoda, G. H., M. Benfield, M. M. Sisak, and P. Anson. 2002. Tracking acoustic transmitters by code division multiple access (CDMA)-based telemetry. Hydrobiologia 483:275–286.

Oppermann, I., M. Hamalainen, and J. Iinatti. 2004. UWB Theory and Applications. Wiley, Chichester, UK.

Rigby, P. R., Y. Andrade, and R. K. O'Dor. 2005. Interpreting behavioural data from Radio-Acoustic Positioning Telemetry (RAPT) systems. African Journal of Marine Science 27:395–400.

Sibert, J. R., and J. L. Nielsen, editors. 2001. Electronic tagging and tracking in marine fisheries. Kluwer Academic Publishers, Dordrecht, The Netherlands.

Stone, R. P., C. E. O'Clair, and T. C. Shirley. 1992. Seasonal migration and distribution of female red king crabs in a southeast Alaskan estuary. Journal of Crustacean Biology 12:546–560.

Taggart, S. J., J. Mondragon, A. G. Andrews, and J. K. Nielsen. 2008. Spatial patterns and movements of red king and Tanner crabs: implications for the design of marine protected areas. Marine Ecology Progress Series 365:151–163.

Thorstad, E. B., F. Okland, D. Rowsell, and R. S. McKinley. 2000. A system for automatic recording of fish tagged with coded acoustic transmitters. Fisheries Management and Ecology 7:281–294.

Welch, D. W., G. W. Boehlert, and B. R. Ward. 2002. POST-the Pacific Ocean salmon tracking project. Oceanologica Acta 25:243–253.

American Fisheries Society Symposium 76:251–264, 2012

Distribution and Movement of Catadromous Fish: Design and Implementation of a Freshwater-Estuarine Acoustic Telemetry Array

Chris T. Walsh*
Industry and Investment NSW, Cronulla Fisheries Research Centre of Excellence
P.O. Box 21, Cronulla, NSW, 2230, Australia
and
University of Wollongong, School of Biological Sciences
Northfields Avenue, Wollongong, NSW, 2522, Australia

Ivars V. Reinfelds
NSW Office of Water, P.O. Box 53, Wollongong, NSW, 2500, Australia

Ron J. West
University of Wollongong, School of Biological Sciences
Northfields Avenue, Wollongong, NSW, 2522, Australia

Charles A. Gray
Industry and Investment NSW, Cronulla Fisheries Research Centre of Excellence
P.O. Box 21, Cronulla, NSW, 2230, Australia

Dylan E. van der Meulen
Industry and Investment NSW, Cronulla Fisheries Research Centre of Excellence
P.O. Box 21, Cronulla, NSW, 2230, Australia

Abstract.—Movement patterns of two co-occurring catadromous fishes, estuary perch *Macquaria colonorum* and Australian bass *Macquaria novemaculeata*, were investigated in a large tidal river in southeastern Australia. Nineteen estuary perch and seventeen Australian bass were captured as mature adults from the Shoalhaven River, surgically implanted with electronic transmitters and released. Forty-nine Vemco (VR2W) acoustic receivers were strategically placed throughout the river from the Tallowa Dam wall downstream to the sea (a distance of 75 km). Between September 2007 and February 2008, a total of 800,263 detection events were recorded with most fish detected in the middle (estuary perch) to upper (Australian bass) estuarine reaches of the river. Both species made extensive use of the estuary, with no estuary perch and only three Australian bass entering the freshwater, indicating that the freshwater residency phase of these catadromous fishes may not be obligatory. The data also suggests that estuary perch and Australian bass exhibit high site fidelity,

*Corresponding author: chris.walsh@industry.nsw.gov.au

which, along with their large scale movements may be influenced by factors such as river discharge and prey availability. This large freshwater-estuarine telemetry array combined with critical abiotic information (river discharge and salinity) has and will provide a greater understanding of catadromous fish movement, particularly in relation to habitat utilization and environmental flows.

Introduction

Environmental flows are recognized as playing an important role in the productivity of freshwater (Bayley 1988; De Graaf 2003) and estuarine systems (Drinkwater and Frank 1994; Loneragan and Bunn 1999; Gillanders and Kingsford 2002; Robins et al. 2005). In particular, the migrations and spawning of diadromous fishes are typically linked to environmental cues such as river discharge, salinity and temperature (McDowall 1988; Pusey et al. 2004). For many Australian fish species, the precise triggers for migration, spawning, and subsequent dispersal are not known, but several species appear to require flooding at seasonally appropriate times (Harris 1988; Gehrke et al. 1999). Likewise, the input of freshwater can affect the salinity profile of an estuary and therefore strongly influence the distribution of fish within it (Robins et al. 2005).

Many rivers in Australia have been modified by human development including those impounded by weirs and dams (www.ozcoasts.org.au). These structures change the natural flow regimes of rivers including seasonal flow cycles, variability of flows, and water quality, resulting in a potential loss of cues for fish spawning and dispersal (Gehrke and Harris 2000). Increases in water extraction have also lead to reductions in flow depths that create barriers to upstream and downstream movement of fish (Whitfield 1994; Gehrke et al. 1999; Reinfelds et al. in press). Recent studies have shown that such impacts have contributed to a dramatic decline in native fish populations throughout rivers in southeastern Australia (Gehrke and Harris 2000).

The use of passive acoustic monitors capable of recording the presence and hence movement, behavior and habitat utilization of animals tagged with acoustic transmitters has become an increasingly popular research tool (Lucas and Baras 2000; Heupel et al. 2006; Hindell 2007). Individually-coded pinger tags used in association with automated acoustic listening stations (receivers) can monitor the movement of animals at fine temporal and spatial scales of resolution. Movement can then be related to environmental variables (e.g., river discharge, salinity and temperature) collected concurrently via co-instrumentation or retrospectively from other sources. Long-term deployment of these receivers enables annual or seasonal patterns of movement to be examined so that factors affecting changes in habitat use or distribution can be elucidated (Heupel and Hueter 2001; Egli and Babcock 2004; Heupel et al. 2004).

Estuary perch *Macquaria colonorum* (Günther, 1863) and Australian bass *Macquaria novemaculeata* (Steindachner, 1866) represent the only two known catadromous percichthyids world-wide. Both these species are distributed throughout the estuarine and freshwater reaches of coastal rivers, lakes and lagoons in southeastern (SE) Australia and are thought to occupy similar habitats (McDowall 1996). There is little quantitative data pertaining to the distribution, residency and habitat utilization of estuary perch and Australian bass, particularly within the estuarine environment. There is also little information available regarding their large scale movements (including both pre- and post-spawning) in relation to environmental flows. The main objectives of this paper were to: 1) outline the design and implementation of a freshwater-estuarine acoustic telemetry array in the Shoalhaven River—a large regulated river system in SE Australia; and 2) provide preliminary results on the instream distribution and movement of two catadromous Percichthyidae fishes—estuary perch and Australian bass.

Methods

Study area

The Shoalhaven River (–34°53'S, 150°45'E) has a total catchment area of approximately 7,300 km^2 (Figure 1). The Tallowa Dam impounds the Shoalhaven River some 75 km upstream of the Pacific Ocean with mean and median annual inflows (1909–2004) being 1,043 GL and 631 GL, respectively. This dam is part of the Sydney and Shoalhaven water supply schemes and was completed in 1976. Due to its small operational capacity relative to annual inflows, the dam frequently overtops during wet years creating unmanaged spills of water to the river downstream that heavily supplement regulated environmental flow releases. The river downstream of Tallowa Dam extends for 25 km to the tidal limit, and is confined in a sandstone gorge. The freshwater reaches are comprised of a series of long pools averaging 830 m in length with a mean maximum depth of 6.5 m, separated by cobble dominated riffles, rapids, and runs. The estuary extends for 47 km upstream of the entrance at Crookhaven Heads, has a tidal range of approximately 1.2 m and is classed as a micro tidal, mature, wave-dominated barrier estuary (Roy et al. 2001).

Equipment used

Vemco V13–1 L coded acoustic transmitters operating on 69 kHz frequencies were inserted into mature estuary perch and Aus-

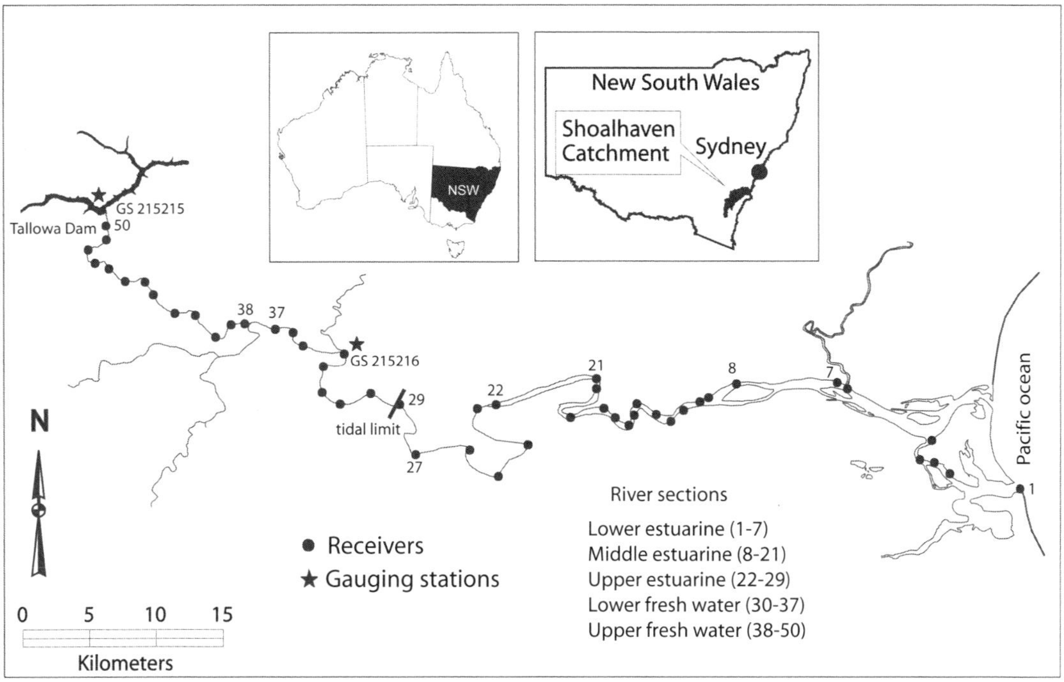

Figure 1. Study area showing receiver locations.

tralian bass. The transmitters, with a random pulse rate of 60–180 s were 36 mm long, 13 mm in diameter and weighed approximately 6 g in water (<2% of the fish's body mass). Battery life within these transmitters is approximately two years. Vemco VR2W acoustic receivers were used to continuously monitor the presence of the unique, digitally-coded transmitter signals emitted by the acoustic tags. The receivers record the date, time, temperature, depth, and identity of tagged fish within the detection range of the unit.

Receiver deployment

A linear array of 49 receivers was deployed in the freshwater and estuarine reaches of the Shoalhaven River below Tallowa Dam (Figure 1). Freshwater receivers were anchored by cobble-filled bags (100 kg) and suspended beneath styrene surface floats with 5 mm steel cable. Deployment locations for freshwater receivers were optimized on the basis of pool-riffle bathymetry modeled from detailed survey data with 1–2 receivers typically deployed in the deepest sections of each major pool. Estuarine receivers were deployed either on anchor chains of existing channel marker buoys or moored independently via a weighted float and anchor system. These receivers were deployed singularly or within arrays (two to three receivers with overlapping detection ranges) from the tidal limit downstream to Crookhaven Heads. The selection of deployment sites for all receivers were based on detailed bathymetric data, salinity profiles and existing knowledge of Australian bass and estuary perch aggregation sites—including spawning and pre/post spawning areas. Detection ranges for freshwater and estuarine receivers were assessed by trolling Vemco V13–1 L range-testing transmitters with a 6 s delay while recording tracks at maximum (10 m) resolution with a Garmin handheld GPS (±5 m positional accuracy). The GPS track time-stamps were related to time stamps for receiver detections to determine detection ranges in ArcView 3.3. Receiver clocks were synchronized to GPS clocks prior to range testing.

Fish capture and surgical procedure

Between September and December 2007 (Austral spring to summer), 19 estuary perch and 17 Australian bass were tagged and released with acoustic transmitters within the estuarine reaches of the Shoalhaven River (Table 1). To reduce possible negative effects on gonad development and spawning behavior, all fish were tagged after peak spawning activity (June–August). Fish were captured via hook and line and were operated on either in situ or transported to the nearby aquaculture facility at the Shoalhaven Marine and Freshwater Centre (West Nowra, NSW, Australia). Fish were removed from their holding tanks and placed in an anesthetic bath containing 50 g L^{-1} of Aqui-S solution for approximately 2–3 min. Fish were weighed (±1 g), measured (fork length, ± 1 mm) and then placed ventral-side up in a V-shaped surgical cradle. During surgery, 25 g L^{-1} of Aqui-S solution was irrigated over the gills of the fish using a battery operated pump. A 20 mm incision was made with a scalpel on the ventral mid-line between the pelvic fins and the anus. An acoustic tag was inserted into the peritoneal cavity of the fish and the incision was closed with three synthetic absorbable sutures (Ethicon Vicryl 3–0). Each fish was injected with oxytetracycline at a dosage of 75 mg kg^{-1} fish weight. All fish were further identified with an external T-bar anchor tag (Hallprint) inserted below the dorsal fin. Fish were allowed to recover in ambient water until they were observed to swim normally and then released close to the point of capture. Recent studies have indicated that survival rates of 100% could be expected (Gehrke et al. 2001; C. T. Walsh, Department of Primary Industries, unpublished data).

Table 1. Summary of individual tracking data for estuary perch (EP) and Australian bass (AB).

Fish ID	Species	sex	Fork length (mm)	Date tagged and released	Last date detected	Number of detections	Total days tracking
1035952	EP	f	360	24-Sep-07	29-Feb-08	36988	159
1035953	EP	f	373	23-Oct-07	29-Feb-08	26456	130
1035954	EP	m	358	21-Sep-07	29-Feb-08	4452	162
1035955	EP	m	366	21-Sep-07	29-Feb-08	4036	162
1035956	EP	f	376	21-Sep-07	29-Feb-08	20584	162
1035957	EP	*	327	14-Dec-07	29-Feb-08	22694	78
1035958	EP	f	403	14-Dec-07	25-Dec-07	2668	12
1035960	EP	*	335	14-Dec-07	29-Feb-08	24082	78
1036284	EP	f	300	21-Sep-07	29-Feb-08	37248	162
1036285	EP	m	321	24-Sep-07	29-Feb-08	53762	159
1036286	EP	f	315	24-Sep-07	29-Feb-08	46222	159
1036287	EP	f	322	24-Sep-07	29-Feb-08	5762	159
1036288	EP	f	312	24-Sep-07	29-Feb-08	43855	159
1036289	EP	f	339	23-Oct-07	29-Feb-08	15114	130
1036290	EP	m	331	24-Sep-07	29-Feb-08	19664	159
1039619	EP	m	312	24-Sep-07	29-Feb-08	51723	159
1039758	EP	m	321	14-Dec-07	29-Feb-08	15169	78
1039759	EP	m	314	14-Dec-07	29-Feb-08	6228	78
1039761	EP	m	314	14-Dec-07	29-Feb-08	44089	78
1036282	AB	f	367	21-Sep-07	29-Feb-08	14780	162
1036283	AB	f	346	21-Sep-07	21-Sep-07	14	1
1039617	AB	*	355	23-Oct-07	23-Oct-07	27	1
1039620	AB	f	364	23-Oct-07	05-Dec-07	2317	44
1039621	AB	m	346	24-Sep-07	22-Oct-07	6011	29
1039623	AB	*	348	23-Oct-07	29-Feb-08	8793	130
1039624	AB	f	328	23-Oct-07	29-Feb-08	57689	130
1039695	AB	f	414	24-Sep-07	29-Feb-08	15926	159
1039696	AB	f	373	23-Oct-07	07-Nov-07	566	16
1039697	AB	f	428	23-Oct-07	29-Feb-08	16946	130
1039698	AB	f	370	23-Oct-07	29-Feb-08	23152	130
1039699	AB	*	375	23-Oct-07	07-Nov-07	1362	16
1039700	AB	f	364	23-Oct-07	06-Feb-08	18354	107
1039701	AB	f	419	24-Sep-07	15-Jan-08	3616	114
1039702	AB	*	445	24-Sep-07	29-Feb-08	100418	159
1039703	AB	*	386	23-Oct-07	23-Oct-07	76	1
1039704	AB	m	429	24-Sep-07	29-Feb-08	49420	159

* Sex could not be differentiated

Data analysis

Fish movements were monitored for a 162 d period between 21 September 2007 and February 29 2008 in the Shoalhaven River. Based on receiver data, the proportion of time that each fish spent within the vicinity of each receiver and within each of the five river reaches (LE, lower estuary; ME, middle estuary; UE, upper estuary; LF, lower freshwater; UF, upper freshwater; see Figure 1) was analyzed. Area use within the river was calculated as the time each tagged fish spent in the vicinity of each acoustic receiver (Cowley et al. 2008). A fish was assumed to be in the vicinity of a particular receiver when two (or more) consecutive detections were recorded within 60 min. Absence times of more than 60 min were assigned in equal proportions to the receivers frequented before and after the absence period. Differences in time spent between the five river reaches for estuary perch and Australian bass were tested using a nonparametric Kruskal-Wallis analysis of variance (ANOVA). Analyses and posthoc multiple comparisons were performed using the Statistica 6 (Statsoft) software package with a P-value of <0.05 considered significant.

Flow rates were continuously monitored at two gauging stations (215215, 215216) downstream of Tallowa Dam (Figure 1) while seven conductivity loggers (Odyssey) located throughout the estuary measured salinity at hourly intervals during the study. The onset, progression, and cessation of fish movement were also examined in relation to river discharge and salinity.

Results and Discussion

Array performance

Freshwater receivers returned a mean detection radius of 190 m (range, 125–255 m). Minimum detection ranges in the freshwater pools under low flow conditions were constrained by pool bathymetry and geometry, but receivers were typically capable of detecting 40% of the length of the pools in which they were deployed. Estuarine detection ranges were highly dependent on the study site and tidal conditions, with maximum detection distances ranging from 280 m to 420 m (mean, 350 m). Detection ranges were greater in the estuarine areas, more than likely due to constraints imposed by the pool-riffle geometry of the freshwater reaches. The likelihood of a fish not being detected when moving past a receiver was only 0.4% for estuary perch and 2.7% for Australian bass, respectively. Therefore, movements by fish through the array system were unlikely to escape detection.

A total of 800,263 detections were recorded on both the freshwater and estuarine receivers from September 2007 to February 2008. Of the 19 estuary perch tagged, 18 were still detected by the end of the study, compared to 10 out of a possible 17 Australian bass. It seems unlikely that undetected Australian bass died as a result of the surgery, as postsurgery rates of survival were near 100% (see above). We consider it more likely that these undetected fish may have been captured by commercial or recreational fishers and not reported, died of natural causes, tag failure, or resided in areas outside receiver detection ranges. Australian bass are known to use habitats that include shallow weedy areas, undercut banks and woody debris (snags), whereas estuary perch prefer to reside in deep channels (Harris 1983; Williams 1970). Hence, habitat preferences by Australian bass may render it less susceptible to receiver detection.

Distribution and movement

During the study period 480,796 detections were recorded by estuary perch within the tidal reaches of the Shoalhaven River. Of these, 90% of the detections were recorded from 85% of the tagged fish in the middle reaches where salinities typically ranged from 10 to 20 ppt. (Figure 2). These results (no. of detections per station) were also mirrored by the proportion of time spent within the vicinity of each receiver (Figure 3). Estuary perch spent significantly more time per fish day in the middle estuarine reaches (median of 92%) of the Shoalhaven River than in the other estuarine environments (ANOVA: $H_{(2,48)} = 28.12, p < 0.001$). There was no significant difference between the time spent in lower and upper estuarine reaches ($p > 0.05$). A total of 319,467 detections were recorded by Australian bass, mainly in the upper estuarine reaches (85% of detections from 75% of fish) of the Shoalhaven River (Figure 2). Australian bass spent significantly more time per fish day in these upper estuarine reaches (median of 66%) than in the other estuarine

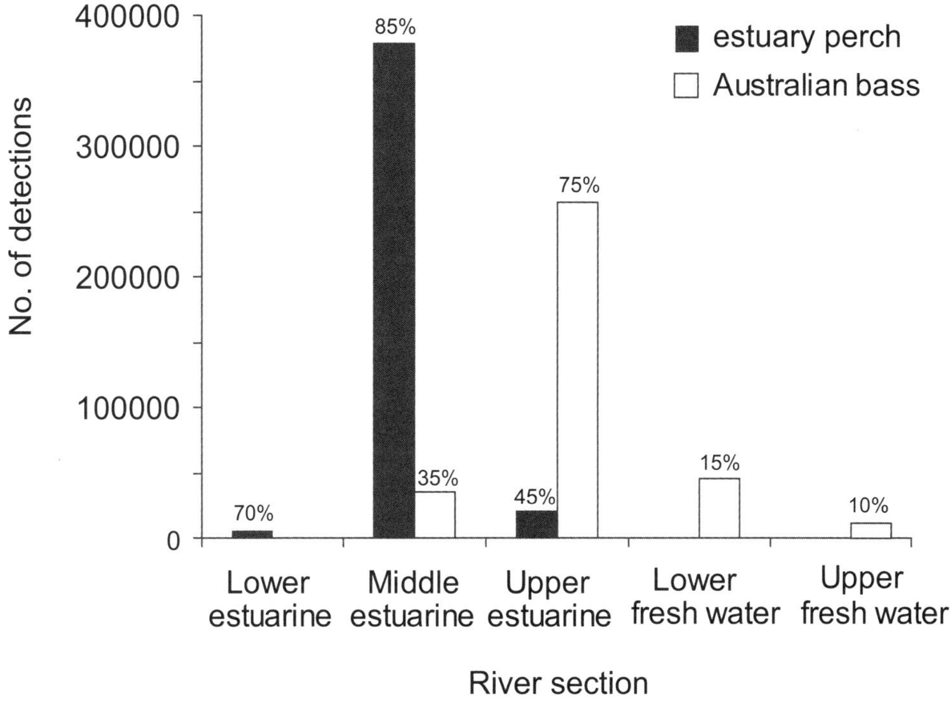

FIGURE 2. Total number of detections recorded for estuary perch *Macquaria colonorum* and Australian bass *Macquaria novemaculeata*. Percentages above bars represent proportion of tagged population detected.

or freshwater environments (ANOVA: $H_{(3,40)} = 17.61$, $p < 0.001$) (Figure 3). There were no significant differences between time spent in the lower freshwater and upper freshwater reaches ($p > 0.05$). Similar to estuary perch, Australian bass spent a considerable proportion of their time resident in similar locations (Figure 3).

Estuary perch made extensive use of the tidal reaches of the estuary and no fish were detected in freshwater above the tidal limit (Figure 4). Likewise, no fish were detected by the receiver closest to the river mouth. Twelve estuary perch (66%) made significant downstream movements (>5 km) which corresponded with minor flood events (December 2007 and February 2008) and decreasing river salinities (Figure 6). Subsequently, when the salinity in the estuary increased after the flood events, these fish returned to their original habitats. Of the Australian bass released, only three out of the surviving 10 tagged Australian bass migrated above the tidal limits into freshwater (Figure 5). Two of these fish (ID 1036282 and 1039695) continued their upstream migration to the dam wall on the initial rise of the December 2007 flood event (Figure 6). These fish completed their 24 km migration to the dam wall in 58 and 94 h, respectively. Five of the remaining eight fish made significant downstream movements associated with this flood event. No fish were detected in the lower estuarine reaches of the Shoalhaven River (Figure 5).

Based on the number of detections and the proportion of time spent in the vicinity of a receiver, there appears to be a distinct difference in habitat preference between estuary perch and Australian bass. The distribution of tagged estuary perch in the Shoalhaven River corresponds with the findings of Williams (1970) and McCarraher and McKenzie

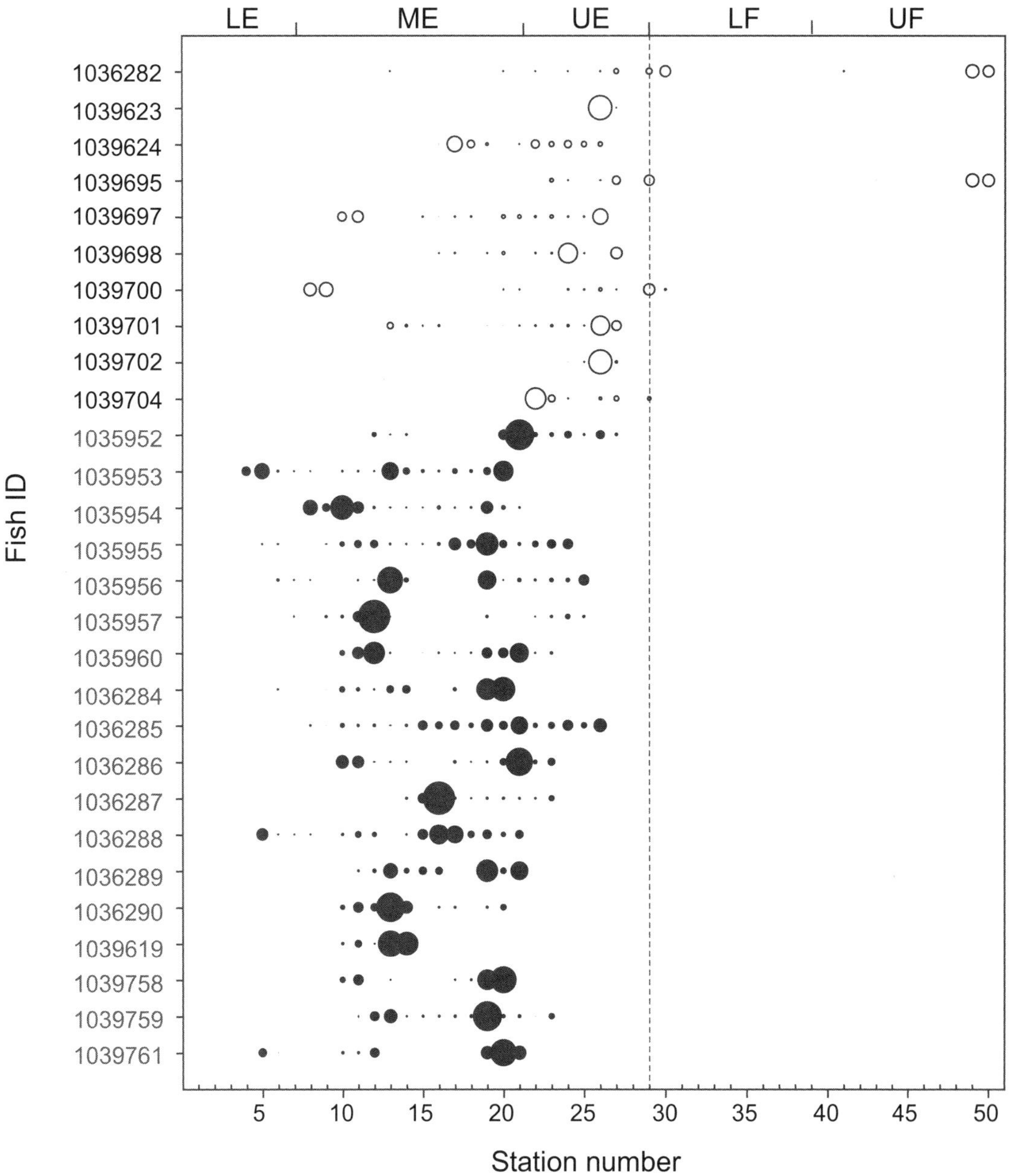

Figure 3. Bubble-plot representation of the proportion of time (percentage frequency) each tagged estuary perch (•, n = 18) and Australian bass (○, n = 10) spent at each station in the Shoalhaven River between September 2007 and February 2008. Abbreviations: LE, lower estuary; ME, middle estuary; UE, upper estuary; LF, lower freshwater; UF, upper freshwater. Note: only fish at liberty for more than 50 d were included.

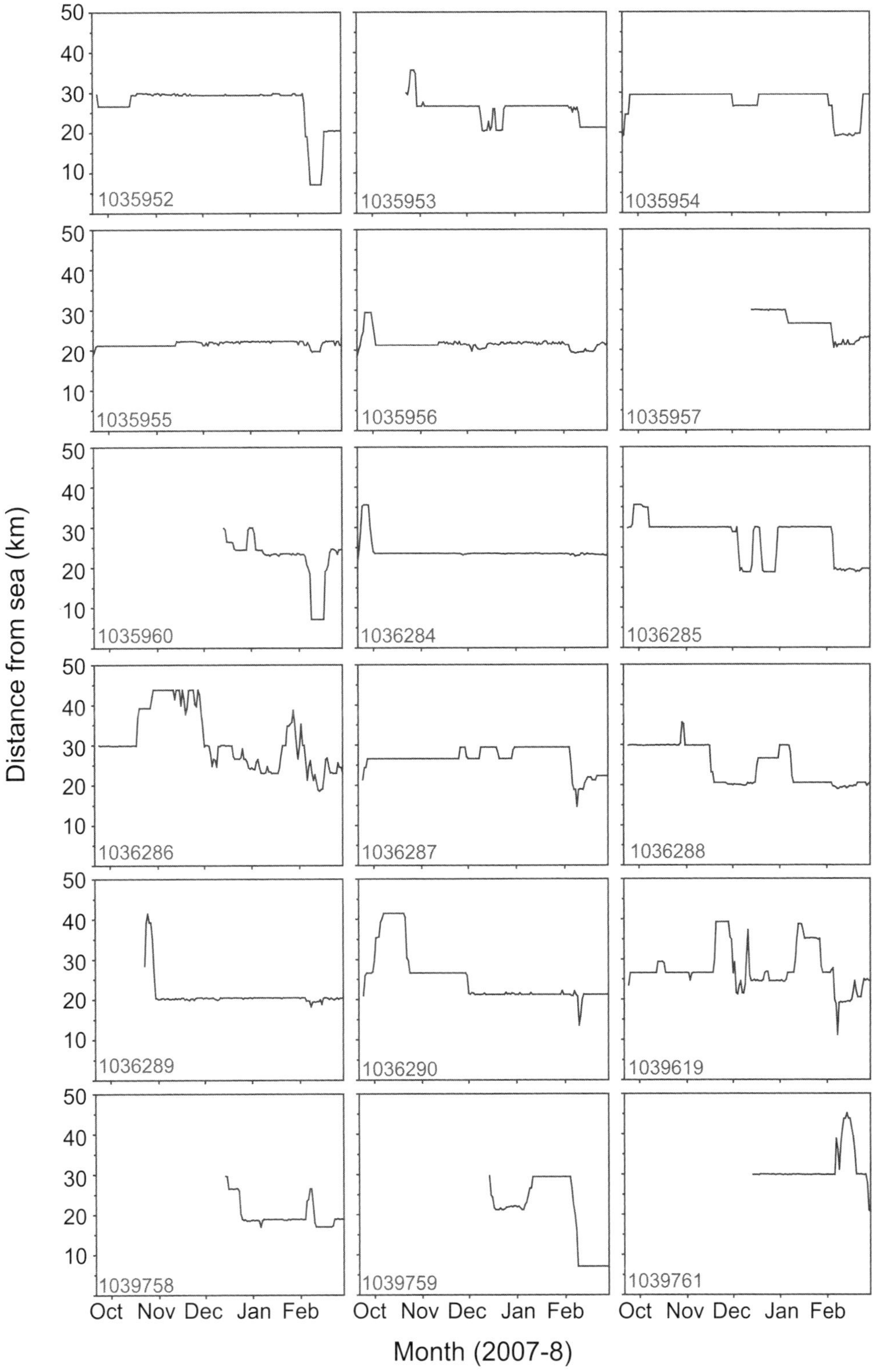

FIGURE 4. Movement of tagged estuary perch between September 2007 and February 2008. Note: only fish at liberty for more than 50 d were included.

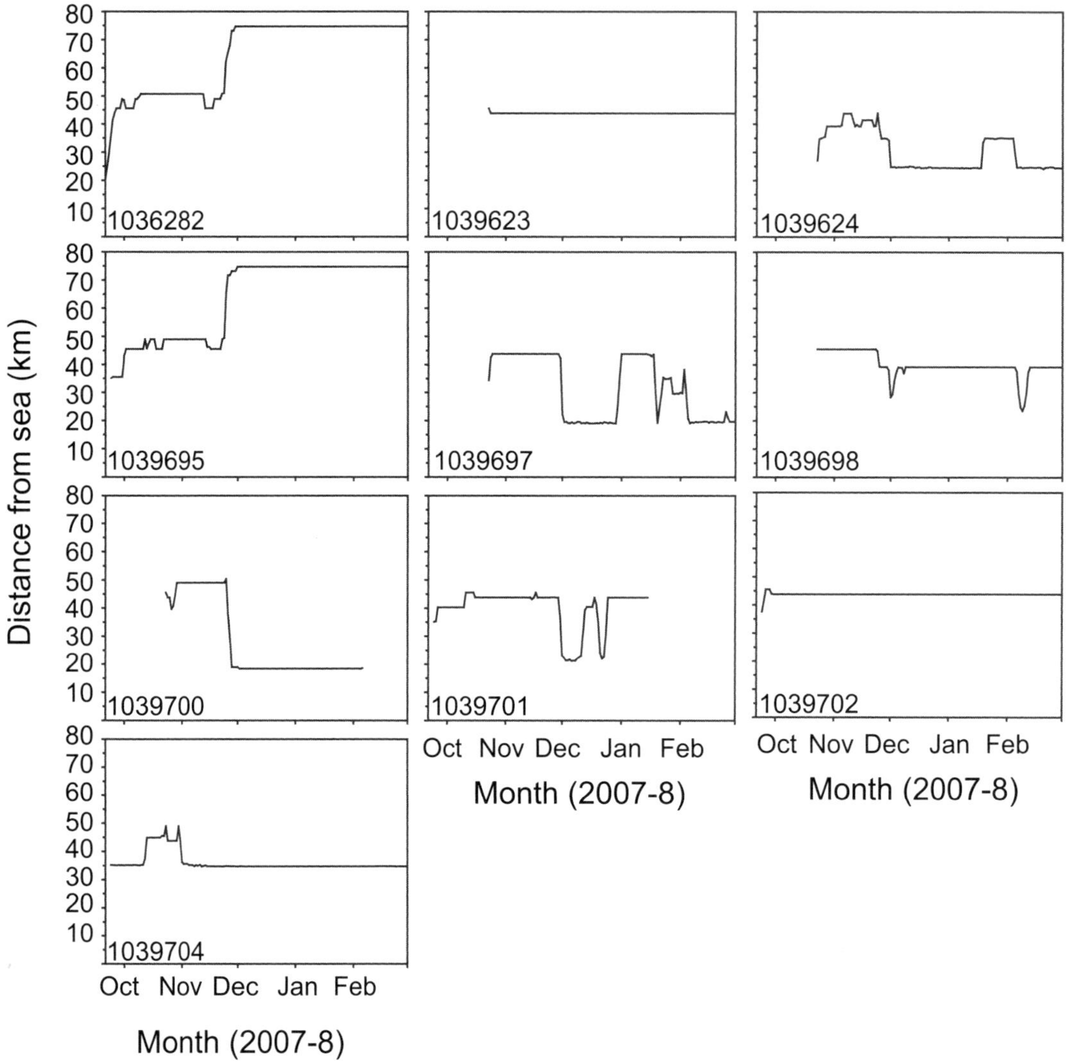

Figure 5. Movement of tagged Australian bass between September 2007 and February 2008. Note: only fish at liberty for more than 50 d were included.

(1986), who showed that significant populations of estuary perch were captured in the middle to upper reaches of tidal rivers during the austral spring to summer periods. However, McCarraher and McKenzie (1986) also found estuary perch residing in the freshwater of several catchments. In this study, none of the tagged estuary perch were recorded above the tidal limits, and similarly, Williams (1970) did not report any captures in freshwater from surveys done on the Shoalhaven River.

Only two tagged Australian bass spent a significant proportion of their time in the upper freshwater habitats of the Shoalhaven River, with only three out of the seven females migrating to freshwater. Previous work has shown that only female adult Australian bass reside in the freshwater reaches of catchments in the spring to autumn months, while male adult fish tend to reside in the estuary (Harris 1986, 1988; McDowall 1996). Mallen-Cooper (2000) postulated that increases in river discharge act as important triggers

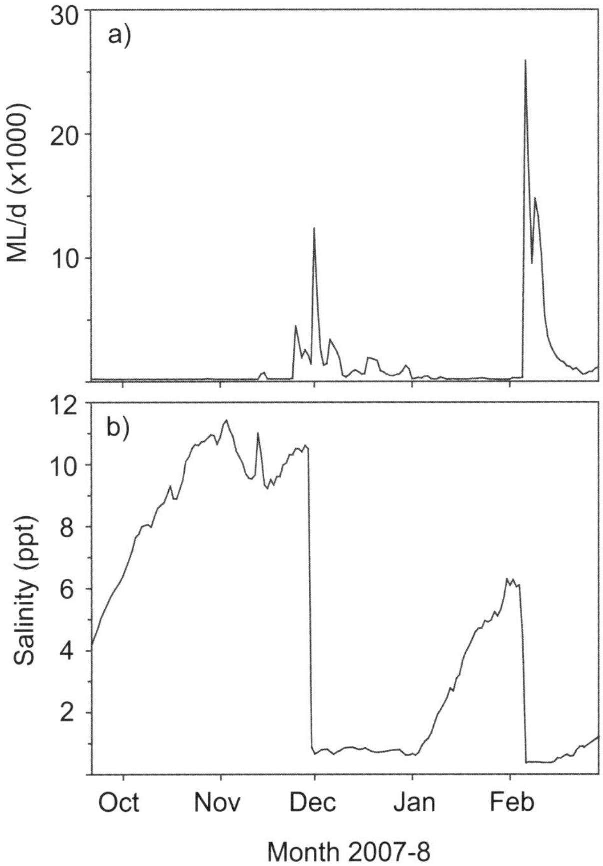

FIGURE 6. a) River discharge (ML/d) measured at Tallowa Dam; and b) Salinity measured at Nowra (middle estuarine) between September 2007 and February 2008.

for these post spawning upstream migrations. Only two fish made significant postspawning freshwater migrations in the Shoalhaven River that correlated with a significant increase in river discharge. Therefore, further work is required to determine the timing and magnitude of freshwater flows required to stimulate post spawning migration for Australian bass. Nevertheless, this study has shown that the freshwater residency phase of these two catadromous species may not be obligatory.

A number of biotic and abiotic factors have been found to induce behavioral responses in fish and impact on their distribution within estuaries (Whitfield 1994). For example, high utilization of specific habitat and long periods of limited movement (stationary behavior) were evident in both tagged estuary perch and Australian bass in the Shoalhaven River. Fish exhibiting a high degree of site fidelity is a common trait found among many estuarine-dependent species (e.g. mulloway *Argyrosomus japonicus*, Taylor et al. 2006; spotted grunter *Pomadasys commesonnii*, Childs et al. 2008). Childs et al. (2008) suggested that the high-use areas of spotted grunter coincided with the highest abundances of their preferred prey. Indeed, Cowley et al. (2008) hypothesized that strong selection exists for estuarine-dependent species which prey on nonmobile benthic organisms to exhibit resident behavior and occupy areas of optimal food availability. Howell (2004) and Harris (1985) showed that adult estuary perch and Australian bass captured in large estuaries feed predominantly on decapods (e.g., shrimp, prawns and crabs). Moreover, freshwater input and the subsequent fluctuations in salinity levels within the estuarine environment have been shown to influence the distribution and abundance of many predatory and prey species (Robins et al. 2005). Therefore, the observed area-use patterns (ie: limited resident behaviour) and movement in response to environmental flows by tagged estuary perch and Australian bass may well be influenced by prey availability.

Conclusion

The acoustic telemetry array deployed in this study is currently the largest freshwater-estuarine monitoring network in Australia and is one of the largest such networks in the world. Results have demonstrated that the design and implementation of this array is highly effective for studying the distribution and movement of two catadromous fishes in a large tidal river. In particular, this study has provided new information on the extent of the instream distribution and movement of estuary perch and Australian bass, indicating that the freshwater residency phase of these catadromous species may not be obligatory. It has also shown that site fidelity and large scale movement of both these species may be influenced by environmental flows and prey availability. Further insight is required into the spatial and temporal usage of habitats by estuary perch and Australian bass, along with information on the magnitude, frequency, duration and seasonality of environmental flows required to stimulate and facilitate spawning and migratory requirements of these two fishes.

Acknowledgments

The authors thank Will Robbins for his constructive comments on the manuscript. Southern Bass and Basin Lure and Fly fishing clubs as well as Steve Starling are gratefully acknowledged for assisting with the capture of fish. We would like to thank Shoalhaven Marine and Freshwater Centre (University of Wollongong) for the use of their facilities and Craig Burnes for providing excellent technical support. This study was funded by Industry and Investment NSW, NSW Office of Water, NSW Saltwater Recreational Fishing Trust and University of Wollongong. This study complied within the requirements of the Animal Research Act 1985 (ACEC # 06/03).

References

Bayley, P. B. 1988. Factors affecting growth rates of young tropical floodplain fishes: seasonality and density-dependence. Environmental Biology of Fishes 21:127–142.

Childs, A. R., P. D. Cowley, T. F. Naesje, A. J. Booth, W. M. Potts, E. B. Thorstad, and F. Økland. 2008. Estuarine use by spotted grunter *Pomadasys commersonnii* in a South African estuary, as determined by acoustic telemetry. African Journal of Marine Science 30(1):123–132.

Cowley, P. D., S. E. Kerwath, A. R. Childs, E. B. Thorstad, F. Økland, and T. F. Naesje. 2008. Estuarine habitat use by juvenile dusky kob *Argyrosomus japonicus* (Sciaenidae), with implications for management. African Journal of Marine Science 30(2):247–253.

De Graaf, G. 2003. The flood pulse and growth of floodplain fish in Bangladesh. Fisheries Management and Ecology 10:241–247.

Drinkwater, K. F., and K. T. Frank. 1994. Effects of river regulation and diversion on marine fish and invertebrates. Aquatic Conservation: Marine and Freshwater Ecosystems 4(2):135–151.

Egli, D. P., and R. C. Babcock. 2004. Ultrasonic tracking reveals multiple behavioural modes of snapper (*Pagrus auratus*) in a temperate no-take marine reserve. ICES Journal of Marine Science 61:1137–1143.

Gehrke, P. C., K. L. Astles, and J. H. Harris, J. H. 1999. Within catchment effects of flow alteration on fish assemblages of the Hawkesbury-Nepean River system, Australia. Regulated Rivers: Research and Management 15:181–198.

Gehrke, P. C., and J. H. Harris. 2000. Large-scale patterns in species richness and composition of temperate riverine fish communities, south-eastern Australia. Marine and Freshwater Research 51(2):165–182.

Gehrke, P. C., D. M. Gilligan, and M. Barwick. 2001. Fish communities and migration in the Shoalhaven River—before construction of a fishway. NSW Fisheries Final Report Series No. 26.

Gillanders, B. M., and M. J. Kingsford. 2002. Impact of changes in flow of freshwater on estuarine and open coastal habitats and the associated organisms. Oceanography and Marine Biology 40:233–309.

Harris, J. H. 1983. The Australian bass, Macquaria novemaculeata. Doctoral dissertation. University of New South Wales, Australia.

Harris, J. H. 1985. Diet of the Australian bass, *Macquaria novemaculeata* (Perciformes: Percichthyidae) in the Sydney Basin. Australian Journal of Marine and Freshwater Research 36(2):219–234.

Harris, J. H. 1986. Reproduction of the Australian bass, *Macquaria novemaculeat*a (Perciformes: Percichthyidae) in the Sydney Basin. Australian Journal of Marine and Freshwater Research 37(2):209–235.

Harris, J. H. 1988. Demography of Australian bass, *Macquaria novemaculeata* (Perciformes, Percichthyidae), in the Sydney Basin. Australian Journal of Marine and Freshwater Research 39:355–369.

Hindell, J. S. 2007. Determining patterns of use by black bream, Acanthopagrus butcheri (Munro, 1949) of re-established habitat in a south-eastern Australian estuary. Journal of Fish Biology 71:1331–1346.

Heupel, M. R., and R. E. Hueter. 2001. Use of a remote acoustic telemetry system to monitor shark movements in a coastal nursery area. Pages 217–236 *in* J. R. Sibert and J. L. Nielsen, editors. Electronic tagging and tracking in marine fisheries. Kluwer Academic Publishers, Dordrecht, Netherlands.

Heupel, M. R., C. A. Simpendorfer, and R. E. Hueter. 2004. Estimation of shark home ranges using passive monitoring techniques. Environmental Biology of Fishes 71:135–142.

Heupel, M. R., J. M. Semmens, and A. J. Hobday. 2006. Automated acoustic tracking of aquatic animals: scales, design and deployment of listening station arrays. Marine and Freshwater Research 57:1–13.

Howell, T., L. J. Laurenson, J. H. Myers, and P. L. Jones. 2004. Spatial, temporal and size-class variation in the diet of estuary perch (*Macquaria colonorum*) in the Hopkins River, Victoria, Australia. Hydrobiologia 515:29–37.

Loneragan, N. R., and S. E. Bunn. 1999. River flows and estuarine ecosystems: implications for coastal fisheries from a review and a case study of the Logan River, southeast Queensland. Australian Journal of Ecology 24(4):431–440.

Lucas, M. C., and E. Baras. 2000. Methods for studying spatial behaviour of freshwater fishes in the natural environment. Fish and Fisheries 1(4):283–316.

Mallen-Cooper, M. (2000). Taking the mystery out of migration. Pages 101–111 *in* D. A., Smith, D. C., Koen, J. D. editors. Fish movement and Migration, Australian Society for Fish Biology Workshop Proceedings. Hancock Bendigo, Victoria.

McDowall, R.M. 1988. Diadromy in fishes: migrations between freshwater and marine environments. Timber Press. London.

McDowall, R. M. 1996. Australian freshwater cods and basses. In Freshwater Fishes of South-eastern Australia: 150–158. Reed, Sydney.

McCarraher, D. B. and J. A. McKenzie. 1986. Observations on the distribution, growth and diet of estuary perch (*Macquaria colonorum*) in Victorian waters. Arthur Rylah Institute for Environmental Research Technical Report Series No. 42, Department of Conservation, Forests and Lands, Melbourne, Australia.

Pusey, B. J., M. J. Kennard, and A. H. Arthington. 2004. Freshwater Fishes of Northeastern Australia. CSIRO Publishing, Victoria, Australia.

Reinfelds, I., Lincoln-Smith, M., Haeusler, T., Ryan, D. and Growns, T. 2009. Hydraulic assessment of environmental flow regimes to facilitate fish passage through natural riffles: Shoalhaven River below Tallowa Dam, New South Wales, Australia. River Research and Applications 26:589–604.

Robins, J. B., I. A. Halliday, J. Staunton-Smith, D. G. Mayer, and M. J. Sellin. 2005. Freshwater-flow requirements of estuarine fisheries in tropical Australia: a review of the state of knowledge and application of a suggested approach. Marine and Freshwater Research 56(3):343–360.

Roy, P. S., R. J. Williams, A. R. Jones, R. Yassini, P. J. Gibbs, B. Coates, R. J. West, P. R. Scanes, J. P. Hudson, and S. Nichol. 2001. Structure and function of south-east Australian estuaries. Estuarine, Coastal and Shelf Science 53:351–384.

Taylor, M. W., S. D. Laffan, S. D. Fielder., and I. M. Suthers. 2006. Key habitat and home range of mulloway *Argyrosomous japonicus* in a south-east Australian estuary: finding the estuarine niche to optimise stocking. Marine Ecology Progress Series 328:237–247.

Whitfield, A. K. 1994. Abundance of larval and 0+ juvenile marine fishes in the lower reaches of three southern African estuaries with differing freshwater inputs. Marine Ecology Progress Series 105:257–267.

Williams, N. J. 1970. A comparison of the two species of the genus *Percalates* Ramsey and Ogilby (Percomorphi: Macquariidae) and their taxonomy. NSW State Fisheries Research Bulletin 11.

American Fisheries Society Symposium 76:265–288, 2012

Differential Habitat Use between Adult European Sea Bass and North Atlantic Cod in the Inner Oslo Fjord: Influence of Abiotic Environmental Variables

Anne-May Ilestad
Norwegian University of Life Sciences (UMB), Ås, Norway
and
Norwegian Institute for Water Research (NIVA), Oslo, Norway

Thrond O. Haugen*
Norwegian Institute for Water Research (NIVA), Oslo, Norway
and
Norwegian University of Life Sciences (UMB), Ås, Norway

Jonathan E. Colman
Norwegian University of Life Sciences (UMB), Ås, Norway
and
Department of Biology, University of Oslo (UiO), Oslo, Norway

Abstract.—Environmental variables and interspecific interactions influence fishes' habitat use. The eurythermic and euryhaline European sea bass *Dicentrarchus labrax* is increasing in Norwegian waters. We compare diel- and seasonal habitat use and home ranges of this species with Atlantic cod *Gadus morhua* in the Oslo-fjord, Norway. Individual depth and horizontal positions were obtained using acoustic telemetry. Fish were tracked using both portable and stationary receivers. Both species displayed high site fidelity from June to October. In late October, the sea bass migrated out of the core area whereas the cod remained. Cod home ranges (11.4 ± 5.5 ha) were smaller than sea bass home ranges (42.6 ± 26.1 ha) throughout the study. Home range increased with individual body size in sea bass, but not cod. The two species had differential diel- and seasonal depth use. Sea bass used depths shallower than 4 m during June through September, with diel vertical migrations rarely exceeding 2 m. After September, sea bass sought deeper waters; stabilizing at 11 m in November with a diel migration amplitude of 7 m. Cod stayed deeper than the 9–10 m thermocline in August with vertical migration amplitude of 7 m and moved into daytime depths of 6.5 m with diel amplitude of 4.5 m in December. Sea bass actively selected warmer temperatures and the lowest salinities available, whereas cod avoided temperatures higher than 13°C and salinities below 24 ppt. The differential temperature preference for these species is likely the main driver for the differential depth utilization. A model using air temperatures as a predictor was constructed

*Corresponding author: throndhaugen@umb.no

predicting daytime and night-time depth use. This study provides novel insights to the habitat use of the two species under Nordic fjord conditions—information to be included in upcoming management programs emerging due to the continuing northward colonization of sea bass along the Norwegian coast.

Introduction

Aquatic life is continuously affected by a changing environment. Among other things, predictions of a wetter and warmer climate in the Northern hemisphere will impact the ocean's temperature and salinity levels (IPCC 2007). Changes in the marine environment ultimately shape available habitat and resources for marine organisms and influence their distribution on a temporal and spatial scale (Stenevik and Sundby 2007). The impact of such changes and relocations is poorly documented. However, recent efforts have focused considerably on the disappearance of species from areas once inhabited by well-established populations. On the other hand, a changing environment will likewise enable species to establish populations in new areas. The arrival of new species may alter ecosystems by influencing trophic networks (food webs). Importantly, new species could potentially out-compete already established species (Mackenzie et al. 2007).

European sea bass (*Dicentrarchus labrax*, hereafter 'sea bass') have become increasingly prevalent in Norwegian coastal waters over the last 10–15 years (Colman et al. 2007). It is a highly eurythermic species, tolerating temperatures between 2–32°C (Jourdan-Pineau et al. 2010) and with a preference for warm water (Pickett and Pawson 1994). Sea bass from the North Sea have been shown to require temperatures higher than 9°C for spawning (Pawson et al. 2000). Pawson et al. (2007) suggest that "...climate warming may have lengthened the duration of residence of adult sea bass in summer feeding areas and ... has allowed sea bass to spawn much further north than previously" in the North Sea (Colman et al. 2007). Furthermore, sea bass is also a euryhaline species that can inhabit a wide range of salinity levels (Conides and Glamuzina 2006; Dalla Via et al. 1998). Thus, the ability of adult bass to tolerate highly variable temperatures and salinity levels could be a competitive advantage over feeding grounds and habitat (Pickett and Pawson 1994).

The diet of sea bass is similar to that of other commercially and ecologically important species in Norwegian waters such as Atlantic cod (*Gadus morhua*, hereafter 'cod'), Atlantic salmon *Salmo salar* and brown trout *Salmo trutta* (Colman et al. 2007). From a young age (approximately 3 months), sea bass are opportunistic predators feeding on a range of different marine species, e.g. small fish, crustaceans, annelid worms (Pickett and Pawson 1994). The type of food depends on the distribution and prey availability. Deep-water sea bass prey mainly on fish. Offshore-living sea bass eat pelagic fish (mackerel (Scombridae) and pilchard (Clupeidae)), while sea bass occupying inshore waters (summer/winter) often eat more shallow-water food types (crustaceans and annelid worms) (Pickett and Pawson 1994). Movement patterns for a predatory fish such as sea bass are likely influenced by the activity and movement patterns of their preferred prey species and resource availability, as well as interspecific interactions with other predatory fish species.

There are positive correlations between sea temperature and activity of fish, where temperature influences the ability of fish to feed (Rankin and Jensen 1993) and to avoid predators (Domenici et al. 2007; Linehan et al. 2001). Pickett et al. (1994) found that feeding of sea bass is much reduced when sea temperatures were below 10°C, and feeding has been shown to stop below 7°C (Pastoureaud 1991). If applied to Norwegian conditions, one can expect low food intake and reduced activity levels and movement for adult sea bass in the period of November to May.

Prior to late 1980s, mature sea bass inhabiting waters surrounding the UK migrated between summer foraging grounds, offshore prespawning, and spawning areas on the south and west coasts of England and Wales (Pawson et al. 1987, 2005). At that time the southern North Sea was too cold in winter to support maturing female bass. The migration patterns out of the southern North Sea had changed by 2000–2004 (Pawson et al. 2007) and spawning sea bass have been observed as far north as the west coast of Norway (Matre) (own observations), and researchers now address whether sea bass along the Norwegian coast is part of a self-supporting stock (Colman et al. 2007).

Despite sea bass being widely distributed in Norwegian waters, it is only recognized by the Norwegian authorities as a 'rare visitor.' Analyses of scale samples have revealed that Norwegian sea bass have a higher growth rate than sea bass sampled from other populations in the North Sea. Furthermore, data indicate that Norwegian sea bass mature one year earlier, as 4-years-olds, compared to 5-year-olds in most other southern European populations (Colman et al. 2007). During the last 5 years, Norwegian fishermen report more sea bass being caught, in both by-catches and targeted fishing. There is also an increasing awareness and interest among commercial fishermen and recreational anglers for catching sea bass in Norway (Colman et al. 2007). However, as to yet, Norway has no management plans for regulating the fishing of sea bass (Colman et al. 2007).

Presently, there is no information as to how sea bass will impact the existing Norwegian marine ecosystems. As a nonindigenous species, its presence could potentially influence the distribution and production of established species through direct mortality and competition. Sea bass may also influence Norwegian marine ecosystem diversity and overall ecological interactions with the entire food chain (Occhipinti-Ambrogi 2007; Parker et al. 1999). It is therefore of utmost importance to study the habitat use of this new species in Norwegian waters and to compare its behavior with indigenous species so as to be in a good position for formulating knowledge-based management programs for this species in the future. Owing to the fjord structuring of the Norwegian shoreline, existing knowledge on sea bass habitat use, based on studies from elsewhere in Europe, may have little relevance for Norwegian conditions. It is therefore exciting to study how sea bass habitat use and behavior in Norwegian waters compares to other sea bass populations, especially from southern areas.

As the first study ever, we compare the habitat use of coexisting sea bass and cod. This study also presents the first estimates on sea bass home ranges, habitat selection ratios, and daily- and seasonal vertical movement patterns in Norwegian waters. The reason why we compare sea bass with the indigenous cod is that cod is a commercially very important species in most of North Atlantic waters (Cote et al. 2004; Swain et al. 1998). Hence, any documentation of interaction effects between the two species may have important commercial implications. We test and quantify whether the daily movement patterns (i.e., diel vertical migration) of sea bass and cod could be attributed to external environmental variables such as temperature and salinity, and also if migration and home range sizes change over seasons. Finally, we test whether the sea bass and cod exhibit differential habitat preferences (i.e., temperature, salinity) and explore whether there is overlapping use or avoidance (segregation in the water column).

Methods

The study area

The core study area is located in the inner Oslo Fjord, in the Bærum basin and comprises 2,200 ha of surface area (Figure 1). The Oslofjord is 107 km long, with a mean annual surface temperature of 7.5°C. Our study area is part of the Bærum basin and has an average depth of 30 m. Three medium sized rivers (Sandvika River, Asker River, Lysaker River with median discharge of 3.0, 0.4 and 1.6 m^3/s respectively) and numerous smaller rivers empty into the Bærum basin. The fjord is a popular recreation area for citizens in Oslo and surrounding areas.

Tagging procedure

All fish used in this study were sampled by angling. A total of 20 fish (average length of 45 cm) were captured within the study area; 11 adult European sea bass, 9 Atlantic cod. Sea bass were caught by casting from land or boat, using a Rapala Original 13 cm lure with the barbs removed from the three treble hooks (to minimize injury to the fish). Cod were caught jigging from a boat and using a blue and sliver 22 g Møresilda with the barbs removed from the one treble hook. Once hooked, fish were retrieved as quickly as possible to minimize "the fight" and reduced loss of energy and accumulation of lactic acid that can be detrimental (Davie and

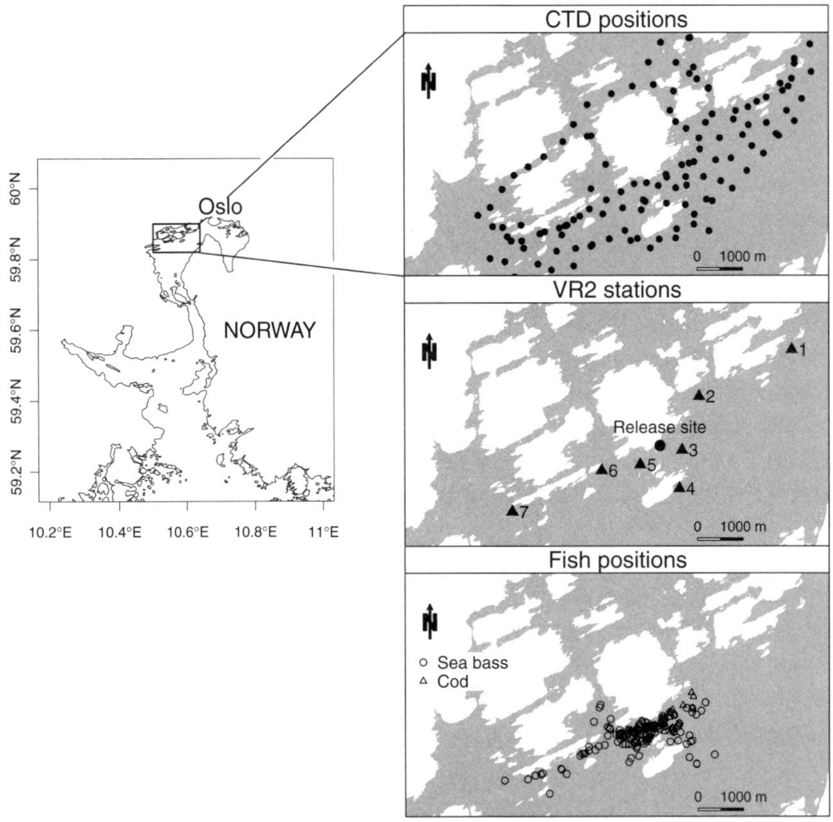

Figure 1. Map of the study area with positions of sites where CTD profiles were measured (top panel) and positions of VR2w stations and the surgery- and release site (mid panel) and manual tracking positions for the sea bass and cod in this study (lower panel).

Kopf 2006). To further minimized damage to the fish, no net was used when landing. Each fish was landed by being pulled out of the water and lowered directly into an 80 L tank of water where the hook was removed and the fish would remain until tagging. All handling of the fish was done using a wet towel. The holding tanks were constructed of hard, black plastic and were covered with a towel to provide darkness. The black color of the tank and darkness provided by the towel was found in earlier fieldwork to be very important for sea bass. Lighter (white) tank/bucket/plastic colors and sunlight stressed sea bass and caused mortality in an earlier study (Colman et al. 2007). New water was provided to the tanks by buckets within every 5 min while there were fish in a tank. No more than 5 fish were in one tank at a time.

The tagging procedure was carried out in two periods; 26 June 2007 and 30 August 2007. Nine sea bass were tagged in the first period and the remaining two sea bass and all the cod were tagged in the latter period. One sea bass was killed by an angler 63 d after its release. Individual lengths were 48.89 ± 6.58 cm (±SD) for sea bass and 46.11 ± 5.46 cm for cod. Following the implantation protocol recommended by Mulcahy (2003), each fish was tagged with acoustic transmitters (ADT-MP13, Thema AS), measuring 7 ± 37 mm and weighing 11 g in air and 6 g in water. The transducer weight to fish weight ratio ranged from 0.4% to 2.6% (0.9 ± 0.4, SD). The transmitter contained a pressure sensor that enabled depth sensoring at a precision of 0.2 m, with a maximum depth of 100 m (i.e., in cases where the individuals dwelled at depth >100 m the tag would provide information of 100 m). The reason for using maximum of 100 m was that the deepest point in the Bærum basin is approximately 85 m.

For sedation, each fish was placed in a 50 L black bucket with seawater and clove oil (concentration of 2 mL per 5 L seawater) (Mylonas et al. 2005). Once anesthetized (after 1–2 min), the fish were placed in a V-shaped polystyrene cradle with the belly pointing upwards. A constant flow of water was provided over the gills from a silicon hose attached to an aquarium pump submerged in a tank with maintenance dosage (0.5 mL per 5 L seawater). Before a small incision was made (12–15 mm) on the peritoneal cavity (Cote et al. 2002; Fabrizio and Pessutti 2007), the surgical area was carefully cleaned with chlorhexidine (0.5 mg/mL). The ethanol-sterilized transmitter was rinsed in sterile physiological salt water before being implanted in the peritoneum of the fish. Using monofilament suture, three stitches sealed the incision using surgical sewing equipment. All the surgery was carried out by Professor Trygve Poppe (Norwegian School of Veterinary Science). The tagging program along with the implantation protocol was approved by the National Animal Research Authority (Licence number 07/45773) before the implantation was initiated.

After surgery, the length and identification number of each fish was noted before they were transferred for overnight recovery in a 1.5 × 2.5 m keep-net located 5 m below sea surface. The full surgery procedure from the moment a fish was removed from its keep tank to the time it was released in the overnight keep-net was less than 2 min per fish. After 24 h of recuperation in the keep-net, each fish was transferred back into the tanks in the boat and released into the study area at their respective capture locations.

Tracking procedure

Tracking data were obtained using two methods; manual tracking and stationary tracking (logging). Daytime manual tracking was conducted from June 2007 to February 2008, and seven stationary VR2w receivers operated from September 2007 (Figure 1). Tracking began the same day as the release, using a THELMA AR-MANUAL portable receiver. We used an omnidirectional hydrophone to detect individuals within the detection range and a directional hydrophone for cross-tracking to obtain a position estimate for the detected individual. Each position was obtained where the signal (RSSI) was strongest, i.e. the highest position resolution. The tracking signals (ID and depth information) were displayed on a screen. The strength of the RSSI was noted as well as the GPS position and depth. With a range of up to 1 km (tested at a location 40 km south of the study area, in the same fjord), the best performance of the THELMA receiver proved to be within 100–300 m from a tagged fish. The transmitters were programmed to ping at the 69 kHz frequency with random intervals at every 20–60 s. Estimated longevity of acoustic transmitters was 12 months (guaranteed to be longer than 7 months).

In addition to random manual tracking from June 2007 through to February 2008, an intensive tracking procedure was carried out in three sequential weeks (in August, October, and December) during daytime. Tracking covered 120 stations based on a grid of 500 × 500 m within the study area (Figure 1). Water temperature, salinity, and depth were measured at each station using a salinity-temperature-depth logger (CTD, SAIV-SD204, Figure 1). At each station, tracking was conducted for 10 min. At stations where fish were detected, cross tracking with the directional hydrophone was conducted in order to assess as precise position as possible (using the RSSI information) and the corresponding depths of the fish were recorded with a precision of 0.2 m. All tracked individuals were assigned the depth-specific salinity and temperature from the nearest CTD–profile matching the depth position of the fish.

In addition to the manual tracking system, seven stationary underwater omni-directional receiver loggers (VR2w, VEMCO, http://www.vemco.com/) were placed within the study area (Figure 1). Within a range of up to 1 km, identification number and depth of detected fish were recorded and stored on flash memory devices. The data were transferred to ordinary laptop computers via bluetooth connection. The seven receivers were moored within the study area (Figure 1). The receivers were attached to the mooring rope at about 2 m depth, with the hydrophone pointing downwards. Solar cell driven buoy lights and radar reflectors were mounted on the top of the 60 L orange buoys. To prevent potential losses of data, the VR2w were emptied every month. This was carried out in situ via the bluetooth connection. Each buoy was also marked with a waterproof note providing information about the project and contacts if necessary.

Statistics

Manual tracking data

A total of 815 individual daytime depth positions were obtained and differences in depth use between sea bass and cod were tested using a Welch two sample *t*-test. During the manual tracking, sea depth for each observation was obtained to describe the maximum sea depth available for the individual fish. This was obtained for 378 individual observations.

Differences in water temperature and salinity use between the sea bass and cod were also tested using a Welch two sample *t*-test, but separately for the early season (August/September) and the late season (October), since the CTD measurements were obtained separately for those two periods. Salinity and temperature were registered for 279 of the observations. By combining the salinity use and temperature use of the two species with the *availability* of these variables in the study area we could assess habitat selection ratios (w_i). These ratios, and their associated Bonferroni-adjusted 95% confidence intervals, were estimated using the FishTel software (version 1.4vi, (Rogers and White 2007). Selection ratios with confidence intervals that did not overlap with one was interpreted as positive habitat selection and ratios lower than one were interpreted as habitat avoidance. For temperature, discrete habitat types were constructed from 1 degree bins, whereas salinity types were grouped into 2 ppt bins. Log-likelihood ratio tests for both within-species and between-species variation in habitat selection were conducted using the same software and following procedures provided in Rogers and White (2007).

To explore horizontal habitat use, we used information from the manual tracking positioning data for estimation of home range (HR) shapes and sizes. The home ranges were estimated as individual-wise polygons (minimum-convex-polygon approach) for individuals for whom we had more than three positions over a defined time slot. We also estimated kernel-based Utilization Distribution (UD) areas. Even though the UDs showed similar patterns as the polygon HRs for the sea bass and cod, the generally few observations resulted in unrealistically large UD estimates and they were therefore not taken into account when interpreting the results. Based on observations of sea bass shoaling and migration behavior during October to November, the home ranges were estimated separately for these two periods. Hence, by exploring shapes and sizes of HRs estimated over the June–September period and for the June–December period, eventual seasonal and species differences in these HR attributes should appear. We also explored to what degree HR size could be attributed to individual status by regressing HR area with individual body size. The home ranges were estimated using the adehabitat/ade4 library in R (R Development Core Team 2009). Differences in habitat sizes between cod and sea bass were tested using Welch two-sample analysis of variance (ANOVA) (i.e., accounting for heteroscedasticity).

VR2w data

Since the VR2w data entailed more or less continuous observations on depth use for many of the individuals, it allowed for more detailed exploration on diel patterns in the vertical aspect of the habitat use compared to the manual tracking data. In particular, we explored depth use patterns for the sea bass and cod during night time and day time. Information on time of sunset and sunrise was retrieved from the sunset computer available from MindSpring (http://www.mindspring.com/~cavu/sunset.html). Species differences in seasonal change in the daily amplitude of vertical migration (maximum depth – minimum depth) were also explored. All these tests were performed fitting various mixed effects linear models (lme). The response variable depth was log transformed to homogenize variation. To control for variation due to repeated measurements of the same individuals in the model, ID was modeled as random effect (Pinheiro and Bates 2000). In order to allow for estimation of nonparallel temporal trajectories, the fixed effects in the models (i.e., species and time) was always tested under a fully factorial design:

$$\ln(\text{depth}) = \text{time} + \text{species} + \text{time} \times \text{species} + \text{individual(rand)} + \text{VR2(rand)} + \text{error} \quad (1)$$

where time was coded as a month effect and individual ID and VR2-station was modeled as a random effects to account for nonindependency of observations. The models were fitted using the nlme library included in R (version 2.10.1, R Development Core Team 2009). Parameters were estimated using the restricted maximum likelihood technique (Pinheiro and Bates 2000).

To explore how temperature affected depth use in the sea bass and cod in more detail than the manual tracking data allowed for, we used temperature data from a local weather station situated about 4 km off the core study area (Asker weather station: 59.8562°N, 10.437°E). These temperature data matched the high temporal resolution of the VR2w data series. However, owing to the high specific heat capacity of water the temperature exchange between air and water constitute a delayed diffusion process. We assessed the most appropriate time lag of this temperature diffusion process using the model selection procedure Akaike Information Criterion (AIC). A number of lme models with the same model structure as model (1), but with air temperature replacing time, was fitted to the individual daily mean depth position data. Importantly, the air temperature effect was time lagged by varying degree of number of days. Technically, the delay process was performed as moving averages of the air temperature and the model receiving the best AIC supported (i.e., the most optimal moving average level) was chosen for further interpretation (Burnham and Anderson 1998).

Results

Nine of the eleven tagged sea bass (81.8%) and seven of the nine cod (77.8%) were recorded during the tracking period (August 2007 to February 2008). One sea bass was caught by an angler during the study period. The remaining fish may have outmigrated from the study area, died, or been caught by anglers (and not reported) or the transmitters may have failed.

During June to October, both the tagged sea bass and cod showed strong site fidelity. This was confirmed by data from the VR2w loggers operated from August 2007. Due to winter migration (away from the core study area) during late October when the water approached isothermal (at 10–11°C, Figure 2), tagged sea bass were not detected by the VR2w station 5, but some of them were detected by stations 6 and 7 (Figure 1). Cod did not show signs of dispersal and were recorded until the last tracking day on 11 February 2008.

In total, 635 CTD profiles were measured over the August to December period (Figure 2). At depths larger than 15 m temperatures were stable at 8–10°C over the entire August to December period. However, for the shallower 10–12 m, there were substantial temporal variations in these profiles. In August and September, there was a pronounced thermocline at 8–10 m and two halinoclines at ~3 m and ~10 m. Isothermal was reached during October at which there was ~10°C all through the water column. In November and December a new thermocline established at 8–10 m with temperatures lower than 8–10°C in the upper layers. During this period the salinity was higher than 25 ppt all through the water column for most profiles.

Daytime home range, habitat use and -selection: manual tracking data

Even though both species displayed strong site fidelity over the June to September period, home ranges (95-level polygons, Figure 3) were significantly larger for sea bass (42.6 ± 26.1 ha, SD) than cod (11.4 ± 5.5 ha) during this period (Figure 3). There was significantly larger among-individual variation in home-range sizes for sea bass than cod (CV_{cod} = 0.45; $CV_{sea\ bass}$ = 0.62). In the sea bass, part of the inter-individual variation could be linked to individual size (BL, body length) as large individuals had larger home ranges than small ones (linear regression: ln(HR) = –32.6 + 9.2ln(BL), $R^2_{adj} = 0.71$, $p = 0.012$, $n = 8$). All cod home ranges had some degree of area overlap, whereas 62% (13 out of 21 comparisons) of the sea bass home ranges did not overlap at all. For the 8 sea bass comparisons that had overlapping areas, there was a significant negative relationship between individual body size difference and percentage overlap

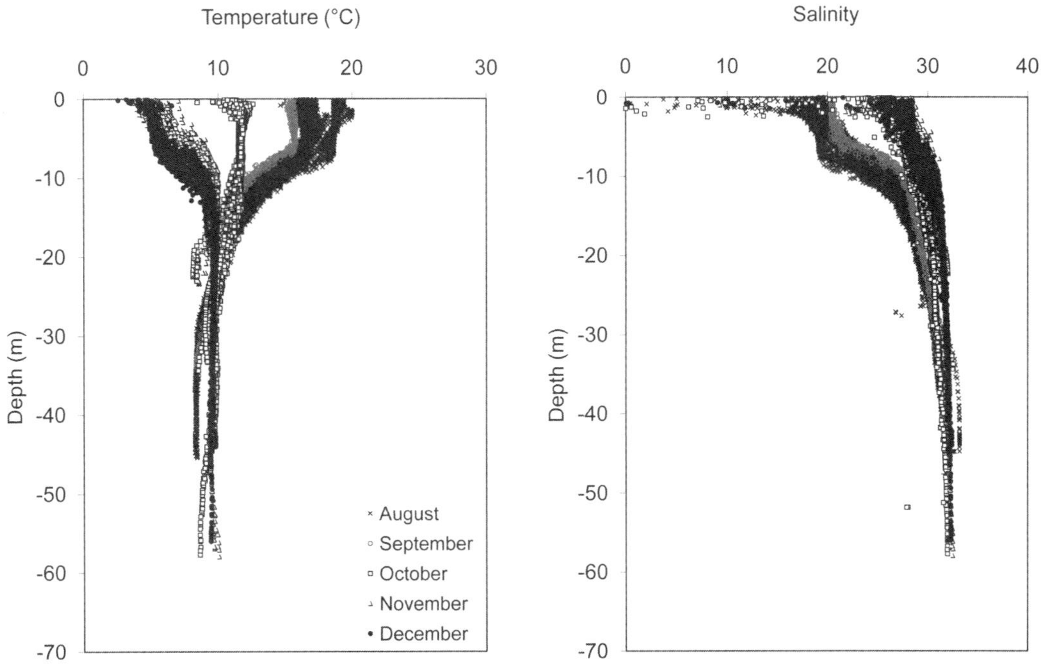

FIGURE 2. Salinity and temperature profiles based on measurements from all CTD profiles (Figure 1) recorded during August to October 2007.

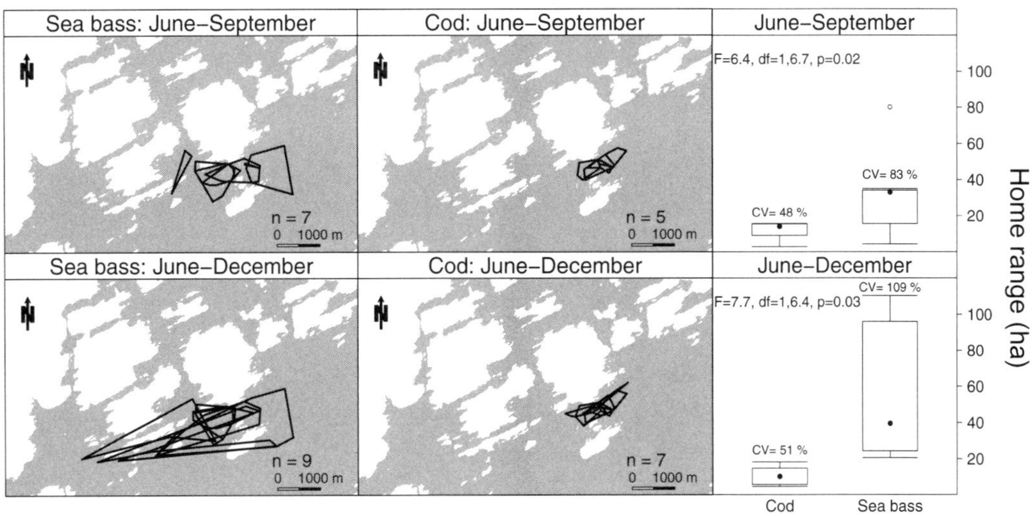

FIGURE 3. Minimum Polygon Home Ranges (HR) for sea bass and cod from the inner part of Oslofjord as estimated over two time periods; June–October (upper row) and June–December (lower row). Corresponding box-plots of HR size distributions for the sea bass and cod and the corresponding Welch-Anova test statistics, testing for differences in HR-sizes between the two species, is also provided (rightmost panels).

($r_p = -0.72$, $p = 0.04$, $n = 8$), i.e., the more different the body sizes, the less overlap in home range. When extending the period over which home range was estimated to the June to December period, a south-west dispersal pattern appeared for the sea bass and mean home range increased greatly to an average of 71.0 ± 82.2 (range: 20.6–261.5 ha). However, the body size effect on home-range size for sea bass was no longer evident for the June–December interval. The cod home ranges remained similar when comparing among the periods.

Daytime depth use for sea bass and cod differed significantly from each other (Figure 4). While sea bass had a preference for shallower water, cod moreover occupied deeper water (Figure 4). Mean depth recorded for sea bass and cod were 2.74 ± 2.21 m (±SD) (range: 0.35–15.68 m) and 9.79 ± 4.70 m (range: 0.39–18.81 m) respectively. Mean daytime depth use for the two fish species also differed between seasons (early; August–September and late; October–December). Sea bass occupied depths of 2.5 ± 1.8 m (early) and 4.09 ± 3.2 m (late), while cod stayed at 9.68 ± 3.2 m earlier in the season compared to 10.10 ± 3.9 m in late season.

Mean daytime salinity of sea bass (23.34 ± 3.67 ppt) differed significantly from cod (25.94 ± 3.49 ppt) (Welch two sample t-test; $t = -5.86$, $df = 131.73$, $p < 0.00001$, n

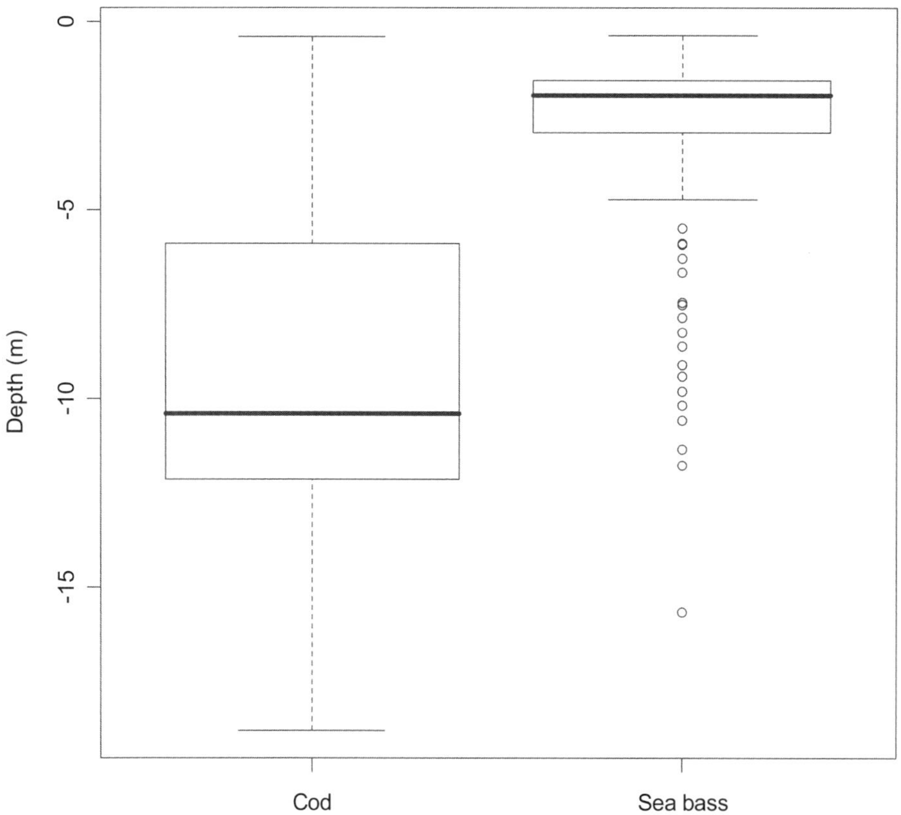

Figure 4. Box plot of daytime depth (m) position of cod and sea bass during June–October 2007. Data were retrieved by manual tracking.

= 298). Using the same test, mean daytime temperature differed significantly between sea bass (14.22 ± 2.39°C) and cod (12.45 ± 3.24°C) (t = 4.59, df = 104.78, p < 0.00001, n = 298).

The habitat selection analyses revealed that sea bass individuals were highly synchronized in their habitat selection as, except for salinity in October, there was no evidence for significant inter-individual variation in habitat selection (Figure 5). Cod individuals displayed significant among-individual variation in August for their degree of habitat selection in relation to both temperature and salinity, whereas they were more synchronized in October. During August, both species performed nonrandom habitat selection, but during October, only sea bass individuals did so. The selection ratios where highly different between the two species as the sea bass always actively selected the warmer temperatures and the lowest salinities available, whereas the cod avoided temperatures above 13°C and salinities below 24 ppt.

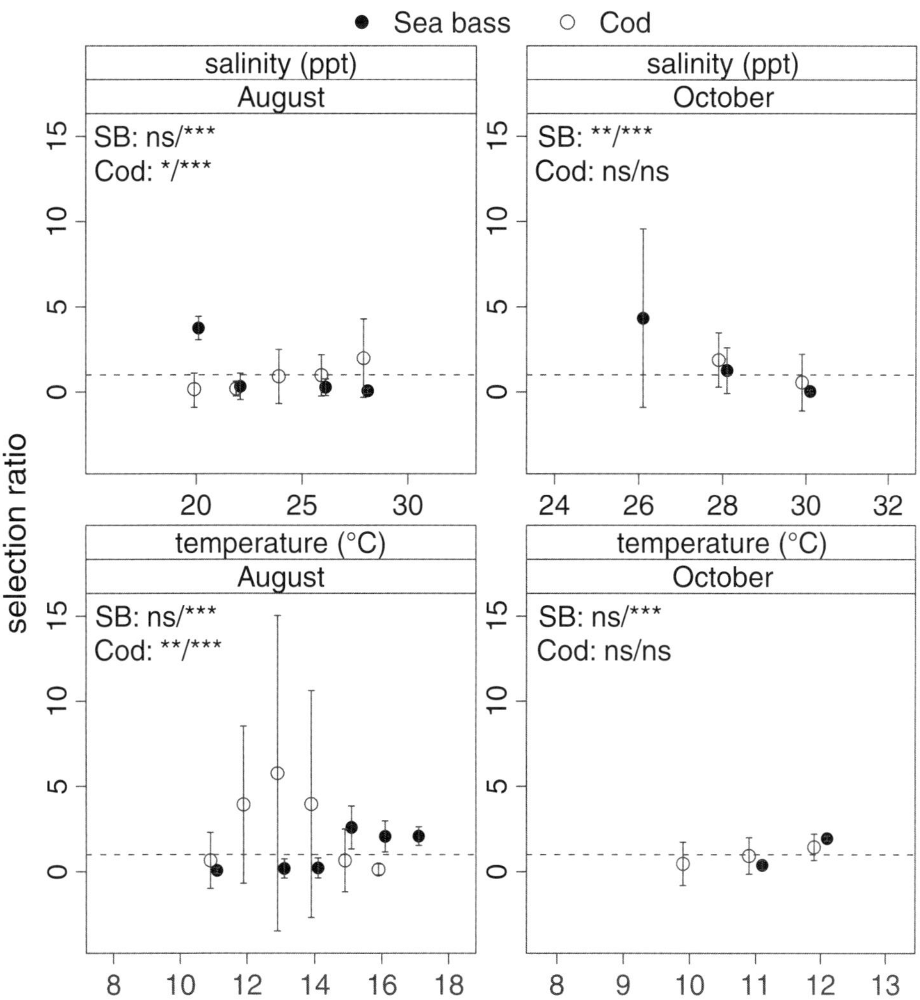

Figure 5. Temperature and salinity habitat selection ratios (w_i) for cod and sea bass from the inner part of Oslofjord during August and October 2007. Points represent species-specific mean w_i values for a given habitat variable level and error bars represent 95% confidence bounds. Dashed horizontal line for w_i = 1 represent the "no-selection" line.

Diel vertical migration pattern: the VR2w-data

The VR2w loggers registered more than 600,000 pings over the August–December 2007 period. 68% of the recorded pings were from station 5 (Figure 6), which is the station closest to the release site (Figure 1).

The cod and sea bass that were registered by the VR2w-loggers had significantly different depth positions both during night and daytime, and they also had different temporal trajectories in their depth use over the August to December period ($p_{time*species} < 0.0001$, lme-analysis). In general, sea bass went from shallow water layers to deeper water layers as the fall period progressed, whereas cod sought more shallow water (Table 1). The diel vertical migration amplitude also changed differently over time ($p_{time*species} < 0.0001$, lme-analysis), where cod vertical migrated 7 m in September and decreased to 4 m when approaching December and sea bass migrated 2 m in September and ended up at 6–7 m in December (Figure 7). There was, however substantial day-to-day within-individual variation in these amplitudes, as the within-individual variance component accounted for 72% of the explained variation. The among-VR2 station variation in daily vertical migration amplitudes accounted for less than 0.3% of the variation. The main reason for the inter-individual variation in diel vertical migration amplitude could be attributed to a substantially higher inter-individual variation in daytime depth use compared to the nocturnal depth use. In the daytime depth use lme-model, the individual variation accounted for 72% of the explained variation, compared to 44% in night-time depth use.

The air temperature was shown to have a very different effect on depth use, both during day- and night-time, between the two species. A model where the air temperature was averaged over 6 d received the highest AIC support (Table 2) and a highly significant interaction effect between this averaged air temperature and species estimated that sea bass go shallower with increasing temperature (0.88 m/°C), whereas cod go deeper (–0.27 m/°C). This model accounted for 73% of the variation in daytime depth use over the September to December period (Figure 8). A similar model applied on night-time data showed the same tendency, except for cod having a shallower intercept at night (–5.6 m versus -6.8 m) whereas sea bass had the same intercept (–12.1 m). The model explained 74% of the variation of nigh-time data.

Discussion

Our study demonstrates that habitat segregation exists between sea bass and cod in our study area. Part of this segregation can be attributed to differential habitat preferences, as temperature and salinity selection ratios differ between the two. However, it is not possible to discern whether competition over (food) resources can explain some of the segregation observed, as neither food nor stomach content data were available.

Habitat segregation

Home range or habitat segregation in wild adult sea bass has not, to our knowledge, been reported before. However, it is well known from sea bass mark–recapture studies in British waters that the sea bass during the summer and autumn season display high site fidelity, where tagged individuals rarely are recaptured more than 16 km off the tag and release site (Pawson et al. 2007). During winter to spring, the UK sea bass migrate to pre spawning and spawning areas that might be located more than 80 km away from the summer feeding areas. Because subsequent recaptures of returning spawners are almost exclusively within the same feeding area as

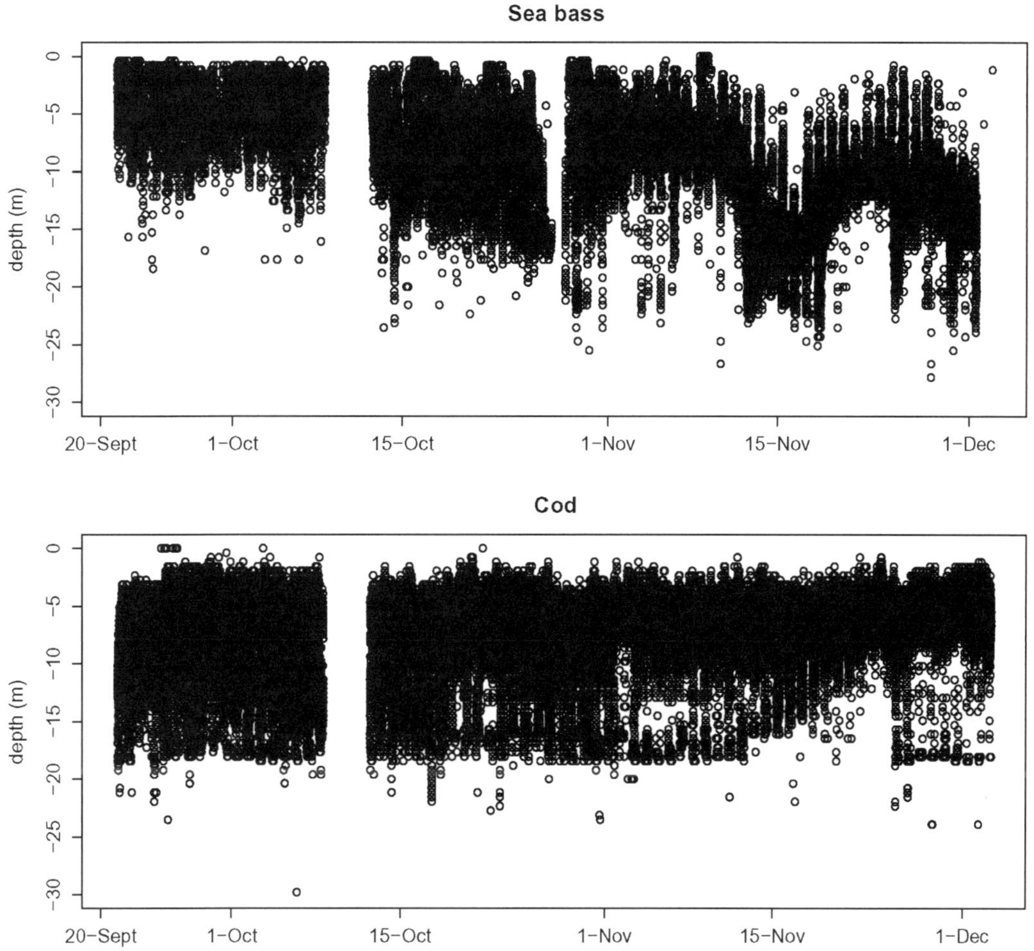

FIGURE 6. Depth positions of cod and sea bass recorded by VR2-stations 3–7 during September to December 2007.

they were tagged, UK sea bass is regarded as a philopatric species (Pawson et al. 2008). A similar pattern is also evident from a sea bass mark–recapture study in Norway. With just one exception, all recaptures have been less than 10 km off the release site (author's unpublished data). Since we know that the sea bass both spawn and stay in the fjord area as post juveniles and adults, it is likely that Norwegian sea bass can be recognized as a philopatric species as well.

In coastal Atlantic cod, more home range estimates have been reported (Espeland et al. 2007; Neat et al. 2006; Robichaud and Rose 2004; Turner et al. 2002). Most of these studies report that coastal cod populations are resident and rarely dispersing more than 15–20 km (Neat et al. 2006). Recently, Espeland and coworkers (Espeland et al. 2007), using acoustic transmitters and triangulation, reported home range sizes of 27.04 ha (range 3.05–77.33 ha) for Atlantic cod located in a Norwegian south-coast fjord. Unfortunately, this study was conducted over a small time slot (44 d) and during the spawning season (February to March). The researchers followed up with some manual tracking showing that many of the cod individuals resided in the particular strait area during summer and autumn, but 50% of the fish tagged dispersed

TABLE 1. Mean and SD of depth observations of cod and sea bass from VR2-stations 3–7 during night and day time over the September to December period in 2007.

Part of day	Species	Month	mean depth	s.d.
Day	Cod	September	–9.40	1.34
Day	Cod	October	–8.70	0.84
Day	Cod	November	–7.41	0.94
Day	Cod	December	–6.59	1.00
Day	Sea bass	September	–4.55	1.06
Day	Sea bass	October	–5.46	1.90
Day	Sea bass	November	–10.50	3.21
Day	Sea bass	December	–8.20	9.93
Night	Cod	September	–8.14	0.74
Night	Cod	October	–7.35	1.02
Night	Cod	November	–5.98	0.63
Night	Cod	December	–5.48	0.68
Night	Sea bass	September	–4.23	0.94
Night	Sea bass	October	–5.89	2.48
Night	Sea bass	November	–10.42	3.02
Night	Sea bass	December	–9.51	5.13

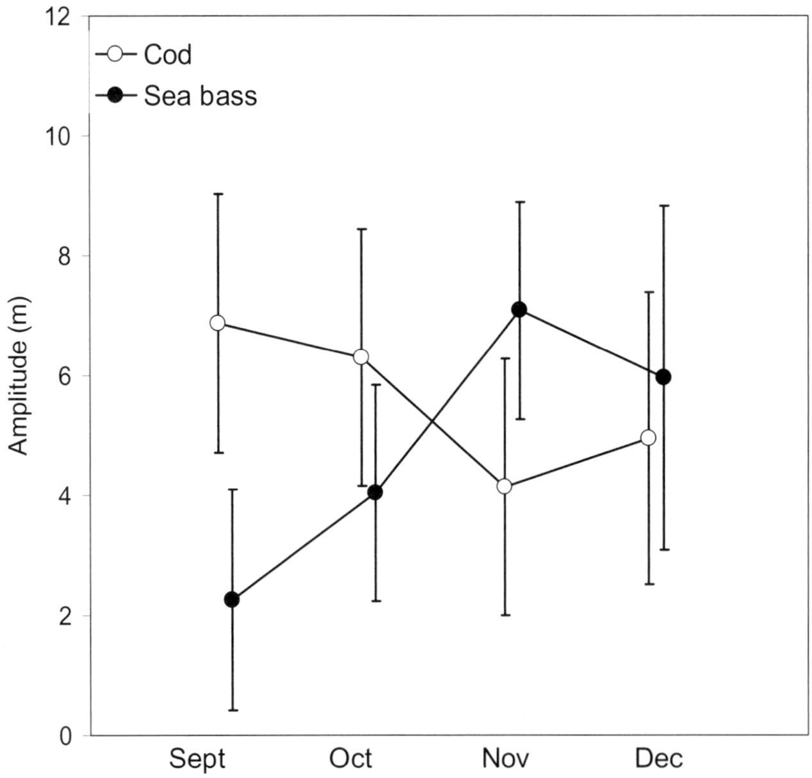

FIGURE 7. Monthly least square means (2S.E.) for diel vertical migration amplitude over the September to December period at VR2-stations 3–7 for cod and sea bass.

TABLE 2. Parameter estimates retrieved from the best lme model estimating the daytime and nighttime depth position of sea bass and cod as function of air temperature (moving average over 6 d) and species as retrieved from the most supported lme model (Table 2). The models have been fitted depth use data from VR2-stations 3–7. The model accounts for individual variation (random effect—not shown) and VR2-station effects (random—not shown) and explains 73 and 76% of the total variation in daytime and nighttime depth use, respectively. The model presented was the most supported among 8 models with varying extent of moving average. The intercept represents the cod's intercept, and the species effect constitutes the sea bass.

Part of day	Term	Estimate	SE	*t*-value	*p*-value
Day	Intercept	–6.84	0.30	–22.51	<.0001
	species	–5.17	0.44	–11.82	<.0001
	ma(6)temperature	–0.27	0.05	–5.40	<.0001
	species* ma(6)temperature	1.15	0.07	–15.88	<.0001
Night	Intercept	–5.60	0.27	–20.74	<.0001
	species	–6.55	0.39	–16.87	<.0001
	ma(6)temperature	–0.25	0.05	–5.49	<.0001
	species* ma(6)temperature	1.13	0.06	–17.47	<.0001

more than 5 km. Two individuals were still in the study area one year after the tagging. Since our study was primarily undertaken during the summer-to-winter period the home range estimates are likely not comparable to the Espeland and coworker study. The habitat use in cod may change over the year (Espeland et al. 2008; Swain et al. 1998) and the habitat use in spawning cod is very likely to differ from nonspawning cod. However, we have reasons to believe that the cod included in our study had very high site fidelity over the entire year since a follow-up tracking trip conducted in February 2009 revealed that 4 out of the 7 cod individuals were still present in the study core area.

Our cod home range area estimates may be smaller than the real areas. The reason for this is that the areas have been estimated using daytime data only. If cod predominantly feed during night-time, as reported by Grant and Brown (1998), the home ranges estimated in our study may be biased small. However, a study conducted by Lokkeborg and Ferno (1999) show that cod may feed all through the day. Hence, our home ranges may still not be severely biased. In future work, activity sensors (e.g., tilt or heart rate sensors) should be included in the transmitter tags to resolve this diel activity pattern issue.

We found a significant positive correlation between body size and June–October home range in sea bass. This finding can be coupled to larger daily food intake demands for larger individuals (Eroldogan et al. 2004; Person-Le Ruyet et al. 2004). Larger individuals will also face lower swimming costs for a given swimming distance (and speed) than smaller individuals (Rankin and Jensen 1993). The home range area overlap results may indicate that sea bass are more territorial than cod, as virtually all cod involved in this study had overlapping areas, whereas only 38% of the sea bass comparisons did so. Interestingly, in the group of sea bass pairwise comparisons that had overlapping areas, individuals with similar body sizes tended to have larger degree of home range area overlap than those that differed in body size. This result indicates that sea bass individuals form groups with equally sized individuals that stay together within the same home range

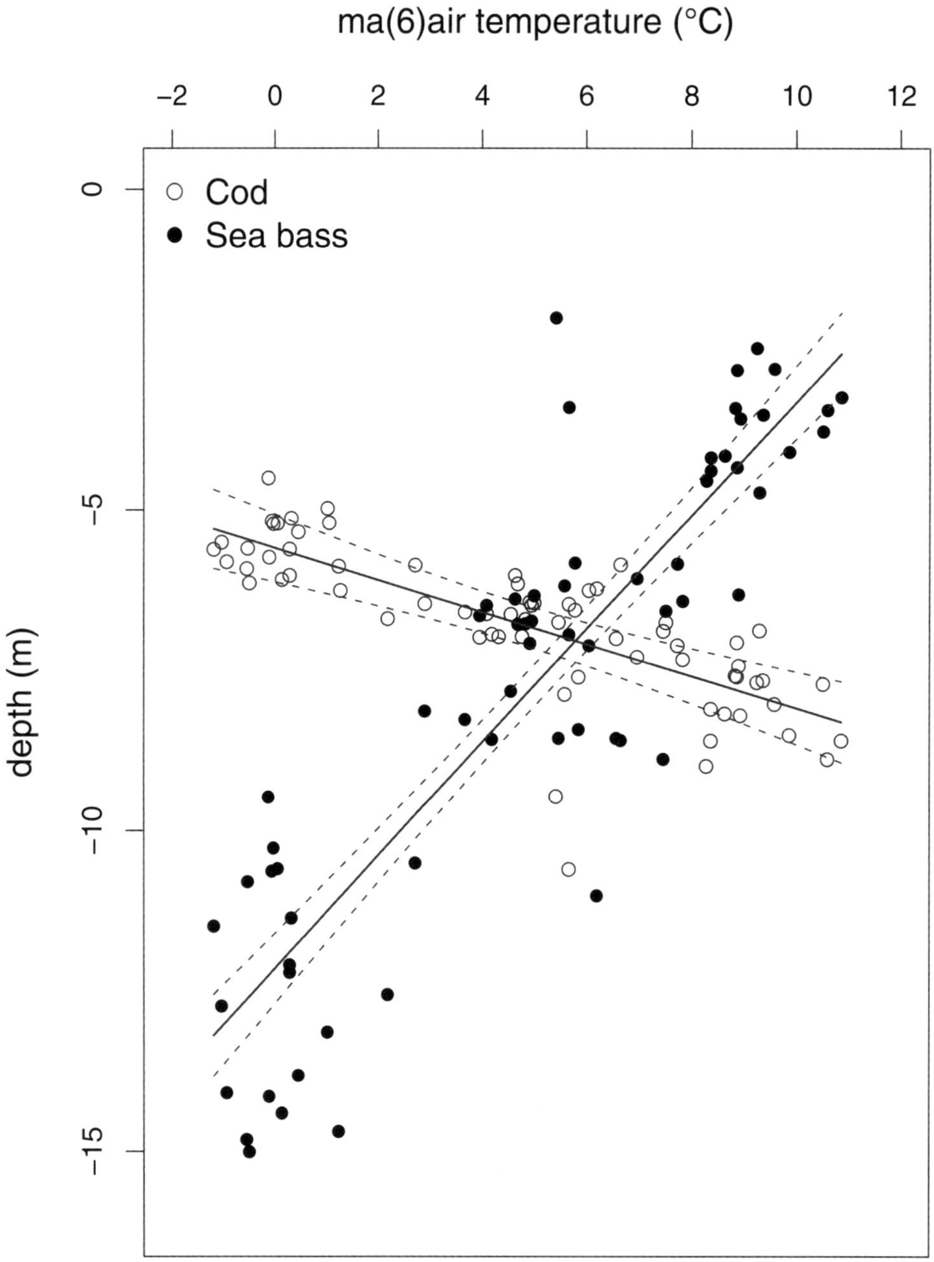

FIGURE 8. Predicted daytime depth use (solid lines), and corresponding 95% confidence bounds (dashed lines) as function of air temperature (moving average over 6 d) and species as retrieved from the most supported lme model (Table 2). The model accounts for individual variation and explains 73% of the total variation in depth use.

area. This kind of grouping behavior is often observed in wild sea bass (own observations, Pickett and Pawson 1994; Quayle et al. 2009). Owing to the low percentage of overall home-range area overlap, these groups seem to interact in an agonistic fashion. We have no direct observations that can support such a hypothesis, but this finding should be explored in greater detail in the future.

We documented that cod and sea bass from the inner part of the Oslo fjord had highly differentiated temperature and salinity preferences (Figure 5). Owing to the spatio-temporal features of the salinity- and temperature profiles in the study area, where the degree of variation was greater in the upper layers than in lower layers (Figure 2), the scope for habitat selection was greater for the upper layers. Knowing that sea bass is a euryhaline and eurythermic species (Pickett and Pawson 1994), it is not unexpected to find that sea bass inhabit the more unstable upper layers, and cod remain deeper in the water column (below the thermocline in August and October, Figure 2) where both temperature and salinity more stable. Even though euryhaline and eurothermic, the sea bass individuals were clearly picky when it came to both temperature and salinity—in fact, more than cod individuals. In general, sea bass selected locations/layers that had the highest temperature available and low salinity, whereas cod individuals selected temperatures at around 12°C and tended to select the higher salinities. These findings are pretty much in accordance with the general tolerance physiology literature for the sea bass and cod. Tolerable temperatures for sea bass range from 5–28°C, and the optimal temperature (for growth, at least in southern populations) is 22–24°C (Sahin et al. 2000; Zanuy and Carrillo 1985). Sea bass are reported not to be able to maintain growth at temperatures below 10°C (Russell et al. 1996). Cod is a cold-stenothermic species that may be found at temperatures ranging from –1 to 15°C (Björnsson et al. 2001), with optimum growth at 8–10°C for individuals larger than 1000 g (Björnsson et al. 2007). By combining the observed habitat selection ratios with the measured temperature- and salinity profiles, the observed vertical distribution for the sea bass and cod could easily be predicted. In fact, even air temperature could be used as a very good predictor for depth use in the two species (Figure 8). Hence, the vertical distribution for the sea bass and cod in our study seems largely determined by temperature.

We suggest that the observed difference in late-autumn behavior between the two species, where sea bass migrate towards deeper waters further south in the Oslo fjord and cod remain, can, at least in part, be explained by the differentiated temperature (and salinity) preferences between the two species. As water temperature in the upper layers cools as autumn progresses, sea bass seek areas with deeper water, i.e., where the warmest water is located. In a particular area located about 4 km southwest of the core study area, there is an 80 ha trough of deep water (maximum at about 85 m), where two sea bass individuals where tracked February 2008. At depths larger than 50 m, the water temperature is stable at 10–10.5°C throughout the winter (December through April, author's unpublished data). Catches by local gillnet and recreational fishermen confirm that sea bass stay in deep areas (>50 m) during winter. Hence, sea bass appear to utilize deeper areas as a winter "refuge" where the temperature is sufficiently high to maintain basic body functioning and to some extent also feed. As cod prefer cold water, they do not have to migrate towards winter refuges, but can remain at the site throughout the year and simply alter their depth use.

The observed difference in seasonal development in the diel depth utilization of the sea bass and cod can also be explained by their differentiated temperature preference. In August/September, cod stay in preferred cold water during daytime and ascend to feed in nonpreferred shallow waters during night-time to feed on crab, shrimps and small fish. As the autumn progresses, the shallow areas get colder (Figure 2), and hence, the cod undertake shorter and shorter vertical migration distances (Figure 7). Sea bass individuals were shown to undertake increasing diel vertical migration distances as the autumn progressed (Figure 7), staring with a two meter amplitude in summer and ending up with an amplitude of 7 m in early winter. Again, this can be linked to temperature preference, as during summer and early autumn optimum temperatures were located in the surface layers where also most of the prey is located. However, as the surface layers cool down, the sea bass seek deeper water in order to find warmer water. In order to maintain growth, sea bass will need to undertake diel vertical migrations, feeding in shallow waters and digesting in warmer, deeper layers. Diel vertical migration has recently also been demonstrated in in-shore sea bass along the British coast. Using miniature electronic data storage tags (DSTs), Quayle et al. (2009) found that the most commonly observed behavior was characterized by sea bass maintaining positions in midwater and making frequent ascents and descents of between two and ten meters. For sea bass dwelling off-shore, the same researchers found that sea bass undertook vertical migrations with amplitudes even exceeding 120 m. The same study also showed that the degree of vertical migration changed over the seasons, and what they termed as in-shore behavior, i.e., minimum depth <3 m and diel amplitude <7 m, dominated in the May to October period. These results are very consisted with our findings for the same part of the season. Unfortunately, our study does not allow for comparisons with Quayle et al. (2009) winter and spring results, but during that period, the UK sea bass were mainly located off-shore in pre spawning areas or in spawning areas.

We documented that the individual diel vertical migration pattern of inner Oslo fjord cod changed over the summer to winter period. Similar seasonal changes in vertical migration have been documented for other North Sea cod populations as well (Hobson et al. 2007; Neat et al. 2006; Turner et al. 2002). The Neat et al. (2006) study shows coastal cod individuals from the Shetland Isles swam deep (>70 m) from late October to late February and ascend to shallower waters during the spawning period in February/March. After the spawning period, the cod split into three groups, where one group resides in shallow waters, another goes deep and a third group stays in intermediate depths. Interestingly, this study demonstrates that there is a tremendous inter-individual variation in summer and autumn diel vertical migration patterns. Some individuals display an extensive vertical movement pattern, while others did not move at all. Among those moving, some displayed a nocturnal ascent pattern whereas others ascended during daytime. The authors draw the conclusion that it is meaningless—for this particular cod population—to generalize the diel vertical movement behavior. They rather suggest that each individual develops its own movement pattern and that there is an underlying individual plasticity in cod that allows for this (Reebs 2002). The cod involved in our study seem far more synchronized in their diel vertical movement patterns. There was variation in both how deep they went during daytime and also how shallow they went during night-time, but all individuals seemed to display nocturnal ascents. Righton et al. (2001) have suggested that differences in bathymetry and habitat among populations may explain among-population differences in diel vertical movement patterns observed in cod. We suggest adding differences in temperature profile to the story. For our particular system, we suggest that the diel vertical movement pattern (for both cod and sea bass) is driven by the temperature profile (stratified during summer and most of autumn, Figure 2), combined with the main food items being located mainly in shallow waters (shore crabs in particular, but also sprat *Sprattus sprattus*). Such a system, where the food is present in a suboptimal (to the cod) warm water layer on top and colder water with little prey in deeper layers, may allow for less inter-individual diel vertical movement variation. This population-level variation in diel vertical movement found in cod warrants further investigation. We suggest that future analyses combine cod depth-use data with prey distribution and temperature/salinity profiles.

Differentiated hunting strategies between sea bass and cod may also influence the differences in depth use (Hobson et al. 2007). While cod is a solitary species attacking prey from below (Jobling 1995), shoaling is common among sea bass (Pickett and Pawson 1994). Sea bass in our study area are often observed in shoals chasing prey fish on the surface, along land and islands, and in shallow coves. The shoaling behavior appears cooperative, but is only visible when they reach the surface and hunt in shallow water. Prey fish have even been observed chased onto beaches.

Sea bass individuals in the size range of 25–30 cm have, under controlled semi-natural conditions, been shown to have a predominantly diurnal feeding activity pattern in summer and autumn and nocturnal activity in winter and spring (Azzaydi et al. 2007; Sanchez-Vazquez et al. 1998). Hence, most of our data would be from periods where we should expect the sea bass to be predominantly feeding in daytime. In cod, feeding in older-than 0-group individuals has been reported to be predominantly nocturnal throughout the year (Grant and Brown 1998; but see Lokkeborg and Ferno 1999). Unfortunately, our data cannot be used for drawing direct inferences about the activity level of the fish, but our own observations during daytime confirms that sea bass, in summer and autumn, are highly active during daytime and with possible peaks at dawn and dusk (crepuscular). Cod are also seen feeding during daytime, thus supporting Lokkeborg and Ferno (1999), but seem to have feeding peaks during night (own observations). This seemingly diel segregation in feeding activity will result in further habitat segregation between the sea bass and cod. Both species have similar, but varied diets. They feed on smaller fish (e.g., Atlantic herring *Clupea harengus* and European sprat), shrimps (e.g., northern krill *Meganyctiphanes norvegica*), crabs (mainly green crab, *Carcinus maenas*), polychaets and mussels (Pickett and Pawson 1994). The similar prey composition implies that exploitative competition may still prevail between the two regardless of the observed habitat segregation. They simply eat the same food at different times and positions.

Distributional consequences

The differences found in thermal and salinity preferences between the sea bass and cod are largely in accordance with what is reported in controlled temperature- and salinity-tolerance literature (Björnsson et al. 2001; Claireaux et al. 1995a, 1995b; Claireaux et al. 2000). If global warming is going to progress as predicted by the IPCC (IPCC 2007) in the near future, this may affect the distribution of the sea bass and cod and may benefit the sea bass at the cost of cod (Bernardo et al. 2007; Calosi et al. 2008; Drinkwater 2005; Portner et al. 2008). According to recent macro-physiological theory, dramatic shifts in cod distribution may be expected (Portner et al. 2008). However, from historical records, we know that even during the warm Atlantic period (7000–3900 BC), when European sea bass was a common species in north Atlantic waters (together with species like common stingray *Dasyatis pastinaca*, black sea bream *Spondyliosoma cantharus* and swordfish *Xiphias gladius*), Atlantic cod was still a common species. It is thus not trivial to forecast what the ultimate outcome will be in terms of fish species composition in coastal north-Atlantic waters under an IPCC-predicted climate regime. Species may change horizontal distribution, but may also change vertical distribution and live in refuge water layers that hold preferred habitat characteristics (e.g., temperatures and salinities) and as such allow for "co-existence" with invading species that occupy other parts of the water column.

Management

With the exception of Colman et al. (2007), no studies on wild European sea bass have been conducted in Scandinavia. Fisheries managers and scientists in the UK are urging for improved management plans, especially due to increasing interest in sea bass among commercial and recreational fishermen (Fritsch et al. 2007). A similar trend and concern for management has been expressed in recent studies in Norway (Colman et al. 2007). The present increase in sea bass numbers in Norwegian waters is coupled with an increasing desire by recreational anglers and commercial fishermen to catch sea bass. However, there has also been an increasing concern for the potential negative influence a growing sea bass population in Norwegian waters might have on the already established fisheries. There will certainly be interactions and alterations in the ecosystem, but it is too early to speculate on whether these are negative. Our study show that habitat segregation exists between sea bass and cod in our study area, but it is not possible to discern whether competition occurs or whether segregation is a result of displacement or the separate influence of environmental variables on the two species independent of each other. Since no data on depth- and habitat selection of cod before the increase of sea bass in the study area are available, it is not possible to determine whether the cod exhibited habitat segregation due to an increasing presence of sea bass.

Since knowledge about habitat use and habitat selection constitute key components in management ecology—and even conservation ecology (Rice 2005)—the results from our study comprise an important first step providing key knowledge that can allow for sound ecological management of the entire European sea bass population in general, and the Norwegian sea bass and their influence on the Norwegian coastal ecosystem in particular. Knowing that the sea bass and cod under study differ in habitat use and also in seasonal horizontal- and vertical distribution, even within such a small area as the inner Oslo fjord, provide good prospects for development of efficient species differentiated protection measures if desired. Our data and tests illustrate that acquiring information pertinent to sound management is possible.

The current study was conducted in a very restricted area during the summer-to-winter period and involved just 17 individuals. To increase inference generality, future acoustic telemetry studies should involve winter-to-summer observations (to fully explore seasonality issues) and more individuals and study sites (to extend large-scale spatial relevance). More indigenous fish species should also be involved in future studies, especially including brown trout *Salmo trutta* as an important coastal recreational fish species, to fully assess interaction effects between these species and the sea bass. In addition, focus on other habitat characteristics, such as substrate type, macrophyte type, and human activity locations should also be incorporated to fully understand the co-occurring fish species' habitat preferences. Finally, prey selectivity ratios will have to be estimated for drawing more ultimate inferences about the degree of habitat segregation observed between cod and sea bass.

Acknowledgments

We would like to acknowledge Stig Aaslund for enthusiastic field assistance both during fish sampling and during the tracking work. Trygve Poppe is thanked for his precise implantation surgery of the transmitters, for fish sampling and for provision of good stories. Gratitude goes to Einar Helsing for kindly lending us his terrace under the implantation process, his beach for the recovery of fish and his buoy used for the keep-net. We are also grateful to Mike Pawson, Alex Mann and Håkon Rekstad for help with sampling of sea bass and cod for tagging. Jan Eyolf Bjørnsen at Thelma AS gave advice under the tracking process. Norwegian Institute of Water Research (O-27074) along with Elbe Normark AS provided funding.

References

Azzaydi, M., V. C. Rubio, F. J. López, F. J. Sánchez-Vázquez, S. Zamora, J. A. Madrid. 2007. Effect of restricted feeding schedule on seasonal shifting of daily demand-feeding pattern and food anticipatory activity in European sea bass (*Dicentrarchus labrax* L.). Chronobiology International 24(5):859–874.

Bernardo, J., R. J. Ossola, J. Spotila, and K. A. Crandall. 2007. Interspecies physiological variation as a tool for cross-species assessments of global warming-induced endangerment: validation of an intrinsic determinant of macroecological and phylogeographic structure. Biology Letters 3(6):695–698.

Björnsson, B., A. Steinarsson, and T. Arnason. 2007. Growth model for Atlantic cod (*Gadus morhua*): Effects of temperature and body weight on growth rate. Aquaculture 271:216–226.

Björnsson, B., A. Steinarsson, and M. Oddgeirsson. 2001. Optimal temperature for growth and feed conversion of immature cod (*Gadus morhua* L.). ICES Journal of Marine Science 58(1):29–38.

Burnham, K. P., and D. R. Anderson. 1998. Model selection and inferences. Springer Verlag, New York.

Calosi, P., D. T. Bilton, and J. I. Spicer. 2008. Thermal tolerance, acclimatory capacity and vulnerability to global climate change. Biology Letters 4(1):99–102.

Claireaux, G., D. M. Webber, S. R. Kerr, and R. G. Boutilier. 1995a. Physiology and behavior of free-swimming Atlantic cod (*Gadus morhua*) facing fluctuating salinity and oxygenation conditions. Journal of Experimental Biology 198(1):61–69.

Claireaux, G., D. M. Webber, S. R. Kerr, and R. G. Boutilier. 1995b. Physiology and behavior of free-swimming Atlantic cod (*Gadus morhua*) facing fluctuating temperature conditions. Journal of Experimental Biology 198(1):49–60.

Claireaux, G., D. M. Webber, J. P. Lagardere, and S. R. Kerr. 2000. Influence of water temperature and oxygenation on the aerobic metabolic scope of Atlantic cod (*Gadus morhua*). Journal of Sea Research 44(3–4):257–265.

Colman, J. E., M. G. Pawson, J. Holmen, T. O. Haugen, and G. D. Pickett. 2007. European sea bass in the North Sea; past, present and future status, use and management challenges. Pages 111–129 *in* Ø. Aas, editor. Global challenges in recreational fisheries: development, management and science. Blackwell, Oxford, UK.

Conides, A. J., and B. Glamuzina. 2006. Laboratory simulation of the effects of environmental salinity on acclimation, feeding and growth of wild-caught juveniles of European sea bass *Dicentrarchus labrax* and gilthead sea bream, *Sparus aurata*. Aquaculture 256(1–4):235–245.

Cote, D., S. Moulton, P. C. B. Frampton, D. A. Scruton, and R. S. McKinley. 2004. Habitat use and early winter movements by juvenile Atlantic cod in a coastal area of Newfoundland. Journal of Fish Biology 64(3):665–679.

Cote, D., L. M. N. Ollerhead, R. S. Gregory, D. A. Scruton, and R. S. McKinley. 2002. Activity patterns of juvenile Atlantic cod (*Gadus morhua*) in Buckley Cove, Newfoundland Hydrobiologia 483(1–3):121–127.

Dalla Via, J., P. Villani, E. Gasteiger, and H. Niederstatter. 1998. Oxygen consumption in sea bass fingerling *Dicentrarchus labrax* exposed to acute salinity and temperature changes: metabolic basis for maximum stocking density estimations. Aquaculture 169(3–4):303–313.

Davie, P. S., and R. K. Kopf. 2006. Physiology, behaviour and welfare of fish during recreational fishing and after release. New Zealand Veterinary Journal 54(4):161–172.

Domenici, P., G. Claireaux, and D. J. McKenzie. 2007. Environmental constraints upon locomotion and predator-prey interactions in aquatic organisms: an introduction. Philosophical Transactions of The Royal Society B: Biological Sciences 362(1487):1929–1936.

Drinkwater, K. F. 2005. The response of Atlantic cod (*Gadus morhua*) to future climate change. ICES Journal of Marine Science 62(7):1327–1337.

Eroldogan, O. T., M. Kumlu, and A. Aktas. 2004. Optimum feeding rates for European sea bass *Dicentrarchus labrax* L. reared in seawater and freshwater. Aquaculture 231(1–4):501–515.

Espeland, S. H., A. F. Gundersen, E. M. Olsen, H. Knutsen, J. Gjøsæter, and N. C. Stenseth. 2007. Home range and elevated egg densities within an inshore spawning ground of coastal cod. ICES Journal of Marine Science 64(5):920–928.

Espeland, S. H., E. M. Olsen, H. Knutsen, J. Gjøsæter, D. Danielssen and N. C. Stenseth. 2008. New perspectives on fish movement: kernel and GAM smoothers applied to a century of tagging data on coastal Atlantic cod. Marine Ecology-Progress Series 372:231–241.

Fabrizio, M. C., and J. P. Pessutti. 2007. Long-term effects and recovery from surgical implantation of dummy transmitters in two marine fishes. Journal of Experimental Marine Biology and Ecology 351(1–2):243–254.

Fritsch, M., Y. Morizur, E. Lambert, F. Bonhomme, and B. Guinand. 2007. Assessment of sea bass (*Dicentrarchus labrax*, L.) stock delimitation in the Bay of Biscay and the English Channel based on mark-recapture and genetic data. Fisheries Research 83(2–3):123–132.

Grant, S. M., and J. A. Brown. 1998. Diel foraging cycles and interactions among juvenile Atlantic cod (*Gadus morhua*) at a nearshore site in Newfoundland. Canadian Journal of Fisheries and Aquatic Sciences 55(6):1307–1316.

Hobson, V. J., D. Righton, J. D. Metcalfe, and G. C. Hays. 2007. Vertical movements of North Sea cod. Marine Ecology Progress Series 347:101–110.

IPCC. 2007. Climate Change 2007: The Physical Science Basis. Contribution of Working Group I to the Fourth Assessment. Pages 996 *in* S. Solomon and 7 coeditors, editors. Report of the Intergovernmental Panel on Climate Change. Cambridge University Press, Cambridge, United Kingdom and New York, NY, USA.

Jobling, M. 1995. Environmental Biology of Fishes, 1st edition, volume 16. Chapman and Hall, London.

Jourdan-Pineau, H., A. Dupont-Prinet, G. Claireaux, and D. J. McKenzie. 2010. An investigation of metabolic prioritization in the European sea bass, *Dicentrarchus labrax*. Physiological and Biochemical Zoology 83(1):68–77.

Linehan, J. E., R. S. Gregory, and D. C. Schneider. 2001. Predation risk of age-0 cod (*Gadus*) relative to depth and substrate in coastal waters. Journal of Experimental Marine Biology and Ecology 263(1):25–44.

Lokkeborg, S., and A. Ferno. 1999. Diet activity pattern and food search behaviour in cod, *Gadus morhua*. Environmental Biology of Fishes 54(3):345–353.

Mackenzie, B. R., H. Gislason, C. Mollmann, and F. W. Koster. 2007. Impact of 21st century climate change on the Baltic Sea fish community and fisheries. Global Change Biology 13(7):1348–1367.

Mulcahy, D. M. 2003. Surgical implantation of transmitters into fish. ILAR Journal 44(4):295–306.

Mylonas, C. C., G. Cardinaletti, I. Sigelaki, and A. Polzonetti-Magni. 2005. Comparative efficacy of clove oil and 2-phenoxyethanol as anesthetics in the aquaculture of European sea bass (*Dicentrarchus labrax*) and gilthead sea bream (*Sparus aurata*) at different temperatures. Aquaculture 246(1–4):467–481.

Neat, F. C., and 7 coauthors. 2006. Residency and depth movements of a coastal group of Atlantic cod (*Gadus morhua* L.). Marine Biology 148(3):643–654.

Occhipinti-Ambrogi, A. 2007. Global change and marine communities: Alien species and climate change. Marine Pollution Bulletin 55(7–9):342–352.

Parker, I. M., and coauthors. 1999. Impact: toward a framework for understanding the ecological effects of invaders. Biological Invasions 1:3–19.

Pastoureaud, A. 1991. Influence of starvation at low-temperatures on utilization of energy reserves, appetite recovery and growth character in sea bass, *Dicentrarchus labrax*. Aquaculture 99(1–2):167–178.

Pawson, M. G., M. Brown, J. Leballeur, and G. D. Pickett. 2008. Will philopatry in sea bass, *Dicentrarchus labrax*, facilitate the use of catch-restricted areas for management of recreational fisheries? Fisheries Research 93(1–2):240–243.

Pawson M. G., D. F. Kelley, G. D. Pickett. 1987. The distribution and migrations of bass, *Dicentrachus labrax* L, in waters around England and Wales as shown by tagging. Journal of the Marine Biological Association of the United Kingdom 67:183–217.

Pawson, M. G., G. D. Pickett, and P. R. Witthames. 2000. The influence of temperature on the onset of first maturity in sea bass. Journal of Fish Biology 56(2):319–327.

Pawson, M. G., G. D. Pickett, J. Leballeur, M. Brown, and M. Fritsch. 2007. Migrations, fishery interactions, and management units of sea bass (*Dicentrarchus labrax*) in Northwest Europe. ICES Journal of Marine Science 64(2):332–345.

Pawson, M. G., G. D. Pickett, and M. T. Smith. 2005. The role of technical measures in the recovery of the UK sea bass (*Dicentrarchus labrax*) fishery 1980–2002. Fisheries Research 76(1):91–105.

Person-Le Ruyet, J., K. Mahe, N. Le Bayon, and H. Le Delliou. 2004. Effects of temperature on growth and metabolism in a Mediterranean population of European sea bass, *Dicentrarchus labrax*. Aquaculture 237(1–4):269–280.

Pickett, G. D., and M. G. Pawson. 1994. Sea bass: biology, exploitation and conservation. Chapman and Hall, London.

Pinheiro, J. C., and D. M. Bates. 2000. Mixed-effects models in S and S-PLUS. Springer, New York.

Portner, H. O., and coauthors. 2008. Cod and climate in a latitudinal cline: physiological analyses of climate effects in marine fishes. Climate Research 37(2–3):253–270.

Quayle, V. A., D. Righton, S. Hetherington, and G. Pickett. 2009. Observations of the behaviour of European sea bass (*Dicentrarchus labrax*) in the North Sea. Pages 103–119 *in* J. L. Nielsen, H. Arrizabalaga, N. Fragoso, A. Hobday, M. Lutcavage, and J. Sibert in Tagging and Tracking of Marine Animals with Electronic Devices, volume 9. Springer, Dordrecht.

R Development Core Team. 2009. R: a language and environment for statistical computing. R Foundation for Statistical Computing, Vienna, Austria.

Rankin, J. C., and F. B. Jensen. 1993. Fish ecophysiology. Chapman and Hall, London.

Reebs, S. G. 2002. Plasticity of diel and circadian activity rhythms in fishes. Reviews in Fish Biology and Fisheries 12(4):349–371.

Rice, J. C. 2005. Understanding fish habitat ecology to achieve conservation. Journal of Fish Biology 67:1–22.

Righton, D., J. Metcalfe, and P. Connolly. 2001. Fisheries—different behaviour of north and Irish sea cod. Nature 411(6834):156–156.

Robichaud, D., and G. A. Rose. 2004. Migratory behaviour and range in Atlantic cod: inference from a century of tagging. Fish and Fisheries 5(3):185–214.

Rogers, K. B., and G. C. White. 2007. Analysis of movement and habitat use from telemetry data. Pages 625–676 *in* M. L. Brown, and C. S. Guy, editors. Analysis and Interpretation of Freshwater Fisheries Data. American Fisheries Society, Bethesda, Maryland.

Russell, N. R., J. D. Fish, and R. J. Wootton. 1996. Feeding and growth of juvenile sea bass: The effect of ration and temperature on growth rate and efficiency. Journal of Fish Biology 49(2):206–220.

Sahin, T., B. Akbulut, and M. Aksungur. 2000. Compensatory growth in sea bass (*Dicentrarchus labrax*), sea bream (*Sparus aurata*) and rainbow trout (*Oncorhynchus mykiss*). Turkish Journal of Zoology 24(1):81–86.

Sanchez-Vazquez, F. J., M. Azzaydi, F. J. Martinez, S. Zamora, and J. A. Madrid. 1998. Annual rhythms of demand-feeding activity in sea bass: Evidence of a seasonal phase inversion of the diel feeding pattern. Chronobiology International 15(6):607–622.

Stenevik, E. K., and S. Sundby. 2007. Impacts of climate change on commercial fish stocks in Norwegian waters. Marine Policy 31(1):19–31.

Swain, D. P., G. A. Chouinard, R. Morin, and K. F. Drinkwater. 1998. Seasonal variation in the habitat associations of Atlantic cod (*Gadus morhua*) and American plaice (*Hippoglossoides platessoides*) from the southern Gulf of St. Lawrence. Canadian Journal of Fisheries and Aquatic Sciences 55(12):2548–2561.

Turner, K., D. Righton, and J. D. Metcalfe. 2002. The dispersal patterns and behaviour of North Sea cod (*Gadus morhua*) studied using electronic data storage tags. Hydrobiologia 483(1–3):201–208.

Zanuy, S., and M. Carrillo. 1985. Annual cycles of growth, feeding rate, gross conversion efficiency and hematocrit levels of sea bass (*Dicentrarchus labrax* L) adapted to 2 different osmotic media. Aquaculture 44(1):11–25.

American Fisheries Society Symposium 76:289–311, 2012

Marine Migratory Behavior of Hatchery-Reared Anadromous and Wild Non-Anadromous Sockeye Salmon Revealed by Acoustic Tags

Chris C. Wood*
Fisheries and Oceans Canada, Pacific Biological Station
3190 Hammond Bay Road, Nanaimo, British Columbia V9T 6N7, Canada

David W. Welch
Kintama Research Corp.
10-1850 Northfield Road, Nanaimo, British Columbia V9T 6N7, Canada

Lyse Godbout
Fisheries and Oceans Canada, Pacific Biological Station
3190 Hammond Bay Road, Nanaimo, British Columbia V9T 6N7, Canada

James Cameron
4743 Francis Peninsula Road, Madeira Park, British Columbia V0N 2H1, Canada

Abstract.—We investigated the marine migratory behavior and survival of Sakinaw Lake sockeye salmon *Oncorhynchus nerka* during their outbound migration as juveniles and return migration as adults two or more years later by tracking individuals that had been implanted with Vemco acoustic tags programmed to have two periods of active transmission. We tracked both hatchery-reared anadromous sockeye salmon ('hatchery sockeye') and wild nonanadromous 'kokanee,' two genetically-distinct, sympatric ecotypes inhabiting Sakinaw Lake, British Columbia. Tagged kokanee were distinguished from wild sockeye by haplotype frequencies at two mitochondrial DNA genes. Migrations were inferred from detections by the Pacific Ocean Shelf Tracking (POST) receivers, and supplemental tracking near the release site and in Sakinaw Lake. We found no significant differences between the ecotypes in the proportion of 'migratory' fish (those detected migrating seaward by POST telemetry in the year of release, 42% of all 254 fish released) or in the proportion of ocean-going fish (those detected at receivers near the open ocean, 20% of all fish released). Seaward migration in both ecotypes was primarily northward through Johnstone Strait in 2 of the 3 years studied (92% of migratory fish in 2004 and 84% in 2006). A significantly higher proportion of fish moved southward in 2005 (45% of migratory fish) than in 2004 or 2006, but this difference could not be attributed to ecotype, body size, or release date. One significant difference observed between the ecotypes was that 6 kokanee but no sockeye migrated back into Sakinaw Lake within 2 weeks of release in 2006. The number of tagged fish detected as returning adults with operational tags was

*Corresponding author: Chris.Wood@dfo-mpo.gc.ca

low (3 sockeye at the release site and 2 kokanee at Sakinaw Creek), but none of these fish had been detected crossing seaward POST lines as juveniles and thus appeared to be nonmigratory. The adult return rate of these nonmigratory tagged fish (3.4% in sockeye, 4.3% in kokanee) was higher than for migratory tagged fish (0% for both ecotypes). This discrepancy suggests that factors outside the Strait of Georgia have caused the poor marine survival that is preventing recovery of the endangered Sakinaw sockeye population (mean <0.2% since 2003).

Introduction

Anadromous sockeye salmon *Oncorhynchus nerka* increasingly face extirpation near the southern part of their range because of human impacts and climate change (Gustafson et al. 2007; Waples et al. 2008). The population of sockeye salmon in Sakinaw Lake in southern British Columbia, hereafter called Sakinaw sockeye, has declined by over 99% from an average of about 5,000 spawning fish annually prior to 1990 to <30 since 2001. This population was designated Endangered by the Committee on the Status of Endangered Wildlife in Canada (COSEWIC) following a citizen's petition for emergency status assessment in 2002 (Irvine et al. 2005). Overexploitation of adults by intensive mixed-stock fisheries undoubtedly contributed to the decline (COSEWIC 2003), but recovery is now constrained by the poor survival of juveniles after they leave Sakinaw Lake as 'smolts.' Reliable counts of both smolts and adults passing a weir at the lake's outlet since 2003 indicate that smolt-to adult survival has averaged <0.2% (range 0–0.7%), even though regulations imposed in 2004 have largely eliminated incidental fishing mortality (see Integrated Fisheries Management Plans, http://www-ops2.pac.dfo-mpo.gc.ca/xnet/content/MPLANS/archive.htm). Smolt abundance is being maintained by good freshwater survival of hatchery-reared fry stocked into Sakinaw Lake as part of a captive breeding program that began in 2003, but the natural spawning population is unlikely to recover unless marine survival improves in the near future.

Research is needed to reveal the presumably climate-related causes of this extremely low marine survival. When our study began in 2004, smolt-to-adult survival of Sakinaw sockeye was an order of magnitude less than that reported for the two sockeye salmon index populations (Chilko and Cultus lakes) in the Fraser River, despite the proximity of their ocean entry (Figure 1), and the larger size of Sakinaw smolts (Figure 2). Large smolt size has been shown to favor marine survival in Chilko sockeye (Henderson and Cass 1991) and in sockeye populations generally (Burgner 1991). Like other Fraser River sockeye populations, Chilko sockeye are thought to migrate quickly northward through the Strait of Georgia towards the open ocean (Groot et al. 1988). We hypothesized that Sakinaw sockeye might experience higher marine mortality than Chilko sockeye if they migrate seaward by a different route or at a different time, perhaps remaining in the Strait of Georgia for a longer period and encountering unfavorable conditions there. However, results from this tagging study refute our initial hypothesis, and it is now evident (see Figure 2) that marine survival has subsequently declined for both Cultus and Chilko sockeye (Grant et al. 2010)

Conditions that adversely affect marine survival can favor the evolution of nonanadromous life histories in sockeye salmon (Wood 1995) and other fishes (Gross 1987; Hendry et al. 2004). Most sockeye populations include some 'residual sockeye' which are nonanadromous progeny of anadromous females, and primarily male (Ricker 1938). In

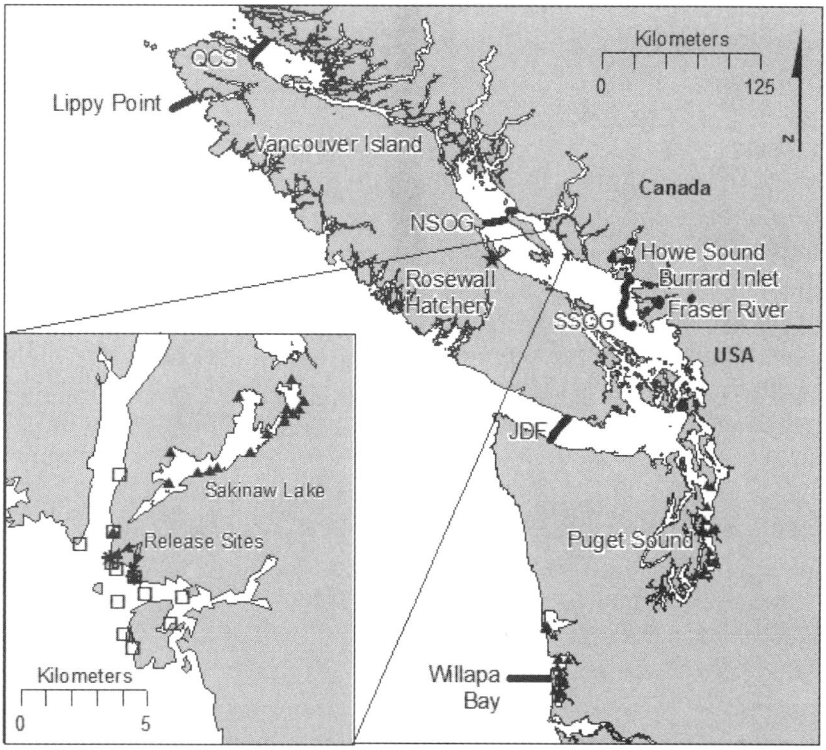

FIGURE 1. Map showing the location of Sakinaw Lake (inset), release sites (asterisks) and mobile tracking (squares), POST receivers (solid circles, often forming lines), and other supplementary fixed receivers (solid triangles). QCS: Queen Charlotte Strait; NSOG: Northern Strait of Georgia; SSOG: Southern Strait of Georgia; JDF: Strait of Juan de Fuca.

contrast, 'kokanee' exist as self-perpetuating populations of nonanadromous individuals that are reproductively isolated from anadromous sockeye salmon. Kokanee evolve from residual sockeye when anadromous populations become land-locked (Ricker 1940; Nelson 1968) but also evolve in sympatry when conditions are suitable (Foote et al. 1989; Wood and Foote 1996). Once isolated from anadromous sockeye salmon, kokanee tend to evolve specific adaptations for nonanadromous life (Ricker 1940; Foote et al. 1989) but limited evidence suggests that kokanee can revert to an anadromous life history if they are recently derived from anadromous sockeye salmon (see Discussion). Kokanee in Sakinaw Lake are genetically distinct from Sakinaw sockeye (Wood et al. 2008), but they remain abundant and share the same spawning sites and rearing habitat. It therefore seems plausible that Sakinaw kokanee could be used to create an anadromous sockeye run if the present population of Sakinaw sockeye becomes extinct.

In this study, we used acoustic tagging technology and Pacific Ocean Shelf Tracking (POST) telemetry to investigate the marine migratory behavior and survival of sympatric populations of sockeye salmon and kokanee released at sea near the outlet of Sakinaw Lake. Our objectives were, first, to investigate the early marine migratory behavior of Sakinaw sockeye in search of clues about the possible causes for poor marine survival; and second, to evaluate the feasibility of forcing the seaward migration of kokanee as a means

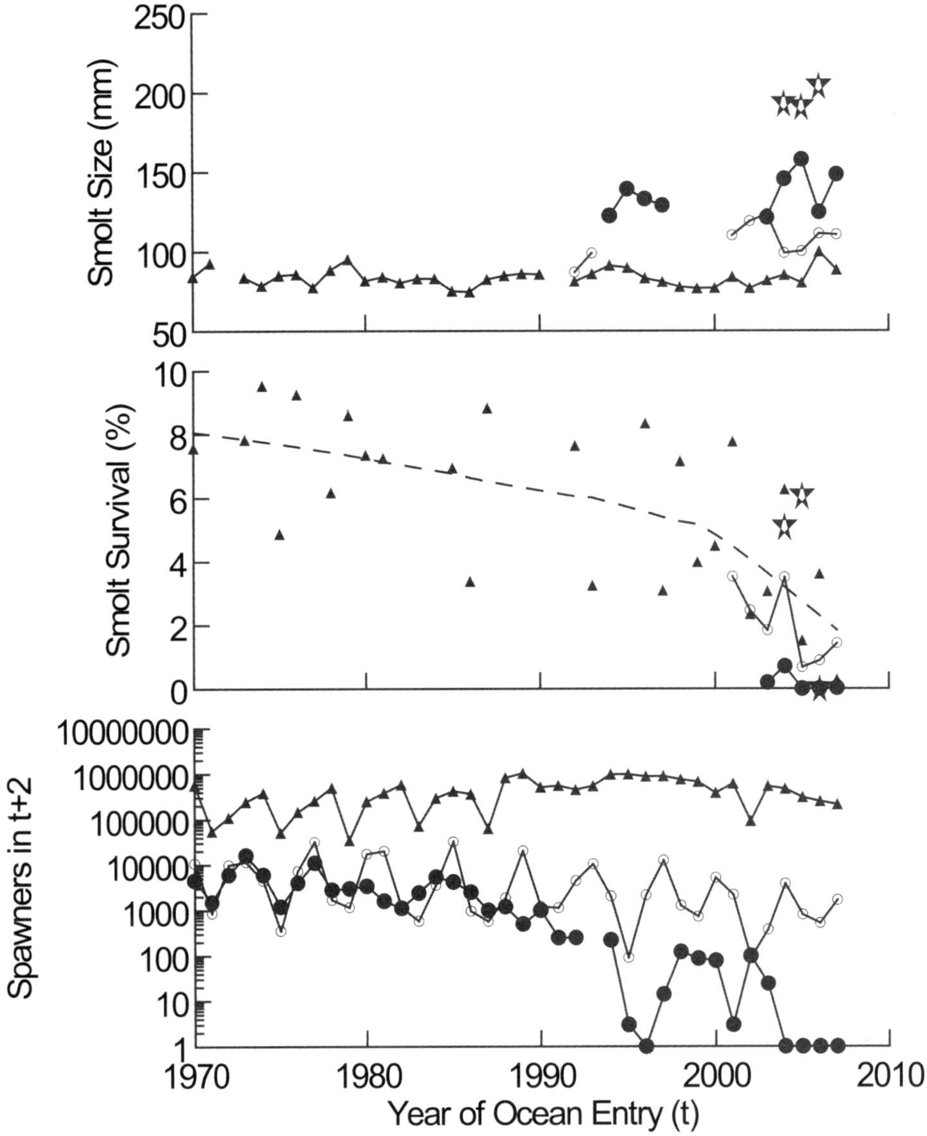

Figure 2. Trends in mean smolt size, smolt-to-adult survival, and number of spawning adults in Sakinaw sockeye (solid circles) compared with two Fraser River populations—Cultus Lake (open circles) and Chilko Lake (triangles). Stars denote mean size at release of all acoustically tagged Sakinaw fish and survival to adult return for the nonmigratory component. Dashed line represents LOWESS smoothing of the trend in Chilko Lake smolt survival.

to recreate an anadromous sockeye salmon run.

Study Site

Sakinaw Lake is a small lake (surface area 6.9 km2) which drains into the east side of the Strait of Georgia 80 km northwest of Vancouver, British Columbia. Access to the Strait of Georgia is through a short creek (<1 km) that is dammed to regulate lake level. A fish ladder allows fish passage and facilitates enumeration of seaward migrating 'smolts' and returning adults. Access to the Pacific Ocean is possible either to the north through Johnstone Strait and Queen Charlotte Strait, or to the south through the Strait of Juan de Fuca.

Methods

POST Acoustic telemetry

The POST project is a permanent continental-scale telemetry research platform designed to track migrations of Pacific salmon and other species along the west coast of North America (Welch et al. 2003, 2004, 2006, 2009; Rechisky et al. 2009). Listening 'curtains' consisting of Vemco VR-2 or VR-3 passive acoustic receivers placed in lines at strategic locations enable individual identification and an almost complete count of acoustically-tagged fish along each possible migration corridor. Detections at successive lines allow inferences about migration route, speed, and survival. In this study, the POST lines of primary interest were those that must be crossed to reach the ocean, *viz*. the Northern Strait of Georgia (NSOG) and Queen Charlotte Strait (QCS) lines across the northern migration route, and the Strait of Juan de Fuca (JDF) line across the southern migration route (Figure 1).

Supplementary POST lines were deployed to the south of Sakinaw Lake in Howe Sound (all years), the Southern Strait of Georgia (SSOG, 2006), Point Atkinson (2005), Burrard Inlet (2006), and the Fraser River (all years), as well as off the northwest coast of Vancouver Island (Lippy Point in 2006 and Brooks Peninsula in 2004 and 2005) (details in Welch et al. 2009). A few detections of Sakinaw fish were recorded on compatible fixed receivers in Puget Sound (deployed by K. Andrews, NOAA Fisheries). Here we refer to the supplementary POST lines south of Sakinaw and the fixed receivers in Puget Sound collectively as the 'inside' lines.

Acoustic tag programming

We surgically implanted Vemco V9–1 L acoustic tags into hatchery-reared sockeye salmon (hereafter hatchery sockeye) and wild kokanee to track their seaward migration following release in the Strait of Georgia near the outlet of Sakinaw Lake. The tags were programmed to transmit acoustic codes at 69 kHz with a random time interval between successive transmissions of 40–100 s for kokanee released in 2006, and 30–90 s for all other release groups. To provide an opportunity to track the return migration of adults, battery life was extended by programming the tags to have two periods of active acoustic transmission, with the first period of transmission ceasing in late summer (by which time smolts were expected to be out of range of the POST arrays; Groot and Cooke 1987), and to begin transmitting again in early summer of the expected year of return migration at age 4.

In 2004, 97 sockeye were released in two groups, 63 on 1 June 2004 18:00 UTC and 34 on 16 June 2004 21:00 UTC, each bearing a V9–1 L tag programmed to stop transmitting its acoustic code 8 September 2004–24 June 2006 (first group) or 1 October 2004–17July 2006 (second group) and then to continue transmitting until battery exhaustion. In 2005, a single group of 47 kokanee was released on

3 June 2005 02:00 UTC with V9–1 L tags programmed to 'sleep' between 28 July 2005 and 13 June 2007. In 2006, a total of 110 tags was released on 3 June 2006 04:00 UTC: 50 V9–1 L tags in age-1 hatchery sockeye were programmed to sleep between 19 July 2006 and 14 May 2008; 47 V9–1 L tags in kokanee (estimated to be age 2 based on length) were programmed to sleep between 29 July 2006 and 14 June 2007 (nearly one year earlier than the sockeye because they were one year older at the time of tagging); and 13 smaller V9–6 L tags, all but 2 implanted in age-1 hatchery sockeye, were programmed to transmit continuously until exhaustion (expected after 1 October 2006).

Fish tagging

Age-1 hatchery-reared sockeye were tagged as a proxy for wild Sakinaw sockeye smolts which were too small to tag with V9–1 L tags and too valuable to risk in experimental studies given the endangered status of the population. Wild Sakinaw sockeye smolts are typically less than 150 mm at age 1, averaging 120–130 mm (Murray and Wood 2002) (Figure 2). The hatchery sockeye were reared in the Rosewall Creek Hatchery on Vancouver Island for 1.5 years, but originated from gametes from wild anadromous parents that had been collected from their spawning site in Sakinaw Lake. For tagging, we selected smolts larger than 170 mm which Welch et al. (2007) recommend as the minimum length to carry the V9–1 L tag. Fork length of tagged fish averaged 193 mm in 2004 and 213 mm in 2006 (Table 1). Average body mass of smolts was estimated from length-weight conversions based on data for sibling hatchery sockeye sampled in 2004 (ln[weight in g] = 3.13 ln[fork length in mm] – 12.18).

V9–1 L tags measure 24x10 mm and weigh 3.6 g in air; the smaller V9–6 L tags measure 21x9 mm and weigh 2.6 g. The V9–1 L tags ranged from 11% (hatchery sockeye in 2006) to 13% (kokanee in 2005) of average body length, and from 4 to 5% of average body weight. Transmitter burden was always less than 8% of body weight (because all fish were over 170 mm and 49 g), which is less than the maximum value of 16% of length and 8% of weight recommended by Lacroix et al. (2004) for juvenile salmonids, or 10.3% of weight judged by Brown et al. (2006) to have only a small effect on critical swimming speed and no detectable effect on survival over 21 d in sockeye salmon. On average, transmitter burden was less than the upper limit of 5.8% of weight recommended by Hall et al. (2009) for Chinook salmon.

Wild kokanee were obtained by seining in the main basin of Sakinaw Lake in 2005 and 2006. When this component of the study began, we were uncertain whether these freshwater resident fish were residual sockeye or kokanee. A small (4-mm^2) section of caudal fin was excised at the time of tagging and preserved in ethanol. Later genetic analysis based on mitochondrial DNA data (Wood et al. 2008) indicated that virtually all the freshwater resident fish tagged in 2005 and 2006 were kokanee (see Results). In both years, we selected the smallest fish from those larger than 170 mm in an effort to target age-2 rather than age-3 or age-4 individuals. Fork length of tagged kokanee averaged 191 mm in 2005 and 192 mm in 2006, similar to the hatchery sockeye tagged in 2004 but smaller than the hatchery sockeye tagged in 2006 (Table 1).

Animal care protocols were approved annually by institutional animal care committees prior to all field work, and the acoustic tags were surgically implanted following standard procedures (Welch et al. 2007). Smolts were sedated with 1-ppm metomidate HCl (Aquacalm) for a minimum of 10 min, then brought under full anesthesia 3–4 min prior to surgery using 70-ppm Tricaine Methane Sulfonate (MS-222) mixed with a 140-ppm $NaHCO_3$ buffer. Gills were continuously irrigated during surgery with recirculating water

TABLE 1. Summary of acoustic tag releases in sockeye and kokanee from Sakinaw Lake and subsequent detections.

		Tagging Data				Number (%) detected					
		Release		Length (mm)	Pre-release	As juveniles in year of release				As adults	
Year	Ecotype	Date (UTC)	Number	mean (SD)	mortality	All sites	migratory[a]	outer lines[b]	Sakinaw	POST array	Release site
2004	sockeye	1 June 18:00	63	191 (14.5)	0	29 (46%)	27 (43%)	12 (19%)	--	0	1 (2%)
	sockeye	16 June 21:00	34	197 (14.0)	0	12 (35%)	12 (35%)	6 (18%)	--	0	2 (6%)
		all	97	193 (14.6)		41 (42%)	39 (40%)	18 (19%)	--	0	3 (3%)
2005	kokanee	3 June 02:00	47	191 (12.3)	24%[c]	26 (55%)	22 (47%)	13 (28%)	1[d] (2%)	0	2 (4%)
2006	sockeye	3 June 04:00	61	213 (11.5)	0	36 (59%)	26 (42%)	10 (16%)	0	0	0
	kokanee	3 June 04:00	49	192 (10.0)	2%	28 (57%)	19 (39%)	11 (22%)	6 (12%)	0	0
		all	110	204 (15.2)	1%	64 (58%)	45 (41%)	21 (19%)	6 (5%)	0	0
all	all	all	254	197 (15.5)	4%	131 (52%)	106 (42%)	52 (20%)	7[b] (3%)	0	5 (2%)

[a] detected at NSOG, QCS, JDF or Lippy Point acoustic arrays

[b] detected at QCS or JDF acoustic arrays

[c] 11.5% for fish collected 5 days before surgery; 90% for those collected the day of surgery.

[d] includes the detection of reactivated tag number 2802 in 2007; possibly underestimated because fixed receivers were not deployed in Sakinaw Lake in 2005

containing 50-ppm MS-222 (and 100-ppm $NaHCO_3$) flowing into the mouth. A tag was inserted into the abdominal cavity through a mid-ventral line incision, which was closed using 2 or 3 polydioxanone monofilament sutures. Fish were then transferred to a recovery tank. Water baths were changed as needed to maintain dissolved oxygen levels between 8 and 11 ppm and water temperatures within 2°C of ambient hatchery tank water. Vidalife, a water conditioner and mucous-protectant, was used in all water baths.

Hatchery sockeye were tagged at the hatchery and allowed to recover for 9–11 d (but only 3 d for the late release group in 2004). Tagged fish were then sedated in 0.2-ppm metomidate HCl and transported by truck and floatplane in oxygenated containers to a marine sea pen constructed of perforated stainless steel panels (1x1x2 m). Because the hatchery water was initially much colder (8–9°C) than the seawater (16–17°C), the water in the transport containers was warmed as gradually as possible over the 3-h period of transport. Kokanee were tagged at Sakinaw Lake (surface water temperature 18°C) and transported to the marine sea pen more directly than the hatchery sockeye (all within 30 h and most within 8 h of surgery) because we believed the wild fish would be reluctant to feed in confinement. All tagged kokanee were held in the marine sea pen for at least 2 d after surgery to allow time for recovery from anesthesia, acclimation to seawater, and to monitor mortality, tag shedding, or abnormal swimming behavior prior to release.

Releases and early tracking

In 2004 and 2005, tagged fish were held and released in a sheltered cove at Daniel Point, 1.3 km south of the outlet of Sakinaw Lake (Figure 1). That site became inaccessible by floatplane in 2006, so the release site was moved to Lees Bay, 2.8 km south of the outlet of Sakinaw Lake. Fish were released during daylight in 2004 (instructions to release at dusk were misunderstood), but at or after dusk in 2005 and 2006 (Table 1).

In 2006, we examined movement from the release site over the week following release by deploying a Vemco VR-100 mobile receiver at each of 12 sites within a roughly 3-km radius of the release site (Figure 1). Each site was monitored from a boat for about 5 min at intervals of 1, 2, 3, 5 and 7 d after release. Vemco VR-2 fixed receivers were deployed at the release site and at the outlet to Sakinaw Lake continuously from 16 August 2006 until 28 November 2007, within Sakinaw Lake from 29 May 2006 until 14 February 2007, and immediately north and south of the outlet of Rosewall Creek from 22 September until 21 November 2006. In subsequent years, two fixed receivers were redeployed only at the release site and the outlet of Sakinaw Lake where they operated continuously from 3 June to 4 December 2008 and 22 May to 23 December 2009.

Data Analysis

Data from fixed receivers included geographical position, date, time, and tag identity for each detection. Prior to analysis, the raw detection data were screened for false detections that sometimes arise from the collision of two or more transmissions or echoes (multipath). Data from receivers were initially screened using an algorithm described by Welch et al. (2009), before being added to the POST Secretariat's on-line database (http://www.postcoml.org/page.php?section = database). We considered a detection to be real only if the tag code was detected more than once at the same location within a 30-min interval.

Segment distance between receiver locations was defined as the shortest possible route by water. Travel time was measured from release to the time of first detection at each location. Segment distance divided by travel time provides a minimum estimate of swimming speed.

Migration routes were categorized as 'northward direct' (ND, if detected at the NSOG or QCS array without earlier detection at a more southern array), 'southward direct' (SD, if detected at a receiver south of the release site without earlier detection at the NSOG array), 'northward following reversal' (NR, if detected at the NSOG or QCS array after detection at a more southern array), or 'southward following reversal' (SR, if detected at a southern array after detection at the NSOG array). Tags were also assigned to the categories of 'migratory' if detected at one of the three primary POST lines that had to be crossed to reach the open ocean (i.e., NSOG, QCS, or JDF but excluding inside lines); 'ocean going' if they crossed the outermost lines (QCS or JDF), and 'reactivated' if they were detected after the end of the programmed sleep period. The 13 V9–6 L tags released in 2006 without programmed sleep periods were excluded from analyses involving reactivated tags.

Frequencies of fish detections in each category were compared among release groups using likelihood ratio or Mantel-Haenszel tests computed by the cross-tabulation procedures in SYSTAT 11 (SYSTAT 2004). Because hatchery sockeye released in 2006 were larger on average than other groups, we classified fish as 'small' (<210 mm) or 'large' (≥210 mm), such that virtually all fish in the other release groups would be considered 'small.' We then looked for an effect of size-class on detection frequency for all the migration route categories, using data for the hatchery sockeye release in 2006. We used the Mantel-Haenszel procedure to test across all release groups for an association between detection frequencies and size-class or ecotype (sockeye or kokanee) while controlling for year of release. After finding no significant differences, we ignored size differences in subsequent tests.

We tested for differences among groups in travel time to and between the three primary lines across migration corridors (NSOG, QCS, and JDF) with the nonparametric Kruskal Wallis procedure in SYSTAT 11 (SYSTAT 2004). We also tested for correlations between fork length at release and travel time, and then compared the relative influence of distance, year, and ecotype on travel time to each line by constructing general linear models (GLM) in SYSTAT 11 (SYSTAT 2004). Similarly, we estimated rate of dispersion from the release site by regressing the log-transformed counts of fish detected after release against days at large with ecotype as a co-variate in the GLM procedure.

Results

Identification of kokanee

Analysis of mitochondrial DNA (mtDNA, specifically a PCR-RFLP assay of the cytochrome b and ND1 genes described by Wood et al. 2008) indicated that most if not all of the freshwater residents tagged in 2005 and 2006 were the progeny of nonanadromous spawners, and thus, kokanee rather than residual sockeye. Three mitochondrial DNA haplotypes (female lineages, designated #1, #3, and #5) occurred in the tagged fish in roughly equal proportions: 29%, 42% and 29% in 2005, and 30%, 41%, and 28% in 2006. The same haplotypes occurred, but in different proportions, in samples of the anadromous (19%, 2%, and 78%) and nonanadromous ecotypes (31%, 36% and 33%) collected from spawning sites in Sakinaw Lake (Wood et al. 2008). This highly significant difference in haplotype proportions ($P < 0.001$) indicates that anadromous and nonanadromous sockeye are reproductively isolated within Sakinaw Lake, and that the nonanadromous ecotype is a self-perpetuating population of kokanee. MtDNA haplotype composition of tagged fish was not significantly different from that of the kokanee sample ($P = 0.75$), but was significantly different from that of the sockeye sample ($P < 0.001$). Haplotype #3, virtually absent in sockeye (2%) but predominant in kokanee (36%), was the predominant haplotype in both the tagged fish groups (41–42%). Thus, it is statistically most probable that the released group comprised only kokanee. There was no significant difference in mean length among haplotypes within either release group ($P > 0.77$), and thus no indication that the smaller fish were residual sockeye.

Survival to release

No hatchery sockeye died prior to release as a result of tagging, transport, temperature shock, or failure to acclimate to seawater, either in 2004 or 2006. All swam away, apparently healthy, following an acclimation period in the sea pen of 20 h (in 2004) or 55 h (in 2006). All hatchery sockeye had been held for an extended period (3–11 d) in freshwater at the hatchery following surgery to implant tags.

Prerelease mortality of wild kokanee averaged 24% in 2005 and 2% in 2006. Most of the mortality in 2005 resulted from an unplanned effort to bolster sample size by capturing additional fish on the day of tagging. Of 62 fish transferred to the sea pen, 15 died within 24 h, but the remaining 47 appeared healthy after 48 h and were released. Later inspection revealed that 9 of 10 fish (90%) captured, tagged, and transferred to seawater the same day (31 May) had died compared with only 6 of 52 fish (11.5%) captured on 26–27 May, tagged on 30 May, and transferred to seawater on 31 May. In 2006, all kokanee were captured on the 29 May and held in a pen in Sakinaw Lake until surgery on 31 May, transferred to the seawater pen later the same day, and released 55 h later. None died before or immediately after surgery, and only 1 of 50 (2%) died in the sea pen before release. Thus, it seems likely that the additional mortality for fish collected on 31 May 2005 was due to repeated sedation and excessive cumulative stress within a single day.

Behavior at the release site

The number of tagged fish detected by mobile tracking within a 3-km radius of the release site declined exponentially from 48 fish (27 sockeye and 21 kokanee, or 44% of all fish released) the morning following release (day 1) to 5 fish (3 sockeye and 2 kokanee) on day 7. This decline corresponds to a rate of emigration from the release site of 30% per day, with no significant difference between ecotypes ($P > 0.13$).

Six tagged kokanee (but no hatchery sockeye) migrated back into Sakinaw Lake between days 8 and 11. These early return migrations were confirmed by detections with the fixed receivers and supplementary mobile tracking in Sakinaw Lake. To gain access to the lake, at least four of these kokanee must have jumped a 1-m vertical dam because the fish ladder was deliberately blocked to count salmon smolts moving downstream until 14 June 2006 (day 11 after release).

Detections by POST lines and supplementary receivers

Of a total of 254 fish released over three years, 131 (52%) were later detected by fixed receivers in the year of release (Table 1). Detected fish were not significantly larger or smaller than undetected fish within any release group ($P > 0.26$), or overall ($P > 0.18$). After controlling for year of release (in Mantel-Haenszel tests), detection probability was not significantly associated with size-class ($P = 0.82$) or ecotype ($P = 0.92$). MtDNA haplotype composition was not significantly different among detected and undetected groups of kokanee ($P > 0.38$). We conclude that neither size nor ecotype influenced migratory behavior or survival in a way that biased overall probability of detection by the fixed receivers.

Migration routes

Most fish detected after release moved directly north across the NSOG line (the ND route), but proportions varied among years, ranging from 23% of the 47 fish released in 2005 to 36% of 97 released in 2004 (Figure 3). The vast majority (90%) of the 80 northward migrants crossed the eastern end of the NSOG line, passing east of Texada Island through Malaspina Strait. This directed migration behavior was evident each year and did not differ significantly between ecotypes ($P > 0.25$). In each year, a few fish (≤5% of releases) crossed the NSOG line after being detected first by inside lines south of the release site, indicating that they had reversed direction (the NR route). Of the 47 fish released in 2005, most moved south—either directly (the SD route, 26%) or indirectly after crossing the NSOG line and reversing direction (the SR route, 4%). Detections at southern inside lines averaged 5% (of fish released) in Howe Sound, 1% at the SSOG line, 2% in Burrard Inlet, 4% in the Fraser River, and 2% in Puget Sound. (Note that these percentages for inside lines cannot be added because individual fish were sometimes detected at more than one line.)

Relative frequencies of detection in each migration route category (ND, NR, SD, SR) were not significantly affected by release date in 2004 ($P = 0.41$) or ecotype in 2006 ($P = 0.32$, after excluding fish that returned directly to Sakinaw Lake). Similarly, absolute frequencies of detection in each migration route category were not significantly associated with release date in 2004 ($P > 0.12$), ecotype in 2006 ($P > 0.10$), or size-class of hatchery sockeye released in 2006 ($P > 0.67$). In contrast, both relative and absolute frequencies of detection were strongly associated with year for the ND and SD routes ($P < 0.002$, ecotypes and release dates combined) but not for the NR or SR routes ($P > 0.31$). After controlling for year effect, we found no significant association with ecotype ($P > 0.19$, size classes combined) or size-class ($P > 0.20$, ecotypes combined). In sum, a significantly higher proportion of fish moved southward in 2005 than in other years, but this behavior could not be attributed to ecotype, body size, or release date.

Migration speed

Travel time to the NSOG, QCS, or JDF lines was not significantly correlated with fork length at release within any release group after correction for multiple comparisons. Travel times to the QCS line (median = 22 d) and JDF line (median = 25 d) were not significantly different among release groups ($P > 0.10$), but travel times to the NSOG line were significantly less ($P = 0.01$) for the 2004 release (sockeye, median = 6 d) than for other groups (median = 11 d). Travel times to the NSOG or QCS lines in 2004 were not significantly different between release dates ($P > 0.18$). These median travel times correspond to minimum sustained swimming speeds of between 6.5 km/d (2004) and 3.5 km/d (other years) to NSOG, 12 km/d to QCS, and 9.4 km/d to JDF.

We also combined data from all fixed receivers (ignoring possible differences in swimming speed along different migration routes) to regress travel time on absolute distance from the release site with year and ecotype as covariates (Figure 4). Absolute distance and year of release were both highly significant variables ($P < 0.001$) but ecotype was not ($P = 0.30$). The slopes of these regressions imply minimum sustained swimming speeds of 17 km/d in 2004, 22 km/d in 2005, and 18 km/d in 2006, after allowing for

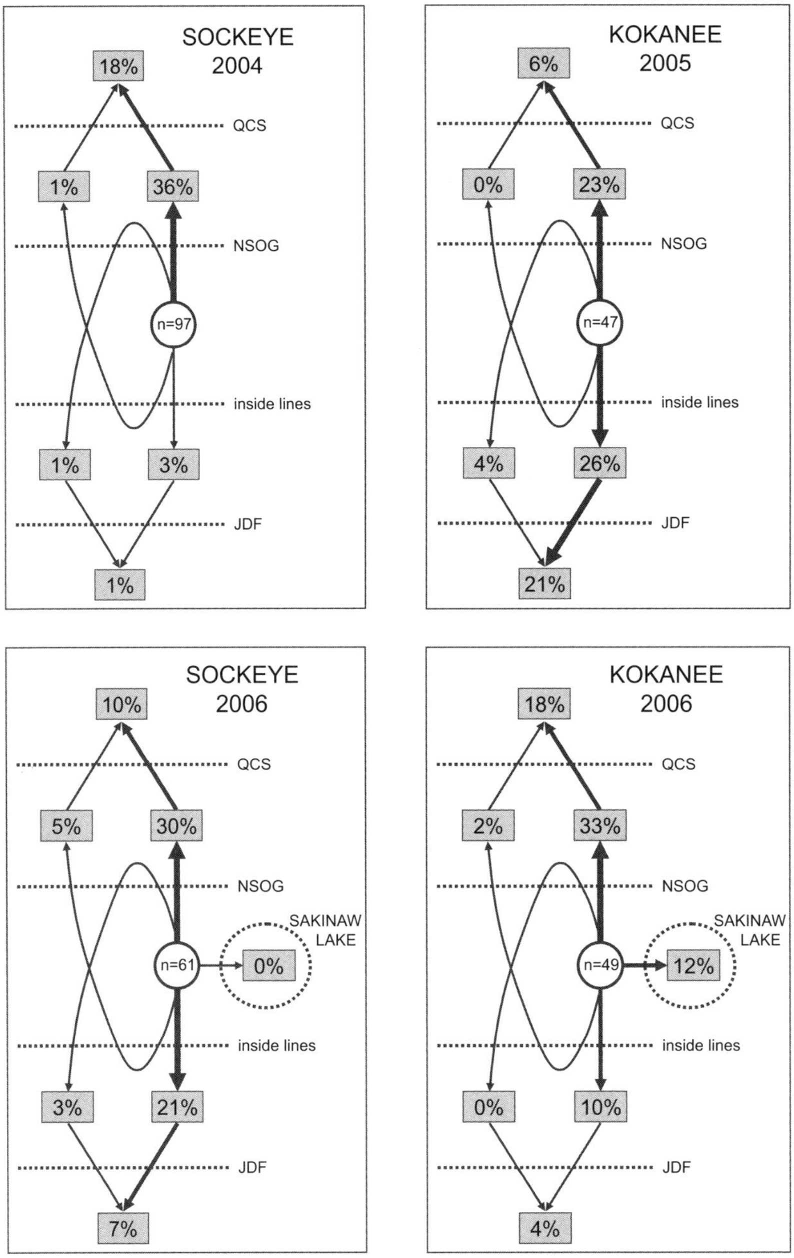

Figure 3. Schematic diagram of migration routes past principal lines of acoustic receivers (dotted lines) by year and ecotype. Arrows indicate the migration route categories: north direct, south direct, north following reversal, south following reversal, and into Sakinaw Lake (measured in 2006 only). The number in the circle indicates the number of tags released (*n*); the number in each rectangle shows the percentage of tags detected along that route (also represented approximately by the thickness of the arrow). Numbers in the uppermost and lowermost rectangles summarize the percentage of tags that were detected reaching the open ocean by the northern or southern route, respectively.

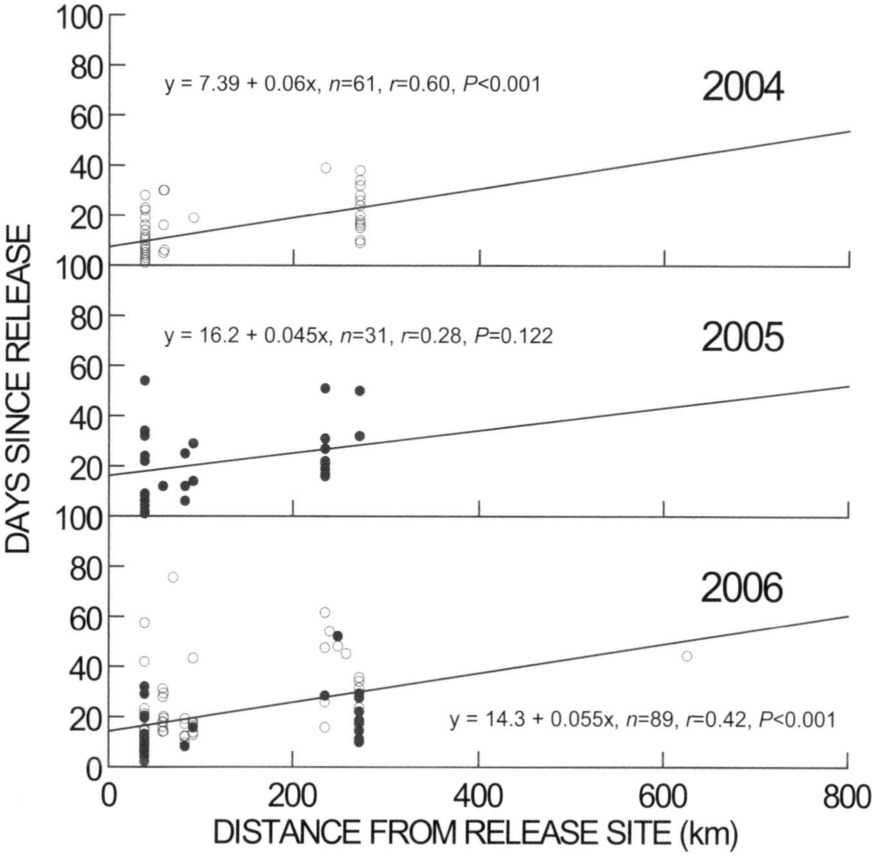

Figure 4. Linear regressions to determine minimum sustained speed of migration. Open circles denote hatchery sockeye; solid circles denote wild kokanee. Estimates of the intercept range from 7 d (2004) to 14 d (2005), and estimates of the slope range from 17 km/d (2004) to 22 km/d (2005).

an initial period of milling near the release site of 7–14 d (based on intercepts of regression lines).

Median travel times between the NSOG and QCS lines did not differ significantly ($P = 0.96$) between sockeye in 2004 (median 11.5 d, $n = 16$) and kokanee in 2005 (10.0 d, $n = 3$), but were significantly less ($P = 0.01$) for kokanee in 2006 (9.0 d, $n = 9$) compared with sockeye in 2006 (median 18.0 d, $n = 5$). Again, these median travel times imply minimum sustained swimming speeds ranging from 13 to 26 km/d. Sample sizes in these comparisons were too small to detect any effect of fish length.

Ocean going tag detections

The total proportion of ocean going fish (i.e., those detected at the outer QCS or JDF lines) averaged 20% across release groups (range 16–28%, Table 1). This proportion was not significantly affected by size-class within hatchery sockeye released in 2006 ($P = 0.54$), ecotype ($P = 0.55$ after controlling for possible year effects, Mantel-Haenzsel test), nor year ($P = 0.24$ after combining ecotypes).

Reactivated tag detections

By autumn 2009, 6 tags have been detected after transmissions were programmed to start. Three of these transmitted for much longer than had been expected—at least 798, 815 and 913 d after subtracting time in sleep mode. Tag number 2802 was detected in Sakinaw Lake during mobile tracking surveys on 1 August, 16 August, and 28 November 2007, and on each occasion, it was located near the main spawning beach in the upper basin. This tag had been placed in a 177-mm long kokanee released in 2005 and was not detected at POST lines in either 2005 or 2007. Fixed receivers were deployed in Sakinaw Lake only between 29 May 2006 and 14 February 2007, and the tag would have been in sleep mode between 28 July 2005 and 13 June 2007. Thus, the kokanee carrying tag 2802 could have returned to Sakinaw Lake any time between its release on 3 June 2005 and its detection there on 1 August 2007. We assume that it returned to Sakinaw Lake soon after release in 2005, because this behavior was observed in 12% of the kokanee released in 2006. Accordingly, tag 2802 was excluded from further analyses involving returning adult fish.

Three reactivated tags were detected among hatchery-reared sockeye originally transported from Rosewall Creek, Vancouver Island, and released in 2004. Two of these were from the late release group with fork lengths at release of 211 mm (tag number 2087) and 194 mm (tag number 2088). Both were detected intermittently over a number of consecutive dates in 2006 by a fixed receiver at the release site. Tag 2087 was detected for a period of 80 d (30 July to 17 October); tag 2088 was detected for a period of 6.5 d (29 August to 4 September). These tags had not been detected at any of the POST lines or other fixed receivers, either as juveniles in 2004 or as adults after reactivation. The detections at the release site could not be explained by early mortality at the release site because the first detections occurred well after 17 July 2006, the date the tags were programmed to begin retransmitting. Moreover, the timing of the first detections (late August–September) is consistent with the natural return migration timing for this sockeye population, and the intermittency of detections suggests that the tagged fish were milling in the vicinity of the release site. The third reactivated sockeye tag (number 2034 from the early 2004 release group, 175 mm at release) was detected repeatedly and intermittently at the release site over two consecutive nights (5–6 September 2008). Like the others, tag 2034 had not been detected previously at POST lines or at any of the supplementary receivers. None of the three sockeye tags were detected by receivers in Sakinaw Lake or near its outlet.

Two more reactivated tags were detected among kokanee released in 2005 (tag 2781, 183 mm at release; tag 2796, 204 mm at release). Both were detected at the outlet of Sakinaw Lake in the summer of 2009, but neither were detected at the release site. Tag 2781 was detected repeatedly over many consecutive days, mostly at night, between 7 June and 7 July 2009. Tag 2796 was detected repeatedly over a shorter period between 15 and 20 June 2009. Tag 2796 might have ascended into Sakinaw Lake without being detected again as access was unobstructed once the smolt weir was removed on 8 June 2009 and video surveillance for returning adult sockeye did not begin until 25 June 2009; no receivers were deployed within Sakinaw Lake in 2009. Tag 2796 had not been detected previously by any other receivers. Tag 2781 had been detected when it crossed the outer POST line in Howe Sound 11 d after release (22 June 2005) and reached the inner (northern) line in Howe Sound the next day; it was last detected at the outer Howe Sound line 15 and 16 d after release, perhaps re-entering the Strait of Georgia.

No reactivated tags were ever detected among fish in the migratory group (65 sockeye and 41 kokanee). In contrast, 5 reactivated tags were detected as returning adults

(3 in sockeye and 2 in kokanee) among the remaining ('nonmigratory') tag group which comprised 98 sockeye and 47 kokanee (excluding the nonsleeper V9–6 L tags and tag 2802 which is presumed to have entered Sakinaw Lake shortly after release). Averaged over all years, estimated survival from release to adult return ranged from 0% for both ecotypes in the migratory group to 3.4% for nonmigratory sockeye and 4.3% for nonmigratory kokanee. Statistical power was low given the low number of adult returns, so the statistical significance of this difference in return rate between migratory and nonmigratory fish is marginal ($P = 0.02$, likelihood ratio chi-square or $P = 0.08$, Fisher's exact test). Still the difference is qualitatively striking, and may be biologically significant (see Discussion). No reactivated tags were detected from releases in 2006, so excluding 2006 yields slightly higher estimates of survival to return (5.2% for nonmigratory sockeye released in 2005 and 8.3% for nonmigratory kokanee released in 2006), and slightly increases statistical signficance.

Discussion

Migratory routes and swimming speed

Several lines of evidence suggest that the tagged fish required a few days following release to begin their directed migration. First, a large proportion were detected by mobile tracking near the release site over the first week after release in 2006; our estimate for rate of dispersion from the release site was only 30% per day. Many of these fish made meandering excursions into a local inlet (Pender Harbour). Others made even longer excursions into more southern inlets (Howe Sound, Burrard Inlet, the Fraser River, and Puget Sound). These unexpected migrations were more apparent in hatchery-reared sockeye than wild kokanee, but the difference was not statistically significant. Second, the intercepts in regressions of travel time on distance were all between 7 and 14 d, suggesting an initial period of milling near the release site. This interpretation is also consistent with the single tag detected at Lippy Point which migrated a distance of at least 626 km in 44 d, a minimum average swimming speed of only 9 km/d from the release site to JDF, but 22 km/d from JDF to Lippy Point; and with the average minimum swimming speed between the NSOG and QCS lines (22 km/d) for the 33 fish observed over all years and release groups. It is unclear whether the inferred delay in migration was required for orientation and route finding or reflects inexperience (hatchery-rearing or nonanadromous tendency), a variable reaction to the stress of tagging and release, or a period of acclimation to seawater.

Once underway, average migration speed (estimated from the slope of regression lines) ranged from 17 km/d to 22 km/d, which corresponds to 1.0–1.3 body lengths per sec (BL/s). This migration speed is similar to estimates for Cultus Lake sockeye salmon (0.46–1.82 BL/s, Welch et al. 2009) and steelhead trout (anadromous rainbow trout *O. mykiss*, 0.5–2.0 BL/s, Welch et al. 2004; Melnychuk et al. 2007), but faster than estimated for Somass River sockeye salmon smolts (0.3 BL/s) migrating across Barkley Sound on the west coast of Vancouver Island (Wood et al. 1993), and marine migrations of post-smolts of Atlantic salmon *Salmo salar* (0.53–0.56 BL/s) and brown trout *S. trutta* (0.06 BL/s) (Thorstad et al. 2007). Within our study, body length at release had no detectable effect on speed of migration or migration route, presumably because length at release varied only narrowly (95% of tagged fish were within 197 mm ± 31 mm).

In 2 of the 3 years studied, most Sakinaw sockeye and kokanee migrated directly northward following release, with 90% migrating through Malaspina Strait, the narrow channel on the mainland side of Texada Island. The same behavior was observed in Cultus Lake sockeye salmon (Welch et al. 2009) and earlier studies have demonstrated that the largest aggregations of Fraser River sockeye generally tend to move northward along the eastern side of the Strait of Georgia and Johnstone Strait (Groot and Cooke 1987; Groot et al. 1988). The consistency of this migration behavior across studies suggests that sockeye smolts might find their way by orienting relative to the mainland coast as they migrate northward, but most tagged smolts migrated multiple kilometers away from the coast. Welch et al. (2009) and Groot and Cooke (1987) also found that significant numbers of Fraser River sockeye salmon enter Howe Sound and other inlets.

A significantly higher proportion of tagged fish (kokanee) moved southward in 2005 than in other years, but this behavior could not be attributed to ecotype, body size, or release date. It seems that migration routes and speeds can be affected by year-to-year variations in unknown climatic factors. Oceanographic conditions and fish ecology were anomalous in 2005 both in the Strait of Georgia (Beamish et al. 2008) and along the west coast of North America (DFO 2006) resulting in generally poor survival of salmon smolts entering the ocean in 2005 (R. Beamish, Fisheries and Oceans, personal communication). On the other hand, migratory behavior and marine survival was not unusual for Cultus Lake sockeye salmon tagged and released in 2005 compared to 2004 or 2006 (Welch et al. 2009).

Migration was directed northward most strongly in 2004 as evident in the higher proportions detected at the NSOG and QCS lines, shorter travel time to the NSOG line, and fewer detections per fish at the NSOG line. We note, however, that the number of tags detected at the JDF line in 2004 (only 1 of 97 tags released) might be biased low because one third of the JDF receivers were lost that year creating gaps in the line (Welch et al. 2009).

Proportion of tags detected after release

On average, 49% of the migratory tags (n = 106) were subsequently classified as ocean going (n = 52), most passing north through Queen Charlotte Strait. This proportion of ocean going tags likely underestimates actual survival in the Strait of Georgia for several reasons. First, some tags might be missed. However, the probability of detection is thought to be high for the marine POST lines, and is estimated at 90% for the NSOG line based on a Cormack-Jolly–Seber capture/recapture model and subsequent detections at the outer lines (Welch et al. 2009). Applying this correction to the QCS and JDF lines would only boost the ocean-going proportion from 49% to 55% of migratory tags.

Second, some tags might have failed, been shed, or entered sleep mode before the fish had completed their migration out of the Strait of Georgia. Less than 10% of Vemco-style tags were shed from salmonids held for up to 2 years in tanks (Rikardsen and Thorstad 2008). In 2004, most detections at the outer lines occurred well before the programmed shut-off date (between 8 September and 1 October). For this reason, the sleeper programming was modified in subsequent years to shut-off earlier (late July) in an attempt to extend the transmission period after reactivation. Thus, the programmed dates of tag activity could not significantly bias estimates of proportion migrating seaward unless migration out of the Strait of Georgia is bimodal with the second mode occurring after September.

Third, some tagged fish might have remained within the Strait of Georgia throughout the marine phase of their lives. None of the five tags detected in adults after reactivation were detected crossing the NSOG, QCS, or JDF lines as seaward-migrating juveniles or returning as adults. The simplest explanation is that these fish remained within the Strait of Georgia during the period that the tags were active and the POST lines operational (until 30 August 2004 and after 17 July 2006 in the case of sockeye released in 2004). Accordingly, we refer to these fish as nonmigratory to simplify our comparison with migratory fish detected at a primary POST line in the year of release. Fish in the migratory group might also have stopped migrating seaward, accounting for the reduced proportion in the ocean-going group, but none were detected returning across POST lines or at receivers near the release site.

Nonmigratory fish survived to be detected as returning adults at an overall rate of >3%, which is within the typical range of marine survival in a viable sockeye salmon population. In contrast, the 0% rate of return for migratory tagged fish closely matched that of the natural (nonviable) anadromous sockeye population in Sakinaw Lake (average <0.2%, range 0–0.7% since ocean-entry year 2003; Figure 2). Overall smolt-to-adult survival of Sakinaw Lake sockeye has been too low for a significant number of smolts to have survived at the measured rate observed for the nonmigratory fish in our study. We speculate that the nonmigratory behavior was an artifact of the study, perhaps induced by rearing smolts to an unnaturally large size before release, but we cannot exclude the possibility that there are two distinct migratory phenotypes with the rarer phenotype now experiencing higher marine survival.

Feasibility of re-anadromization

Our findings also lend credibility to speculation that residual sockeye or kokanee could be used to recreate anadromous populations. Kokanee survived the transition to full-strength seawater just as well as hatchery sockeye. This was not surprising given similar results in previous experiments to compare the seawater adaptability of sympatric sockeye and kokanee (Foote et al. 1992) and land-locked kokanee (Foote et al. 1994). Sockeye and kokanee both dispersed from the release site at the same rate, and showed the same tendency to migrate northward along the mainland coast through Malaspina Strait when released in the same year (2006). Despite repeated tests, we found no differences that could be attributed to ecotype in the choice of migration route, travel times to POST lines, or rates of survival to the open ocean, although small sample sizes sometimes limited statistical power.

One significant difference between the migratory behavior of hatchery sockeye and kokanee was that (at least) 9 kokanee, but no hatchery-sockeye, migrated back to the outlet of Sakinaw Lake, and 7 of these are known to have entered the lake. These observations demonstrate that kokanee transported from the lake in buckets and released at sea 2.8 km from the lake outlet, are able to home back to their natal lake. In fact, tag 2802 was detected after reactivation in 2007 on the spawning ground in the upper basin of Sakinaw Lake. On the other hand, we were not surprised that hatchery sockeye did not migrate into Sakinaw Lake, because they had no prior exposure to water from the Sakinaw watershed; they had been collected as gametes from wild spawners and reared in well water at the Rosewall Creek hatchery on the east coast of Vancouver Island, 60 km distant. No sockeye were detected at either of two fixed receivers deployed near the outlet of Rosewall Creek in 2006, but this result was not surprising either because the sockeye were reared in well water and never exposed to water from Rosewall Creek.

Perhaps most importantly, two tagged kokanee returned to Sakinaw Lake after spending four years at sea. This confirms that Sakinaw kokanee can survive a fully anadromous migration, consistent with scanty evidence documented elsewhere. Foerster (1947) describes an experiment in which 63,874 hatchery-reared age-1 kokanee from Kootenay Lake, BC, were released into Sweltzer Creek which drains Cultus Lake into the lower Fraser River. The kokanee population in Kootenay Lake has been isolated from anadromous sockeye salmon for about 10,000 years by a natural waterfall (Northcote 1973). One might expect these kokanee to have lost many adaptations for anadromous migration, yet they still exhibit a circannual cycle of seawater adaptability (Foote et al. 1994). Foerster (1947) reports that no kokanee were observed in the expected year of return (at age 4), either in marine fisheries or in Sweltzer Creek, but the following year, 25 fish bearing the same marks (pelvic fin clips) were caught in the fishery, and 17 others returned to Cultus Lake. Those caught in the fishery were judged to be age 5 (implying that they were sea-run kokanee), but the age of fish returning to Cultus Lake could not be confirmed. Interestingly, both tagged kokanee that returned to Sakinaw Lake as adults were also older than expected (age 6 rather than age 4).

In another experiment documented by Rounsefell (1958), hatchery-reared progeny of kokanee from Wenatchee Lake, Washington State were marked and planted back into Wenatchee Lake, and 0.5% were later recovered as anadromous adults. Ricker (1972) speculates that these kokanee had been recently derived from anadromous sockeye in Wenatchee Lake. Kokanee or residual sockeye are now absent or very rare in Wenatchee Lake (A. Murdoch, Washington Department of Fish and Wildlife, personal communication).

Two hatchery-reared kokanee were captured in the Pacific Ocean after being planted in Lake Toro, Hokkaido, and migrating to sea of their own volition (Kaeriyama et al. 1982). When hatchery-reared kokanee from the same population were forced to migrate to sea, as many as 2.8% survived to return as anadromous adults (Urawa et al. 1999). This kokanee population had been transplanted from Lake Sikotshu, which in turn had originated from anadromous sockeye salmon that were transplanted from Lake Akan as early as 1893, and from the Kuril Islands as recently as 1925–1940 (Urawa et al. 1999).

These studies all suggest that kokanee can revert to an anadromous life history under some conditions. In most of these cases (but excluding the Cultus Lake experiment), the 'sea-run kokanee' were thought to have been recently derived from an anadromous sockeye population. Therefore, we should not expect anadromy to re-evolve quickly in all situations. As a counter-example, the anadromous population of sockeye salmon in Cowichan Lake on Vancouver Island BC is thought to have been extirpated thousands of years ago when the outflow shifted from the west coast to the east coast of Vancouver Island (Beamish 1982). Despite the high density and abundance of kokanee in Cowichan Lake, and relatively easy access upstream from the Strait of Georgia (the lake is used by anadromous Chinook *O. tshawytscha* and coho salmon *O. kisutch*), no anadromous sockeye salmon run has become re-established within historic records (Rutherford et al. 1988). Furthermore, records from annual smolt trapping to enumerate Chinook salmon smolts in the Cowichan River between 1991 and 2002 confirm that kokanee rarely if ever migrate downstream from Cowichan Lake (Nagtegaal et al. 2004 and earlier references therein).

Acoustic tag technology

This research was possible through recent advances in acoustic technology, and the extraordinary opportunities provided by large-scale telemetry array infrastructures (POST). Surgical procedures for implanting miniaturized acoustic tags in juvenile salmonids have been perfected to the point that little or no short-term mortality was observed in this study, similar to the results from surgical trials on steelhead (Welch et al. 2007), coho (Chittenden et al. 2009), and Chinook (Rechisky and Welch 2010). Prerelease mortality was unacceptably high in wild kokanee in 2005, but this was attributed to excessive sedation during capture and handling, not to the surgery; learning from this experience, we were able to almost eliminate prerelease mortality of wild kokanee in 2006. The proportion of fish surviving to be detected following release was generally sufficient for meaningful statistical comparisons of migration route, speed, and rate of survival to the ocean among release groups.

To our knowledge, this was the first study (when it began in 2004) to rely on programming of Vemco tags to extend tag life so as to cover both the start and end of the marine migration period; similar programming was later adopted in 2005–2007 to investigate the return migrations of Cultus Lake sockeye (Welch et al. 2009). Although we recommend further study to determine the overall reliability of sleeper tags, six reactivated Sakinaw tags were detected over two years, and three, up to four years after release. These observations confirm that sophisticated programming can work in the field, and that at least some individuals (both sockeye salmon and kokanee) were able to survive to maturity with the tags in place. Extending tag life by programming a sleep period provides obvious benefits, but at the cost of losing information on fish behavior during the sleep period.

Our study also illustrates the potential synergies of mixing technologies to explore different spatial and temporal scales. The POST array and Vemco tags provided consistent large-scale and long-term monitoring, which was supplemented by mobile tracking and tactical deployment of additional fixed receivers to meet specific local and short-term objectives. By replicating release groups within a season, we obtained additional evidence that hatchery-reared sockeye salmon can be useful as a proxy for understanding the behavior of wild sockeye salmon. The relatively consistent pattern of early marine migration observed among migratory fish from all hatchery-reared release groups suggests an underlying propensity (presumably genetic) regardless of the details of release. On the other hand, we attribute the surprising proportion of nonmigratory fish among both hatchery-reared sockeye and wild kokanee to their unnaturally large size at release, or perhaps to hatchery rearing or release conditions; their smolt-to-adult survival was an order of magnitude higher, >3% overall, compared with an average <0.2% observed for coded-wire tagged smolts in the wild Sakinaw sockeye population.

Although not without problems, acoustic tags can be especially useful for behavioral studies of aquatic animals that are rare, difficult to capture, or belong to endangered populations such that only a limited number of individuals can be recaptured, and where every tag must be informative. Our tagging efforts allowed us to refute our initial hypothesis—that endangered Sakinaw sockeye experience higher marine mortality than nonendangered Fraser River sockeye because they remain for a longer period in the Strait of Georgia and encounter unfavorable conditions there. In 2 out of 3 years, migration route and timing through the Strait of Georgia were similar to that reported for upper Fraser River sockeye populations (Groot and Cooke 1987; Groot et al. 1988). The unexpectedly high survival of nonmigratory tagged fish further suggests that the extremely poor survival of wild Sakinaw sockeye smolts has been caused by factors operating outside of the Strait of Georgia. Our study also provides some evidence that nonanadromous kokanee can survive forced anadromous migration to return as adults. However, further study is needed to determine the extent to which reversion to anadromy is possible, and whether a sympatric kokanee population could be used as a last resort to restore an extirpated anadromous sockeye run.

Acknowledgments

This work is a contribution to the Census of Marine Life. M. Jacobs and A. Ladouceur performed all the surgery; L. Clint and L. Willott reared the Sakinaw sockeye and helped with their transport; C. Murray, J. Irvine, G. McBain, and the Sechelt Indian Band assisted with field work at Sakinaw Lake; A. Porter provided summaries of POST detection data and prepared Figure 1; and S. Grant, M. Bradford, and S. Baillie provided unpublished size and survival data in Figure 2 for Chilko, Cultus, and Sakinaw sockeye, respectively. We thank F. Whoriskey and M. Bradford for loaning us receivers; K. Andrews, C. Chittenden and M. Melnychuk for sharing tag detection data from their own research; and three anonymous reviewers for many excellent suggestions. Funding was provided by Fisheries and Oceans Canada (SARA program), with matching funding in 2005 from the Pacific Salmon Commission Southern Endowment Fund; the Pacific Salmon Foundation provided additional funding to JC in 2005; most of the marine POST sub-arrays cited in this study were funded with a grant from the Gordon and Betty Moore Foundation.

References

Beamish, R. J. 1982. *Lampetra macrostoma*, a new species of freshwater parasitic lamprey from the west coast of Canada. Canadian Journal of Fisheries and Aquatic Sciences 39:736–747, 1982.

Beamish, R. J., R. M. Sweeting, K. L. Lange, and C. M. Neville. 2008. Changes in the population ecology of hatchery and wild coho salmon in the Strait of Georgia. Transactions of the American Fisheries Society 137:503–520.

Brown, R. S., D. R. Geist, K. A. Deters, and A. Grassell. 2006. Effects of surgically implanted acoustic transmitters >2% of body mass on the swimming performance, survival and growth of juvenile sockeye and Chinook salmon. Journal of Fish Biology 69:1626–1638.

Burgner, R.L. 1991. The life history of sockeye salmon (*Oncorhynchus nerka*). Pages 3–117 in C. Groot and L. Margolis editors. Pacific salmon life histories. UBC Press, Vancouver.

Chittenden, C.M., K.G. Butterworth, K.F. Cubitt, M.C. Jacobs, A. Ladouceur, D.W. Welch & R.S. McKinley. 2009. Maximum tag to body size ratios for an endangered coho salmon (*O. kisutch*) stock based on physiology and performance. Environmental Biology of Fishes: 84:129–140.

COSEWIC. 2003. COSEWIC assessment and status report on the sockeye salmon *Oncorhynchus nerka* Sakinaw population in Canada. Committee on the Status of Endangered Wildlife in Canada. Ottawa. Available: http://www.sararegistry.gc.ca/virtual_sara/files/COSEWIC/sr%5Fsockeye%5Fsalmon%5Fe%2Epdf (May 2010).

DFO (Department of Fisheries and Oceans Canada). 2006. State of the Pacific Ocean 2005. DFO Science. Ocean Status Report. 2006/001. Available: http://www.pac.dfo-mpo.gc.ca/science/psarc-ceesp/osrs/StateofOceans2005fnl.pdf (Accessed May 2010).

Foerster, R. E. 1947. Experiment to develop sea-run from land-locked sockeye salmon (*Oncorhynchus nerka kennerlyi*). Journal of the Fisheries Research Board of Canada 7:88–93.

Foote, C. J., I. Mayer, C. C. Wood, W. C. Clarke, and J. Blackburn. 1994. On the developmental pathway to nonanadromy in sockeye salmon, *Oncorhynchus nerka*. Canadian Journal of Zoology 72:397–405.

Foote, C. J., C. C. Wood, W. C. Clarke, and J. Blackburn. 1992. Circannual cycle of seawater adaptability in *Oncorhynchus nerka*: genetic differences between sympatric sockeye salmon and kokanee. Canadian Journal of Fisheries and Aquatic Sciences 49:99–109.

Foote, C. J., C. C. Wood, and R. E. Withler. 1989. Biochemical genetic comparison of sockeye salmon and kokanee, the anadromous and nonanadromous forms of *Oncorhynchus nerka*. Canadian Journal of Fisheries and Aquatic Sciences 46:149–158.

Grant, S.C.H., C.G.J. Michielsens, E.J. Porszt, and A. Cass. 2010. Pre-season run size forecasts for Fraser River Sockeye Salmon (*Oncorhynchus nerka*) in 2010. Canadian Science Advisory Secretariat 2010/042. Available: http://www.dfo-mpo.gc.ca/csas-sccs/publications/resdocs-docrech/2010/2010_042-eng.htm (November 2010).

Groot, C., R. E. Bailey, L. Margolis, and K. Cooke. 1988. Migratory patterns of sockeye salmon (*Oncorhynchus nerka*) smolts in the Strait of Georgia, British Columbia, as determined by analysis of parasite assemblages. Canadian Journal of Zoology 67:1670–1678.

Groot, C., and K. Cooke. 1987. Are the migrations of juvenile and adult Fraser River sockeye salmon (*Oncorhynchus nerka*) in nearshore waters related? Pages 53–60 *in* H. D. Smith, L. Margolis, and C. C. Wood, editors. Sockeye salmon (*Oncorhynchus nerka*) population biology and future management. Canadian Special Publication of Fisheries and Aquatic Sciences 96.

Gross, M. R. 1987. Evolution of diadromy in fishes. Pages 14–25 *in* M. J. Dadswell, R. J. Klauda, C. M. Moffitt, R. L. Saunders, R. A. Rulifson, and J. E. Cooper, editors. Common strategies of anadromous and catadromous fishes. American Fisheries Society, Symposium 1, Bethesda, Maryland.

Gustafson, R., R. S. Waples, J. M. Myers, J. J. Hard, G. J. Bryant, O. W. Johnson, and L. A. Weitkamp. 2007. Pacific salmon extinctions: quantifying lost and remaining diversity. Conservation Biology 21:1009–1020.

Hall, J. E., J. Chamberlin, A. N. Kagley, C. Greene, and K. L. Fresh. 2009. Effects of gastric and surgical insertions of dummy ultrasonic transmitters on juvenile Chinook salmon in seawater. Transactions of the American Fisheries Society 138:52–57.

Henderson, M. A., and A. J. Cass. 1991. Effect of smolt size on smolt-to-adult survival for Chilko Lake sockeye salmon (*Oncorhynchus nerka*). Canadian Journal of Fisheries and Aquatic Sciences 48:988–994.

Hendry, A. P., T. Bohlin, B. Jonsson, and O. K. Berg. 2004. To sea or not to sea? Anadromy versus non-anadromy in salmonids. Pages 92–125 *in* A. P. Hendry and S. C. Stearns, editors. Evolution illuminated: salmon and their relatives. Oxford University Press, Oxford UK.

Irvine, J. R., M. R. Gross, C. C. Wood, L. B. Holtby, N. D. Schubert, and P. G. Amiro. 2005. Canada's Species At Risk Act: an opportunity to protect "endangered" salmon. Fisheries 30:11–19.

Kaeriyama, M., S. Urawa, and T. Suzuki. 1982. Anadromous sockeye salmon (*Oncorhynchus nerka*) derived from nonanadromous kokanees: Life history in Lake Toro. Scientific Reports of the Hokkaido Salmon Hatchery 46:157–174.

Lacroix, G. L., D. Knox, and P. McCurdy. 2004. Effects of implanted dummy acoustic transmitters on juvenile Atlantic salmon. Transactions of the American Fisheries Society 133:211–220.

Melnychuk, M. C., Welch, D. W., Walters, C. J., and Christensen, V. 2007. Riverine and early ocean migration and mortality patterns of juvenile steelhead trout (*Oncorhynchus mykiss*) from the Cheakamus River, British Columbia. Hydrobiologia 582:55–65.

Murray, C., and C. Wood. 2002. Status of Sakinaw Lake sockeye salmon (*Oncorhynchus nerka*). DFO Canadian Science Advisory Secretariat Research Document 2002/088. Available: http://www.dfo-mpo.gc.ca/csas/Csas/DocREC/2002/RES2002_088e.pdf (May 2010).

Nagtegaal, D. A., E. W. Carter, N. K. Hop Wo, and K. E. Jones. 2004. Juvenile chinook production in the Cowichan River, 2002. Canadian Manuscript Reports of Fisheries and Aquatic Sciences 2679.

Nelson, J. S. 1968. Distribution and nomenclature of North American kokanee, Oncorhynchus nerka. Journal of the Fisheries Research Board Canada 25:409–414.

Northcote, T.G. 1973. Some impacts of man on Kootenay Lake and its salmonids. Great Lakes Fishery Commission Technical Report No. 25.

Rechisky, E. L., D. W. Welch, A. D. Porter, M. C. Jacobs, A. Ladouceur. 2009. Experimental measurement of hydrosystem-induced mortality in juvenile Snake River spring Chinook salmon using a large-scale acoustic array. Canadian Journal of Fisheries and Aquatic Sciences. 66:1019–1024.

Rechisky, E.L., and Welch, D.W. 2010. Surgical implantation of acoustic tags: Influence of tag loss and tag-induced mortality on free-ranging and hatchery-held spring Chinook (*O. tschawytscha*) smolts. Pages 71–96 *in* K. S. Wolf and J. S. O'Neal, editors. PNAMP Special Publication: tagging, telemetry and marking neasures for monitoring fish populations—A compendium of new and recent science for use in informing technique and decision modalities: Pacific Northwest Aquatic Monitoring Partnership Special Publication 2010–002.

Ricker, W. E. 1938. "Residual" and kokanee salmon in Cultus Lake. Journal of the Fisheries Research Board, Canada 4:192–218.

Ricker, W. E. 1940. On the origin of kokanee, a fresh-water type of sockeye salmon. Proceedings and Transactions of the Royal Society of Canada 34:121–135.

Ricker, W.E. 1972. Heredity and environmental factors affecting certain salmonid populations. Pages 19–160 *in* R.A. Simon and P.A. Larkin editors. The stock concept in Pacific salmon. H.R. MacMillan Lectures in Fisheries, University of British Columbia, Vancouver, B.C.

Rikardsen, A. H. and E. B. Thorstad. 2012. Choosing the best attachment methods for telemetry tags: experiences from northern salmonids. Proceedings of the symposium on advances in fish tagging and marking technology, Auckland, New Zealand, 24–28 February 2008. American Fisheries Society, Bethesda, Maryland.

Rounsefell, G. A. 1958. Anadromy in North American Salmonidae. U.S. Fish and Wildlife Service, Fishery Bulletin 58:171–185.

Rutherford, D. T., C. C. Wood, and J. R. Candy. 1988. Stock assessments and biological investigations of kokanee (*Oncorhynchus nerka*) in Cowichan Lake, British Columbia 1984–1987. Canadian Manuscript Report of Fisheries and Aquatic Sciences 1995.

SYSTAT. 2004. SYSTAT 11. SYSTAT Software, Inc. Richmond, California. Available: http://www.systat.com (May 2010).

Thorstad, E. B., Okland, F., Finstad, B., Sivertsgard, R., Plantalech, N., Bjorn, P. A., and McKinley, R. S. 2007. Fjord migration and survival of wild and hatchery-reared Atlantic salmon and wild brown trout post-smolts. Hydrobiologia 582:99–107.

Urawa, S., M. Ban, M. Fukuwara, T. Suzuki, and M. Kaeriyama. 1999. Progressive technologies for the artificial production of anadromous sockeye salmon in Japan. Bulletin of Tohoku Regional Fisheries Research Laboratory 62:141–150.

Waples, R. S., G. R. Pess, and T. Beechie. 2008. Evolutionary history of Pacific salmon in dynamic environments. Evolutionary Applications 1:189–206.

Welch, D. W., S. D. Batten, and B. R. Ward. 2007. Growth, survival and tag retention of steelhead trout (*O. mykiss*) surgically implanted with dummy acoustic tags. Hydrobiologia 582:289–299.

Welch, D. W., Boehlert, G. W., and Ward, B. R. 2003. POST–the Pacific Ocean Salmon Tracking Project. Oceanologia Acta 25:243–253.

Welch, D. W., M. C. Melnychuk, E. L. Rechisky, A. D. Porter, M. J. Jacobs, A. Ladouceur, R. S. McKinley, and G. D. Jackson. 2009. Freshwater and marine migration and survival of endangered Cultus Lake sockeye salmon smolts using POST, a large-scale acoustic telemetry array. Canadian Journal of Fisheries and Aquatic Sciences 66:736–750.

Welch, D. W., S. Turo, S. D. Batten. 2006. Large-scale marine and freshwater movements of white sturgeon. Transactions of the Amercian Fisheries Society 135:140–143.

Welch, D. W., B. R. Ward, and S. D. Batten. 2004. Early ocean survival and marine movements of hatchery and wild steelhead trout (*O. mykiss*) determined by an acoustic array: Queen Charlotte Strait, British Columbia. Deep-Sea Research 51:897–909.

Wood, C. C. 1995. Life history variation and population structure in sockeye salmon. Pages 195–216 *in* J. L. Nielsen, editors. Evolution and the aquatic ecosystem: defining unique units in population conservation. American Fisheries Society Symposium 17, Bethesda, Maryland.

Wood, C. C., J. W. Bickham, R. J. Nelson, C. J. Foote, and J. C. Patton. 2008. Recurrent evolution of life history ecotypes in sockeye salmon: implications for conservation and future evolution. Evolutionary Applications 1:107–221.

Wood, C. C., and C. J. Foote. 1996. Evidence for sympatric genetic divergence of anadromous and non-anadromous morphs of sockeye salmon (*Oncorhynchus nerka*). Evolution 50:1265–1279.

Wood, C. C., N. B. Hargreaves, D. T. Rutherford, and B. T. Emmett. 1993. Downstream and early marine migratory behaviour of sockeye salmon (*Oncorhynchus nerka*) smolts entering Barkley Sound, Vancouver Island. Canadian Journal of Fisheries and Aquatic Sciences 50:1329–1337.

American Fisheries Society Symposium 76:313–323, 2012

Potential Effects of Acoustic Tag Strength Variation on Detection Probabilities of Migrating Fish: Recommended Measurements Prior to Tagging

Michael C. Melnychuk*

Department of Zoology, and Fisheries Centre, University of British Columbia 2202 Main Hall, Vancouver, British Columbia V6T 1Z4, Canada

Abstract.—A common assumption in acoustic or radio telemetry studies is that tag transmission strength is homogeneous for a particular tag type, which in turn supports the assumption that detection ranges or mark–recapture detection probabilities are homogenous among tagged fish. Variation among tags in acoustic intensity could reduce precision in detection probability estimates that do not account for it, and therefore possibly in the precision of survival or abundance estimates. Simple methods are suggested for quantifying variation in tag strength prior to tagging fish and incorporating these measurements into mark–recapture models. At little extra effort to the researcher, these measurements could explain part of the variation in detection probability estimates and therefore could increase the precision of survival or abundance estimates of migrating fish. This potential source of variation in detection probabilities was investigated in a case study with migrating salmon smolts. An index of tag strength was quantified while coded acoustic tags were activated prior to tagging fish. Detection and survival probabilities were estimated with standard mark–recapture methods for the downstream and early ocean migration. A model that included the tag strength index as an additive covariate of detection probabilities had a reasonable level of support compared to a model without the index, suggesting that this source of variation should not be ignored.

Introduction

Tagging studies are often used to estimate the abundance and/or important demographic parameters of a population such as survival rates. Estimates of these quantities of interest must account for various nuisance parameters, the most common of which are capture or detection probabilities (p; the proportion of live, tagged animals that are captured or detected in a sampling period or location). Mark–recapture or tag-recovery models often make the simplifying assumption that in a given capture period or at a detection location, tagged animals have the same probability of being captured or detected (i.e., homogeneity in p).

In capture-mark–recapture studies that take place over time periods, animal behavior often plays a role in capture probabilities. The chance of being captured may vary among individual animals. Further, the probability of capture may change for a particular individual animal contingent upon previous

*Corresponding author: mmel@uw.edu
Current address: School of Aquatic Fishery Sciences, University of Washington, Box 355020, Seattle, Washington, USA

captures. Biases caused by such heterogeneity or behavioral change in capture probabilities on survival or abundance estimates have long been recognized and methods have been developed to deal with them (e.g., Pollock 1982; Pledger et al. 2003).

In capture-tag-detection studies for migrating animals that take place over detection locations, which are similar from a statistical point of view (e.g., Burnham et al. 1987; Melnychuk 2009), animals are not re-captured following the initial tagging period. Instead, they are detected as they pass telemetry receiver stations (e.g., acoustic, radio, or PIT-tag detectors) deployed along migration routes. It is commonly assumed that heterogeneity in p at each station can be ignored in these cases as long as tag types, swimming speeds, and specific travel routes are similar among tagged animals. Also, since previous detection is not expected to affect the chance of detection at a particular station, 'trap-dependent' behavior can be ignored.

Contrary to these commonly held assumptions, the acoustic intensity of tags might vary among the tagged animals of a population. A tag with a stronger output will transmit its signal further through water than one with a weaker output. For example, the maximum variation among VEMCO acoustic tags is about ± 4 dB around the average power output for a particular tag type, which could lead to about a 15–20% gain or reduction in expected detection range under normal ocean conditions (Doug Pincock, VEMCO/Amirix, personal communication). If such variation exists and does affect range, stronger tags might be more likely to be detected by acoustic receivers. If this variation is consistent over time, with stronger tags consistently transmitting signals of greater intensity throughout their lifespan, they should consistently be detected with greater chance at detection stations along a migration route. Variation in tag strength could be due to variation in battery strength or electronic components of tags.

Here, I describe simple methods for quantifying variation in acoustic intensity of transmitters before they are implanted into study animals. I show how this information can later be incorporated as a covariate of detection probabilities in capture-tag-detection (hereafter, mark–recapture) analyses, which may lead to less biased and/or more precise estimates of survival and detection probabilities. In a case study, I tested the assumption of homogeneity in p among tagged fish that migrated past successive receiver stations. Prior to tagging fish, I measured a relative index of tag strength. After fish migrated through the study area and detection data were downloaded, I used mark–recapture models to estimate survival. In some models, I used the tag strength index as a covariate of p, and evaluated the relative support for models with and without this covariate.

Methods

Acoustic tags sold by VEMCO often come with a small magnet attached which acts on a reed switch inside the tag. When the magnet is removed the tag begins to transmit pulse trains, which each consist of 6–8 pulses (typically, 16 pulse trains are transmitted in rapid succession, after which pulse train intervals fall into a programmed delay period, e.g., 30–90 s between pulse trains). Tags are generally left deactivated until immediately before tagging fish in order to prolong battery life. As these individually-coded tags are activated, researchers will often verify that tag IDs are properly decoded using either a receiver with display panel (e.g., VEMCO VR-100) or a data-logging receiver (e.g., VEMCO VR-2).

Measuring a tag strength index prior to tagging

Even if a batch of tags are all of the same model, manufactured at the same time, and contain the same components, the acoustic intensity of transmissions can still vary among individual tags. This variation can be quantified while activating tags, prior to attaching or implanting them into fish. Two types of tag strength indices may be quantified:

(1) a continuous index of tag strength can be assessed by directly measuring the acoustic intensity of tag pulses with a high-frequency oscilloscope. At some fixed distance from tag to hydrophone or probe, the voltage or decibel level (dB) of pulses can be measured. If an oscilloscope is not available, a VR-100 receiver has a signal strength meter that can provide relative voltage levels of tag pulses.

(2) A discrete index of tag strength can be assessed by quantifying the proportion of pulse train transmissions that are successfully decoded by a VR-100 or VR-2 receiver. A known number of pulse trains is transmitted from a single tag over some period (e.g., the 16 initial pulse trains in rapid succession), and at some fixed distance from tag to hydrophone, some proportion of these will be decoded.

In both cases, tags with a greater recorded voltage (or dB) or a higher proportion of transmissions successfully decoded represent a greater acoustic output. The distance from tag to hydrophone and the gain setting on the oscilloscope or receiver should be determined so as to provide a high level of contrast in strength index values among tags. For the continuous index, as wide a range as possible of voltage or dB levels is desirable. For the discrete index, a wide range of proportions of decoding success is desirable.

When quantifying tag strength indices, several factors should be taken into account; many of the following recommendations were suggested by Dale Webber (VEMCO/Amirix, personal communication). Most importantly, the above measurements should be taken in water in a controlled environment (e.g., a tank in a laboratory setting). This ensures that the proper ambient pressure is placed on the tag transducer so it transmits at the proper resonance. The tag should be consistently aligned and placed in the same position on a cradle, designed to keep the tag vertical with its transducer end as free as possible from the cradle. The same hydrophone should be used for all tags in a batch. The hydrophone can be suspended from the ceiling and the water tank can be placed on sponge foam or other insulation, both of which can minimize physical transmission of background noise. Neither the cradle nor the hydrophone should be moved during the period when all tags in a batch are measured.

Maintaining a consistent level of background noise among activated tags is crucial, and reducing the level of background noise is better still. Fluorescent lights and electromagnetic interference can both create noise at the frequencies at which acoustic tags operate (often 69 kHz for VEMCO tags). Sound-absorbing walls in the water tank are important for reducing echoes, thereby isolating the intensity of each tag pulse.

The discrete tag strength index that relies on decoding success may not be possible to assess with a VR-2 in water, since nearly all transmissions would likely be decoded in typical tank sizes (a water tank several hundreds of meters long would be necessary to provide enough contrast in detection success). If an oscilloscope or VR-100 are not available for measuring tag strength, then in-air measurements with a VR-2 may be taken (see case study below). It is important to note, however, that in-air measurements are not necessarily correlated with in-water measurements due to the medium affecting electromechanical aspects of the transducer and its mounting; it is even possible that they are completely uncorrelated (Doug Pincock,

VEMCO/Amirix, personal communication). In-water measurements should thus be taken whenever possible.

Incorporating tag strength measurements into mark–recapture analyses

Variation in signal strength among tags is expected to result in variation in the probability of detecting a tag in a tracking study. In migrating fish especially, implanted tags may not be within detection range of a receiver for much time, so only a few transmissions while within range may be expected. If tag strength is measured prior to tagging fish, then since tags are individually-coded, this information can be used retrospectively to potentially explain variation among tags in detection probabilities at receiver stations.

A commonly-used mark–recapture model for migrating fish is the open-population Cormack-Jolly–Seber model (CJS; Cormack 1964; Jolly 1965; Seber 1965), which allows survival probabilities (ϕ) in each segment of the migration (with segments delimited by receiver stations) and detection probabilities (p) at each station to be estimated simultaneously. Optimization routines of mark–recapture models are often run in a "link-space" for numerical estimation purposes; one commonly used is the logit-link. Detection probabilities are typically modeled as being separate for each station i, and shared among all tags of some release group of migrating fish:

$$(1)\ \text{logit}(p_i) = \beta_0 + \beta_i \times \text{station}_i\ .$$

This is the classic CJS sub-model for p, where "station$_i$" terms are simply dummy variables. Coefficients represent an overall intercept (β_0, which is not necessary to represent explicitly), and station-specific terms (β_i) that allow for independence among stations in p estimates, $\hat{p}$.

The tag strength index can be incorporated as an individual covariate of p, which constrains $\hat{p}$ for a given tag j to be a function of its index value:

$$(2)\ \text{logit}(p_{i,j}) = \beta_0 + \beta_i \times \text{station}_i + \beta_S \times \text{strength index}_j\ .$$

As formulated in equation (2), the strength index is an additive covariate (rather than multiplicative, or interacting with station i), assuming that its effect on p is consistent across all stations. The coefficient β_S represents this effect, i.e., the slope of the relationship between p and the tag strength index in logit space. As with any time-invariant individual covariate in a mark–recapture model, it is also assumed that relative tag strength measured during activation has a lasting effect throughout the migration. That is, the louder tags at activation are still the louder ones as tags pass receiver stations. Either of these assumptions could be invalid if minor variation in tag pulses is random over time or if the ranking among tags of measured strength at activation changes at some time after activation.

Measured variation in tag signal strength may or may not directly relate to variation in p. Different hypothesized models like equations (1) and (2) can be compared in an information-theoretic framework (e.g., Akaike's Information Criterion, AIC), using the detection data to arbitrate between which model(s) have a better balance between goodness of fit to the data and the number of parameters required to achieve that fit. Such comparisons may reveal that the tag strength index is an important source of variation in p, so should be accounted for when estimating p and ϕ for fish release groups.

Case study: tag strength index

In 2007, prior to tagging coho salmon smolts from Tenderfoot Creek Hatchery in British Columbia, I quantified a simple index of in-air tag strength while VEMCO V7–2 L tags (136 dB re 1μPa at 1 m) were being activated. In-water measurements of tag strength were not feasible in this study, and neither an oscilloscope nor a VR-100 was available. Although in-air indices are sub-optimal, they were measured in hopes of revealing at least some correlation with what the tag strength would be in water, as evidenced by detection patterns in the field.

A small cradle was built from electrical tape to ensure that orientation was consistent among tags. One tag at a time was placed in the cradle with the transducer end of the tag always facing the same way. A VEMCO VR-2 receiver was placed flat on a folded towel, with the hydrophone tip on the same level vertically and at a consistent distance from the tag (80 cm), perpendicular to the long axis of the tag. The hydrophone tip was aligned so it was tilted towards the cradle about 15° away from parallel with the tag.

A tag was placed adjacent to the hydrophone tip to ensure proper decoding, and the magnet was removed. Immediately after the first pulse train was transmitted and received by the VR-2, the tag was placed in the cradle for the remaining 15 rapid-succession pulse trains. After these 16 pulse trains were transmitted the tag was moved out of range of the receiver. The number of decoded transmissions out of the last 15 for each tag was counted after downloading the VR-2. This index ranged from 0 to 15, with higher values representing more frequent detection.

Ambient noise wherever tag activation occurs will affect signal:noise ratios of tag transmissions and thus decoding success. Some tags were activated at a hatchery immediately prior to tagging fish rather than the day before in a quieter environment. At the hatchery, separation distances were reduced.

It is important that measurements are consistent among tags, but the particular distances or angles between hydrophone tip and tag used in this case study are not meaningful in themselves. The distance from tag to hydrophone was set to achieve a reasonable level of contrast in the index. If separation distances are too short or too far, decoding success will be too high or too low, respectively, so contrast would be poor. Of the 199 tags tested, 97 were not decoded (Figure 1). The remaining 102 tags were spread across the index values, with fairly frequent occurrence at high index values. Note that if a continuous measure of tag strength had been taken rather than the relative, discrete measure used, the distribution of tag strength index values would more likely be normal in shape.

Case study: mark–recapture models

Tagged fish were released at a hatchery and migrated past seven VR-2 receiver stations in rivers and reliably past three stations arranged in listening lines of multiple VR-2 receivers after ocean entry. Detection on up to two more ocean lines was possible, but this coho salmon population exhibited consistent migration behavior only past the first three (Melnychuk et al. 2010). Ocean lines were deployed under the Pacific Ocean Shelf Tracking Project (POST; Welch et al. 2003, in press).

Variants of the CJS model were fit to detection data to estimate survival (ϕ) and detection (p) probabilities. In all models, the survival sub-model was "time-varying," with ϕ estimated separately for each segment of the migration, i.e., [ϕ_{Seg}]. Detection sub-models varied, the simplest also being "time-varying," with p estimated separately for each detection station, i.e., [p_{Station}], as in equation (1). Three other models were fit that involved the tag strength index as an additive individual covariate, i.e., p was constrained to be a linear function (in logit space) of the strength index. The first model involved station-specific p and a common covariate across all stations, [$p_{\text{Station + tag strength}}$], as in equation (2). A second model involved station-specific p, where the slope of the covariate was permitted to vary among river and ocean habitats,

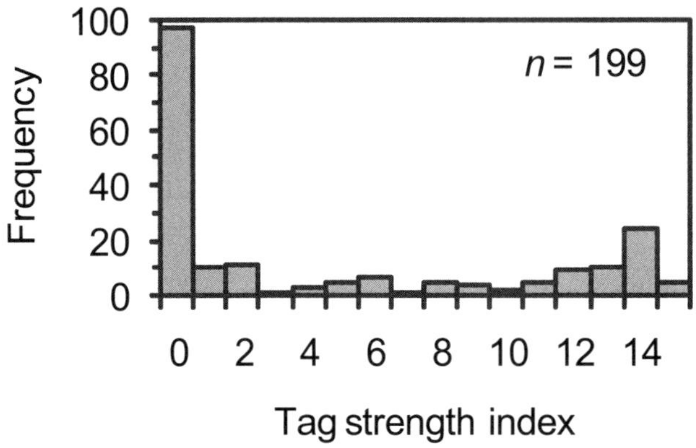

Figure 1. Frequency distribution of tag strength index values, defined as the number of detections out of 15 transmissions. Tag indices were assessed prior to tagging fish.

[$p_{\text{Station + habitat} \times \text{tag strength}}$]. The third model was more constrained: it replaced station-specific variation in p with habitat-specific variation, where a common p was assumed for all stations within the same habitat, [$p_{\text{habitat + tag strength}}$].

Models were fit to detection data using Program MARK (White and Burnham 1999) implemented through RMark (Laake and Rexstad 2009). AIC scores were used to compare candidate models (Burnham and Anderson 2002). First, a variance inflation factor ($\hat{c}$) was estimated to compensate for extra-binomial variation in estimated parameters (Burnham et al. 1987). I estimated $\hat{c}$ assuming the general CJS model, [ϕ_{Seg}, p_{Station}], using the deviance ratio bootstrapping method in Program MARK ($\hat{c}$ = 1.239). This $\hat{c}$ estimate was used for model comparison, with computed QAICc values corrected for both extra-binomial variation and small sample sizes.

After comparing candidate models, a p sub-model with reasonable support and involving the strength index covariate was compared to the model not involving the index as a covariate. First, the estimate and confidence limits of the coefficient for the strength index (β_S in equation (2) was extracted to assess whether slopes of logit(p) versus the strength index differed from zero.

Second, predicted fits of p across the range of the index covariate were calculated for each receiver station. Third, estimates and SEs of ϕ and p were compared to assess how inclusion of the strength index covariate affected the precision of model parameters.

Results

Case study: model selection results

Maintaining a common ϕ sub-model across candidate models allowed the effect of each p sub-model to be isolated. The most parsimonious p sub-model was the simple station-varying (CJS) model that did not involve the tag strength covariate (Table 1). However, the sub-model that involved the additive covariate as well as station-specific p also had a reasonable level of support (ΔQAICc = 0.6). Including the strength index as a covariate reduced twice the negative log-likelihood by nearly 2, indicating an overall improvement in goodness-of-fit (even if the gain in fit was not fully compensated for by including the extra parameter, in terms of AIC scores). The other two p sub-models had little support within the model set (Table 1).

Another form of tag strength index was also evaluated. The X/15 discrete tag

TABLE 1. Model selection results for detection probability (p) models.

Model	k[a]	–2•ln(**L**)	QAICc [b]	ΔQAICc	Akaike weight
$\phi_{Seg,}\ p_{Station}$	24	1613.8	1351.7	0.0	0.55
$\phi_{Seg,}\ p_{Station + tag\ strength}$	25	1612.0	1352.3	0.6	0.40
$\phi_{Seg,}\ p_{Station + habitat \times tag\ strength}$	27	1611.8	1356.4	4.7	0.05
$\phi_{Seg,}\ p_{habitat + tag\ strength}$	15	1967.3	1618.3	266.5	0.00

[a] The parameter count is adjusted to include the number of potentially estimated parameters including those at boundaries of 0 or 1.

[b] QAICc values are adjusted for small sample sizes and extra-binomial variation with $\hat{c} = 1.239$.

strength index was reduced to a binary index of whether the tag was heard 0 or 1^+ times out of the 15 transmissions. This binary index was used as a group covariate of p similar to the p sub-models involving individual covariates. In general it did not perform as well as the X/15 discrete covariate and is not shown in Table 1.

Case study: effect of tag strength index on detection probability

In the second model listed in Table 1, p was specified as in equation (2). If there was no effect of the tag strength index on p, the value of β_S would be zero. Estimated β_S and 95% confidence limits were 0.020 (–0.009–0.050). Confidence intervals included zero, although about 5/6 of the interval was positive.

After models are fit to data, p and its standard error can be estimated for a specified value of a covariate. These predicted values of p over the range of covariate values (again, for the second model listed in Table 1) are shown in Figure 2. There was considerable variation in p among successive receiver stations, attributable to site-specific depth, receiver array geometry, and environmental noise conditions. At each station, there was at least some effect of the tag strength index on predicted $\hat{p}$. Since the modeled covariate was additive, the same effect was assumed for all stations (slopes are equal in logit-space, but were back-transformed to a probability scale so appear to vary among stations). Predicted $\hat{p}$ differed by about 0.05–0.08 between low and high values of the index. In some cases when overall $\hat{p}$ was low (e.g., station 4), this difference in percentage points amounts to a ~25% increase in $\hat{p}$ from low to high values of the strength index.

For the two candidate models with reasonable levels of support, estimates $\hat{p}$ and $\hat{\phi}$ were compared. These models differed by only one parameter (the tag strength index, as an individual covariate), so the effect of incorporating the strength index on values of $\hat{p}$ and $\hat{\phi}$ and their precision can be assessed. Estimates of p and ϕ at each receiver station or in each segment were very similar between models, with <1% difference in all estimates (Figure 3). The small differences that did occur were not consistent in direction, so it appears that the relative bias of incorporating (or perhaps not incorporating) the tag strength index is negligible. Similarly, the precision of $\hat{p}$ and $\hat{\phi}$ was similar between the two models. For all paired estimates, differences in coefficients of variation were <0.003. With the exception of lower precision of $\hat{\phi}$ for the [$p_{Station + tag\ strength}$] model at one station (8, Figure 3), differences between models in 95% confidence limit widths were <0.006 for all estimates.

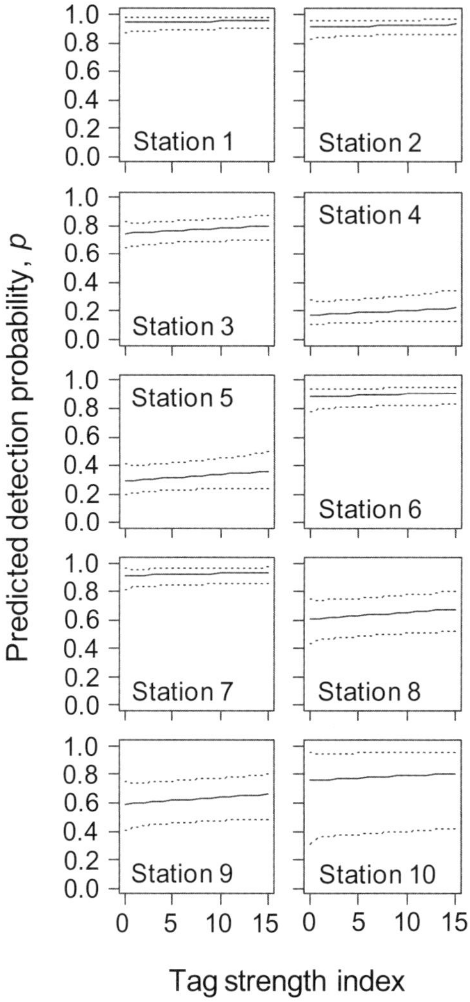

FIGURE 2. Predicted fits of mark–recapture detection probabilities at each receiver station over the range of measured tag strength index values. The model assumed is $[p_{\text{Station + tag strength}}]$. Dotted lines show 95% confidence bands around the best predicted fits.

Discussion

Effect of tag strength variation on detection probability

When conducting tagging studies, researchers usually assume that tags of the same type will have the same transmission range and probability of being detected. This assumption of homogeneity in p is crucial for mark–recapture models where inferences are made about abundance or survival (especially if individual covariates are used to constrain survival estimates). While heterogeneity in p is unlikely to cause bias in $\hat{\phi}$ when survival is homogeneous (Pollock and Raveling 1982), it could affect the precision. The model selection results from the case study suggest some degree of variation in p as a result of variation in acoustic strength among tags of the same type. The effect was not large, as the 95% confidence limits of β_S did not exclude zero, and predicted fits of p over the covariate range showed only a slight positive trend. The

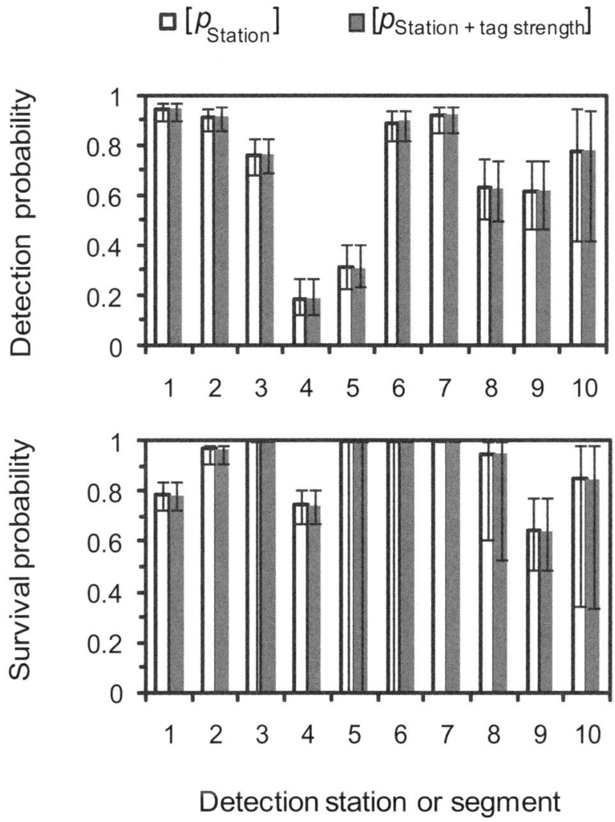

FIGURE 3. Estimated detection probabilities at each receiver station and survival probabilities in each segment of the migration for Tenderfoot Creek coho. White bars represent estimates for the Cormack–Jolly–Seber model, [$p_{Station}$], and gray bars show estimates for model [$p_{Station + tag\ strength}$]. Error bars represent 95% confidence intervals around the estimates.

covariate did, however, have some explanatory power and improved the goodness-of-fit to the data, just not enough in terms of AIC scores to compensate for the extra parameter. The effect was perhaps large enough that the assumption of homogeneity in p should not be taken for granted in tagging studies. That is not to say that such a strength index should replace station-dependence in p, but that it could be used in addition to station-dependent model specifications.

The effect of tag strength variation on p is likely minor compared with effects of spacing of receivers in an array, tag type, and river or oceanographic conditions that affect background noise and therefore signal:noise ratios. Tag types are generally identical within a study population, and station-dependence (which incorporates receiver spacing) is generally accounted for in mark–recapture models, so these factors are already often used to explain much of the total variation in observed detection data. A minor effect of tag strength was observed here despite tag strength measurements being sub-optimal; it is unknown how often such variation occurs and how large it may be. Simple measurements of tag strength prior to tagging fish, preferably in water, may help to explain some additional part of the variation in p within a population.

A trade-off in bias and precision is often inherent in mark–recapture models. Incorporating additional parameters into models often improves the accuracy of parameter estimates, but may reduce the precision of parameter estimates because an additional source(s) of uncertainty (i.e., way of explaining variation in the data) is admitted. On the other hand, precision may increase when an important source of variation in p is accounted for. In this study, incorporating the discrete tag strength index as an individual covariate of p slightly improved the goodness-of-fit of the model to data (ΔQAICc <2, Table 1), but this was not associated with much of a reduction in precision in estimates of either p or ϕ (Figure 3). There was very little and no consistent difference between models in p or ϕ, suggesting that negligible relative bias exists from incorporating (or perhaps failing to incorporate) the strength index as a covariate. Under a more optimal measure of tag strength, reduction in bias and/or increase in precision of p and ϕ may be greater.

The small effect of the strength index on p was in the expected direction. Tags that were detected more frequently out of 15 transmissions after activation (high covariate values) tended to be the ones more likely to be detected by acoustic receivers. The variation in values of the index was likely attributable to variation in acoustic output of the tag, with some tags transmitting signals slightly louder than others, possibly as a result of subtle differences in battery strength, electrical components, or mounting of the transducer.

Further recommendations for measuring tag strength

The experimental set-up used for measuring the discrete tag strength index in the case study was rudimentary. Some tags were activated at a hatchery and others were activated in a quieter room the day before tagging fish. The distance between the tag cradle and hydrophone was decided after only a few tags were measured; spending more time assessing an appropriate distance would have likely provided better contrast in the index, with relatively more large values (Figure 1). Most importantly, the measurements were performed in air. Since these are not necessarily correlated with in-water measurements of tag strength, hypothetically there could be no discernable effect of an in-air index on p even if there was variation in tag strength in water. A 'cleaner' set-up and a tag strength index measured in water, as described earlier, could potentially have more explanatory power for p.

Whatever index of tag strength is used, it is important that it involves as much contrast as possible across its range. A 50:50 split of tags falling into the lowest and highest tag strength values is statistically optimal in terms of precision, while a distribution closer to uniform would allow the form of the relationship between p and the index to be investigated. A larger number of programmed rapid-pulse intervals when tags are first activated could be requested by the user so that proportions of transmissions detected for a discrete index are based on a larger sample, thereby improving accuracy. All tags within a group should be activated in the same location with the same set-up over as short a time period as possible so as to maximize consistency. Several indices (e.g., both a continuous and a discrete index) can be assessed for each tag to find a statistically preferable covariate, and the same index can be measured more than once on each tag to quantify random variation in the index value of each tag, to ensure consistency.

The time cost of quantifying a tag strength index prior to tagging is not overly prohibitive. Depending on the index used, measurements may take about two or three minutes per tag. Indices measured in water are highly preferable, and would create little further time cost. Some manufacturers may routinely perform tag strength measurements as part of their quality control, and the researcher could also ask whether this information is available. Considering that such an index may help to explain some of the variation in estimated detection probabilities in tagging studies, it may be worthwhile to spend the time assessing this potential source of variation before tagging fish.

Acknowledgments

I thank Doug Pincock, Carl Schwarz, Carl Walters, and two anonymous reviewers for useful comments on the manuscript, and Dale Webber and Doug Pincock for useful discussions. I thank Kintama Research Corp. and POST members for general support and field work involving fish tagging and receiver deployments. Funding for the tags in this study was provided by the Pacific Salmon Foundation, CN Railway Co. through the Cheakamus Ecosystem Restoration Technical Committee, and the B.C. Pacific Salmon Forum.

References

Burnham, K. P., and D. R. Anderson. 2002. Model selection and multimodel inference: a practical information-theoretic approach. Springer-Verlag, New York.

Burnham, K. P., D. R. Anderson, G. C. White, C. Brownie, and K. H. Pollock. 1987. Design and analysis methods for fish survival experiments based on release-recapture. American Fisheries Society, Monograph 5, Bethesda, Maryland.

Cormack, R. M. 1964. Estimates of survival from the sighting of marked animals. Biometrika 51:429–438.

Jolly, G. M. 1965. Explicit estimates from capture-recapture data with both death and immigration-stochastic model. Biometrika 52:225–247.

Laake, J., and E. Rexstad. 2009. RMark—an alternative approach to building linear models in MARK. Appendix C *in* E. Cooch and G. White, editors. Program MARK: a gentle introduction, 7th edition. RMark version 1.8.9.

Melnychuk, M. C. 2009. Estimation of survival and detection probabilities for multiple tagged salmon stocks with nested migration routes, using a large-scale telemetry array. Marine and Freshwater Research 60:1231–1243.

Melnychuk M. C., D. W. Welch, and C. J. Walters. 2010. Spatio-temporal migration patterns of Pacific salmon smolts in rivers and coastal marine waters. PLoS ONE 5(9):e12916. doi:10.1371/journal.pone.0012916

Pledger, S., K. H. Pollock, and J. L. Norris. 2003. Open capture-recapture models with heterogeneity: I. Cormack-Jolly-Seber model. Biometrics 59:786–794.

Pollock, K. H. 1982. A capture-recapture design robust to unequal probability of capture. Journal of Wildlife Management 46:752–757.

Pollock, K. H., and D. G. Raveling. 1982. Assumptions of modern band-recovery models, with emphasis on heterogeneous survival rates. Journal of Wildlife Management 46:88–98.

Seber, G. A. F. 1965. A note on the multiple recapture census. Biometrika 52:249–259.

Welch, D. W., Boehlert, G. W., and Ward, B. R. 2003. POST - the Pacific Ocean salmon tracking project. Oceanologica Acta 25:243–253.

Welch, D. W., M. C. Melnychuk, J. C. Payne, E. L. Rechisky, A. D. Porter, G. D. Jackson, B. R. Ward, S. P. Vincent, C. C. Wood, and J. Semmens. In press. In situ measurement of coastal ocean movements and survival of juvenile Pacific salmon. Proceedings of the National Academy of Sciences.

White, G. C. and K. P. Burnham. 1999. Program MARK: Survival estimation from populations of marked animals. Bird Study 46 Supplement:120–139.

American Fisheries Society Symposium 76:325–346, 2012

Estimating Detection Probabilities for Linear Acoustic Monitoring Arrays

Alistair J. Hobday*
CSIRO Marine and Atmospheric Research, Wealth from Oceans Flagship
Castray Esplanade, Hobart, Tasmania, Australia

Doug Pincock
Amirix Systems—VEMCO Division
Nova Scotia, Canada

Abstract.—Acoustic monitoring is a widely used low-cost technique for studying the movements of aquatic animals. One common deployment configuration is a linear array of receivers such that passage of a tagged animal across a line of receivers is detected. The typical goal is to estimate the fraction of a population that moves across the line. Receivers in an array can have nonoverlapping or overlapping detection probability envelopes. It has been assumed in the case of nonoverlapping arrays that the detection rate of tagged animals is proportional to the coverage of the line. Unfortunately, the estimation process is not as simple as previously believed. In fact the probability of detection is more Gaussian-shaped rather than uniform with distance from the receiver, and varies over time due to biotic and abiotic noise. Wind-generated noise in particular can have a major influence on the performance of receivers. Range testing, while important, will not solve the problem. In fact, the temporal complexity of this variable detection probability envelope renders a statistical solution improbable. Simulation modeling provides a potential alternative for estimating detection probability for nonoverlapping arrays, however, for most situations, the best solution is to design an overlapping array, with the detection range estimated for the worst environmental conditions.

Introduction

The study of animal movements in marine environments remains a core focus for marine scientists. For example, movement information is used in fisheries stock assessments to set stock boundaries (e.g., Hampton and Gunn 1998), and a recent need is in the design and location of fisheries management zones (Hobday and Hartmann 2006; Pecl et al. 2006) and marine protected areas (Louzao et al. 2006). For many species traditional mark–recapture studies have been used to obtain this information, but this approach is limited by the scale and the need for recapture of tagged animals. The real-time monitoring of animals in the ocean has been undertaken by a variety of labor intensive methods, including active acoustic tracking. An additional constraint in active tracking is that a few animals are only tracked for short periods of time immediately after tagging when behavior may

*Corresponding author: Alistair.Hobday@csiro.au

also be modified (e.g., Gunn et al. 1999; Lutcavage et al. 2000; Davis and Stanley 2002), or there is discontinuity in tracking of species with more site fidelity.

One option for increasing the efficiency of acoustic telemetry is the use of passive acoustic receivers to detect tagged study animals. Relatively low-cost acoustic listening receivers are being used widely throughout the world (for review see Heupel et al. 2006). Animals tagged with acoustic tags emitting an acoustic pulse are detected when they pass close to these receivers, i.e. within the detection range. The technology is robust and simple, and is one of a suite of new tools being used for a number of new ocean monitoring studies (OTN, www.oceantrackingnetwork.org; CoML-POST www.coml.org/descrip/post, Welch et al. 2003; IMOS-AATAMS-Australia; http://www.imos.org.au/facilities/tagging-monitoring.html). Some of the study goals, as with other historical studies, are to discover the migration routes taken by the tagged animals, estimate the fraction of animals surviving, and determine site fidelity.

When receivers are arranged in groups or arrays, two patterns are common; clusters and lines (for review, see Domeier 2005; Heupel et al. 2006). A recent survey involving 20 large projects with a total of over 1300 receivers deployed showed an even split between these two patterns (Heupel et al. 2005). Lines of receivers are usually employed in migration studies, where the goal is to estimate the fraction of tagged animals that cross a line of receivers (e.g., Comeau et al. 2002; Stark et al. 2005, Pecl et al. 2006). A common configuration is a cross-shelf array (e.g., Welch et al. 2003; Hobday 2003). These linear configurations have been variously described as lines, gates, or curtains (Heupel et al. 2006). The arrangement of the receivers can be such that detection ranges are overlapping (receivers closely spaced), or nonoverlapping (receivers widely spaced).

In this paper, we discuss the issues associated with the technology incorporated in the VR1, VR2, VR2W, VR3 and VR4 family of receivers manufactured by Vemco/Amirix (www.vemco.com)—henceforth referred to simply as the receiver—but the findings can also be extended to other types of receivers such as the Vemco VR100 and products from suppliers such as Sonotronics (www.sonotronics.com) and Lotek (www.lotek.com). Most receivers detect a train of coded pulses which encode in the time intervals between pulses an 8, 12, or 16 bit ID Code (or 8 bits of sensor data plus an 8 bit ID code). If all the pings in a coded sequence are not received correctly, the tag will not be registered at the receiver. When receivers are recovered, the set of tag detections (tag ID, data from any sensors on the tags, date and time) can be downloaded, and the results used to address the study question.

In this paper, detection probability for a particular distance is the proportion of coded pulse trains that arrive at the acoustic receiver from that distance that are correctly decoded. This can only be correctly measured from data collected in detection range tests conducted under the full range of anticipated conditions. Our overall goal is to provide guidance to users designing arrays and/or attempting to interpret data from arrays by:

1. Demonstrating the underlying factors which lead to variations in a receivers' detection probability envelope size and shape (geometry) and how these will vary depending on environmental (usually weather) conditions. This is supported by the results from extensive field tests.

2. Illustrating the biases and errors that can be introduced in estimates of total number of fish passing a receiver or line of receivers based on actual number of detections by a nonoverlapping array.

3. Providing guidance on situations where these biases can be kept to a manageable level.

4. Providing guidance on the factors determining the detection performance of an array (both overlapping and non overlapping designs).

Estimating Animal Movements—Uniform versus Gaussian Detection Probability

The proportion of animals crossing a line of receivers may be simply reported as the number detected divided by the number of animals tagged, however, this assumes that all animals that crossed the line were detected. If this assumption is not valid, for example because the line is not completely covered as in a nonoverlapping line of acoustic receivers, or if the interval between pulses is too long given the swimming speed of the tagged animal (Welch et al. 2003), then the true number of animals crossing will be underestimated (Hobday et al. 2009). Thus, one can consider correcting the number of animals detected for the proportion of the line in which detections were actually possible.

For example Comeau et al. (2002) estimated Atlantic cod *Gadus morhua* crossing rates based on a uniform model of detection probability (Figure 1a). For that study, detection probability (RE) (different from our definition of detection probability) was calculated as:

$$RE_1 = \frac{\sum_1^N \frac{2D_r}{M_r}}{N} \quad (1)$$

where N is the number of receivers, D_r is the average detection range, and M_r is the distance between adjacent receivers. The probability of detection inside the 'detection range' is assumed to be 100%, although corrections could be made to this formula to account for a detection rate of less than 100% at the receiver. A conclusion based on this calculation for receivers that do not have overlapping detection ranges, would be that if RE is 0.5 and 20% of animals were detected at the line, that 40% of animals must have crossed the line. That is, the total percentage of the study population that crossed the line (T%) is

$$T_{\%} = \frac{D_{\%}}{RE} \quad (2)$$

where D% is the percent of tagged animals detected at the line, and RE is the receiver efficiency as defined in equation (1).

The above method assumes that the detection range is constant, and animals are only detected at a single receiver and are moving in only one direction with distance from the receiver uniformly distributed. Any deviation from these assumptions biases the estimate. In most cases, large biases will also be introduced by the fact that, as this paper will demonstrate, the detection probability envelope differs greatly from the uniform rectangular shape (Figure 1). The decline in detection probability at longer ranges is caused by diminished receiver range during periods when acoustic conditions are less than ideal, most commonly due to environmental noise. We will refer to this type of nonuniform detection probability envelope as Gaussian as it can be approximated with a Gaussian distribution (see below).

Influences on Detection Range and Probability

The maximum distance at which a receiver can detect signals is influenced significantly by environmental noise conditions (authors' personal observations). A longer discussion of the environmental factors influencing detection range is provided in Heupel et al. (2006); here we focus on the factors with temporal variation that influence esti-

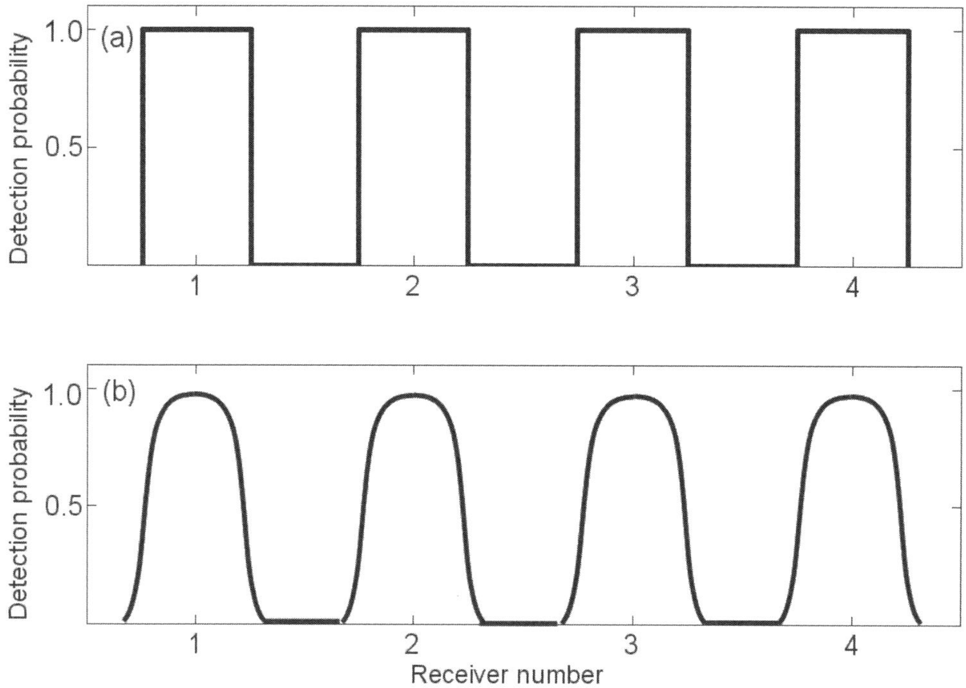

FIGURE 1. Schematic illustration of two detection probability models as a function of distance from four receiver locations. **A.** Uniform detection probability envelope. **B.** Gaussian-shaped detection probability envelope.

mates of detection probability envelope size and shape. Most attention is given to abiotic variables, which are more commonly the limiting factor for signal detection, although biological or manmade sources can be significant in some locations.

Biotic factors that lead to variation in the detection range, defined as the maximum detection distance, and detection probability (detection at any distance), over time include biological growth and biological noise. Biological growth can occur on the receivers and may lead to a gradual long-term decline in detection range (Heupel et al. 2008). Deep placement (>25 m) or regular cleaning/replacement of receivers minimizes this impact (authors' personal observations; Heupel et al. 2008). A daily cycle (short-term) in detection range can also occur due to nocturnal cycles in biological noise generated by fishes (Families Batrachoididae, Gadididae, Holocentridae, Sciaenidae, Pomacentridae, and Gobiidae) (e.g., Fine 1997) or snapping shrimp (e.g., *Alpheus*) (Johnson et al. 1947; Knowlton and Moulton 1963).

Abiotic factors that influence detection include environmental changes such as wave action ("sea state" noise)—a major cause of variation in detection probability in open ocean arrays. Deterioration in detection range/probability also occurs with heavy rainfall, which may be the most important factor for some embayed settings (Simpfendorfer et al. 2008). The mooring design can also impact detection: detection range and probability can be impacted by noisy moorings, shadow areas due to floats or nearby structures, or daily/seasonal changes in boat traffic (Clements et al. 2005; Heupel et al. 2006; Simpfendorfer et al. 2008).

Receiver Characteristics and Shape of the Detection Probability Envelope

The VR2 detection range is approximately omni–directional. Thus, detection range for a bottom mounted receiver in water deeper than the receiver's range will form a semicircular detection probability envelope (Figure 2a), while the detection probability envelope in more typical deployments in shallower water are more constant with depth (Figures 2b and 2c). Therefore, in theory the envelope shape for a shallow deployment will be close to cylindrical in the water column, moving towards a half sphere as deployment depth increases. In practice, one is interested in the range at the same depth as the fish is swimming which, unless the receiver and the fish are at the same depth, will be less than the maximum detection range.

The maximum detection range of a receiver is strongly influenced by ambient noise, the most common influence being weather-related. We have thus restricted our analysis to this example, however, the analysis is no different if other noise sources are significant. Wind-generated noise leads to a decrease in maximum detection range (Figure 3) (Wenz 1962; Urick 1983). Wind data can be combined with the detection pattern of the receiver to determine the size and shape of the detection probability envelope. Because detection range varies with noise levels (e.g., Figure 3) receiver detection range is not constant, but will increase or decrease in all directions according to local conditions at the time. Therefore, in a long term test, varying acoustic conditions will cause detection rates to decline as distance increases and as the percentage of time that acoustic conditions are suitable for detection decreases. A number of other users have also noted this behavior in their field studies (personal communication to authors; Simpfendorfer et al. 2008), and their observations in part motivated this paper.

When designing an array, observations of a decrease in detection probability need to be interpreted with care. In particular, long-term testing showing a detection probability of *X* percent at a particular range does not mean that *X* percent of the transmissions from a particular tagged fish passing will be detected. Rather, it means that virtually all transmissions from a fish will be detected for *X* percent of the time that fish is within that range; and conversely, few or no detections at 100 – *X*% of times. As a consequence, at distances for which the long term detection probability is below 100%, the probability of a particular fish being detected by that receiver is unknown. This imposes the requirement that the detection envelopes of adjacent receivers need to overlap significantly to ensure detection of passing fish. The issue of the amount of overlap required will be addressed in a later section. The actual shape of the detection probability envelope depends on how wind (or other environmental factor) varies throughout the test. For example, wind speed

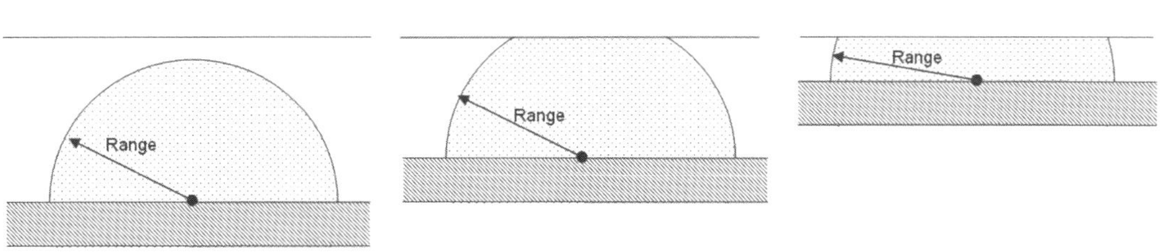

Figure 2. Detection range for a bottom-mounted receiver. A. Depth greater than detection range. B. Moderate depth. C. Shallow depth.

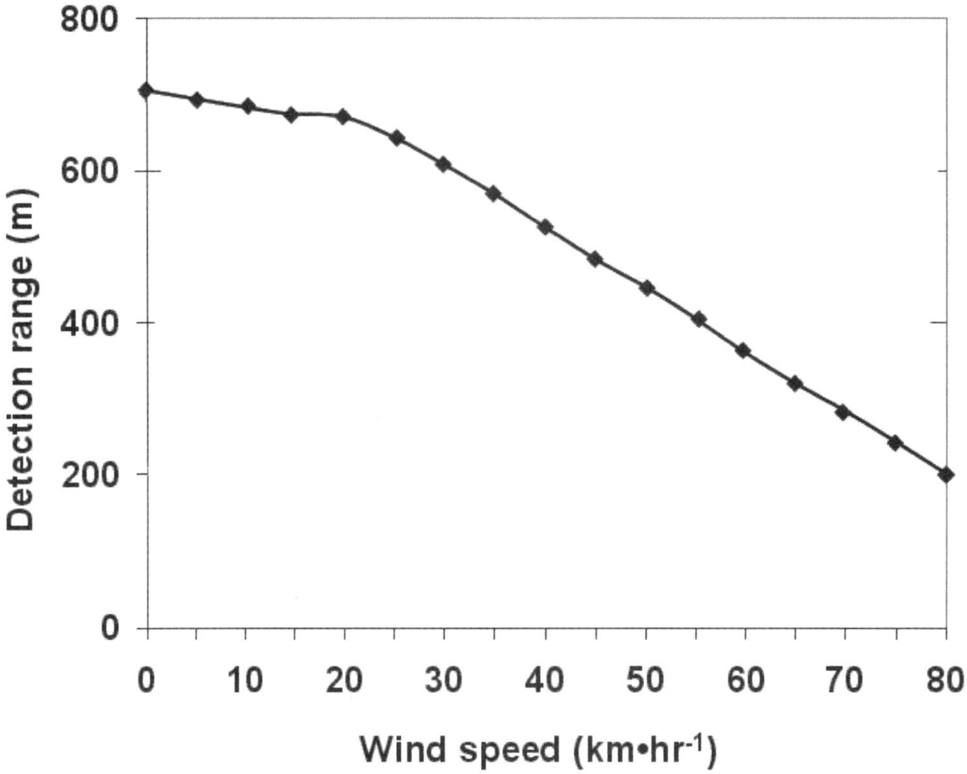

FIGURE 3. Theoretical VR2 detection range as a function of wind speed in normal ocean conditions (Transmitter Power: 153 dB re 1 uPa @ 1 m). Results are calculated by bringing together noise level for a particular wind speed from Wenz curves (Wenz 1962) and the signal level at various ranges from the sonar equation (Urick 1983). An online calculator (www.vemco.com) can be used to develop similar detection range relationships for other transmitter powers and receivers.

distributions can often be approximated by a Gaussian distribution, as used in this paper.

Detection Tests to Determine Detection Range

Clearly, knowledge of the effective range of the receivers and the shape of the detection probability envelope is essential for interpreting the results from an acoustic array. Most users now recognize that detection tests can define the envelope shape and range, however, detection tests results are only applicable to the whole experiment if the conditions for the test either match those encountered in the experiment or if allowance is made for differences in environmental conditions compared to when detection tests were performed. In stable and known ocean conditions (i.e., deep water), one can estimate performance reasonably well based on the manufacturer's receiver and tag specifications by bringing together noise level for a particular wind speed from Wenz curves (Wenz 1962) and the signal level at various ranges from the sonar equation (Urick 1983) so that and detailed range testing may be unnecessary (Pincock 2009). However, if conditions are suspected to be variable or unknown, especially in shallow, estuarine, or obstruction-rich waters, the number of factors that can influence range and their variability from one location to another dictate that in situ range testing be done.

The most rigorous—but unfortunately the most time consuming—procedure involves installing a receiver in an actual anticipated or representative deployment situation, placing a tag a fixed distance away, and determining from the receiver log the percent of transmissions that were detected (Heupel et al. 2006). These controlled detection tests would be repeated for a number of ranges with sufficient time at each range for all representative conditions to be encountered—this may require many days at each location to experience the full range of conditions (e.g., consider infrequent winter storms or seasonal tidal signals). If this procedure is considered to be too time consuming or unrealistic, a variety of other methods can be used to determine range, and the method depends on the isolation of the study area (the number of times the array can be visited to conduct experiments); further discussion is provided in Heupel et al. (2006).

1. Towed tag(s). Tags can be deployed behind a vessel and towed/drifted past the receiver. If the positions of the receiver and the vessel-tag are known, usually via GPS, then linking the timestamp of the detection at the receiver with the vessel location at the time of signal transmission can be used to determine the distance between the two objects, and hence the detection range. Issues with this approach include vessel noise interfering with the test—this should be ruled out in advance with testing designed for this purpose if the vessel is available—and variable depth of the towed tag. Using a depth tag for this test ensures that one knows the depth of the detected tag. The time interval between tag acoustic pulses must also be short to allow sufficient detections, particularly when vessel speed is relatively fast. This towing method is usually employed in remote locations, where the array is only visited a few times. Tests should be carried out in a range of environmental conditions; however, this is rarely possible, particularly for the worst conditions. Repeated experiments over a number of years may increase the sample size of detection tests in a range of conditions. An example of comprehensive testing of this type is presented in the next section.

2. Sentinel tag(s). This involves simply mooring one or more low repetition rate (e.g., 1 transmission per hour) tags at known distances from the receiver throughout the study time period and analyzing the detection rate of these tags over time. For example, if one places three sentinel tags at varying distances from the receiver (such that one tag is within a range at which 100% detection is anticipated, with the other two further out) examination of the receiver log will show the times when reception at the longer ranges is good and when it is not. It is possible that fouling on the sentinel tag(s) without change in the abiotic environment may also decrease the detection range over time, but this is also an issue for the receiver (Heupel et al. 2008). Ideally, receivers and sentinel tags would be raised and cleaned before this becomes a significant issue. Using sufficient tags to describe the envelope may also be prohibitive.

3. Direct noise and sound propagation measurements. The general approach involves collection of ambient noise data with subsequent off-line analysis, which takes into account the observed attenuation of the signals, noise levels present and the detection threshold and bandwidth of the receiver. This approach has the potential to provide direct insight into issues surrounding expected

[maximum] detection range as well as factors limiting it and may even point the way to improving the situation. However, the equipment and analysis tools required are not available to most users. Attempting this approach for all receivers in a large array in a remote location would be a challenging under taking and likely impractical.

Environmental Influences in a Controlled Detection Range Test

Results from comprehensive detection tests at a depth of 20 m in Shad Bay, Nova Scotia (a sheltered bay) illustrate receiver performance in variable conditions. Tests were conducted by mooring tags for one week at set distances on the same horizontal plane as the receiver. The detection characteristics observed (Figure 4a) are similar to those illustrated in Figure 1b (with the exception that no data were collected at longer ranges where the detection rate would have been less than 50%). The increase in performance at 550 m illustrates one of our concerns with variable weather: ocean conditions were calmer during the time the transmitter was at this distance, and so range was greater than at 475 m. Results from a distance of 275 m illustrate the influence of environmental variation (Figure 4b). Detection probability was commonly over 95%, but there were periods when performance was poorer. While in situ weather observations were not made, examination of regional online weather data showed that the two points of greatly diminished performance occurred during the two most extreme wind/rain conditions in the area during this test.

Data for this receiver at all distances over the two month test period showed that, with the exception of the results at 550 m discussed above, the general trend as distance increased was a decrease in the percentage of time in which detection probability was high (Table 1).

Environmental Influences in Experimental Array Detection Tests

Cross-shelf arrays have been deployed by the first author since 2002 in southern Western Australia to determine the migration speed and cross-shelf position of juvenile southern bluefin tuna *Thunnus maccoyii* (Figure 5) (Hobday 2003; Hobday et al. 2009). The study area is located where a fishery-independent survey for juvenile tuna is carried out, and estimates of migration speed and timing and time spent inside the survey area are required to correct the observed biomass levels.

Detection tests using towed tags have been undertaken for these cross-shelf arrays over a number of years in a range of sea states [Beaufort 0–7, 0–35 knots]. Receivers are typically moored at a depth of 25 m below the sea surface at a depth of 40 m (coast) to 140 m (shelf break). When receivers are recovered, the time of detection is matched with the position of the vessel to determine the maximum detection distance (Figure 6).

Data from these detection tests shows a decline in detection frequency with distance (e.g., Heupel et al. 2006, Figure 7) that is consistent with a Gaussian-shaped, rather than uniform detection probability envelope. The decline in frequency very close to the receiver is due to limited detection tests where the vessel drifted within 50 m of the receiver. Thus, for the conditions under which these tests were conducted, the equation provided by Comeau et al. (2002) would underestimate the proportion of detected animals, as coverage is in fact less than assumed using a uniform detection probability envelope. The observed detection probability envelope shape (Figure 7) is consistent with both the theoretical shape (Figure 1b) and the results from controlled field measurements (Figure 4 and Table 1).

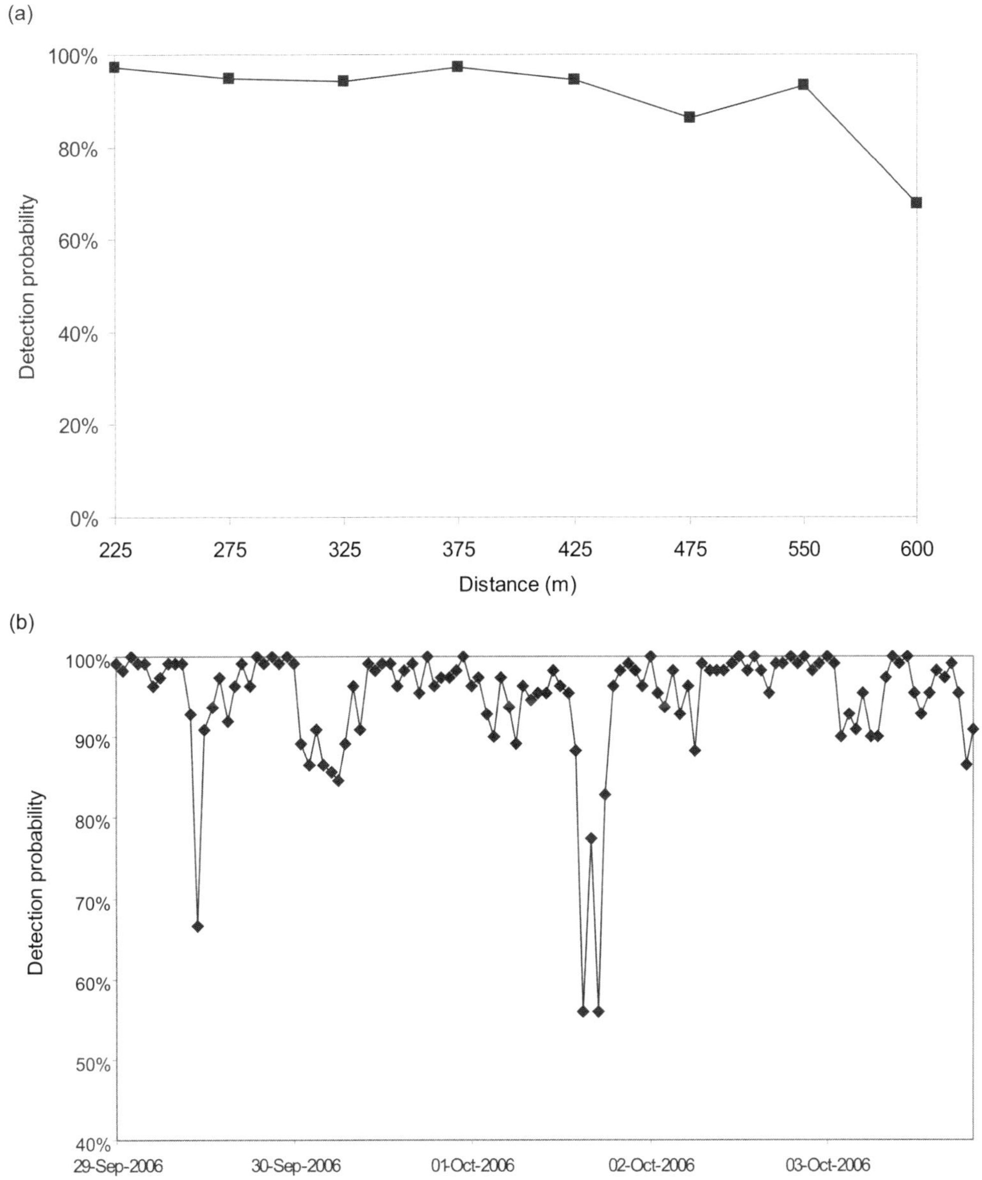

FIGURE 4. A. Controlled detection test results from Shad Bay, Nova Scotia, September–November 2006. Test tags were moored at increasing distances from the receiver for 1 week at each distance. B. Detection probabilities at a distance of 275 m from the receiver with each data point representing one hour during which 110 tag transmissions took place.

Table 1. Detection probability summary for a range of distances in the Shad Bay detection test (25 September to 25 November, 2006).

Range (meters)	Percent of test period with detection probability > 0.95	Minimum detection probability observed in a one hour segment
225	84%	0.85
275	71%	0.65
325	67%	0.5
375	88%	0.85
425	62%	0.65
475	59%	0.1
550	70%	0.3
600	16%	0.05

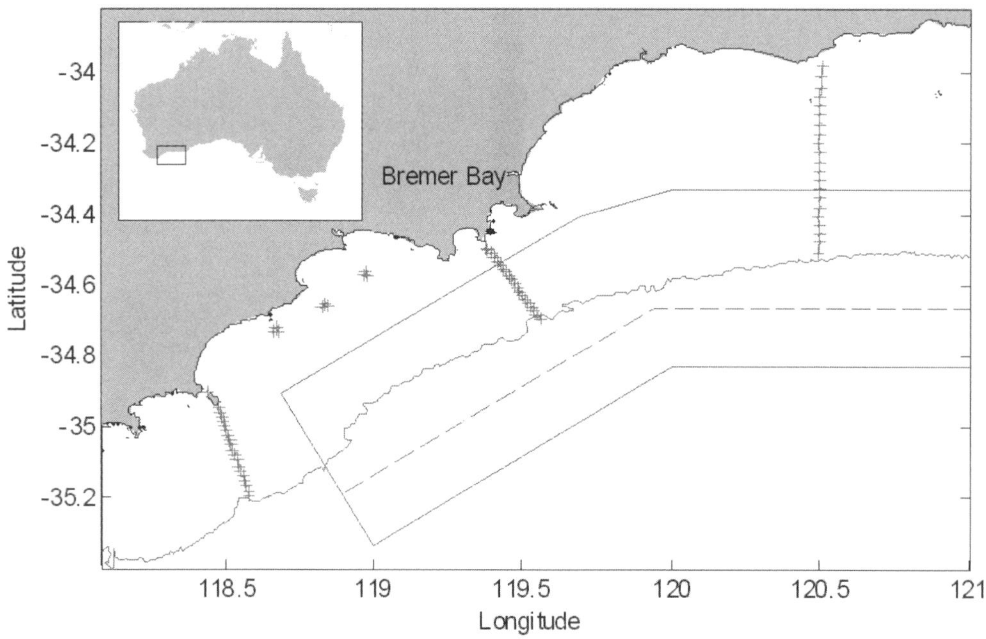

Figure 5. Acoustic monitoring study area in southern Western Australia. Three lines of receivers run from the coast (depth ~40 m) to the 200 m contour, a distance of 30 km (middle line) to 50 km (eastern line). The box centered on the 200 m contour is the fishery-independent survey region; the dashed line divides a subsection of the survey region.

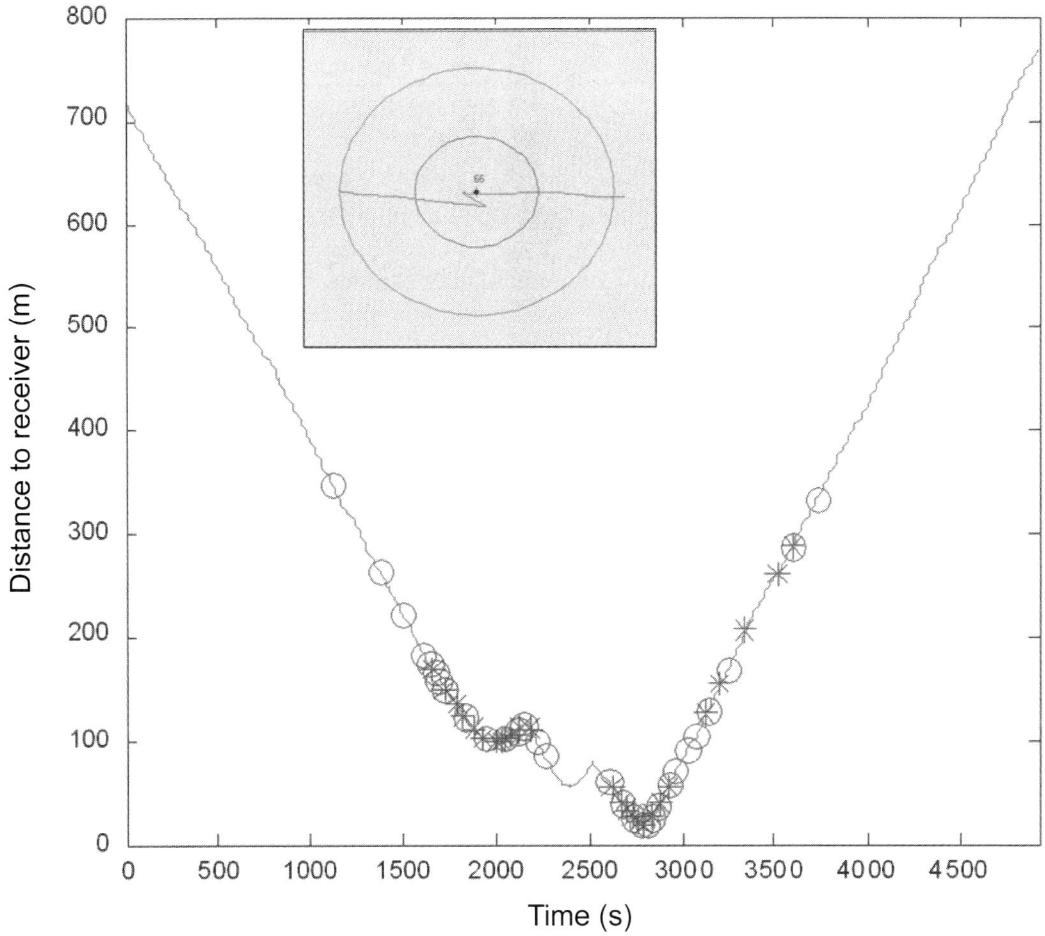

Figure 6. Detection range test example (southern Western Australia 2006–07). **Inset.** Drift track of two towed tags (V16, Output Level 153dB re 1 uPa @ 1 m) past one receiver in a linear array. The inner circle around the receiver represents the maximum observed detection distance in the test (360 m) and the outer circle is a radius of 800 m. **Main figure:** Pulse time (circles and stars) for each tag as a function of tag distance from the receiver. Initially vessel and tags drift toward the receiver and distance decreases, then distance increases again as the tags drift away. The discontinuity close to the receiver was when the engines were started to reposition the vessel.

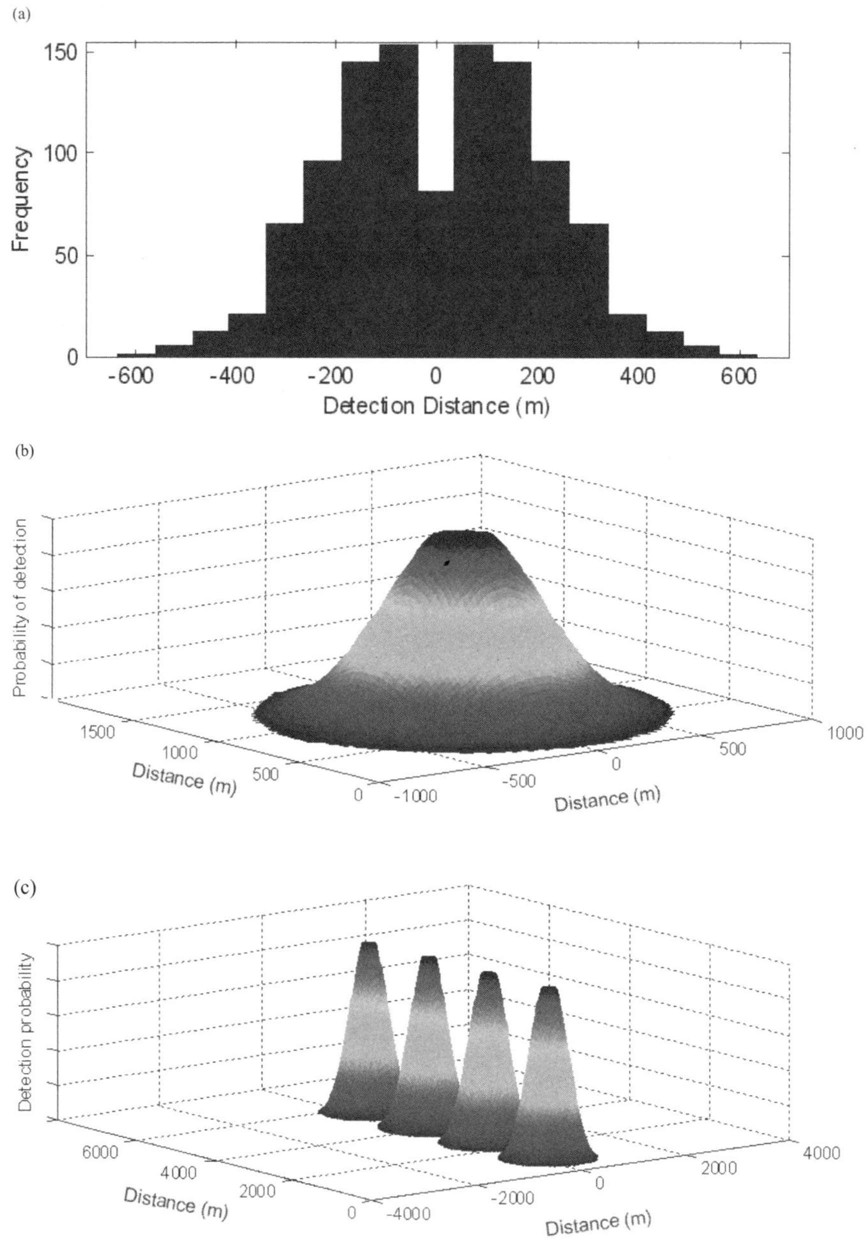

Figure 7. (a). Number of times that a towed acoustic tag was detected as a function of distance from the acoustic receiver combined from a large set of range test experiments in southern West Australia. The tag was towed at the same depth as the receiver (25 m). The histogram data are mirrored around the zero-distance point to form a detection probability envelope. The decrease in detection frequency close to the receiver is due to the difficulty of judging the drift of the vessel during the detection test from 600 m away, thus it does not drift the tag right over the receiver, and so the count of detections very close to the receiver is somewhat lower. **(b).** The detection probability envelope in two dimensions approximates a cone, with the maximum probability at the center and then decreasing in all directions from the receiver. **(c).** When arranged in lines, the cones form a leaky curtain, which will detect only a portion of the passing tagged animals. The top of the cones may be "flattened" in good conditions, indicating near-uniform detection to some range before a decline occurs.

Detection Probability Estimation Using a Simulation Model

In order to illustrate the difference in detection probability that can result from assumptions about the shape of the detection probability envelope we developed an individual based model (IBM). The IBM simulates tagged individuals migrating under a range of specified conditions (swimming speed, pulse rate, number of other transmitters present and weather related noise conditions) past a linear acoustic array. In the discussion below, we will focus on wind as being the dominant factor determining the noise at the receiver and hence the dominant influence on detection range at any given time. The approach is exactly the same if another noise source is significant.

The first step for the model is forming the detection probability envelope. This can be determined for particular wind statistics (mean wind speed and standard deviation) by first determining the relationship between range and wind speed for the transmitter being used (see Figure 3). Then, using the assumed wind statistics, one can determine detection range for different wind speeds, and hence plot the detection probability at different distances (Figure 8). The greater the wind variability, the more the envelope deviates from the uniform assumption (Figure 1a). The Gaussian approximation for the detection probability envelope (Figure 8) is obtained from

$$P_D(R) = 1 - \Phi(M, \sigma) \quad (3)$$

where P_D is the detection probability and $\Phi(M, \sigma)$ the cumulative Gaussian probability density function with mean, M, and standard deviation, σ. M and σ are chosen to fit the calculated envelope with M moving towards the maximum detection range and

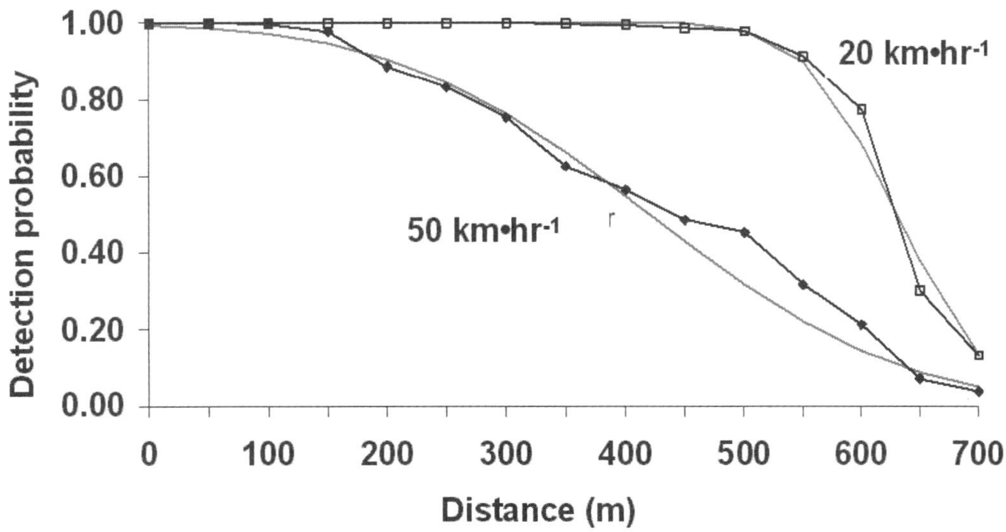

FIGURE 8. Calculated detection probability envelopes for two different mean wind speeds (standard deviation = 0.5 × mean wind speed). Black lines are the calculated results obtained using the on-line range calculator at www.vemco.com and are for a transmit level of 153 db re 1 μPa @ 1 m (the same as used for the tests in south-western Australia as in Figure 3). Gray continuous lines are Gaussian approximations of the type used in the model. The values of mean detection range (M) and SD to fit the data for (i) 20 km × hr^{-1} were M = 630 m and SD = 10% of M and (ii) 50 km × hr^{-1} were M = 420 m and SD = 40% of M.

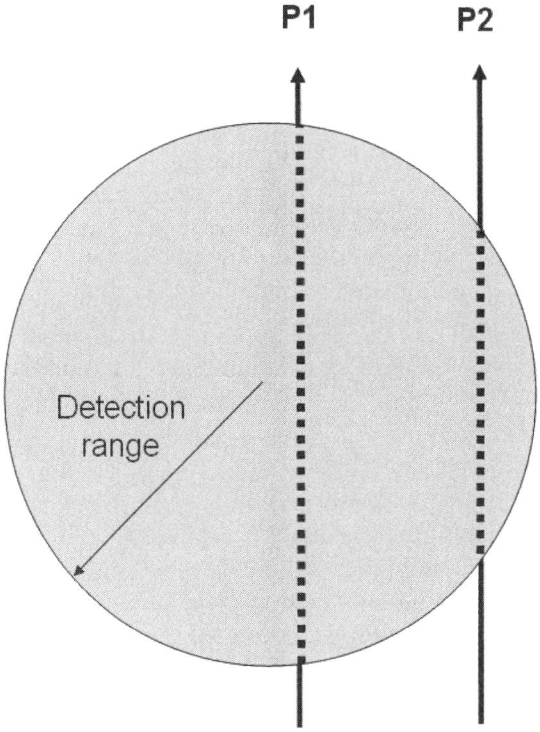

Figure 9. Two tagged fish passages (P1 and P2) perpendicular to an array. Path length (dashed line segment) for a particular fish passage increases as distance from receiver decreases and as detection range increases.

σ decreasing for less severe conditions (Figure 8). P_D is also a function of the difference in depth between the fish and receiver but, in the following, we assume that receiver and fish are at the same depth to simplify the discussion.

The second step is to determine the probability of detection of each tagged fish coming within range of a receiver. The model assumes that fish swim perpendicular to the array (Figure 9). More meandering paths would lead to a longer time within the detection probability envelope and to periods where the fish is closer to a receiver and hence more likely to be detected. The probability of a fish being detected crossing the line was the probability of at least one of n tag transmissions being detected, which is

$$P(D_{1 \text{ or more pulses}}) = 1 - P(U_{all}), \text{ where } P(U_{all}) = P(U_1) \times P(U_2) \times \ldots P(U_n) \quad (4)$$

where $P(U_i)$ is the probability of transmission i being undetected. This depends on the distance of the fish from the receiver, the probability of detection at that range as determined by the detection envelope and the number of transmissions potentially detectable by the receiver. The number of possible transmissions (n) is determined by the path length (Figure 9), fish swimming speed and tag transmission rate.

When the probability of more than one transmitter being within range of the receiver becomes significant, determination of $P(U_i)$ needs to take "collisions" into account—a collision being the occurrence of transmissions from different transmitters overlapping in time which leads to failure to detect one or both tags (Pincock 2009).

This model was first parameterized to reflect the tuna migration study in southern Western Australia which used tags transmitting at random intervals between 20 and 69 s at an output level of 153 dB re 1 μPa @ 1 m, giving a range in calm ocean conditions of from 700 m to just over 200 m in very severe conditions (Figure 3). Wind speed for this study can be reasonably approximated as Gaussian with a mean of 20 km·hr^{-1} and standard deviation of 10 km·hr^{-1} (Figure 10).

Applying the model to some representative scenarios (for wind conditions similar to those of the study area as well as more severe conditions) for situations in which there are no collisions (i.e., no more than one transmitter is within the detection range of the receiver at any time) shows that detection decreases with swimming speed and increased time between tag transmissions (Table 2). As these results summarize detection performance for tags that come within the maximum detection range of a single receiver, it is no surprise that there is never a situation in which 100% of transmitters are detected. For this model, this would only occur if the wind were constant and known for the entire period. The significant fall off in detection probability for stronger wind conditions (50 km·hr^{-1}) is due to the fact that many of the fish swimming within the maximum range of the receiver are out of the effective detection range in these conditions (Table 2).

These results also illustrate the biases that will be introduced if one attempts to estimate the total number of fish passing a receiver based on a uniform detection probability envelope assumption (equation (1) and (2) using a range determined under a fixed set of conditions (Table 2). The exact value of these errors depends on the wind statistics, which will likely differ from the Gaussian assumption, but the trend is clear. Depending on conditions, predictions can range from serious over-estimation to serious under-estimation. Results for two different uniform envelopes (Table 2) demonstrate that, while in certain situations, it might be feasible to choose the right distance, great care is required. This is discussed further in the next section.

When multiple transmitters are present, a number of each tag's transmissions will not be detected and this detection probability decreases as the tag density (number of tags present within the receiver detection range) increases. This effect is the same as increasing the mean time between transmissions (Table 2). For example, the decline in detection probability for fish swimming at 4 m ·s^{-1} with an average tag transmission interval of 120 s at a wind speed of 20 km·hr^{-1} when tag density increases to 3 tags, is from 84% to 67%. Note that it is not realistic to predict detection rates over a study period since this would require a meaningful estimate of the statistics of tag density and its correlation with detection range – an unlikely possibility. These calculations simply illustrate how performance degrades with tag density (Table 2). Since collisions are random, the resulting detection rates for a single trial are also variable and, therefore, the detection rates shown for a tag density of 3 tags are the minimum that will be observed 95 times out of 100. Average detection rates would be higher than these minimums, but design to achieve a minimum detection rate is recommended. While the preceding analysis concentrates on a single receiver, the biases will be the same for a nonoverlapping array if one tries to make a determination of the total number of tagged animals crossing the whole array.

Discussion—Implications for Detection Probability

Given the results and various confounding variables, is there a solution for estimating detection probability, and thus the number of animals that actually cross a line of non-

(a)

(b)

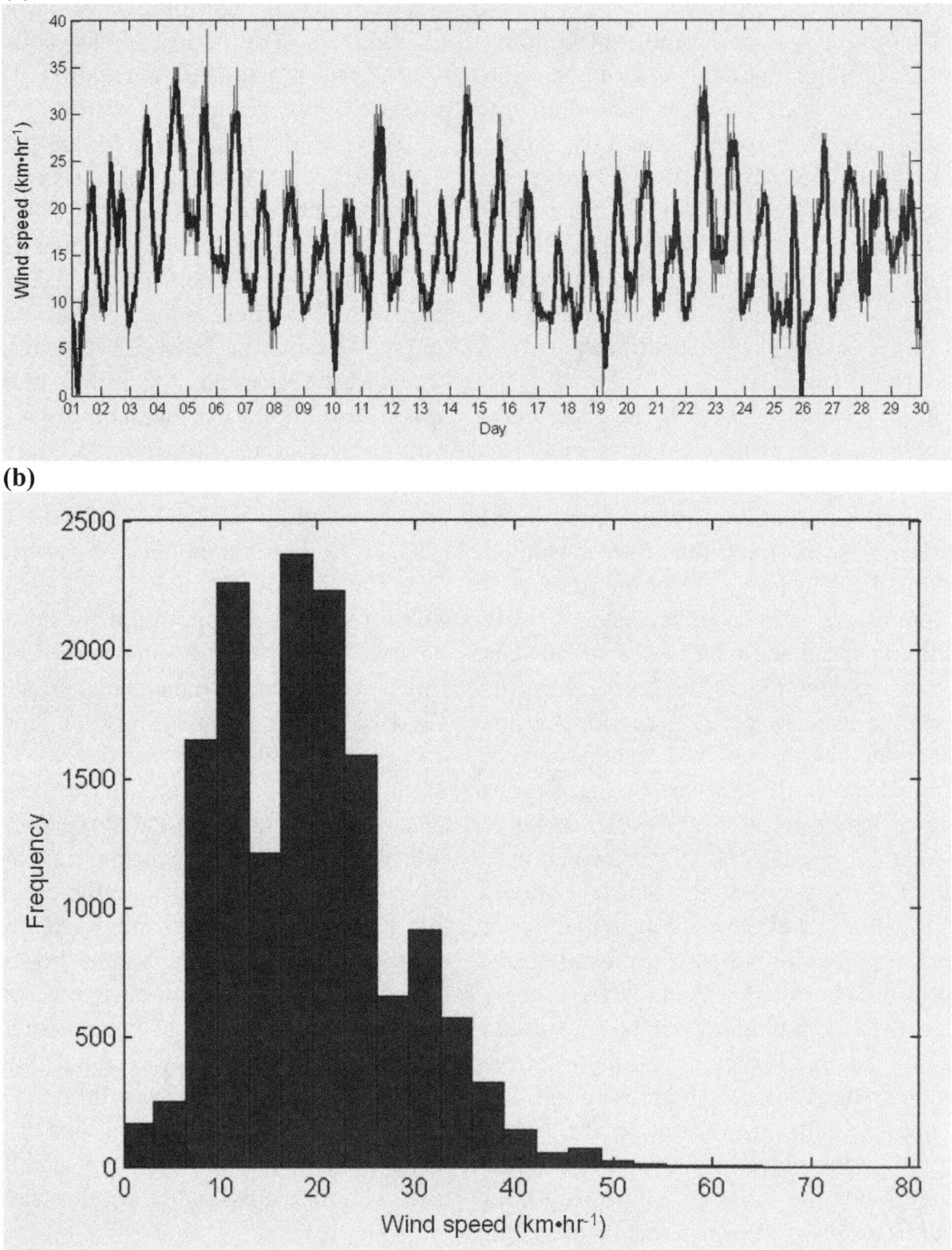

FIGURE 10. (a). Example time series of winds in January 2006 close to the array of receivers in southern Western Australia. Note the diurnal signal, as well as lower frequency cycles. The 30 min record has been smoothed with a 5-point centered running mean. **(b).** Frequency distribution of wind speed at Hopetown, Western Australia (close to the acoustic array) measured every 20 min for the months January to May for the years 2003 to 2006.

TABLE 2. Summary of results from analysis of model simulation using acoustic transmitters with output 153 dB re 1 μPa @ 1 m for two wind conditions (20 $km{\cdot}hr^{-1}$ and 50 $km{\cdot}hr^{-1}$), a range of swimming speeds and mean transmission times. The percent of tags detected when 1 or 3 acoustic tags are present are shown. The predicted percent of tags detected is based on a tag density of 1 using an assumption of a uniform envelope of 500 m and 300 m.

Wind conditions mean ± SD (km·hr-1)	Swimming speed ($m{\cdot}s^{-1}$)	Mean time between tag transmissions (s)	Percent of tags detected from simulation using Gaussian model (tag density)		Predicted percentage of fish using uniform envelope assumption (compared to number predicted from Gaussian model)	
			1 tag	3 tags	500 m envelope	300 m envelope
20 ± 10	0.5	45	90%	90%	126%	210%
	4.0	45	89%	88%	125%	208%
	0.5	120	90%	90%	126%	210%
	4.0	120	84%	67%	118%	196%
	0.5	240	90%	89%	126%	210%
	2.0	240	84%	77%	118%	196%
	4.0	240	59%	11%	83%	138%
50 ± 25	0.5	45	58%	59%	81%	135%
	4.0	45	57%	42%	80%	133%
	0.5	120	57%	59%	80%	133%
	4.0	120	47%	25%	66%	110%
	0.5	240	59%	57%	83%	138%
	2.0	240	47%	37%	66%	110%
	4.0	240	19%	2%	27%	44%

overlapping acoustic receivers? Clearly, as the results indicate, a deterministic solution as proposed for a simple uniform detection probability envelope by Comeau et al. (2002) is usually impractical given the complexities of the marine environment that impact the shape of the Gaussian envelope.

Is it possible to use a Gaussian envelope (i.e., time varying uniform envelope whose size is determined by wind or other range limiting factor) to predict fish passage? Theoretically it is possible, but this requires that the actual conditions for the study be identical to the assumed noise conditions. An example of variation in number of fish predicted when the mean wind deviates just 5 km·hr^{-1} from the assumed value illustrates the problem (Table 3). While the results might be acceptable in some cases, these are optimistic as other factors such as standard deviation of winds differing from the assumed value and deviations from standard Gaussian shape could introduce significant additional errors.

To this point, we have focused entirely on wind or other noise sources as the cause of errors in predictions of fish passing a nonoverlapping array. Large as these might be, even larger errors could arise from the fact that predictions of this type are based on an assumption that the actual passage of fish is uniformly distributed both in space and time. In summary, then, all the uncertainty of such predictions leads to the conclusion that acceptable results with regard to estimating total fish passage are usually only possible with an overlapping array, such that detection is as close as possible to 100%.

Finally, with regard to nonoverlapping arrays, and assuming that fish passage is uniform in space and time (or alternately that the statistics are known), a nonoverlapping array could still be feasible if measures are taken to determine the detection probability envelope characteristics throughout the study. Possibilities include:

1. The use of a set of "sentinel" tags at known and increasing ranges from the receiver. At any given time, the envelope applied to the detection data would be based on which of the sentinel tags were being detected. Even two or three such tags would significantly improve predictions.

2. The use of a receiver which can detect the ambient noise level and log it with detections. One can use knowledge of the noise level to obtain reasonably accurate estimates of the detection probability envelope over time. The underlying technology of the VR2 cannot support this capability but future, likely more expensive products, could.

Table 3. Example of errors in the predicted number of fish passing a linear array due to differences in mean wind speed from the assumed value using a Gaussian envelope. For this example swimming speed is 2 m·s^{-1} and mean time between transmissions is 60 seconds.

Mean wind speed (km·hr-1)	Percent of tags detected	Predicted number of tags per 100 passing with actual mean wind speed varying by ± 5 km·hr-1
10	93%	98 to 103
20	85%	94 to 104
30	75%	93 to 105
40	64%	92 to 108
50	53%	91 to 110

3. Artificially shortening the receiver's detection range so that there is little or no variability of detection probability envelope. Successful implementation of this approach would require a high level of confidence on the distribution of fish passage in space and time as it has the effect of ignoring many detectable transmissions.

Thus, we suggest that overlapping arrays are best for estimating total fish passage because of variation in the detection range with environmental variability, and hence the shape of the detection probability envelope over time. Criteria and formulae for the design of overlapping arrays as well as the performance as receiver separation and sample size (i.e., the number of tagged fish actually crossing the space between the receivers) are presented elsewhere (e.g., Welch et al. 2003; Pincock 2009). For example, the formulae provided by Welch et al. (2003) could be adapted for the expected minimum distance based on expected environmental conditions. In designing overlapping arrays, we provide a few overall comments.

The general objective is to place receivers so that in the entire space between them, one or both receivers have a detection probability close to 100%. For example, a separation of 1000 m (2 times reliable detection range) would be appropriate if moderate wind conditions shown will be experienced during the experiment (20 km·hr^{-1}), but less than 400 m between receivers is needed for the more severe conditions (50 km·hr^{-1}) (Figure 8). Pincock (2009) suggests the separation be twice the distance at which the receiver detection probability falls to 50% under the worst anticipated conditions and shows that few if any fish will be missed with such a strategy.

Since the amount of overlap is chosen to ensure detection in the worst conditions anticipated, many tags passing the linear array will be detected by more than one receiver. This helps improve the quality of the data. It is tempting to shorten the time between tag transmissions so that fast-moving fish are more likely to be detected; however, the potential for collisions when multiple tags are passing complicates this potential solution. As the number of tags within the detection range increases, higher transmission rates will actually lead to a reduction in detection rate (Table 4). Therefore, if a large number of tags within the detection range are likely, the time between transmissions needs to be relatively long. Of course, there are limits to this and it may not be realistic to set parameters appropriately to deal with the simultaneous arrival of a large number of fast moving fish. Such scenarios, if anticipated, require more complex tag coding schemes, or multiple overlapping lines of receivers (e.g., Pecl et al. 2006).

In conclusion, the often observed (and theoretically predicted) trend of declining detection rate with distance is not a simple single variable issue, but rather results from a combination of receiver characteristics and the surrounding noise and propagation conditions. Users need to carefully consider whether the conditions for their receiver locations permit array design based on stable and known ocean conditions or if they need to conduct in situ testing. In most cases they will not be able to spend the time and effort to produce results comparable to the two examples described here. Therefore, it is important to take into account the extent to which environmental conditions for the study might be worse than those for the measurements and place the receivers accordingly.

A simulation and modeling approach as outlined above to estimate the detection probability can be very useful. While the analysis presented here used a combination of tools and software developed by CSIRO and Vemco, we are investigating the possibility of making an integrated capability available to other researchers. This would support vari-

TABLE 4. Example of the time taken to detect all tags present within a detection array for a range of numbers of tags present and two mean transmission intervals (30 s and 120 s).

Number of tags	Time (minutes) to detect all tags	
	30 seconds	120 seconds
2	1.1	2.5
4	2.9	4.5
6	5.5	6.5
8	9.5	8.5
10	15.7	10.7
20	151	25.5

ables such as tag characteristics, fish swimming speed, fish residency, receiver separation, (environmental) noise, and propagation conditions (either for stable and known ocean conditions or modified according to the result of site specific range or other testing).

The methods described here can be modified for most applications including acoustic receiver clusters. In the case of acoustic clusters, animal movements and habitat use may also be underestimated if the detection range/probability is subject to variation. Armed with a more realistic understanding of the behavior of these widely-used low cost receivers, the quality of information from this technology will only improve. The major programs that plan on using this technology will no doubt further improve the quantitative and statistical treatment of the data collected.

Acknowledgments

AH's research on southern bluefin tuna using acoustic receivers has been financially supported by CSIRO, AFMA, FRA, JFA and the School of Zoology at the University of Tasmania. The involvement of my collaborators Ryo Kawabe, Yoshi Takao, and Kazu Miyashita is appreciated, as is the assistance of a number of field staff over the years. The experience of the CSIRO electronics and mechanical workshops has led to improved mooring design over the years. The logistical, moral and culinary support of Geoff Campbell (FV Quadrant) and his crew in the field is gratefully acknowledged. DP's research is a contribution to the Census of Marine Life and the Pacific Ocean Tracking Project (POST), and he acknowledges the contributions of colleagues at Amirix Systems including Andy Reid, Dale Webber, Denise King, Terry Fraser and Bruce Oakley for their help in defining issues as well as providing data and feedback on the results presented here. The controlled detection range testing results presented are only a part of the extensive and valuable work done by Chad Murphy and Tim Stone. Many of Vemco's customers including David Welch, Kim Holland, Laurent Dagorn and Ron O'Dor have provided helpful input on their experience in the field. Russ Bradford and two anonymous reviewers improved the clarity of the manuscript.

References

Clements, S., D. Jepsen, M. Karnowski, and C. B. Schreck. 2005. Optimization of an acoustic telemetry array for detecting transmitter-implanted fish. North American Journal of Fisheries Management 25:429–436.

Comeau, L. A., S. E. Campana, and M. Castonguay. 2002. Automated monitoring of a large-scale cod (*Gadus morhua*) migration in the open sea. Canadian Journal of Fisheries and Aquatic Sciences 59:1845–1850.

Davis, T. L. O., and C. A. Stanley, 2002. Vertical and horizontal movements of southern bluefin tuna (*Thunnus maccoyii*) in the Great Australian Bight observed with ultrasonic telemetry. Fishery Bulletin 100: 448–465.

Domeier, M. L. 2005. Methods for the deployment and maintenance of an acoustic tag tracking array: an example from California's Channel Islands. Marine Technology Society Journal 39:74–81.

Fine, M. L. 1997. Endocrinology of sound production in fishes. Marine Freshwater Behaviour and Physiology 29:23–45.

Gunn, J. S., J. D. Stevens, T. L. O. Davis, and B. M. Norman. 1999. Observations on the short-term movements and behaviour of whale sharks (*Rhincodon typus*) at Ningaloo Reef, Western Australia. Marine Biology 135:553–559.

Hampton, J., and J. Gunn. 1998. Exploitation and movements of yellowfin tuna (*Thunnus albacares*) and bigeye tuna (*T. obesus*) tagged in the north–western Coral Sea. Marine and Freshwater Research 49:475–489.

Heupel, M. R., C. Simpfendorfer, and C. Lowe. 2005. Passive acoustic telemetry technology: current applications and future directions, Mote Technical Report Number 1066, Mote Marine Laboratory, Sarasota, Florida.

Heupel, M. R., J. M. Semmens, and A. J. Hobday. 2006. Automated acoustic tracking of aquatic animals: scales, design and deployment of listening station arrays. Marine and Freshwater Research 57:1–13.

Heupel, M. R., K. L. Reiss, B. G. Yeiser, and C. A. Simpfendorfer. 2008. Effects of biofouling on performance of moored data logging acoustic receivers. Limnology and Oceanography: Methods 6:327–335.

Hobday, A. J. 2003. Nearshore migration of juvenile southern bluefin tuna in southern Western Australia, 14th Recruitment Monitoring Program Workshop, RMWS14–05. CSIRO Marine Research, Hobart, Australia.

Hobday, A. J., and K. Hartmann. 2006. Near real-time spatial management based on habitat predictions for a longline bycatch species. Fisheries Management and Ecology 13:365–380.

Hobday, A. J., R. Kawabe, Y. Takao, K. Miyashita and T. Itoh. 2009. Correction of an abundance index using acoustic tag data for juvenile southern bluefin tuna in southern Western Australia. Pages 405–422 *in* J. R. Nielsen, A. Sibert, J. Hobday, M. E. Lutcavage, H. Arrizabalaga, N. Fragosa, editors. Tagging and Tracking of Marine Animals with Electronic Devices II. Reviews: Methods and Technologies in Fish Biology and Fisheries. Netherlands, Springer.

Johnson, M. W., F. A. Everest, and R. W. Young. 1947. The role of snapping shrimp (*Crangon* and *Synalpheus*) in the production of underwater noise in the sea. Biological Bulletin 93: 122–138.

Knowlton, R. E., and J. M. Moulton. 1963. Sound production in the snapping shrimps *Alpheus* (*Crangon*) and *Synalpheus*. Biological Bulletin 125:311–331.

Louzao, M., K. D. Hyrenbach, J. M. Arcos, P. Abello, L. G. De Sola, and D. Oro. 2006. Oceanographic habitat of an endangered Mediterranean Procellariiform: implications for Marine Protected Areas. Ecological Applications 16:1683–1695.

Lutcavage, M. E., R. W. Brill, G. B. Skomal, B. C. Chase, J. L. Goldstein, and J. Tutein. 2000. Tracking adult North Atlantic bluefin tuna (*Thunnus thynnus*) in the northwestern Atlantic using ultrasonic telemetry. Marine Biology 137:347–358.

Pecl, G. T., S. R. Tracey, J. M. Semmens, and G. D. Jackson. 2006. Use of acoustic telemetry for spatial management of southern calamary *Sepioteuthis australis*, a highly mobile inshore squid species. Marine Ecology Progress Series 328:1–15.

Pincock, D. G. 2009. Detection performance of lines of VR2W/VR3 Receivers. Available: www.vemco.com/pdf/line_performance.pdf (January 2010).

Simpfendorfer, C. A., M. R. Heupel, and A. B. Collins. 2008. Variation in the performance of acoustic receivers and its implication for positioning algorithms in a riverine setting. Canadian Journal of Fisheries and Aquatic Sciences 65:482–492.

Stark, K. E., G. D. Jackson, and J. M Lyle. 2005. Tracking arrow squid movements with an automated acoustic telemetry system. Marine Ecology Progress Series 299:167–177.

Urick, R. J. 1983. Principles of Underwater Sound, 3rd edition, McGraw-Hill, New York.

Welch, D. W., G. W. Boehlert, and B. R. Ward. 2003. POST—the Pacific Ocean salmon tracking project. Oceanologica Acta 25:243–253.

Wenz, G. M. 1962. Acoustic ambient noise in the ocean: Spectra and sources. Journal of the Acoustical Society of America 34:1936–1956.

Section IV

Radio Telemetry

American Fisheries Society Symposium 76:349–369, 2012

Is Radio Telemetry Getting Washed Downstream? The Changing Role of Radio Telemetry in Studies of Freshwater Fish Relative to Other Tagging and Telemetry Technology

Steven J. Cooke*
Fish Ecology and Conservation Physiology Laboratory, Department of Biology Carleton University, 1125 Colonel By Drive, Ottawa K1S 5B6, Canada

Eva B. Thorstad
Norwegian Institute for Nature Research, NO-7485 Trondheim, Norway

Abstract.—Radio telemetry is now considered a standard tool for fisheries professionals studying fish in freshwater systems. However, interest in radio telemetry technology may be waning in popularity relative to other technologies such as acoustic telemetry and passive integrated transponders, which have recently become popular for tracking studies in freshwater. The purpose of this paper is to identify the unique characteristics of radio telemetry that continue to make it an extremely powerful tool for elucidating the fundamental biology of aquatic organisms, particularly those living in shallow, fluvial habitats, and in providing information to enable effective management and conservation of aquatic resources. By evaluating the peer-reviewed literature on radio telemetry in aquatic systems over a ten year period, we summarized the advances in fisheries science owing to radio telemetry and identified trends related to study design, tagging techniques, and tracking strategies. We also identified research questions and management needs that cannot be effectively addressed using technologies other than radio telemetry, with main emphasis on hydropower development, microhabitat use and movements, mortality, predation, winter biology, natural history studies of imperiled fish in developing countries, and studies in habitats where physical characteristics restrict the usefulness of other techniques. We emphasize that it is important for radio telemetry practitioners to not dismiss simple manual tracking studies where individual tagged fish are discriminated by using discrete frequencies. Indeed, such an approach can yield insight that is not possible with fixed stations alone, and is particularly useful for developing countries where capacity and finances may not exist for deployment of fixed stations. By characterizing and clarifying the actual and potential role for radio telemetry within the broader landscape of tagging and telemetry technologies, we hope to enable telemetry practitioners to make informed decisions regarding the optimal choice of tagging technology when designing and executing telemetry studies in freshwater systems. We submit that radio telemetry is not getting washed downstream, but there are now simply more technological options for researchers interested in the study of freshwater fish.

*Corresponding author: Steven_Cooke@carleton.ca

Introduction

Tagging and telemetry techniques have become standard tools for fisheries professionals studying topics such as the spatial ecology and migration of fish (reviewed in Lucas and Baras 2000). There are many technologies that can be used to track fish, although perhaps the most widely known is radio telemetry (Cooke et al., in press). However, interest in radio telemetry technology appears to be waning in popularity relative to other technologies such as acoustic telemetry and passive integrated transponders (PIT tags). Our objective was therefore to clarify the changing role of radio telemetry for the study of freshwater fish by exploring three questions: 1) What is the current state of radio telemetry for the study of fish in freshwater systems? 2) Is interest for radio telemetry waning and if so, why? and 3) What are the specific applications and environments where radio telemetry is particularly useful?

By reviewing the peer-reviewed literature on radio telemetry in aquatic systems over a ten year period, we summarized the advances in fisheries science owing to radio telemetry and identify trends related to study design, tagging techniques, tracking strategies, and data analysis. This analysis will enable us to characterize and clarify the actual and potential role for radio telemetry within the broader landscape of tagging and telemetry technologies so that telemetry practitioners can make informed decisions regarding the optimal choice of tagging technology when designing and executing telemetry studies in freshwater systems. We conclude by identifying the unique characteristics of radio telemetry that continue to make it an extremely powerful tool for elucidating the fundamental biology of organisms, particularly those living in fluvial habitats, and in providing information to enable effective management and conservation of aquatic resources.

Background on Radio Telemetry

Radio telemetry is a form of biotelemetry and relies upon the transmission of information between a tag placed in or on the fish and a remote receiver with a receiving antenna. Radio transmitters emit electromagnetic energy in the radio frequency (RF) range, usually in the very high frequency (VHF) band between 30 and 300 MHz (Sisak and Lotimer 1998). Radio transmitters use a quartz crystal to control the transmitting frequency, and the transmitter itself requires a transmitting antenna. The antenna typically trails from the transmitter although transmitters where the antenna is coiled and sealed within the body of the transmitter are also available. Radio signals propagate omni-directionally in the water, but only those wave vectors almost perpendicular (<6° from the vertical) to the air–water interface emerge into the air and can be detected by an aerial receiving antenna. Key variables influencing range of radio transmitters include water depth and conductivity (Velle et al. 1979). Signal strength of the transmitter also influences the range. However, increased signal strength, longer battery lifetime and increased pulse rate are all variables that require larger batteries, which will increase the overall size of transmitters. Radio tracking has become the preferred method for use in shallow (usually <5 m), low conductivity (usually <500 μS cm-1) lakes, ponds, rivers and streams (Cooke et al., in press). Most electromagnetic frequencies are rapidly attenuated by seawater and are usually undetectable within a fraction of a meter from the source, explaining why normal VHF radio telemetry is useless for marine applications (see Cooke et al., in press for discussion).

There are a number of different transmission types that can be used in radio telemetry studies. The most common is the pulsed system (i.e., pingers) where individuals are discriminated through frequency separation (e.g., individual fish on different frequencies) or different pulse rates (e.g., time between pulses varies among fish on a single frequency). These methods require significant bandwidth, which means that much time is needed to scan for fish on different frequencies. Digitally coded transmitters emit a unique numerical code that discriminate them from other tags on the same frequency. Although most frequently used for automated telemetry systems, coded transmitters are also useful for manual tracking because the coded signal can more easily be detected audibly during levels of high background noise, and the scan time can be reduced when searching for a large number of fish spread over a large geographical area.

Radio telemetry technology was first applied to the study of fish behavior in freshwater systems in the mid 1900s. Before radio transmitters were used to track fish, they were used for tracking terrestrial wildlife. Even today, radio transmitters or hybrid technologies that include using radio signals to communicate with satellites are considered the standard in wildlife biology and continue to be the focus of entire books on the topic (Millspaugh and Marzluff 2001). Radio telemetry tools enable wildlife biologists to track animals at a variety of spatial scales and from a variety of tracking platforms (planes, trucks, snowmobile, boat, foot). Since the first use of radio telemetry with fish, there have been many technological innovations (e.g., coded transmitters, field-programmable transmitters, fixed stations; e.g., Eiler 1995) and improvements in tagging (Bridger and Booth 2003; Jepsen et al. 2002) and tracking techniques, and data analysis (Millspaugh and Marzluff 2001). Radio telemetry studies now require that the practitioner have specialized training and knowledge. In parallel to the developments in radio telemetry technology have been innovations in acoustic telemetry and passive integrated transponder (PIT) tags. Unlike radio telemetry, acoustic telemetry is quite effective in deep waters and areas of high conductivity (e.g., marine systems) where radio signals are attenuated (Voegeli and Pincock 1996). PIT tagging systems have been developed whereby small fish can be implanted with PIT tags and then and then their movements tracked if they move past checkpoints (i.e., antennas with dataloggers; Gibbons and Andrews 2004). The advantage of PIT tags is their small size and duration over many years, but the disadvantage is the short detection range.

Trends from the Fish Telemetry Conference Held in Europe

Within a popular fish telemetry conference held in Europe, which is arguably the single most authoritative source of information on this topic, there have been a series of proceedings produced and published. The conferences have been arranged in Liège, Belgium (1995), La Rochelle, France (1997), Lowestoft, UK (1999), Trondheim, Norway (2001), Ustica, Italy (2003), Sesimbra, Portugal (2005) and Silkeborg, Denmark (2007). Each symposium has yielded a module in the journal Hydrobiologia or a book (Baras and Philippart 1996; Moore and Russell 2000; Spedicato et al. 2005). The most recent proceedings (from the 2007 conference) were published as a module in the journal Fisheries Management and Ecology. We mined the proceedings to evaluate temporal trends in the use of radio telemetry for the study of fish in freshwater. Studies focused solely on marine systems were excluded as this was not a relevant comparison. Two of the books (Baras and Philippart 1996; Moore and Russell 2000) included extended abstracts which were also summarized. The proceedings from the 2007 conference were published only recently so we relied on the book of abstracts from the conference. Studies that used acoustic transmitters and acoustic receivers, but where recordings were transmitted from the acoustic receiver to a land based station as a radio signal, were defined as acoustic studies. The approach adopted in this review was more reasonable than sorting through every single telemetry paper published in the broader peer reviewed literature, and it is our assertion that the trends exemplified at the conference reflect the trends in the broader literature given that the conference has recently changed its name to the "International Fish Telemetry Conference" because of the geographic diversity of presenters and research sites.

Use of Radio Telemetry Relative to Other Tracking Technologies

At the first conference held in 1995, all of the published papers covering studies in freshwater were based on radio telemetry platforms. Through time, however, the proportion of radio telemetry papers has steadily declined to the point where in 2007 they constituted less than 50% of all papers on fish telemetry in freshwater (Figure 1). Beginning in 1997 and continuing through to 2007, there has been a steadily increasing trend towards the use of acoustic telemetry in freshwater to the point where it currently constitutes about 36% of all telemetry studies on freshwater fishes. Also PIT tagging and other passive technologies were increasing in use to the point where they now constitute 26% of the freshwater papers published in the telemetry proceedings. Clearly, radio telemetry is no longer the only option for research on fish in freshwater. A number of high profile acoustic marine tracking systems have been extended into estuarine and freshwater environments (e.g., Crossin et al. 2007; Melnychuk et al. 2007) and these have led to a greater appreciation for acoustic techniques in freshwater systems. The same has not occurred for radio telemetry in marine systems for obvious reasons,

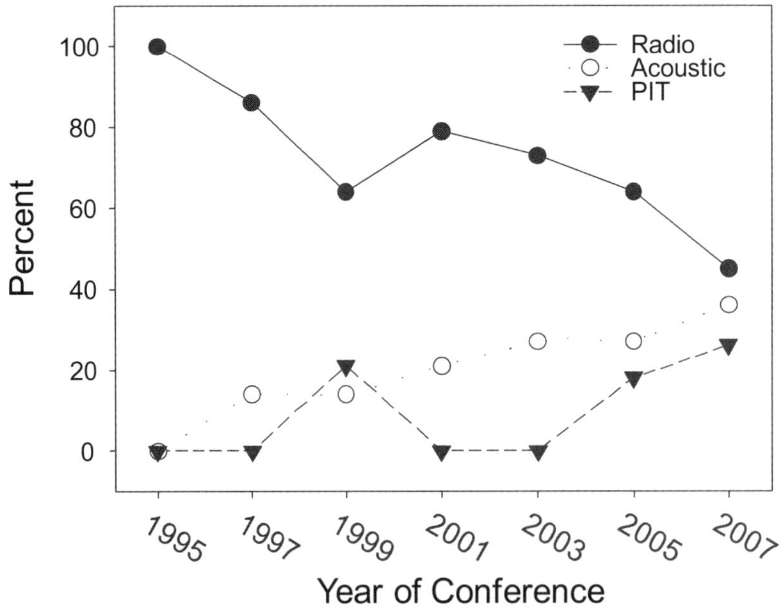

Figure 1. Trends in radio telemetry, acoustic telemetry, and PIT/transponder techniques for studies of fish in freshwater as determined from papers and abstracts (only for years 1996, 2000, and 2007) published in proceedings from the Fish Telemetry Conference Held in Europe.

although radio technology is used to transmit signals to satellites (e.g., popup satellite tags; Block 2005) or to transmit data from acoustic receiving systems to a shore based station (O'Dor et al. 1998; Lembo et al. 2002).

Trends from the Peer Reviewed Literature over a 10 year Period

We systematically evaluated all published peer reviewed literature using ISI Web of Science for 10 year period (January 1 1998 to December 31 2007) focused on use of radio telemetry for the study of freshwater fish. All searches were conducted on January 3 2008 using the string "fish* and radio* and (track* or tag* or telem*)" and were then manually filtered to exclude erroneous papers. As above, we also filtered any studies where radio signals were used to transmit data collected from archival loggers or receivers. We also excluded papers published as part of the Fish Telemetry Conference held in Europe to minimize biases associated with these "pulses" of publishing that would influence temporal trends. After filtering (we excluded about 20% of total hits), we had a list of 269 papers dealing with fish radio telemetry in peer reviewed outlets. For some elements of the review, we also excluded papers that used dummy transmitters or were synthetic (review) in nature (31 papers fell into this category). From each paper we extracted information to identify trends in radio telemetry use through time. We also acknowledge that many telemetry studies are conducted by government agencies, utilities and environmental consultants and reasonably few of these would end up in the peer reviewed literature. When appropriate, we discuss where this may have relevance to the trends that we report.

Publishing Trends in Radio Telemetry Studies

The number of radio telemetry studies published (excluding syntheses and dummy transmitter studies as above) over a ten year period has generally increased ($r = 0.76$, $P < 0.05$) over time (Figure 2). The fewest number of papers was 10 in 1998 and the high was 43 in 2004. The majority (63%) of articles on radio telemetry and fish were published in fish-centric journals such as Journal of Fish Biology, North American Journal of Fisheries Management, Transactions of the American Fisheries Society, Ecology of Freshwater Fish, Environmental Biology of Fishes, and Canadian Journal of Fisheries and Aquatic Sciences (Figure 3). However, there are several studies that have used radio telemetry on fish, all within the last few years, that have been published in leading general ecology and biology journals such as Oecologia (Crook et al. 2001), Ecology (Cooke et al. 2006), Proceedings of the Royal Society of London B (Fleming et al. 2000), Biometrics (Cowen and Schwarz 2005) and Aquatic Toxicology (Scott et al. 2005). Often these studies are more interdisciplinary than those published in traditional fisheries outlets, such as linking behavior with physiology and genetics (see Cooke et al. 2008). For example, Cooke et al. (2006) applied radio telemetry transmitters to adult sockeye salmon *Oncorhynchus nerka* and obtained a blood sample and gill biopsy prior to releasing the fish. *A posteriori*, the authors were able to evaluate the physiological status of fish that failed to reach spawning grounds versus those that were successful. Radio telemetry studies of terrestrial wildlife are more frequently featured in general ecology and conservation biology journals, and it would benefit fisheries science and radio telemetry to publish some of the best studies on fish and radio telemetry in outlets that would

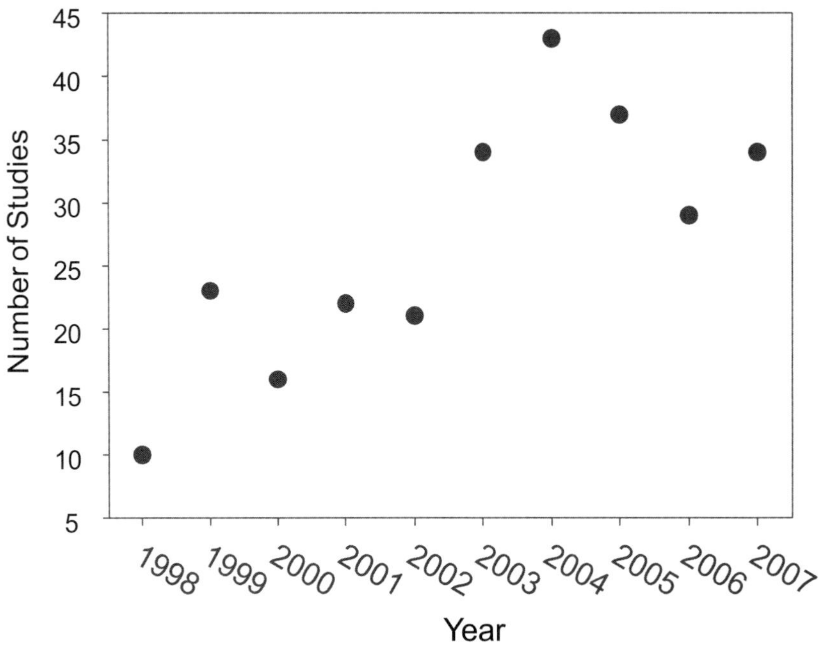

Figure 2. Temporal patterns in the annual number of radio telemetry studies published in peer reviewed journals over a ten year period.

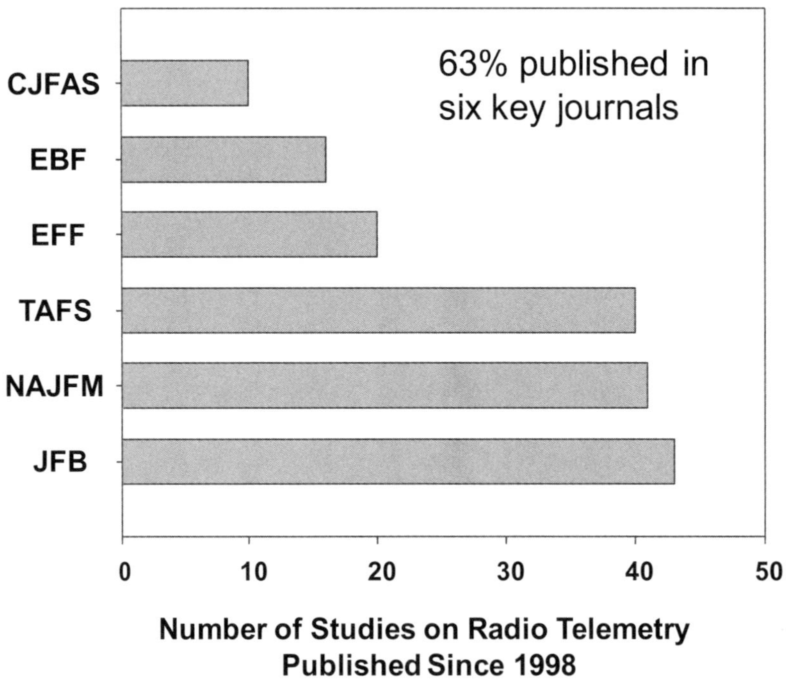

Figure 3. Trends in use of primary journal outlets for fish radio telemetry studies published in peer reviewed journals over a ten year period. Abbreviations are as follows: Canadian Journal of Fisheries and Aquatic Sciences (CJFAS), Ecology of Freshwater Fish (EFF), Environmental Biology of Fishes (EBF), Transactions of the American Fisheries Society (TAFS), North American Journal of Fisheries Management (NAJFM), and Journal of Fish Biology (JFB).

be more accessible to nonfish-centric ecologists. Papers at the fringe of a discipline can yield insight into opportunities for creative uses of technology.

The majority of radio telemetry studies were conducted in developed countries with the USA, Canada, Australia and several European countries being the sites of the most studies (Figure 4). However, there were also several radio telemetry studies that have been conducted in less developed countries such as Kenya (Mlewa et al. 2007), Namibia (Thorstad et al. 2001b), Peru (Cohen et al. 1999), and Thailand (Mitamura et al. 2007).

Roughly equal numbers of studies were conducted on anadromous/catadromous fish ($N = 122$, 48% of studies) and resident freshwater species ($N = 131$, 52% of studies). Half of the papers published were focused on salmonids ($N = 134$, 50% of the studies; 20 studies on either parr or smolts) and in particular, Atlantic salmon *Salmo salar*, rainbow trout *Oncorhynchus mykiss* and brown trout *Salmo trutta*. Other popular study taxa included centrarchids ($N = 20$), cyprinids ($N = 18$), acipenserids ($N = 14$), percids ($N = 12$), esocids ($N = 7$), percycthids ($N = 6$), and petromyzontids ($N = 4$). The majority of studies ($N = 212$, 82% of studies) focused on a single species, forty studies included tagging two species (15% of studies) and three studies covered three species (1% of studies). Only four studies covered four or more species (2% of studies). The most species studied in a single published paper was seven (Weitkamp et al. 2003).

The duration of radio tagging studies was median 5 months, with a mode of one month (Figure 5). The longest study period was 40 months. Studies of short duration (two months or less) tended to focus on smolt out-migration, fishway evaluations, examinations of catch-and-release fishing mortality, or short term habitat use of small fish. Longer term studies (multiple years) tended to focus on seasonal habitat use and migrations

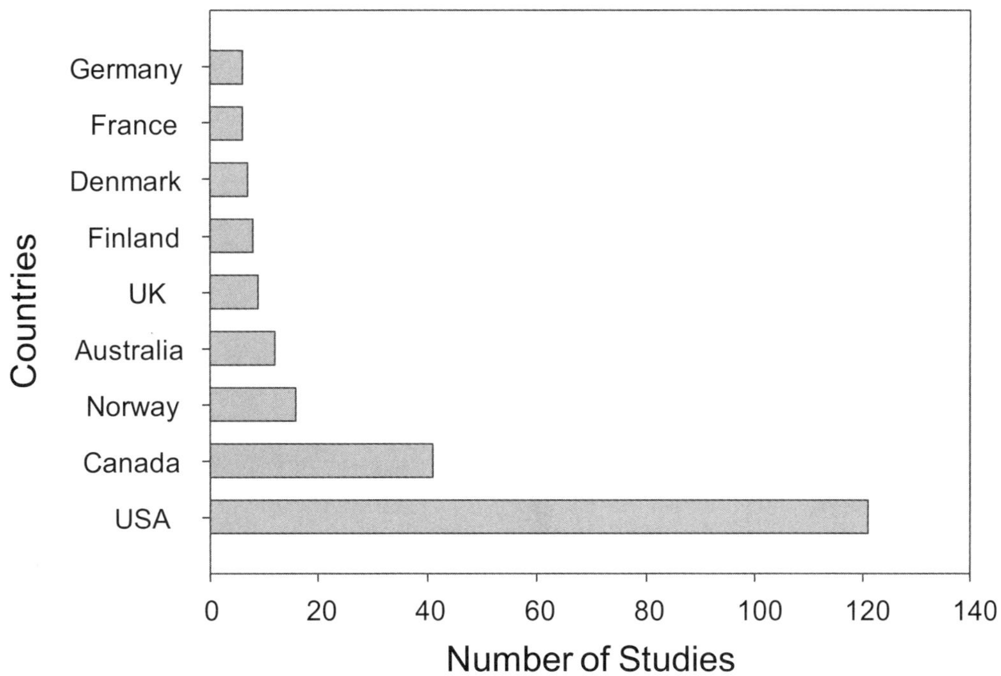

FIGURE 4. Regional patterns of study locations for fish radio telemetry studies published in peer reviewed journals over a ten year period.

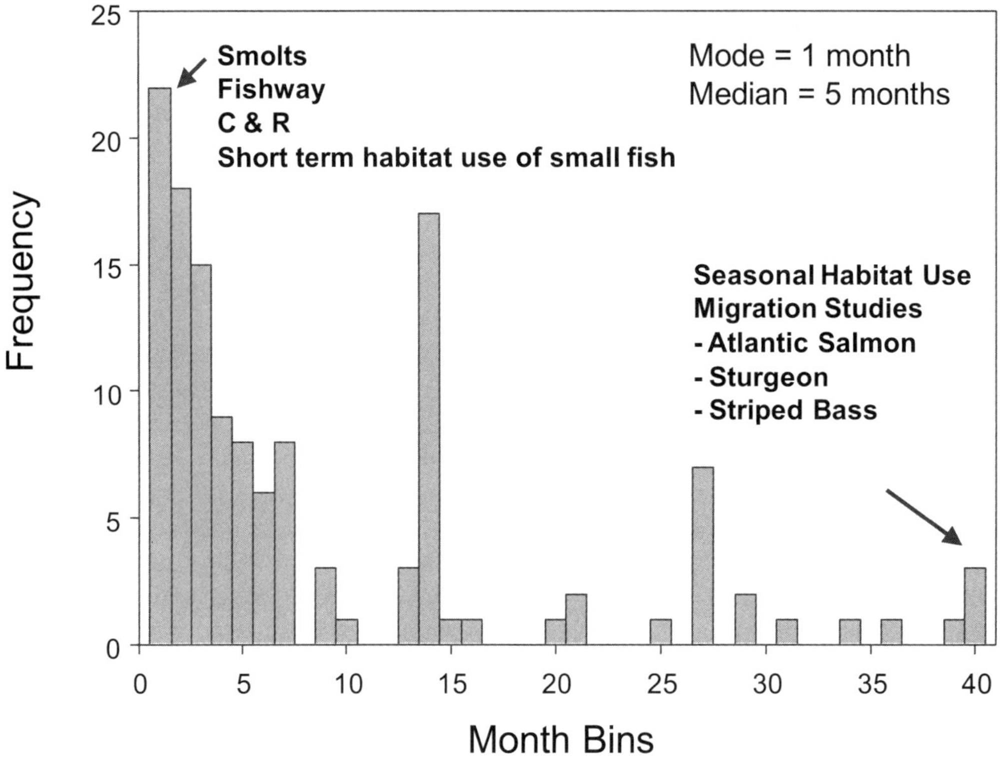

Figure 5. Distribution of tracking durations (in months) for fish radio telemetry studies published in peer reviewed journals over a ten year period.

of large-bodied fish species such as Atlantic salmon, sturgeon (Acipenseridae) and striped bass *Morone saxatilis*.

The most common radio transmitter attachment technique was intracoelomic surgery (N = 154). Of these studies where information was reported (i.e., N = 131), 124 (95%) had the antenna exit the body cavity while 7 (5%) coiled the antenna within the cavity or used transmitters that had a precoiled antenna within the transmitter. It is worth noting that internal antenna configurations can suffer reductions in signal strength. For acoustic transmitters, there is no external antenna so this eliminates infection and other problems that are inherent when a radio transmitter antenna exits the body. External attachment (backpack style) was the second most popular method (N = 42). Gastric methods were also used (N = 13) but this approach was largely restricted to studies of adult Pacific salmonids (*Oncorhynchus* spp.) during spawning migrations (when feeding ceased). A single study used urogenital tagging methods (i.e., Colavecchia et al. 1998) and eight studies combined methods.

The number of radio tags used in each study ranged from 1 to 18,286 (mean 373 ± 1833 [Std. Dev.], median 35, mode 20; Figure 6). The four studies with the largest number of tags were all focused on the evaluation of mortality rates among outmigrating smolts or returning adults in the Columbia River (Keefer et al. 2004; Schreck et al. 2006; Caudill et al. 2007; Naughton et al. 2007). In total, 90,360 radio tags were used in these studies published over a ten year period, with an estimated transmitter cost of $22.59 million USD (assuming $250USD/tag). Clearly this is a gross underestimate as many studies

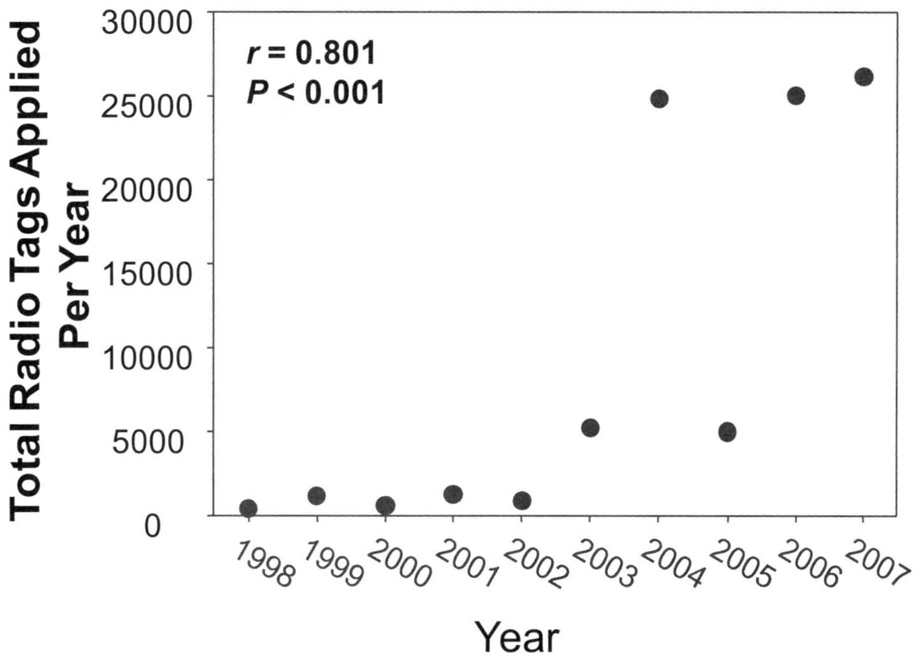

FIGURE 6. Trends in the total number of radio tags applied annually in fish radio telemetry studies published in peer reviewed journals over a ten year period.

are not published in peer reviewed literature, including studies with large numbers of individuals in the Pacific northwest. In addition, we excluded tag numbers from papers published as part of the Fish Telemetry Conference in Europe. The total number of tags used annually increased with year during 1998–2007 ($r = 0.80$, $P < 0.001$). Also the average and median number of tags deployed in each study increased in this period (average: $r = 0.89$, $P < 0.001$; median: $r = 0.50$, $P < 0.05$), suggesting some positive movement towards increased sample sizes (Figure 7). However, there was no significant relationship between the year and the minimum sample size in a given study. This may represent an artifact of management briefs/short communications publishing unique findings on one or a few individuals (Herke and Moring 1999; Thorstad et al. 2001a).

The majority of studies ($N = 230$) used simple radio transmitters. Several studies combined technologies, either in the transmitter itself (e.g., use of a combined acoustic radio tag, CART tag; $N = 3$; see Deary et al. 1998) or by using a combination of radio and acoustic tags ($N = 12$), a combination of radio and PIT tags ($N = 11$), or one study where radio, PIT and acoustic tags were all used (Collins et al. 2000). However, typically PIT tags were used for identifying unique individuals if fish were recaptured rather than as a telemetric monitoring tool for free-swimming fish. Several studies combined radio and archival techniques ($N = 2$). For example, Mitamura et al. (2007) used radio telemetry to locate a data logger that was temporarily attached to a Mekong giant catfish *Pangasianodon gigas* to elucidate depth distribution. In marine mammal research, this approach is used frequently to locate animals for the retrieval of time-depth recorders.

In most studies, individual fish were recognized by using discrete frequencies ($N = 128$, 62% of studies). Some studies used varied pulse intervals, or pulse rates, to discrimi-

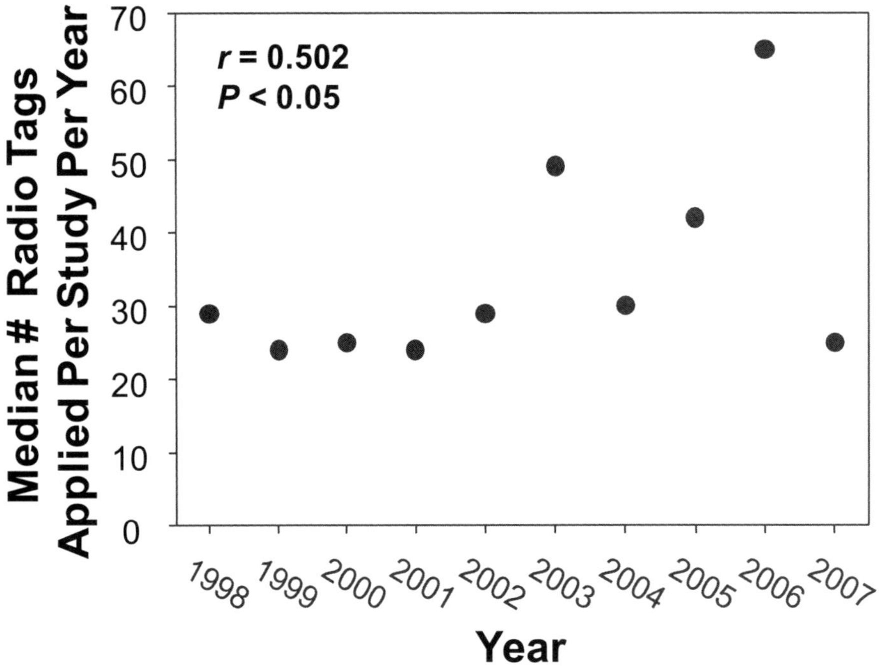

Figure 7. Trends in the median number of radio tags applied annually in fish radio telemetry studies published in peer reviewed journals over a ten year period.

nate between individuals (N = 28, 13% of studies). Interest in this approach using different frequencies and pulse rates to recognize individuals seems to be waning. Digitally coded transmitters are now frequently used (N = 52, 25% of studies) and in fact, some commercial manufacturers are moving away from producing other tag types. Coded tags allow one to tag multiple fish with the same frequency, and the digital coding scheme enables the researcher to discriminate between individual fish using a purpose-built receiver.

Of the studies that reported tracking details, the majority used manual tracking techniques such as tracking by foot, car, boat, aircraft, bike, kayak, snowmobile (N = 136). Automated telemetry arrays were also popular (N = 60), as were studies that combine the two techniques (e.g., supplementing fixed stations with manual tracking, N = 24). Although it was our perception that studies involving manual tracking were becoming less common, a temporal analysis revealed that the proportion of studies that either used arrays or manual tracking did not change over the ten year period (Figure 8). The reason for predicting a greater emphasis on automated systems reflected our perspective that the increasing sample sizes makes manual tracking quite difficult. Furthermore, telemetry arrays represent the "state of the art," and most of the commercial manufacturers of telemetry equipment appear to be emphasizing fixed stations over manual tracking. Clearly, not every telemetry study requires an elaborate array with automated receivers and this depends on the study question, system (environment, organism and scale), and funding levels. There are trade-offs between the expense of fixed stations, including maintenance and potential for vandalism, and the labor costs associated with tracking. One of the benefits of manual tracking is the value-added knowledge that can be obtained from being in the field constantly. Further, information on detailed behavior patterns when the fish

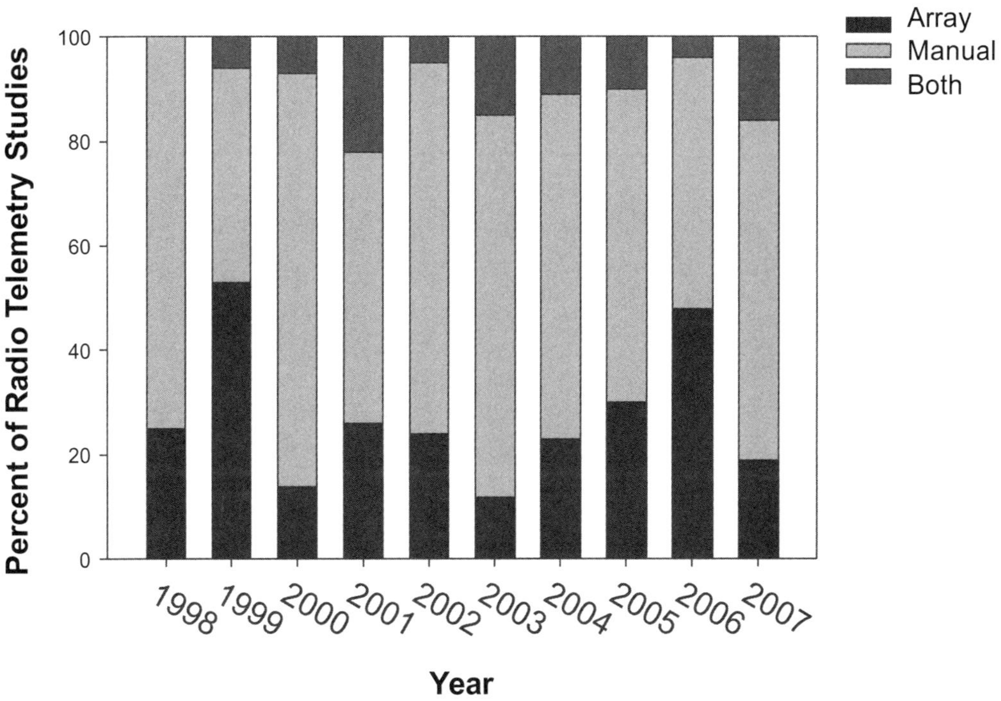

FIGURE 8. Trends in the use of different radio tracking methods in fish radio telemetry studies published in peer reviewed journals over a ten year period.

is outside the range of fixed stations can only be collected by manual tracking. Combining arrays with manual tracking seems to be an effective approach, but only 11% of studies have done this.

Although we did not quantify antenna configurations, there seems to be growing use of underwater antennas. These can be set-up in an array to obtain detailed positions of fish near industrial infrastructure such as dam forebays and canals (Beeman et al. 2004), or through use of "dip" antennas to manually locate transmitters to within several centimeters (Niemela et al. 1993). There have also been attempts to use a downrigger technique to deliver an underwater antenna to depth and track fish in lake systems (Bunt et al. 2003). Use of underwater antennas is ineffective at large spatial scales due to signal attenuation in water.

Over the ten year period examined, 2/3 of the studies were basic in focus ($N = 149$, 66%, relative to $N = 77$, 34% applied; Figure 9). The mean sample size for applied studies (median = 55, mean = 692) was significantly ($P < 0.05$) larger than basic studies (median = 29, mean = 223). However, the study duration (in months) did not differ between applied (median = 5, mean = 8) and basic (median = 5, mean = 7.5) studies. Overall, studies on habitat use/preference ($N = 173$), movement ($N = 127$), migration ($N = 109$), and fine scale estimates of swimming speeds/activity ($N = 20$) were the most common (note that an individual study may have fallen into several categories and was counted as such). A surprising number of studies were specifically focused on understanding winter biology ($N = 54$). In terms of more applied work, common topical areas included hydropower flow issues ($N = 35$), mortality studies ($N = 19$), fish passage studies ($N = 11$), research on endangered species ($N = 6$), studies dealing with aquaculture ($N = 6$) and studies associ-

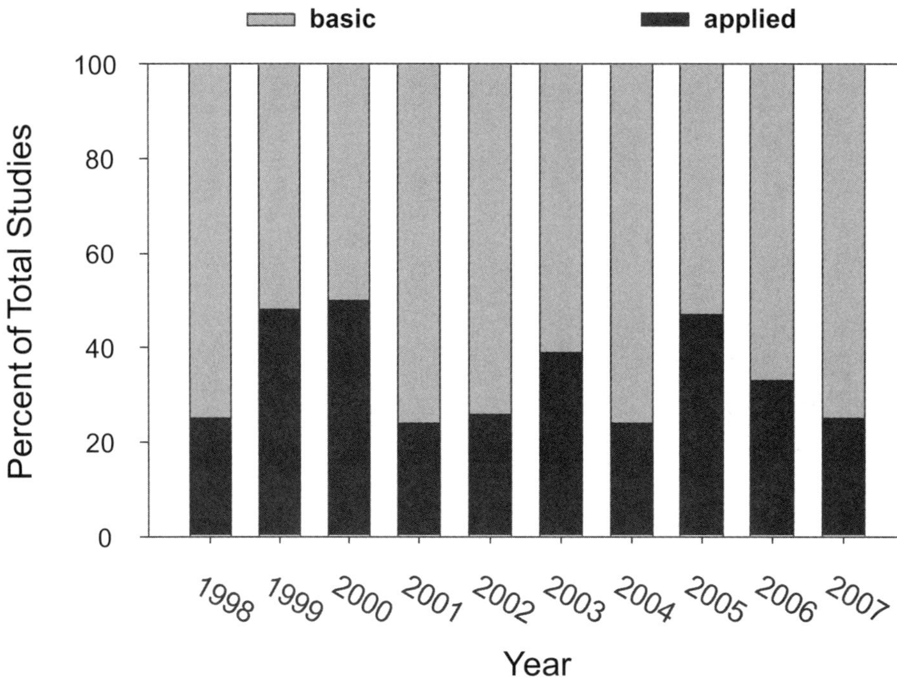

Figure 9. Trends in the focus (i.e., applied versus basic) of fish radio telemetry studies published in peer reviewed journals over a ten year period.

ated with catch-and-release fishing ($N = 4$). We conducted an analysis to determine if the sample sizes differed between basic and applied realms. Indeed, the mean sample size for applied studies (median = 55, mean = 692) was significantly ($P < 0.05$) larger than basic studies (median = 29, mean = 223). However, the study duration (in months) was similar between applied (median = 5, mean = 8) and basic (median = 5, mean = 7.5) studies.

Where does Radio Telemetry Excel?

All telemetry and tagging technology have a series of inherent capabilities and limitations that make them differentially appropriate for research and management applications. Based on our own experience as well as insight from the literature and colleagues, we developed a list of eight specific applications where radio telemetry is ideal and usually the only or best means of achieving the desired outcome. The list is not exhaustive as there are many objectives that can be achieved with radio telemetry.

1. Fishway, entrainment or bypass assessments—Hydropower development and other anthropogenic activities including the development of irrigation channels and locks all have the potential to alter fish behavior, habitat use, and survival, and there is increased interest in documenting these impacts. Radio telemetry is a powerful tool for the assessment of the interactions between fish and humans in or near this infrastructure. Some innovations in acoustic telemetry (e.g., 3 dimensional systems) are quite useful in open areas such as forebays or tailraces (provided that noise is not too great), but these technologies do not work well within engineered structures. Fixed radio telemetry stations involving many antennas and combined digital switching and tracking systems can track rapidly moving fish as they transit through engineered structures. For example, Bunt et al. (1999a) tracked fish below a low head barrier dam equipped with two Denil fishways and identified that fish were having difficulty locating the fishway entrance. Such knowledge led to a retrofit of the fishway to enhance passage. Radio telemetry can also provided detailed information on time to traverse a structure, passage efficiency, attraction efficiency, as well as areas that fish are using for resting or are difficult to pass/negotiate (Bunt et al. 1999a). Manual tracking is less frequently used for these applications, particularly on larger systems were there are many tagged fish (e.g., Keefer et al. 2004). However, focused radio tracking of few individuals can also be effective.

2. Manual tracking studies in riverine or shallow systems that require detailed evaluations of microhabitat use—The classic application of radio telemetry, and one that will continue to be important, is the study of fish microhabitat use in fluvial or shallow systems. For example, Matheney and Rabeni (1995) monitored the movement and habitat use of northern hog suckers *Hypentelium nigricans* in an Ozark stream over an entire year. More recently, Bunt et al. (1999b) used detailed manual tracking to evaluate the microhabitat use of brown trout during periods of high and low flow in a hydro-peaking river. Some of the fish were located in "subterranean" environments (deep in unembedded cobble), an observation that would not have been possible with acoustic telemetry. Such observations can be facilitated by using coaxial cable with the end exposed rather than an antenna (Niemela et al. 1993) or an extended-reach technique, which uses a small loop antenna attached to a lightweight telescopic pole (Broadhurst and Ebner 2007) to obtain more accurate positions. If field observations are collected around the clock, it is also possible to evaluate diel habitat use of fish as has been done for brown trout (Young 1995) and dace *Leuciscus leuciscus* (Clough and Ladle 1997). With increasingly small radio tags under development, it will soon be possible to study some small freshwater species or early life stages, data which tend to be currently obtained using mark–recapture techniques (e.g., creek chub, *Semotilus atromaculatus*; Belica and Rahel 2008). Data obtained from detailed microhabitat studies can be used for hydraulic modeling for the evaluation of various hydropower operations on fish (Scruton et al. 2002).

3. Studies that require external application of tags on relatively small fish to assess mortality or behavior at fine temporal scales—Small radio tags can be externally attached to fish. The entire procedure requires several seconds and can be done without anesthetizing fish while they are submerged in water. There is more flexibility in the size and shape of radio transmitters than acoustic ones based on the lack of a ceramic (the acoustic output module) in the former. Radio transmitters can be laterally compressed so that they hug the side of the fish and cause minimal drag. This type of deployment (i.e., external) is ideal for the evaluation of mortality or behavior and is often the preferred method for quantifying nonharvest fishing mortality (Donaldson et al. 2008). For example, to evaluate the consequences of air exposure on the postrelease behavior and survival of angled largemouth bass, Thompson et al. (2008) used small radio tags and tracked fish for five days. Mortality was negligible and there was little evidence of behavioral impairments. Conversely, a similar study with smallmouth bass revealed that fish which suffered barotrauma after being captured as part of a live release fishing tournament died at higher rates and took longer to leave the release site than fish without barotrauma (Gravel and Cooke 2008). Radio telemetry has the potential to be used for studies of inland bycatch survival and external tagging is also effective for evaluating fishway performance without having to perform invasive surgery.

4. Studies on threatened/important species in developing countries where there is interest in building capacity and using simple technology—The current crisis in the loss of freshwater fish diversity (Dudgeon et al. 2006) extends beyond the developed world and includes many inland systems in developing countries. Indeed, in many of these countries fish serve as the primary protein source for much of the population. Radio telemetry has much to offer the study of inland fishes, including those species that are endangered (Cooke 2008). However, there is need to develop regional capacity within the local research communities and ensure effective transfer of fish tracking technology (Marmulla and Bènech 1999). To this end, the Food and Agriculture Organization of the United Nations has sponsored a number of workshops (see Baras et al. 2002) designed to teach researchers in developing countries about radio telemetry technology. There are also increasingly more examples of information exchange among researchers in different regions through active collaborations, often funded by international development agencies (Thorstad et al. 2001b; Økland et al. 2005). Radio telemetry is relatively inexpensive in terms of capital, especially when using manual tracking (labor costs also tend to be relatively low in developing countries). In most cases, manual tracking studies are sufficient for obtaining critical information on the movement, habitat use, and distribution of fish. Studies that use manual tracking are excellent means of engaging the local community, which in fact also applies to other stakeholder groups in the developed world (angling groups, school children, conservation organizations, native peoples).

5. Studies where avian/terrestrial predation, stranding or human harvest are issues—A distinct advantage of radio telemetry over acoustic telemetry is the potential to be used to locate transmitters that have been removed from or left the aquatic environment. For example, mammalian or avian predators can prey on fish and remove the carcass (and transmitter) from the water. Thorstad et al. (2004) were able to locate a radio transmitter from a fish that was tagged to evaluate postrelease fishing mortality of cichlids in Africa, and located the transmitter in a tree frequented by fish eagles. In a long term study of smallmouth bass *Micropterus dolomieu* demographics and behavior in Ontario, Barthel et al. (2008) located a transmitter in a mink den at a depth of more than 1 m in the soil. Radio telemetry is also extremely effective for locating stranded fish in fluvial systems, either in temporary pools or on land (Berland et al. 2004.). Strategic placement of telemetry receiving stations can also yield important information on fish harvest. For example, Spoelstra (2001) placed a logging telemetry receiver at a parking lot in a tailrace fishery for saugeye (*Sander canandensis* × *vitreus*) in Ohio that was known to have low levels of tag reporting to quantify harvest. In British Columbia, telemetry receivers were placed in four key fish processing plants, which collectively handled the majority of all salmon sockeye processing in the province to determine levels of commercial harvest and to retrieve tags without damaging the processing equipment (English et al. 2004).

6. Winter biology in fluvial systems—Despite the fact that winter is a critical period in the life of many fish species, particularly in temperate environments, little is known about the distribution and behavior of fish during this season. Radio telemetry techniques have provided one of the only tools for the study of fish during winter. Early work involved snorkeling surveys (e.g., Cunjak and Power 1986) but this technique did not enable the continuous or repeated monitoring of individually-known fish, and cannot be used during ice cover. Brown et al. (2001) used radio telemetry to study the winter movement of tagged brown trout *Salmo trutta*, white sucker *Catostomus commersoni* and common carp *Cyprinus carpio* in the Grand River, Ontario. In particular, the researchers focused on studying fish behavior during periods of ice break-up and flooding. Indeed, both white sucker and common carp moved the longest distances (downstream) during flooding and ice break-up, whereas brown trout moved less during these times. The authors also revealed that backwater habitats were important areas of refuge for riverine fishes during periods of flooding and ice break-up. Numerous other studies have quantified fish position, corresponding habitat variables, and fish activity in winter, typically for comparisons with other seasons and usually in lotic systems. Radio telemetry is also effective in winter in lentic systems. Bauer and Schlott (2004) used radio telemetry in winter to evaluate common carp behavior relative to dissolved oxygen conditions in a pond environment in Germany. The carp moved to more oxygen-rich habitats during hypoxic events. An advantage of radio telemetry for studying fish movement and habitat use is that positions can be obtained through the ice. In fact, near exact positions can be obtained by walking on the ice. Using acoustic telemetry, this could only be accomplished by drilling holes in the ice to deploy a hydrophone. PIT tagging systems can be used by walking on or near ice, but range detection is limited and this approach only works in very shallow, small streams (Roussel et al. 2004).

7. Studies in areas with entrained air such as discharges or tailwaters—Because acoustic telemetry requires the detection of sound energy in water, the signal to noise ratio must be high in order for a transmitter to be detected. In regions such as dam tailwaters, fishways, discharge canals, and rapids entrained air, turbulence, and high flows can obscure acoustic telemetry transmissions (Thorstad et al. 2000). However, radio telemetry techniques excel in these conditions. For example, DePhilip et al. (2005) used radio telemetry to identify spawning locations of walleye *Sander vitreus* in the Ausauble River, Michigan in the vicinity of a tailrace. At a thermal effluent canal on the north shore of Lake Erie, Cooke and McKinley (1999) monitored the residency patterns of common carp and channel catfish. In the same canal, radio telemetry was also effective for monitoring the activity of fish in and around tempering pumps, which is an inherently noisy environment (Cooke et al. 2000). Typically, studies near or around dams or other industrial/utility infrastructure use fixed stations with a combination of aerial and underwater antennas (Cooke et al. 2004). However, it is important to acknowledge that acoustic telemetry technology is becoming more robust for positioning fish in noisy environments, in particular with three dimensional acoustic telemetry systems (Ehrenberg and Steig 2003). Nonetheless, radio telemetry is still the most effective and utilized technology for studies in noisy environments.

8. Studies of fish where there is potential for large migrations in vast watersheds—Radio telemetry technology, unlike PIT tags or acoustic telemetry, is well suited for the use of aircraft or truck in situations where telemetered fish need to be tracked over long distances. For example, Collins et al. (2000) tracked Atlantic sturgeon *Acipenser oxyrinchus oxyrinchus* in the Combahee and Edisto rivers, South Carolina, primarily by using airplane. Although this tracking approach is popular, Roberts and Rahel (2005) noted that the accuracy of aerially determined locations can limit the types of conclusions one can draw about fish habitat use. They revealed that the overall mean location error of aquatic telemetry studies that used aircraft tracking was 158 m, a level of error much greater than most other radio tracking techniques. Nonetheless, the ability to track fish via aircraft is a powerful technique that is appropriate for determining crude (within 100 s of meters) positions in remote areas and rugged terrain where other tracking techniques are impractical.

The Future of Radio Telemetry, or, Is Radio Telemetry Getting Washed Downstream?

Radio telemetry is no longer the only option for research on the spatial ecology of freshwater fishes. Acoustic telemetry arrays are being used frequently in lakes or large rivers, but the disadvantage is that manual tracking is crude and awkward. PIT tagging studies that use fixed antennas or mobile reader systems are effective for small fluvial systems, but are otherwise limited. However, radio telemetry has distinct advantages for some applications as outlined above. Beyond the opportunities outlined above, there are several other unique characteristics of radio telemetry that are worthy of discussion. For example, the smallest radio transmitters available are far smaller than the smallest acoustic transmitters, enabling tagging and tracking smaller fish with radio than acoustic transmitters, notwithstanding the low power output and short lifespan of small radio tags. Further, radio transmitters typically cost less to produce than acoustic transmitters. In fact, we would like to emphasize a "back to basics" approach with radio telemetry. Indeed, there are still many questions that are best addressed using simple single frequency tags and manual tracking. There is no question that fixed stations do add more opportunities, but there are also often issues associated with security and power. There is also immense value of being at one's study site frequently. Fixed stations alone mean infrequent visits and the opportunity to miss out on intriguing biological or hydrological events. One of the challenge with radio transmitters is that the antenna typically exits the body cavity and can serve as a source of infection and biofouling. Future research is needed to reduce the potential consequences of the antennas on fish.

Our objective here was to characterize and clarify the actual and potential role for radio telemetry within the broader landscape of tagging and telemetry technologies, so that telemetry practitioners can make informed decisions regarding the optimal choice of telemetry technology when designing and executing studies in freshwater systems. So is radio telemetry getting washed downstream? We would submit that radio telemetry continues to be a very effective tool for research on freshwater fish. There is no doubt that there are now many options available for tracking studies in freshwater, and many of these tools are in their infancy. As fisheries practitioners, it is not our job to advocate for a specific technology. Instead, we should become familiar with all options that are available and then use whatever technique is most appropriate for addressing the desired study objective (which may not be telemetry at all!). Given that, we believe that radio telemetry will not fade in popularity but will simply remain one of the tools available for fisheries practitioners. Here we provided a list of specific situations in which radio telemetry excels. There are also other situations where radio telemetry is useful, and there are instances in which other technologies can accomplish the same objectives as radio telemetry. We hope that this critical appraisal of the current state of radio telemetry will help to clarify the actual and potential role of radio telemetry for the study of freshwater fish.

Acknowledgments

We thank Larry Hildebrand for encouraging us to develop this presentation and for his support with this focus. Lisa Thompson kindly assisted with extracting data from the literature. Kyle Hanson and two anonymous referees provided comments on an earlier draft of this paper. We thank our many collaborators, mentors, and students for teaching us about radio telemetry. In particular, we thank Finn Økland, Chris Bunt, Rich Brown, Scott Hinch, and Martyn Lucas. Cooke and his research using radio telemetry are supported by the Canada Foundation for Innovation, the Ontario Research Fund, the Ontario Ministry of Research and Innovation, the Canada Research Chairs Program, BC Hydro, and the Natural Sciences and Engineering Research Council. The contents of this paper were presented as the "plenary" in the radio telemetry session at the Tagging Symposium held in Auckland, New Zealand.

References

Baras, E., and J.-C. Philippart. 1996. Underwater biotelemetry. Proceedings of the First Conference and Workshop on Fish Telemetry in Europe. University of Liège, Belgium.

Baras, E., V. Benech, and G. Marmulla. 2002. Outcomes of a pilot fish telemetry. Workshop for developing countries. Hydrobiologia 483:9–11.

Barthel, B. L., S. J. Cooke, J. H. Svec, C. D. Suski, C. M. Bunt, F. J. S. Phelan, and D. P. Philipp. 2008. Divergent life histories among smallmouth bass inhabiting a connected river-lake system. Journal of Fish Biology 73:829–852.

Bauer, C., and G. Schlott. 2004. Overwintering of farmed common carp (*Cyprinus carpio* L.) in the ponds of a central European aquaculture facility—measurement of activity by radio telemetry. Aquaculture 241:301–317.

Beeman, J. W., C. Grant, and P. V. Haner. 2004. Comparison of three underwater antennas for use in radiotelemetry. North American Journal of Fisheries Management 24:275–281.

Belica, L. A. T., and F. J. Rahel. 2008. Movements of creek chubs, Semotilus atromaculatus, among habitat patches in a plains stream. Ecology of Freshwater Fish 17:258–272.

Berland, G., T. Nickelsen, J. Heggenes, F. Økland, E. B. Thorstad, and J. Halleraker. 2004. Movements of wild Atlantic salmon parr in relation to peaking flows below a hydropower station. River Research and Applications 20:957–966.

Block, B. A. 2005. Physiological ecology in the 21st Century: advancements in biologging science. Integrative and Comparative Biology 45:305–320.

Bridger, C. J., and R. K. Booth. 2003. The effects of biotelemetry transmitter presence and attachment procedures on fish physiology and behavior. Reviews in Fisheries Science 11:13–34.

Broadhurst, B., and B. Ebner. 2007. An improved technique for small-scale radio-tracking of crayfish and benthic fishes in upland streams. Transactions of the American Fisheries Society 136:423–427.

Brown, R. S., G. Power, and S. Beltaos. 2001. Winter movements and habitat use of riverine brown trout, white sucker and common carp in relation to flooding and ice break-up. Journal of Fish Biology 59:1126–1141.

Bunt, C. M., C. Katopodis, and R. S. McKinley. 1999a. Attraction and passage efficiency of white suckers and smallmouth bass by two Denil fishways. North American Journal of Fisheries Management 19:793–803.

Bunt, C. M., S. J. Cooke, and R. S. McKinley. 1999b. Movement and summer habitat of brown trout (*Salmo trutta*) below a pulsed-discharge hydroelectric generating station. Regulated Rivers: Research and Management 15:395–403.

Bunt, C. M., S. J. Cooke, and D. P. Philipp. 2003. A modified downrigger for detecting radio transmitters in deep water. North American Journal of Fisheries Management 23:258–263.

Caudill, C. C., W. R. Daigle, M. L. Keefer, C. T. Boggs, M. A. Jepson, B. J. Burke, R. W.

Zabel, T. C. Bjornn, and C. A. Peery. 2007. Slow dam passage in adult Columbia River salmonids associated with unsuccessful migration: delayed negative effects of passage obstacles or condition-dependent mortality? Canadian Journal of Fisheries and Aquatic Sciences 64:979–995.

Clough, S., and M. Ladle. 1997. Diel migration and site fidelity in a stream-dwelling cyprinid, *Leuciscus leuciscus*. Journal of Fish Biology 50:1117–1119.

Cohen, M. A., W. R. C. Beaumont, and N. C. Thorp. 1999. Movement and activity patterns of the black piranha. Environmental Biology of Fishes 54:45–52.

Colavecchia, M., C. Katopodis, R. Goosney, D. A. Scruton, and R. S. McKinley. 1998. Measurement of burst swimming performance in wild Atlantic salmon (*Salmo salar* L.) using digital telemetry. Regulated Rivers: Research and Management 14:41–51.

Collins, M. R., T. I. J. Smith, W. C. Post, and O. Pashuk. 2000. Habitat utilization and biological characteristics of adult Atlantic sturgeon in two South Carolina rivers. Transactions of the American Fisheries Society 129:982–988.

Cooke, S. J. 2008. Biotelemetry and biologging in endangered species research and animal conservation: relevance to regional, national, and IUCN Red List threat assessments. Endangered Species Research 4:165–185.

Cooke, S. J., and R. S. McKinley. 1999. Winter residency and activity patterns of channel catfish, Ictalurus punctatus (Rafinesque), and common carp, *Cyprinus carpio* L., in a thermal discharge canal. Fisheries Management and Ecology 6:515–526.

Cooke, S. J., S. G. Hinch, M. C. Lucas, and M. Lutcavage. In press. Biotelemetry and biologging. *In* A. Zale, et al., editors. Fisheries Techniques, 3rd edition. American Fisheries Society, Bethesda, Maryland.

Cooke, S. J., C. M. Bunt, and R. S. McKinley. 2000. Winter residency of smallmouth bass in a thermal discharge canal: implications for tempering pump operation. North American Journal of Fisheries Management 20:288–295.

Cooke, S. J., C. M. Bunt, and J. F. Schreer. 2004. Understanding fish behavior, distribution and survival in thermal effluents using fixed telemetry arrays: a case study of smallmouth bass in a discharge canal during winter. Environmental Management 33:140–150.

Cooke, S. J., S. G. Hinch, A. P. Farrell, D. A. Patterson, K. Miller-Saunders, D. W. Welch, M. R. Donaldson, K. C. Hanson, G. T. Crossin, I. Olsson, M. S. Cooperman, M. T. Mathes, K. A. Hruska, G. N. Wagner, R. Thomson, R. Hourston, K. K. English, S. Larsson, J. M. Shrimpton, and G. Van Der Kraak. 2008. Developing a mechanistic understanding of fish migrations by linking telemetry with physiology, behavior, genomics and experimental biology: an interdisciplinary case study on adult Fraser River sockeye salmon. Fisheries 33: 321–338.

Cooke, S. J., S. G. Hinch, G. T. Crossin, D. A. Patterson, K. K. English, M. C. Healey, J. M. Shrimpton, G. Van Der Kraak, and A. P. Farrell. 2006. Mechanistic basis of individual mortality in Pacific salmon during spawning migrations. Ecology 87:1575–1586.

Cowen, L., and C. J. Schwarz. 2005. Capture-recapture studies using radio telemtry with pre-mature radio-tag failure. Biometrics 61:657–664.

Crook, D. A., A. I. Robertson, A. J. King, and P. Humphries. 2001. The influence of spatial scale and habitat arrangement on diel patterns of habitat use by two lowland river fishes. Oecologia 129:525–533.

Crossin, G. T., S. G. Hinch, S. J. Cooke, D. W. Welch, S. D. Batten, D. A. Patterson, G. Van Der Kraak, J. M. Shrimpton, and A. P. Farrell. 2007. Behaviour and physiology of sockeye salmon homing through coastal waters to a natal river. Marine Biology 152:905–918.

Cunjak, R. A., and G. Power. 1986. Winter habitat utilization by stream resident brook trout (*Salvelinus fontinalis*) and brown trout (*Salmo trutta*). Canadian Journal of Fisheries and Aquatic Sciences 43:1970–1981.

Deary, C., D. A. Scruton, G. H. Niezgoda, S. McKinley, D. Cote, K. D. Clarke, D. Perry, T. Lindstrom, and D. White. 1998. A dynamically switched combined acoustic and radio transmitting (CART) tag : An improved tool for the study of diadromous fishes. Marine Technology Society Journal 32:63–69.

DePhilip, M. M., J. S. Diana, and D. Smith. 2005. Movement of walleye in an impounded reach of the Au Sable River, Michigan, USA. Environmental Biology of Fishes 72:455–463.

Donaldson, M. R., R. Arlinghaus, K. C. Hanson, and S. J. Cooke. 2008. Enhancing catch-and-release science with biotelemetry. Fish and Fisheries 9:79–105.

Dudgeon, D., A. H. Arthington, M. O. Gessner, Z. Kawabata, D. Knowler, C. Loque, R. J. Naiman, A-H. Prieur-Richard, D. Soto, M. L. J. Stiassny, and C. A. Sullivan. 2006. Freshwater biodiversity: importance, status, and conservation challenges. Biological Reviews 81:163–182.

Ehrenberg, J. E., and T. W. Steig. 2003. Improved techniques for studying the temporal and spatial behavior of fish in a fixed location. ICES Journal of Marine Science 60:700–706.

Eiler, J. H. 1995. A remote satellite-linked tracking system for studying Pacific salmon with radio telemetry. Transactions of the American Fisheries Society 124:184–193.

English, K., C. Sliwinski, M. Labelle, W. R. Koski, R. Alexander, A. Cass, and J. Woodey. 2004. Migration timing and in-river survival of Late-run Fraser River sockeye using radiotelemetry techniques. Report prepared by LGL Limited, Sidney, B.C. for the Pacific Biological Station of Fisheries and Oceans Canada.

Fleming, I. A., K. Hindar, I. B. Mjølnerød, B. Jonsson, T. Balstad, and A. Lamberg. 2000. Lifetime success and interactions of farm salmon invading a native population. Proceedings of the Royal Society, London B: Biological Sciences 267:1517–1524.

Gibbons, J. W., and K. M. Andrews. 2004. PIT tagging: Simple technology at its best. BioScience 54:447–454.

Gravel, M. A., and S. J. Cooke. 2008. Severity of barotrauma influences the physiological status, post-release behaviour, and fate of tournament-caught smallmouth bass. North American Journal of Fisheries Management 28:607–617.

Herke, S. W., and J. R. Moring. 1999. Soft harness for external attachment of large radio transmitters to northern pike (*Esox lucius*). Fisheries Research 39:305–312.

Jepsen, N., E. B. Thorstad, E. Baras, and A. Koed. 2002. Surgical implantation of telemetry transmitters in fish: how much have we learned? Hydrobiologia 483:239–248.

Keefer, M. L., C. A. Peery, T. C. Bjornn, M. A. Jepson, and L. C. Stuehrenberg. 2004. Hydrosystem, dam, and reservoir passage rates of adult Chinook salmon and Steelhead in the Columbia and Snake Rivers. Transactions of the American Fisheries Society 133:1413–1439.

Lembo, G., M. T. Spedicato, F. Økland, P. Carbonara, I. A. Fleming, R. S. McKinley, E. B. Thorstad, M. Sisak, and S. Ragonese. 2002. A wireless communication system for determining site fidelity of *Epinephelus marginatus* (Lowe, 1834) using coded acoustic transmitters. Hydrobiologia 483:249–257.

Lucas, M. C., and E. Baras. 2000. Methods for studying spatial behaviour of freshwater fishes in the natural environment. Fish and Fisheries 1:283–316.

Marmulla, G., and V. Bènech. 1999. Necessity for an information exchange network for fisheries projects using biotelemetry in developing countries. Pages 197–205 *in* A. Moore and I. Russell, editors. Advances in fish telemetry. Proceedings of the Third Conference on Fish Telemetry in Europe, Norwich, UK, 20–25 June 1999.

Matheney, M. P., and C. F. Rabeni. 1995. Patterns of movement and habitat use by northern hog suckers in an Ozark stream. Transactions of the American Fisheries Society 124:886–897.

Melnychuk, M. C., D. W. Welch, C. J. Walters, and V. Christensen. 2007. Riverine and early ocean migration and mortality patterns of juvenile steelhead trout (*Oncorhynchus mykiss*) from the Cheakamus River, British Columbia. Hydrobiologia 582:55–65.

Millspaugh, J., and J. Marzluff. 2001. Radiotelemetry and Animal Populations. Academic Press, San Diego, CA.

Mitamura, H., Y. Mitsunaga, N. Arai, Y. Yarnagishi, M. Khachaphichat, and S. T. Viputhanuma. 2007. Vertical movements of a Mekong Giant catfish (*Pangasianodon gigas*) in Mae Peum Reservoir, Northern Thailand, monitored by a multi-sensor micro data logger. Zoological Science 24:643–647.

Mlewa, C. M., J. M. Green, and R. Dunbrack. 2007. Are wild African lungfish obligate air breathers? Some evidence from radio telemetry. African Zoology 42:131–134.

Moore, A., and I. Russell. 2000. Advances in Fish Telemetry. CEFAS, Lowestoft, U.K.

Naughton, G. P., C. C. Caudill, C. A. Peery, T. S. Clabough, M. A. Jepson, T. C. Bjornn, and L. C. Stuehrenberg. 2007. Experimental evaluation of fishway modifications on the passage behaviour of adult Chinook salmon and steelhead at Lower Granite Dam, Snake River, USA. River Research and Applications 23:99–111.

Niemela, S. L., J. B. Layzer, and J. A. Gore. 1993. An improved radiotelemetry method for determining use of microhabitats by fishes. Rivers 4:30–35.

O'Dor, R. K., Y. Andrade, D. M. Webber, W. H. H. Sauer, M. J. Roberts, M. J. Smale, and F. M. Voegeli. 1998. Applications and performance of Radio-Acoustic Positioning and Telemetry (RAPT) systems. Hydrobiologia 372:1–8.

Økland, F., E. B. Thorstad, C. J. Hay, T. F. Næsje, and B. Chanda. 2005. Patterns of movement and habitat use by tigerfish (*Hydrocynus vittatus*) in the Upper Zambezi River (Namibia). Ecology of Freshwater Fish 14:79–86.

Roberts, J. J., and F. J. Rahel. 2005. Accuracy of aerial telemetry in fisheries studies. North American Journal of Fisheries Management 25:660–666.

Roussel, J-M., R. A. Cunjak, R. Newbury, D. Caissie, and A. Haro. 2004. Movements and habitat use by PIT-tagged Atlantic salmon parr in early winter: the influence of anchor ice. Freshwater Biology 49:1026–1035.

Schreck, C. B., T. P. Stahl, L. E. Davis, D. D. Roby, and B. J. Clemens. 2006. Mortality estimates of juvenile spring-summer Chinook salmon in the Lower Columbia River and estuary. Transactions of the American Fisheries Society 135:457–475.

Scott, D. M., M. C. Lucas, and R. W. Wilson. 2005. The effect of high pH on ion balance, nitrogen excretion and behaviour in freshwater fish from an eutrophic lake: a laboratory and field study. Aquatic Toxicology 73:31–43.

Scruton, D. A., R. S. McKinley, N. Kouwen, W. Eddy, and R. K. Booth. 2002. Use of telemetry in the development and application of biological criteria for habitat hydraulic modeling. Hydrobiologia 483:83–94.

Sisak, M. M., and J. S. Lotimer. 1998. Frequency choice for radio telemetry: the HF vs. VHF conundrum. Hydrobiologia 371/ 372:53–59.

Spedicato, M. T., G. Lembo, and G. Marmulla. 2005. Aquatic telemetry: advances and applications. Proceedings of the Fifth Conference on Fish Telemetry held in Europe, Ustica, Italy, 9–13 June 2003. FAO/COISPA, Rome.

Spoelstra, J. A. 2001. Movement of reservoir-stocked saugeye between tailwaters, streams, and rivers. Master's thesis, Ohio State University, Columbus.

Thompson, L. A., M. R. Donaldson, K. C. Hanson, R. Arlinghaus, and S. J. Cooke. 2008. Physiology, behavior and survival of angled and air exposed largemouth bass. North American Journal of Fisheries Management 28:1059–1068.

Thorstad, E. B., F. Økland, and T. G. Heggberget. 2001a. Are long term negative effects from external tags underestimated? - Fouling of an externally attached telemetry transmitter. Journal of Fish Biology 59:1092–1094.

Thorstad, E. B., C. J. Hay, T. F. Næsje, and F. Økland. 2001b. Movements and habitat utilization of three cichlid species in the Zambezi River, Namibia Ecology of Freshwater Fish 10:238–246.

Thorstad, E. B., F. Økland, D. Rowsell, and R. S. McKinley. 2000. A new system for automatic recording of fish tagged with coded acoustic transmitters. Fisheries Management and Ecology 7:281–294.

Thorstad, E. B., C. J. Hay, T. F. Næsje, B. Chanda, and F. Økland. 2004. Effects of catch-and-release angling on large cichlids in the subtropical Zambezi River. Fisheries Research 69:141–144.

Velle, J. I., J. E. Lindsay, R. W. Weeks, and F. M. Long. 1979. An investigation of the loss mechanisms encountered in propagation from a submerged fish telemetry transmitter. Pages 228–237 *in* Proceedings of the 2nd International Conference on Wildlife Biotelemetry. Laramie, Wyoming.

Voegeli, F. A. and D. G. Pincock. 1996. Overview of underwater acoustics as it applies to biotelemetry. Pages 23–30 *in* E. Baras and J. C. Philippart, editors. Underwater Biotelemetry, Proceedings of the First Conference and Workshop on Fish Telemetry in Europe, University of Liege, Belgium.

Weitkamp, D. E., R. D. Sullivan, T. Swant, and J. DosSantos. 2003. Behavior of resident fish relative to total dissolved gas supersaturation in the lower Clark Fork River. Transactions of the American Fisheries Society 132:856–864.

Young, M. K. 1995. Telemetry-determined diurnal positions of brown trout (*Salmo trutta*) in two south-central Wyoming streams. American Midland Naturalist 133:264–273.

American Fisheries Society Symposium 76:371–378, 2012

Using Radio Telemetry to Evaluate the Success of a Hatchery Steelhead Recycling Program

Eva Schemmel*

Department of Zoology, University of Hawaii at Manoa
2538 McCarthy Mall, Edmondson 152, Honolulu, Hawaii 96822, USA

Shaun Clements

Oregon Department of Fish and Wildlife, Fish Division
28655 Highway 34, Corvallis, Oregon 97333, USA

Carl Schreck

Oregon Cooperative Fish and Wildlife Research Unit Leader, U.S. Geological Survey
Department of Fish and Wildlife, Oregon State University
104 Nash Hall, Corvallis, Oregon 97330, USA

David Noakes

Department of Fisheries and Wildlife and Oregon Hatchery Reseach Center
Oregon State University, 104 Nash Hall, Corvallis, Oregon 97330, USA

Abstract.—Many salmonid hatcheries in the Pacific Northwest recycle adult fish. This involves the transport and release of adult salmon or steelhead (*Oncorhynchus* spp.) back downstream following their collection at the hatchery. The goal is to increase recreation along the river by providing additional opportunities for anglers to harvest fish. However, the effects of such programs have seldom been critically evaluated. We implanted radio-tags in adult hatchery origin steelhead captured at a hatchery and released them downstream to estimate the number of recycled radio-tagged steelhead that were caught in the recreational fishery, returned to the hatchery, and estimate the number that strayed. The fish tended to hold at the release site or outside of the hatchery entrance. Anglers reported catching 10.3% of the recycled fish, one of these was caught in a tributary. Due to hatchery trap closures, only approximately 10.3% of the recycled fish were caught again at the hatchery. Although more fish were available for recreational harvest, this recycling program increased the potential for fish from hatchery and wild origin to interact.

*Corresponding author: eschemmel@gmail.com

Introduction

Hatcheries were established in the Pacific Northwest to mitigate for the decline of salmon and steelhead (*Oncorhynchus* spp.) populations (Hilborn 1992). Traditionally, the majority of hatcheries focused on increasing harvest opportunities and supplementing wild populations. To increase harvest opportunities, many hatcheries "recycle" adult salmonids. Recycled fish are sometimes termed "re-runs." Recycling involves the downstream transport and release of adult hatchery fish that have returned to a hatchery or sorting facility. The goal of a recycling program is to provide anglers with additional opportunities to catch returning fish that were not captured during their original migration back to the hatchery (Lindsay et al. 2001). The success of such programs is often measured in terms of increased angler harvest. However, other factors, such as the incidence of straying, are becoming increasingly important. Despite this, there have been few efforts to evaluate the effects of most recycling programs on angler harvest or the behavior of un-caught adult salmonids.

The studies that have been reported have found only modest catch rates, with increased potential for recycled fish to stray. There is evidence that homing fidelity is decreased in fish that are recycled (Lindsay et al. 2001). Evenson and Cramer (1984) reported that only 9.5% of recycled Chinook salmon *O. tshawytscha* were caught by anglers. During their 3 years study, the return rate of recycled fish to the hatchery was 27–62%. They also estimated that 4% of the recycled fish strayed into other rivers (Evenson and Cramer 1984). Conversely, Dyson and Apperson (2005) found higher catch rates (35%) of recycled summer Chinook salmon in a recreational fishery and only 1% returned to the hatchery. Recycled fish that are not recaptured back at the hatchery may lead to an increase in competition, interbreeding, and density dependent interactions with wild salmonids.

We used radio-telemetry to evaluate the behavior and catch rates of adult hatchery steelhead after they have returned to the hatchery and been recycled back downstream to provide for the recreation fishery. Our objectives were to determine if recycling hatchery fish increases angler harvest, reduces homing fidelity, and increases stray rates.

Methods

Study Site

We conducted the study on the Clackamas River, a tributary of the lower Willamette River, Oregon, USA (Figure 1). The Clackamas River Basin contains both hatchery and wild populations of salmon and trout and is a popular fishing destination. The Clackamas River contains a wild population of winter steelhead listed as threatened under the Endangered Species Act (1998). Oregon Department of Fish and Wildlife (ODFW) introduced Skamania stock hatchery summer steelhead into the Clackamas River in the 1970s (Kostow et al. 2003). Today, the ODFW Clackamas Hatchery acclimates 175,000 summer steelhead smolts for release into the Clackamas River in April. The Clackamas River hatchery recycling program began in 1998 when hatchery summer steelhead were no longer passed above a barrier dam, North Fork Dam, to protect native winter steelhead spawning grounds. We monitored the behavior of radio-tagged adult steelhead in the 45 river kilometer (rkm) reach of the Clackamas River, from downstream of North Fork Dam to the mouth of the Clackamas River.

Fish Collection

We obtained 39 summer steelhead adults (2.9 ± 0.1 kg, 15 males, 24 females) from ODFW Clackamas Hatchery 36 rkm upstream of the mouth of the Clackamas River

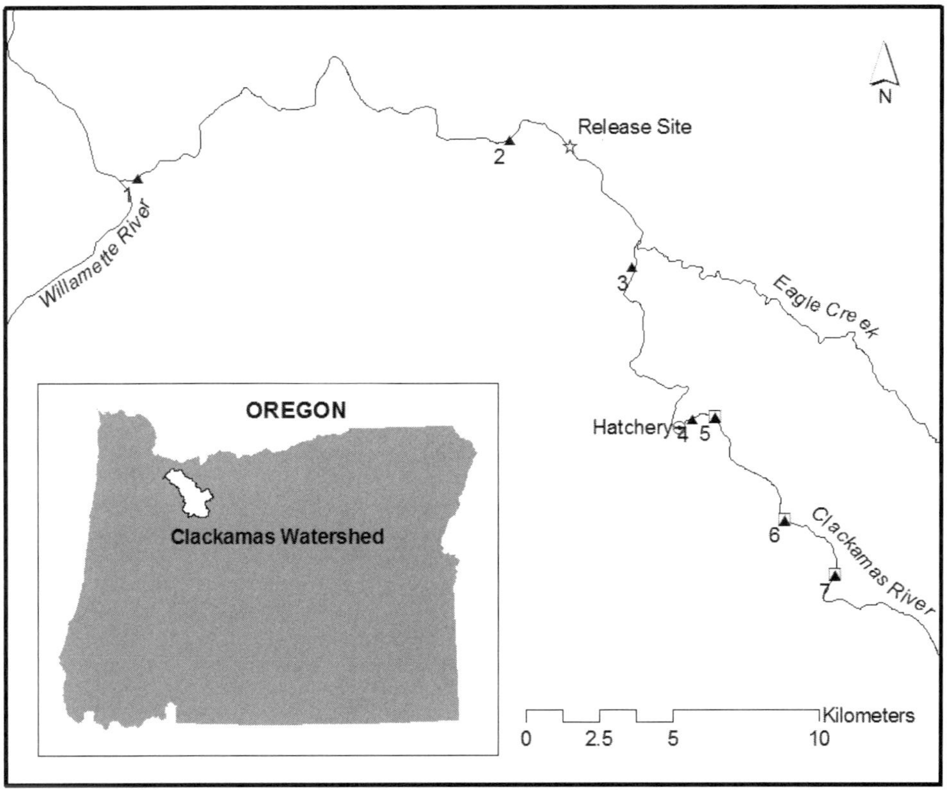

Figure 1. Locations of the fixed monitoring stations (1–7). Station 1 is at the confluence of the Clackamas and Willamette Rivers. Stations 5–7 were located on dams; River Mill Dam, Diversion Dam, and Faraday Dam, respectively. Tagged steelhead were released at Barton Park (Release Site).

(Figure 1). The fish were collected between 22 July 2007 and 1 August 2007 and held in a raceway (16–19°C) prior to tagging.

All fish were internally tagged with one of two coded radio transmitters (Lotek NTC-6–2: 9.1 × 30.1 mm, 4.5 g in air or Lotek CEMG2-R16–25: 16 × 45 mm, 11.9 g in air). Transmitters had a burst rate of 3 s with a tag life of six months. We anesthetized each fish individually in a darkened container containing a buffered solution of 0.05% (50 mg/L) tricaine methanesulfonate (MS-222, Argent Chemical Laboratories, Redmond, Washington) until the fish lost its equilibrium with a maximum exposure time of 8 min. The fish was placed ventral side up on a surgical table. The gills were continuously perfused with a diluted solution (0.025%) of MS-222 though a perforated tube placed in the buccal cavity. A 6–8 cm ventral incision was made between the pectoral and pelvic fins (Okland et al. 1997). Sex was determined from visual observation of the gonads. A radio transmitter was then inserted into the body cavity and the incision was closed using absorbable monofilament in a simple interrupted suture pattern. We also marked the fish externally with individually numbered Floy tags to identify them for recognition by anglers and to mark that these fish were not fit for human consumption.

Following surgery we held the fish in a flow-through raceway at the hatchery for 24 h to monitor their condition before release and to validate that all of the transmitters were functional. The fish were then transported in aerated river water to Barton Recreational Park, 21 rkm upstream from the confluence with the Willamette River (Figure 1), and released into the Clackamas River. We released tagged steelhead during the morning, with four releases occurring between 23 July 2007 and 2 August 2007.

Tracking and Analysis

We used a combination of fixed monitoring stations and mobile tracking to record the location of radio-tagged, recycled steelhead. We used one of two receivers (Lotek SRX 600 and Lotek SRX 400, Lotek, Newmarket, Ontario, Canada) and two four-element yagi antennas (150 Mhz, Cushcraft, Manchester, New Hampshire, USA) at our fixed stations. The fixed monitoring stations were located at seven sites throughout the study reach (Figure 1). We measured the detection efficiency at each fixed station by floating submerged transmitters past each site using a rubber bouy. Detection efficiency was 100% at all sites. Mobile tracking included river rafting and foot surveys. When river conditions allowed, we surveyed the entire study reach every 3 d during the summer (August) and fall (November). For mobile tracking, the location of each radio-tagged steelhead was recorded using a Lotek SRX 600 receiver with an integrated global positioning system (GPS). The receiver recorded GPS positions when there were at least four satellites in range. The receiver recorded the signal strength, time, tag code, and GPS location. The location of each fish was plotted based on the highest signal strength at a given gain.

Homing Fidelity

Recycled steelhead homing fidelity was estimated as the number of fish that returned to the hatchery trap and to the hatchery entrance. Fish that were re-caught in the Clackamas Hatchery trap were recorded and subsequently recycled back downstream by ODFW personnel to the original release location, following standard operating procedures. The Clackamas Hatchery trap was closed between 29 September 2007 and 01 March 2008. The number of fish that re-entered the trap is therefore based on the current operating procedure and may have been higher had the trap been open continuously. Due to trap closures, we estimated the time of return to the hatchery locale based on the first detection at the confluence of the Clackamas River and Dog Creek (a tributary connecting the hatchery to the Clackamas River), hereafter termed the "hatchery entrance." Thus, it does not represent recapture in the hatchery trap, but we felt that it was important to monitor as an estimate of recycled hatchery steelhead homing fidelity.

Straying

The number of recycled hatchery steelhead that strayed out of the Clackamas River was determined from detections of out migrating steelhead at the fixed station (Station 1) at the mouth of the Clackamas River. In addition, Eagle Creek, the main tributary along the lower Clackamas River contains wild populations of winter steelhead and coho salmon (USFWS 2007). This tributary was monitored by motor vehicle mobile tracking. Radio tagged steelhead that were found in this tributary were considered strays.

In addition to strays, the disappearance of radio tagged fish was estimated by the nondetection of fish that had previously been recorded several times within the study reach, and had neither migrated outside the study reach nor been recovered. The time of disappearance was based on the last known record from the fixed monitoring stations or the manual surveys. Fish disappearances may represent tag failure, unreported angler harvest, animal predation, or fish straying into tributaries along the Clackamas River.

Angler Success

Angler catch of recycled fish was determined by volunteer reporting. We used a variety of methods to encourage voluntary angler reporting. Each fish was externally tagged with an individually numbered tag that provided contact details to report the catch. The study was also widely publicized among local angling groups and was featured on television, in press releases, and on the ODFW website. We also posted information on the study at major fishing areas along the river. Catch reports were collected at the ODFW Clackamas office and at the ODFW Clackamas Hatchery. The catch reports included information on fish id number, catch location, time, gear used, and fish condition.

Results

Homing Fidelity

We recaptured only four (2 males, 2 females) of the radio-tagged steelhead (10.3%). These fish entered the trap in September 2007 and March 2008. The first fish re-entered the trap on 9 September, 36 d after release. Three of the four fish were recycled for a second time; however, the last returning steelhead re-entered the trap in March and so it was not recycled. All three of the fish that were recycled for a second time returned to the hatchery trap for a third time.

The majority of fish moved upriver following release but tended to reside in the river at the hatchery entrance. In total, between July and March, we estimated that 87.5% (n = 21) of females and 80.0% (n = 12) of males returned to the hatchery entrance after being recycled downstream. These estimates include the fish that entered the hatchery trap. A large percentage of the recycled steelhead moved upriver to the hatchery entrance during August, one month after release (females: 45.8%, n = 11; males: 33.3%, n = 5) (Figure 2). However, most of the fish took between 2 and 4 months to cover the 15 rkm between the release site and the hatchery entrance (Figure 2).

Straying

None of the radio-tagged fish passed the lower-most receiver placed at the confluence of the Willamette and Clackamas River. In addition, we did not observe any of the radio-tagged fish entering into the tributaries of the Clackamas River during the period in which we were actively tracking the fish (July–December). However, two fish were never detected post release and six (15.4%) of the recycled fish disappeared during the study period. Four of these fish disappeared between Station 2 and Station 3, the reach containing the tributary Eagle Creek, one of which was reported by an angler as being caught in Eagle Creek. The remaining 2 fish disappeared near the hatchery. In addition, two of the radio-tagged fish were recovered dead from the river during the study period. They did not appear to have died from hooking- or tagging-related injuries.

Angler Success

Anglers reported catching 10.3% (n = 4) of the radio-tagged steelhead. They were caught throughout the study period. In addition, we recovered a radio-tag in shallow water at a popular fishing location near the hatchery creek entrance, possibly discarded by an angler after capture of the fish. Most of the reported landings occurred at the release location and the entrance to the Clackamas Hatchery. This matches the general distribution of fish that we observed during the radio tracking.

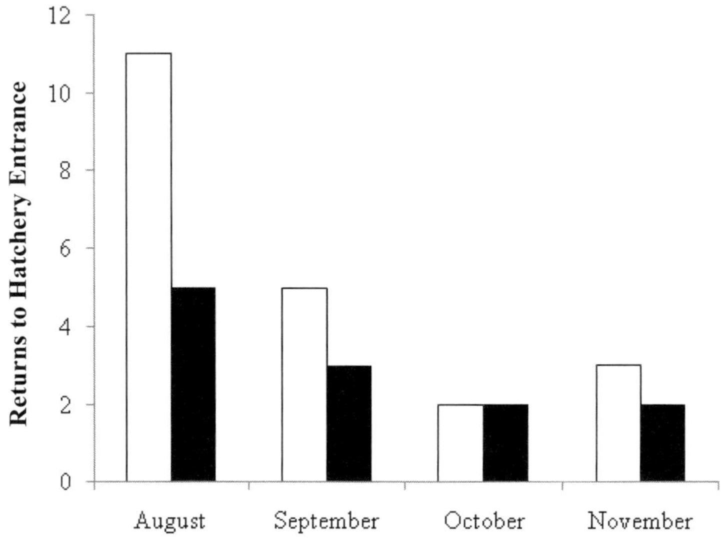

Figure 2. Monthly return of female and male radio-tagged adult summer steelhead to the entrance of the Clackamas Fish Hatchery. Returns were based on the first date of detection at the mouth of Dog Creek (the tributary that connects Clackamas hatchery to the Clackamas River). We released 15 males (black bars) and 24 females (white bars).

Discussion

Our results show that recycled hatchery fish are caught in a recreational fishery. Anglers caught a minimum of 10.3% of recycled steelhead during our study, based on surveys from recreational anglers. However, a significant number (15.4%) of radio-tagged fish disappeared during the study period. Given that detection efficiency at the fixed monitoring stations was close to 100%, the disappearance of tagged fish must be explained by tag failure, natural predation, fishing mortality, or straying into tributaries. There are relatively few natural predators within the study area, and the range of the radio transmitter would have allowed for detection of fish within 100 m of the river. Because of the intensive fishing effort on the river over a period of several months and the high number of radio-tagged fish that could not be accounted for, we speculate that many of these nondetections are the result of unreported angler harvest of radio-tagged steelhead or are fish that strayed into tributaries. Despite our efforts, the volunteer reporting rate was probably low, although we have no way to estimate the exact rate. During the study, we received several anecdotal reports from anglers who witnessed other anglers catching and keeping radio-tagged fish. Oregon State law prohibits the retention of a radio-tagged fish so it is possible that some anglers chose not to report their catch.

We were able to monitor the homing fidelity of recycled steelhead back to the hatchery. Our results suggest that fish tended to move upstream to the hatchery after being recycled. The majority tended to reside near the entrance to the hatchery. The return rates of recycled fish to the hatchery entrance were above 80%. Due to hatchery trap operating procedures very few fish (~10%) were actually recaptured in the hatchery trap. However, our results suggest that if the hatchery trap remained open recapture rates of recycled fish may have been higher. We suggest that hatchery managers could reduce the potential of for straying of recycled fish by keeping the hatchery trap operational throughout the season.

We found that while this recycling program increased the number of fish available to anglers, it also increase the incidence of straying. Straying of hatchery fish may pose threats to wild salmon and steelhead populations (Quinn 1993). We did not detect any recycled steelhead moving out of the Clackamas River past the radio receiver at the mouth, however, we found evidence that recycled fish strayed into tributaries along the river. We had one incidence of a recycled steelhead straying into a tributary, Eagle Creek, closer to the time of spawning. This fish was caught and reported by an angler in December. The last detection of this fish before it was reported caught was at the mouth of Eagle Creek in mid November. Three additional fish disappeared along the stretch of river containing the Eagle Creek tributary. Our radio tracking surveys did not detect any of these fish in Eagle Creek. We suggest that our vehicle tracking method may not have been sensitive enough to detect the radio transmitted signals in the Eagle Creek tributary, especially in areas where it flows through steep valleys.

In addition to recycling programs potentially increasing straying, they may also result in an increased number of hatchery fish in the river of release. The Clackamas River and Eagle Creek tributaries that the recycled fish were most likely to stray into both contain wild salmon and steelhead populations (USFWS 2007). Increased numbers of hatchery fish may pose conservation risks to wild salmonids, especially when nonnative hatchery stocks are released (Waples 1991; Currens et al. 1997; Scientific Review Team 1998). These concerns include potential negative competitive interactions (Flagg et al. 2000; Kostow and Zhou 2006; Kostow 2009), disease transfer (Currens et al. 1997; Amos and Thomas 2002), and interbreeding with wild salmonids (Waples 1991; Kostow et al. 2003; Hayes et al. 2004; Araki et al. 2007).

We used radio telemetry to determine the benefits and risks of recycling programs. The benefit of recycling programs will likely depend upon the river system, fish stock, and the fishery management goals in the region. Our results suggest that the recycling program for adult summer steelhead on the Clackamas River increases the number of fish available for the recreational fishery. However, the large number of recycled fish that were not caught by anglers and were not recaptured at the hatchery raises concerns over the impacts that these hatchery fish could have on wild stocks within the Clackamas basin. However, our study suggests that the majority of the recycled fish return to the hatchery vicinity and we speculate that it may be possible to reduce the incidence of straying in this program by managing the operation of the hatchery trap to ensure that a greater proportion of the recycled fish are recaptured. We suggest that recycling programs that operate in areas that contain wild stocks should be closely monitored for straying and potential negative impacts on wild populations.

Acknowledgments

Technical support was provided by Lotek through a Memorandum of Agreement with the Oregon Hatchery Research Center. We thank Mike Vandentillaart from Lotek for technical advice and assistance. Logistic support was provided by the Clackamas Office of ODFW, with special thanks to Todd Alsbury, Danette Ehler, and Isaac Sanders of the Clackamas Office of ODFW, and the ODFW Clackamas Hatchery, with special thanks to Hatchery Manager, Brian Zimmerman, and his staff. Field assistance was provided by a number of volunteers from ODFW, OSU, and Portland General Electric (PGE). This study was conducted under permission of the IA-

CUC Project Number 3587 of Oregon State University. Funding support was provided by the Oregon Department of Fish and Wildlife, the Fisheries and Wildlife Department of Oregon State University, the Oregon Hatchery Research Center, and the Thomas G. Scott and Henry Mastin Scholarships.

References

Amos, K., and J. Thomas. 2002. Disease interaction between wild and cultured fish: observations and lessons learned in the Pacific Northwest. Bulletin of the European Association of Fish Pathologists 22:95–102.

Araki, H., W. R. Ardren, E. Olsen, B. Cooper, and M. S. Blouin. 2007. Reproductive success of captive-bred steelhead trout in the wild: evaluation of three hatchery programs in the Hood River. Conservation Biology 21:181–190.

Currens, K. P., A. R. Hemmingsen, R. A. French, D. V. Buchanan, C. B. Schreck, and H. W. Li. 1997. Introgression and susceptibility to disease in a wild population of rainbow trout. North American Journal of Fisheries Management 17:1065–1078.

Dyson, N. W., and K. A. Apperson. 2005. Summer chinook salmon sport fishery on the south fork of the Salmon River, Idaho 2003. Fisheries Management Investigations #IDFG 05-57:1–25.

Evenson, M. D., and S. P. Cramer. 1984. An evaluation of recycling hatchery spring chinook salmon through the sport fishery in the upper Rogue River. Report Series 84–10, Oregon Department of Fish and Wildlife, Corvallis.

Flagg, T. A., and nine coauthors. 2000. Ecological and behavioral impacts of artificial production strategies on the abundance of wild salmon populations. NOAA Technical Memorandum NMFS-NWFS-41.

Hayes, S. A., M. H. Bond, C. V. Hanson, and R. B. Macfarlane. 2004. Interactions between endangered wild and hatchery salmonids: can the pitfalls of artificial propagation be avoided in small coastal streams? Journal of Fish Biology 65:101–121.

Hilborn, R. 1992. Hatcheries and the future of salmon in the Northwest. Fisheries 17:1–8.

Kostow, K. E., A. R. Marshall, and S. R. Phelps. 2003. Naturally spawning hatchery steelhead contribute to smolt production but experience low reproductive success. Transactions of the American Fisheries Society 132:780–790.

Kostow, K. E., and S. Zhou. 2006. The effect of an introduced summer steelhead hatchery stock on the productivity of a wild winter steelhead population. Transactions of the American Fisheries Society 135:825–841.

Kostow, K. 2009. Factors that contribute to the ecological risks of salmon and steelhead hatchery programs and some mitigating strategies. Reviews in Fish Biology and Fisheries 19:9–31.

Lindsay, R. B., K. R. Kenaston, and R. K. Schroeder. 2001. Reducing impacts of hatchery steelhead programs. Oregon Department of Fish and Wildlife, Portland.

Okland, F., B. Finstad, R. S. McKinley, E. B. Thorstad, and R. K. Booth. 1997. Radio-transmitted electromyogram signals as indicators of physical activity in Atlantic salmon. Journal of Fish Biology 51:476–488.

Quinn, T. P. 1993. A review of homing and straying of wild and hatchery-produced salmon. Fisheries Research 18:29–44.

Scientific Review Team. 1998. Review of artificial production of anadromous and resident fish in the Columbia River Basin. Report to the Northwest Power Planning Council, Portland, Oregon.

USFWS Columbia Basin Hatchery Review Team. 2007. Eagle Creek NFH assessments and recommendations report. p. 1–102.

Waples, R. S. 1991. Genetic interactions between hatchery and wild salmonids: lessons from the Pacific Northwest. Canadian Journal of Fisheries and Aquatic Sciences 48:124–133.

American Fisheries Society Symposium 76:379–384, 2012

An Improved Method for Obtaining Fine-Scale Location of Radio Tags when Tracking by Boat

John D. Koehn*
Arthur Rylah Institute for Environmental Research, Department of Sustainability and Environment, 123 Brown Street, Heidelberg 3084, Australia

John H. Eiler
Auke Bay Laboratories, Alaska Fisheries Science Center, National Marine Fisheries Service 17109 Point Lena Loop Road, Juneau, Alaska 99801, USA

John A. McKenzie
Arthur Rylah Institute for Environmental Research, Department of Sustainability and Environment, 123 Brown Street, Heidelberg 3084, Australia

William G. O'Connor
Biodiversity and Ecosystem Services, Department of Sustainability and Environment 8 Nicholson Street, East Melbourne 3000, Australia

Abstract.—This study describes an improved method that decreases tracking time but increases accuracy and precision for the fine-scale location of radio tags by reducing the receiving antenna efficiency. Test tags (48–49 MHz) were hidden in a lowland river and independently located by boat tracking using both the antenna reduction method and standard triangulation techniques. The antenna reduction method was found to be faster and more accurate, locating transmitters to within 0.19 ± 0.13 m (mean ± SE) of their actual location (error range 0–1.50 m) compared to 19.53 ± 4.81 m (error range 0–70.00 m) using triangulation. This simple approach can be applied to a wide range of radio telemetry studies and avoids many of the major sources of location error associated with triangulation including minimising the distance between the transmitter and receiver and avoiding the potential need for mapping.

Introduction

Radio tracking is a widely used method for studying the movements and habitat use of fishes at a range of scales (e.g., Eiler et al. 2000). Radio telemetry is undertaken using a variety of equipment and different personnel, in a wide range of conditions and locations, with tracking undertaken by air, land, or boat. In order to be efficient, radio tracking needs to be undertaken with both speed and accuracy, with minimal disturbance to the study animal (Kenward 2001). While these

*Corresponding author: John.Koehn@dse.vic.gov.au

attributes are presumably considered in most studies, potential errors associated with the techniques used are rarely reported (Simpkins and Hubert 1998), even though they can cause limitations in the accuracy of data collected (White and Garrott 1990).

Whenever fish positions are estimated using radio-tracking, some error will inevitably occur, either in locating the fish, mapping their positions, or both (Brenden et al. 2004). Total location error dictates the scale at which reliable inferences can be made, and failure to account for these errors can lead to inaccurate interpretations from the data (Simpkins and Hubert 1998). These errors are of greater concern when the data are being used to determine fine-scale movements or habitat use (Niemela et al. 1993). Triangulation is a common method used to locate radio-tagged fish (White and Garrott 1990), however, this approach can be time consuming and may introduce substantial error with respect to location accuracy (Stasko and Pincock 1977; White and Garrott 1990; Simpkins and Hubert 1998).

This paper describes an improved method for locating radio tags when boat tracking in a lowland river by physically reducing the receiving antenna and its efficiency to increase fine-scale tracking accuracy. Controlled tests compared this 'antenna reduction' method to standard triangulation techniques to quantify the relative speed and accuracy of the two methods.

Methods

Study site

The study was conducted over a 1 km, single-channelled, reach of the lower Ovens River in southeastern Australia (146 00 E, 36 00 S) from 1 to 14 July 1992. The Ovens River in this reach has an average width of 50 m, maximum depth under nonflood flows of 11 m, is slow-flowing with a soft substratum, contains considerable amounts of woody habitat and had a water conductivity of about 50 EC (Koehn 2009b).

Field trials

This study used Model 9 48–49 MHz radio transmitters (Advanced Telemetry Systems, Isanti, Minnesota, USA). 'Hide and seek' trials were conducted to determine the speed and accuracy with which transmitters could be located using triangulation and the antenna reduction methods. Ten different transmitters were used during the trials with transmitters being randomly selected and placed within the study area and then independently located. Site, transmitter depth (1.0–3.9 m), and tracking method were randomly selected for each trial, with pairs of sites with similar microhabitat characteristics used for each technique to minimize confounding factors. The transmitters were connected to a float submerged >40 cm below the water surface so that the actual location of the radio tag could not be visually identified.

Tracking was undertaken by boat using a loop antenna together with a Model 2100 scanning receiver (Advanced Telemetry Systems). Loop antennae are relatively compact (i.e., suitable for field use) and bidirectional, with maximum signal strength occurring when either edge is pointed at the transmitter. A faint or 'null' signal occurs when the plane of the antenna is perpendicular to the transmitter direction. As the 'null' is easier to distinguish than the maximum signal (Winter 1983), we used this technique to take directional bearings for triangulated locations. Two bearings, with a minimum angle of 60°, were used to determine each location, with markers placed on each bank and the transmitter location determined to be the point of intersection.

The antenna reduction method consisted of incrementally reducing the antenna and hence lowering its ability to detect the transmitter signal. This was accomplished with the following steps:

1. Reducing receiver gain to lowest level that signal was readily detected; removing the receiving antenna from the coaxial cable when a strong signal was encountered and using the coaxial cable as an antenna; removing the coaxial cable from the receiver when a strong signal was encountered, and detecting the signal only with the receiver; then

2. Reducing the receiver gain to determine the position when the receiver was positioned directly above the transmitter.

Reducing the receiver gain was also used in step 1 to reduce signal intensity and reception range.

Tracking for each trial was initiated at a distance of 1 km downstream from the hidden transmitter and conducted by moving through the study reach until it was considered that the transmitter had been successfully located. The time taken from the start of the tracking procedure to estimating the transmitter location was recorded using a stopwatch. The estimated location was then marked with another float attached to a weighted marker. The linear distance between the estimated and actual locations was measured to the nearest 0.01 m and defined as the location error (Zimmerman and Powell 1995).

Results

Fifty-eight transmitter location trials were conducted during this study, comprising 32 trials using triangulation method and 26 trials using the antenna reduction method. Tags were located more quickly using the antenna reduction method (26.54 ± 3.92 min per location) (mean ± SE) compared to the triangulation method (43.00 ± 4.28 min) (Figure 1a). Location accuracy was greater using the reduction method with transmitters located to within 0.19 ± 0.13 m (mean ± SE), of their actual position (error range 0–1.50 m) compared to 19.53 ± 4.81 m (error range 0–70.00 m) using triangulation (Figure 1b).

Discussion

This paper describes a simple method that can improve speed and accuracy of boat radio tracking. This antenna reduction method was quicker and more accurate at locating 48–49 MHz transmitters than the triangulation method. The accuracy observed during this study provides sufficient resolution when fine-scale location data are required. As fine-scale tracking can be time consuming, the increased efficiency of this method also allows more data to be collected in any given tracking period. The reduced speed and lesser location accuracy of the triangulation method suggest that it has limitations for collecting fine-scale movement and habitat use data. Similar concerns have also been expressed by other authors (Stasko and Pincock 1977; White and Garrott 1990; Simpkins and Hubert 1998).

Negating the need for triangulation and any associated mapping avoids the greatest causes of error in location determination (Stasko and Pincock 1977; White and Garrott 1990; Simpkins and Hubert 1998). Restricting the range at which the signal is audible minimizes the distance between the receiver and the transmitter, and hence error in the determination of the transmitter location (Simpkins and Hubert 1998). With any directional tracking there is a potential 'tracking angle error' which causes location error to increase with distance from the receiver. By placing the receiver as close as possible to the transmitter, this distance is not only minimized but cannot be affected by distance variables such

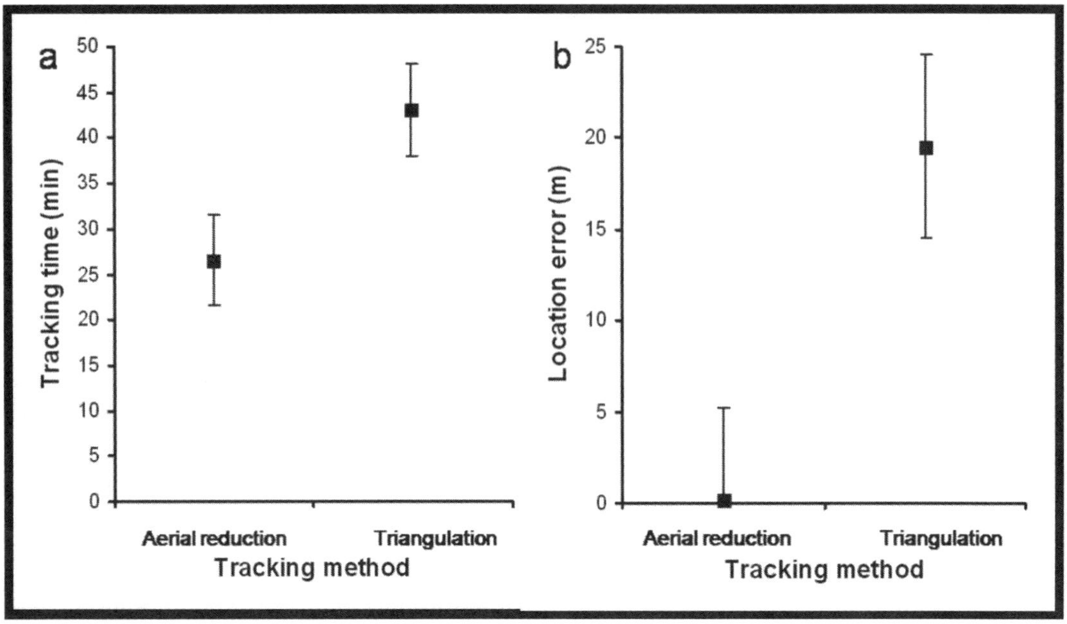

Figure 1. a. tracking time (mean ± SE); **b.** location error (mean ± SE), for aerial reduction and triangulation tracking methods.

as river width or distance from the bank. This makes this boat-based location method even more attractive in large rivers and lakes.

Our approach is simple, utilizes only one receiver, and provides improvements on fine-scale location accuracy over triangulation methods. Using a similar method with coaxial cable as an omni-directional antenna, Niemela et al. (1993) reported a mean error of 2.9 m with a 95% CL defined by a circle of 6.37 m. Bruno et al. (1990) reported accuracy to within 1 m using a similar method in a lake habitat. The findings from this study provide confidence in the data collected in associated studies on the microhabitats of fish determined using this radio-tracking method (Nicol et al. 2007; Koehn 2009a, 2009b) which used a 12 m x 12 m habitat grid with 3 m x 3 m cells (Nicol et al. 2007). The total location error range of <1.5 m indicates that fish locations documented should fall within the confines of a habitat cell.

The antenna reduction method, however, does have limitations as it is most effective when the boat is positioned over the top of the fish, something which is not always feasible. It is worth noting however that this technique was applicable among habitats with substantial amounts of large woody debris, even if the tracker had to leave the boat and climb onto the wood above the transmitter. This was sometimes the case when tracking Murray cod *Maccullochella peelii* in this area (Koehn 2009a, 2009b). Another potential limitation is the possibility of disturbing the fish when approaching by boat causing aberrant behavior. This may be more problematic with shallow or mid-water species that may be frightened when approached, and methods that minimize disturbance may be necessary. Although not quantified, such disturbance has not been considered of concern in a range of studies on a variety of species that have used this tracking method in this area (Nicol et al. 2007; Koehn 2009a, 200b; J. Koehn, S. Nicol, J. O'Connor, J. Lyon, personal observations).

Triangulation, however, is often necessary for shore-based tracking and has the advantage of maintaining a distance between the tagged animal and the tracker, hence limiting disturbance caused by an approaching boat. A modified approach may be to use a combination of methods. For example, the use of an efficient, omni-directional antenna such as a vertical dipole (Beeman et al. 2004), in conjunction with a directional antenna which makes it possible to distinguish the direction of the signal at close range. Broadhurst and Ebner (2007) reported improved location accuracy (1.00 ± 0.21 m) using a loop antenna on a telescopic pole in a small stream compared to triangulation. Telemetry studies on salmon in Alaska and northwestern Canada (Eiler et al. 1992, 2006) also used an adjustable attenuator pad incorporated into the coaxial line between the antenna and receiver to reduce signal reception. This pad was used to vary the amount of signal loss, providing additional control over the amount of signal received, while still utilizing directional antennas.

While directional tracking of high (140–160 MHz) frequency transmitters, is somewhat easier than low (30–50 MHz) frequency transmitters, all signals can become essentially omni-directional depending on the distance from the source. Water depth, signal bounce, and site specific factors (e.g., geomorphology) (Winter 1983), can all potentially cause additional location error. Although our study focused on low (48–49 MHz) frequency transmitters in lowland rivers, the antenna reduction method described here has also been successfully used to extensively radio track salmon and retrieve high frequency transmitters in the large river drainages of Alaska and northwestern Canada (Eiler et al. 1992, 2006).

The success and versatility of the antenna reduction method suggests that this approach can be applicable to a wide range of telemetry studies. Increased tracking efficiency can result in the collection of additional data, the elimination of triangulation and any need for mapping, and the minimization of errors associated with the distance between the transmitter and the receiver. Such improvements contribute to improved accuracy in determining fine-scale transmitter locations. Unfortunately, details of location accuracy and tracking efficiency are rarely reported. We would encourage other researchers to conduct and report on site specific trials and techniques so that measurement error can be assessed and others may benefit from these improved techniques.

Acknowledgments

The authors thank Damien O'Mahony, Justin O'Connor and Pat O'Leary who assisted with the trials and others who assisted the Murray cod radio tracking project. This work was funded by the Murray-Darling Basin Commission and Department of Sustainability and Environment, Victoria. Advice on statistics was provided by Dave Ramsey and valuable comments on manuscript drafts were kindly provided by Matt Jones, Jed MacDonald and the two anonymous reviewers.

References

Brenden, T. O., B. R. Murphy, and E. M. Hallerman. 2004. Estimating total error in locating radiotelemetry transmitters by homing in a riverine environment. Journal of Freshwater Ecology 19:295–304.

Beeman, J. W., C. Grant, and P. V. Haner. 2004. Comparison of three underwater antennas for use in radiotelemetry. North American Journal of Fisheries Management 24:275–281.

Broadhurst, B., and Ebner, B. 2007. An improved technique for small-scale radio-tracking of crayfish and benthic fishes in upland streams. Transactions of the American Fisheries Society 136:423–427.

Bruno, N. A., R. W. Gregory, and H. L. Schramm. 1990. Nest sites used by radio-tagged largemouth bass in Orange Lake, Florida. North American Journal of Fisheries Management 10:80–84.

Eiler, J. H., B. D. Nelson, and R. F. Bradshaw. 1992. Riverine spawning by sockeye salmon in the Taku River, Alaska and British Columbia. Transactions of the American Fisheries Society 121:701–708.

Eiler, J. H., D. J. Alcorn, and M. R. Neuman, editors. 2000. Biotelemetry 15. Proceedings of the 15th international symposium on biotelemetry. 10–15 May 1999. Juneau, Alaska.

Eiler, J. H., T. R. Spencer, J. J. Pella, and M. M. Masuda. 2006. Stock composition, run timing, and movement patterns of Chinook salmon returning to the Yukon River basin in 2004. U.S. Department of Commerce, NOAA Technical Memorandum NMFS-AFSC-165.

Kenward, R. E. 2001. A manual for wildlife radio tagging. Academic Press, London.

Koehn, J. D. 2009a. Using radio telemetry to evaluate the depths inhabited by Murray cod *Maccullochella peelii peelii*. Marine and Freshwater Research 60:317–320.

Koehn, J. D. 2009b. Multi-scale habitat selection by Murray cod (*Maccullochella peelii peelii*) in two lowland rivers. Journal of Fish Biology 75:113–129.

Nicol, S. J., R. J. Barker, J. D. Koehn, and M. A. Burgman. 2007. Structural habitat selection by the critically endangered trout cod, *Maccullochella macquariensis*, Cuvier. Biological Conservation 138:30–37.

Niemela, S. L., J. B. Layzer, and J. A. Gore. 1993. An improved radiotelemetry method for determining use of microhabitats by fishes. Rivers 4:30–35.

Simpkins, D. G., and W. A. Hubert. 1998. A technique for estimating the accuracy of fish locations identified by radiotelemetry. Journal of Freshwater Ecology 13:263–268.

Stasko, A. B., and D. G. Pincock. 1977. Review of underwater biotelemetry, with emphasis on ultrasonic techniques. Journal of the Fisheries Research Board of Canada 34:1261–1285.

White, G. C., and R. A. Garrott. 1990. Analysis of wildlife radio-tracking data. Academic Press, San Diego, California.

Winter, J. D. 1983. Underwater biotelemetry. Pages 371–395 *in* L. A. Nielsen, and D. L. Johnson, editors. Fisheries techniques. American Fisheries Society, Bethesda, Maryland.

Zimmerman, J. W., and R. A. Powell. 1995. Radio-telemetry error: location error method compared with error polygons and confidence ellipses. Canadian Journal of Zoology 73:1123–1133.

Section V

Chemical and Biological Markers

Foreword to Chemical and Biological Markers Section

R. Keller Kopf

Although chemical and biological marking techniques have been in use for over 30 years, recent advances in molecular science and mass spectrometry technology now facilitates their routine application by fisheries scientists and biologists. An exponential rise in peer-reviewed publications using these methods shows the ubiquity with which this technology has been adopted and suggests a productive future (Figure 1). Growth in this area of research has focused largely on otolith chemistry, genetic markers, and mass chemical marking of calcified structures.

Today, researchers use otolith chemistry, DNA sequencing, stable isotope or fatty acid analyses, artificial mass marking, parasite diversity or other unique 'signatures' to answer a variety of questions about the life history of fish and the environments they inhabit. These techniques have been instrumental to answering questions about stock discrimination, population origin, habitat use, dispersal and migration, age and growth, trophic status and ecotoxicology. Manuscripts published in these proceedings use artificial marks and natural signatures to retrace the life history of Eastern Australian salmon *Arripis trutta*, spotted chub mackerel *Scomber australasicus*, European eel *Anguilla anguilla* and golden perch *Macquaria ambigua*.

Because chemical and biological techniques often utilize natural signatures, unlike other tagging methods, this approach typically does not require individual animals to be marked. More traditional tagging approaches are often subject to low tag-recapture rates

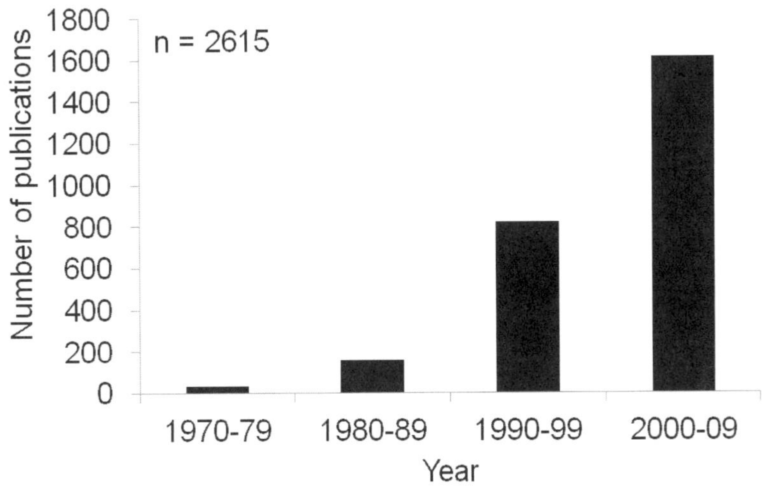

Figure 1. Number of peer-reviewed publications using chemical or biological marking techniques between 1970 and 2009. ProQuest Aquatic Science & Fisheries Abstracts; search terms: Abstract "fish" OR "otolith chemistry" AND "chemical tag/marker OR biological tag/marker."

and are constrained by post-release effects and low sample size. Natural signature tags make use of existing heterogeneity within or among populations and therefore all individuals are effectively marked prior to sampling. Natural markers used in these proceedings include elemental composition of otoliths, parasite diversity, and mtDNA which are used to discriminate stocks and to assess the effectivness of hatchery re-stocking programs (Hughes et al.; Lin et al.; Schmarr et al.). The application of all three techniques to chub mackerel provides a good example of how multiple lines of evidence can, and should, be used to test putative hypotheses (Schmarr et al.). As an alternative to natural signatures, mass chemical marking of hatchery reared fish has great potential to advance our knowledge of fish biology and to evaulatue the success of stocking and environmental restoration programs. Previously hindered by lethal and subjective identification methods, a non-lethal and objective method of identifying artificial flourescent marks in fish is presented by Crook et al. and tested in the field by marking and releasing approximately 20,000 fingerlings.

American Fisheries Society Symposium 76:389–396, 2012

Quantitative Measurement of Calcein Fluorescence for Non-lethal, Field Based Discrimination of Hatchery and Wild Fish

David A. Crook*, Damien J. O'Mahony
Arthur Rylah Institute for Environmental Research, Department of Sustainability and Environment, 123 Brown Street, Heidelberg, 3084, Victoria, Australia

Bronwyn M. Gillanders, Andrew R. Munro[a]
Southern Seas Ecology Laboratories, Darling Building DX 650 418, School of Earth and Environmental Sciences, University of Adelaide, Adelaide, 5005, South Australia, Australia

Andrew C. Sanger
New South Wales Department of Primary Industries 3/556 Macauley Street, Albury, 2640, New South Wales, Australia

Abstract.—Recent studies have described nonlethally detectable techniques for externally marking fish with fluorescent compounds such as calcein. We describe the use of a portable fluorometer for the nonlethal quantitative measurement of fluorescence in calcein marked golden perch *Macquaria ambigua*. The results demonstrate that calcein marked fish could be unambiguously identified more than 2 years after marking when held in the laboratory; the body mass of the fish had increased from 0.8 to 1.3 g (median 1.0 g) to 6–44 g (median 11.1 g) over this period. Similarly, marked fish released into the wild at approximately 1 g were detected at up to 18 months postrelease and 66 g in weight. However, some deterioration in the detectability of the external marks was apparent in the larger fish recaptured from the wild. Based on these findings, we suggest that the technique has the potential to provide a practical and objective means of discriminating hatchery and wild fish under field conditions.

Introduction

Chemical marking of fish for identification purposes has become an important tool in fisheries management and research (Nielsen 1992). Although chemical marks have been used in fisheries research for several decades, a major drawback of commonly used techniques is the requirement to either kill the fish (e.g., otolith marks) or remove a sample of tissue (e.g., fin spine marks) to determine whether or not the fish has been marked (Munro et al. 2008; Taylor et al. 2005). While examination and chemical analysis of otoliths

*Corresponding author: David.Crook@dse.vic.gov.au
[a]Current address: Commercial Fisheries Division, Alaska Department of Fish and Game, 333 Raspberry Road, Anchorage, Alaska 99518, USA

can provide detailed information about fish (e.g., age, migration history, identification of artificial chemical marks [see review by Campana 1999]), it is undesirable to kill fish where nonlethal alternatives exist, particularly when dealing with threatened or recreationally important species. Several recent studies have described techniques for marking fish with externally detectable fluorescent compounds such as calcein (Mohler 1997, 2003; Leips et al. 2001; Frenkel et al. 2002; Negus and Tureson 2004; Stubbing and Moss 2007) and alizarin red S (Bashey 2004; Crook et al. 2007) which can be detected on live fish without the need for destructive sampling.

Fluorescent compounds absorb light of a given wavelength and then emit light at longer wavelengths due to shifts in the energy states of electrons. Many compounds become fluorescent only when combined with a specific target substance: for example, calcium in the bony structures of animals (Pautke et al. 2005). One of the main advantages of using fluorescence as a marking technique is that the excitation energy levels required to generate a fluorescence signal are low and living tissue can be examined with no adverse physiological effects. To date, the detection of fluorescent chemical marks on fish has involved the use of a light source of specific wavelength to excite the fluorophore and barrier filters to remove extraneous wavelengths and allow for visual detection of the fluorescence (Mohler 1997, 2003; Leips et al. 2001; Bashey 2004; Negus and Tureson 2004; Stubbing and Moss 2007). A portable unit for the visual detection of fish marked with calcein (marketed in the USA as SE-MARK) is commercially available (Western Chemical Inc., Ferndale, Washington, USA).

Although visual detection of fluorescent marks has proved effective in several previous studies, a disadvantage is that it relies upon a subjective judgment by the operator as to the presence of a mark. This can be particularly problematic if the tissue under examination exhibits autofluorescence; i.e. wavelengths within the emission spectrum that are emitted in the absence of the fluorescent mark due to excitation of endogenous fluorophores naturally present within the tissue. For example, fish tend to exhibit autofluorescence with the excitation and barrier filters used to detect calcein marks (Bashey 2004) and this has the potential to cause difficulties for the discrimination of marked and unmarked fish, particularly for field staff that may not have experience in identifying fluorescent marks. In addition, visual detection of fluorescent marks is best performed under darkened conditions to reduce incidental light that can obscure the marks. This is often difficult to achieve under field conditions. Here, we describe a technique for the nonlethal quantitative measurement of fluorescence in calcein marked fish that has the potential to provide a practical and objective means of discriminating hatchery and wild fish under field conditions.

Methods

Study species

Golden perch *Macquaria ambigua* is a wholly freshwater member of the Family Percichthyidae that inhabits lowland rivers and floodplain lakes in the inland regions of temperate south-eastern Australia (Harris and Rowland 1996). It is a popular angling species and hatcheries produce approximately two million fingerlings (30–50 mm TL) per annum for stocking to enhance recreational fisheries.

Laboratory study—fish marking

Fifty golden perch fingerlings of approximately 1 g in weight were marked with calcein using the osmotic induction technique described by Crook et al. (2009) at an aquarium facility located in Melbourne, Australia.

A 5% salt solution was prepared by dissolving 100 g of natural salt in 2 L of aquarium water. A 1% solution of calcein was prepared by adding 10 g of calcein powder (2,4-bis-[N,N′-{dicarbo methyl}-aminomethyl] fluorescein) to 1 L of aquarium water. The pH of this solution was then gradually adjusted to 7.0 by adding 1.0 M sodium hydroxide. The fish were placed in a 200-mL plastic container with a fine mesh bottom and immersed in the salt solution for 10 min, rinsed in aquarium water for 5 s, and then immersed in the calcein solution for 10 min. All solutions were aerated during the marking procedure. The fish were held individually for the first 100 d after marking in 5-L containers on a recirculating, bio-filtered water system. After this period, 12 of the marked fish were moved to a 200-L tank and held together until the fluorometer measurements were taken approximately 26 months after marking. Unmarked control fish from the same hatchery as the marked fish were held in a separate 200-L tank until the fluorometer measurements were taken. All fish were fed live blackworms (approximately 0.1 g per fish) 3–5 times per week, water temperature was kept at 19–20°C, and a 12-h light/dark regime was maintained.

Calcein detection

A hand-held GFP-Meter (Opti-Sciences, Inc., Hudson NH USA) was tested for measuring fluorescence in the marked fish. The unit is powered by four AA batteries, weighs 1 kg and is housed in a weather resistant plastic case. The cost was approximately $US 4,940 (purchased January 2007). It is configured with a flexible fiber-optic probe (1 m length, 4 mm diameter) to bring light energy to and from the sample. The end of this probe is held against the sample surface and a reading is recorded after 3 s. The unit uses a solid state LED as a low energy excitation source and a PIN diode detector for detection. Thin film interference filters are incorporated into the optics to ensure signal separation and the source is optically monitored for drift. The unit uses a modulated detection system to minimize the effects of temperature drift, stray light, and photokinetic effects such as fluorescence from external light sources or luminescence. A wide variety of source/filter combinations are available for detection ranges from ultraviolet to infrared depending on the specific filter requirements. For the detection of calcein marked fish, we used a unit equipped with a 465 nm excitation filter and a 530 nm emission filter. The unit can collect data in un-calibrated "raw" signal units or can be normalized to standard blank and signal reference samples. The unit has data logging capability, with a capacity of up to 2,730 sample points, and stored data can be viewed immediately or downloaded to a computer via a serial port.

To examine the distribution of fluorescence and determine the optimal positions on the fish to take measurements with the fluorometer, the marked and unmarked golden perch were examined and photographed in May 2007 using a fluorescence stereomicroscope (model MZ16 F, Leica, Heerbrugg, Switzerland) linked to a digital camera (Infinity Capture 3.5.1, Lumenera Corp., Ottawa, Canada). A GFP3 filter set (Leica, Heerbrugg, Switzerland) was fitted to the microscope to allow detection of the calcein marks (470 nm excitation filter with 40 nm half bandwidth, 525 nm band pass barrier filter with 50 nm half bandwidth). Based on an examination of these photographs and initial trials with the fluorometer, it was decided to take measurements from each fish at the ventro-posterior edge of the preoperculum, as this area had the most consistent high quality fluorescent marks (Figure 1).

Calibration of the fluorometer was conducted by using a flat black surface as the blank and an unmarked dead fish as the reference standard. Unmarked dead fish were noted to have fluorescence values between

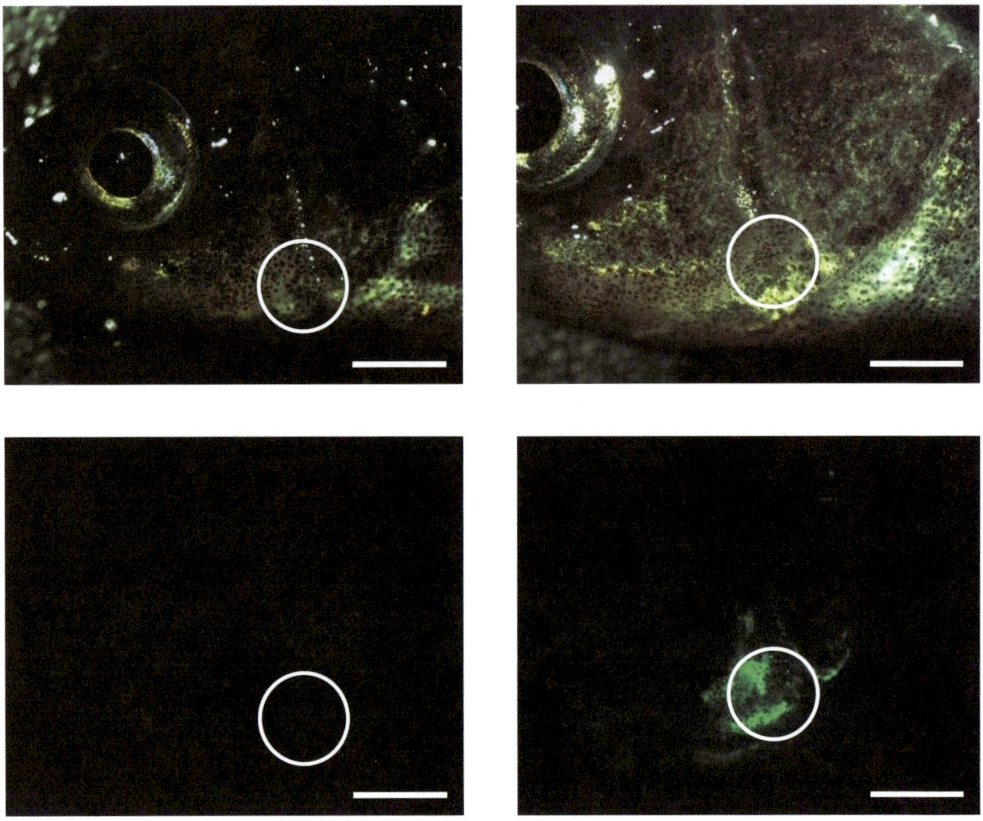

FIGURE 1. Photographs of the heads of a 77 mm TL unmarked golden perch (left) and a 104 mm TL calcein marked golden perch (right) taken 26 months after marking. Photographs of both fish were taken using identical settings on a fluorescence stereomicroscope with unfiltered white light (top) and the GFP3 filter set (bottom) as the light source. The sample:reference fluorescence ratios of the unmarked and marked fish were 0.292 and >1.8 (signal overload) respectively. The circles represent the locations of the fluorescence measurements taken using the GFP hand held fluorometer. The scale bars represent 4 mm length.

those of unmarked and marked live fish and thus were deemed a suitable reference standard for our purposes. An intermediate reference sample was used to provide a dynamic measurement range that allowed for readings both below and above the reference standard. Using this calibration set-up, the blank value was 68 A-D tics (standard measure of fluorescence signal intensity), the reference standard value was 2,695 A-D tics and the system gain was 159 units (system gain ranges from 0 to 255 units). Values in A-D tics were converted to ratios of the sample to reference values.

Measurements from live fish were taken from the 12 calcein marked fish (61–115 mm TL, median 81 mm) and 12 unmarked control fish (54–96 mm, median 65 mm) held in the aquarium, as well as from 52 additional unmarked golden perch collected by electrofishing from two rivers in southern New South Wales, Australia (Billabong Creek, 35.12.926 S, 144.48.251 E, n = 32, range 37–436 mm, median 129 mm; Edward River, 35.25.772 S, 144.45.512 E, n = 20, range 170–429 mm, median 314 mm). The highest of six sample: reference values taken for each fish (three from each side) was used as the final value. Measurements ceased if the unit recorded a "signal overload" (i.e., signal: reference value >1.8).

Field evaluation

To examine mark retention and detectability under field conditions, approximately 20,000 golden perch fingerlings (30–50 mm TL) were marked at the Narrandera Fisheries Centre, Narrandera, New South Wales, Australia, on 17th March 2008 using the osmotic induction technique described above. A calcein concentration of 0.5% was used to mark these fish, however, as use of this concentration produces similar mark quality to 1.0% calcein at a much reduced cost (see Crook et al. 2009). The marked fish were held in 2000 L tanks for one day after marking and were then transported by vehicle to Billabong Creek, NSW, where they were released at a single location known as "Zara" (35.09.224 S, 144.43.952 E).

To recapture marked fish, sampling was conducted using a boat-mounted electrofishing unit (Smith-Root 5 GPP, Smith-Root, Vancouver, Washington, USA) in June and October 2008, and March, July and October 2009. The stocking site at Zara and three additional sites were sampled: Wanganella, Caroonboon and Windouran, located 5 km upstream, 30 km downstream and 60 km downstream of Zara respectively. The reaches sampled ranged from 0.8 to 3 km in length. The GFP fluorometer was used as described above to collect fluorescence measurements from the preoperculums of golden perch below 300 mm TL (encompassing the possible maximum size of marked fish). The fish were subsequently killed and their otoliths removed, mounted in resin and sectioned, and then examined under the fluorescence microscope for calcein marks.

Results

Calcein detection

Using the calibration settings described above, the unmarked fish had sample:reference fluorescence ratios ranging from 0.016 to 0.300 (unmarked hatchery fish, range 0.101–0.298, median 0.212; Billabong Creek fish, range 0.016–0.300, median 0.068; Edward River fish, range 0.019–0.136, median 0.060). In contrast, all 12 of the calcein marked fish had sample: reference fluorescence ratios of >1.8 (i.e., the unit recorded "signal overload"). These results demonstrate that calcein marked fish held in the laboratory could be unambiguously and objectively identified more than 2 years after marking; the body mass of the fish had increased from 0.8 to 1.3 g (median 1 g) to 6–44 g (median 11.1 g) over the 2-year grow-out period.

Field evaluation

A total of 30 fish with calcein marks on their otoliths were recaptured up to 583 d postrelease, including 26 from the release site at Zara, three from Wanganella and one from Caroonboon (Table 1). The recaptured fish ranged from 37 to 157 mm TL (2–66 g wet weight). Of these fish, 28 recorded "signal overloads" (sample:reference >1.8), while the remaining two fish had sample: reference values of >1 (Table 1). Thus, all 30 recaptured fish could be readily discriminated from unmarked fish (see above) based on the fluorometer readings. The two fish that did not register signal overloads were 100 mm TL (10 g) and 157 mm TL (66 g) respectively (Table 1). Visual examination of these fish under the fluorescence microscope showed some evidence of deterioration in mark quality (particularly in the larger individual), with the intensity and area of fluorescence relative to body size reduced compared with smaller fish.

Discussion

This study has shown that external calcein marks can be objectively detected on live

Table 1. Details of calcein marked golden perch recaptured from Billabong Creek. Fish were marked and released as fingerlings of 25–40 mm TL. Sample:reference values of >1.8 represent signal overload readings on the fluorometer. Nonmarked fish previously collected from Billabong Creek had sample:reference values of 0.016–0.300.

Site	Recapture date	Days post-release	TL (mm)	Sample: reference
Zara	19/06/2008	93	37	>1.8
Zara	19/06/2008	93	40	>1.8
Zara	19/06/2008	93	40	>1.8
Zara	19/06/2008	93	41	>1.8
Zara	19/06/2008	93	42	>1.8
Zara	19/06/2008	93	43	>1.8
Zara	19/06/2008	93	45	>1.8
Zara	19/06/2008	93	49	>1.8
Zara	19/06/2008	93	49	>1.8
Zara	10/10/2008	206	46	>1.8
Zara	10/10/2008	206	65	>1.8
Zara	5/03/2009	352	71	>1.8
Zara	5/03/2009	352	72	>1.8
Zara	5/03/2009	352	74	>1.8
Zara	5/03/2009	352	74	>1.8
Zara	5/03/2009	352	75	>1.8
Zara	5/03/2009	352	76	>1.8
Zara	5/03/2009	352	76	>1.8
Zara	5/03/2009	352	77	>1.8
Zara	5/03/2009	352	83	>1.8
Zara	5/03/2009	352	87	>1.8
Zara	5/03/2009	352	100	1.046
Wanganella	1/07/2009	470	83	>1.8
Wanganella	3/07/2009	472	98	>1.8
Wanganella	1/07/2009	470	108	>1.8
Zara	4/08/2009	504	89	>1.8
Zara	4/08/2009	504	96	>1.8
Zara	3/08/2009	503	83	>1.8
Zara	4/08/2009	504	87	>1.8
Caroonboon	22/10/2009	583	157	1.100

golden perch using the GFP meter both in the laboratory and under field conditions for more than 18 months and with more than a 60-fold increase in fish body mass postmarking. Furthermore, the sensitivity of the fluorometer and the large differences in fluorescence values between marked and unmarked fish suggest that discrimination may be possible at larger sizes and over longer periods than examined here. However, some of the largest marked fish recaptured in the field did have reduced fluorescence values, suggesting gradual loss of mark quality. In particular, the reduction in the area of fluorescence relative to the size of the fish made it more difficult to locate the mark when taking fluorescence measurements from the larger fish. A longer term study is required to determine the ultimate longevity of detectable marks in golden perch under field conditions.

Previous studies have shown that detectability of fluorescent marks is highly dependent on the growth rates of the fish and the morphological characteristics of the species (Negus and Tureson 2004; Frenkel et al. 2002). Detailed trials for each species under consideration are therefore required to determine the optimal locations for measurement and appropriate fluorescence values for discriminating between marked and unmarked fish. The longevity of external calcein marks also appears to be particularly influenced by exposure to sunlight, which can cause rapid deterioration of mark intensity (Bashey 2004; Honeyfield et al. 2008). Our field study was conducted in a relatively turbid river (turbidity [NTU]: median = 63, range = 18–260), and golden perch are a mainly benthic species that typically occupy habitats of >1.5 m depth and are strongly associated with woody debris and other types of cover (Crook et al. 2001). Thus, the fish in our study were probably exposed to minimal amounts of sunlight. It is possible that external marks on golden perch would be detectable for shorter amounts of time in less turbid water, although the species' behavior and habitat preferences would still limit the amount of exposure to sunlight. Nonetheless, it is apparent that the habits of particular species and ambient environmental conditions need to be taken into account when assessing the potential longevity of calcein marks.

The unit used in this study was not specifically designed to measure fluorescence in fish, with its primary application the measurement of fluorescence in plant tissues and extracts. It is possible that optimization of this type of technology specifically for measuring fluorescence in fish may further improve the detection of chemical markers. Furthermore, although the unit used here was configured to the excitation and emission spectra of calcein, a wide range of filters are available and it is possible that units could be developed to detect other chemical markers that exhibit fluorescence, such as alizarin red S (Bashey 2004; Crook et al. 2007). Although the quantitative measurement of fluorescent marks on live fish in the field is still in the early stages of development, we believe that with further refinement and development, such an approach has considerable potential for use in fisheries research and management.

Acknowledgments

We gratefully acknowledge Larry Taitelbaum from Opti-Sciences for his assistance and advice regarding the use of the GFP unit. Stephen Thurstan is acknowledged for his advice on fish husbandry and for supplying the fingerlings. Jed Macdonald, Karl Pomorin, Justin O'Mahony and Terry Glenane assisted in the aquarium, and Larry Taitelbaum and Jeremy Hindell provided instructive comments on earlier versions of the manuscript. Financial support for this research was provided by the Murray-Darling Basin Authority (Native Fish Strategy project MD741), the Australian Research Council (LP0348611), Industry and Investment New South Wales Department and the Victorian Department of Sustainability and Environment. The research was carried out under animal ethics permit ARI AEC 07/03.

References

Bashey, F. 2004. A comparison of the suitability of alizarin red S and calcein for inducing a nonlethally detectable mark in juvenile guppies. Transactions of the American Fisheries Society 133:1516–1523.

Campana, S. E. 1999. Chemistry and composition of fish otoliths: pathways, mechanisms and applications. Marine Ecology Progress Series 188:263–297.

Crook, D. A., A. I. Robertson, A. J. King, and P. Humphries. 2001. The influence of spatial scale and habitat arrangement on diel patterns of habitat use by two lowland river fishes. Oecologia 129:525–33.

Crook, D. A., D. J. O'Mahony, B. M. Gillanders, A. R. Munro, and A. C. Sanger. 2007. Production of external fluorescent marks on golden perch fingerlings through osmotic induction marking with alizarin red S. North American Journal of Fisheries Management 27:670–675.

Crook, D. A., D. J. O'Mahony, A. C. Sanger, A. R. Munro, B. M. Gillanders, and S. Thurstan. 2009. Development and evaluation of methods for osmotic induction marking of golden perch (*Macquaria ambigua*) with calcein and alizarin red S. North American Journal of Fisheries Management 29:279–287.

Frenkel, V., G. Kindschi, and Y. Zohar. 2002. Noninvasive, mass marking of fish by immersion in calcein: evaluation of fish size and ultrasound exposure on mark endurance. Aquaculture 214:169–183.

Harris, J. H. and S. J. Rowland 1996. Family Percichthyidae: Australian freshwater cods and basses. Pages 150–163 *in* R. M. McDowall, editor. Freshwater Fishes of South-Eastern Australia. Reed, Chatswood, New South Wales, Australia.

Honeyfield, D. C., T. Kehler, J. W. Fletcher, and J. W. Mohler. 2008. Effect of artificial sunlight on the retention of external calcein marks on lake trout. North American Journal of Fisheries Management 28:1243–1248.

Leips, J. C., T. Baril, F. H. Rodd, D. N. Reznick, F. Bashey, F., G. J. Visser, and J. Travis. 2001. The suitability of calcein to mark poeciliid fish and a new method of detection. Transactions of the American Fisheries Society 130:501–507.

Mohler, J. W. 1997. Immersion of larval Atlantic salmon in calcein solutions to induce a nonlethally detectable mark. North American Journal of Fisheries Management 17:751–756.

Mohler, J. W. 2003. Producing fluorescent marks on Atlantic salmon fin rays and scales with calcein via osmotic induction. North American Journal of Fisheries Management 23:1108–1113.

Munro, A. R., B. M. Gillanders, T. S. Elsdon, D. A. Crook, and A. C. Sanger. 2008. Enriched stable isotope marking of juvenile golden perch *Macquaria ambigua* otoliths. Canadian Journal of Fisheries and Aquatic Sciences 65:276–285.

Negus, M. T., and F. T. Tureson. 2004. Retention and nonlethal external detection of calcein marks in rainbow trout and Chinook salmon. North American Journal of Fisheries Management 24:741–747.

Nielsen, L. A. 1992. Methods of marking fish and shellfish. American Fisheries Society. Bethesda, Maryland.

Pautke, C., S. Vogt, T. Tischer, G. Wexel, H. Deppe, S. Milz, M. Schieker, and A. Kolk. 2005. Polychrome labelling of bone with seven different fluorochromes: enhancing fluorochrome discrimination by spectral image analysis. Bone 37:441–445.

Stubbing, D. N., and R. D. Moss. 2007. Success of calcein marking via osmotic induction in brown trout fry, *Salmo trutta*. Fisheries Management and Ecology 14:231–233.

Taylor, M. D., D. S. Fielder, and I. M. Suthers. 2005. Batch marking of otoliths and fin spines to assess the stock enhancement of *Argyrosomus japonicus*. Journal of Fish Biology 66:1149–1162.

American Fisheries Society Symposium 76:397–417, 2012

Discriminating Stocks of *Scomber australasicus* Using a Holistic Approach: a Pilot Study

David W. Schmarr*
Marine Parasitology Laboratory, School of Earth and Environmental Sciences
The University of Adelaide, Adelaide 5005, Australia
and
South Australian Research and Development Institute (Aquatic Sciences)
P.O. Box 120, Henley Beach 5022, Australia

Ian D. Whittington
Marine Parasitology Laboratory, School of Earth and Environmental Sciences
The University of Adelaide, Adelaide 5005, Australia
and
Monogenean Research Laboratory, Parasitology Section
The South Australian Museum, Adelaide 5000, Australia

Jennifer R. Ovenden
Molecular Fisheries Laboratory, Queensland Department of Primary Industries and Fisheries, Floor 6, North Tower, Queensland Biosciences Precinct
University of Queensland 4072, Australia

Tim M. Ward
South Australian Research and Development Institute (Aquatic Sciences)
P.O. Box 120, Henley Beach 5022, Australia

Abstract.—This study assesses the suitability of genetic approaches, parasitology and otolith microchemistry for determining the stock structure of spotted chub mackerel *Scomber australasicus* in Australasian waters and establishes protocols for using these techniques to determine variability within and among putative stocks. Seventy-five fish from three locations across the geographical distribution of *S. australasicus* in Australian waters (SE Queensland, South Australia and SE Western Australia) and one location in New Zealand were examined. Genetics and parasite assemblage were analyzed for all fish; otolith microchemistry of Australian fish was also examined. Techniques were successfully developed to extract and amplify a segment of the mtDNA control region, and results showed significant genetic heterogeneity among fish from Western Australia, Queensland, and New Zealand. Parasite analysis identified several taxa that are suitable for use as biological tags and enabled discrimination of fish collected from the four locations. Studies of otolith microchemistry using LA-ICP-MS had sufficient power to distinguish fish from the three Australian

*Corresponding author: david.schmarr@sa.gov.au

locations. This study suggests that there are multiple stocks of *S. australasicus* within Australian waters, proposes protocols for future studies of finer scale stock structure, and discusses the efficacy of each technique for stock discrimination.

Introduction

Knowledge of stock structure is critical for the management of pelagic fisheries (Ihssen et al. 1981). Different stocks of pelagic fishes, such as sardine (Gaughan et al. 2002), horse mackerel *Trachurus trachurus* (Abaunza et al. 2003), Atlantic herring *Clupea harengus* (Ruzzante et al. 2006), and Atlantic mackerel *Scomber scombrus* (Villamor et al. 2004), often have distinct biological characteristics (e.g., abundance, fecundity, recruitment, growth rate, mortality) that affect their responses to fishing pressure. Managing multiple stocks on the basis of stock assessments that do not consider stock structure can lead to localized depletion. For example, the Californian sardine fishery in the middle of last century did not account for four separate stocks, resulting in the sequential collapse of stocks from north to south, and eventual collapse of the entire fishery (Radovich 1982). The North Sea herring fishery comprises several seasonally mixing spawning stocks that are managed as one unit. Overfishing in the 1970s led to the collapse of the fishery and in the subsequent recovery, some stocks failed to return to their spawning grounds (Corten 2001).

A range of techniques have been used, with varying success, for stock discrimination of pelagic fishes (Begg and Waldman 1999; Cadrin et al. 2004; Ihssen et al. 1981; Pawson and Jennings 1996). The combination of two or more stock discrimination techniques, preferably for the same samples, has been termed a "holistic approach" (Begg and Waldman 1999) and can increase the likelihood of discerning stock structure. Although the value of using several stock discrimination techniques has been emphasized repeatedly, this approach has been applied to only a few species, including Narrow-barred Spanish mackerel *Scomberomorus commerson* (Moore et al. 2003; Newman et al. 2006; Ovenden and Street 2006) and horse mackerel (Abaunza et al. 2003; Karaiskou et al. 2004; MacKenzie et al. 2004).

In both the *S. commerson* and *T. trachurus* studies, parasite assemblages, genetics and otolith microchemistry were used for stock discrimination. These techniques provide information at varying temporal scales. In doing so, these techniques can be used to discriminate stocks based on the definition of a stock proposed by Ihssen et al. (1981): an intraspecific group of randomly mating individuals with temporal or spatial integrity.

Genetics provides stock structure information at an evolutionary timescale. Genetic spatial heterogeneity indicates group integrity over large time scales, but due to the inherited nature of genetic markers, only a small level of mixing is required to prevent population homogeneity. Analysis of mtDNA provides stock resolution at higher levels of geneflow than other genetic methods. This has enabled genetic approaches for investigating the population structure of many marine fish species (Ovenden 1990) including highly mobile species such as *Scomber* (Nesbo et al. 2000). Changes in the nucleotide sequence of mtDNA accumulate at the population or stock level and are regulated by evolutionary forces, such as genetic drift, migration, mutation and selection (Stepien 1995). The mtDNA 'control' or 'D-loop' region promises to be one of the most useful genetic markers. Properties that make it useful for stock structure studies include high sequence variability, protein noncoding, selectively neutrality and random accumulation of mutations (Stepien 1995). Failure to detect genetic stock structure may be explained by other hypotheses such as recent stock separation or contemporary geneflow. In this instance, alternative stock discrimination techniques may be more effective according to the required outcomes of the stock analysis.

Parasite assemblages can provide information about the past location and movement of the host fish over a timescale ranging from days to years, depending upon the biology and distribution of the parasite species studied. The principle underpinning the use of parasites for stock discrimination is that fish from different regions may harbor different parasite assemblages and analysis of these assemblages can be used to define different stocks (Lester 1990; Lester and MacKenzie 2009; MacKenzie 2002; MacKenzie and Abaunza 1998). The utility of parasite species for stock discrimination is directed by several general criteria, the most desirable of which are: 1) parasites should have significantly different levels of infection in the different locations sampled; 2) parasites should persist in the host for a sufficient period of time; 3) infection levels should be constant over time; 4) parasites should be easily detected, removed and identified; and 5), parasites that affect the behavior of the fish should be avoided, especially if that behavior increases the catchability of the fish (Lester and MacKenzie 2009; MacKenzie and Abaunza 1998).

Otoliths provide a permanent record of individual growth and aspects of the physical and chemical environment encountered by a fish over its entire life. Spatial heterogeneity in otolith microchemistry data can be investigated in relation to life history providing information about the temporal integrity of a stock over the lifetime of the fish. The analysis of chemical composition of otoliths, or otolith microchemistry, is based on three general assumptions about the nature of otoliths: 1) otolith material is not subject to dissolution or resorption once it has been laid down on the otolith surface; 2), the trace element uptake of otoliths reflects the chemical and physical environment (Campana et al. 2000); and 3), otolith composition is stable on removal and storage, and is not contaminated during storage or analysis (Thresher 1999). Variation in water chemistry among locations results in different otolith composition which can be used to distinguish stocks (Campana 1999; Campana and Thorrold 2001).

Spotted chub mackerel *Scomber australasicus* is distributed anti-tropically in the Pacific Ocean, the southeastern Indian Ocean and the Red Sea (Collette 1999). S. australasicus is an epi-pelagic, neritic species that schools by size (Collette and Nauen 1983) and is an important component of pelagic ecosystems off southern Australia. Recent moves to expand the commercial fishery in Australia have been constrained by a justifiably precautionary approach to management of the resource in the absence of information about stock size, distribution and structure.

The stock structure of *S. australasicus* in Australasian waters is uncertain. One study found significant differences between Australia and New Zealand in the morphology of monogenean parasites (Rohde 1987). However, Scoles et al. (1998) found no genetic differences between *S. australasicus* from Australia and New Zealand using mtDNA RFLP analysis and cytochrome *b* sequencing, despite identifying distinct North and South Pacific populations and a discrete Red Sea population. However, the methods used and sample sizes of Scoles et al. (1998) may not have been suitable for distinguishing between these populations.

There is no information on the stock structure of *S. australasicus* in Australian waters. As several oceanographic factors, such as Bass Strait, the East Australia Current, the Flinders Current and the Leeuwin Current, may influence the distribution and movement patterns of *S. australasicus* (Cresswell and Griffin 2004; Godfrey et al. 1980; Middleton and Cirano 2002), there is significant potential for structure within the Australian population. Such structure has significant implications for the management of *S. australasicus* in Australia, and warrants investigation.

This study had the following three aims. (1) To conduct a preliminary assessment of the potential level of stock structure of *S. australasicus* in Australasia. (2) To establish specific protocols for the use of each technique in a detailed future study of the stock structure of *S. australasicus* in Australasia. These protocols will involve recommendations regarding sample sizes based on power analysis (Peterman 1990), which are critical for ensuring that type-two errors are avoided in stock discrimination studies (Mapstone 1995). (3) To assess the efficacy of genetic approaches, parasitology and otolith microchemistry for investigating *S. australasicus* stock structure in Australasia.

Methods

Samples were collected between June and November 2003 at three locations within Australia and one in New Zealand, which were selected to comprise the widest distribution of *S. australasicus* as possible (Figure 1). All fish were frozen immediately then transported to Adelaide, Australia, for processing. In the laboratory, each fish was defrosted, then the length, weight, sex, gonad development and mature gonad weight of each fish was recorded. The gills and viscera were removed for further examination for parasites. Unless otherwise stated, all statistical analyses were performed using SPSS v.13.0.

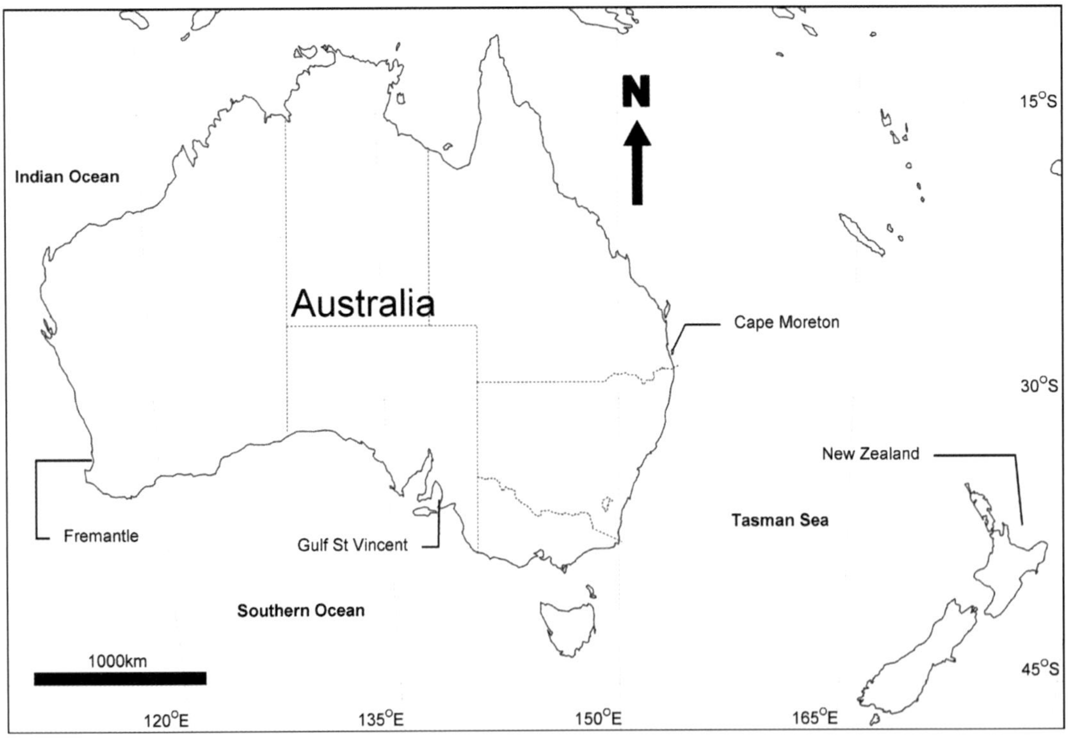

FIGURE 1. Collection locations of Scomber australasicus samples in Australia and New Zealand.

Genetic analysis

Raw DNA was extracted from muscle tissue samples stored in 10% DMSO. Several methods of extraction were used, but the most practical was a modified salting out method (Miller et al. 1988). For Australian fish, a segment of mtDNA was amplified using a PCR protocol (Ovenden et al. 2002) and primers Pro889U20 (CCW CTA ACT CCC AAA GCT AG) and TDKD1291L21 (CCT GAA ATA GGA ACC AAA TGC) designed to amplify approximately 375 bp of the left domain of *Scomberomorus commerson* mtDNA control region (Ovenden and Street 2006). The PCR products were purified then sequenced using forward and reverse primers. Sequences were obtained with an ABI automated sequencer using the chain-termination method with BigDye terminators. New Zealand fish were extracted using a Gentra Puregene DNA purification protocol. The PCR conditions were: 50–100 ng of DNA, 0.1 μM of each primer, 200 μM of each dNTP, Finnzymes Dynazyme buffer equivalent to 1.5 μM MgC12, and 0.5U of DyNAzyme EXT DNA Polymerase in a reaction volume of 25 μL. PCR cycling conditions on an Eppendorf Thermal Cycler were: 94°C for 2 min followed by 35 cycles of 45 s at 94°C, 45 s at 48° and 1 min at 72°C with a final extension of 72°C for 6 min. PCR products were purified then sequenced using the forward primer.

Sequences were edited using Bioedit v5.0.9 (Hall 1999), aligned using Clustal X v1.83 (Thompson et al. 1997) and compared by analysis of molecular variance (AMOVA) using Arlequin v2.0 (Schneider et al. 2000). To determine the sample size required for a future stock discrimination study, the binomial sampling equation was calculated with 95% confidence using the minimum observed haplotype frequency (Bartley et al. 1995).

Parasite analysis

Each fish was examined thoroughly for parasites. The skin, gills, and oral surfaces were examined for ectoparasites under a dissecting microscope. The stomach, pyloric caeca, and intestine were opened longitudinally and washed with saline solution. Parasites were removed, sorted under a dissecting microscope to the lowest taxonomic group possible and by their location on the fish, and preserved according to the conventions of Pritchard and Kruse (1982). Parasite material collected in this study is deposited in the Australian Helminth Collection and the Marine Invertebrate Collection at the South Australian Museum. The measures of parasite infection used in this study were prevalence (the number of fish in each sample infected by each parasite species) and abundance (the average number of parasites per fish in each sample) as defined by Bush et al. (1997). Conventional multivariate analyses usually require normally distributed data, but the distribution of abundance data were nonnormal in this study. Typically, there were two common features of parasite distribution: a large component in the zero category and a negative binomial distribution. To bring the distribution of parasite abundance closer to normality, it was transformed by the natural logarithm plus one. Due to instances of parasites being absent from all fish in some areas, an insignificant random number (between –0.005 and 0.005) was added to each count enabling matrix inversion in discriminant function analyses (DFA) (Lester et al. 1985). The results of the DFA were displayed graphically as plots of the first versus second canonical discriminant functions. Confidence limits (95%) were represented by circles around the group centroids with radius $\sqrt{5.99/n}$, where 5.99 is the critical value in a χ^2 test where $\alpha = 0.05$, df = 2 and n = sample size (Mardia et al. 1979). Classification rates of the discriminant functions were calculated based on group classification. Multivariate analyses of prevalence data were performed using hierarchical cluster analyses displayed graphically as dendrograms, where the distance between clusters was measured by Euclidean distance and clusters were linked using Ward's method. Estimates of power were calculated for univariate and multivariate data using multivariate analysis of variance (ANOVA) (MANOVA).

Otolith microchemistry

The sagittal otoliths were dissected with adhering tissue removed. They were rinsed in Milli-Q water and placed in a sterile plastic multi-well plate. Otoliths were embedded in Struers Epofix resin and transversely sectioned through the nucleus of the otolith to a thickness of 1–1.5 mm using a Buehler Isomet low-speed diamond saw. Otolith sections were hand-polished with 10 and 3 μm lapping film lubricated with Milli-Q water. Polished otolith sections were mounted on glass slides with Crystalbond Thermoplastic adhesive (28 per slide in two rows of 14 sections). Slide-mounted otoliths were ultrasonicated in Milli-Q water for 5 min. These slides were air-dried and stored in clean individual zip-lock bags.

The laser ablation ICP-MS (LA-ICP-MS) system was used as described by McCulloch et al. (2005). To remove potential surface contaminants, a transect along each otolith was preablated by laser using a 180 μm diameter spot. Several otoliths were spot-analyzed to screen for detectable elements. These elements were manganese (Mn), magnesium (Mg), rubidium (Rb), strontium (Sr), barium (Ba) and lead (Pb). Calcium (Ca) was analyzed for use as an internal standard in a ratio to other elements.

Otoliths were analyzed in series of seven at a time. Two reference standards (a powdered coral standard and NIST 614 (National Institute of Standards and Technology, Gaithersburg, Maryland, USA)) were measured before and after every seven otoliths to correct for instrument drift. Each otolith was ablated along a continuous transect from ventral to dorsal edge using a 22 μm diameter laser spot pulsing at 50 Hz and a scan rate of 0.6 $mm{\cdot}min^{-1}$. Between each otolith ablation, the system was allowed to return to background levels. Each otolith transect was standardized by subtracting background readings and correcting against the NIST 614 reference standard. The readings were then transformed into a ratio to calcium and smoothed using a 10 point running median and an 11 point running mean to remove spiked outputs. Readings from each of the elements analyzed were averaged from 100 μm at the edge of the otolith and from 100 μm across the core of the otolith. These two measures were used to compare fish from different regions. Sample statistics were analyzed using univariate and multivariate data, from which power was calculated. Multivariate data were analyzed using the same statistical methods as the parasite abundance data. Discriminant functions were plotted with 95% confidence intervals and classification rates were calculated. In order to determine whether juvenile fish had developed in different water bodies, elemental composition at the core of the otolith was analyzed by hierarchical cluster analysis and presented as a dendrogram, where the distance between clusters was measured by Euclidean distance and clusters were linked using Ward's method.

Results

A total of 75 fish were captured for this study. Twenty fish from Cape Moreton in SE Queensland (Qld) (mean FL ± SE = 207 ± 4 mm) and 19 fish from Gulf St. Vincent in South Australia (SA) (mean FL ± SE = 239 ± 6 mm) were line caught, while 23 fish from

Fremantle in Western Australia (WA) (mean FL ± SE = 241 ± 10 mm) and 13 fish from the north of New Zealand (NZ) (mean FL ± SE = 403 ± 6 mm) were captured by purse seine.

Genetic analysis

A 345 bp region at the 5' end of the mtDNA D-loop was examined. Complete high quality sequences were obtained from 14 Qld fish, 23 WA fish and 13 NZ fish. DNA from the SA sample was degraded and unsuitable for further analyses. Queensland fish had 11 different haplotypes with 27 polymorphic sites, a haplotype diversity of 0.956 ± 0.045 and nucleotide diversity of 0.029 ± 0.016. Western Australian fish had 21 different haplotypes with 42 polymorphic sites, haplotype diversity of 0.992 ± 0.015 and nucleotide diversity of 0.028 ± 0.015. New Zealand fish had 10 different haplotypes with 34 polymorphic sites, haplotype diversity of 0.961 ± 0.041 and nucleotide diversity of 0.027 ±0.015. Each haplotype has been lodged on GenBank (accession numbers JN827256-JN827305; http://www.ncbi.nlm.nih.gov/). A comparison of sequence data using AMOVA showed a significant difference between these populations (F_{ST} = 0.0898, p = 0.00684). Pairwise differences showed that the WA population was significantly different from both the Qld and NZ populations (F_{ST} = 0.0915, p = 0.0198 and F_{ST} = 0.1389, p = 0.003), but the Qld population was not significantly different from the NZ population (F_{ST} = –0.0063, p = 0.3722). The least common haplotypes had a frequency of 2.7%. These preliminary F_{ST} estimates were based upon the samples available and may have reflected stochasticity or unequal sample size. Binomial sampling theory (Bartley et al. 1995) was used to estimate the sample size required to draw each unique haplotype from the entire population with 95% confidence. This was calculated as a total sample size of at least 55 specimens.

Parasite analysis

Thirty-one parasite species were identified: 15 digenean, 2 nematode, 2 acanthocephalan, 4 cestode, 4 monogenean, 2 isopod and 2 copepod species (Table 1). The prevalence of parasites ranged from 1 to 96% (Table 1). Of these parasites, 16 species were prevalent enough (>10%) for comparison of populations (Bush et al. 1990). The first two transformed parasite abundance canonical discriminant functions allowed 97.5% of the samples to be assigned to their capture location (Table 2). The overall correct assignment rate was 95.3%, with 100, 90, 84.2 and 100% of fish from NZ, Qld, SA, and WA correctly classified respectively. The plot of the first two canonical discriminant functions shows the large extent of differentiation between each location (Figure 2). Fish from NZ were strongly differentiated from Australian fish by the first discriminant function. Australian fish were differentiated along the second discriminant function.

Hierarchical cluster analyses of parasite prevalence data (Figure 3) showed three distinct clusters: one containing only NZ fish (1), another containing 19 Qld fish with two SA fish (2), and the last containing one Qld, 17 SA and 23 WA fish (3A and 3B). The last cluster was broken down into two further clusters: one small cluster containing one Qld, eight SA and four WA fish (3A); and another cluster containing 19 WA fish and nine SA fish (3B). Cluster analyses used parasite prevalence data, which is based on the presence or absence of a parasite species on each fish. This approach eliminates the affect of environmental or stochastic effects that may bias abundance or infection intensity measures. Cluster analysis indicates that there may be a large amount of exchange between WA and SA populations, a small amount of exchange between Qld and SA fish and a very small exchange of fish between WA and Qld fish. It also shows that there is no exchange of fish between Australia and NZ.

TABLE 1. Prevalence (P) and abundance (A) of parasites from *Scomber australasicus* sampled at four locations in 2003.

Parasite group	Parasite species	Total (n = 75) P (%)	A (± S.E.)
Isopoda			
	Ceratothoa imbricata	12	0.37 ± 0.13
	Nerocila macleayi	1	0.01 ± 0.01
Trematoda			
	Lecithocladium sp.	77	21.16 ± 3.56
	Hemiurid sp.	13	0.63 ± 0.28
	Dinuris sp.	1	0.57 ± 0.57
	Opechona sp.1	32	18.52 ± 6.39
	Opechona sp.2	3	0.49 ± 0.48
	Allocreadid sp.1	59	14.27 ± 3.74
	Allocreadid sp.2	16	0.33 ± 0.11
	Allocreadid sp.3	1	0.08 ± 0.08
	Nematobothrid type 1	8	0.61 ± 0.46
	Nematobothrid type 2	17	1.31 ± 0.48
	Nematobothrid type 3	29	2.75 ± 1.13
	Nematobothrid type 4	29	2.91 ± 0.73
	Didymocystis sp.1	5	0.16 ± 0.08
	Didymocystis sp.2	1	0.03 ± 0.03
	Neodiplotrema sp.	15	0.25 ± 0.08
Copepoda			
	Clavellisa scombri	19	0.27 ± 0.09
	Caligus sp.1	4	0.04 ± 0.02
Monogenea			
	Kuhnia scombri	31	1.97 ± 0.64
	Kuhnia scombercolias	9	0.15 ± 0.06
	Pseudokuhnia minor	15	0.77 ± 0.33
	Grubea australis	4	0.05 ± 0.03
Cestoda			
	Trypanorhynch metacestode type 1	5	0.09 ± 0.05
	Trypanorhynch metacestode type 2	1	0.03 ± 0.03
	Tetraphyllid metacestode type 1	1	0.01 ± 0.01
	Cestode scolex polymorphus type 1	3	0.49 ± 0.41
Acanthocephala			
	Rhadinorhynchus sp.1	15	0.25 ± 0.10
	Rhadinorhynchus sp.2	16	0.69 ± 0.21
Nematoda			
	Ascarid nematodes[1]	96	20.60 ± 3.36

[1] *Hysterothylacium* sp. and *Contracaecum* sp.

TABLE 2. Results of discriminant function analysis of S. australasicus samples from Australia and New Zealand with standardized coefficients of the first and second canonical discriminant functions for transformed parasite abundances. Parasite species with prevalence below 10% were excluded from these analyses.

	Function 1	Function 2
Eigenvalue	16.575	4.511
% of Variance	76.6	20.9
Canonical Correlation	0.971	0.905
Parasite species		
Ceratothoa imbricata	–0.149	–0.163
Lecithocladium sp.	–0.538	–0.559
Opechona sp. 1	–0.090	–0.358
Allocreadid sp. 1	–0.061	0.129
Hemiurid sp.	–0.008	0.297
Allocreadid sp. 2	0.111	0.312
Nematobothrid type 2	0.011	0.226
Nematobothrid type 3	–0.085	0.143
Nematobothrid type 4	0.014	0.693
Neodiplotrema sp.	0.582	–0.331
Clavellisa scombri	0.125	0.051
Kuhnia scombri	–0.049	–0.411
Kuhnia scombercolias	0.213	0.094
Pseudokuhnia minor	0.000	0.349
Rhadinorhynchus sp. 1	0.004	0.049
Rhadinorhynchus sp. 2	0.832	–0.293
Ascarid nematodes	–0.310	–0.172

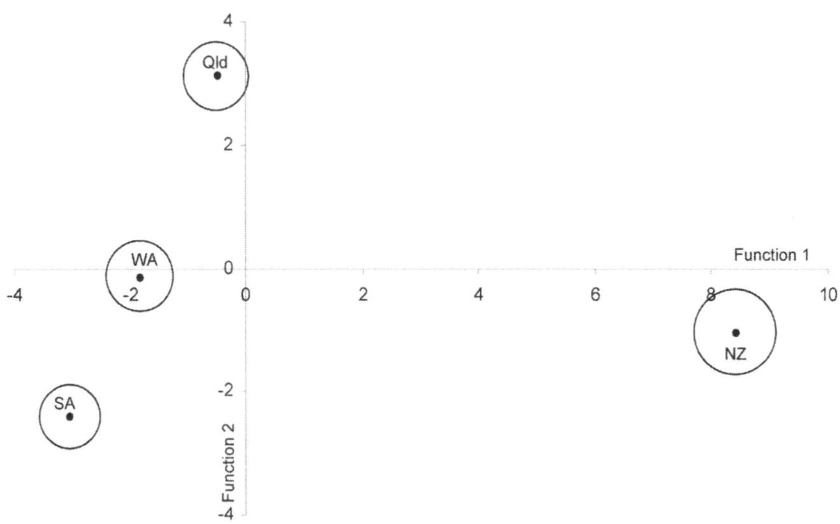

FIGURE 2. Results of DFA of parasite abundance plotting the first two functions at group centroids with 95% confidence rings.

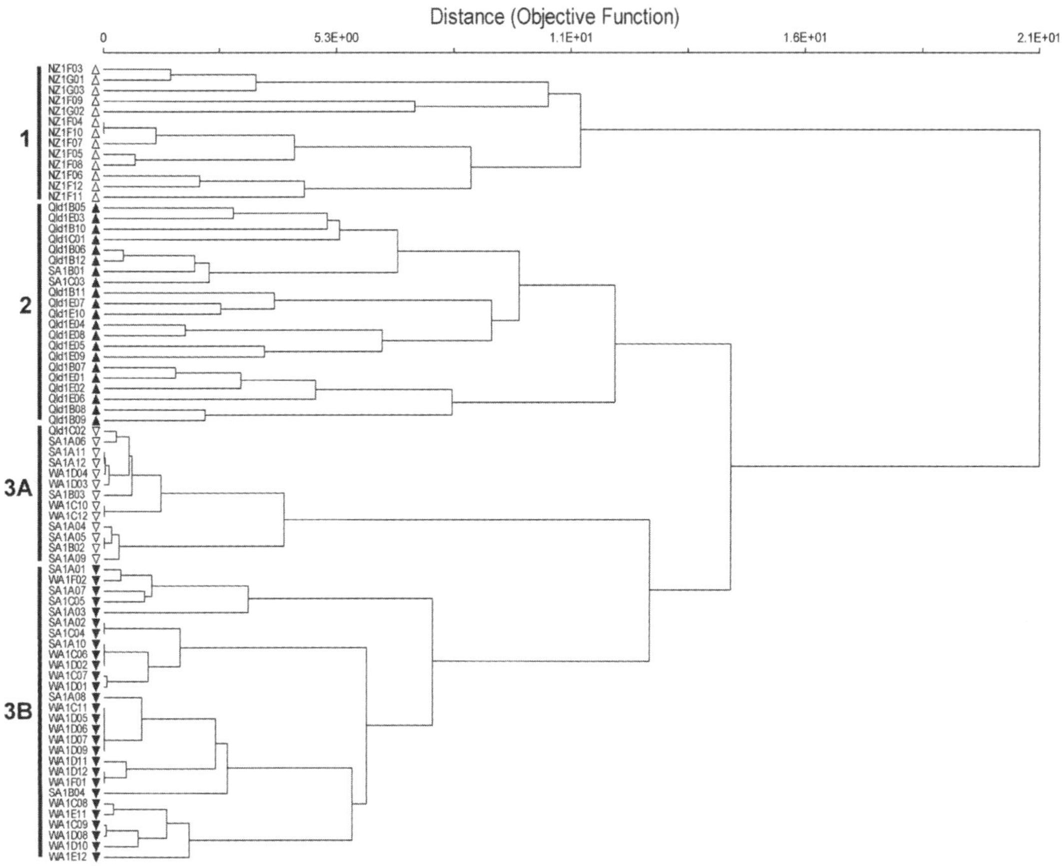

FIGURE 3. Hierarchical cluster analysis dendrogram using parasite prevalence data. Triangles indicate clusters assigned to groups. Labels indicate fish identity and location.

Results from MANOVA showed a statistically significant difference between populations of fish collected across Australasia (Wilk's $\lambda = 0.024$, $F = 7.848$, df = 51, $P < 0.001$), and a dataset with a high degree of power (>0.99 at $\alpha = 0.05$). Univariate data confirmed these results with a majority of the parasites having a high degree of power to discriminate populations (Table 3). Results of power analyses indicate that sample sizes in this study are sufficient for further stock structure study.

Otolith microchemistry

Otoliths from 19 Qld fish, 18 SA fish and 23 WA fish were sampled for the elements Manganese (Mn), Calcium (Ca), Magnesium (Mg), Rubidium (Rb), Strontium (Sr), Barium (Ba) and Lead (Pb).

The first and second canonical discriminant functions of element concentrations from the core of the otolith contributed 100% to the classification of samples (Table 4). The overall correct classification rate was 68.3%, with 63.2, 77.8 and 65.2% of fish from Qld, SA, and WA correctly classified. The plot of the two canonical discriminant functions from the otolith core shows that each location is significantly differentiated from the other locations (Figure 4). SA fish were discriminated from Qld fish along the second discriminant function, with Mg and Mn contributing most to this result (Table 4). The first

TABLE 3. Results of univariate analyses of parasite abundance from *S. australasicus* collected from Australian and New Zealand waters with observed statistical power at $\alpha = 0.05$.

Parasite species	*F*	df	*P*	Observed Power
Ceratothoa imbricata	10.347	3	<0.001	0.998
Lecithocladium sp.	66.401	3	<0.001	>0.999
Opechona sp. 1	3.038	3	0.035	0.690
Allocreadid sp. 1	0.195	3	0.900	0.085
Hemiurid sp.	3.986	3	0.011	0.816
Allocreadid sp. 2	9.294	3	<0.001	0.995
Nematobothrid type 2	4.567	3	0.006	0.869
Nematobothrid type 3	3.475	3	0.020	0.754
Nematobothrid type 4	24.137	3	<0.001	>0.999
Neodiplotrema sp.	13.399	3	<0.001	>0.999
Clavellisa scombri	0.751	3	0.525	0.203
Kuhnia scombri	4.202	3	0.009	0.837
Kuhnia scombercolias	8.511	3	<0.001	0.992
Pseudokuhnia minor	3.634	3	0.017	0.775
Rhadinorhynchus sp. 1	6.179	3	0.001	0.954
Rhadinorhynchus sp. 2	21.058	3	<0.001	>0.999
Ascarid nematodes	5.724	3	0.001	0.938

TABLE 4. Results of discriminant function analysis of *S. australasicus* samples from WA, SA, and Qld with standardized coefficients of the first and second canonical discriminant functions for trace element concentrations at the core, edge and core+edge of the otolith.

	Core		Edge	
	Function 1	Function 2	Function 1	Function 2
Eigenvalue	0.361	0.267	0.992	0.117
% of Variance	57.5	42.5	89.5	10.5
Canonical Correlation	0.515	0.459	0.705	0.324
Element				
Mg	0.368	0.700	0.375	–0.387
Mn	–0.571	0.730	–0.807	0.373
Rb	0.077	0.322	0.268	0.653
Sr	0.407	–0.342	–0.267	0.427
Ba	0.699	0.261	0.897	0.127
Pb	0.147	–0.600	0.024	–0.504

and second canonical discriminant functions of element concentrations from the edge of the otolith contributed 100% to the classification of samples (Table 4). The overall correct classification rate was 76.7%, with 73.7, 77.8 and 78.3% of fish from Qld, SA and WA correctly classified. The plot of the two canonical discriminant functions from the otolith edge shows significant differentiation of WA fish from SA and Qld fish, but does shows some overlap in confidence intervals for SA and Qld fish (Figure 5). WA fish were discriminated from SA and Qld fish along the first discriminant function, with Ba contributing most to this result (Table 4).

A similar pattern is illustrated by hierarchical cluster analyses (Figure 6). Based on otolith-core element concentration, most fish were assigned to their capture location, a small number of fish were assigned to their closest neighboring sample region, but none was assigned to their farthest neighboring sample region. Cluster analysis and DFA of otolith-core element concentration indicate that most fish originated in the same location as the fish that they were captured with, but a small number possibly originated in other locations. Results from otolith-edge element concentration show that more fish were assigned to their capture location, although

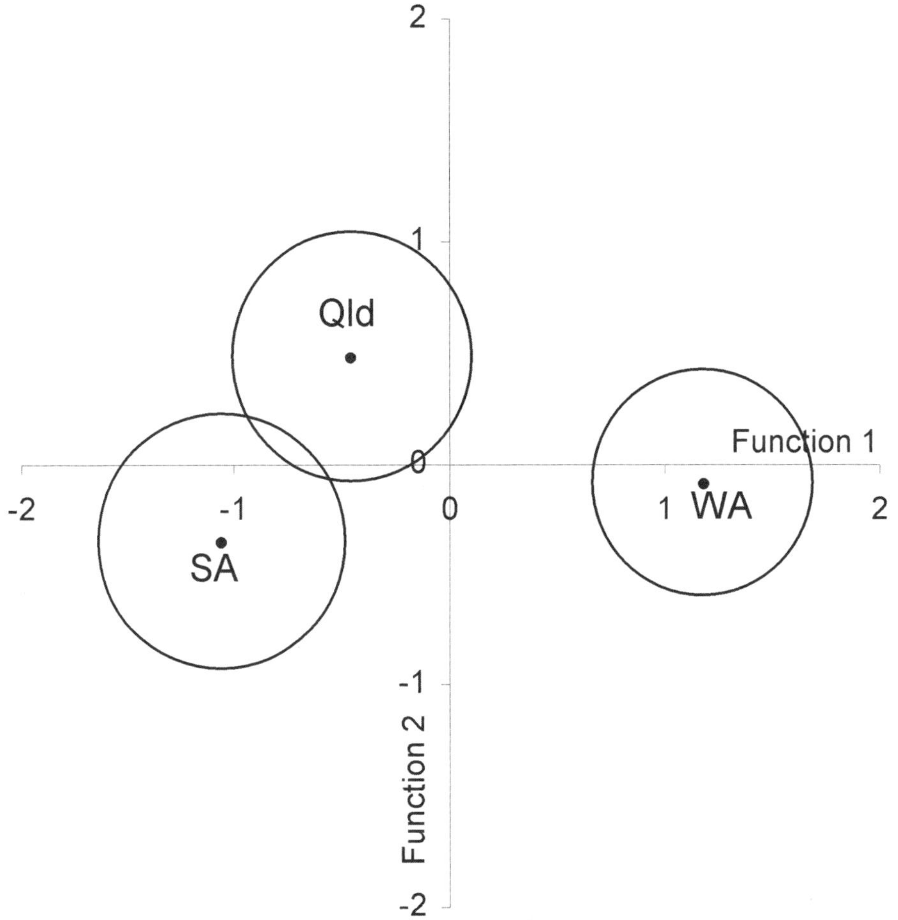

Figure 5. Results of DFA of elemental composition from the edge of the otolith plotting the two functions at group centroids with 95% confidence rings.

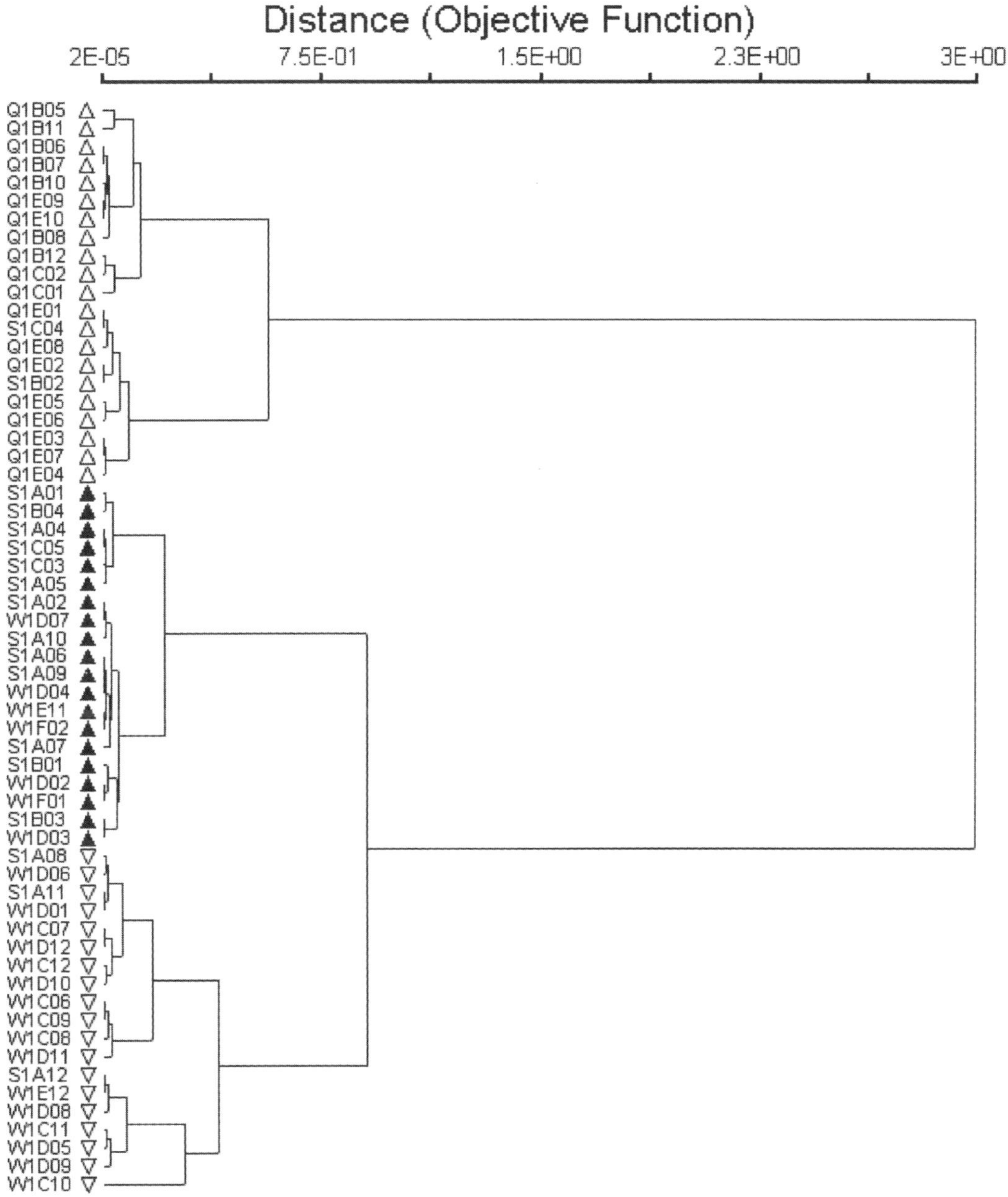

Figure 6. Hierarchical cluster analysis dendrogram using elemental composition from the core of the otolith. Triangles indicate clusters assigned to groups. Labels indicate fish identity and location (Q – Queensland, S – South Australia, W – Western Australia).

with slightly less confidence than otolith-core element concentration.

Results from MANOVA on otolith edge and core element concentrations show that there is a statistically significant difference between populations of fish collected across Australia (Edge: Wilk's $\lambda = 0.449$, $F = 4.264$, df = 12, $P < 0.001$. Core: Wilk's $\lambda = 0.58$, $F = 2.717$, df = 12, $P = 0.003$), and both datasets had a high degree of power (>0.99 and 0.975 at $\alpha = 0.05$). Univariate data confirmed the importance of Ba and Mn, which were the only elements showing high (>0.8) statistical power (Table 5). The three elements that contributed most towards the DFA were Ba, Mn and Mg. For Mg to have high statistical power in future stock structure studies would require a sample size of at least 40 fish per location.

Discussion

Each of three stock discrimination techniques used in this study detected significant variation between sample locations. This heterogeneity warrants further fine-scale stock structure analyses. Protocols were established in this study that with small modifications, will be appropriate for further fine-scale stock structure studies using the sample sizes and methods discussed below. The efficacy of each technique is assessed according to three factors: (1) The technique's capability to discriminate stocks; (2) The technique's practicality in terms of sample size and design, sample processing speed and certainty, and availability of equipment and materials; and (3), The type of information that each technique provides and whether that information is meaningful for testing stock structure within the framework of the stock concept. Each of the aims is addressed below in relation to stock discrimination method.

Genetic analysis

This study has identified at least two genetically different populations of *S. australasicus* in Australasian waters. There appears to be a separation between western (WA) and eastern populations (Qld and NZ), but no genetic subdivision within fish from eastern populations. Scoles et al. (1998) found distinct North and South Pacific populations and

Table 5. Results of univariate analyses of element concentration at the core and edge of otoliths from S. australasicus collected from WA, SA, and Qld waters with observed statistical power at $\alpha = 0.05$.

Element		*F*	df	*P*	Observed Power
Edge	Mg	1.040	2	0.360	0.223
	Mn	5.471	2	0.007	0.830
	Rb	2.464	2	0.094	0.476
	Sr	0.346	2	0.709	0.103
	Ba	12.178	2	>0.001	0.994
	Pb	0.211	2	0.810	0.081
Core	Mg	3.704	2	0.031	0.657
	Mn	2.832	2	0.067	0.535
	Rb	0.314	2	0.732	0.098
	Sr	1.148	2	0.324	0.243
	Ba	5.114	2	0.009	0.803
	Pb	1.732	2	0.186	0.349

a discrete Red Sea population, but no difference between Australian and New Zealand populations. Their study used mtDNA RFLP analysis and cytochrome *b* sequencing. Both markers are less sensitive to differences between populations than control region sequences due to their slower rate of change. Additionally, their sample sizes may have been insufficient to detect differences. In contrast, Nesbo et al. (2000) could differentiate multiple spawning stocks in eastern Atlantic mackerel *S. scombrus* using control region sequences, while cytochrome *b* sequences only detected transatlantic differentiation. The haplotype diversity and nucleotide diversity in the present study were consistent with those found in a stock structure study of Atlantic *S. scombrus* and Pacific chub mackerel *S. japonicus* using mtDNA control region sequences (Zardoya et al. 2004), while the FST was higher and significant in this study.

Reliable high quality mtDNA control region sequences were obtained using the protocols developed for this study. These sequences were highly variable. Despite the fact that the FST for this *S. australasicus* population was above average for marine species (Waples 1998), and differentiated some populations, it is highly unlikely that the sample is representative of the genetic diversity within the entire population. Conclusive evidence of stock composition was hindered by small and unequal sample sizes and the relatively small amount of information contained in the short segment of mtDNA that was sequenced. As a result, the genetic stock structure of *S. australasicus* in Australian waters is still uncertain. Further studies should aim to increase the sample size, sample more locations, increase the length of the sequence that is amplified, and importantly, replicate samples over time (Waples 1998).

The efficacy of genetic analysis for investigating *S. australasicus* stock structure was confirmed in this study. The genetic marker employed here was able to discriminate stocks of *S. australasicus*. The sample sizes used in this study were sufficient to detect heterogeneity in the stock, and subsequent stock structure analyses will require only small increases in sample size. Compared with other techniques, genetic analyses required time-consuming and complex laboratory techniques and data processing. However, this was mitigated by the amount of information provided by molecular sequence data. Genetic stock structure indicates restricted gene flow between populations over an evolutionary timescale. There is likely to be very little interchange between stocks, and consequently if overfishing depleted one stock, other fish in the population would not replenish it. Due to the sensitivity of genetic heterogeneity to small amounts of gene-flow, this technique should be considered the most basic measure of stock structure and the minimum unit for spatial management.

Parasite analysis

Parasites from Australian and New Zealand *S. australasicus* are suitable for use in discriminating stocks. Using multivariate analyses, both parasite abundance and prevalence accurately predicted fish capture location. The parasites found in this study compare with those found in three previous studies of parasite assemblages of *S. australasicus*. Korotaeva (1974) found 18 species of parasite in a sample of 54 S. australasicus, 11 of which are likely to be in common with the present study. In a study of 15 *S. australasicus* ectoparasites, Perera (1993) found no evidence of seasonality in parasite infection intensities and very little evidence for a correlation between infection intensity and fish size. It is likely that at least 10 of those species were found in the present study. However, several parasites were only identified to genus or family. Perera (1993) identified four or five types of didymozoid, which based on her descriptions, are the same as the nematobrothrid types 1–4 found in this study. Due to the difficulty in removing three of these species from the head, fins and mouth, only those infecting the gill should be considered for further studies. Using the same dataset, Hayward et al. (1998) concluded that the parasite assemblage for each fish is more likely to be influenced by extrinsic factors such as capture location, environmental conditions and intermediate host availability, rather than the influence of other parasites in the host population or from other fish. This confirmed the findings of a study of over 100 species of fish, including *S. japonicus*, in which Rohde et al. (1995) found that parasite assemblages were influenced less by school behavior and host size, and more by temperature. This is important because it suggests that parasite assemblage is a signature related to the waters the fish has inhabited in the past and may be used to discriminate fish stock.

There have been several studies of *Scomber* spp. stock structure using parasite data, although most work has been on *S. japonicus* and *S. scombrus*. A study by Rohde (1987) found significant differences between Australian and New Zealand *S. australasicus* on the basis of differences in sclerite size and shape in the monogeneans *Kuhnia* spp. and *Pseudokuhnia* sp.. However, the stock structure of *S. australasicus* within Australian waters was not examined. Both of these parasite species were observed in the present study. Pozdnyakov and Vasilenko (1994) found that parasitological data complemented their discrimination of two stocks of S. japonicus in the northwest Pacific using reproductive parameters. Cremonte and Sardella (1997) were able to discriminate populations of *S. japonicus* in the Argentine Sea using multivariate analysis of parasite data. MacKenzie (1990) used trypanorhynch cestode plerocerci effectively as biological tags for Atlantic mackerel *S. scombrus* in the northeast Atlantic. Unfortunately, cestodes were rare parasites in the present study and are unlikely to be of any further use in this region. The success of the present study and previous studies using similar methods for multivariate analysis of parasite data may warrant further application of this technique. Future studies should have similar sample sizes to those used here, but cover a greater number of sample locations and, as with genetic analyses, replicate samples over time. Future studies should also focus on those parasite groups and habitats found to be informative in this study.

Parasite analyses proved to be very powerful at discriminating populations of *S. australasicus*. This technique requires standard laboratory equipment and parasite identification keys, however sorting and identifying parasites is a time consuming and labor intensive process. Sample sizes for future stock structure studies will not need to be much larger than those used in this study. Parasite assemblage data may indicate previous movement of fish through the area where parasites are endemic. Alternatively, in incidences where the parasite has a direct life cycle, the stock may harbor a population of the parasite through direct transmission between fish. Therefore, parasite assemblage data can provide information about stock identity and movement over the lifespan of the fish. However, to determine the origin of stocks, future studies should only use permanent parasites, that is those that were acquired by juvenile fish (Lester and MacKenzie 2009).

Otolith microchemistry

Otolith microchemistry is a feasible technique for stock discrimination of *S. australasicus*. Multivariate analyses confirmed that broadly separated populations could be discriminated, although there was some overlap in the classification of fish. Compared with genetic and parasite analyses, there are few studies on stock discrimination of small pelagic species using otolith microchemistry, and no studies of *Scomber*. Several studies have investigated stock structure of pelagic species, although most relate to different sampling instruments or they are based on larger pelagic species. For instance, Edmonds and Fletcher (1997) used stable isotope ratios to discriminate stocks of Pacific sardine *Sardinops sagax* in southwestern Australian waters. Proctor et al. (1995) studied the stock structure of the southern bluefin tuna *Thunnus maccoyii* using probe microanalysis of otolith composition, and attributed their findings of no population differentiation to homogeneity of the pelagic environment and low instrument sensitivity.

Examination of ontogenetic variation in otolith elemental composition is one of the most powerful uses of otolith microchemistry (Campana and Thorrold 2001). Analysis of the individual chemical life history was not possible in this study because of poor ontogenetic resolution. However, it was possible to sample areas of the otolith attributed to the juvenile and most recent periods in the fish's life. This enabled to test whether fish that were captured at the same location had the same juvenile elemental signature and possibly the same origin, and whether the elemental signature for the period prior to capture was also able to distinguish between capture locations. Future studies should concentrate on these two sections of the otolith, replicate samples over time, and if possible, compare the elemental signatures of fish from common year-classes. Sample sizes were sufficient for stock discrimination using some elements in this study, but should be increased for future studies to enable the use of a greater range of elements. Given that alternative elements may be detected at other locations due to differences in water chemistry, future studies should also examine a larger suite of elements.

The efficacy of otolith microchemistry in this study lies in the ability to discriminate stocks with greater resolution than genetic techniques. Due to the small effect size for detecting significant differences in elemental concentrations between pelagic fish populations, sample sizes may need to be larger. However, because otolith samples can be processed in large batches at a time, the sample processing rate for otolith microchemistry is faster than other stock discrimination techniques. This technique could be limited by the availability of analytical equipment such as the LA-ICP-MS in some countries. Otolith microchemistry enables stock discrimination based on the past history of the fish. This information is vital for establishing the temporal integrity of fish stocks. Data from the core of the otolith can be used to confirm whether fish collected from different locations have a common or independent origin. Due to the permanency of otoliths, analysis of microchemistry should also be regarded as a contemporary measure of stock structure, although it may be less sensitive than other techniques due to low variability in element concentrations in the open ocean (Campana 1999).

Summary and Conclusions

The three different stock discrimination techniques found consistent patterns in the analyses. There were significant genetic differences between distant populations (WA versus Qld and NZ) while closer populations (Qld versus NZ) had no significant difference. While populations were significantly different for both parasite and otolith analyses, there was a small amount of overlap between adjacent sampling locations (WA and SA, SA and Qld) and less between distant locations (WA and Qld).

In general, these patterns show that a majority of *S. australasicus* are more closely associated with *S. australasicus* from the same capture location than those from other areas.

The genetic differences between *S. australasicus* populations separated by larger distances, the potential barriers to movement, and the similarities between closer populations could be interpreted as localized movement of fish within the study area. However, it is unlikely that these patterns provide evidence of *S. australasicus* movement between areas. There are several oceanographic and biogeographic features that may influence the spatial heterogeneity of *S. australasicus* such as Bass Strait, the East Australia Current, the Flinders current and the Leeuwin Current (Cresswell and Griffin 2004; Godfrey et al. 1980; Middleton and Cirano 2002). Given the wide geographic scale of this study, these significant biogeographic influences, and the fact that only a small number of fish migrating per generation will extinguish genetic differences between populations, it is likely that the number of misclassified fish were a result of processes other than movement between the locations sampled. It may be due to movement from unsampled locations with overlapping parasite assemblages and otolith chemistry. Further *S. australasicus* stock structure studies should be designed to detect movement between closer locations with consideration of the above oceanographic and biogeographical features.

This pilot study has demonstrated the effectiveness of three stock discrimination techniques to discriminate between four broadly separated populations of *S. australasicus* across Australia and New Zealand. The future use of these three techniques for stock discrimination studies should be governed by the costs, benefits and guidelines outlined above. Further studies of *S. australasicus* stock structure will require samples collected on a finer spatial scale, larger sample sizes to account for smaller differences between closer sample sites, and temporal replication to establish the stability of each marker over time.

Acknowledgments

We thank Paul Lewis (WA Fisheries), Peter Smith (National Institute of Water and Atmospheric Research), Jonathan Staunton-Smith (QDPI-Southern Fisheries Centre), and David Fleer and Richard Saunders (SARDI) for collection of samples. Thanks are due to Damien Broderick and Raewyn Street at QDPI-SFC Molecular Laboratory, and Leanne Wheaton, Ralph Foster and Steve Donnellan at SA Museum Evolutionary Biology Unit for help with genetic work. We also thank Les Kinsley at the Australian National University Research School of Earth Sciences for assistance with otolith microchemistry. Rissa Williams and Kate Hutson provided valuable comments on drafts of this paper.

References

Abaunza, P., L. Gordo, C. Karlou-Riga, A. Murta, A.-T.-G.-W. Eltink, M.-T. Garcia-Santamaria, C. Zimmemann, C. Hammer, P. Lucio, S.-A. Iversen, J. Molloy, and E. Gallo. 2003. Growth and reproduction of horse mackerel, *Trachurus trachurus* (Carangidae). Reviews in Fish Biology and Fisheries 13:27–61.

Bartley, D., D. Kent, and M. Drawbridge. 1995. Conservation of genetic diversity in a white seabass hatchery enhancement program in southern California. Pages 249–258 *in* H. L. Schramm Jr., and R. G. Piper, editors. Uses and effects of cultured fishes in aquatic ecosystems. American Fisheries Society, Symposium 15, Betheseda, Maryland.

Begg, G. A., and J. R. Waldman. 1999. An holistic approach to fish stock identification. Fisheries Research 43:35–44.

Bush, A. O., J. M. Aho, and C. R. Kennedy. 1990. Ecological Versus Phylogenetic Determinants of Helminth Parasite Community Richness. Evolutionary Ecology 4:1–20.

Bush, A. O., K. D. Lafferty, J. M. Lotz, and A. W. Shostak. 1997. Parasitology meets ecology on its own terms: Margolis et al. revisited. Journal of Parasitology 83:575–583.

Cadrin, S. X., K. D. Friedland, and J. R. Waldman, editors. 2004. Stock identification methods: applications in fishery science. Elsevier Academic Press, Amsterdam.

Campana, S. E. 1999. Chemistry and composition of fish otoliths: pathways, mechanisms and applications. Marine Ecology Progress Series 188:263–297.

Campana, S. E., G. A. Chouinard, J. M. Hanson, A. Frechet, and J. Brattey. 2000. Otolith elemental fingerprints as biological tracers of fish stocks. Fisheries Research 46:343–357.

Campana, S. E., and S. R. Thorrold. 2001. Otoliths, increments, and elements: keys to a comprehensive understanding of fish populations? Canadian Journal of Fisheries and Aquatic Sciences 58:30–38.

Collette, B. 1999. Mackerels, molecules, and morphology. Pages 149–164 *in* Proceedings of the 5th Indo-Pacific Fisheries Conference, Noumea, 1997. Société Française d'Ichtyologie.

Collette, B. B., and C. E. Nauen. 1983. Scombrids of the world. An annotated and illustrated catalog of tunas, mackerels, bonitos and related species known to date, volume 2. FAO, Rome.

Corten, A. 2001. The role of "conservatism" in herring migrations. Reviews in Fish Biology and Fisheries 11:339–361.

Cremonte, F., and N. H. Sardella. 1997. The parasita Fauna of Scomber japonicus Houttuyn, 1782 (Pisces: Scombridae) in two zones of the Argentine Sea. Fisheries Research 31:1–9.

Cresswell, G. R., and D. A. Griffin. 2004. The Leeuwin Current, eddies and sub-Antarctic waters off south-western Australia. Marine and Freshwater Research 55:267–276.

Edmonds, J. S., and W. J. Fletcher. 1997. Stock discrimination of pilchards *Sardinops sagax* by stable isotope ratio analysis of otolith carbonate. Marine Ecology Progress Series 152:241–247.

Gaughan, D. J., W. J. Fletcher, and J. P. McKinlay. 2002. Functionally distinct adult assemblages within a single breeding stock of the sardine, *Sardinops sagax*: management units within a management unit. Fisheries Research 59:217–231.

Godfrey, J. S., G. R. Cresswell, T. J. Golding, A. F. Pearce, and R. Boyd. 1980. The separation of the East Australian Current. Journal of Physical Oceanography 10:430–440.

Hall, T. A. 1999. BioEdit: a user-friendly biological sequence alignment editor and analysis program for Windows 95/98/NT. Nucleic Acids Symposium Series 41:95–98.

Hayward, C. J., K. M. L. Perera, and K. Rohde. 1998. Assemblages of ectoparasites of a pelagic fish, slimy mackerel (*Scomber australasicus*), from south-eastern Australia. International Journal for Parasitology 28:263–273.

Ihssen, P. E., H. E. Booke, J. M. Casselman, J. M. McGlade, N. R. Payne, and F. M. Utter. 1981. Stock identification: materials and methods. Canadian Journal of Fisheries and Aquatic Sciences 38:1838–1855.

Karaiskou, N., A. Triantafyllidis, and C. Triantaphyllidis. 2004. Shallow genetic structure of three species of the genus *Trachurus* in European waters. Marine Ecology Progress Series 281:193–205.

Korotaeva, V. D. 1974. Helminths of some commercial marine fish of the sub-order Scomberoidei from the Australian region. Izvestiya Tikhookeanskogo Nauchno Issledovatel'skogo Instituta Rybnogo Khozyaistva i Okeanografii TINRO Parazity morskikh zhivotnykh 88:61–66 [In Russian].

Lester, R. J. G. 1990. Reappraisal of the use of parasites for fish stock identification. Australian Journal of Marine and Freshwater Research 41:855–864.

Lester, R. J. G., A. Barnes, and G. Habib. 1985. Parasites of Skipjack Tuna, *Katsuwonus pelamis* - Fishery Implications. Fishery Bulletin 83:343–356.

Lester, R. J. G., and K. MacKenzie. 2009. The use and abuse of parasites as stock markers for fish. Fisheries Research 97:1–2.

MacKenzie, K. 1990. Cestode parasites as biological tags for mackerel (*Scomber scombrus* L.) in the Northeast Atlantic. Journal Du Conseil International Pour L'exploration De La Mer 46:155–166.

MacKenzie, K. 2002. Parasites as biological tags in population studies of marine organisms: an update. Parasitology 124:S153-S163.

MacKenzie, K., and P. Abaunza. 1998. Parasites as biological tags for stock discrimination of marine fish: a guide to procedures and methods. Fisheries Research 38:45–56.

MacKenzie, K., N. Campbell, S. Mattiucci, P. Ramos, A. Pereira, and P. Abaunza. 2004. A checklist of the protozoan and metazoan parasites reported from the Atlantic horse mackerel, *Trachurus trachurus* (L.). Bulletin of the European Association of Fish Pathologists 24:180–184.

Mapstone, B. D. 1995. Scalable decision rules for environmental impact studies: effect size, type I, and type II errors. Ecological Applications 5:401–410.

Mardia, K. V., J. T. Kent, and J. M. Bibby. 1979. Multivariate analysis, volume xv. Academic Press, London.

McCulloch, M., M. Cappo, J. Aumend, and W. Müller. 2005. Tracing the life history of individual barramundi using laser ablation MC-ICP-MS Sr-isotopic and Sr/Ba ratios in otoliths. Marine and Freshwater Research 56:637–644.

Middleton, J. F., and M. Cirano. 2002. A northern boundary current along Australia's southern shelves: The Flinders Current. Journal of Geophysical Research C—Oceans 107:3129–3143.

Miller, S. A., D. D. Dykes, and H. F. Polesky. 1988. A simple salting out procedure for extracting DNA from human nucleated cells. Nucleic Acids Research 16:1215.

Moore, B. R., R. C. Buckworth, H. Moss, and R. J. G. Lester. 2003. Stock discrimination and movements of narrow-barred Spanish mackerel across northern Australia as indicated by parasites. Journal of Fish Biology 63:765–779.

Nesbo, C. L., E. K. Rueness, S. A. Iversen, D. W. Skagen, and K. S. Jakobsen. 2000. Phylogeography and population history of Atlantic mackerel (*Scomber scombrus* L.): a genealogical approach reveals genetic structuring among the eastern Atlantic stocks. Proceedings of the Royal Society of London Series B: Biological Sciences 267:281–292.

Newman, S., R. Buckworth, M. Mackie, P. Lewis, T. Bastow, and J. Ovenden. 2006. Spatial subdivision of adult assemblages of Spanish mackerel, *Scomberomorus commerson* (Pisces: Scombridae) from western, northern and eastern Australian waters through stable isotope ratio analysis of sagittal otolith carbonate. In R. Buckworth, S. Newman, J. Ovenden, R. Lester, and G. McPherson, editors. The stock structure of Northern and Western Australian Spanish mackerel. Final Report, Fisheries Research & Development Corporation Project 1998/159. Fisheries Group, Department of Business Industry and Resource Development, Northern Territory Government, Darwin, Australia.

Ovenden, J., and R. Street. 2006. Genetic population structure of Spanish mackerel. *In* R. Buckworth, S. Newman, J. Ovenden, R. Lester, and G. McPherson, editors. The stock structure of Northern and Western Australian Spanish mackerel. Final Report, Fisheries Research & Development Corporation Project 1998/159. Fisheries Group, Department of Business Industry and Resource Development, Northern Territory Government, Darwin, Australia.

Ovenden, J. R. 1990. Mitochondrial DNA and marine stock assessment: a review. Australian Journal of Marine and Freshwater Research 41:835–853.

Ovenden, J. R., J. Lloyd, S. J. Newman, C. P. Keenan, and L. S. Slater. 2002. Spatial genetic subdivision between northern Australian and southeast Asian populations of *Pristipomoides multidens*: a tropical marine reef fish species. Fisheries Research 59:57–69.

Pawson, M. G., and S. Jennings. 1996. A critique of methods for stock identification in marine capture fisheries. Fisheries Research 25:203–217.

Perera, K. M. L. 1993. No evidence for seasonality in the ectoparasitic fauna of slimy mackerel, *Scomber australasicus*. Australian Journal of Marine and Freshwater Research 44:709–719.

Peterman, R. M. 1990. Statistical power analysis can improve fisheries research and management. Canadian Journal of Fisheries and Aquatic Sciences 47:2–15.

Pozdnyakov, SE, and A. V. Vasilenko. 1994. Distribution, migration routes and helminth fauna in *Scomber japonicus* from the northwestern Pacific. Journal of Ichthyology 34:74–91.

Pritchard, M. H., and G. O. W. Kruse. 1982. The collection and preservation of animal parasites. University of Nebraska Press, Lincoln.

Proctor, C. H., R. E. Thresher, J. S. Gunn, D. J. Mills, I. R. Harrowfield, and S. H. Sie. 1995. Stock structure of the southern bluefin tuna *Thunnus maccoyii*: an investigation based on probe microanalysis of otolith composition. Marine Biology 122:511–526.

Radovich, J. 1982. The Collapse of the California sardine fishery: what have we learned? California Cooperative Oceanic Fisheries Investigations Reports 23:56–80.

Rohde, K. 1987. Different populations of *Scomber australasicus* in New Zealand and southeastern Australia, demonstrated by a simple method using monogenean sclerites. Journal of Fish Biology 30:651–657.

Rohde, K., C. Hayward, and M. Heap. 1995. Aspects of the ecology of metazoan ectoparasites of marine fishes. International Journal for Parasitology 25:945–970.

Ruzzante, D. E., S. Mariani, D. Bekkevold, C. André, H. Mosegaard, L. A. W. Clausen, T. G. Dahlgren, W. F. Hutchinson, E. M. C. Hatfield, E. Torstensen, J. Brigham, E. J. Simmonds, L. Laikre, L. C. Larsson, R. J. M. Stet, N. Ryman, and G. R. Carvalho. 2006. Biocomplexity in a highly migratory pelagic marine fish, Atlantic herring. Proceedings of the Royal Society of London B: Biological Sciences 273:1459–1464.

Schneider, S., D. Roessli, and L. Excoffier. 2000. Arlequin: A software for population genetics data analysis. Genetics and Biometry Laboratory, Department of Anthropology, University of Geneva, Geneva, Switzerland.

Scoles, D. R., B. B. Collette, and J. E. Graves. 1998. Global phylogeography of mackerels of the genus *Scomber*. Fishery Bulletin 96:823–842.

Stepien, C. A. 1995. Population genetic divergence and geographic patterns from DNA sequences: examples from marine and freshwater fishes. Pages 263–287 *in* J. L. Nielsen, editor. Evolution and the aquatic ecosystem: defining unique units in population conservation. American Fisheries Society, Symposium 17, Bethesda, Maryland.

Thompson, J. D., T. J. Gibson, F. Plewniak, F. Jeanmougin, and D. G. Higgins. 1997. The CLUSTAL-X windows interface: flexible strategies for multiple sequence alignment aided by quality analysis tools. Nucleic Acids Research 25:4876–4882.

Thresher, R. E. 1999. Elemental composition of otoliths as a stock delineator in fishes. Fisheries Research 43:165–204.

Villamor, B., P. Abaunza, and A. C. Farina. 2004. Growth variability of mackerel (*Scomber scombrus*) off north and northwest Spain and a comparative review of the growth patterns in the northeast Atlantic. Fisheries Research 69:107–121.

Waples, R. 1998. Separating the wheat from the chaff: patterns of genetic differentiation in high gene flow species. Journal of Heredity 89:438–450.

Zardoya, R., R. Castilho, C. Grande, L. Favre-krey, S. Caetano, S. Marcato, G. Krey, and T. Patarnello. 2004. Differential population structuring of two closely related fish species, the mackerel (*Scomber scombrus*) and the chub mackerel (*Scomber japonicus*), in the Mediterranean Sea. Molecular Ecology 13:1785–1798.

American Fisheries Society Symposium 76:419–437, 2012

An Examination of the Population Structure of the Eastern Australian Salmon (*Arripis trutta*) Stock in Southeastern Australia

Julian M. Hughes*
Industry & Investment New South Wales, Cronulla Fisheries Research Centre of Excellence
P.O. Box 21, Cronulla, New South Wales 2230, Australia
and
Fisheries & Marine Environmental Research Facility, School of Biological, Earth and Environmental Sciences
University of New South Wales, Sydney, New South Wales 2052, Australia

John Stewart
Industry & Investment New South Wales, Cronulla Fisheries Research Centre of Excellence
P.O. Box 21, Cronulla, New South Wales 2230, Australia

Bronwyn M. Gillanders
Southern Seas Ecology Laboratories, School of Earth and Environmental Sciences
University of Adelaide, South Australia 5005, Australia

Iain M. Suthers
Fisheries & Marine Environmental Research Facility, School of Biological, Earth and Environmental Sciences
University of New South Wales, Sydney, New South Wales 2052, Australia

Abstract.—The population structure of the eastern Australian salmon *Arripis trutta* stock in the waters of southeastern (SE) Australia was examined using information provided by historical as well as current data sources. An extensive tag-recapture program and aging study undertaken during the 1960s demonstrated widespread mixing of the *A. trutta* population in SE Australian waters and established a robust model of general movement of fish from Tasmania north to Victoria and NSW with the approach of sexual maturity at ~four years of age. However, this work also hypothesized that the portion of the stock at Flinders Island in Tasmanian waters was resident and did not undergo this northward migration. Otolith chemistry analyses were therefore used as a tool in a 'weight of evidence' approach to further examine the population structure of the *A. trutta* stock in SE Australia. Samples of five year old *A. trutta* for analysis of otolith chemistry were collected over seven weeks from two sites (10 per site) within each of four locations: northern NSW, southern NSW, Victoria and Tasmania. The cores and edges of otoliths were analyzed using laser ablation in-

*Corresponding author: Julian.Hughes@industry.nsw.gov.au

ductively coupled plasma mass spectrometry. Univariate analyses did not find spatial differences for any of the elements Li, Na, Mg, Mn, Ba or Sr between locations. Multivariate analyses however, did find differences between the multi-element 'fingerprints' of fish from Tasmania compared to each of the other locations (which were similar). This difference was driven by a group of fish collected from Flinders Island in north-eastern Tasmanian waters. The fish collected at this site were also significantly smaller at five years of age than fish from all other sites, indicating reduced growth rates. The lack of consequential and definitive differences in otolith chemistry data combined with the highly migratory nature of *A. trutta* in this region demonstrated by tagging studies confirm that the most likely stock structure model for *A. trutta* in SE Australia is of a single well mixed biological stock spanning Tasmania, Victoria and NSW with fish moving north from Tasmania to mainland Australia with the approach of sexual maturity. However, the reduced growth rates and distinct elemental signature for *A. trutta* from Flinders Island highlights the need for further work to examine the preexisting hypothesis of a potential resident sub-population there.

Introduction

The eastern Australian salmon *Arripis trutta* (Forster, 1801) is an abundant, pelagic species of fish distributed in coastal marine waters from southern Queensland to western Victoria and Tasmania (Paulin 1993). *Arripis trutta* is a member of the teleost family Arripidae, a single genus family containing four species, which are all found in the temperate waters of southern Australia and New Zealand (Paulin 1993). Arripids are a highly mobile family of fishes all of which undertake extensive migrations of up to thousands of kilometres in length motivated by life history, reproductive or seasonal cues (Malcolm 1959; Stanley 1978; Cappo et al. 2000; Ayvazian et al. 2004).

Arripis trutta has a strong schooling habit and at times may form vast shoals of up to thousands of metric tons per school around ocean beaches and areas of exposed coast (Malcolm 1966a). It sustains high volume commercial fisheries in New South Wales (NSW), Victoria and Tasmania (2287 t landed in 2005–2006; Australian Bureau of Agricultural and Resource Economics 2007), as well as being an enormously popular recreational sportfish due to its excellent fighting qualities. Owing to its substantial commercial fishery *A. trutta* has been the subject of considerable past scientific investigation in southeastern (SE) Australia (Malcolm 1959, 1966a, 1966b, 1966c, Nicholls 1973; Stanley and Malcolm 1977; Stanley 1978).

The distribution, movements, and age composition of *A. trutta* have been examined via an extensive tag-recapture program and aging study (Stanley 1978). Over 5,200 *A. trutta* were aged using scales, and then tagged and released in the waters of NSW, Victoria and Tasmania. A large proportion (36%) of these fish were later recaptured and, importantly, not a single tagged fish released in NSW or Victoria was recaptured in Tasmanian waters. These results, combined with the lack of spawning activity observed in Tasmanian waters (Stanley and Malcolm 1977), suggested that the movement of fish from Tasmanian to mainland Australian waters across Bass Strait was a one-way migration and occurred with the onset of sexual maturity at ~four years old. These results, in addition to the results of other tagging programs on *A. trutta* (Stanley 1988a, 1988b), also showed that bidirectional movement of fish between NSW and Victorian coastal waters occurred regularly and it was concluded that *A. trutta* in SE Australia was a single, well-mixed biological stock (Stanley 1978). However, due to the presence of unusually old fish in the Flinders Island area of northeastern Tasmanian waters (20% > four years old), both Stanley (1978) and Malcolm (1966c) hypothesized the possible existence of a resident sub-population of fish which did not undergo the migration from Tasmanian to mainland waters with the onset of sexual maturity.

The physical and chemical characteristics of the different water bodies encountered by fish may vary at a range of spatial scales. Specific trace elements present in the environment of fish are incorporated into the growing surface of their otoliths and these can reflect the ambient environmental conditions of the water in which fish reside (Fowler et al. 1995a, 1995b). Since fish that spend at least part of their lives in different water masses often produce otoliths of different elemental composition, the otolith elemental composition (or 'elemental fingerprint') can serve as an environmentally induced tag of groups of fish (Campana et al. 2000). This technique has become increasingly popular and extensively used over the past 20 years for examining movement patterns and distinguishing among populations and stocks of many fish species throughout the world (e.g., Edmonds et al. 1989; Campana & Gagné 1995; Campana et al. 2000; Daverat et al. 2005; McCulloch et al. 2005; Tzeng et al. 2005). Since the existing population structure data for the SE Australian *A. trutta* stock were not conclusive, we chose to use otolith chemistry analysis as a tool in a 'weight of evidence' approach to further examine *A. trutta* population structure in this region. Otolith chemistry analysis was chosen because 1), the physical and chemical characteristics of the different water bodies encountered by *A. trutta* throughout SE Australia are likely to vary, and 2), it had the potential to provide dissimilar but complementary information to that from existing data sources (i.e., tagging and aging studies).

The stock structure of *A. trutta* has been of considerable interest to fishery managers in the three states (NSW, Victoria, Tasmania) where it commonly occurs in SE Australia as a species which could benefit from cross-jurisdictional management arrangements because of 1), its broad geographical distribution, 2), its long distance migrations and 3), its commercial and recreational importance. Our aim therefore, was to use a 'weight of evidence' approach which utilized information provided by historical tagging and aging studies in addition to analyses of otolith chemistry and size-at-age to examine the potential for population structure within the *A. trutta* stock in SE Australia.

Methods

Sample Collection and Experimental Design

Arripis trutta for otolith chemistry analyses were collected over seven weeks in 2007 (March 30 to May 16) by commercial fishers from NSW, Victoria and Tasmania using beach haul and purse seine nets. A subsample ranging from 12 to 89 fish were taken from each catch and frozen. Ages of fish were estimated by counting opaque zones in otolith sections. Ten fish, each estimated to be five years old were then randomly chosen from the catches from each of two sites (separated by kilometers to tens of kilometers) from each of four locations on the SE Australian coast: northern NSW, southern NSW, Victoria and Tasmania (Figure 1; Table 1). Selecting five year old fish was done to ensure that the fish had been alive for the same period and minimized the possibility of a confounding temporal influence on spatial comparisons (Fowler et al. 2005).

Sample Preparation and Analysis

All fish collected were frozen for at least 24 h before being defrosted, measured (FL to

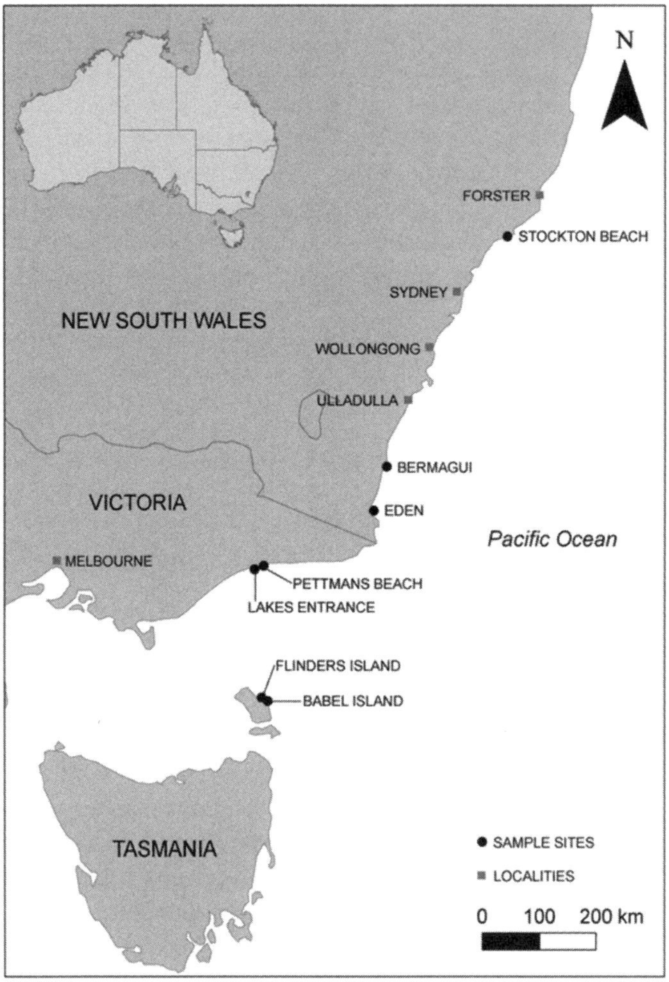

Figure 1. Map of southeastern Australia showing collection locations of *Arripis trutta* samples taken for analysis of otolith chemistry.

the nearest 0.1 cm) and dissected. The sagittal otoliths were removed using Teflon-coated forceps and cleaned in ultrapure deionized water. Otoliths were air dried overnight and stored in plastic 1.5 mL microcentrifuge tubes. One otolith from each fish was embedded in indium spiked (~30 ppm) two-part epoxy resin (Struers Epofix) and a ~300 μm section was taken through the core perpendicular to the long axis of the otolith using a slow speed saw (Buehler Isomet) and two spaced diamond blades. The blades were lubricated in ultrapure water and each otolith section was then polished using 9 μm lapping film and ultrapure water. Sections were then fixed to a glass microscope slide in random order (up to 14 sections per slide) using indium spiked (~200 ppm) thermoplastic glue (Crystalbond 509) and sonicated for 5 min in an ultrasonic cleaner (Unisonics Australia FXP4). Slides were allowed to dry overnight in a laminar flow cabinet before being transferred into individual clean sealable plastic bags prior to analysis.

Table 1. Details of capture sites and dates, fishing gears used and lengths (mean ± SE) for 5 year old *Arripis trutta*. n is sample size, FL is fork length, * denotes significantly different at $p < 0.05$.

Location	Site	Latitude, Longitude	Date caught	Gear	n	FL (cm)
Northern NSW	Stockton Beach 1	32° 49' S, 151° 54' E	May	Beach haul	10	44.84 ± 0.62
	Stockton Beach 2	32° 49' S, 151° 54' E	May	Beach haul	10	46.90 ± 0.32
Southern NSW	Bermagui	36° 25' S, 150° 05' E	April	Purse seine	10	46.65 ± 0.75
	Eden	37° 04' S, 149° 54' E	April	Purse seine	10	49.09 ± 0.71*
Victoria	Lakes Entrance	37° 52' S, 147° 59' E	May	Purse seine	10	45.24 ± 0.58
	Pettmans Beach	37° 49' S, 148° 11' E	April	Purse seine	10	45.69 ± 0.66
Tasmania	Babel Island	39° 56' S, 148° 19' E	May	Purse seine	10	45.06 ± 0.42
	Flinders Island	39° 56' S, 148° 12' E	April	Purse seine	10	40.65 ± 0.97*

Arripis trutta otoliths were analyzed using a New Wave Nd Yag 213 nm UV laser operated in Q-switch mode connected to an Agilent 7500cs inductively coupled plasma-mass spectrometer (ICP-MS). The laser was operated at a frequency of 5 Hz and 65% power. To determine background concentrations, the chamber gases were analyzed without any sample present for 30 s prior to the commencement of ablation. Otolith sections were ablated in a sealed chamber and ablated otolith material was transported in an argon and helium carrier gas stream to the ICP-MS. A reference standard (NIST 612, National Institute of Standards and Technology) was analyzed at the beginning and end of each session and periodically throughout the session to correct for mass bias and instrument drift. Concentrations of each element analyzed (Lithium-7 (Li), Sodium-23 (Na), Magnesium-25 (Mg), Manganese-55 (Mn), Strontium-88 (Sr) and Barium-138 (Ba)) were standardized to Calcium-43 (Ca), by expressing concentrations of elements as ratios to Ca (mmol/mol). For all otolith sections, both the core region and outer edge were ablated by the laser using a 30 μm diameter spot. The core region represented juvenile otolith growth and the outer edge represented otolith growth immediately prior to collection.

Statistical Analyses

Univariate analyses of variance (ANOVAs) were used to test whether elemental compositions (Li, Na, Mg, Mn, Sr, Ba) differed between the core and edge of the otolith as well as between sites and among locations. A three-factor design treated 'Core/Edge' as fixed, and 'Site' as a random factor nested within 'Location.' Heterogeneous data (Cochran's C-test, $p < 0.05$) were Ln(x + 1) transformed. Where significant differences were detected in ANOVA, means were compared using Student-Newman-Keuls (SNK) tests to determine where the differences occurred.

The discrimination between geographic areas in SE Australian waters based on multiple elements in *A. trutta* otoliths was tested using discriminant analysis. Discriminant analysis has two steps: (1) an *F*-test (Wilks' lambda) is used to test if the discriminant model as a whole is significant, and (2); if the *F*-test shows significance, then the individual independent variables are assessed to see which differ significantly in mean by group and these are used to classify the dependent variable. Quadratic discriminant function analysis (DFA) was then used to determine the ability of elemental signatures to correctly assign fish to their collection locations. Backward stepwise quadratic functions were determined in SYSTAT (Point Richmond, California) and then used to classify fish to groups using a jackknifed method with prior probabilities of group membership being equal. The accuracy of classifications to these groups was determined as the percentage calculated to be correct. This analysis was done separately for both the core region and outer edge of the otoliths.

A single-factor ANOVA was also used to test whether mean lengths of fish differed between sites. Heterogeneity of variances and significant differences detected in ANOVA were examined using the methods described above.

Results

Univariate Analyses

Core-edge comparisons—Significant differences in elemental concentrations were detected between the core region and outer edge of the otolith for all elements analyzed using ANOVA (Li:Ca, Na:Ca, Mg:Ca, Sr:Ca, Ba:Ca) except Mn:Ca (Table 2). The otolith core region consistently contained significantly more Li:Ca, Na:Ca, Mg:Ca and Ba:Ca than the otolith edge al-

Table 2. ANOVA results comparing elemental concentrations (Li:Ca, Na:Ca, Mg:Ca, Mn:Ca, Sr:Ca, Ba:Ca) of core regions and outer edges of otoliths from 5 year old fish collected from Stockton Beach 1 and 2 (northern NSW), Bermagui and Eden (southern NSW), Lakes Entrance and Pettmans Beach (Victoria), and Babel and Flinders Islands (Tasmania). $^{*}p < 0.05$, $^{**}p < 0.01$, $^{***}p < 0.001$, df is degrees of freedom.

Source	df	F
Li:Ca		
Core-Edge	1	9.65*
Location	3	1.70
Site (Location)	4	1.02
Core-Edge x Location	3	0.87
Core-Edge x Site(Location)	4	0.55
Residual	144	
Na:Ca		
Core-Edge	1	1003.59***
Location	3	4.38
Site(Location)	4	1.13
Core-Edge x Location	3	1.58
Core-Edge x Site(Location)	4	0.60
Residual	144	
Mg:Ca		
Core-Edge	1	46.55**
Location	3	0.09
Site(Location)	4	3.58**
Core-Edge x Location	3	1.24
Core-Edge x Site(Location)	4	1.19
Residual	144	
Mn:Ca		
Core-Edge	1	1.18
Location	3	5.34
Site(Location)	4	1.20
Core-Edge x Location	3	14.44
Core-Edge x Site(Location)	4	0.19
Residual	144	
Sr:Ca		
Core-Edge	1	53.61**
Location	3	0.36
Site(Location)	4	1.85
Core-Edge x Location	3	3.62
Core-Edge x Site(Location)	4	0.88
Residual	144	

Table 2. Continued.

Source	df	F
Ba:Ca		
Core-Edge	1	36.47**
Location	3	1.55
Site(Location)	4	0.57
Core-Edge x Location	3	1.40
Core-Edge x Site(Location)	4	0.65
Residual	144	

though the magnitude of this variation differed by location (Figure 2a, b, c, f; Table 2). In contrast, the otolith edge consistently contained significantly more Sr:Ca than the core region (Figure 2e; Table 2), but once again the magnitude of the variation differed by location.

Spatial comparisons—No spatial differences were detected by ANOVA for the elements Li:Ca, Na:Ca, Mn:Ca, Sr:Ca and Ba:Ca (Table 2). The only significant difference detected was for variation in Mg:Ca between sites for two of the four locations (Table 2). Otoliths of fish collected from Bermagui had higher Mg:Ca concentrations than fish from Eden in southern NSW (Figure 2c). Similarly, the otoliths of fish collected from Pettmans Beach in Victoria had higher Mg:Ca than in fish from Lakes Entrance (Figure 2c).

Multivariate Analyses

A significant difference between locations was detected for the multi-element signatures for both the otolith core region (Wilks' lambda = 0.682, $F_{9, 180} = 3.416$, $p < 0.001$) and outer edge (Wilks' lambda = 0.565, $F_{12, 193} = 3.878$, $p < 0.0001$) of five year old fish (Figure 3). The difference in core regions appeared to be driven by the influence of fish from southern NSW being distinct from fish from the other three locations sampled and Victorian fish being distinct from those from Tasmania (Figure 3a). With the exception of Tasmania and Victoria the ellipses from all locations overlapped one another (Figure 3a). For the outer edge of the otolith, Tasmania separated from the other three locations (Figure 3b). This trend was driven by the influence of fish from Flinders Island (Figure 4). There was however, considerable overlap between southern NSW and Victoria, and southern NSW and northern NSW (Figure 3b). The 95% confidence ellipses around the mean values for each location suggests that there was less separation among the four locations for the otolith core region because the ellipses overlapped for all locations (Figure 3a), than for the outer edge where the ellipse around the mean value for Tasmania was separate to those of the other three locations (Figure 3b).

A low percentage of five year old fish were correctly classified on the basis of multi-element signatures in the core region (44%) and outer edge (38%) of the otolith (Table 3). Correct classification of fish to their sampling locations based on elements in the core region ranged from 20% for fish from northern NSW up to 65% for fish from Victoria (Table 3a). For the outer edge of the otolith, correct classification of fish to their sampling locations ranged from 20% for fish from southern NSW up to 65% for Tasmania (Table 3b).

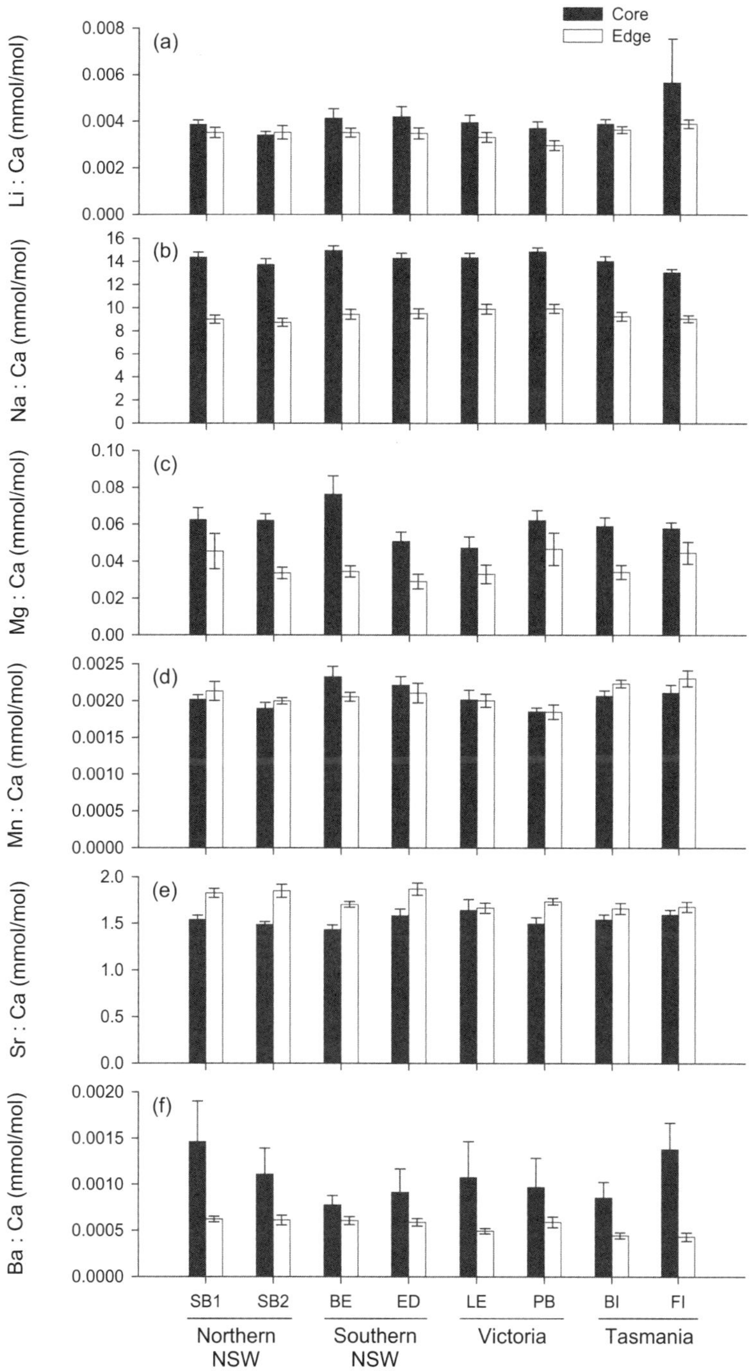

FIGURE 2. Mean (±SE) concentration ratios (mmol/mol) of **a)** Li:Ca, **b)** Na:Ca, **c)** Mg:Ca, **d)** Mn:Ca, **e)** Sr:Ca and **f)** Ba:Ca from the core region (■) and outer edge (□) of 5 year old *Arripis trutta* otoliths collected from northern NSW (SB1 = Stockton Beach 1, SB2 = Stockton Beach 2), southern NSW (BE = Bermagui, ED = Eden), Victoria (LE = Lakes Entrance, PB = Pettmans Beach) and Tasmania (FI = Flinders Island, BI = Babel Island). n = 10 for each site.

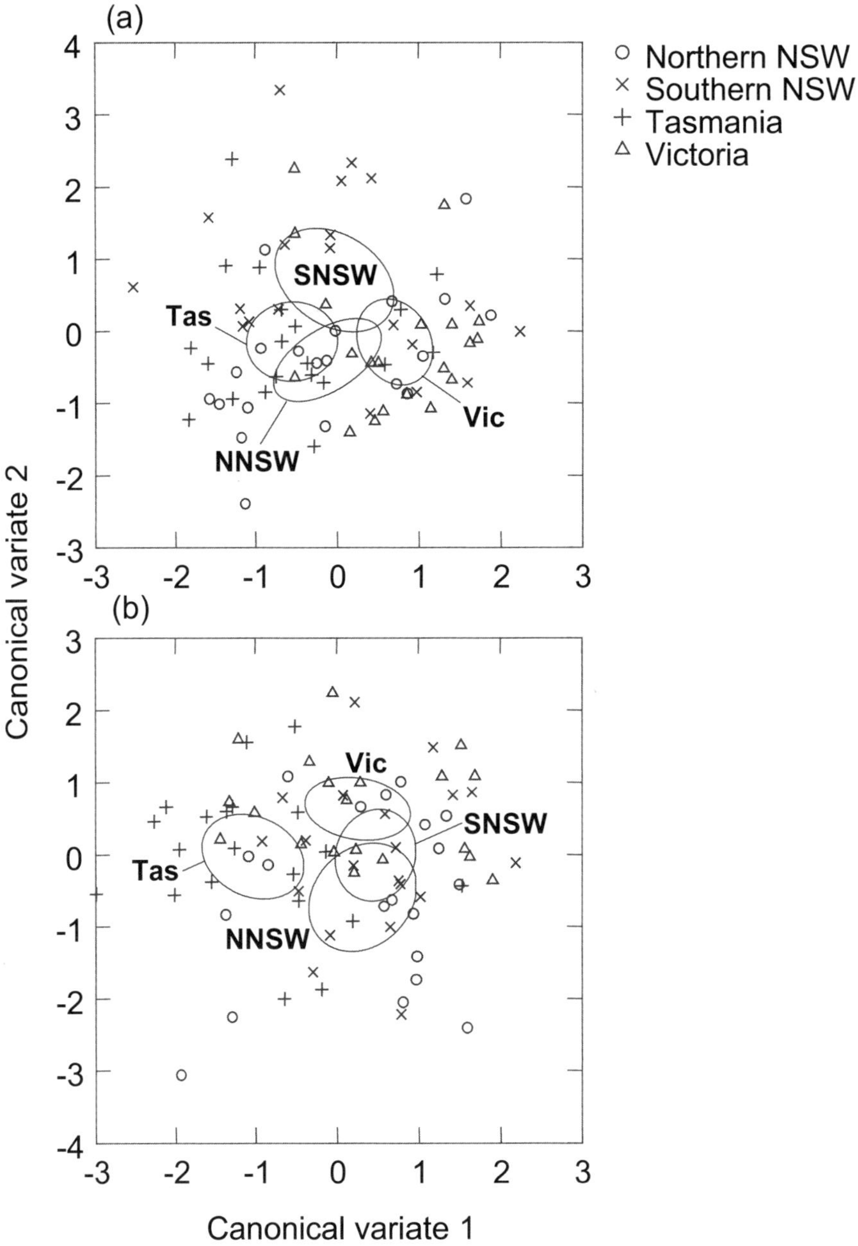

Figure 3. Canonical variate plot summarising variation in otolith elemental signatures for 5 year old *Arripis trutta* from northern NSW (NNSW- ○), southern NSW (SNSW- ×), Victoria (Vic- Δ) and Tasmania (Tas- +) for a) the core region, and b) the outer edge. Ellipses represent 95% confidence intervals around the means for each location. n = 20 for each location.

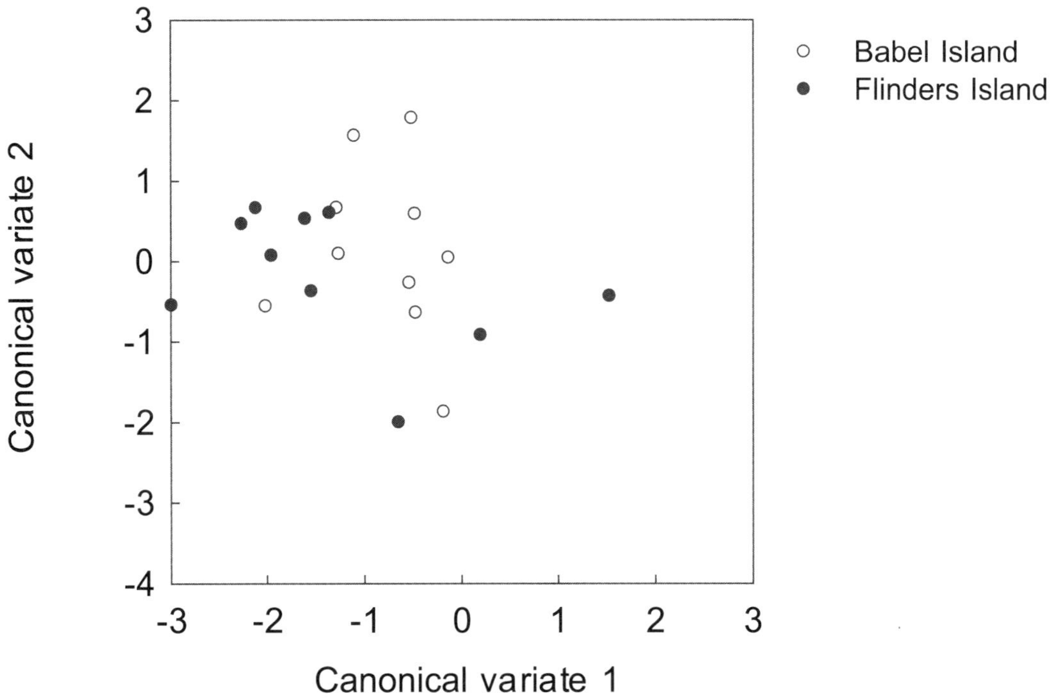

FIGURE 4. Canonical variate plot summarising variation in otolith elemental signatures for 5 year old *Arripis trutta* from sites in Tasmanian waters (Babel Island- ○, Flinders Island- ●). n = 10 for each location.

TABLE 3. Results of quadratic discriminant function analysis (DFA) where 5 year old fish from each location were classified to each group based on otolith elemental signatures (Li:Ca, Na:Ca, Mg:Ca, Mn:Ca, Sr:Ca, Ba:Ca). The percentage of fish classified to each group is shown for **a)** the core region, and **b)** the outer edge of the otolith. Overall, 44% (core) and 38% (edge) were correctly classified. Values in **bold** indicate fish which were correctly classified. n = 20 for each location.

		Northern NSW	Southern NSW	Victoria	Tasmania
a)	Northern NSW	**20**	15	25	40
	Southern NSW	10	**40**	30	20
	Victoria	15	20	**65**	0
	Tasmania	20	10	20	**50**
b)	Northern NSW	**35**	15	25	25
	Southern NSW	30	**20**	35	15
	Victoria	25	20	**30**	25
	Tasmania	15	10	10	**65**

Length-at-age

ANOVA detected a significant difference in the mean length of fish at age five years among sites (ANOVA; $F_{7, 72} = 13.42$, $p < 0.0001$). SNK tests showed that fish collected from Eden (southern NSW) were significantly larger than fish from all other sites and fish collected from Flinders Island (Tasmania) were significantly smaller than fish from all other sites (Table 1). There were no differences in fish length between any of the other sites (Table 1). These results indicate that fish from Eden had grown faster, and similarly fish from Flinders Island had grown slower, than fish from all other sampled sites.

Discussion

Single Elements

Differences in elemental concentrations were detected between the core region and outer edge of *Arripis trutta* otoliths for all elements except Mg:Ca. Spot analyses of the core region represent the juvenile part of the otolith, while edge ablations reflected the elements present during the life of the fish immediately prior to being caught. Such differences may be a result of the changing uptake and/or incorporation of elements onto the otolith surface with the age or life history stage of the fish as a result of ontogeny (Sadovy and Severin 1992). Ontogenetic changes such as growth variation from larval to juvenile and juvenile to adult can result in morphological and physiological modifications within fish. Ontogeny has been shown to affect otolith chemistry for several fish species (Toole et al. 1993; Fowler et al. 1995a, 1995b), but not for others (Daverat et al. 2005; Elsdon and Gillanders 2005).

Variation in the environment of the fish (i.e., the waterbody the fish lives in) also has the potential to produce the variation seen between the core and edge regions of the otolith (Elsdon and Gillanders 2003, 2005; Daverat et al. 2005; Tzeng et al. 2005). The outer edge of *A. trutta* otoliths contained significantly more Sr:Ca than the core region. Conversely, the core region contained significantly more Ba:Ca that the outer edge. Correlations between ambient and otolith Sr:Ca and Ba:Ca concentrations have been investigated in several studies and a strong positive relationship exists for both elements (e.g., Elsdon and Gillanders 2003, 2005). The concentrations of both Sr:Ca (Bath Martin et al. 2004; Daverat et al. 2005) and Ba:Ca (Elsdon and Gillanders 2005) in otoliths have also previously been shown to be related to environmental salinity. The low Sr:Ca and high Ba:Ca concentrations found in core regions of otoliths suggest that as juveniles *A. trutta* are found in lower salinity environments than the environment they inhabit as adults. Indeed, juvenile *A. trutta* are often found in estuaries in both Australia (Stanley and Malcolm 1977) and New Zealand (Hartill and Walsh 2005).

With the exception of Mg:Ca, which has been shown to vary independent of environmental concentrations (Campana 1999), there were no significant differences in the spatial variation of individual elements in the otoliths of *A. trutta* collected from different sites. A lack of important single-element differences among locations on a similar spatial scale (tens to hundreds of kilometers) has previously been shown for several other studies on marine fish species (e.g., Kalish et al. 1996; Thorrold et al. 1997; Edmonds et al. 1999; Gillanders et al. 2001; Stransky et al. 2005). In this study, the fact that single-element analyses failed to find significant differences between locations, whereas examination of the multi-element signature did, also serves to highlight the limitations of univariate analyses of single element data in otolith chemistry analysis.

The small variation in elemental composition of otoliths of *A. trutta* collected from

SE Australian waters may reflect the relatively homogeneous environmental conditions present in this region. The comparatively homogeneous environment of the Mediterranean in the vicinity of the southwest Iberian Peninsula was suggested as a possible cause of the lack of significant otolith chemistry differences found between locations for the sparid *Diplodus vulgaris* (Gillanders et al. 2001). These authors hypothesized that this environmental homogeneity was due to a lack of major rivers in proximity to study locations, the low annual rainfall and the similar influence of Atlantic currents in the region. In SE Australia however, there is considerable freshwater input into the nearshore marine environment via substantial rainfall and numerous large river and estuary systems. In addition, the influence of the prevailing East Australian Current (EAC) is highly seasonal (Ridgway and Dunn 2003) likely resulting in a relatively heterogeneous environment with high potential for differences in water chemistry over the ~850 km spatial scale of the study.

Alternatively, fish may move among different areas within this region. It has been conclusively shown by past tagging studies that *A. trutta* undertake extensive movements between the coastal waters of Tasmania, Victoria and NSW (Malcolm 1959; Stanley 1978, 1988a, 1988b). This highly migratory behavior may therefore obscure any consistent differences in otolith chemistry that exist between the water bodies at each collection location. In SE Australia, elemental concentrations in ambient water are highly variable over small (within estuary) to large (between estuary) spatial scales (Elsdon and Gillanders 2006; Hamer and Jenkins 2007), and between estuarine and coastal marine waters (Hamer et al. 2006). Water chemistry in this region also has the potential to vary over a range of temporal scales from tidal cycles to years (Hatje 2003; Elsdon and Gillanders 2006; Hamer et al. 2006; Hamer and Jenkins 2007).

Multi-element Signatures

The multi-element signatures showed more separation among locations for analyses of the outer edge of the otolith compared to that for analyses of the otolith core. Similarly for the differences between the cores and edges for the single-element analyses discussed above, this pattern may reflect similarities in the environment of juveniles or may be a result of ontogenetic factors, or a combination of both (Gillanders and Joyce 2005). It is not possible to entirely discount the influence of ontogeny on the otolith elemental signatures seen in this study without controlled experiments in the laboratory (Elsdon and Gillanders 2005) or knowledge of environmental concentrations. The poorer separation seen in the core analyses compared with the edge may be because the fish come from a common, or a few common origins. *Arripis trutta* juveniles have been recorded from estuaries throughout SE Australia from Botany Bay in NSW to Port Philip Bay in Victoria (Stanley 1978) and also Tasmania (Malcolm 1966a, 1966b). The ability of otolith elemental signatures in tracking movements and/or habitat shifts throughout the life history of a fish has been an often employed tool in many previous studies on many fish species which utilize estuarine (Gillanders and Kingsford 1996, 2000; Gillanders 2002; McCulloch et al. 2005) or freshwater (Daverat et al. 2005; Tzeng et al. 2005) environments as juveniles. Alternatively there may be a significant amount of population mixing when the fish are young. The greater separation of locations for edge elemental fingerprints may therefore reflect the dispersal of *A. trutta* from estuarine nursery areas to offshore marine areas of the SE Australian coast with age. A similar increased degree of spatial segregation during the adult phase has been previously inferred from otolith chemistry for many other marine fish species (e.g., Stransky et al. 2005; Fowler et al. 2005).

A significant difference was detected for the multi-element signature of otolith edges among locations where *A. trutta* were collected. This difference was driven by a distinct multi-element signature from the otolith edges of Tasmanian fish which separated them from fish from the other three mainland locations (i.e., northern NSW, southern NSW and Victoria). Results from the DFA using multi-element signatures to classify fish to a location showed that attempting to classify fish using either the core (overall 44%) or edge (overall 38%) signature was extremely difficult and resulted in a low degree of accuracy, but importantly, the highest percentage of correctly classified fish using the edge signatures came from Tasmanian waters (65%). A similar overall low classification rate, but better separation by edge chemistry than core signatures, has previously been reported for deepwater fish species in the Atlantic (Swan et al. 2003; Stransky et al. 2005).

Significant differences in otolith chemistry similar to that found in this research have been previously used to infer potential population or stock differentiation in many other published studies (e.g., Edmonds et al. 1989; Campana and Gagné 1995; Campana et al. 2000). Our current understanding of the life cycle of *A. trutta* in this region is that following spawning in southern NSW and eastern Victoria between October and March there is a southward movement of early life history stages to Tasmania (Malcolm 1966a). Evidence of spawning activity has never been recorded from Tasmanian waters and all *A. trutta* there are immature (Stanley and Malcolm 1977). On the basis of the dispersal of the larval and juvenile stages, the progeny of spawning in mainland waters are likely to be well mixed. The primary nursery areas for juveniles are the bays and estuaries of northeastern (NE) Tasmania, although a smaller proportion is consistently found in eastern Victoria and southern NSW (Malcolm 1966c). As they reach maturity (~four years), the juveniles undertake a one-way migration from Tasmania across Bass Strait via the Flinders Island chain and return to the waters of eastern Victoria and southern NSW as adults (Stanley 1978). Following migration, extensive inter-mixing of fish and large-scale movements occur throughout mainland waters (Stanley 1978, 1988a, 1988b). The migration across Bass Strait from Tasmanian to mainland waters may represent a significant change in water chemistry and this may be reflected by the significantly different multi-element fingerprints for fish from Tasmania. In fact, it appears as though the only two groups of *A. trutta* that are spatially separate enough to be exposed to the sufficiently different water masses required to produce different otolith elemental signatures are Tasmanian fish prior to migration, and fish already present in mainland waters.

Notably, it was the group of fish collected from the Flinders Island site that were responsible for the difference between the multi-element signature of the otolith edges of Tasmanian fish which separated them from fish from the other three mainland locations. The two Tasmanian sites were separated by just ~20 km and sampling of both sites occurred within a seven week period. The distinct elemental signature from fish collected from Flinders Island is therefore unlikely related to small scale variation in the water chemistry found in NE Tasmania as there are no major rivers in close proximity to the two sites from which pulses of freshwater may be discharged, thus influencing ambient water chemistry in terms of salinity, temperature and trace elements. The influence of the EAC is also likely to affect both sites in a similar manner (Ridgway and Dunn 2003) and therefore unlikely to vary enough over such small spatial and temporal scales to produce the different elemental signatures evident for the two sites.

Alternatively, Malcolm (1966c) and Stanley (1978) proposed the hypothesis that some *A. trutta* from the waters around Flinders Island may be a contingent (*sensu* Secor 1999) nonreproductive resident population which do not undertake this northerly migration across Bass Strait. This hypothesis was based on aging data demonstrating an unusually large number (20% >four years old) of *A. trutta* present in catches sampled from Flinders Island which were older than their reported age-at-maturity (Stanley and Malcolm 1977). If there is a nonmigratory sub-population of *A. trutta* in the coastal waters of NE Tasmania, then their continuous residence in these waters could very well produce the distinct multi-element signature we see for Tasmanian fish. The significantly smaller size (and resultant slower growth) of five year old *A. trutta* sampled from Flinders Island compared with all other sampling locations further supports a 'weight of evidence' approach which serves to highlight the need for further work to examine this resident sub-population hypothesis. Many species of fish have migratory and resident forms, particularly anadromous species (Northcote 1978; Elliot 1994; Morinville and Rasmussen 2006). Nonmigratory individuals of such species often exhibit slower growth than migratory individuals (Jonsson 1985; Elliot 1988; Jonsson and Jonsson 1993). Slower growth may be a result of factors like adverse environmental conditions or food quality or quantity (Nordeng 1983; Elliot 1988; Olsson et al. 2006) that migratory forms are better able to exploit or avoid as a result of their migratory habit. It is now being increasingly shown that many marine fish species also have resident and migratory forms (e.g., Secor 1999; Dunn and Pawson 2002; Fukumori et al. 2008). Of particular relevance to *A. trutta*, recent work using stable isotope ($\delta^{13}C$ & $\delta^{18}0$) analysis of otolith carbonate has suggested migratory and resident sub-populations within the stock of another member of the family Arripidae in Australia, the Australian herring *A. georgiana*; the population of which has been shown to consist primarily of migratory animals along with a smaller proportion of resident individuals (Ayvazian et al. 2004). Analysis of otolith chemistry of yellowtail scad *Trachurus novaezelandiae* in Jervis Bay (NSW) has also recently shown that the population consists of resident and migratory sub-populations; the resident animals being significantly smaller at the same age than migratory members of the population (Heagney 2010).

Overall however, the lack of consequential and definitive differences in otolith chemistry data presented here (i.e., no spatial differences in single element analyses or among multi-element signatures for the otolith edges of fish collected from mainland locations) in combination with the results of historical tagging studies (Stanley 1978, 1988a, 1988b), lead us to conclude that the most likely population model for *A. trutta* in SE Australia is of a single biological stock spanning NSW, Victoria and Tasmania. Widespread mixing of individuals occurs both as larvae and juveniles prior to recruitment, in addition to as adults once reproductive maturity is reached and the migration of Tasmanian fish across Bass Strait occurs. In fact, *A. trutta*'s extensive migratory capabilities in SE Australian waters demonstrated by past research have led previous researchers to conclude that *A. trutta* in SE Australia are indeed a single well-mixed biological stock in this region (Stanley 1978).

Conclusions

The demonstration of consistent core-edge differences in otolith chemistry using both uni- and multi-variate analyses is consistent with observations of juvenile estuarine residency of *A. trutta* in SE Australia. In addition, multi-element signature data suggests some differences between fish collected from Tasmanian waters compared with fish collected from other locations in SE Australia.

We suggest that the distinct otolith chemistry of Tasmanian fish is most likely reflective of differences in water chemistry between the waters of Tasmania and mainland Australia. The highly migratory nature of *A. trutta* in this region demonstrated by tagging data combined with the lack of differences in otolith chemistry between mainland locations confirm that the most likely stock structure model for *A. trutta* in SE Australia is of a single well mixed biological stock spanning Tasmania, Victoria, and NSW with fish moving north from Tasmania to mainland Australia with the approach of sexual maturity. However, the cause of the distinct otolith elemental signature for fish from Tasmania were those fish collected from Flinders Island, which when considered along with the reduced growth rates for this group of fish lends more support to the 'weight of evidence' approach which highlights the need for further work to examine the hypothesis of a resident sub-population in this area (Malcolm 1966c; Stanley 1978). Further work should include the use of otolith transect analysis from adult fish collected throughout SE Australia to provide age-specific information on migration patterns in this region. Additional analysis of the otoliths of juvenile fish from multiple locations within SE Australia could also assist in the identification of possible important nursery areas or habitats. These approaches have the potential to provide substantial benefits to the management of this important species via our understanding of the population structure, life history, and migratory behavior of *A. trutta* in the waters of SE Australia.

Acknowledgments

We thank D. Rotherham and two anonymous reviewers for comments on earlier drafts of this manuscript. We are also grateful to the commercial fishers who collected the fish to be used in this study: Harry Mitchelson, Alf Patane, Stuart Richey, and Gary Warren. This study was funded by the Fisheries Research and Development Corporation and the NSW Recreational Saltwater Fishing Trust (FRDC Project No. 2006/018).

References

Ayvazian, S. G., T. P. Bastow, J. S. Edmonds, J. How, and G. B. Nowara. 2004. Stock structure of Australian herring (*Arripis georgiana*) in southwestern Australia. Fisheries Research 67:39–53.

Australian Bureau of Agricultural and Resource Economics. 2007. Australian Fisheries Statistics 2006. Australian Bureau of Agricultural and Resource Economics and Fisheries Research and Development Corporation, ABARE Project 2983, Canberra, Australia.

Bath Martin, G. E., S. R. Thorrold, and C. M. Jones. 2004. Temperature and salinity effects on strontium incorporation in otoliths of larval spot (*Leiostomus xanthurus*). Canadian Journal of Fisheries and Aquatic Sciences 61:34–42.

Campana, S. E. 1999. Chemistry and composition of fish otoliths: pathways, mechanisms and applications. Marine Ecology Progress Series 188:263–297.

Campana, S. E., and Gagné, J. A. 1995. Cod stock discrimination using ICPMS elemental assays of otoliths. Pages 671–691 *in* D. H. Secor, J. M. Dean and S. E. Campana, editors. Recent developments in fish otolith research. University of South Carolina Press, Columbia, South Carolina.

Campana, S. E., G. A. Chouinard, J. M. Hanson, A. Fréchet, and J. Brattey. 2000. Otolith elemental fingerprints as biological tracers of fish stocks. Fisheries Research 46:343–357.

Cappo, M., C. J. Walters, and R. C. Lenanton. 2000. Estimation of rates of migration, exploitation and survival using tag recovery data for western Australian "salmon" (*Arripis truttaceus*: Arripidae: Percoidei). Fisheries Research 44:207–217.

Daverat, F., J. Tomas, M. Lahaye, M. Palmer, and P. Elie. 2005. Tracking continental habitat shifts of eels using otolith Sr/Ca ratios:

validation and application to the coastal, estuarine and riverine eels of the Gironde-Garonne-Dordogne watershed. Marine and Freshwater Research 56:619–628.

Dunn, M. R., and M. G. Pawson. 2002. The stock structure and migration of plaice populations on the west coast of England and Wales. Journal of Fish Biology 61(2): 360–393

Edmonds, J. S., M. J. Moran, N. Caputi, and M. Morita. 1989. Trace element analysis of fish sagittae as an aid to stock identification: Pink snapper (*Chrysophrys auratus*) in western Australia waters. Canadian Journal of Fisheries and Aquatic Sciences 46:50–54.

Edmonds, J. S., R. A. Steckis, M. J. Moran, N. Caputi, and M. Morita. 1999. Stock delineation of pink snapper and tailor from Western Australian by analysis of stable isotope and strontium/calcium ratios in otolith carbonate. Journal of Fish Biology 55:243–259.

Elliot, J. M. 1988. Growth, size, biomass and production in contrasting populations of trout *Salmo trutta* in two Lake District streams. Journal of Animal Ecology 57:49–60.

Elliot, J. M. 1994. Quantitative Ecology of the Brown Trout. Oxford University Press, Oxford, UK.

Elsdon, T. S., and B. M. Gillanders. 2003. Relationship between water and otolith elemental concentrations in juvenile black bream *Acanthopagrus butcheri*. Marine Ecology Progress Series 260:263–272.

Elsdon, T. S., and B. M. Gillanders. 2005. Consistency of patterns between laboratory experiments and field collections in otolith chemistry: an example and applications for salinity reconstructions. Marine and Freshwater Research 56:609–617.

Elsdon, T. S., and B. M. Gillanders. 2006. Identifying migratory contingents of fish by combining otolith Sr:Ca with temporal collections of ambient Sr:Ca concentrations. Journal of Fish Biology 69:643–657.

Fowler, A. J., S. E. Campana, C. M. Jones, and S. R. Thorrold. 1995a. Experimental assessment of the effect of temperature and salinity on elemental composition of otoliths using solution-based ICPMS. Canadian Journal of Fisheries and Aquatic Sciences 52:1421–1430.

Fowler, A. J., S. E. Campana, C. M. Jones, and S. R. Thorrold. 1995b. Experimental assessment of the effect of temperature and salinity on elemental composition of otoliths using laser ablation ICPMS. Canadian Journal of Fisheries and Aquatic Sciences 52:1442–1449.

Fowler, A. J., B. M. Gillanders, and K. C. Hall. 2005. Relatioship between elemental concentration and age from otoliths of adult snapper (*Pagrus auratus* Sparidae): implications for movement and stock structure. Marine and Freshwater Research 56:661–678.

Fukumori, K., N. Okuda, H. Hamaoka, T. Fukumoto, D. Takahashi, and K. Omori. 2008. Stable isotopes reveal life history polymorphism in the coastal fish *Apogon notatus*. Marine Ecology Progress Series 362:279–289.

Gillanders, B. M. 2002. Temporal and spatial variability in elemental composition of otoliths: implications for determining stock identity and connectivity of populations. Canadian Journal of Fisheries and Aquatic Sciences 59:669–679.

Gillanders, B. M., and M. J. Kingsford. 1996. Elements in otoliths may elucidate the contribution of estuarine recruitment to sustaining coastal reef populations of a temperate reef fish. Marine Ecology Progress Series 141:13–20.

Gillanders, B. M., and M. J. Kingsford. 2000. Elemental fingerprints of otoliths of fishmay distinguish estuarine 'nursery' habitats. Marine Ecology Progress Series 201:273–286.

Gillanders, B. M., P. Sanchez-Jerez, J. Bayle-Sempere, and A. Ramos-Espla. 2001. Trace elements in otoliths of the two-banded bream from a coastal region in the south-west Mediterranean: are there differences among locations? Journal of Fish Biology 59:350–363.

Gillanders, B. M., and T. C. Joyce. 2005. Distinguishing aquaculture and wild yellowtail kingfish via natural elemental signatures in otoliths. Marine and Freshwater Research 56:693–704.

Hamer, P. A., G. P. Jenkins, and P. Coutin. 2006. Barium variation in *Pagrus auratus* (Sparidae) otoliths: A potential indicator of migration between an embayment and ocean waters in south-eastern Australia. Estuarine, Coastal and Shelf Science 68:686–702.

Hamer, P. A., and G. P. Jenkins. 2007. Comparison of spatial variation in otolith chemistry of two fish species and relationships with water chemistry and otolith growth. Journal of Fish Biology 71:1035–1055.

Hartill, B. and C. Walsh. 2005. Characterisation of the kahawai fisheries of New Zealand and a review of biological knowledge. Final Research Report for Ministry of Fisheries Research Project KAH2004–01 Objective 1.

Hatje, V. 2003. Particulate trace metal and major elemental distributions over consecutive tidal cycles in Port Jackson Estuary, Australia. Environmental Geology 44:231–239.

Heagney, E. 2010. Pelagic fish in coastal waters: management of their fine-scale population structure within hydrographic habitats. Doctoral dissertation. University of New South Wales, Australia

Jonsson, B. 1985. Life history patterns of freshwater resident and sea-run migrant brown trout in Norway. Transactions of the American Fisheries Society 114:182–194.

Jonsson, B., and N. Jonsson. 1993. Partial migration: niche shift vs. sexual maturation in fishes. Reviews in Fish Biology and Fisheries 3:348–365.

Kalish, J. M., M. E. Livingston, and K. A. Schofield. 1996. Trace elements in the otoliths of New Zealand Blue Grenadier (*Macruronus novaezelandiae*) as an aid to stock discrimination. Australian Journal of Marine and Freshwater Research 47:537–42.

Malcolm, W. B. 1959. The populations of Australian 'salmon', *Arripis trutta* (Bloch & Schneider), in Australian waters. Australian Journal of Marine and Freshwater Research 10:22–29.

Malcolm, W. B. 1966a. Synopsis for F.A.O. species and stocks thesaurus of data on *Arripis trutta* (Bloch and Schneider). Commonwealth-States Fisheries Conference Southern Pelagic Project Committee Technical Session Volume 3, SPP(T)66/1/4, Cronulla, Australia.

Malcolm, W. B. 1966b. Distribution of juvenile salmon in Tasmanian coastal waters. Commonwealth-States Fisheries Conference Southern Pelagic Project Committee Technical Session Volume 3, SPP(T)66/44A, Cronulla, Australia.

Malcolm, W. B. 1966c. Area of distribution and movement of the eastern subspecies of the Australian 'salmon'. Commonwealth-States Fisheries Conference Southern Pelagic Project Committee Technical Session Volume 3, SPP(T)66/44A, Cronulla, Australia.

McCulloch, M., M. Cappo, J. Aumend, and W. Müller. 2005. Tracing the life history of individual barramundi using laser ablation MC-ICP-MS Sr-isotopic and Sr/Ba ratios in otoliths. Marine and Freshwater Research 56:637–644.

Morinville, G. R., and J. B. Rasmussen. 2006. Does life-history variability in salmonids affect habitat use by juveniles? A comparison among streams open and closed to anadromy. Journal of Animal Ecology 75:693–704.

Nicholls. A. G. 1973. Growth in the Australian "salmon," *Arripis trutta* (Bloch and Schneider). Australian Journal of Marine and Freshwater Research 24:159–176.

Nordeng, H. 1983. Solution to the 'char problem' based on Arctic char (*Salvelinus alpinus*) in Norway. Canadian Journal of Fisheries and Aquatic Sciences 40:1372–1387.

Northcote, T. G. 1978. Migratory species and production in freshwater fishes. Pages 326–359 *in* S. D. Gerking, editor. Ecology of Freshwater Fish Production. Blackwell Scientific Publications, Oxford, UK.

Olsson, I. C., L. A. Greenberg, E. Bergman, and K. Wysujack. 2006. Environmentally induced migration: the importance of food. Ecology Letters 5:645–651.

Paulin, C. 1993. Review of the Australasian fish family Arripididae (Percomorpha), with a description of a new species. Australian Journal of Marine and Freshwater Research 44:459–471.

Ridgway, K. R., and J. R. Dunn. 2003. Mesoscale structure of the mean East Australian Current System and its relationship with topography. Progress in Oceanography 56:189–222.

Sadovy, Y., and K. P. Severin. 1992. Trace elements in biogenic aragonite: correlation of body growth rate and strontium levels in the otoliths of the white grunt, *Haemulon plumieri* (Pisces: Haemulidae). Bulletin of Marine Science 50:237–257.

Secor, D. H. 1999. Specifying divergent migrations in the concept of stock: the contingent hypothesis. Fisheries Research 43:13–34.

Stanley, C. A. 1978. Area of Distribution, Movements, Age Composition and Mortality Rates of the Australian Salmon Population in Tasmania, Victoria and New South Wales. Australian Journal of Marine and Freshwater Research 29:417–433.

Stanley, C. A. 1988a. Tagging experiments on Australian salmon (*Arripis trutta*): recapture data for New South Wales releases, 1939–1965, and Victorian releases 1939–1964. CSIRO Marine Laboratories Report 197.

Stanley, C. A. 1988b. Tagging experiments on Australian salmon (*Arripis trutta*): recapture data for Tasmanian releases, 1949–1964. CSIRO Marine Laboratories Report 193.

Stanley, C. A., and W. B. Malcolm. 1977. Reproductive cycles in the eastern subspecies of the Australian salmon, *Arripis trutta marginata* (Cuvier & Valenciennes). Australian Journal of Marine and Freshwater Research 28:287–301.

Stransky, C., C. -D. Garbe-Schönberg, and D. Günther. 2005. Geographic variation and juvenile migration in Atlantic redfish inferred from otolith microchemistry. Marine and Freshwater Research 56:677–692.

Swan, S. C., J. D. M. Gordon, and T. Shimmield. 2003. Preliminary investigations on the uses of otolith microchemistry for stock discrimination of the deep-water black scabbardfish (*Aphanopus carbo*) in the North East Atlantic. Journal of Northwest Atlantic Fisheries Science 31:221–231.

Thorrold, S. R., C. M. Jones, and S. E. Campana. 1997. Response of otolith microchemistry to environmental variations experienced by larval and juvenile Atlantic croaker (*Micropogonias undulatus*). Limnology and Oceanography 42:102–111.

Toole, C. L., D. F. Markle, and P. M. Harris. 1993. Relationships between otolith microstructure, microchemistry, and early life history events in Dover sole, *Microstomus pacificus*. Fishery Bulletin 91:732–753.

Tzeng, W. N., K. P. Severin, C. H. Wang, and H. Wickstom. 2005. Elemental composition of otoliths as a discriminator of life stage and growth habitat of the European eel, *Anguilla anguilla*. Marine and Freshwater Research 56:629–636.

American Fisheries Society Symposium 76:439–452, 2012

Otolith Sr:Ca Ratios as a Natural Mark to Discriminate Restocked and Naturally Recruited European Eels in Latvia

YU-JIA LIN
Institute of Fisheries Science, National Taiwan University, Taipei 10617, Taiwan, R.O.C.

JEN-CHIEH SHIAO
Institute of Oceanography, National Taiwan University, Taipei 10617, Taiwan, R.O.C.

MARIS PLIKSHS
Latvian Fish Resources Agency, State Agency, Ministry of Agriculture LV-1048, Republic of Latvia

ATIS MINDE
Latvian Fish Resources Agency, State Agency, Ministry of Agriculture LV-1048, Republic of Latvia

YOSHIYUKI IIZUKA
Institute of Earth Sciences, Academia Sinica, Nankang, Taipei 11574, Taiwan, R.O.C.

ISAAK RASHAL
Institute of Biology, University of Latvia, Salaspils, LV-2169, Republic of Latvia

WANN-NIAN TZENG*
Institute of Fisheries Science, National Taiwan University, Taipei 10617, Taiwan, R.O.C.
and
Department of Environmental Biology and Fisheries Science
National Taiwan Ocean University, Keelung 20224, Taiwan, R.O.C.

*Corresponding author: wnt@ntu.edu.tw

Abstract.—Restocking programs of the European eel *Anguilla anguilla* have been conducted for nearly one century in Latvia. To evaluate the efficiency of the eel restocking program and reveal the migratory life histories of European eels in Latvian waters, a total of 75 individuals was collected from the mouth of River Daugava (Daugavgrīva, brackish), a nearby lake (Lake Ķīšezers, freshwater), and a coastal site (Mērsrags, brackish). The naturally-recruited eels consisted of two saltwater types: eels that lived in saltwater and did not enter freshwater (SW, 0–7%) and eels that experienced both freshwater and saltwater, referred to as inter-habitat-shifter (IHS, 60–85%). Restocked eels consisted of purely freshwater types (FW, 7–36.7%) without any exposure to saltwater. Average Sr:Ca ratios at the edge of otoliths were different among sites and origins, and corresponded to the salinities of sampling sites. The mean age at first freshwater entry of IHS was 4.8 ± 2.3 years and was similar to previous studies in Lithuania. The growth rates of FW eels in the river mouth and coast were significantly slower than that of IHS eels ($p < 0.001$ and 0.012). The use of otolith Sr:Ca ratios as a natural mark to distinguish restocked eels here enabled the assessment of restocking efficiency.

Introduction

European eel *Anguilla anguilla* is a facultative catadromous fish spawning in the Sargasso Sea and living in the freshwater, brackish, and costal waters in Europe (Tsukamoto et al. 1998; Tzeng et al. 2000; Tesch 2003). In Latvia, *A. anguilla* was previously one of the most valuable fish sought by fishermen. To enhance local eel production in Latvia, a restocking plan was established in 1927. Approximately 30 million glass eels imported from France were released by the government in 51 Latvian lakes between 1960 to 1988. From 1990 to the present, numerous glass eels were released in inland freshwaters by private fishing enterprises.

Although the restocking plans have been conducted for a long time in Latvia, restocking efficiency was seldom evaluated and the contribution of restocked eels to the local population have not been examined. This might be due to a lack of appropriate marks or tags to discriminate restocked eels from naturally-recruited eels. Recent studies have indicated that otolith Sr:Ca ratio are a useful natural mark to discriminate fish from freshwater and saltwater habitats (Campana 1999; Campana and Thorrold 2001), and have been used to reveal the migratory histories of *A. anguilla* (e.g., Tzeng et al. 1997; Tsukamoto et al. 1998; Tzeng et al. 2000).

Restocked glass eels were collected from estuaries in France and released in freshwater lakes in the middle reaches of the river Daugava. Naturally-recruited eels entered the brackish Baltic Sea and migrated to rivers of southeastern Baltic countries. The differences in route to Latvian waters resulted in different saltwater exposure between restocked and natu-rally-recruited eels. Otolith Sr:Ca ratios, therefore, have been used to examine the restocking efficiency of eels in Lithuania (Shiao et al. 2006) and examine growth differences between restocked or naturally-recruited eels (Lin et al. (2007b). However, restocking efficiency and differences in growth rate have not been evaluated in Latvian waters.

The objectives of this study were: (1) to identify the origins of eels (restocked or naturally-recruited) from three Latvian waters: river mouth (Daugavgrīva), nearby lakes (Lake Ķīšezers), and Baltic coastal region (Mērsrags),and (2) to examine the growth differences of eels with different migratory histories and origins. These results are compared with previous studies in Lithuania (Shiao et al. 2006; Lin et al. 2007b).

Methods

Study site

European eels *A. anguilla* were collected by fyke-nets from Daugavgrīva, the mouth of River Daugava, during August to September, 2005, as well as from nearby Lake Ķīšezers and a coastal site, Mērsrags, during August to September 2006 in Riga Gulf (Figure 1). Riga Gulf is located in the eastern part of the Baltic with the salinity ranging from 2 to 7%. The Latvian coastline of Riga Gulf is about 250 km and the average depth of the coastal zone (2 mi) is around 10 m. Bottom vegetation in the coastal waters of the gulf is unevenly distributed. Brown algae are dominant, followed by the filamentous green algae and red algae. In general, Riga Gulf can be considered highly eutrophic.

Lake Ķīšezers is located close to the coast and connected with the Baltic Sea via the lower reach of River Daugava. It is a freshwater lake, but strong western winds can push brackish water from the Gulf of Riga into the lake for short periods. The area of this lake is 1,730 ha with an average depth of 2.4 m and a maximum depth of 4.5 m in the central part. Lake Ķīšezers is also characterized by the water's brown color and low transparency, which is around 0.5 m in the summer.

The River Daugava is the largest river in Latvia. The width of the lower reach of River Daugava is around 500 m and the depth is around 15 m. The sampling site in the river mouth is brackish with a mean salinity of 4.2%. Three hydroelectric plants were constructed on Daugava and one located 28 km away from the river mouth is not equipped with a fish passage. Therefore, the upstream migration of diadromous fish in the River Daugava is cut off, and downstream migration is limited.

The total length (TL) and weight (TW) of eels were measured to the nearest 1.0 mm and 1.0 g, respectively. Sexes were determined macroscopically from the gross mor-

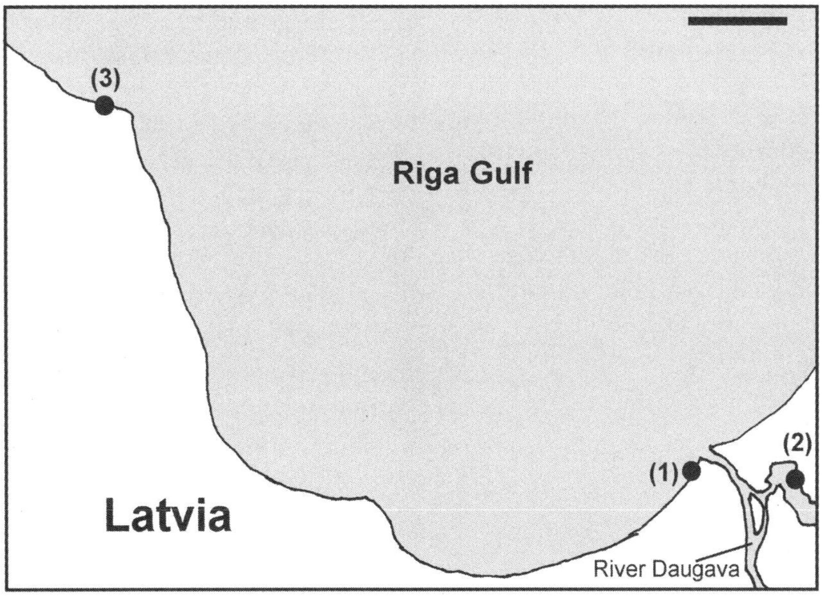

Figure 1. Sampling sites of European eel *Anguilla anguilla* in Gulf of Riga in Latvia. (1) Mouth of River Daugava (Daugavgrīva), (2) Lake Ķīšezers and (3) Mērsrags. Scale bar = 5 km.

phology of gonads. Eels with thin, regularly lobed organs (Syrski's organ) were considered males and individuals with more broad and folded curtain-like gonads were females (Tesch 2003).

Otolith Sr:Ca ratio measurement and age determination

The largest pair of otoliths (sagittae) was removed, dried in air, embedded in epoxy resin, ground and polished until the premordium was exposed. The polished otoliths were coated with carbon under a high-vacuum evaporator. Sr and Ca concentrations in the otolith (wt %) were measured from the primordium to the edge at 10 μm intervals using an electron probe microanalyzer (EPMA, JEOL JXA-8900R); the same conditions and standards described by Lin et al. (2007b). The critical otolith Sr:Ca ratio indicating freshwater residency was set at 2.24×10^{-3} (Shiao et al. 2006), and the region where two consecutive points were lower than the critical value was regarded as freshwater experience (Jessop et al. 2007). If the EPMA transect passed through the vaterite inclusion, it was not used to avoid misidentification of the migratory group (Tzeng et al. 2007; Jessop et al. 2008). The otolith edge was deposited when the fish was caught, thus containing environmental information about the sampling site. Therefore, otolith ratios of the last three points to the otolith edge were averaged. To evaluate whether the different salinities among sites and eel origins can affect average otolith Sr:Ca ratios in the edge, the average values were compared by sites and origins.

After EPMA analysis, the otoliths were re-polished to remove the carbon layer and etched with 5% EDTA for 1–2 min to reveal annual rings. The age of the fish was estimated by counting the number of annual rings and the growth rate (GR) was calculated as $(TL - L_0) \times Age_{-1}$ where L_0 is mean length of elvers (Wang and Tzeng 2000). The age of first freshwater entry was estimated by counting the number of rings prior to the first otolith Sr:Ca ratio meeting freshwater criterion.

Data Analysis

The proportion of restocked eels between sites were compared using a chi-square test. The average of otolith Sr:Ca ratios at the edge were compared by sites and origins using two-way analysis of variance (ANOVA) with Turkey's multiple comparison. Total length, age, and growth rate were log-transformed to meet the assumption of normal distribution (KS Test, all $p > 0.08$) and homogeneity of variance among groups (Levene's Test, all $p > 0.29$). They were then compared among sites and migratory history types by two-way ANOVA. The effects of sites, origins and site-origin interaction on log-transformed growth rates were compared by two-way ANOVA. Although the distribution of the age of first freshwater entry was not normally distrib¬uted for eels from Latvia and Lithuania (KS test, all $p < 0.001$), they were homogeneous to each other (KS test, $p = 0.64$). Therefore, the age at first freshwater entry of the eels between Latvia and Lithuania were compared by nonparametric Wilcoxon Rank Sum test. All the statistical computation were completed in SAS® (Version 8.01) and the significance level α was at 0.05.

Results

Migratory history types and origins of the eels

A total of 75 eels were collected, including 31 in Daugavgrīva, 16 in Lake Ķīšezers, and 28 eels in Mērsrags. All eels were female except one caught in Daugavgrīva, which was excluded from subsequent analysis. The Sr:Ca ratios of eels after the elver stage were classified into one of three categories including: 1) saltwater types (SW) where the otolith Sr:Ca ratios were consistently higher than the critical value of 2.24×10^{-3}; indicating that eels lived exclusively in saltwater (Figure 2a), 2) inter-habitat- shifter (IHS) where the ratios were initially higher and then varied between the critical value; indicating movement between saltwater and freshwater (Figure 2b) and 3) freshwater types (FW) where the ratios were consistently lower than the critical value; indicating exclusive freshwater exposure (Figure 2c). The eels examined were mainly IHS, followed by FW with very few SW individuals. Eighteen IHS, 12 FW, and 1 SW eels were found in Daugavgrīva, 11 IHS and 5 FW eels were found in Lake Ķīšezers, and 26 IHS, 0 FW, and 2 SW eels were identified in Mērsrags (Table 1).

Naturally-recruited (Nat) eels experienced saltwater when migrating across the brackish Baltic Sea while restocked (Rst) eels did not experience saltwater. Therefore, Nat eels included SW and IHS eels and Rst eels included FW types only. However, two IHS types were considered Rst eels, because they entered freshwater after metamorphosis. At radius of about 750 and 1500 μm, they moved to saltwater and showed elevated Sr:Ca ratios representative of brackish coast water (Figure 3). The proportions of Rst eels was 36.7% in Daugavgrīva, 31.2% in Lake Ķīšezers, and 7.1% in Mērsrags, significantly different among sites χ^2 Test, $p = 0.015$) (Table 1). The temporal changes of otolith Sr:Ca ratios differed substantially between Nat and Rst eels. The average otolith Sr:Ca ratios for the Nat eels were approximately 5×10^{-3} at the elver stage and gradually decreased to 2×10^{-3} at the edge. The average ratios of Rst eels remained at a low value of around 1.5×10^{-3} from the elver stage until the edge. The 95% confidence interval for the mean did not overlap until at radius of around 1000 μm (Figure 4), indicating their different migratory histories.

Average otolith Sr:Ca ratio in the edge among sites and origins

In Nat eels, the average (±SD) otolith Sr:Ca ratio at the edge was 2.69 ± 1.16, 1.35 ± 0.92 and $2.33 \pm 1.26 \times 10^{-3}$ in Daugavgrīva, Ķīšezers and Mērsrags, respectively. In Rst eels, they were 0.70 ± 0.41, 0.82 ± 0.74 and $2.29 \pm 0.02 \times 10^{-3}$ (Table 2). The average edge ratios of eels were significantly different by sampling sites and origins (Two-way ANOVA, $p = 0.03$ and <0.001, respectively). When compared between origins by sites, the Nat eels had significantly higher average edge ratios than Rst eels in Daugavgrīva ($p < 0.001$). However, the edge ratios were not significantly different between origins in Ķīšezers and Mērsrags ($p = 0.28$ and >0.5, respectively). When compared among sites by origins, the edge ratios of Nat eels from Daugavgrīva and Mērsrags were significantly higher than those of Nat eels from Ķīšezers ($p = 0.01$), corresponding to the salinity of sampling sites. The edge ratios of Rst eels from Mērsrags were significantly higher than those from Daugavgrīva and Ķīšezers ($p = 0.003$), because the two Rst eels moved to the coast (Figure 3).

Otolith annulus patterns and age at first freshwater entry

The otolith annulus patterns of Nat and Rst eels is shown in Figue 5a and 5b. The region from the primordium (P) to the elver check (EC), which corresponded to the marine larval period from leptocephalus to the elver stage, was dark in appearance after etching by HCl. After EC, clear circular annuli were deposited in the otolith and some less-clear pseudo-annuli were found between the success annuli. The otolith annulus patterns were generally similar between the Nat (Figure 5a) and Rst eels (Figure 5b) except that more annuli were found in Rst eels because of older ages. The mean ±SD

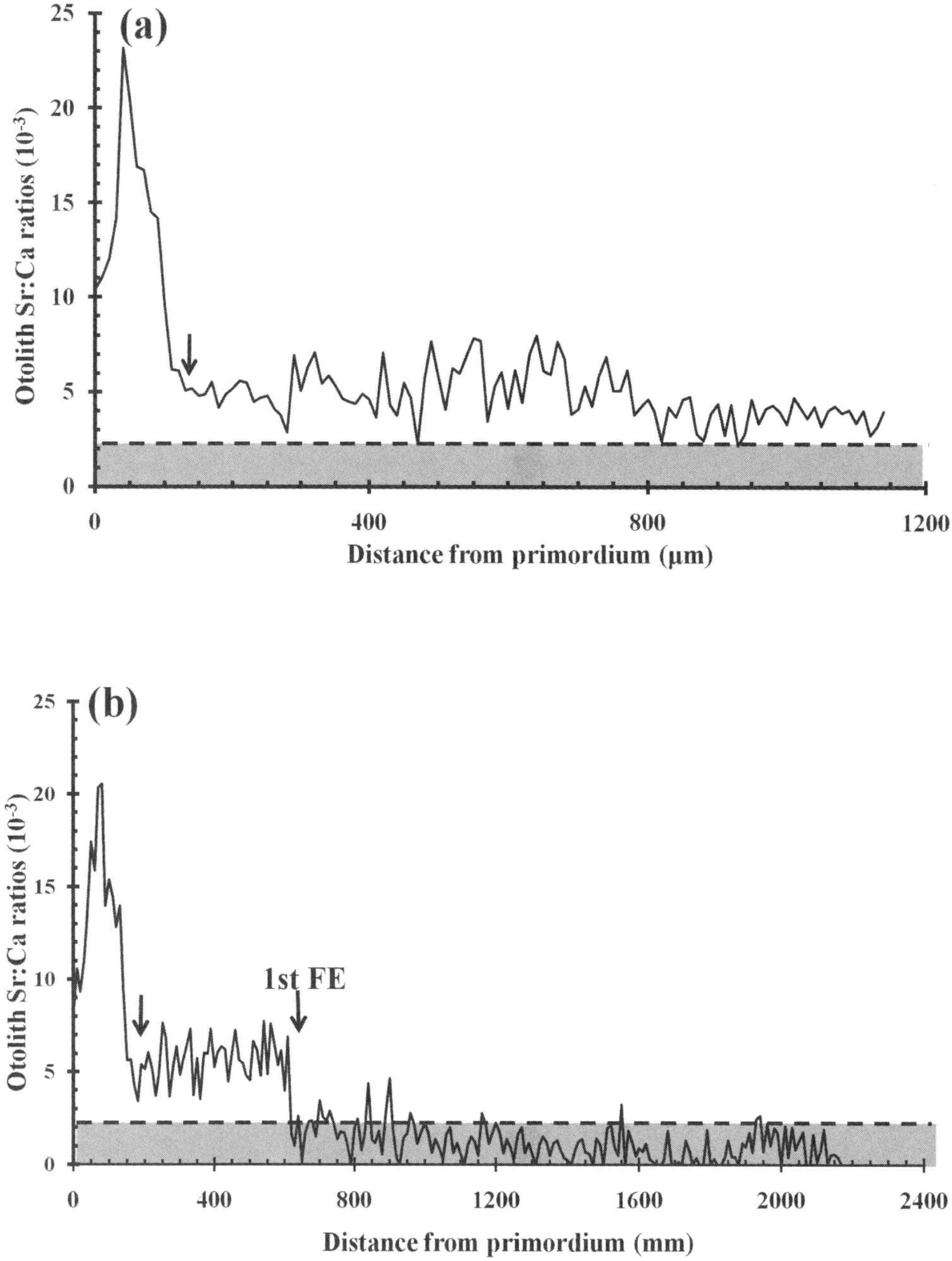

FIGURE 2. Otolith Sr:Ca ratios of (a) saltwater type (SW), (b) inter-habitat-shifter (IHS), and (c) freshwater-type (FW) eels. The grey area under the broken line (Sr:Ca ratio = 2.23×10^{-3}) indicates the freshwater experiences. Arrows without headings = elver check. 1st FE = the first freshwater entry.

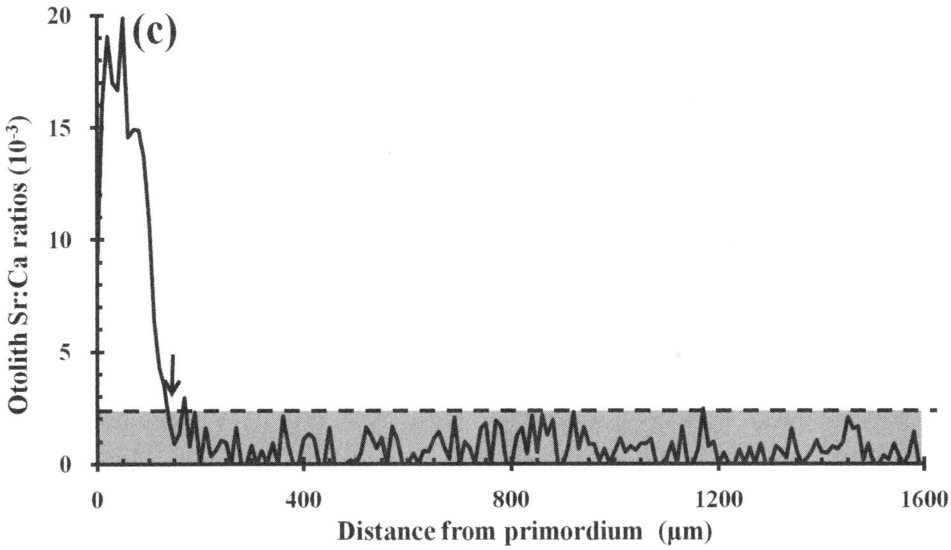

Figure 2. Continued.

age at first freshwater entry of the IHS, determined by otolith Sr:Ca ratios (Figure 2b), was 4.8 ± 2.3 years and ranged from 1 to 11 years and was not significantly different from eels in Lithuania (Shiao et al. 2006; mean ±SD = 5.2 ± 2.1 year, Wilcoxon Rank Sum test, $p = 0.19$). There were two peaks in the age at first freshwater entry in both countries: 55–57% at age 3–5 years and 23–28% at age of 7–9 years (Figure 6).

Growth rate comparisons between sampling sites and migratory history types

FW (equivalently Rst) eels were significantly smaller total length and older age than the IHS (Nat) eels (two-way ANOVA, $p = 0.028$ and <0.001) in Daugavgrīva, but the dif¬ferences in total length and age between FW and IHS eels were not significant in Mērsrags ($p = 0.09$ and 0.13) or Lake Ķīšezers ($p = 0.40$ and 0.60). The growth rates were significantly influenced by sampling sites, migratory history types, and their interaction (two-way ANOVA, all $p < 0.001$). When the sites were separated, the growth rate of FW eels was significantly slower than that of IHS eels ($p < 0.001$) in Daugavgrīva and Mērsrags ($p = 0.01$). The growth rate was not significantly different between FW and IHS eels in Lake Ķīšezers, ($F = 0.85$ $p = 0.37$) (Table 1).

Discussion

Implication of different migratory history types

The predominance of IHS eels indicated that movements from brackish Baltic waters into freshwater habitats was frequent. This result was not consistent with eels in Lithuanian waters (Shiao et al. 2006). Since sampling sites in the present study were either near or in the coastal waters of Latvia, freshwater habitats entering the Baltic Sea appear to be a critical habitat for European eels, which have flexible salinity tolerances (Daverat et al. 2006).

Table 1. European eel *Anguilla anguilla* collected in Daugavgrīva, Lake Ķīšezers and Mērsrags in Latvia with mean salinities (‰) in the brackets. Percentage by origins (Nat: natural-recruited, Rst: restocked eels) and numbers of eels by migratory history types (Ntypes, IHS: inter-habitat- shifters, SW: saltwater type, FW: freshwater type eels). Mean (± SD, ranges in parenthesis) total length (TL, cm) and weight (TW, g), age (years) and growth rate (GR, cm year-1). ND = data were not available.*: All were females except one male (TL = 55.9 cm).

Site	Daugavgriva (4.2)			Lake Ķīšezers (0)		Mērsrags (5.6)		
Origin	Nat (63.3 %)		Rst (36.7 %)	Nat (68.8 %)	Rst (31.2 %)	Nat (92.9 %)		Rst (7.1 %)
Type	IHS	SW	FW	IHS	FW	IHS	SW	IHS
N_{type}	18	1	11(1)*	11	5	24	2	2
TL	58.7 ± 9.9	63.2	69.3 ± 9.9	80.3 ± 7.0	83.6 ± 7.0	78.7 ± 6.3	80.1 ± 10.9	68.3 ± 11.3
			(58.5 ~ 91)	(69.4 ~ 91.7)	(73.8 ~ 90.5)	(66.2 ~ 90.7)	(72.4 ~ 87.8)	(60.3 ~ 76.3)
TW		ND		949 ± 236	1047 ± 291	1021 ± 244	1153 ± 576	691 ± 389
				(521 ~ 1280)	(712 ~ 1330)	(543 ~ 1510)	(745 ~ 1560)	(416 ~ 966)
Age	10 ± 2	13	22 ± 5	14 ± 2	13 ± 4	14 ± 2	14 ± 1	18 ± 4
	(6 ~ 13)		(17 ~ 31)	(10 ~ 17)	(8 ~ 17)	(10 ~ 20)	(13 ~ 15)	(15 ~ 20)
GR	5.5 ± 1.1	4.4	3.0 ± 0.6	5.4 ± 0.8	6.2 ± 2.4	5.5 ± 0.9	5.3 ± 0.2	3.7 ± 1.4
	(4.1 ~ 9.0)		(2.0 ~ 4.0)	(4.5 ~ 7.0)	(4.6 ~ 10.4)	3.7 ~ 7.4)	(5.1 ~ 5.5)	(2.7 ~ 4.7)

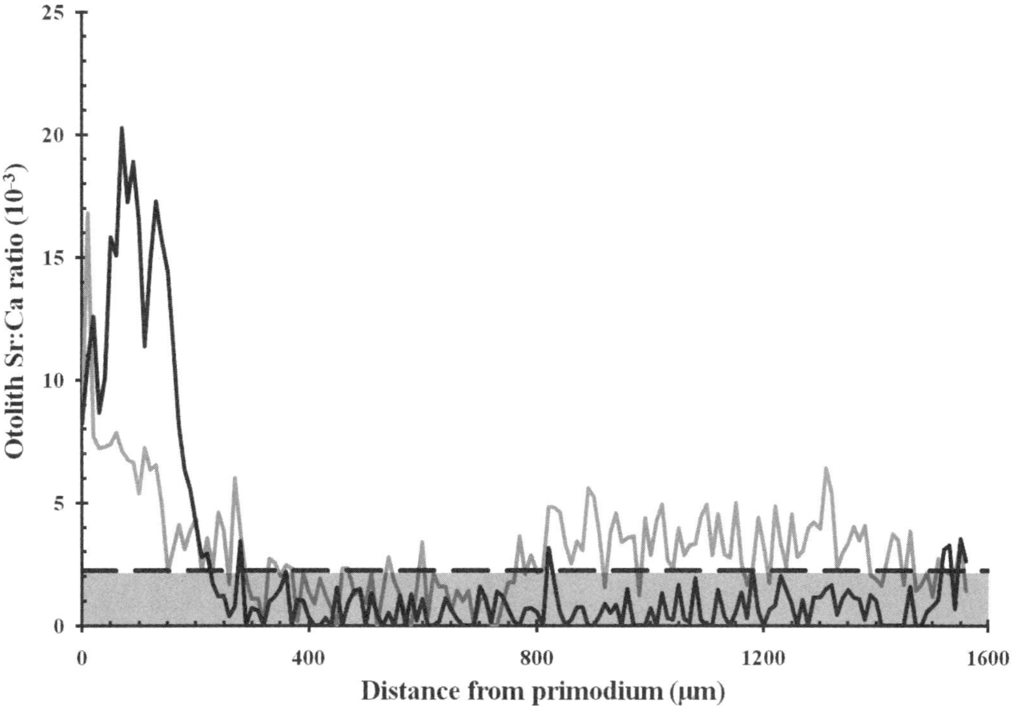

Figure 3. The otolith Sr:Ca ratios of two restocked eels from Mērsrags. The grey area under the black broken line (Sr:Ca ratio = 2.23 × 10^{-3}) indicated the freshwater experiences.

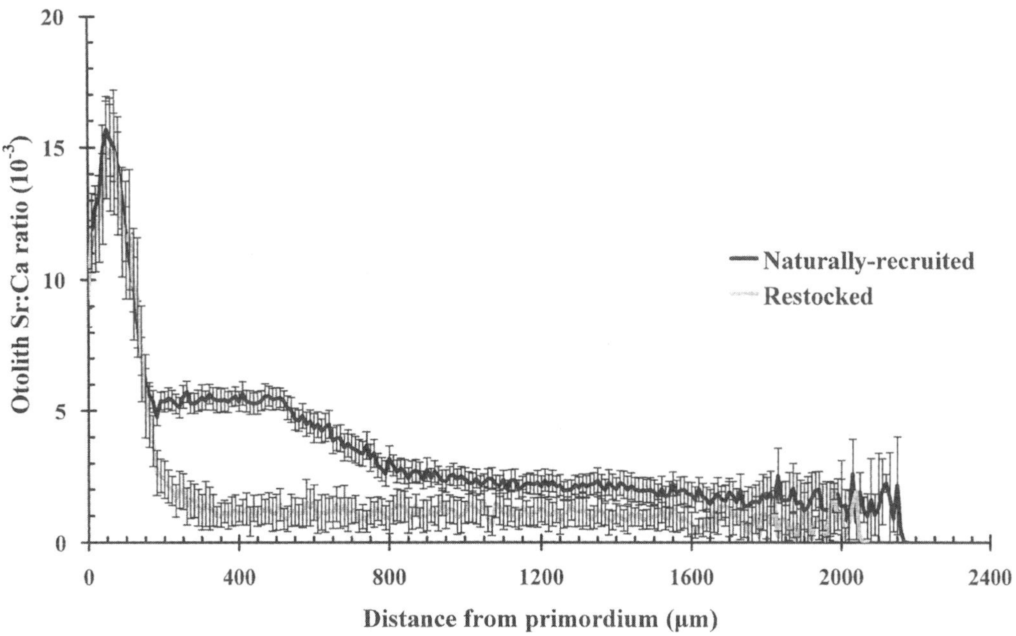

Figure 4. Mean (± 95 % CI) otolith Sr:Ca ratios of naturally-recruited (black line, n = 56) and restocked (grey line, n = 18) eels.

TABLE 2. Mean (± SD, ranges in parenthesis, × 10^{-3}) of otolith Sr:Ca ratio of last three points in the edge of naturally-rectuited (Nat) and restocked eels (Rst) by sites. Different superscripts indicate significant difference in otolith Sr:Ca ratios among sites. The asterisk in Daugavgrīva indicate that the average edge ratios of Nat eels significantly differed from the ratios of Rst eels.

Site	Daugavgrīva*	Ķīšezers	Mērsrags
Nat	2.69 ± 1.16[a] (0.00 ~ 4.24)	1.35 ± 0.92[b] (0.13 ~ 3.03)	2.33 ± 1.26[a] (0.50 ~ 4.59)
Rst	0.70 ± 0.41[a] (0.13 ~ 1.44)	0.82 ± 0.74[a] (0.07 ~ 2.05)	2.29 ± 0.02[b] (2.28 ~ 2.3)

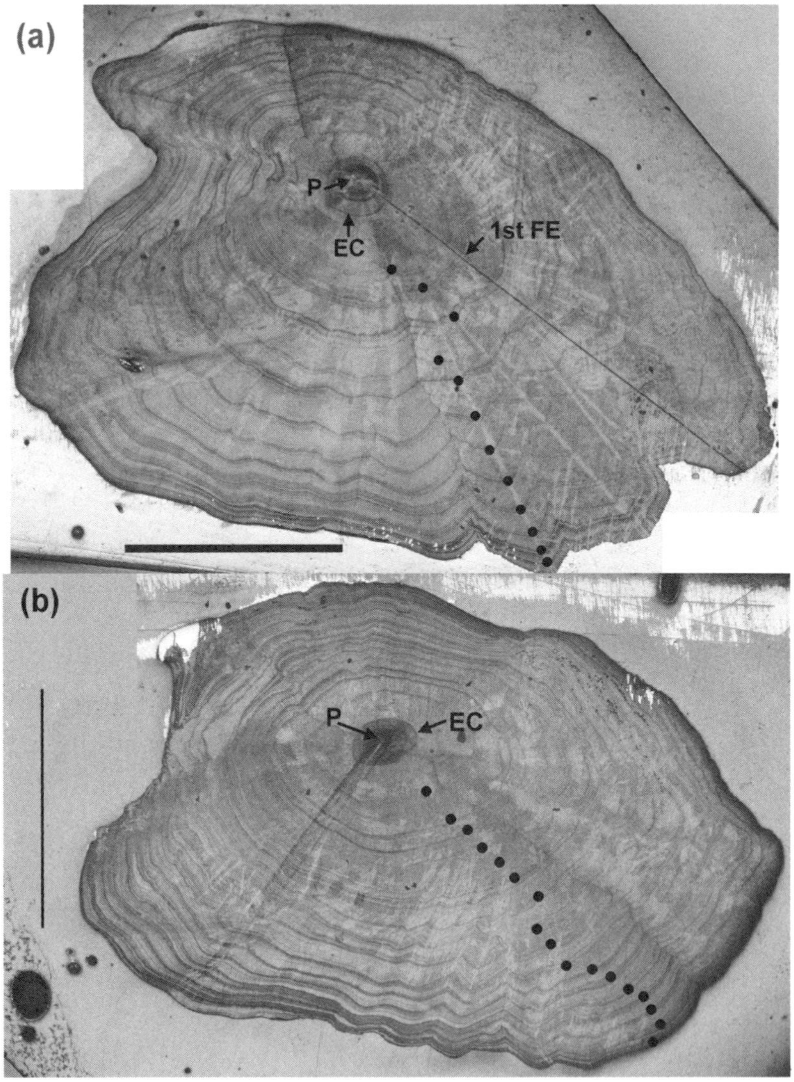

FIGURE 5. Sagittal plan of otoliths of (a) naturally-recruited with Sr:Ca profiles in Figure 2b and (b) restocked eels in Figure 2c. P = primordium, EC = elver check. 1st FE = 1st freshwater entry. Black cycles are annuli. The white line from the primordium to edge was the track where Sr:Ca ratios were measured. Scale bar = 1 mm.

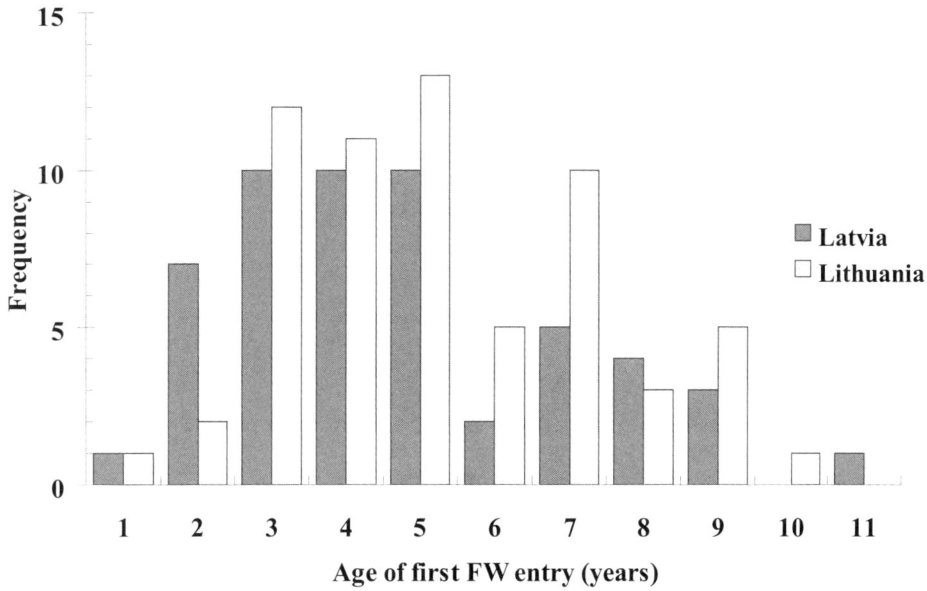

FIGURE 6. Age of first freshwater entry for naturally-recruited eels from Latvia (grey, mean (± SD) = 4.8 ± 2.3 years) and Lithuania (white, 5.2 ± 2.1 years; compiled from Shiao et al. 2006).

Proportion of Rst eels

If some of the Rst eels in Latvian waters came from other countries, such as Lithuania, they had to move though the brackish Baltic, leaving elevated otolith Sr:Ca ratios. All but two eels in Mērsrags entered freshwater and remained there until being caught. Therefore, the majority of Rst eels in Latvia are probably not from other countries. The higher proportion of Rst eels in Lake Ķīšezers and Daugavgrīva than in Mērsrags was expected because Rst elvers were released in some midland lakes of the River Daugava. The outlets of these lakes were equipped with eel weirs that covered most escapements. Downstream migration of eels that have may have overcome the weirs was further limited by the hydroelectric power station dams without fish ladders.

The existence of Rst eels in Daugavgrīva and Lake Ķīšezers implied that at least some Rst eels navigated these obstructions successfully as they were caught in the waters in the lower reach of the River Daugava and Baltic coast. Sampled eels were composed of both yellow and silver eels and therefore, the spawning migration of silver Rst eels (Limburg et al. 2003; Westin 2003) was not limited entirely by physical obstructions to movement. It is unclear how eels passed through these obstructions although it probably occurred during periods of water discharge. Although eels can pass through power plants, they suffer mortality ranging from 15 to 38% (ICES 2007). Therefore, the contribution of Rst eels to downstream sites in Latvia is still constrained.

On the other hand, the proportions of Rst eels in river mouth (36.7%) and coastal lake (31.2%) in Latvia were substantially higher than that in the Curonian Lagoon in Lithuania (20%, Shiao et al. 2006). It was also slightly higher in coastal waters in Latvia (7.1%) than in Lithuania (2%). Numerous factors might explain this phenomenon, such as the number and frequency of glass eels released, days, months, seasons and years of releasing, the geographic features, current systems, behavioral differences, productivity of the waters where the eels were released, and the degree of connectivity between upper and lower reaches.

Average otolith Sr:Ca ratios in the edge

The average Sr:Ca ratio at the edge of Nat eel otoliths corresponded to the salinities of sampling sites. This indicated that the average Sr:Ca ratio of the last three points was useful to distinguish brackish sites from the freshwater sites (Lin et al. 2007a). Insignificant differences in the average edge ratios between Nat and Rst eels in Lake Ķīšezers and Mērsrags also suggests that the average edge ratios can be used to represent the salinity of sampling site. However, in Daugavgrīva, the edge ratios of Rst eels were significantly lower than those of Nat eels. It is suggested that other factors, such as eel origins, should be considered when applying the otolith Sr:Ca ratio in the edge as an indicator of habitat salinity. This might also imply that Rst eels have stayed in midland freshwaters for a certain period and were caught soon after moving to Daugavgrīva.

Age at first freshwater entry

The similarity in age at first freshwater entry between eels sampled in Latvia and Lithuania suggested similar migration behaviors from the the Baltic Sea to freshwater. Because the eels were sampled in different years (2003–2004 in Lithuania and 2005–2006 in Latvia), it also indicated that the age at first entry to freshwater was consistent over the investigated period. However, there was a relatively high degree of individual variation, as indicated by the range from 1 to 11 years. Around the 1910s to 1920s, the elvers were found to enter freshwater in the northeastern and eastern regions of Sweden and southern regions of Lithuania (Figure 6 in Dekker 2003a), suggesting that the eels were able to cross the Baltic Sea at a younger age. But the drastic decrease in recruitment of European eels in recent years (Dekker 2003b) might discourage the eels from migrating eastward (Shiao et al. 2006). Two modes at age of 3–5 and 7–9 implied that the migration of the eels toward freshwaters might be composed of two subtypes; one moving to freshwater quickly and the other moving slowly. This bimodality may be explained, in part, by the different types of movement exhibited by freshwater eels; namely founders, pioneers, home range dweller, and nomads (Feunteun et al. 2003).

Differences in growth rates between migratory history types and eel origins

Because few SW eels were sampled, the growth comparison focused on IHS and FW eels, and Nat versus Rst eels. The growth rates of FW (Rst) eels were smaller than those of IHS (Nat) eels in Daugavgrīva and Mērsrags, but they did not differ in Lake Ķīšezers. The productivity was similar among the three studied sites, and the slower growth rate of FW eels in Daugavgrīva might be due to that the FW eels having escaped from midland lakes with low productivity, in which growth of eels is slow (Lin et al. 2007b). Previous studies in Lithuania suggested that the sampling sites might be the major factor influencing the growth of the eels rather than their origins (Lin et al. 2007b). The significant site-type interaction in this study suggested a more complicated relationship associated with sampling site and natal origin.

In conclusion, the eels collected from three Latvian waters were composed of mainly IHS eels, followed by the FW eels, and the SW eels being the least common. The Rst eels con-tributed around 30% of the total number of eels caught which suggested a substantial contribution of the restocking program to the eel production in three sampling sites.

The average otolith Sr:Ca ratios of last three measuring points were useful to represent the salinity of the sampling site. The otolith annulus pattern was similar between Nat and Rst eels, except more annuli were visible in the otoliths of Rst eels. The age at first freshwater entry of the eels was similar between sites from Latvia and Lithuania, indicating that migration across the Baltic Sea was spatially and temporally consistent over the period of investigation. The growth of FW (Rst) eels was not consistently slower than the IHS (Nat) eels between sampling sites which indicated a complicated site–origin interaction on growth. This study further highlights the importance of otolith Sr:Ca ratio's as a natural mark to discriminate restocked European eels from naturally-recruited eels.

Acknowledgment

We are grateful to Lithuania-Latvia-Taiwan (Republic of China) Mutual Fund for the financial support of the collaborative research project between the countries.

References

Campana, S. E. 1999. Chemistry and composition of fish otoliths: pathways, mechanisms and applications. Marine Ecology Progress Series 188:263–297.

Campana, S. E., and S. R. Thorrold. 2001. Otoliths, increments, and elements: keys to a comprehensive understanding of fish populations? Canadian Journal of Fisheries and Aquatic Sciences 58:30–38.

Daverat, F., K. E. Limburg, I. Thibault, J. C. Shiao, J.J. Dodson, F. Caron, W. N. Tzeng,, Y. Iizuka, and H. Wickström. 2006. Phenotypic plasticity of habitat use by three temperate eel species, *Anguilla anguilla*, *A. japonica* and *A. rostrata*. Marine Ecology Progress Series 308:231–241.

Dekker, W. 2003a. On the distribution of the European eel and its fisheries Canadian Journal of Fisheries and Aquatic Sciences 60:787–799.

Dekker, W. 2003b. Epilogue: worldwide decline of eel resources necessitates immediate action Fisheries 28:28–30.

Feunteun, E., P. Laffaille, T. Robinet, C. Briand, A. Baistez, J. M .Oliver, A. Lagault, and A. Acou. 2003. Movements of anguillid eels in freshwater habitats. Pages 191–213 *in* K. Aida, K. Tsukamoto, and K. Yamauchi, editors. Advances in eel biology. Springer, Tokyo.

ICES. 2007. Report of the 2007 Session of the Joint EIFAC/ICES Working Group on Eels. ICES CM 2007/ACFM:23.

Jessop, B. M., J. C. Shiao, Y. Iizuka, and W. N. Tzeng. 2007. Effects of inter-habitat migration on the evaluation of growth rate and habitat residence of American eels *Anguilla rostrata*. Marine Ecology Progress Series 342:255–263.

Jessop, B. M., J. C. Shiao, Y. Iizuka, and W. N. Tzeng. 2008. Prevalence and intensity of occurrence of vaterite inclusions in aragonite otoliths of American eel *Anguilla rostrata*. Aquatic Biology 2:171–178.

Limburg, K. E., H. Wickström, H. Svedäng, M. Elfman, and P. Kristainsson. 2003. Do stocked freshwatr eels migrate? Evidence from the Baltic suggests "Yes." Pages 275–284 *in* D. A. Dixon, editor. Biology, management, and protection of catadromous eels. American Fisheries Society, Symposium 33. Bethesda, Maryland.

Lin, S. H., C. W. Chang, Y. Iizuka, and W. N. Tzeng. 2007a. Salinities, not diets, affect strontium/calcium ratios in otoliths of *Anguilla japonica*. Jounal of Experimental Marine Biolology and Ecology 341:254–263.

Lin, Y. J., L. Ložys, J. C. Shiao, Y. Iizuka, and W. N. Tzeng. 2007b. Growth differences between naturally recruited and stocked European eel *Anguilla anguilla* from different habitats in Lithuania. Journal of Fish Biology 71:1–15.

Shiao, J. C., L. Ložys, Y. Iizuka, and W. N. Tzeng. 2006. Migratory patterns and contribution of stocking to the population of European eel in Lithuanian waters as indicated by otolith Sr:Ca ratios. Journal of Fish Biology 69:749–769.

Tesch, F.W. (2003) The eel, 3rd edition. Blackwell Scientific Publications, Oxford, U.K.

Tsukamoto, K., I. Nakai, W. V. Tesch. 1998. Do all freshwater eels migrate? Nature 396:635–636.

Tzeng, W. N., K. P. Severin, H. Wickström. 1997. Use of otolith microchemistry to investigate the environmental history of European eel *Anguilla anguilla*. Marine Ecology Progress Series 149:73–81.

Tzeng, W. N., C. H. Wang, H. Wickström, M. Reizenstein. 2000. Occurrence of the semi-catadromous European eel *Anguilla anguilla* (L.) in Baltic Sea. Marine Biology 137:93–98.

Tzeng, W. N., C. W. Chang, C. H. Wang, J. C. Shiao, Y. Iizuka, Y. J. Yang, C. F. You, L. Ložys. 2007. Misidentification of the migratory history of anguillid eels by Sr/Ca ratios of vaterite otoliths. Marine Ecology Progress Series 348:285–295.

Wang, C. H., and W. N. Tzeng. 2000. The timing of metamorphosis and growth rates of American and European eel leptocephali: A mechanism of larval segregative migration. Fisheries Research 46:191–205.

Westin, L. 2003. Migration failure in stocked eels *Anguilla anguilla*. Marine Ecology Progress Series 254:307–311.

Section VI

Archival and Pop-up Satellite Tags

Foreword to Archival and Pop-up Satellite Tags Section

ANDREW C. SEITZ

Archival and pop-up satellite tags are sophisticated electronic fish marking devices with onboard computers that record and store data measured by a variety of electronic sensors. To acquire data from archival tags, the tag must be physically recovered by recapturing the marked fish. In contrast, data recovery for pop-up satellite tags is independent of fish recapture as these tags release from fishes on a pre-programmed date and archived data summaries are transmitted to the investigator via orbiting satellites. Because neither tag transmits real-time data while deployed on a fish, both types of tags have freed researchers from arduously following study organisms in inhospitable places and during inopportune times to obtain sensor data. The tags can remain in or on animals for several months and have provided unprecedented insight into the ecology of many fish species, particularly those that migrate.

First generation archival and pop-up satellite tags were large, had limited battery life and contained only a few basic sensors, such as temperature and depth. In contrast, the latest generation of archival and satellite tags are highly miniaturized, contain a wide variety of sensors, are capable of storing high volumes of data and have greatly extended battery lives. Because of their increasing versatility and the fact that the cost of the technology has become cheaper, archival and satellite tags are being increasingly applied to a wide variety of freshwater and marine organisms.

The call for papers for the archival and pop-up satellite tagging sessions requested presentations focusing on the development of tag technology, the diversification of their use, innovative approaches to data utilization, and biological/ecological discoveries gleaned from their use. The sessions contained a broad diversity of fish studies from Earth's three major oceans, plus the Laurentian Great Lakes, by speakers from eight countries.

Included in the published proceedings are four studies where the use of archival and pop-up satellite tags has served to advance our knowledge of tag technology as well as provided detailed descriptions of fish behavior through the innovative use of tagging data. The archival tagging session keynote speakers provided a useful analysis of the components necessary for a successful tagging program and how to maximize its benefits, illustrated with examples from their own archival tagging studies on North Sea plaice *Pleuronectes platessa* and Atlantic cod *Gadus morhua* as well as studies by other investigators (Righton and Metcalfe). Another paper describes how "traditional" data (i.e., depth and temperature) from pop-up satellite tags attached to a very "non-traditional fish" were used to describe the diving behavior of whale sharks *Rhincodon typus* off eastern Africa (Brunnschweiler and Sims). Also in the proceedings are descriptions of two innovative tagging studies. The first describes how archival tags were to used to measure body angles of Greenland halibut *Reinhardtius hippoglossoides* to describe their swimming behavior (Albert et al.), while the second describes the use of archival tags to provide area-under-the curve (AUC) estimates of spawning escapement in Pacific salmon *Oncorhynchus* spp. (Shardlow and Hyatt).

American Fisheries Society Symposium 76:457–469, 2012

Diel Oscillations in Whale Shark Vertical Movements Associated with Meso- and Bathypelagic Diving

Juerg M. Brunnschweiler*
ETH Zurich, Raemistrasse 101, Zurich CH-8092, Switzerland

David W. Sims
Marine Biological Association of the UK, The Laboratory
Citadel Hill, Plymouth PL1 2PB, UK
and
School of Marine Science and Engineering, Marine Institute, University of Plymouth,
Drake Circus, Plymouth PL4 8AA, UK

Abstract.—A female whale shark *Rhincodon typus* satellite tagged off the coast of Mozambique and moving to the east coast of Madagascar where the tag popped off after 87 d showed increased meso- and bathypelagic diving while crossing the bathymetrically nonconstraining southern part of the Mozambique Channel. After three weeks with daily maximum diving depths in the epipelagic zone at the beginning of the track, when the whale shark stayed close to the Mozambique coast, 44 and 19 d with maximum diving depths in the meso- and bathypelagic zones, respectively, were recorded. On a total of 10 d the whale shark dived to the maximum depth (1,285.7 m) the pop-off tag could record. Quantitative analysis of the diving time series using Fast Fourier Transform identified a diel cycle of vertical movements, with deeper depths occupied during the day and shallower depths during daytime. Using a single bandpass filter to identify where in the time series the diel oscillations were most pronounced showed that they were mostly associated with periods of meso- and bathypelagic diving during the oceanic phase of the long-distance movement. These data indicate that meso- and bathypelagic diving is common for whale sharks in bathymetrically nonconstraining habitat, expanding the habitat of this circumglobal species into the bathypelagic realm, and support the hypothesis that these were dives related to foraging behavior when crossing less productive deep oceanic waters.

Introduction

Deep diving behavior appears to be a ubiquitous search strategy for many large epipelagic vertebrates including both planktivores and macropredators (Carey and Scharold 1990; Block et al. 1992, 2001; Hooker and Baird 1999; Panigada et al. 1999; Boustany et al. 2002; Sims et al. 2003; Hays et al. 2006). However, while deep diving behavior is known to occur, it remains a relatively unexplored activity for the majority of species, and its exact function remains controversial (Carey and Scharold 1990; Bonfil et al. 2005).

Frequent vertical movements of epipelagic fish through the water column have

*Corresponding author: juerg@gluecklich.net

been suggested as a strategy to (1) detect concentrations of prey (Josse et al. 1998) or the olfactory clues leading to them (Carey and Scharold 1990; Gunn et al. 1999); (2) to conserve energy, i.e. swim up and glide down strategy (Weihs 1973); (3) as a form of behavioral thermoregulation (Carey and Scharold 1990; Holland et al. 1992), or (4) as a way of orientation or navigation (Klimley et al. 2002). For example, vertical movements of plankton-feeding sharks have been shown to closely reflect the diel movements of zooplankton prey (Sims et al. 2005) while rapid ascents in some pelagic sharks and tuna may be undertaken as part of thermoregulatory behavior to regain heat loss at depth, or to re-oxygenate the gills following time spent in low oxygen layers in the water column, stratifications that are known to occur in tropical waters. Marine vertebrate predators may also exhibit plasticity in their diving and foraging behaviors in response to prey availability (Hays et al. 2006; Miller and Trivelpiece 2008). For example, sperm whales *Physeter macrocephalus* target epipelagic prey but also dive to very deep depths (>1000 m) to search for benthopelagic prey. It appears they feed on more sparsely distributed prey items during shallow dives and more densely distributed prey items during deep dives (Teloni et al. 2008).

One reason for distinct gaps in our knowledge of deep diving among species is that obtaining behavioral information when animals are in remote locations far from land is difficult. Much of the most detailed data on deep diving has been derived from animal-attached (external or implanted internal) data loggers that were physically recovered facilitating downloading of data (e.g., for Atlantic bluefin tuna *Thunnus thynnus thynnus*; Block et al. 2005). However, this is not a viable method for species studied in remote locations, where fishery recoveries are unlikely or where fisheries for certain species are not well developed. Consequently, one important goal is to obtain behavioral information from diving animals in remote locations where data loggers cannot be recovered, for example, during long-distance migrations into the open ocean. The Argos system of receivers aboard polar-orbiting satellites allows remote collection of data from transmitter-tagged animals over many weeks or even months regardless of the remoteness of movements (Sims 2010).

As apex predators, elasmobranchs occupy a wide range of marine trophic niches from shallow coastal waters to meso- and bathypelagic depths in the open oceans. For fish, there is no constraint on foraging imposed by the need to surface to breathe, so the strategy on where to forage in the water column may be made simply on the basis of prey concentration (Sims and Quayle 1998; Sims 2000; Sims et al. 2003). The whale shark *Rhincodon typus* is a planktivorous epipelagic species that spends the majority of its time above 100 m and often at the water's surface feeding on plankton (Clark and Nelson 1997; Gunn et al. 1999; Eckert and Stewart 2001; Eckert et al. 2002; Graham et al. 2006; Wilson et al. 2006; Motta et al. 2010). But whale sharks are also known to show meso- and even bathypelagic diving when in bathymetrically nonconstraining habitats (Graham et al. 2006; Wilson et al. 2006; Brunnschweiler et al. 2009). Whale shark deep dives have been proposed to represent foraging dives that specifically target prey seeking refuge at depth and to exploit organisms of the deep scattering layer (Wilson et al. 2006). Nevertheless, deep diving by whale sharks is not known from all locations where they are known to occur so it is arguable how regular this behavior is. So, in addition to recent studies that have advanced the understanding of whale shark feeding behavior in relation to available prey resources and oceanographic factors (Nelson and Eckert 2007; Motta et al. 2010), there is a great need to better understand how whale sharks utilize deep water pelagic habitats in poorly studied locations.

In February 2006, a female whale shark was equipped with a pop-off satellite archival tag off the coast of Mozambique and, based on light-level longitude estimates that are typically accurate and robust (Teo et al. 2004), spent the first three weeks in shallow coastal habitat before moving into oceanic waters crossing the southern part of the Mozambique Channel and around the southern tip of Madagascar where the tag prematurely popped off after 87 d (Brunnschweiler et al. 2009). In bathymetrically nonconstraining habitat, this shark performed dives to depths well into the meso- and bathypelagic zones including the deepest dives so far recorded for the species. Here, detailed vertical movement data are provided that support the hypothesis that these were dives related to foraging behavior.

Methods

The female whale shark (estimated length 6–7 m) was equipped with a PTT-100 pop-off satellite archival tag (Microwave Telemetry Inc.) that archived pressure readings at a resolution of ±5.4 m every 15 min to a maximum depth of 1,285.7 m. An emergency release feature was enabled that detached it automatically when the shark was deeper than ~1,250 m for more than 15 min. After pop off, the tag relayed archived data via Argos, including daily maximum depths, archived pressure readings, and longitude estimates that were used for analysis. Due to the size limit of a single Argos transmission, the organization of depth data and their transmission to Argos satellites is optimized by the manufacturer so that a single Argos message can hold 24 readings. To achieve this, only the data recorded on the hours 0000, 0600, 1200 and 1800 and the 15 min readings on those hours are actual recorded values. Depth data recorded at any other time are transmitted as the difference between that time and the previous hour. This procedure results in a limit to the maximum range the tag can transmit from one depth reading to the next reading an hour later which is 166.8 m for descents and 172.2 m for ascents (Microwave Telemetry Inc., personal communication).

A total of 7,230 archived pressure readings (=87.2%) were transmitted for the time between 19 February and 16 May 2006. Full archival depth datasets (pressure readings every 15 min = 96 depth readings per 24 h) are available for 49 of the 87 consecutive days and, except for Fast Fourier Transform (FFT) analysis (see below), only those were used for analysis of individual dive profiles. Of all archived pressure readings received, 609 are excluded because the whale shark changed depth more than 166.8 m (descent) and 172.2 m (ascent) between hourly readings, respectively, and therefore pressure values do not represent true depth records (see previous paragraph). Overall, the whale shark spent a considerable amount of time in the first 10 m of the water column (Brunnschweiler et al. 2009). Throughout the entire track, 47.6% of the pressure readings were 0 and 5.4 m readings. These readings were defined as the whale shark being at the water surface. An individual dive was defined as the time between the whale shark descending from the water surface until reaching the water surface again. Based on the recorded maximum depth, individual dives were assigned to one of three categories: epipelagic dive (>0 m, <200 m), mesopelagic dive (>200 m, <1,000 m) and a bathypelagic dive (>1,000 m).

All times reported here are local times (GMT + 2–3 h). The whale shark moved in a south-westerly direction along the Mozambique coast before heading eastwards crossing the Mozambique Channel (Brunnschweiler et al. 2009). Time zones were assigned to locations by adding 0.5 h of time for each 7.5 degree increment in longitude. This gives local times of GMT + 2 h for positions between 30°E and 37.5°E, GMT + 2.5 h for positions between 37.5°E and 45°E, and + 3 h for positions between 45°E and 52.5°E, respectively. Day was defined as the time between 0600 hours and 1759 hours and night as the time between 1800 hours and 0559 hours.

The frequency of dives during the long-distance horizontal movement of the whale shark was examined using FFT analysis (for description see Shepard et al. 2006). Briefly, FFT operates by approximating a function with a sum of different sine and cosine terms. The influence of each periodic component is indicated by the magnitude of the corresponding spectral peak in the periodogram. FFTs can identify periodicities up to the Nyquist frequency, which is half the sampling rate, which in this study was one cycle per 7.5 min (Chatfield 1996).

An FFT was applied to the full 87 (consecutive) day depth record using programming routines in MATLAB (Shepard et al. 2006). Due to Argos data transmission constraints some individual swimming depth data points (at 15 min intervals) were not received. We interpolated between missing data by assigning to the gap the mean of depth readings either side of the gap. The gaps never exceeded 45 min in length (2 consecutive data points missing). These gaps made up 1.3% of the total number of data points, with single missing data points making up a further 10% of the total. However, these were randomly distributed through the depth-time series so were unlikely to confound identification of broader temporal-scale frequency components. The finite, nonstationary nature of the data set means the FFT spectral components can interfere with each other, generating frequency peaks that do not represent true periodicity in dive behavior. Therefore, a 'hamming' window function was applied to reduce effects of adjacent spectral components and to help identify biologically meaningful peaks. In addition, a single band-pass filter was applied to each identified spectral frequency peak to indicate the strength (amplitude) of an identified frequency in the time domain of the dive series (Chatfield 1996). For example, a spectral frequency of 1 cycle per day (24 h) was used to filter the entire dive series to locate where in the time series a diel rhythm in dive activity was most pronounced.

Results

Epi-, meso- and bathypelagic dive performance

Regular diving behavior was recorded throughout the entire track (Figure 1). A total of 24 d with maximum diving depths in the epipelagic zone were recorded at the beginning of the track (19 February to 11 March) when the whale shark stayed close to the Mozambique coast at usually less than 10 m depth, and on 11 April and 9 and 11 May when the shark was in deep water (Figures 1 and 2a, b). The first mesopelagic dive to a maximum depth of 468.1 m was recorded on 12 March 2006. From then on, a total of 44 and 19 d with maximum diving depths in the meso- and bathypelagic zones, respectively, were recorded (Figure 1).

Epipelagic dives were performed all along the track with the majority of dives performed between 34°E and 37°E while the shark was in shallow coastal water and 47°E and 49°E while it was in the deep Madagascar Basin southeast of Madagascar, respectively (Figure 2a, 2b). Of a total of 129 epipelagic dives to depths >50 m, 79.1% were performed during the night and 14.7% during the day (descent and ascent between 1800 hours and 0559 hours and between 0600 hours and 1759 hours, respectively), indicating high epipelagic diving activity during the night (Figure 3a). Meso- and bathypelagic dives were recorded between longitudes 38°E and 51°E (Figure 2a) in deep water with bottom depths >4,000 m when the shark was either crossing the southern

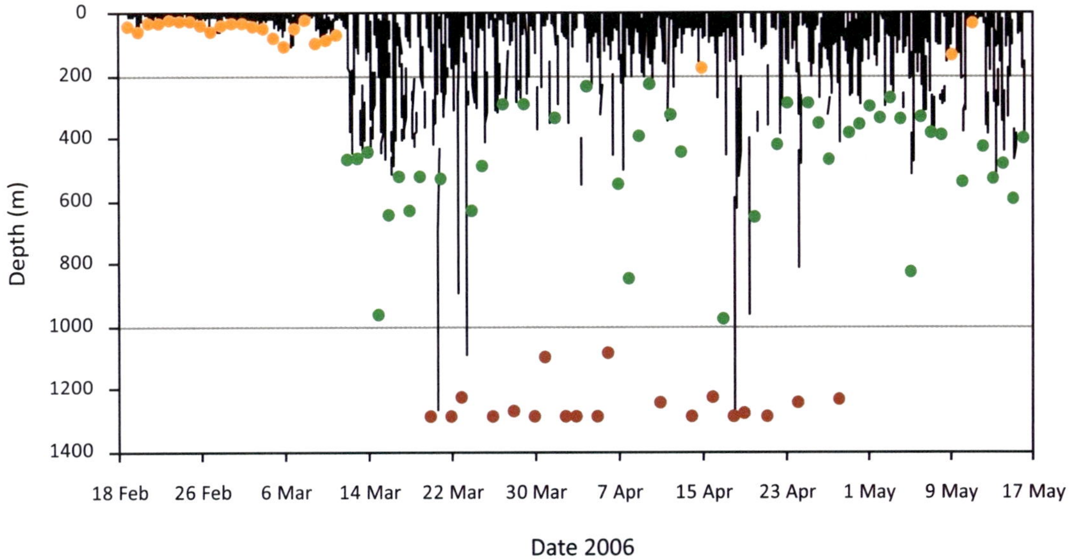

FIGURE 1. Time-depth series of the entire 87 d track (6,621 pressure readings; see Methods for details) and daily maximum depths recorded by the tag in the epipelagic (yellow), mesopelagic (green) and bathypelagic (red) zones. The two horizontal lines at 200 m and 1,000 m indicate the defined zones of diving depth.

part of the Mozambique Channel or was in the deep Madagascar Basin southeast of Madagascar (Figure 2b), near or in cyclonic (spiraling in clock-wise direction) or anti-cyclonic eddy systems (Figure 2c).

Fifty dive profiles included maximum depths in the meso- and bathypelagic zones. Of these, 44% were performed during the day and 10% during the night, indicating a higher deep diving activity during the day (Figure 3b). Except for four mesopelagic dives, all profiles showing the whale shark diving to the meso- or bathypelagic zones had pressure values of unknown depth (see Methods for details; maximum pressure readings unknown = 70%; mean ± SD = 33.2 ± 19%). In the majority of cases, readings that indicate the shark to be deeper than the transmitted depth value were recorded at the beginning of a dive whereas readings that indicate the fish to be less deep were recorded at the end of individual dives. Three dive profiles included actual recorded depths in the bathypelagic zone: on 20 March the whale shark was recorded at 1,264.2 m at 2045 hours, on 23 March at 1,086.7 m at 1515 hours, and on 18 April at 1,285.7 at 0315 hours. Two of these profiles and one mesopelagic dive profile show that the whale shark dived from the epipelagic zone straight to or into the bathypelagic zone at an estimated swimming speed of ~0.9 m s^{-1} (Figure 4).

Dive periodicity

The periodogram calculated for the whale shark had large spectral peaks near zero cycles per day (Figure 5a). This was a consequence of the dive data being offset by the mean dive depth, so only peaks following this initial slope represented periodic signals in the diving time series. Hence, those at or very near zero were ignored as they were not the product of biological signals. For the full dive record there was a pronounced spectral peak at a frequency of 24 h (1 cycle per day) (Figure 5a, 5b). The once daily (diel) signal was not strongly

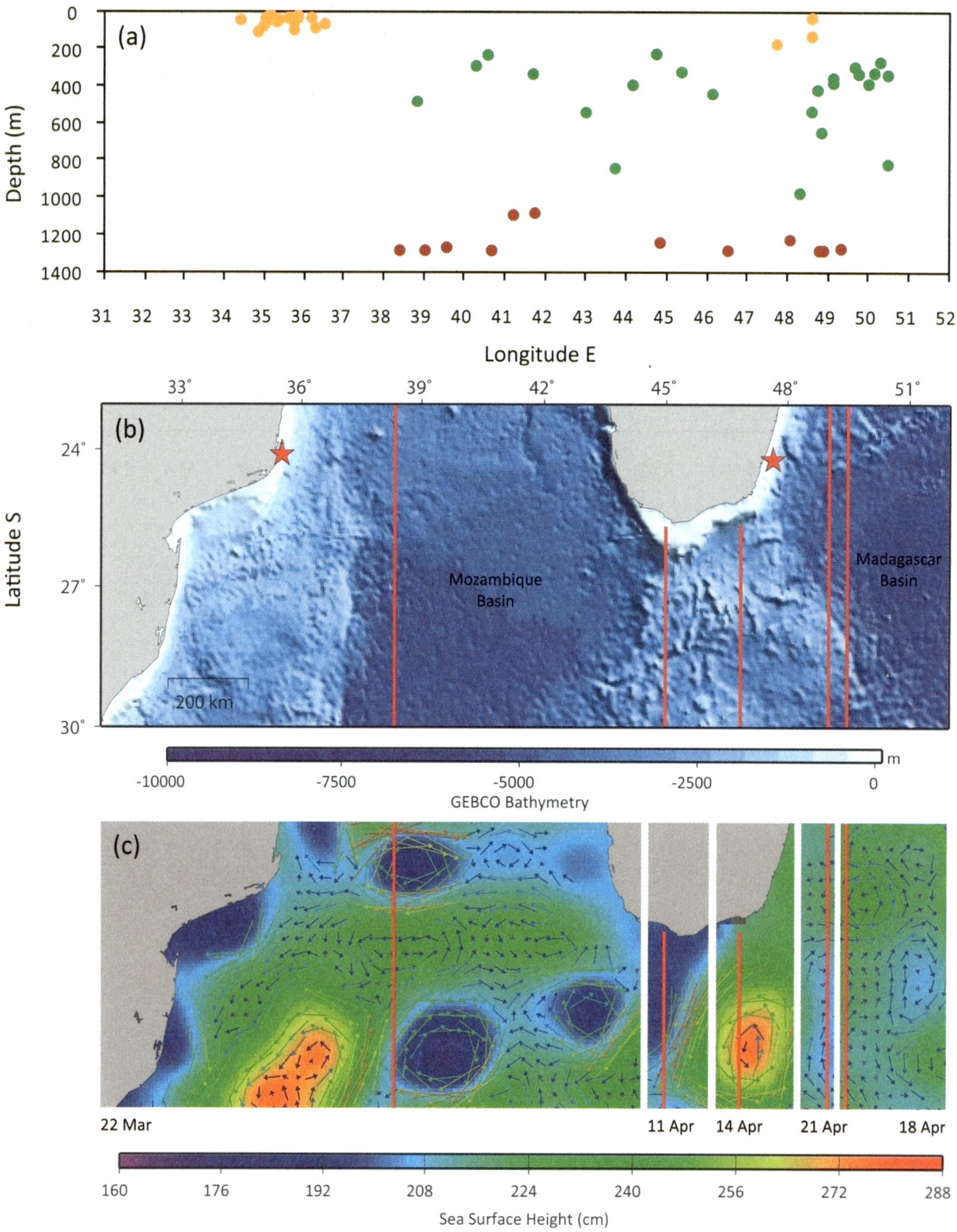

FIGURE 2. (a) Maximum depths in the epipelagic (yellow), mesopelagic (green) and bathypelagic (red) zones at daily longitude estimates (estimated error ± 0.5°). (b) Bathymetric map of the geographic area where the whale shark crossed the Mozambique Channel. The red stars off the coast of Mozambique and south-east of Madagascar indicate the positions where the shark was tagged and from where the first satellite uplink was received. (c) Longitudinal positions (red lines) on sea surface height and geostrophic current maps of five days with maximum diving depths of 1,285.7 m (22 March, 14, 18 and 21 April) and 1,242.8 m (11 April).

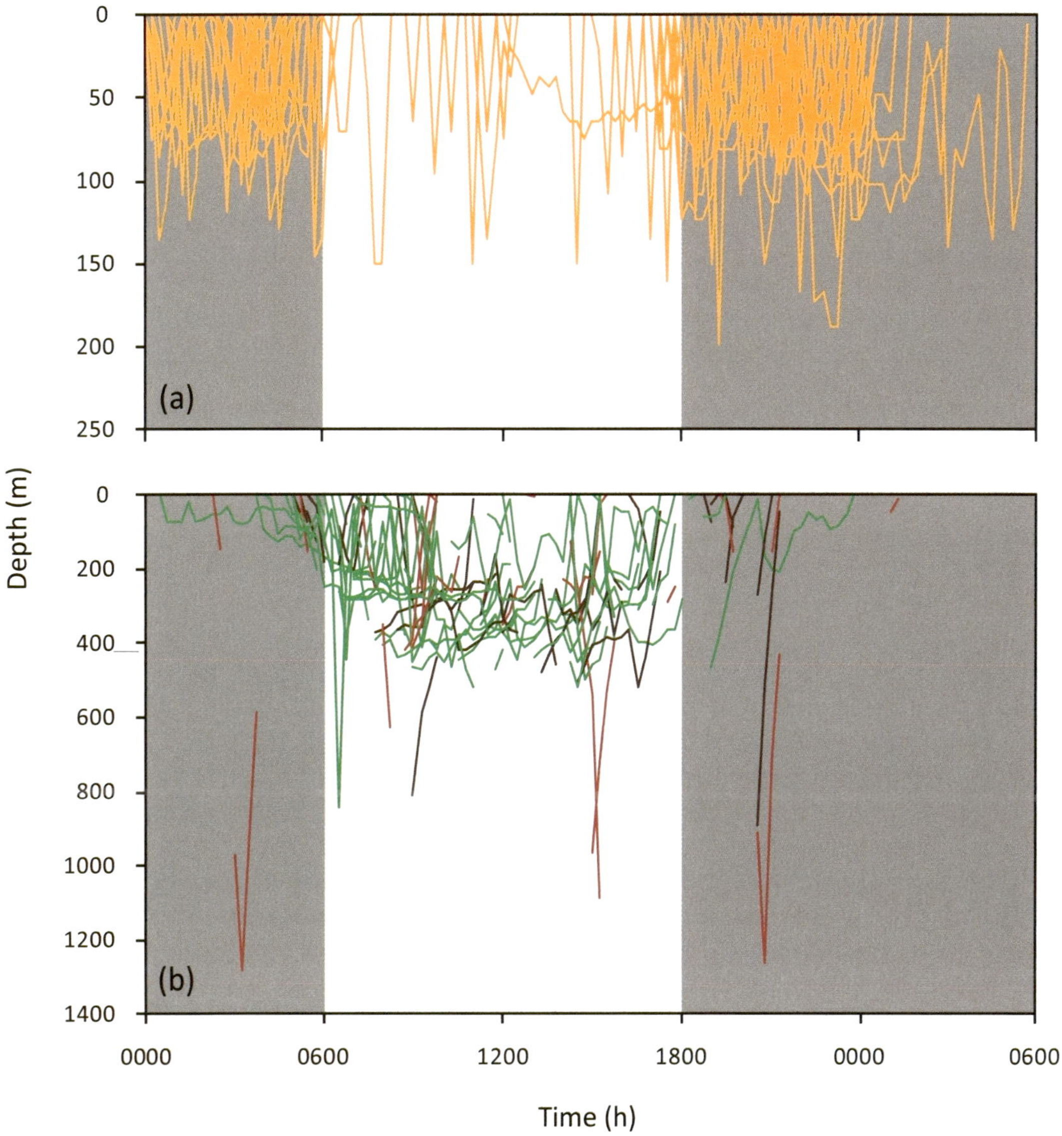

Figure 3. Depth profiles of (a) 129 epipelagic dives and (b) 28 mesopelagic (green), 10 bathypelagic (red) and 12 dives with maximum depths >200 m (brown). Gray shaded areas denote night.

present in the dive depths when the shark was on the Mozambique shelf, but was very strong when it moved into deep water and started the meso- and bathypelagic diving (Figure 5b, 5c). In support of this, the periodicity declined slightly in strength before becoming stronger when the second phase of bathypelagic diving occurred (Figure 5b, 5c).

Discussion

Detailed vertical movement data for sharks in deep oceanic waters are scarce. The deep diving profiles presented here are the first such data available for any shark species, including the deepest ever recorded dives to date for this circumglobal species, and therefore clearly expanding its habitat into the

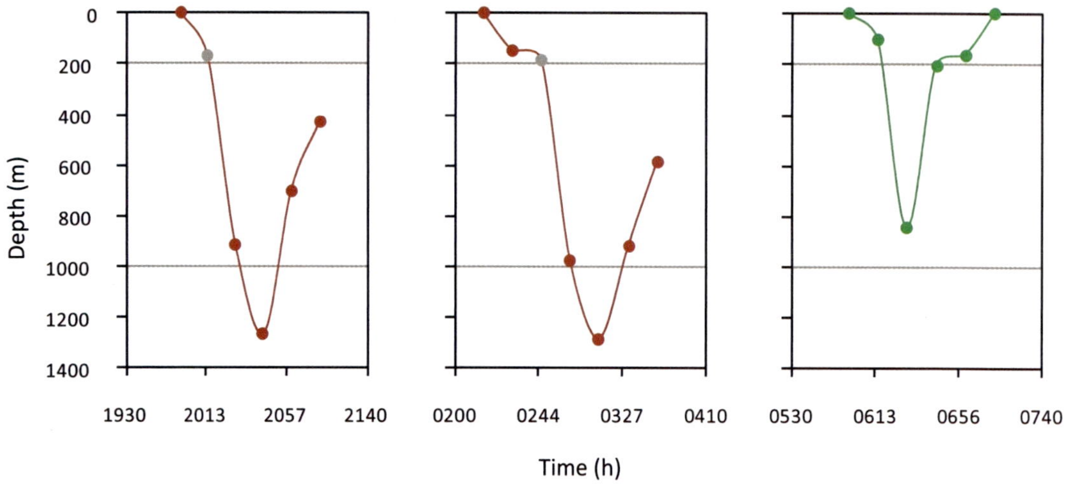

Figure 4. Two bathypelagic (red) dive profiles showing the descent to the maximum depth and one full mesopelagic (green) dive profile. Red and green dots represent archived depth readings every 15 min. Gray dots are minimum depth estimates and do not represent actual depth values (see Methods for details). The two horizontal lines at 200 m and 1,000 m indicate the defined zones of diving depth.

bathypelagic realm. The finding that on 22% of days the maximum dive depths were below 1,000 m indicates that such bathypelagic dives are common for whale sharks in bathymetrically nonconstraining habitat.

Overall, epipelagic dives were mostly observed during the night when plankton is generally closer to the water surface while considerably more meso- and bathypelagic dives were performed during the day (descent and ascent between 0600 hours and 1759 hours) when plankton can be found at greater depths compared to nighttime. This diel pattern was supported quantitatively by FFT analysis, and the pattern is consistent with other studies showing vertical behavior of planktivorous shark species to be a response to diel vertical migrations of their prey (Nelson et al. 1997; Sims et al. 2005; Wilson et al. 2006). The analysis of dive profiles can yield valuable information about the probable purpose of deep dives in marine vertebrates (Amano and Yoshioka 2003; Miller et al. 2004; Watwood et al. 2006, Bost et al. 2007), but because of missing actual depth values, it was not possible to reconstruct the majority of meso- and bathypelagic dive profiles with confidence in this study. Nevertheless, the distribution of missing depth values within individual dive profiles, as well as three reconstructed profiles (Figure 4) indicate that the whale shark dived directly from the epipelagic zone to maximum dive depths at or in the bathypelagic zone at swimming speeds similar to that of surface feeding whale and basking sharks *Cetorhinus maximus* (Sims 2000; Motta et al. 2010). Additionally, the fact that, despite 19 d with maximum depths in the bathypelagic zone being recorded, the tag's emergency release feature was not triggered indicates that the shark only briefly dived into the bathypelagic zone.

The purpose of deep dives to mesopelagic habitats and beyond in this planktivorous shark species remains unknown but is consistent with a general pattern for behavioral plasticity known from a wide range of marine vertebrate predators (Bonfil et al. 2005; Hays et al. 2006; Wilson et al. 2006). Whale sharks are well known to feed on seasonal plankton-rich surface waters and predictable food sources close to coasts and/or reefs (Clark and Nelson 1997; Gunn et al. 1999; Heyman et al. 2001; Wilson et al. 2001; Graham et al.

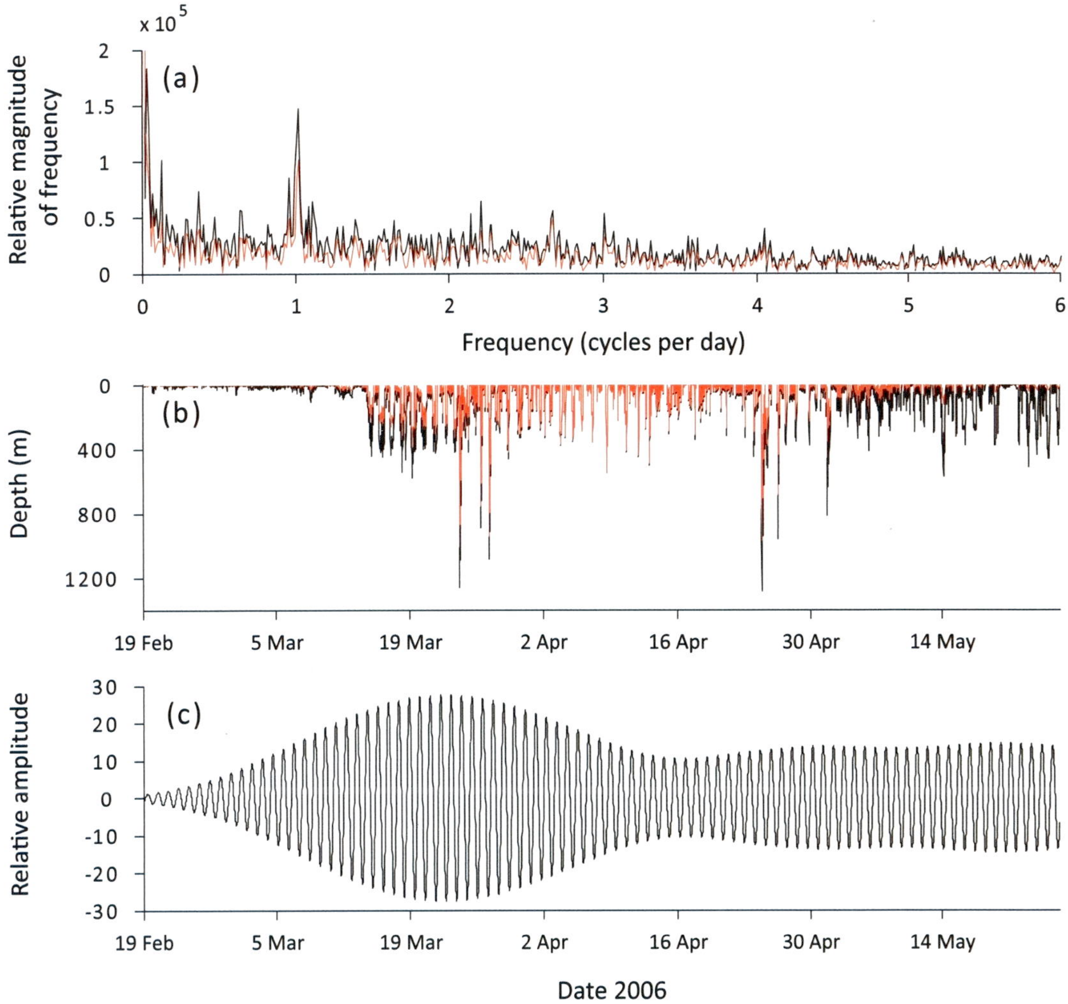

FIGURE 5. Diel oscillations during the oceanic phase of the whale shark's movement. (a) Fast-Fourier Transform generated periodogram for swimming depth indicates vertical movements with a period of 24 h (1 cycle per day). (b) The time series of swimming depths of the female whale shark. (c) Application of a single band-pass filter in the time domain for a periodicity of 1 cycle per day indicate diel dive patterns were most pronounced (greater y-axis amplitude) when meso- and bathypelagic diving occurred in the oceanic phase of long-distant movement. Red lines are results using the hamming window (see Methods for explanation).

2006; Nelson and Eckert 2007; Taylor 2007; Motta et al. 2010), but it remains largely unknown how whale sharks behave when moving between distant seasonal food patches. A possible hypothesis is that they cross less productive deep oceanic waters at relatively high speed to reach distant food-rich patches (Brunnschweiler et al. 2009; Sleeman et al. 2010). Crossing less productive deep oceanic surface waters might trigger deeper dives that aim to explore for feeding opportunities at greater depths. For example, the region south of Madagascar is a source of mesoscale eddies which drift westwards across the southern end of the Mozambique Channel reaching to the channel bottom and in which core water masses with different geophysical properties mix at various depths (de Ruijter et al. 2002, 2004; DiMarco et al. 2002; Quartly and Srokosz 2004). Such physical processes are known to have a structuring role on spatial dynamics and foraging strategy, influence movements and modify the diving behavior of marine predators (Cotté et al. 2007; Lambardi et al. 2008; Bailleul et al. 2010; Kai and Marsac 2010; Mencacci et al. 2010). Additionally, dissolved oxygen concentrations found at the southern end of the Mozambique Channel are generally higher compared to the northern tropical region (DiMarco et al. 2002). For example, at 500 m, an oxygen maximum with high nutrient concentration levels (Wyrtki 1971) spanning the full width of the southern part of the channel can be observed. Such nutrient rich waters might attract planktivorous species resulting in increased deep diving by feeding whale sharks.

The clear diel oscillation in the depth record supports results found previously in quantitative analyses of dive patterns in whale sharks (Graham et al. 2006) and for basking sharks (Shepard et al. 2006). However, in this study it was evident that the diel rhythm was most pronounced when the female whale shark conducted deep diving to meso- and bathypelagic depths during the oceanic phase of the long-distance movement. The fact that the diving behavior of only one, presumably immature (Norman and Stevens 2007), individual was available for analysis precludes us from firmly concluding that foraging within the water column to great depths is a typical behavior of the whale shark. Intraspecific plasticity in behavior are known from a range of oceanic fishes (e.g., Pade et al. 2009; Sims et al. 2009; Queiroz et al. 2010) and we therefore strongly encourage future studies to analyze, where possible, individual deep diving profiles from whale sharks of different size classes and from different geographic areas to determine whether or not deep foraging dives constitute 'normal' behavior.

Acknowledgments

We thank G. Adkison, J.-P. Botha, A. Cumming and F. Gnaegi for their efforts in the field. We also thank E. Shepard, M. Z. Ahmed and V. Wearmouth for their assistance with the Fast Fourier Transform analysis; V. Wearmouth also prepared Figure 5. Use of the Maptool program (http://www.seaturtle.org/) is acknowledged for producing Figure 2. R. Tolentino from Microwave Telemetry Inc. is acknowledged for his technical assistance. We also thank two anonymous reviewers for helpful comments on the manuscript. Juerg M. Brunnschweiler was supported by grants from the Save Our Seas Foundation, Shark Foundation Switzerland, Project AWARE and grant No. 11 9305/1 from the Swiss National Science Foundation. David W. Sims was supported by the Natural Environment Research Council Oceans 2025 Strategic Research Programme (Theme 6 Science for Sustainable Marine Resources) and a Marine Biological Association Senior Research Fellowship.

References

Amano, M., and M. Yoshioka. 2003. Sperm whale diving behavior monitored using a suction-cup-attached TDR tag. Marine Ecology Progress Series 258:291–295.

Bailleul, F., C. Cotté, and C. Guinet. 2010. Mesoscale eddies as foraging area of a deep- diving predator, the southern elephant seal. Marine Ecology Progress Series 408:251–264.

Block, B. A., D. Booth, and F. G. Carey. 1992. Direct measurement of swimming speeds and depth of blue marlin. Journal of Experimental Biology 166:267–284.

Block, B. A., H. Dewar, S. A. Blackwell, T. D. Williams, E. D. Prince, C. J. Farwell, A. Boustany, L. H. Teo, A. Seitz, A. Walli, and D. Fudge. 2001. Migratory movements, depth preferences, and thermal biology of Atlantic bluefin tuna. Science 293:1310–1314.

Block, B. A., S. L. H. Teo, A. Walli, A. Boustany, M. J. W. Stokesbury, C. J. Farwell, K. C. Weng, H. Dewar, and T. D. Williams. 2005. Electronic tagging and population structure of Atlantic bluefin tuna. Nature 434:1121–1127.

Bonfil, R., M. Meÿer, M. C. Scholl, R. Johnson, S. O'Brien, H. Oosthuizen, S. Swanson, D. Kotze, and M. Paterson. 2005. Transoceanic migration, spatial dynamics, and population linkages of white sharks. Science 310:100–103.

Bost, C. A., Y. Handrich, P. J. Butler, A. Fahlman, L. G. Halsey, A. J. Woakes, and Y. Ropert-Coudert. 2007. Changes in dive profiles as an indicator of feeding success in king and Adélie penguins. Deep-Sea Research II 54:248–255.

Boustany, A. M., S. F. Davis, P. Pyle, S. D. Anderson, B. J. Le Boeuf, and B. A. Block. 2002. Expanded niche for white sharks. Nature 415:35–36.

Brunnschweiler, J. M., H. Baensch, S. J. Pierce, and D. W. Sims. 2009. Deep diving behaviour of a whale shark *Rhincodon typus* during long-distance movement in the western Indian Ocean. Journal of Fish Biology 74:706–714.

Carey, F. G., and J. V. Scharold. 1990. Movements of blue sharks (*Prionace glauca*) in depth and course. Marine Biology 106:329–342.

Chatfield, C. 1996. The analysis of time series, 6th edition. Chapman and Hall, London.

Clark, E., and D. R. Nelson. 1997. Young whale sharks, *Rhincodon typus*, feeding on a copepod bloom near La Paz, Mexico. Environmental Biology of Fishes 50:63–73.

Cotté, C., Y.-H. Park, C. Guinet, and C.-A. Bost. 2007. Movements of foraging king penguins through marine mesoscale eddies. Proceedings of the Royal Society of London, Series B: Biological Sciences 274:2385–2391.

de Ruijter, W. P. M., H. Ridderinkhof, J. R. E. Lutjeharms, M. W. Schouten, and C. Veth. 2002. Observations of the flow in the Mozambique Channel. Geophysical Research Letters 29 1502, doi:10.1029/2001GL013714. Available: http://www.agu.org/journals/ABS/2002/2001GL013714.shtml.

de Ruijter, W. P. M., H. M. van Aken, E. J. Beier, J. R. E. Lutjeharms, R. P. Matano, and M. W. Schouten. 2004. Eddies and dipoles around South Madagascar: formation, pathways and large-scale impact. Deep-Sea Research I 51:383–400.

DiMarco, S. F., P. Chapman, W. D. Nowlin Jr., P. Hacker, K. Donohue, M. Luther, G. C. Johnson, and J. Toole. 2002. Volume transport and property distributions of the Mozambique Channel. Deep-Sea Research II 49:1481–1511.

Eckert, S. A., L. L. Dolar, G. L. Kooyman, W. Perrin, and R. Abdul Rahman. 2002. Movements of whale sharks (*Rhincodon typus*) in Southeast Asian waters as determined by satellite telemetry. Journal of Zoology 257:111–115.

Eckert, S. A., and B. S. Stewart. 2001. Telemetry and satellite tracking of whale sharks, *Rhincodon typus*, in the Sea of Cortez, Mexico, and the north Pacific Ocean. Environmental Biology of Fishes 60:299–308.

Graham, R. T., C. M. Roberts, and J. C. R. Smart. 2006. Diving behaviour of whale sharks in relation to a predictable food pulse. Journal of the Royal Society Interface 3:109–116.

Gunn, J. S., J. D. Stevens, T. L. O. Davis, and B. M. Norman. 1999. Observations of the short-term movements and behaviour of whale sharks (*Rhincodon typus*) at Ningaloo Reef, Western Australia. Marine Biology 135:553–559.

Hays, G. C., V. J. Hobson, J. D. Metcalfe, D. Righton, and D. W. Sims. 2006. Flexible foraging movements of leatherback turtles across the North Atlantic Ocean. Ecology 87:2647–2656.

Heyman, W. D., R. T. Graham, B. Kjerfve, and R. E. Johannes. 2001. Whale shark *Rhincodon typus* aggregate to feed on fish spawn in Belize. Marine Ecology Progress Series 215:275–282.

Holland, K. N., R. W. Brill, R. K. C. Chang, J. R. Sibert, and D. A. Fournier. 1992. Physiological and behavioural thermoregulation in bigeye tuna (*Thunnus obesus*). Nature 358:410–412.

Hooker, S. K., and R. W. Baird. 1999. Deep-diving behaviour of the northern bottlenose whale, *Hyperoodon ampullatus* (Cetacea: Ziiphidae). Proceedings of the Royal Society of London, Series B: Biological Sciences 266:671–676.

Josse, E., P. Bach, and L. Dagorn. 1998. Simultaneous observations of tuna movements and their prey by sonic tracking and acoustic surveys. Hydrobiologia 371/372:61–69.

Kai, E. T., and F. Marsac. 2010. Influence of mesoscale eddies on spatial structuring of top predators' communities in the Mozambique Channel. Progress in Oceanography 86:214–223.

Klimley, A. P., S. C. Beavers, T. H. Curtis, and S. J. Jorgensen. 2002. Movements and swimming behavior in three species of sharks in La Jolla Canyon, California. Environmental Biology of Fishes 63:117–135.

Lambardi, P., J. R. E. Lutjeharms, R. Mencacci, G. C. Hays, and P. Luschi. 2008. Influence of ocean currents on long-distance movement of leatherback sea turtles in the Southwest Indian Ocean. Marine Ecology Progress Series 353:289–301.

Mencacci, R., E. De Bernardi, A. Sale, J. R. E. Lutjeharms, and P. Luschi. 2010. Influence of oceanic factors on long-distance movements of loggerhead sea turtles displaced in the southwest Indian Ocean. Marine Biology 157:339–349.

Miller, A. K., and W. Z. Trivelpiece. 2008. Chinstrap penguins alter foraging and diving behavior in response to the size of their principle prey, Antarctic krill. Marine Biology 154:201–208.

Miller, P. J. O., M. P. Johnson, and P. L. Tyack. 2004. Sperm whale behaviour indicates the use of echolocation click buzzes 'creaks' in prey capture. Proceedings of the Royal Society of London, Series B: Biological Sciences 271:2239–2247.

Motta, P. J., M. Maslanka, R. E. Hueter, R. L. Davis, R. de la Parra, S. L. Mulvany, M. L. Habegger, J. A. Strother, K. R. Mara, J. M. Gardiner, J. P. Tyminski, and L. D. Zeigler. 2010. Feeding anatomy, filter-feeding rate, and diet of whale sharks *Rhincodon typus* during surface ram filter feeding off the Yucatan Peninsula, Mexico. Zoology 113:199–212

Nelson, D. R., J. N. McKibben, W. R. Strong Jr., C. G. Lowe, J. A. Sisneros, D. M. Schroeder, and R. J. Lavenberg. 1997. An acoustic tracking of a megamouth shark, *Megachasma pelagios*: a crepuscular vertical migrator. Environmental Biology of Fishes 49:389–399.

Nelson, J. D., and S. A. Eckert. 2007. Foraging ecology of whale sharks (*Rhincodon typus*) within Bahía de Los Angeles, Baja California Norte, México. Fisheries Research 84:47–64.

Norman, B. M., and J. D. Stevens. 2007. Size and maturity status of the whale shark (*Rhincodon typus*) at Ningaloo Reef in Western Australia. Fisheries Research 84:81–86.

Pade, N. G., N. Queiroz, N. E. Humphries, M. J. Witt, C. S. Jones, L. R. Noble, and D. W. Sims. 2009. First results from satellite-linked archival tagging of porbeagle shark, *Lamna nasus*: area fidelity, wider-scale movements and plasticity in diel depth changes. Journal of Experimental Marine Biology and Ecology 370:64–74.

Panigada, S., M. Zanardelli, S. Canese, and M. Jahoda. 1999. How deep can baleen whales dive? Marine Ecology Progress Series 187:309–311.

Quartly, G. D., and M. A. Srokosz. 2004. Eddies in the southern Mozambique Channel. Deep-Sea Research II 51:69–83.

Queiroz, N., N. E. Humphries, L. R. Noble, A. M. Santos, and D. W. Sims. 2010. Short-term movements and diving behaviour of satellite-tracked blue sharks *Prionace glauca* in the northeastern Atlantic Ocean. Marine Ecology Progress Series 406:265–279.

Shepard, E. L. C., M. Z. Ahmed, E. J. Southall, M. J. Witt, J. D. Metcalfe, and D. W. Sims. 2006. Diel and tidal rhythms in diving behaviour of pelagic sharks identified by signal processing of archival tagging data. Marine Ecology Progress Series 328:205–213.

Sims, D. W. 2000. Filter-feeding and cruising swimming speeds of basking sharks compared with optimal models: they filter-feed slower than predicted for their size. Journal of Experimental Marine Biology and Ecology 249:65–76.

Sims, D. W. 2010. Tracking and analysis techniques for understanding free-ranging shark movements and behavior. Pages 351–392 *in* J. C. Carrier, J. A. Musick and M. R. Heithaus, editors. Sharks and their relatives II: biodiversity, adaptive physiology, and conservation. CRC Press, Boca Raton, Florida.

Sims, D. W., and V. A. Quayle. 1998. Selective foraging behaviour of basking sharks on zooplankton in a small-scale front. Nature 393:460–464.

Sims, D. W., E. J. Southall, A. J. Richardson, P. C. Reid, and J. D. Metcalfe. 2003. Seasonal movements and behaviour of basking sharks from archival tagging: no evidence of winter hibernation. Marine Ecology Progress Series 248:187–196.

Sims, D. W., E. J. Southall, G. A. Tarling, and J. D. Metcalfe. 2005. Habitat-specific normal and reverse diel vertical migration in the plankton-feeding basking shark. Journal of Animal Ecology 74:755–761.

Sims, D. W., N. Queiroz, T. K. Doyle, J. D. R. Houghton, and G. C. Hays. 2009. Satellite tracking of the World's largest bony fish, the ocean sunfish (*Mola mola* L.) in the North East Atlantic. Journal of Experimental Marine Biology and Ecology 370:127–133.

Sleeman, J. C., M. G. Meekan, S. G. Wilson, J. J. Polovina, J. D. Stevens, G. S. Boggs, and C. J. A. Bradshaw. 2010. To go or not to go with the flow: Environmental influences on whale shark movement patterns. Journal of Experimental Marine Biology and Ecology 390:84–98.

Taylor, J. G. 2007. Ram filter-feeding and nocturnal feeding of whale sharks (*Rhincodon typus*) at Ningaloo Reef, Western Australia. Fisheries Research 84:65–70.

Teloni, V., P. M. Johnson, J. O. P. Miller, and T. P. Madsen. 2008. Shallow food for deep divers: Dynamic foraging behavior of male sperm whales in a high latitude habitat. Journal of Experimental Marine Biology and Ecology 354:119–131.

Teo, S. L. H., A. Boustany, S. Blackwell Scientific Publications, A. Walli, K. C. Weng, and B. A. Block. 2004. Validation of geolocation estimates based on light level and sea surface temperature from electronic tags. Marine Ecology Progress Series 283:81–98.

Watwood, S. L., P. J. O. Miller, M. Johnson, P. T. Madsen, and P. L. Tyack. 2006. Deep-diving foraging behaviour of sperm whales (*Physeter macrocephalus*). Journal of Animal Ecology 75:814–825.

Weihs, D. 1973. Mechanically efficient swimming techniques for fish with negative buoyancy. Journal of Marine Research 31:194–209.

Wilson, S. G., J. J. Polovina, B. S. Stewart, and M. G. Meekan. 2006. Movements of whale sharks (*Rhincodon typus*) tagged at Ningaloo Reef, Western Australia. Marine Biology 148:1157–1166.

Wilson, S. G., J. G. Taylor, and A. F. Pearce. 2001. The seasonal aggregation of whale sharks at Ningaloo Reef, Western Australia: currents, migrations and the El Niño/Southern Oscillation. Environmental Biology of Fishes 61:1–11.

Wyrtki, K. 1971. Oceanographic Atlas of the International Indian Ocean Expedition, National Science Foundation Publication, OCE/NSF 86–00-001, Washington, D.C.

American Fisheries Society Symposium 76:471–492, 2012

Using Archival Tags to Infer Habitat Use of Central California Steelhead and Coho Salmon

Sean A. Hayes*
NOAA Southwest Fisheries Science Center
110 Shaffer Road, Santa Cruz, California 95060, USA

Morgan H. Bond
NOAA Southwest Fisheries Science Center
110 Shaffer Road, Santa Cruz, California 95060, USA
and
School of Aquatic and Fishery Sciences University of Washington

Brian K. Wells
NOAA Southwest Fisheries Science Center
110 Shaffer Road, Santa Cruz, California 95060, USA

Chad V. Hanson
NOAA Southwest Fisheries Science Center
110 Shaffer Road, Santa Cruz, California 95060, USA
and
Department of Forest Science, Oregon State University

Andrew W. Jones
NOAA Southwest Fisheries Science Center
110 Shaffer Road, Santa Cruz, California 95060, USA
and
Department of Ecology and Evolutionary Biology, Yale University

R. Bruce MacFarlane
NOAA Southwest Fisheries Science Center
110 Shaffer Road, Santa Cruz, California 95060, USA

*Corresponding author: sean.hayes@noaa.gov

Abstract.—We deployed archival temperature loggers on juvenile and adult coho salmon *Oncorhynchus kisutch* and steelhead (anadromous rainbow trout) *O. mykiss* over both the freshwater and marine portions of their lifecycle in order to study their movements and thermal preferences. Beginning in 2003, loggers were deployed on juvenile coho salmon and juvenile and adult steelhead in a small central California coastal stream. A tag recovery from a coho salmon indicates the fish experienced variable temperatures on a daily to weekly basis in the marine environment (mean 13.3°C, range 10–18°C). Tags recovered from steelhead indicate use of a cooler, more stable, thermal habitat window in the marine environment (mean 11.0°C, range 8–14°C), often with little fluctuation over a period of weeks to months, and most thermal changes occurring at the seasonal time scale. Comparisons of steelhead data with sea surface temperature data suggest a northern migration out of the California Current to a narrow band of habitat that fluctuates between the southern boundary of the Bering Sea and north of the 40th parallel. In the shallow freshwater environment, steelhead appeared to be at the mercy of stream temperatures. However, in the estuary, where thermally variable habitats were available, steelhead used a surprisingly broad temperature range, including entering water thought to challenge their thermal tolerances (>20°C) even when cooler waters were available. Temperature loggers recovered on a local beach and island indicate tagged fish were consumed in the estuary by warm-blooded predators. All of these data coupled with a larger number of passive integrated transponder (PIT) tags, are helping to identify discrete habitats fish are using, exact dates of ocean entry and return, and enhance our understanding of marine survival and predation. Finally, archival tags may be useful in understanding habitat use of pelagic long-migrating species like steelhead, by tracking individuals in areas where other tagging technologies are poorly suited.

Introduction

What is known of the ocean distribution of salmon comes largely from commercial and research fisheries (Hartt and Dell 1986; Pearcy 1992). More recently, the results of tagging and marking studies have provided information on point to point locations and movements of individual fish (e.g., Myers et al. 1996; Weitkamp and Neely 2002). This information is biased to maturing or adult age classes due to the nature of gear types and the focus on adult fisheries. Therefore little information on early marine survival or habitat use of juveniles is available. Importantly, early marine habitat use may encompass a critical survival window (Pearcy 1992). Consolidating the above research with growing oceanographic data sets has provided insight into marine habitat use and distributions of Pacific salmon relative to oceanographic features (Burgner et al. 1992; Welch et al. 1998b; Welch et al. 1998c). However, our understanding of the marine component still pales in comparison to the data available on the freshwater life history of salmonids.

Recent advances in electronic tag technology have resulted in tags small enough to be carried by salmonids and yet resilient to the stresses of the riverine and marine environments. This has resulted in two specific research directions: acoustic tags which emit an individual specific code that can be detected by receivers in riverine, estuarine and marine habitats (results reviewed extensively in this volume), and archival tags (or data storage tags) which collect data on the environmental conditions an individual fish encounters, some of which can be used to infer at-sea locations as well (Friedland et al. 2001; Hinke et al. 2005a; Hinke et al. 2005b; Reddin et al. 2006; Reddin et al. 2004; Walker et al. 2005; Walker et al. 2001; Walker et al. 2000). Of course, tradeoffs exist between the two technologies. Acoustic tags can provide population-level estimates of survival and mortality past receiver arrays and fascinating insights into the life history (e.g., the detection of tagged individuals outside their previously expected home range), all without the need to recover the tags. On the other hand, these tags can only provide data where arrays are located, and detection rates are not 100%, resulting in potential biases of knowing only if fish passed a given receiver but nothing of their fate if undetected. Where acoustic arrays provide discrete temporal and spatial data points on multiple fish, archival tags have the advantage of providing continuous data of an individual's movements and habitat use throughout its migration. This of course comes with the necessity of recovering the tag and typically involves a bias of only having data from fish that were successful and survived long enough for the tag to be recovered. In addition, all tagging may be biased toward larger individuals in the population capable of carrying tags. Finally, tagged individuals could be more susceptible to complications associated with tagging that increase mortality or predation, especially acoustic tags which transmit in the ultrasonic frequency band used by many marine mammal predators.

For Pacific salmon populations along the California coast, most data on ocean distributions come from the recovery of coded wire tags (CWT) in ocean and research fisheries on coho *Oncorhynchus kisutch* and Chinook salmon *O. tshawytscha* and consist mostly of catch data for maturing adults (Weitkamp 2010; Weitkamp 2012, this volume; Weitkamp and Neely 2002). To date, almost no data exist on steelhead (anadromous rainbow trout) *O. mykiss* distributions in the California Current ecosystem, despite a much larger coastal breeding range than Chinook or coho salmon (Busby et al. 2000). Recent work by Hinke et al. (2005a, 2005b) used archival tags recording temperature and depth data to study habitat use and distribution potential of subadult and adult Chinook salmon and found that these fish limited their ocean habitat use to a thermal band between 7.6 and 16°C, with 52% of the time spent between 9 and 12°C (Hinke et al. 2005b). Using this information and seasonal data on when tagged fish were known to be in the surface layer, they used sea surface temperature (SST) data from satellites to identify regions of the California Current with temperature ranges matching archival tag data to refine where a given fish was likely to have been. This provided varying precision of location, limited by times when fish were known to be using the surface layer and by the total surface area within the defined thermal limits. Despite lacking high resolution locations at all times, Hinke et al. (2005b) also determined how much surface habitat was available to Chinook salmon at any given time and how much that habitat varied between years, providing the potential for new insights between oceanographic data sets and Chinook salmon growth and survival. Similar work studying habitat use in the California Current ecosystem with archival tags has not been reported for steelhead or coho salmon to this point.

Due to higher densities of salmon stocks in northern Pacific latitudes, similar insights on thermal limits and distribution have been achieved through more traditional net and long-line surveys. In particular, sockeye salmon *O. nerka* have been shown to use a broader but ultimately cooler temperature window than Chinook salmon with ranges of 4–15°C, and seasonal periods rarely exceeding 7°C (Welch et al. 1998a). Their conforming distribution and areas of capture span the central Bering Sea, Japan to the west, Alaska to the east, and the 40th parallel to the south. Steelhead are found in a more restricted band of the North Pacific, both in terms of temperature and geography, with a northern limit south of the Alaskan coast, Bering Sea, and

Aleutian Islands, but a similar southern limit as sockeye salmon. Steelhead captures are reported from waters with a similar range of temperatures from about 5–15°C, but with less seasonal variation than sockeye and a central tendency that fluctuates between roughly 8 and 11°C (Burgner et al. 1992; Welch et al. 1998b). There are a few tag recoveries from northern California steelhead populations that suggest they are migrating to this same ocean region, but in general, data on this are lacking (Burgner et al. 1992; Myers et al. 1996).

Our understanding of steelhead and coho salmon ocean habitat use lags behind that for other Pacific salmon, presumably due to limited fisheries and lower abundance. There are some data on location from at-sea captures of steelhead, but like studies using CWTs, these are limited by where and when surveys were conducted relative to when fish populations overlapped with the survey area. High resolution chronological data are still lacking in the literature and are necessary to develop a clear understanding of the movements and habitat use of these species, as well as how these movements are influenced by environment and food resources.

In this paper, we investigate the application of archival tags that store temperature data to track the habitat use of juvenile steelhead and coho salmon migrating to sea (smolts), as well as adult steelhead performing second ocean migrations (kelts). We used archival tags to address several objectives. The first objective was to study ocean habitat use and likely migration patterns. Secondly, we sought to characterize freshwater habitat use and thermal environment experienced by juvenile steelhead that delayed or skipped their ocean migration and remained in the river for extended periods of time. Finally, we used thermal profiles from tags that were recovered following predation events to characterize the location of some juvenile salmonid predation and the likely predators. We conclude with an evaluation of archival tag technologies for addressing the above questions.

Study Site

Scott Creek is a small coastal watershed, approximately 70 km^2, in central California, 100 km south of San Francisco. Steelhead and coho salmon spawn there naturally, and both populations are subsidized by a small conservation hatchery in the watershed that spawns wild broodstock and releases juveniles at smolt stage (Hayes et al. 2004). Several aspects of the study site make it ideal for archival tagging. The hatchery provided large juvenile fish and served as a tagging laboratory. An adult fish trap and relatively small adult populations (~300 returning adults/species) increased the probability of data logger recovery and enabled the collection of adult steelhead during their return migration to sea for tagging purposes.

Methods

Tag deployment

We began archival tagging in 2003 with one of the smallest archival loggers available. The tag consisted of a temperature logging system, marketed as the I-button (Dallas Semiconductor Corp, subsidiary of Maxim, Sunnyvale California), which collected up to 2,048 data points at 0.5°C resolution (–5–26°C range), with programmable sampling rate. The tag was repackaged in a rubberized coating and distributed for long term marine operation by Alpha mach (Mont-St-Hilaire, QC, Canada) and sold as the "iB-Krill," approximately 25.4 × 13.2 mm and 3.2 g in size. All tags were programmed to maximize sampling duration, by sampling at 4 h intervals, beginning at midnight, sampling six times per day.

Due to concerns about tags interfering with growth and causing drag on juvenile fish, an external attachment method was not considered feasible. Therefore, the tags were surgically implanted into the body cavity of juvenile fish greater than 200 mm fork length and 100 g mass. While it was possible to collect wild steelhead in this size range, wild juvenile coho rarely exceed 140 mm prior to ocean entry. Hatchery fish of both species were used for tagging in place of the smaller wild fish. Hatchery fish were typically scheduled for release between the last week of March and first week of April. Tagging surgeries were typically conducted at least one week before this. In 2003 and 2006, hatchery steelhead had not reached the minimum size threshold, therefore some fish were held back and reared under low density, high food ration conditions, and eventually tagged and released during May or June.

During the surgeries, fish were anesthetized and placed dorsal side down in a V-board lined with moist foam. A 2–3 cm incision was made in the ventral surface approximately 1 cm forward of the pelvic girdle extending towards the head. Tags were presterilized in 80% ethanol, dipped in sterile water and then inserted into the body cavity along with a Passive Integrated Transponder (PIT) tag. The incision was closed with 2–3 surgical staples or polypropylene sutures. The entire procedure typically took 2–3 min and the fish was allowed to recover and observed for at least one week prior to release in the lower section of Scott Creek, approximately 1 km from the ocean.

Adult steelhead kelts received a tag through surgical implantation or external attachment. External attachment used a 2-point attachment similar to the methods of Hinke et al. (2005a, 2005b) and Walker et al. (2000). Briefly, adult female steelhead were captured after spawning during their downstream migration and lightly anesthetized. A pair of 10 cm 12 gauge needles were pushed through the dorsal musculature approximately 3 cm below the dorsal fin and 2.5 cm apart. Two nickel pins were pushed through attachment holes on the tag, then into the exposed ends of the 12 gauge needles and through the dorsal part of the fish. The needles were then removed, leaving the ends of the nickel pins exposed on the opposite side of the fish. Small plastic washers were placed over the pins and the ends of the pins were clipped to leave about 1 cm of pin exposed. The ends were then rolled down with needle-nose pliers to prevent the washer from slipping off. In 2007 and 2008, 14 tags were also surgically implanted in adults, using similar procedures to those used for the aforementioned juveniles. Surgeries were conducted streamside at the Scott Creek weir trapping station. Kelts were allowed to recover in a flow-through live box placed in the stream and were released once they appeared to be completely recovered and ready for rapid swimming (typically after 20 min). Adult tagging was restricted to females, which typically have higher marine survival rates than males in this system (Hayes unpublished data).

Tag recovery and calibration

Tags were recovered through several methods, including some unexpected ones. The plan was to recover tags from fish on return spawning migrations as they entered the adult fish trap on the lower section of the Scott Creek watershed. Fish carrying external temperature loggers were restrained while a pair of small wire cutters was used to clip the pins holding the tag to the dorsal surface of the fish. The tag and pins were then removed, the fish was measured for length and weight and scale samples were taken, followed by recovery and release. All fish returning through the fish trap were scanned for PIT tags and cross referenced with a list of fish carrying internal data loggers. Fish recovered with internal data loggers were transported to the hatchery, anesthetized and the tag removed surgically. This required a 3–5 cm incision on the ventral midline, just forward of the pelvic girdle. There was typically little tissue attached to the tag, and it was removed with hemostats, followed by suturing the incision. After a 2–3 d recovery period, fish were released from the hatchery to continue spawning in the wild. All tags were stamped with a contact number, and in several cases tags were returned from unexpected sources, including three that were found in bird roosting and resting areas and one recovered by a fisherman.

Upon recovery, tags were downloaded using the iButton Viewer 32 TMEX software (Dallas Semiconductor). In most cases, tags were run through a temperature calibration in a water bath from 0–25C to account for potential sensor inaccuracy. A regression curve was established and applied to the raw downloaded data, yielding an accuracy and precision estimated to within 0.25°C. In several situations, there was significant clock drift. To compensate for this, the download file records current clock time of the tag and the downloading computer in the same file. While it was possible to correct for this, it was not clear if drift rates were linear. In several situations tags were analyzed for diurnal temperature changes at sea and/or compared with diurnally changing stream temperatures. In these situations, only tags whose clock drift was less than the time between two sampling points (4 h) at recovery were used.

Environmental measurements

Stream water temperatures were measured on an hourly basis at two locations in the watershed, upstream and at the estuary, (Figure 1) initially with IB-Cod temperature loggers (Alpha Mach, Mont St-Hilaire, Quebec, May 2002 to June 2003) switching to Onset Tidbits (Onset Computer Corp. Pocasset, Massachusetts) in the upper watershed site and YSI 600 XLM data loggers (YSI Inc. Yellow Springs, Ohio) in the estuary (July 2003 through December 2008).

Sea surface temperature (SST) data from the Pacific Ocean north of 25° were obtained from NOAA Coast Watch (http://coastwatch.pfel.noaa.gov/data.html) at a 0.1° resolution. We used the blended sea surface temperature product, which incorporated data from numerous sources: Aqua, NOAA GOES-10, NOAA GOES-12, NOAA POES-17 and POES-18. SST data across the ocean were averaged for each pixel for each month except for during entry and return months during which the fish did not spend the entirety of the month at sea. During these times, the SST temperature data were averaged over the period of the month during which the fish was known to be at sea as determined from the archival temperature data pattern.

Data Analysis

For constructing figures with potential ocean travel range estimates, point to point swim speed times for steelhead were collected from the literature (Burgner et al. 1992; McKinnell et al. 1997; Walker et al. 2000). A total of 77 estimates were found that enabled a calculation of swim speed in km/d. We restricted time between tagging and recapture to events less than 365 d, yielding 57 data points. The maximum estimated speed was 85 km/d and the mean was 24 km/d. For conservative purposes of making maximum range estimates, we used the speed that was two standard deviations above the mean, or 64 km/d.

For ocean temperature records, the point of stream departure or reentry was determined from overlaying temperature data from the estuary on the fish tag data record. The ocean was typically several degrees cooler than the stream in March and April and warmer in February. In addition the stream was subject to daily temperature fluctuations of at least

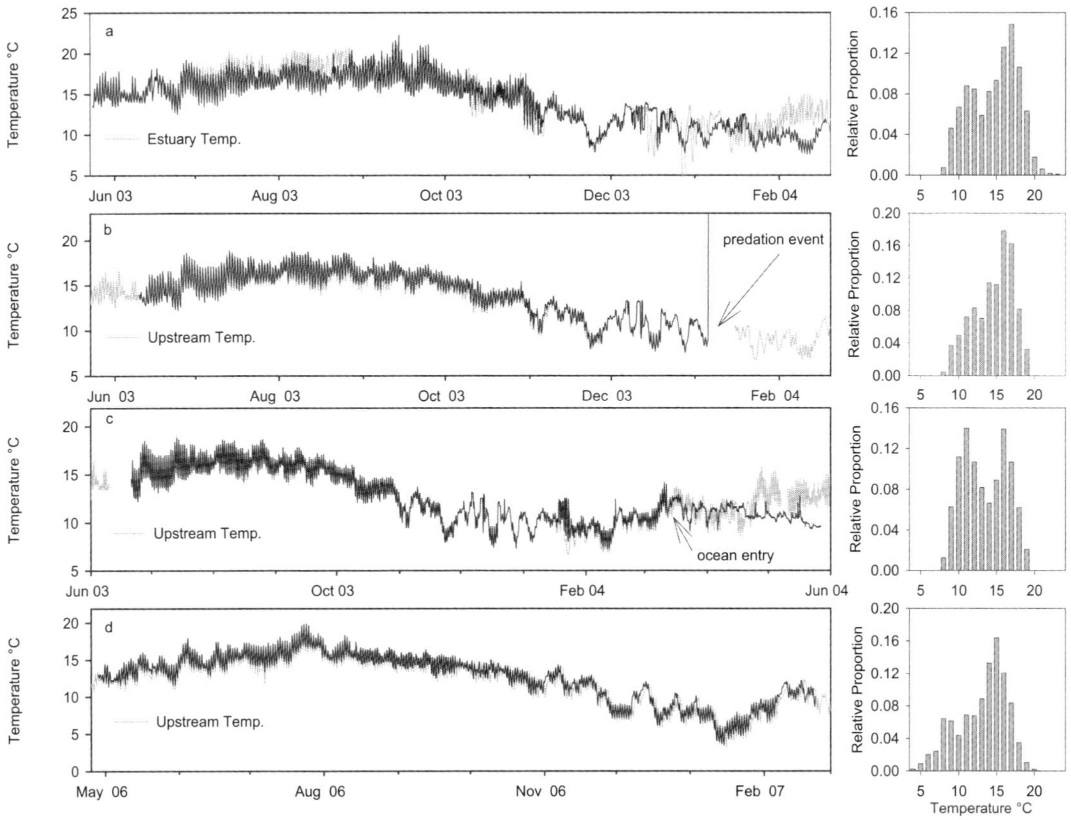

FIGURE 1. Temperature data logger (black line) and corresponding frequency distributions for juvenile steelhead *Oncorhynchus mykiss* that reared in Scott Creek. Estuary or stream temperature data are underlain in gray to demonstrate the relative temperatures of the watershed. Box 1a. shows data for a wild steelhead juvenile from the estuary beginning in May 2003. Box 1b. shows a hatchery steelhead smolt that was released in June 2003 and consumed by a predator in January 2004. Box 1c. shows a hatchery steelhead smolt that was released in June 2003, staying in the stream until March 2004, then entering the ocean, and tag memory filled in late May 2004. Box 1d shows a hatchery steelhead smolt that was tagged in April 2006, resighted several times in the lower watershed throughout the year and tag recovered in February 2007.

two degrees, corresponding to early morning lows and mid afternoon highs that were not present in the ocean portion of fish records. This usually enabled a clear determination of when fish left or re-entered the river to within a 4 h point in time. Stream temperature data were then removed from tag records for which ocean data were available. Extensive evidence from the literature indicates steelhead are very surface oriented, and the vast majority of the data were likely to reflect sea-surface temperatures in the epipelagic zone (Burgner et al. 1992; Nielsen and Zimmerman 2008; Welch et al. 1998b). Therefore monthly frequency distributions of temperature were generated for each archival tag record and were presumed to reflect measurements made at the sea surface. Based on the literature (Walker et al. 2004; Walker et al. 2005; Walker et al. 2001; Walker et al. 2000; Welch et al. 1998a), it did not seem reasonable to make similar assumptions for coho salmon.

We used the archival data from recovered steelhead combined with observed SST data to determine probable habitat within the Pacific Ocean, similar to the methods of Hinke et al. (2005a). Temperature data from each recovered steelhead archival tag were binned into 0.25°C bins, and frequency distributions were determined for each month at sea. We then compared the frequency distributions to the Pacific Ocean SST data to determine possible areas of suitable habitat. The Pacific Ocean SST data were averaged for each month and represented on a gray scale from black (–3°C) to white (34°C), with a color spatial resolution of roughly 0.1 degree (~11 km). Sea surface temperatures that corresponded to temperatures recorded by archival tags were color coded on a rainbow scale with dark blue representing a minimum probability of 0.1, and red matching the highest probability location in a given month. Independent frequency distributions were generated for each month relative to SST data. The full color scale was used in each figure to maximize the potential range contrast, and due to variations in peak probability values between months, was not scaled consistently across figures. We then restricted the identified area of available habitat to the most probable areas occupied by the fish by applying a 64 km/d migration rate. From the origin of ocean entry, we applied increasingly large circles representing maximum distance possible from origin until the radius of the circles reached a distance at which the fish would have had to return to arrive back at the origin by the return date. In the end, this analysis included all the data from the Pacific Ocean within the radius of possible migration distances from Scott Creek for each month at sea. By late summer months, nearly the entirety of the ocean was within the possible range of steelhead salmon (~12,000 km radius).

Results

A total of 579 archival tags were deployed on coho salmon (n = 260) and steelhead (n = 319). Of these, 9 tags were recovered through several methods (Tables 1 and 2). Three tags, one coho and two steelhead, were found deposited on beaches where predators reside and returned to us. One tag was returned to us by a fisherman who captured a coho salmon. Four steelhead were recovered with internal tags requiring surgical removal. Of these, only one had gone to sea, while the other three remained in the stream. One of the three stream fish had apparently expelled the tag, as it could not be found during the surgery and there was a 2 × 3 cm patch of scar tissue on the ventral body wall near where the tag had been implanted. For the hatchery steelhead smolts, where tagging and release were delayed to increase size, an increased residualization rate was observed with many fish choosing not to leave the stream in the year they were released.

Sixty-nine tags were deployed on steelhead female kelts. Two attachment methods were used, with the majority of tags deployed with an external mount during 2004 and 2007. Four tags were implanted in the body cavity in 2007 and an additional 10 tags were implanted in adult kelts in 2008. Of these, two external tags were recovered.

Steelhead juvenile freshwater thermal environment

Four juvenile steelhead tag records (1 wild, 3 hatchery) showed extensive freshwater water residency periods (Figure 1). One wild smolt was collected from and released back to the estuary. This fish was recovered 10 months later at the weir. During the surgery, it was apparent that it was female and was sexually mature with eggs ready for spawning. The tag data indicated that the fish had spent much of its time in the estuary (Figure 1a). A direct comparison to estuary temperatures showed the fish spent 59.6% of its time within 1°C of the estuary surface temperature sensor. In addition the fish spent 32.7% of its

TABLE 1. Summary of tag deployments and recoveries by species. Both internal (in) and external (ex) attachments were used with kelts. Some steelhead *Oncorhynchus mykiss* smolt deployments were delayed (d) in some years to increase fish size.

		# deployed	mean length (mm)	# returned from ocean	# recovered in stream	# recovered elsewhere	# residualized
Coho hatchery smolts							
	2003	30	212	0	0	0	
	2006	214	240	1*	0	2	
	2007	16	258	0	0	0	
Steelhead							
hatchery smolts	2003	28	223	0	0	0	
	2003d	78	246	1	1	1	10
	2006	59	208	0	0	1	
	2006d	49	220	0	1	0	1
wild smolts	2003	4	287	0	1	0	1
	2005	30	219	0	0	0	1
kelts	2004	17 ex	671	1	1	0	
	2007	38 ex, 4 in	665	1	1	0	
	2008	10 in	667	0	0	0	

*not recovered but detected by instream PIT reader reentering river

TABLE 2. Meta data of recovered archival tags.

Species	Age at deployment	Origin	Deployment date	Recovery date	# of days	Fork Length (cm) deployment	Fork Length (cm) recovery	Attachment	Data result
Coho	smolt	hatchery	03/02/06	09/07/06	189	26.9	~45	internal	ocean
Coho	smolt	hatchery	03/14/06	05/31/06	78	21.3		internal	predation
Steelhead	smolt	wild	05/14/03	02/18/04	280	32.7	38.0	internal	stream
Steelhead	smolt	hatchery	06/09/03	02/10/04	246			internal	stream & predation
Steelhead	smolt	hatchery	06/18/03	02/04/05	597	22.3	55.0	internal	stream & ocean
Steelhead	smolt	hatchery	04/25/06	02/23/07	304	22.2	23.5	internal	stream
Steelhead	smolt	hatchery	03/29/06	06/08/06	71	20.3		internal	predation
Steelhead	Kelt	hatchery	04/20/07	02/07/08	293	66.0	72.0	external	ocean
Steelhead	Kelt	wild	03/05/04	02/15/05	347	49.5	56.0	external	ocean

time in water more than 1°C cooler than estuary temperatures, and 7.5% of its time in water more than 1°C warmer than estuary temperatures. During July and August of 2004, the fish was residing in cooler waters, corresponding more to temperatures from a sensor on the bottom of the estuary (data not shown). During September, the fish spent a great deal of time in water that was greater than 20°C, peaking at 23°C, despite having cooler refuge habitat available upstream (Figure 1b). During January, the fish appeared to move into cooler temperatures corresponding to upstream water temperatures (Figure 1b) and was eventually collected at the upstream fish trap on February 18th.

Three additional data loggers were recovered from hatchery steelhead smolts that showed extensive freshwater residence periods (Figure 1b–d). In each case, these were fish that experienced a delayed release. All three fish experienced temperatures that corresponded to upstream temperatures. These temperatures were lower than estuary temperatures during summer and fall months but were similar during the winter/spring high flow season. One fish (Figure 1b) spent seven months in the stream. A comparison to a stream temperature sensor indicated the fish spent 92% of its time in water temperatures within 1°C of the stream sensor and deviated by more than 2°C only once. Another fish (Figure 1c) spent 10 months instream before entering the ocean in March 2004. This assumption was based on the cessation of diurnal temperature fluctuations in the temperature record, which would have been very unlikely in the stream environment. A third fish, tagged in 2006 (Figure 1d), was recaptured three times throughout the year within a 300 m stretch of stream. This fish showed a near perfect correspondence with stream temperature from the sensor in that region. Exact comparisons were not made due to a 25 h drift in the tag clock.

Predation insights

Three loggers were recovered with temperature profiles indicating that fish had been consumed by a warm blooded animal. It is possible these fish died and were eaten by scavengers or were captured by predators. One tag from a steelhead was recovered on Scott Creek beach 7 months after it was deployed (Figure 1b, Figure 2a). The other two tags, one from a steelhead and one from a coho salmon were recovered on Año Nuevo Island, approximately 12 km north of the watershed. An additional seven PIT tags from smolts tagged with temperature loggers were found on the island as well, for a total of 5 coho salmon and 4 steelhead. The upper temperature limit of the archival tag was only 25°C, limiting the ability to measure core temperatures of the predator. In all three cases, the temperatures of the tag corresponded to estuary temperatures for several days leading up to the moment of predation (Figure 2).

Steelhead and coho ocean thermal environment

Archival tags with ocean temperature data were recovered from two steelhead kelts (Figure 3), one steelhead smolt (Figure 1c; March–May 2004) and one coho smolt (Figure 3). An additional tagged coho salmon return was detected by instream PIT tag readers, but the fish evaded recapture. The two steelhead from 2004 experienced temperatures between 7.8 and 14.1°C. The steelhead kelt tagged in 2007 experienced temperature ranges from 7.2 to 14.8°C. Despite differences in age-class, hatchery and wild origin, and years between tag recoveries, the steelhead showed very similar temperature patterns. The coho smolt tagged in 2007 experienced temperature ranges from 9.8 to 17.7°C during 6 months at sea. The coho salmon was recaptured approximately 25 km to the southeast

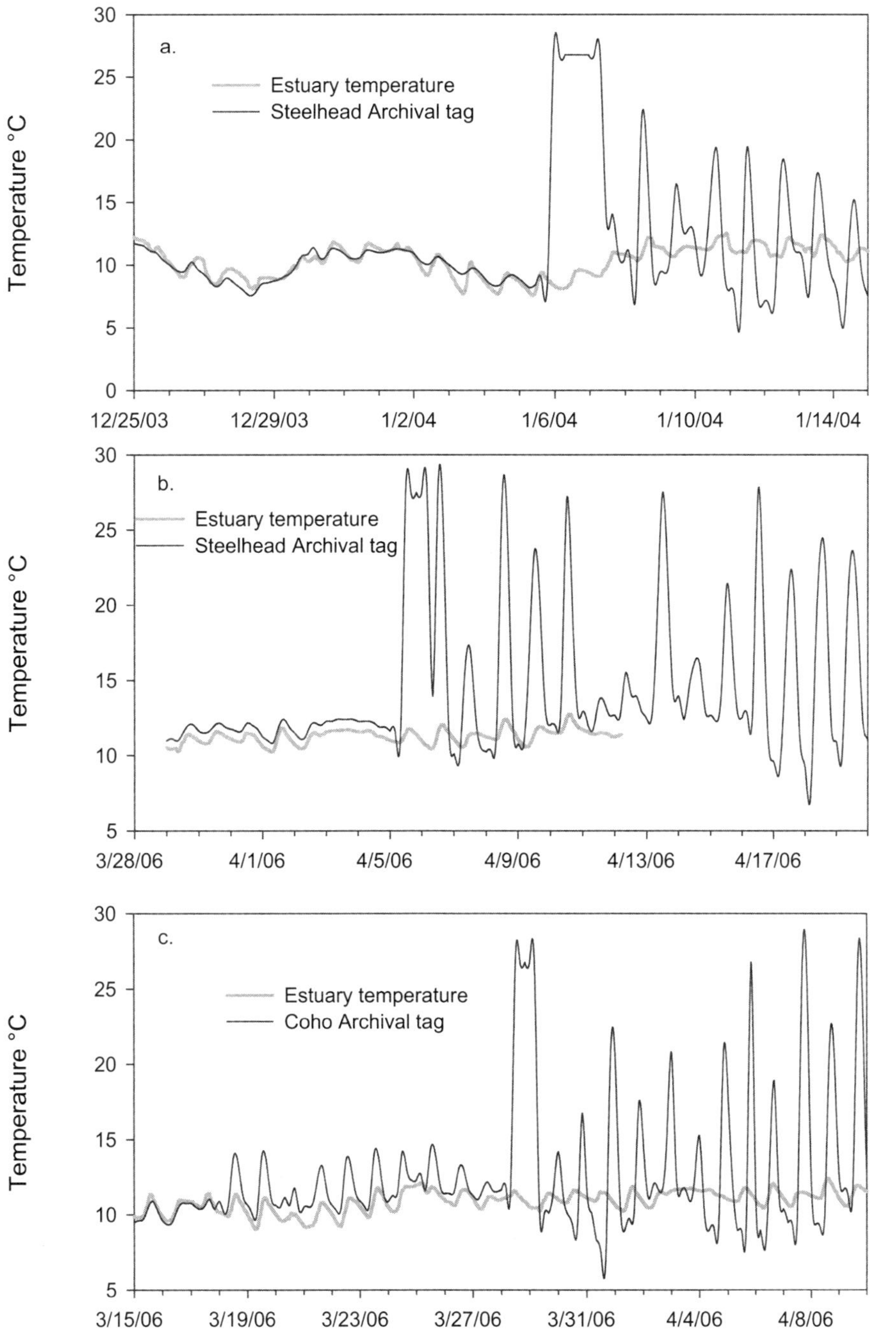

FIGURE 2. Temperature traces of fish consumed by warm blooded predators. Trends show fish tag data overlaid on matching estuary temperature data, followed by predation event, and subsequent deposit of tag onto a terrestrial environment, including (a) a blow up of the hatchery steelhead from Fig 1b with the tag deposited on Scott Creek beach in 2004, (b) a hatchery steelhead and (c) a hatchery coho tagged in March 2006, and tags were recovered on Año Nuevo Island later that summer.

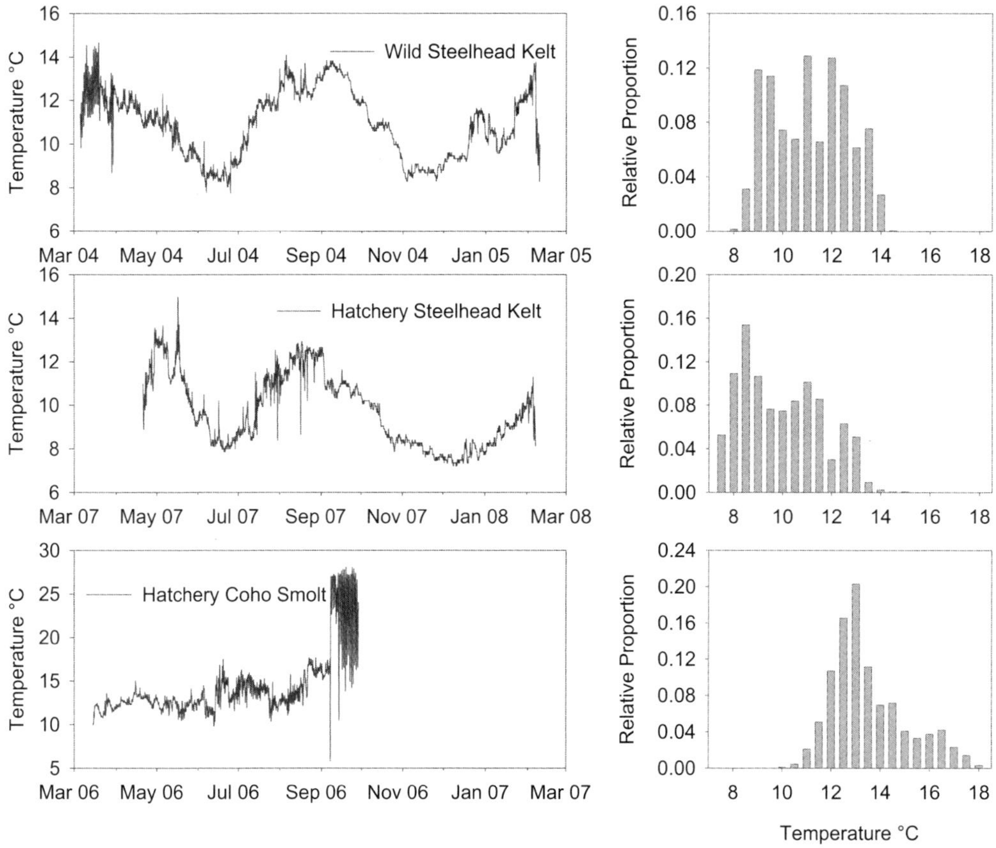

FIGURE 3. Raw temperature logger data and frequency distributions of two steelhead kelts and one coho smolt that went to sea.

of Scott Creek, off of the Santa Cruz Wharf, which was an unexpected location for a coho salmon to be caught in September.

The steelhead data were examined for diurnal changes in temperature, indicative of vertical migration patterns in the water column. The coolest temperatures were typically observed during the 0800 hours reading and the warmest temperatures typically experienced during the 1600 hours reading (Figure 4). There are reports of North American steelhead being found as far west as the Kuril Islands (Myers et al. 2005). If such assumptions were applied to these data, its possible that fish could be in a time zone up to 7 h west of where they were released, which would significantly shift the dawn and dusk times. While the afternoon record tended to be slightly warmer, there were no differences of biological significance, with many portions of the record showing variations of less than 1°C for weeks at a time, providing little insight into any east-west movement that might otherwise be present if daily vertical migrations occurred.

Working with the assumption that steelhead temperature data were from the surface layer and corresponded to satellite SST data, probable habitat plots were generated based on monthly temperature frequency distributions and potential migration speeds to indicate the likely regions of the ocean where steelhead would be found (Figure 5). For all three steelhead, temperatures at ocean entry

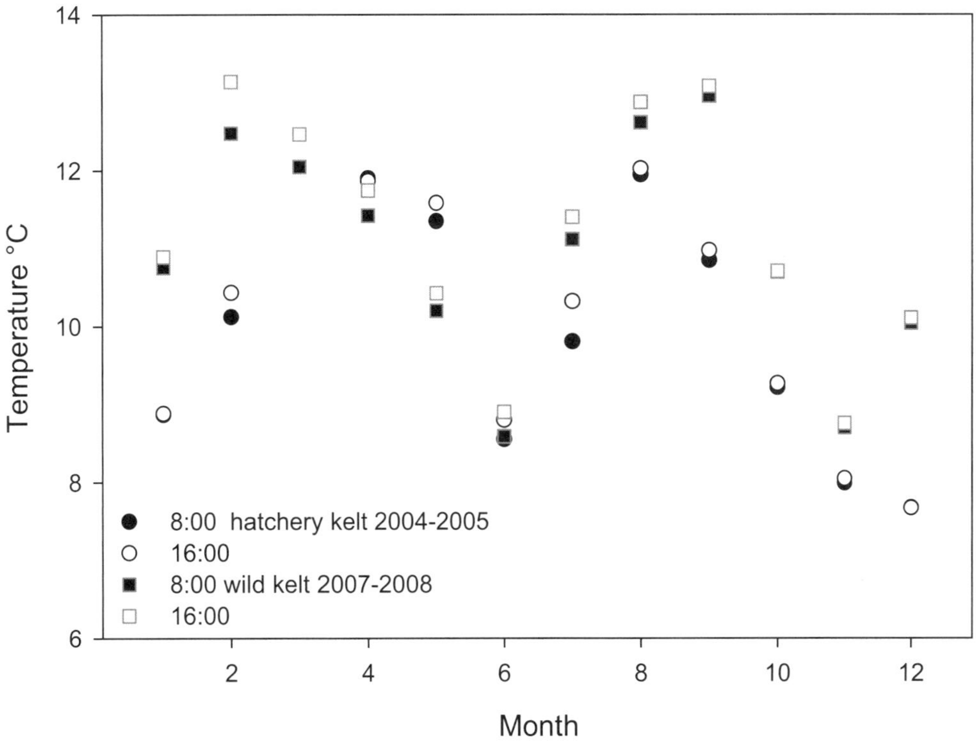

Figure 4. Comparison of monthly mean daily temperatures experienced by the two steelhead kelts for 800 and 1600 during their migration showing the relative lack of diurnal temperature fluctuations in the two records.

corresponded to temperatures along the central coast. Both 2004 fish entered the ocean in March, while the 2007 fish entered in April. The tracks between the three fish were quite similar, with the 2007 fish initially appearing to lag by approximately one month in geographic location. All three fish experienced temperatures that suggest rapid northwesterly movements, and by the third month to areas corresponding to regions outside the California Current, matching coastal regions between Vancouver Island and Southeast Alaska and extending west into the Gulf of Alaska. Suggested temperature ranges continued pushing north to the Aleutian Islands and west to the coasts of Japan and Russia by September in both 2004 and 2007. Ranges began retracting due to travel-time limitations in October, with southward shifting movements beginning in November. Temperatures indicated fish were still outside the California Current in December but began corresponding with California temperatures in January and tightly with temperatures in Central California, just prior to stream reentry in February of both years.

Discussion

These data represent some of the first archival tag recoveries for juvenile Pacific salmonids and the first year long records for Pacific salmonids. Despite low tag return rates of less than 1% for smolts and 2% for steelhead kelts, the data acquired provide insights that could not be produced by any other methodology for comparable costs. From just a few tag returns we were able to document the wide freshwater thermal range that juvenile steelhead and, presumably, coho

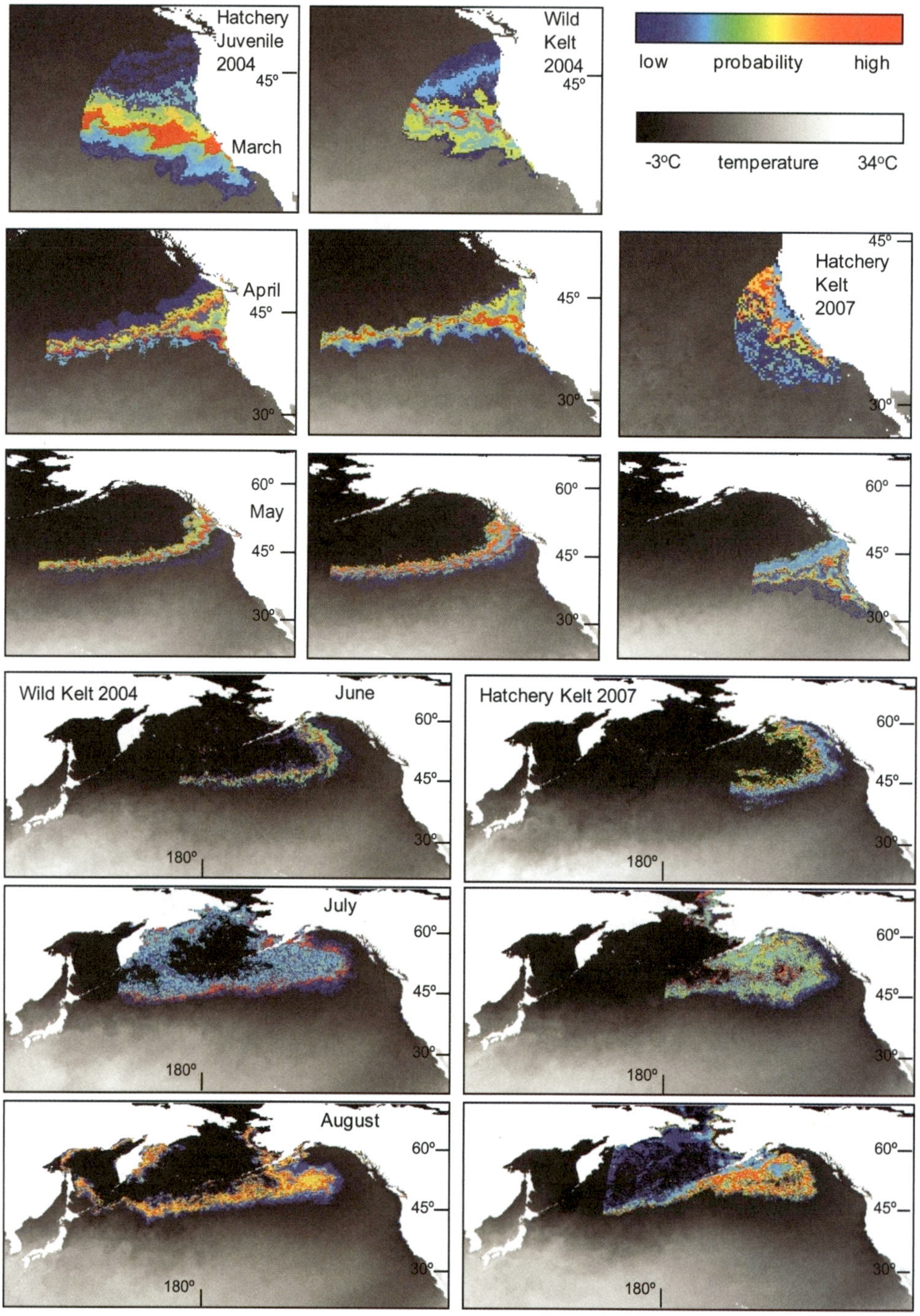

Figure 5. Graphical analysis of available steelhead ocean temperature data for one juvenile hatchery steelhead (March–May 2004), one wild female kelt (March 2004–February 2005) and one hatchery female kelt (April 2007–February 2008). The gray scales represent 0.1° (~11 km) SST values averaged for each month. The minimum SST value represented across the figure is –3°C (black) and the maximum value represented is 34°C (white). The color scale represents monthly probability distributions for the SST experienced by the steelhead during each month. The minimum probability value is 0.01 (blue) and the maximum probability value is given in red, with the upper limit varying between months. Pixels without color represent probability values of zero.

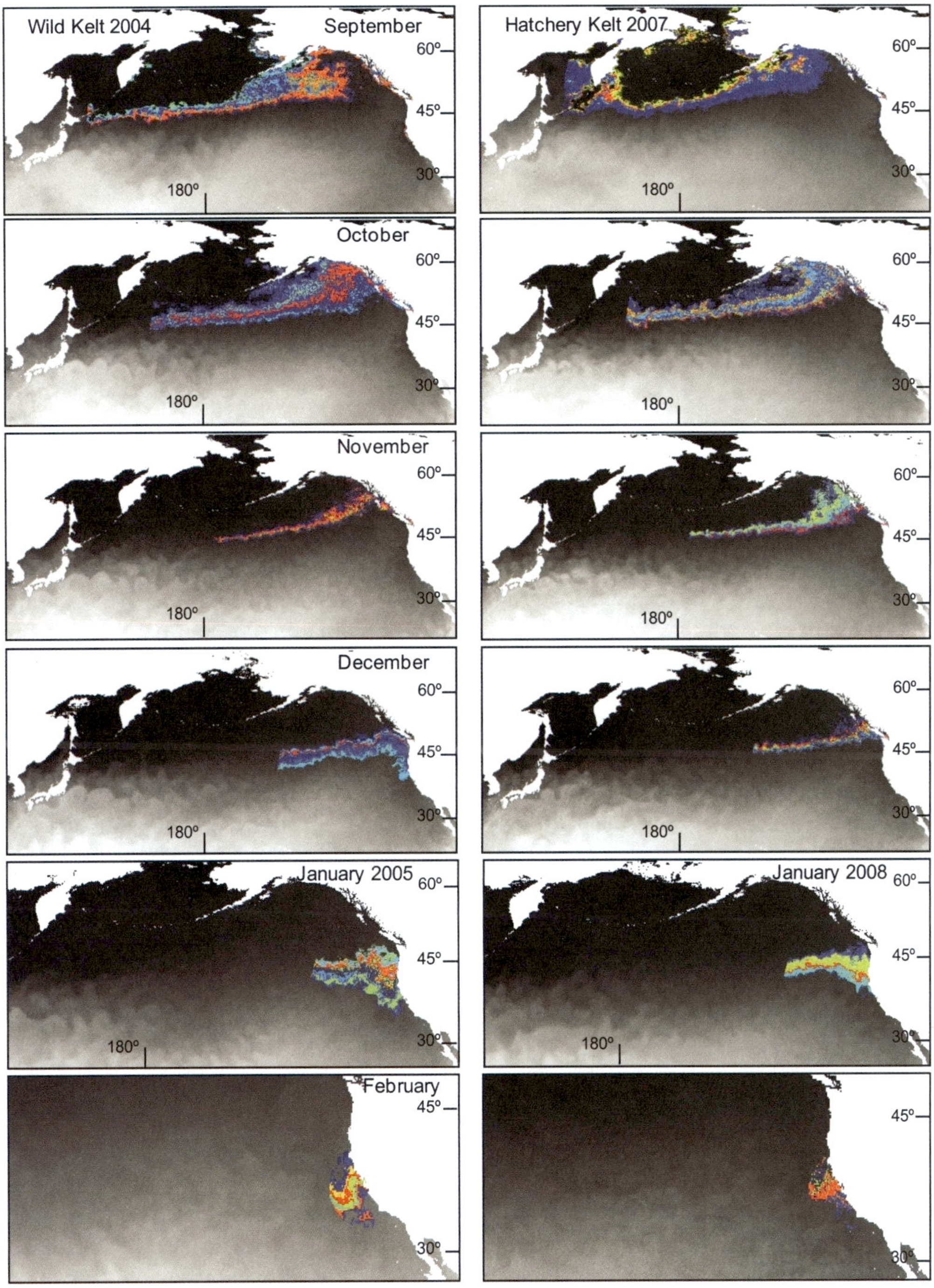

Figure 5. Continued.

experience and contrast it with the extremely narrow thermal range of waters maturing steelhead apparently seek out in the marine environment. New data on coho salmon ocean thermal niche were also provided by a tag return, for comparison with data from other researchers across the range studying coho, steelhead and other species. In addition this work shed new light on the habitats in which juvenile steelhead and coho were susceptible to predation. Finally, the data from several tag returns suggest that central California steelhead are seeking out the same regions of the North Pacific as steelhead from other northern populations and are potentially undertaking one of the longest marine migrations known.

Documenting the freshwater thermal habitat of steelhead was an unexpected but valuable result of this study. Fish carrying archival tags in the upstream habitat showed temperatures with high correspondence to the temperatures recorded by temperature loggers in that habitat. While not surprising, it provides evidence that fish are potentially forced to experience any temperatures that occur in shallow coastal streams like Scott Creek, and that there is very little cool refuge habitat should extreme temperatures occur. This implies that juvenile coho salmon are probably experiencing similar thermal environments. Alternatively, steelhead in this population may simply be indifferent to the stream temperature conditions that occurred during the course of this study and didn't bother to seek out thermal refuge habitat. If fish are truly at the mercy of limited variability in stream habitat conditions, it emphasizes the need to maintain healthy riparian canopy to provide thermal cover for potentially more temperature-sensitive species like coho salmon.

In contrast to the upstream habitat, some of the archival data indicated that steelhead also make use of the estuary habitat. This has already been well documented (Bond et al. 2008; Hayes et al. 2008; Hayes et al. 2004), but it was remarkable to observe that steelhead did not move to accessible cool water habitat even when temperatures were quite warm (>22°C) in the estuary, an indication that central California steelhead may be more tolerant of high temperatures than northern stocks. It also serves as a potential explanation for why coho salmon are not typically observed in the estuary during the warmer summer and fall months in this system (Hayes et al. 2004); they may be restricted to upstream habitat.

Recovery rates of stream fish from the freshwater environment were high. An actual percentage is hard to define, since these recoveries were purely opportunistic and it's likely very high rates could be achieved with concerted collection efforts in shallow streams. This suggests that archival tagging could be a valuable tool for studies of freshwater habitat use. When used in conjunction with instream PIT tag readers (Bond et al. 2007) and high-resolution stream temperature mapping, one could gain tremendous insights into habitat preferences and movements for fish. This could be used for comparative studies of natural habitat, impacted habitats and for evaluating the effectiveness of restoration efforts.

The recovery of archival tags that were consumed by predators was unexpected and serves as a reminder that predators are a component of the ecosystem. While using archival tags to study predation behavior is probably not an ideal strategy, the archival tag recoveries coincided with the recovery of a larger number of PIT tags on Año Nuevo Island (Hayes, unpublished data). Several species of sea birds and pinnipeds used the island as roosting and breeding habitat and this opened up speculation as to which predator was responsible. A higher thermal limit on the tag to measure predator core temperature would have been useful. In either case, all three archival tag recoveries showed tight correspondence to estuary temperatures right up until the moment of predation, suggesting the fish was captured before entering the ocean. Only one species, the Western Gull *Larus occidentalis*, is found on both the island and in the Scott Creek beach/estuary, which identifies it as the most likely predator, or scavenger in the event that the fish were already dead.

The application of the archival tag temperature data to remotely sensed sea surface temperature data indicates that steelhead from central California use the same region of ocean habitat as reported for all other steelhead stocks to date (Burgner et al. 1992; Myers et al. 2005; Myers et al. 1996; Welch et al. 1998b). The method served to produce surprisingly restricted regions of likely distribution. It did make the assumption that all data were from the surface. But this is well supported by both the literature (Burgner et al. 1992; Nielsen and Zimmerman 2008; Welch et al. 1998b) and limited daily variability in temperature profiles; the fish did not migrate below a thermocline. If some of the data were from depths with cooler temperatures, this would serve to only tighten the temperature distributions and reduce the regions identified by our habitat analysis (Figure 5). The alternative explanation that steelhead are staying at cooler depths in local waters is unlikely. Aside from what would likely be major changes in diet associated with different habitat, adult, and maturing age classes of steelhead are rarely captured in the Chinook salmon fisheries along the California coast, and only a few juvenile steelhead (typically <30 cm) have been captured during 10 years of central California juvenile Chinook salmon research surveys (MacFarlane, unpublished data), but are captured with increasing frequency by surveys off Oregon and Washington (Miller 2006; NWFSC et al. 2009).

While not surprising, it is intriguing that steelhead, which are reported to have the most diverse array of freshwater life history strategies (Satterthwaite et al. 2009; Shapovalov and Taft 1954), have the most conserved marine life history strategy of all Pacific salmonids. Steelhead from streams from Alaska to northern California have been recovered in similar regions off the Washington and Canadian coast, extending across the gulf of Alaska all the way to Asia (Myers et al. 2005; Myers et al. 1996). The ocean regions we identified for steelhead from central California suggest they may have the longest migration of any Pacific salmon. Compared to fish originating from Oregon streams, fish from Scott Creek have an additional migration distance of 650 km just to reach the latitude of the California border. Even at an upper swim speed of 64 km/d, this would add an additional 21 d to their migration compared to northern California and Oregon populations. This may contribute to the low marine survival (Bond et al. 2008) and coastal gradient of smaller size fish returning to lower latitude rivers (Withler 1966). It is possible that central California steelhead are more limited in the westward extent of their migration, but insufficient data were collected in this study to examine this idea. Addressing this question would require the use of archival tags with light sensors to perform geolocation analysis, although this might be of limited value during winter months (Teo et al. 2004). Alternatively, it may require increased tagging and recapture efforts or genetic stock identification of fish caught at sea.

Of the three California salmon species, Chinook and coho salmon have migration patterns that are coastal and distributed only limited distances relative to the source of their native river mouth (Weitkamp 2010; Weitkamp 2012, this volume; Weitkamp and Neely 2002). Within the California Current, the population dynamics of both species appear to be tied to local, physical, and biological oceanographic processes (Botsford and Lawrence 2002; Wells et al. 2008b; Wells et al. 2008c; Wells et al. 2007). However, it is likely that California steelhead populations are relatively independent of local conditions, and their populations may track better with oceanographic variables that influence survival of more pelagic Alaskan salmon populations.

It has been suggested that both steelhead and sockeye salmon have energetic optima that restrict their growth potential to certain thermal windows in the marine environment (Welch et al. 1998a; Welch et al. 1998b). In contrast, steelhead from the Scott Creek population have been shown to grow rapidly in an estuarine environment (Hayes et al. 2008), and the present study has shown that the estuarine environment includes temperatures well above the marine thermal window. Yet, these estuarine growth rates are less than the marine growth rates observed for Scott Creek steelhead (Hayes, unpublished data), suggesting estuarine growth potential could be limited by the elevated metabolic rates associated with the warm estuary temperatures.

Steelhead restrict themselves to a narrower but warmer band than sockeye (Welch et al. 1998a) and a cooler band than Chinook (Hinke et al. 2005b) or coho salmon (this study, Walker et al. 2001). It may be that steelhead attempt to reduce potential competition with other salmonids by using habitats that other salmon do not, which is their suggested strategy in freshwater (Atcheson 2010; Cederholm et al. 1997; Fraser 1969; Harvey and Nakamoto 1996). Steelhead marine diet may also be tied to a prey item that is dependant upon a specific temperature band, or they are utilizing a niche that involves going offshore to eat fish, squid and plankton rather than compete with juvenile Pacific salmon species for more nearshore resources (Burgner et al. 1992; Miller 2006).

It is interesting to note the timing of steelhead entry and exit from the river was tied to the time of year when the temperatures in their narrow thermal window had shifted far enough south to be present in central California. It may also be that ocean entry and migration is timed to a period before the California Current has begun its accelerated movement south during the spring upwelling season. Welch et al. (1998b) discussed potential impacts of climate change causing ocean warming. In addition to shifting populations north, changing climatic conditions may result in localized warming of ocean water near river mouths and changes in current patterns. If steelhead are physiologically restricted to a narrow range of ocean temperatures, this warming may prevent or alter the timing of their seaward or return migrations.

Several lessons were learned from this study that would enhance future efforts. The first was that delaying the migration of steelhead in order to give them additional growth opportunity tended to increase their potential to residualize (not leave the stream). While this may be a useful tool for some questions, if one's goal is to characterize the marine portion of a life history, delaying migrations should be avoided. During a concurrent study, a group of hatchery steelhead were raised in lower densities in a separate tank at the hatchery facility. At release, those fish were larger and showed little evidence of fin deterioration commonly associated with steelhead reared in higher densities. As expected, these fish experienced a marine survival rate (3.4%) that was much greater than that of their siblings reared at higher densities (0.05%, Hayes, unpublished data), providing a better rearing method for future tagging studies. In addition, providing fish with several weeks to recover following tagging probably increases their chances of marine survival and allows for extra time to observe potential tag rejection. However fish should be tagged early enough to allow time for release during the normal migration window.

In this study, most of the coho salmon were tagged and released in the spring of 2006, unfortunately coinciding with an unexpected California-wide drop in marine survival associated with fluctuating oceanographic conditions (Lindley et al. 2009; MacFarlane et al. 2008; Wells et al. 2008a). While unforeseeable at the time, it suggests one should plan tagging efforts to distribute tags across several years to buffer against such fluctuations.

Unfortunately, even under the best conditions, marine survival and consequent tag returns of marine fish are low. These studies require large tag budgets to deploy enough tags to ensure some level of success. At this point, archival tag technology is not limiting research on salmonids; rather the issue is more one of cost. Most tag companies have been driven by a research market working on commercial stocks of larger fish or marine mammals where tag recovery rates are higher. In addition, many tag technologies were prototype-tested on pinnipeds over the past two decades due to the high return rates (Boehlert et al. 2001; Boehlert et al. 2002; Costa 1993). As tag technology has evolved, manufacturers have phased out older technologies to keep up with the marine mammal and large-fish market demand for new tags with more memory and new sensors. This has weakened the supply of small, simple and economical tags. In addition, it is difficult to convince funding sources to finance the tagging of fish with an expected return rate of 5% or less. These agencies, often willing to fund $100,000 for 10 marine mammal tags with an expected return of 10 tags, need to be persuaded that there is equal merit in deploying 500 fish tags for the same budget to get 10 returns. While the tag return rates in this study were low, the data yield for a tag budget of approximately $25,000 far outweighed the cost of any off-shore ship-based net survey, and nonship based acoustic receiver arrays are only in early developmental stages and limited in their availability to track fish in habitats off continental shelf breaks.

Acknowledgments

Thanks to the NMFS SWFSC Environmental Research Division for the donation of extra data loggers. Fisheries Restoration Grant Program funding from the California Department of Fish and Game helped support staff that participated in stream tagging and trapping operations. Pat Morris and Guy Oliver recovered the temperature loggers found on Año Nuevo Island. And a special thanks goes out to the field assistants who always took the time when a PIT tagged fish was recovered to stop and check the list for potential archival tag returns.

References

Atcheson, M. 2010. Interannual variation in steelhead trout (*Oncorhynchus mykiss*) diet, growth and consumption in North Pacific marine ecosystems. Master's Thesis. University of Washington, Seattle.

Boehlert, G. W., D. P. Costa, D. E. Crocker, P. Green, T. O'Brien, S. Levitus, and B. J. Le Boeuf. 2001. Autonomous pinniped environmental samplers: using instrumented animals as oceanographic data collectors. Journal of Atmospheric and Oceanic Technology 18(11):1882–1893.

Boehlert, G. W., P. Green-Jessen, D. P. Costa, D. E. Crocker, T. O'Brien, and S. Levitus. 2002. Marine animals as oceanographic data collectors. Earth System Monitor 12(4):1–2.

Bond, M. H., C. V. Hanson, R. Baertsch, S. A. Hayes, and R. B. MacFarlane. 2007. A new low cost in-stream antenna system for tracking passive integrated transponder (PIT) tagged fish in small streams. Transactions of the American Fisheries Society 136:562–566.

Bond, M. H., S. A. Hayes, C. V. Hanson, and R. B. MacFarlane. 2008. Marine survival of steelhead (*Oncorhynchus mykiss*) enhanced by a seasonally closed estuary. Canadian Journal of Fisheries and Aquatic Sciences 65:2242–2252.

Botsford, L. W., and C. A. Lawrence. 2002. Patterns of co-variability among California Current Chinook salmon, coho salmon, Dungeness crab, and physical oceanographic conditions. Progress in Oceanography 53(2–4):283–305.

Burgner, R. L., J. T. Light, L. Margolis, T. Okazaki, A. Tautz, and S. Ito. 1992. Distribution and Origins of Steelhead Trout Oncorhynchus-Mykiss in Offshore Waters of the North Pacific Ocean. International North Pacific Fisheries Commission Bulletin (51):1–92.

Busby, P. J., T. C. Wainwright, and G. J. Bryant. 2000. Status review of steelhead from Washington, Idaho, Oregon, and California. Pages 119–132 *in* E. E. Knudson, C. R. Steward, D. D. MacDonald, J. E. Williams, and D. W. Reiser, editors. Sustainable Fisheries Management: Pacific Salmon. CRC Press, Boca Raton, Florida.

Cederholm, C. J., R. E. Bilby, P. A. Bisson, T. W. Bumstead, B. R. Fransen, W. J. Scarlett, and J. W. Ward. 1997. Response of juvenile coho salmon and steelhead to placement of large woody debris in a coastal Washington stream. North American Journal of Fisheries Management 17(4):947–963.

Costa, D. P. 1993. The secret life of marine mammals. Oceanography 6:120–127.

Fraser, F. J. 1969. Population density effects on survival and growth of juvenile coho salmon and steelhead trout in experimental stream channels. T. G. Northcote, editor. Symposium of salmon and trout in streams. University of British Columbia, Vancouver, B.C.

Friedland, K. D., R. V. Walker, N. D. Davis, K. W. Myers, G. W. Boehlert, S. Urawa, and Y. Ueno. 2001. Open-ocean orientation and return migration routes of chum salmon based on temperature data from data storage tags. Marine Ecology Progress Series (216):235–252.

Hartt, A. C., and M. B. Dell. 1986. Early oceanic migrations and growth of juvenile Pacific salmon and steelhead trout, volume 46. International North Pacific Fisheries Commission, Bulletin 46:1–105.

Harvey, B. C., and R. J. Nakamoto. 1996. Effects of steelhead density on growth of coho salmon in a small coastal California Stream. Transactions of the American Fisheries Society 125(2):237–243.

Hayes, S. A., M. H. Bond, C. V. Hanson, E. V. Freund, J. J. Smith, E. C. Anderson, A. Ammann, and R. B. MacFarlane. 2008. Steelhead growth in a small central California watershed: upstream and estuarine rearing patterns. Transactions of the American Fisheries Society 137:114–128.

Hayes, S. A., M. H. Bond, C. V. Hanson, and R. B. MacFarlane. 2004. Interactions between endangered wild and hatchery salmonids; can the pitfalls of artificial propagation be avoided in small coastal streams? Journal of Fish Biology 65(Supplement A):101–121.

Hinke, J. T., D. G. Foley, C. Wilson, and G. M. Watters. 2005a. Persistent habitat use by Chinook salmon (*Oncorhynchus tshawytscha*) in the coastal ocean. Marine Ecology Progress Series 304:207–220.

Hinke, J. T., G. M. Watters, G. W. Boehlert, and P. Zedonis. 2005b. Ocean habitat use in autumn by Chinook salmon in coastal waters of Oregon and California. Marine Ecology Progress Series 285:181–192.

Lindley, S. T., C. B. Grimes, M. S. Mohr, W. Peterson, J. Stein, J. T. Anderson, L. W. Botsford, D. L. Bottom, C. A. Busack, T. K. Collier, J. Ferguson, J. C. Garza, A. M. Grover, D. G. Hankin, R. G. Kope, P. W. Lawson, A. Low, R. B. MacFarlane, K. Moore, M. Palmer-Zwahle, F. B. Schwing, J. Smith, C. Tracy, R. Webb, B. K. Wells, and T. H. Williams. 2009. What caused the Sacramento River fall Chinook stock collapse? NOAA SWFSC FED Pre-publication report to the Pacific Fishery Management Council. ftp://library.swfsc.noaa/lab_pubs/00920.pdf.

MacFarlane, R. B., B. Wells, and S. A. Hayes. 2008. Coho and Chinook Salmon Decline in California during the spawning seasons of 2007/08. 2 February 2008, NOAA SWFSC FED.

McKinnell, S., J. J. Pella, and M. L. Dahlberg. 1997. Population-specific aggregations of steelhead trout (*Oncorhynchus mykiss*) in the North Pacific Ocean. Canadian Journal of Fisheries and Aquatic Sciences 54(10):2368–2376.

Miller, T. W. 2006. Trophic dynamics of marine nekton and zooplankton in the northern California current pelagic ecosystem. Doctoral dissertation. Oregon State University, Corvallis.

Myers, K., N. Davis, A. Celewycz, E. J. Farley, J. Morris, M. Trudel, M.-A. Fukuwaka, S. Kovalenko, and A. Shubin. 2005. High seas salmonid Coded-Wire Tag Recovery Data, 2005. School of Aquatic and Fishery Science, Fisheries Research Institute, Washington University, Seattle.

Myers, K. W., K. Y. Aydin, R. V. Walker, S. Fowler, and M. L. Dahlberg. 1996. Known ocean ranges of stocks of Pacific salmon and steelhead as shown by tagging experiments, 1956–1995. North Pacific Anadromous Fish Commission Doc 192, University of Washington, NPAFC Doc 192, Seattle.

Nielsen, J. L., and C. Zimmerman. 2008. Acoustic and archival tags describe freshwater to marine migrations and ocean behavior of post-spawn steelhead, Ninilchik River, Alaska. Symposium on advances in fish tagging and marking technology, 24–28 February, Auckland, New Zealand, American Fisheries Society.

NWFSC, CIMRS, and OGI. 2009. Annual Report to Bonneville Power Administration on "Columbia River basin juvenile salmonids: survival and growth in the columbia river plume and northern California current" April 2008–February 2009. NWFSC.

Pearcy, W. G. 1992. Ocean ecology of north Pacific Salmonids. University of Washington Press, Seattle.

Reddin, D. G., P. Downton, and K. D. Friedland. 2006. Diurnal and nocturnal temperatures for Atlantic salmon postsmolts (*Salmo salar* L.) during their early marine life. Fisheries Bulletin 104(3):415–428.

Reddin, D. G., K. D. Friedland, P. Downton, J. B. Dempson, and C. C. Mullins. 2004. Thermal habitat experienced by Atlantic salmon (*Salmo salar* L.) kelts in coastal Newfoundland waters. Fisheries Oceanography 13(1):24–35.

Satterthwaite, W. H., M. P. Beakes, E. M. Collins, D. R. Swank, J. E. Merz, R. G. Titus, S. M. Sogard, and M. Mangel. 2009. Steelhead life history on California's central coast: insights from a state-dependent model. Transactions of the American Fisheries Society 138(3):532–548.

Shapovalov, L., and A. C. Taft. 1954. The life histories of the steelhead rainbow trout (*Salmo gairdneri gairdneri*) and silver salmon (*Oncorhynchus kisutch*). California Department of Fish and Game, Fish Bulletin 98:375.

Teo, S. L. H., A. Boustany, S. Blackwell, A. Walli, K. C. Weng, and B. A. Block. 2004. Validation of geolocation estimates based on light level and sea surface temperature from electronic tags. Marine Ecology Progress Series 283:81–98.

Walker, R., N. Davis, K. Myers, S. Urawa, and K. Hirasawa. 2004. Releases and recoveries of U.S. and NPRB salmonid data storage tags, and recoveries of high seas tags in North America, 2004. School of Aquatic and Fishery Science, Fisheries Research Institute, Washington University Technical Report 406:23.

Walker, R. V., N. D. Davis, K. W. Myers, and J. H. Helle. 2005. New Information from archival tags from Bering Sea tagging, 1998–2004. NPAFC Tech. Rep 6:38–40.

Walker, R. V., N. D. Davis, K. W. Myers, J. M. Murphy, and M. Fukuwaka. 2001. Releases and recoveries of U.S. salmonids data storage tags, and recoveries of high seas tags in North America, 2001. School of Aquatic and Fishery Sciences, University of Washington, (NPAFC Doc. 556.) SAFS-UW-0112, Seattle.

Walker, R. V., K. W. Myers, N. D. Davis, K. Y. Aydin, K. D. Friedland, H. R. Carlson, G. W. Boehlert, S. Urawa, Y. Ueno, and G. Anma. 2000. Diurnal variation in thermal environment experienced by salmonids in the North Pacific as indicated by data storage tags. Fisheries Oceanography 9(2):171–186.

Weitkamp, L. A. 2010. Marine distributions of Chinook salmon from the West Coast of North America determined by coded wire tag recoveries. Transactions of the American Fisheries Society 139(1):147–170.

Weitkamp, L. A. 2012. Marine distributions of coho and Chinook salmon inferred from coded wire tag recoveries. Pages 191–214 *in* J. McKenzie, B. Parsons, A. C. Seitz, R. K. Kopf, M. Mesa, and Q. Phelps, editors. Advances in fish tagging and marking technology. American Fisheries Society, Symposium 76, Bethesda, Maryland.

Weitkamp, L. A., and K. Neely. 2002. Coho salmon (*Oncorhynchus kisutch*) ocean migration patterns: insight from marine coded-wire tag recoveries. Canadian Journal of Fisheries and Aquatic Sciences 59:1100–1115.

Welch, D. W., Y. Ishida, and K. Nagasawa. 1998a. Thermal limits and ocean migrations of sockeye salmon (*Oncorhynchus nerka*): Long-term consequences of global warming. Canadian Journal of Fisheries and Aquatic Sciences 55(3).

Welch, D. W., Y. Ishida, K. Nagasawa, and J. P. Eveson. 1998b. Thermal limits on the ocean distribution of steelhead trout (*Oncorhynchus mykiss*). North Pacific Anadromous Fish Commission Bulletin 1(396–404).

Welch, D. W., B. R. Ward, B. D. Smith, and J. P. Eveson. 1998c. Influence of the 1990 ocean climate shift on British Columbia steelhead (*Oncorhynchus mykiss*) and coho (*O. kisutch*) populations. NPAFC Doc. 310. Department of Fisheries and Oceans, Pacific Biological Station, Nanaimo, B.C.

Wells, B. K., J. C. Field, J. Thayer, C. B. Grimes, W. Bograd, W. Sydeman, F. Schwing, and R. Hewitt. 2008a. Untangling the relationships among climate, prey, and top predators in an ocean ecosystem. Marine Ecology Progress Series 364:15–29.

Wells, B. K., J. C. Field, J. Thayer, C. B. Grimes, W. S. Bograd, W., F. Schwing, and R. Hewitt. 2008b. Untangling the relationships among climate, prey, and top predators in an ocean ecosystem. Marine Ecology Progress Series 364:15–29.

Wells, B. K., C. B. Grimes, J. G. Sneva, S. McPherson, and J. B. Waldvogel. 2008c. Relationships between oceanic conditions and growth of Chinook salmon (*Oncorhynchus tshawytscha*) from Alaska, Washington, and California, USA. Fisheries Oceanography 17:101–125.

Wells, B. K., C. B. Grimes, and J. B. Waldvogel. 2007. Quantifying the effects of wind, upwelling, curl, turbulence, and sea surface temperature on growth and maturation of a California Chinook salmon (*Oncorhynchus tshawytscha*) population. Fisheries Oceanography 16(363–382).

Withler, R. E. 1966. Variability in life history characteristics of steelhead trout (*Salmo gairdneri*) along the Pacific coast of North America. Journal of the Fisheries Research Board of Canada 23(3):365–393.

American Fisheries Society Symposium 76:493–506, 2012

The Tilt-Tag: a Micro-Processor Controlled Tag for Determining Lifespan in Fish and Improving Survey Efficiency in Area-under-the-curve Estimates of Salmon Escapement

Tom F. Shardlow* **and Kim D. Hyatt**

Pacific Biological Station, Department of Fisheries and Oceans Canada
3190 Hammond Bay Road, Nanaimo, British Columbia V9T 6N7, Canada

Abstract.—A micro-controlled fish tag which records post-tagging lifespan was developed, tested as a prototype, and then evaluated in field applications for measuring survey life. The method of constructing the Tilt-Tag and the results of tank test trials on Chinook salmon *Oncorhynchus tshawytscha*, preliminary field trials on chum salmon *Oncorhynchus keta* are reviewed, and full-scale field applications on sockeye salmon *Oncorhynchus nerka* are presented. Survey life (SL) is an essential component for area-under-the-curve (AUC) estimation of Pacific salmon *Oncorhynchus* spp. spawning escapements. However, direct estimates of SL are often unavailable because the estimates mostly require extensive and costly tag-recapture programs. In this study, the Tilt-Tag was used to estimate SL by measuring the elapsed time from tagging until the fish came to rest permanently on its lateral or dorsal surface. Tilt-Tag derived estimates of SL, combined with specification of survey rules that were based on historical run-timing and stream temperature, reduced survey costs by approximately 50% when compared to conventional tag-recapture methods. Abridged details on how to construct the Tilt-Tag are provided so that researchers will be able to make their own tags.

Introduction

Determining lifespan in fish is time consuming and expensive. We present the Tilt-Tag as an alternative to current tagging methods of estimating lifespan for Pacific salmon *Oncorhynchus* spp. and compare survey costs, in person-days, between conventional and Tilt-Tag derived estimates. In salmon spawner surveys, the estimate of the time mature salmon spend in a stream prior to death is often called survey life (SL) (Ames and Phinney 1977), or survey area residence time (Perrin and Irvine 1990; English et al. 1992). Estimates of SL are critically important for the widely used AUC method of determining the number of mature salmon entering freshwater spawning areas i.e. when determining salmon escapement. (English et al. 1992). When using the AUC method, SL is defined as the average number of days a live spawner resides in the AUC survey area. As in:

$$\hat{S} = AUC/SL$$

*Corresponding author: ShardlowT@pac.dfo-mpo.gc.ca

with $\hat{S}$: number of spawners, AUC: area-under-the-curve in spawner days, SL: survey life in days. Estimates of SL have been conducted in several ways using tags. In most cases numbered external tags were applied to fish downstream of the survey area and later recovered on carcasses (Perrin and Irvine 1990). The average number of days between the application and recovery of tags was taken as the SL estimate. Other techniques using tags involve the development of a tag depletion curve from the number of tags recovered on dead fish plotted against time (Bocking et al. 1988). In both of these approaches and in other methods of estimating SL, extensive tag-recovery programs involving daily stream surveys were required to obtain reliable estimates of SL (Fukushima and Smoker 1997; Lady and Skalski 1998; Manske and Schwarz 2000; Korman et al. 2002; Parken et al. 2003). Moreover, one source of uncontrolled error in SL estimates, based on the carcass tag recovery method, is the unknown period of time between the death and recovery of the tagged spawner. Conventional tag studies using external tags are unable to determine this time period because the exact time of death is unknown. Concerns regarding biases in both AUC and SL estimates derived from visual surveys of salmon spawning areas prompted researchers to develop methods that minimize the impact of these biases on the resulting escapement estimates by deriving the estimates for SL and AUC from data collected from the same surveys (English et al. 1992; Irvine et al. 1992). In these studies, SL was estimated from tag depletion curves where the number of tags observed on live fish were plotted against time. Biotelemetry methods using radio-tags, which do not need to rely on daily carcass recoveries, have been used to measure stream life (Koski et al. 1996). The radio-tag signals, recorded remotely, were used to fix the position of the tagged fish. The duration of a constant geographic position of the tag on the spawning ground was taken as the SL for Chinook salmon *Oncorhynchus tshawytscha*. However, it was not always possible to determine if a stationary signal was from a live or a dead spawner. A review of biotelemetry methods in general (Cooke et al. 2004) found no existing technology that could be easily or inexpensively adapted for determining survey life. Methods involving archival data loggers and accelerometers have been used to record swimming activity in whales (Nowacek et al. 2001), flatfish (Kawabe et al. 2003) and tuna (Block et al. 2005) for example. However, these devices could not easily be adapted for use on populations of spawning salmon where hundreds to thousands of tags are needed because the vast majority of the tags in survey life studies would remain un-recovered. More recently, radio tags designed for internal use on salmon have been equipped with motion sensors where the absence of motion indicates the fish is dead. However, daily surveys are still required to detect the radio signals.

The Tilt-Tag is a micro-processor controlled tag using a simple tilt switch and a timer system. It was developed and tested for the single purpose of determining lifespan of spawning fish and is a lower cost alternative to existing methods. The new tag records the elapsed time between application of the tag and the death of the fish. Because the exact time of death is stored electronically, the tagged carcasses do not need to be recovered daily or detected daily as with radio tags, thus reducing survey effort and associated costs. The methods of construction of the tag and the results of test trials (Shardlow et al. 2007) demonstrated that the tag is relatively simple to assemble and can detect time of death within one hour or less. The results of the first full scale field applications of the tag on free-ranging Okanagan River sockeye adults in 2006 and 2007 are presented to illustrate the tags utility. The Tilt-Tag estimates of SL will be compared to conventional tagging estimates that used depletion analysis in the Okanagan sockeye.

The Tag

Concept

The micro-controlled tag records the elapsed time from the time of tag application to the time the fish dies. This duration is determined by sensing the angle or attitude of the fish. An upright position, or 0°, indicates the fish is alive. A tilt beyond a specified angle laterally for more than a specified amount of time, indicates the fish is dead. The angle and required time duration, or lag time, of the tilt can be varied. These parameters are set to assure that the fish is dead by requiring a set maximum tilt angle to last for the specified lag time before the timer stops. The expected durations and angles about a vertical axis assumed for various behavioural events in spawning salmon are listed in Table 1 (Shardlow et al. 2007). Thus, any spawner in the 315° to 45° arc would be considered alive, while a spawner continuously in the 45° to 315° arc for more than 3 min is assumed to be dead. All other behaviors, even though they can occur at the angles assumed by dead fish, will fall far short of the lag time of 3 min required to stop the timer.

Construction

Complete documentation of tag construction is given (Shardlow et al. 2007) in sufficient detail so that others can construct this tag. The following summary provides an overview.

The Tilt-Tag is comprised of four basic components: tilt switches, battery, timer system, and platform. Numerous tilt sensor technologies are commercially available. In this application, a nonmercury, environmentally benign mechanical switch that detects tilt angles of 90° or greater was used. However, when mounted in an offset position, this switch can detect any angle of tilt. Each tag contained two tilt switches, one each for a right or a left rotation of the fish. The tag is powered from coin-type lithium manganese dioxide cell (CR) batteries. Battery life is expected to be several years allowing for re-use of recovered tags.

The timer system consists of a microcontroller, 3-wire serial communication port, a reed switch and a light emitting diode (LED). The timer system is controlled by a reduced instruction set computer (RISC) microcontroller from Texas Instruments. Thin 30-gauge wires protrude from the tags housing to provide a 3-wire serial communication link. The wire is malleable and requires only a few inches for multiple deployments, reducing the possible stress to the fish caused by these protrusions. A magnetic sensor, called a reed switch, monitors for the presence of a magnetic field. The tag's controls such as "deploy" and "upload data" were activated by moving a magnet close to the switch. The tag provides confirmation of the deployment request by a blinking LED. When in deployment mode, the tag can be checked at any time for correct operation. Each time the magnet is

Table 1. Expected duration and angles of behaviors exibited by spawning salmon.

Behavioral event	Expected duration	Angles assumed
Swimming	continuous	315° to 45°
Fighting	< 20 s	0° to 360°
Jumping	< 5 s	0° to 360°
Digging	< 10 s	90° to 270°
Dead	> 3 min	45° to 315°

brought close, the tag sends a series of LED pulses.

The tilt switch, timer system, and battery are mounted on a 32 mm Petersen disc tag. Two mounting pins are placed at the sides of the disk instead of the usual placement of a single central pin. Using two pins secures the tag and prevents rotation that would interfere with the accurate operation of the tilt switches. The tags, when attached to the fish, were not perpendicular but were offset by the dorsal surface angle of the fish. This angle, called the offset angle, needed to be accounted for when mounting the tilt switches into the tag housing. A predetermined offset angle of 45° was adopted as the standard setting although the actual angle after placement on a fish varies with fish species and sex. Once the offset angle was set, the tilt switches were aligned such that a tilt of either 45° or 70° triggered a contact that stopped the timer. An assembled tag weighs approximately 25 g.

Test Results and Trials

Tilt-Tags were subjected to bench tests and field trials during their initial development in 2005 and 2006 (Shardlow et al. 2007). Bench testing, where the tags were vibrated mechanically in a 40 cm circular pattern for 24 h, showed the tags recorded elapsed time with an accuracy of 3 min. SD. Tank trials, where Chinook salmon in holding tanks were repeatedly anesthetized to simulate death, were conducted over a 22 d period, showed the tags recorded the time of "death" to within ± 6 min. The tag recorded time of death correlated strongly to the elapsed time, $r^2 = 0.99$, $n = 16$. There were no mortalities during the tank trials.

In the first field trials chum salmon *O. keta* were tagged and released into a spawning channel (length = 4 km, width = 7.5 m, Avg. depth = 1 m) at the Little Qualicum River, B.C., Canada. Twenty-six operative tags were applied with 20 being recovered during daily walks of the channel. The fact that 6 tags were never recovered is not surprising. In our experience (Shardlow et al. 1985) many tagged carcasses go un-recovered during foot surveys. What was surprising though was that in ideal recovery conditions, like the spawning channel, it took on average 1.6 d to find tagged chums after they had died. Tag recorded SL ranged from 1.4 to 7.9 d with a mean of 6 d. The number of days to dead recovery by walk survey, on the other hand ranged from 2.9 to 11.1 d with a mean of 7.6 d. This bias could have a serious effect on AUC derived estimates of escapement where tags are recovered during walk surveys. To our knowledge this type of recovery bias has gone unreported in previous studies.

Field Applications

Study location and methods

Okanagan River sockeye salmon migrate up the Columbia River and into the Okanagan basin. Spawning takes place in British Columbia, Canada in a 19-km stretch of river between Vaseux Lake and the town of Oliver. (Figure 1). The majority of spawning occurs in a 7-km section of river between McIntyre Dam and vertical drop structure 12 (VDS 12) located at the Oliver Bridge. This section, where counts of spawners are conducted annually, is known as the index section of the river. Spawning typically takes place from the end of September to early November each year with the peak spawn occurring near mid-October (Stockwell and Hyatt 2003)

In October of 2006 and 2007, sockeye salmon were tagged before the peak of spawning in the index section of the river. Spawners were caught using a beach-seine or captured from a holding pond at a temporary fish fence. Once captured, spawners were

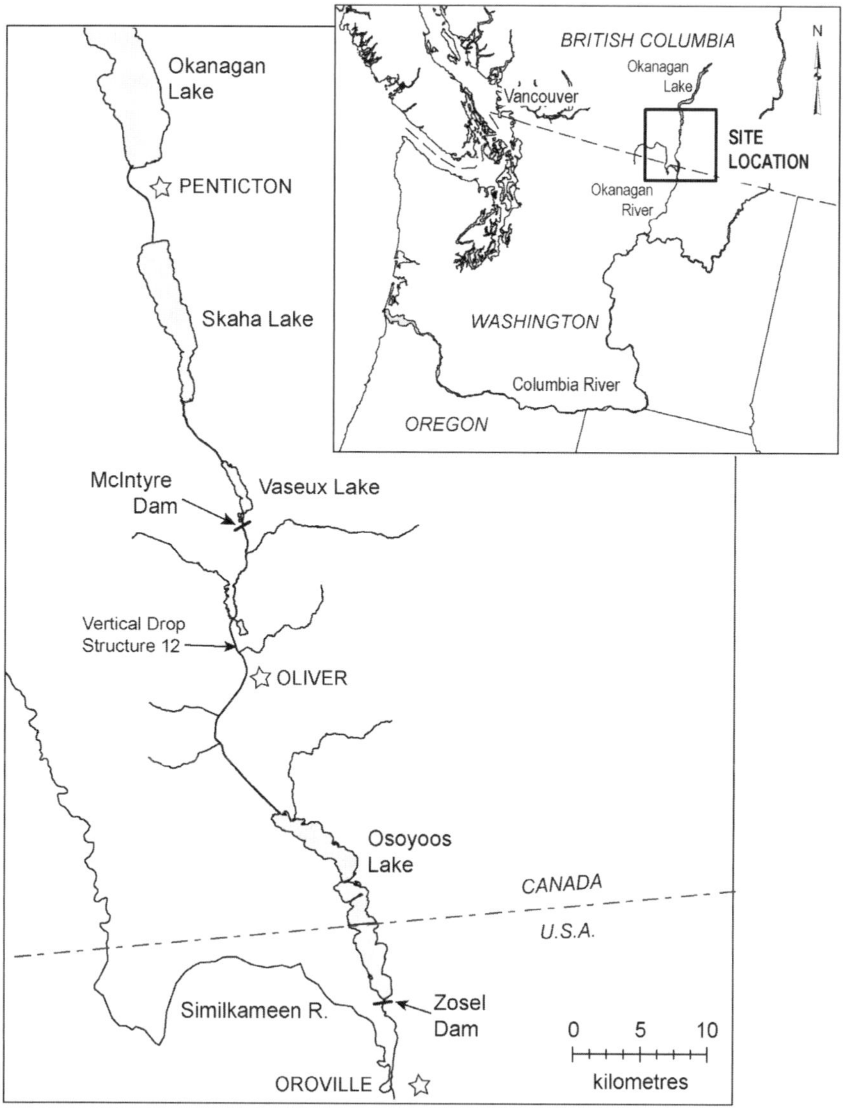

FIGURE 1. Map of study area.

randomly sampled then stratified by sex to obtain equal numbers of males and females. The fish were placed on a tagging board fitted with a holding cradle and color coded Tilt-Tags were attached near the dorsal fin. After tagging, the fish were retained for a minute or two to assure they were swimming in an upright position before release into the river. In 2006, 70 Tilt-tags were applied on October 4th, 5 on October 6th, and 12 on October 14th. In 2007, 79 tags were applied on October 5th and 76 on October 16th.

Four observers sitting in an inflatable boat surveyed the index section while drifting downstream. Observers counted live, dead, and tagged sockeye during survey trips conducted on average two times per week during the spawning season. Separate surveys on foot were conducted once per week later in the season once die-off commenced.

Crews of three walked downstream counting sockeye carcasses and recovering tags from dead fish (Figure 2). Post tagging lifespan data were recovered via a serial port connection from the Tilt-Tags.

Tagging results

A total of 233 tags were applied and 39 tagged carcasses were recovered giving an overall recovery rate of 0.17 (Table 2). The sex ratio of fish with tags applied and of recovered carcasses did not depart significantly from 50:50. In 2006, 14 tags were recovered (Table 3) which showed a mean SL of 13.6 d. The average number of days between application and recovery was 16.9 which means the tagged fish were dead but remained unrecovered for an average of 3.3 d (*t*-test, one-tail, $p < 0.05$). In 2007, 25 tags were recovered overall (Table 4) with a mean SL of 13.9 d and carcasses remained un-recovered for an average of 6.6 d. The early tag group from October 5th application showed a slightly longer SL of 15.2 compared to the SL of 12.2 d for fish tagged on October 16 (*t*-test, left tail $p = 0\ 0.06$). The SL by sex was similar (Table 2) for the 2006–07 recoveries.

For the depletion analysis, live tagged fish were released and later counted visually in the river. We followed the depletion method used by Tschaplinski and Hyatt (1990) and English et al. (1992) where SL was calculated by dividing the area under the tag depletion curve by the total number of tags initially applied. Less than half of the tags released were subsequently counted alive between the time of release to the time of the first count. Mortality during this period is expected to be negligible. Therefore, the remaining fish either went undetected during visual surveys or migrated down-stream of the study area. A low rate of detection is not uncommon in visual surveys and may be influenced greatly by observer efficiency (Shardlow et al. 1987). Fortunately, observer efficiency can be assumed constant because flows in the Okanagan River remained very constant during the October survey period at $10.56 \pm 0.60\ m^3/s$ in 2006 and $10.97 \pm 1.34\ m^3/s$ in 2007.

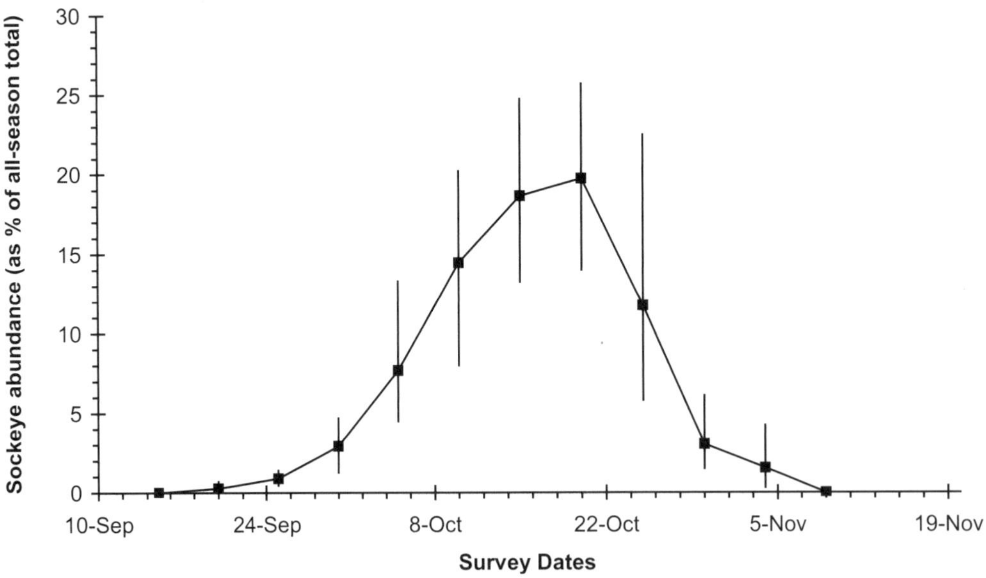

FIGURE 2. Average percent sockeye spawner abundance in the index section of the Okanagan River for the years 2001 to 2007. The bars show maximum and minimum range.

TABLE 2. Sex ratio of tags applied and recovered[a] for sockeye salmon in the Okanagan River index section, 2007 & 2006.

	Total	Male	Female
2007			
No. Applied	155	88	67
Sex ratio applied		0.57	0.43
No. Recovered	25	15	10
P recovered	0.16	0.17	0.15
Sex ratio recovered		0.60	0.40
Survey life (SL)		11.9	14.3
SD, SL		3.0	4.8
2006			
No. Applied	78	40	38
Sex ratio applied		0.51	0.49
No. Recovered	14	4	10
P recovered	0.18	0.10	0.26
Sex ratio recovered		0.29	0.71
Survey life (SL)		13.7	14.1
SD, SL		4.4	6.1
2006-07 Combined			
No. Applied	233	128	105
Sex ratio applied		0.55	0.45
No. Recovered	39	19	20
P recovered	0.17	0.15	0.19
Sex ratio recovered		0.49	0.51
Survey life (SL)		13.4	14.2
SD, SL		4.2	5.3

[a] recovered as carcasses.

Depletion analysis produced an estimate of SL = 14.5 d for 2006 (Table 5). In 2007, depletion estimates of SL were 16.1 d for the October 5th tags and 11.6 d for the October 16th tag group. The average for both tag groups combined was 14.2 d (Table 6).

When estimating SL by using either Tilt-Tags or conventional tags, fish are captured after entering the survey area. Since tagged individuals will have spent some unknown amount of time in the area prior to tagging, SL estimates will be biased low. However in this study the conventional tags were simply the Tilt-Tags without the electronic data recovered. Therefore any time bias will be identical when comparing the two tag types.

The finding that SL is shorter for the October 16th tag group than for the October 5th tags is consistent with the generally observed result that late timed spawners are shorter lived (Perrin and Irvine 1990).

Comparing SL estimates from the Tilt-Tag with depletion estimates.

For comparative purposes we considered a reduced set of Tilt-Tag recoveries used to produce the estimate of SL. Table 6B shows

Table 3. Elapsed time between tag application and dead recovery compared to time recorded on tags for sockeye salmon released in the Okanagan River index section, October, 2006.

Tag no.	Date applied	Date recovered	Days to dead recovery	Recorded days on tag	Days dead but unrecovered
1	4-Oct	20-Oct	16	13.1	2.9
2	4-Oct	20-Oct	16	11.7	4.3
3	4-Oct	18-Oct	14	13.8	0.2
4	4-Oct	26-Oct	22	21.2	0.8
5	4-Oct	25-Oct	21	7.5	13.5
6	4-Oct	25-Oct	21	17.0	4.0
7	4-Oct	26-Oct	22	13.7	8.3
8	4-Oct	20-Oct	16	15.7	0.3
9	4-Oct	18-Oct	14	13.9	0.1
10	4-Oct	26-Oct	22	17.9	4.1
11	6-Oct	25-Oct	19	18.2	0.8
12	14-Oct	26-Oct	12	9.5	2.5
13	14-Oct	19-Oct	5	5.0	0.0
14	14-Oct	30-Oct	16	12.3	3.7
Mean			16.9	13.6	3.3
SD			4.8	4.4	3.8

SL from carcass recovery dates restricted to (a) October 30th and November 2nd, or (b) November 2nd recoveries only. Table 7 summarizes the comparative results and shows that the depletion estimate of SL, in all but one case, fell within ± 1 SD of SL determined by the Tilt-Tag. However, there was a large difference in the survey effort required to measure SL. Depletion analysis required more person-days than did Tilt-Tag determined SL. This is true even when samples sizes are reduced. Combining the October 5th tag application with the October 16th application to form the 2007 combined SL, we found that three trips (i.e. two days of tag application followed by one day of carcass recovery on November 2nd) was sufficient to produce a usable estimate of 14.3 d ± 4.8 SD. Under this sampling effort, 8 person-days were expended compared to the 17 person-days required to produce the depletion estimate of SL = 19.2 d. For this comparison the float trip samples for the depletion analysis were reduced from seven to four survey trips which we consider the minimum effort required to produce a usable estimate. We also note that the depletion estimate, using minimal samples, falls outside of the range of estimates produced by all other methods.

Low-effort versus high-effort AUC escapement estimates

As noted above, reliable estimates of SL based on Tilt-Tag application require the expenditure of fewer person-days of effort than depletion-based SL estimates to support AUC determinations of spawner abundance. However, the savings in effort between the two approaches can be increased greatly if survey effort required to determine the shape of the annual spawner-abundance curve could also be reduced to the three to four trip minimum required for Tilt-Tag determinations of SL. Sur-

TABLE 4. Elapsed time between tag application and dead recovery compared to time recorded on tags for sockeye salmon released in the Okanagan River index section, October-November 2007.

Tag no.	Date Applied	Date Recovered	Days to dead recovery	Recorded days on tag	Days dead but unrecovered
1	5-Oct	21-Oct	16	14.4	1.6
2	5-Oct	21-Oct	16	7.6	8.4
3	5-Oct	22-Oct	17	13.7	3.3
4	5-Oct	30-Oct	25	21.5	3.5
5	5-Oct	30-Oct	25	20.2	4.8
6	5-Oct	30-Oct	25	16.5	8.5
7	5-Oct	30-Oct	25	21.2	3.8
8	5-Oct	30-Oct	25	9.5	15.5
9	5-Oct	30-Oct	25	18.3	6.7
10	5-Oct	30-Oct	25	16.3	8.7
11	5-Oct	2-Nov	29	12.0	17.0
12	5-Oct	2-Nov	29	7.2	21.8
13	5-Oct	2-Nov	29	8.8	20.2
14	5-Oct	2-Nov	29	25.2	3.8
15	16-Oct	21-Oct	16	11.1	4.9
16	16-Oct	30-Oct	14	11.0	3.0
17	16-Oct	30-Oct	14	8.7	5.3
18	16-Oct	22-Oct	6	4.1	1.9
19	16-Oct	30-Oct	14	10.2	3.8
20	16-Oct	2-Nov	18	15.8	2.2
21	16-Oct	2-Nov	18	14.8	3.2
22	16-Oct	2-Nov	18	14.3	3.7
23	16-Oct	2-Nov	18	13.9	4.1
24	16-Oct	2-Nov	18	14.6	3.4
25	16-Oct	2-Nov	18	16.2	1.8
Mean			20.5	13.9	6.6
S.D.			6.2	5.0	5.8

Tilt Tags	Mean SL (days)	S.D.
Overall	13.9	5.0
Oct 5th applicaton	15.2	5.7
Oct 16th application	12.2	3.6

TABLE 5. Depletion estimate from time series counts of tags on live fish in the index section of the Okanagan River from 70 tags released on October 4, 2006.

Date October	Days since release	Count live tags
6	2	32[1]
13	9	36
16	12	35
19	15	5
22	18	1
25	21	2
27	23	0
	Depletion estimate of survey life = 14.5 days	

[1]For the depletion estimate the October 6 count was assumed to be 36 since that many were alive on October 13.

TABLE 6. A. Depletion estimate from time series counts of tags on live fish in the index section of the Okanagan River from October 5th and October 16th releases, 2007. **B.** Stream life (SL) estimates based on 2007 sub-sampled Tilt-Tag recoveries from survey dates restricted to either (a) October 30 and November 2, or (b) November 2 only.

A. Depletion estimate.

	Oct 5		Oct 16	
	79 tags released		76 tags released	
Survey date	Days since release	Live tags counted	Days since release	Live tags counted
17-Oct	12	36		30
19-Oct	14	17		28
24-Oct	19	8	8	26
26-Oct	21	0	10	16
28-Oct	23	2	12	17
29-Oct	24	0	13	7
2-Nov	28	0	17	1
		5-Oct	16-Oct	Average
SL estimate		16.5 d	11.9 d	14.2 d

B. Tilt-Tag estimate.

	SL	*n*	SD
(a)			
Overall	14.8	20	4.8
Oct. 5 application	16.1	11	5.9
Oct. 16 application	13.3	9	2.6
(b)			
Nov. 2 application	14	10	4.8

Table 7. Comparison of survey-life estimated by Tilt Tags vs depletion analysis for 2006 and 2007 Okanagan sockeye.

Year	Method	Tags applied	Tags recovered	Survey life (d), ± SD	Survey effort (person-days)
All	Tilt Tag	78	14	13.6 ± 4.4	17
2006	Depletion	78	36[a]	14.5	24
Early run[b]	Tilt Tag	80	11	16.1 ± 5.9	11
2007	Depletion	80	36[a]	16.5	26
Late run[c]	Tilt Tag	76	9	13.3 ± 2.6	11
2007	Depletion	76	30[a]	11.9	26
Combined	Tilt Tag[d]	76	10	14.3 ± 4.8	8
2007	Depletion[e]	76	66[a]	19.2	17

[a] Max. No. live tags observed
[b] October 5th tag application, and recoveries for Oct. 30 & Nov. 2 only
[c] October 16th tag application, and recoveries for Oct. 30 & Nov. 2 only
[d] Oct.5 & 16th application, and recoveries for Nov. 2 only
[e] Float surveys reduced from 7 to 4 trips.

Table 8. A comparison of high-effort versus low-effort AUC estimates of sockeye salmon spawner abundance in the index section of the Okanagan River for the years 2002, 2004, 2006, and 2007.

	High-effort		Low-effort		
Brood year	No. of surveys	AUC[1] estimate	No. of surveys	AUC estimate	Difference as a % of high-effort AUC
2002	11	3,991	4	3,888	+2.58
2004	13	40,103	4	42,493	–5.95
2006	14	15,070	4	15,295	–1.49
2007	12	10,487	4	10,339	+1.41

[1] An all-year historic average SL of 11 days was used for the 2002 and 2004 estimates. The Tilt-Tag determined SL of 14 days was used for 2006 and 2007.

vey effort during 2006 and 2007 to determine the shape of the spawner-abundance curves amounted to 12 and 10 stream-inspection dates of the Okanagan River spawning area respectively. However, review of background information on water temperature, and start, peak, and end dates for spawning by sockeye suggests a stepwise approach may be used to approximate AUC estimates from just four stream inspections as follows. Step-1: Recent observations (Figure 2) indicate it is safe to assume that few, if any, sockeye spawn either before or after September 25th or November 5th respectively in a given year so these dates are used as AUC start and end dates on which the spawner population equals zero. Step-2: Historic observations (n = 25) also indicate the all-year average date of peak spawning occurs on October 14th (range October 1–October 25) coincident with the interval when average weekly water temperature declines to 12°C (Stockwell and Hyatt 2003). Thus, if the first escapement enumeration survey is scheduled for October 7th, it should serve to identify the slope of the rise between the date of zero abundance and the average date of peak abundance. A notable exception to this fixed-date scheduling would occur if mean weekly water temperature was predicted to approach 12°C earlier than October 7th in which case the first survey would be executed in the week coinciding with average water temperature of 12°C. Step-3: More than 75% of the AUC occurs within the interval extending from one week before to two weeks after the occurrence of peak adult abundance (Figure 2). Thus, an adequate AUC description should emerge from additional surveys scheduled to coincide with the average date of peak abundance (October 14th), one week after peak abundance (October 21st) and then two weeks after peak abundance (October 28th). The latter two survey dates should not only identify the slope of the decline from peak abundance to the assumed November 5th date of zero-abundance but should also provide ample opportunity to retrieve Tilt-Tags for SL estimates because the peak of salmon carcass availability generally occurs between the 7th and 20th day following the peak of spawning activity (Stockwell and Hyatt 2003).

We used the stepwise, rules-based approach for survey scheduling described here to extract subsets of observations from the more extensive series obtained by field crews during 2002, 2004, 2006, and 2007 (Figure 3). These smaller sets of observations, retrieved on the basis of our rules-based approach, were then used to generate "low-effort" AUC estimates for comparison with previously completed "high-effort" AUC-based estimates of spawner abundance in a given year (Table 8). These results confirm that our rules-based, low-effort survey approach has the potential to yield AUC estimates that vary little from those derived in recent years from much higher expenditures of effort. The principal sources of error in AUC estimates originate from low resolution, specification of spawner abundance curves and from inaccurate determinations of SL. Our results indicate that the large reduction in survey effort (68%) associated with a rules-based determination of AUC introduces a small increase in error (maximum difference of 5.95% between low and high-effort AUC estimates, Table 8).

By contrast, tag-recovery estimates of SL based on recovery of carcasses would introduce much larger errors (e.g., 24.2–47.4%, Tables 3 and 4) into AUC determinations which vary in direct proportion to errors in a given SL estimate. In addition, depletion estimates of SL based on low-effort recoveries, i.e. 4 trips, indicated a 34% error (19.2 d, Table 8). Consequently, we conclude that our rules-based approach to scheduling survey effort combined with Tilt-Tag technology for SL determinations allows not only a minimum 50% reduction in field effort but also a reduction in the overall error associated with annual AUC estimates of spawner abundance.

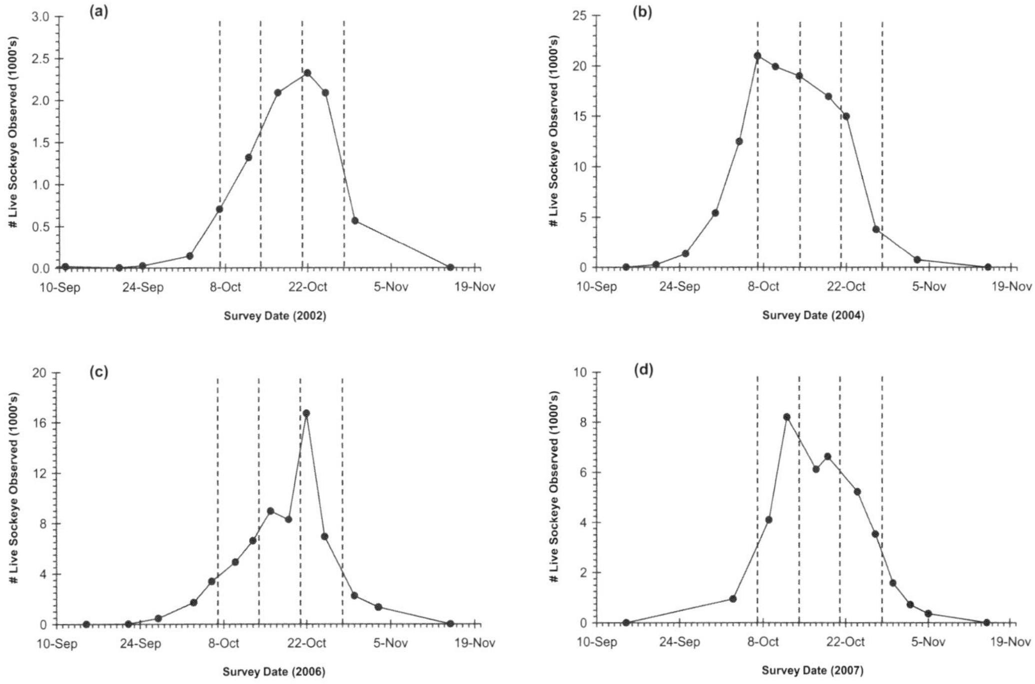

FIGURE 3. Sockeye spawner counts in the index section of the Okanagan River for the years 2002, 2004, 2006, and 2007. The vertical bars indicate the dates chosen for low-effort sampling.

Summary

Application of the Tilt-Tag for determinations of SL can reduce survey (person-day) costs by approximately 50% when compared to conventional tag-recovery methods. The Tilt-Tag determined SL is more precise than tag-recapture estimates because the time of death can be determined to within hours by the Tilt-Tag rather than to within days by conventional tags. Moreover, if used in combination with known run timing and the step-wise approach to survey scheduling, Tilt-Tags reduce total effort required to execute AUC-based escapement estimates by more than 50% with little loss of between-year resolution of changes in salmon abundance. Finally, the Tilt-Tag is relatively easy to manufacture and the cost per tag applied is expected to be low when compared to alternative archival tags. Our aim is not to manufacture the tags, but to make the technology freely available to whomever may wish to make them. We estimate material costs per tag were approximately \$55.00 CAN, but this would be reduced by more than half if the tags were made on a production basis.

Acknowledgments

We thank Margot Stockwell for helping to prepare this manuscript and members of the Okanagan Nation Fisheries Department for their able assistance in the field.

References

Ames, J., and D. E. Phinney. 1977. 1977 Puget Sound summer-fall Chinook methodology: escapement goals, run size forecasts, and in-season run size updates. Technical Report No. 29, Washington Department of Fisheries.

Block, B. A., S. L. H. Teo, A. Walli, A. Boustany, M. J. W. Stokesbury, C. J. Farwell, K. C.

Weng, H. Dewar, and T. D. Williams. 2005. Electronic tagging and population structure of Atlantic bluefin tuna. Nature 434:1121–1127.

Bocking, R. C., J. R Irvine, K. K. English, and M. Labelle. 1988. Evaluation of random and indexing sampling designs for estimating coho salmon (*Oncorhynchus kisutch*) escapement to three Vancouver Island streams. Canadian Technical Report Fisheries and Aquatic Sciences No. 1639.

Cooke, S. J., S. G. Hinch, M. Wikelski, R. D. Andrews, L. J. Kuchel, T. G. Wolcott, and P. J. Butler. 2004. Biotelemetry: a mechanistic approach to ecology. Trends in Ecology Evolution 19:334–343.

English, K. K., R. C. Bocking, J. R. Irvine. 1992. A robust procedure for estimating salmon escapement based on the area-under-the-curve method. Canadian Journal of Fisheries and Aquatic Sciences 49:1982–1989.

Fukushima, M., and W. W. Smoker. 1997. Determinants of stream life, spawning efficiency, and spawning habitat in pink salmon in the Auke Lake system. Alaska. Canadian Journal of Fisheries and Aquatic Sciences 54:96–104.

Kawabe, R., K. Nashimoto, T. Hiraishi, Y. Naito, and K. Sato. 2003. A new devise for monitoring the activity of freely swimming flatfish, Japanese flounder (*Paralichthys olivaceus*). Fisheries Science 69:3–10.

Koski W. R., M. R. Link, and K. K. English. 1996. Distribution, timing, fate and numbers of Chinook salmon returning to the Nass River watershed in 1992. Canadian Technical Report of Fisheries and Aquatic Sciences No. 2129.

Korman, J. R., N. M. Ahrens, P. S. Higgins, and C. J. Walters. 2002. Effects of observer efficiency, arrival timing, and survey life on estimates of escapement for steelhead trout (*Oncorhynchus mykiss*) derived from repeat mark-recapture experiments. Canadian Journal of Fisheries and Aquatic Sciences 59:1116–1131.

Lady, J. M., and J. R. Skalski. 1998. Estimators of stream residence time of Pacific salmon (*Oncorhynchus* spp.) based on release-recapture data. Canadian Journal of Fisheries and Aquatic Sciences 55:2580–2587.

Manske, M., and C. J. Schwarz. 2000. Estimates of stream residence time and escapement based on capture-recapture data. Canadian Journal of Fisheries and Aquatic Sciences 57:241–246.

Nowacek, D. P., M. P. Johnson, P. L. Tyack, K. A. Shorter, W. A. McLellan, and D. A. Pabst. 2001. Buoyant balaenids: the ups and downs of buoyancy in right whales. Proceedings of the Royal Society of London B: Biological Sciences 268:1811–1816.

Parken, C. K., R. E. Bailey, and J. R. Irvine. 2003. Incorporating uncertainty into area-under-the curve and peak counts salmon escapement estimation. North American Journal of Fisheries Management 23:78–80.

Perrin C. J., and J. R. Irvine. 1990. A review of survey life estimates as they apply to the area-under-the-curve method for estimating the spawning escapement of Pacific salmon. Canadian Technical Report of Fisheries and Aquatic Sciences No. 1733.

Shardlow T., R. Hilborn, and D. Lightly. 1985. Components analysis of instream escapement methods for Pacific salmon (*Oncorhynchus* spp.). Canadian Journal of Fisheries and Aquatic Sciences 44:1031–1037.

Shardlow, T. F., K. D. Hyatt, and S. R. Pearson. 2007. A new microcontrolled fish tag for accurately estimating life span and survey life of spawning salmon. Aquatic Living Resources 20:197–203.

Stockwell, M. M., and K. D. Hyatt. 2003. A summary of Okanagan sockeye salmon (*Oncorhynchus nerka*) escapement survey observations by date and river segment from 1947 to 2001. Canadian Data Report of Fisheries and Aquatic Sciences 1106.

Tschaplinski, P. J. and K. D. Hyatt. 1990. Abundance, migration timing, and biological characteristics of sockeye salmon (*Onchorhynchus nerka*) returning to Henderson Lake, Vancouver Island during 1988. Canadian Technical Report of Fisheries and Aquatic Sciences 1758.

American Fisheries Society Symposium 76:507–527, 2012

Distinguishing Pelagic and Demersal Swimming of Deepwater Flatfish by Recording of Body Angles

Ole Thomas Albert*

Institute of Marine Research, P.O. Box 6404, NO-9294 Tromsø, Norway

Yvan Lambert

Pêches et Océans, Institut Maurice-Lamontagne 850, route de la Mer Mont-Joli, Québec G5H 3Z4, Canada

Tone Vollen, Carla Freitas, and Lise Heggebakken

Institute of Marine Research, P.O. Box 6404, NO-9294 Tromsø, Norway

Abstract.—Based on previous investigations of flatfish swimming performance, three hypotheses regarding frequency distribution of body angles (pitch and roll) during periods of pelagic excursions and periods when the fish swim along the bottom were formulated and evaluated. A tagging experiment on Greenland halibut *Reinhardtius hippoglossoides* was made with archival tags (DST) recording body angles each 15 min together with ambient depth. A total of 19 tags were recovered, representing 80,000 h of recording. The paper shows that for periods when pelagic or demersal swimming behavior can be deduced from the depth trajectories alone, the angular spectra were consistent with the hypothesized patterns for pelagic and demersal phases, respectively. Based on this distinction, the percentage of pelagic occupancy was estimated for each month of the year. With almost no occurrences of vertical swimming positions, the common perception that Greenland halibut adopts such position during pelagic phases was rejected.

Introduction

Archival tags recording depth and temperature, and sometimes other parameters, at preprogrammed time intervals, have been increasingly important in studying distribution and behavior of marine species (Metcalfe & Arnold 1997; West and Stevens 2001; Hunter et al. 2003; Solmundsson et al. 2003; Hunter et al. 2004b; Neat et al. 2006; Hobson et al. 2007; Nichol et al. 2007; Teo et al. 2007). The depth recorded by these tags is the ambient depth of the tagged animal and bears no direct information of the bathymetry. A common challenge with analyses of recorded time series from demersal fish is therefore to distinguish periods when the fish are distributed in close proximity to the bottom, from periods when they are more pelagic in their distribution. This distinction may be of great importance for understanding the behavioral ecology of the species (e.g., foraging times and areas, energy budgets, migrations), as well as the availability of the fish to sampling and fishing gears. For many species, a general understanding of the vertical distribution

*Corresponding author: ole.thomas.albert@imr.no

may be gathered acoustically, but flatfish generally have very low acoustic target strength due to the lack of swim bladder, and acoustic measurements are presently not feasible.

The common approach to distinguishing pelagic from demersal distribution of groundfish from archival tag recordings is to consider the size and the suddenness of a change in depth (West and Stevens 2001; Hunter et al. 2003; Hunter et al. 2004a). This practice may be well founded for species inhabiting flat areas but may not be equally useful for species living on the continental slope. Also, selection of a value to represent a sudden or large depth change may be arbitrary and dependant on the sampling rate.

Our approach was based on theoretical behavioral differences in swimming performance of flatfish when they are pelagic, as opposed to on or near the bottom (Olla et al. 1972; Kawabe et al. 2003, 2004). The purpose of this paper was to evaluate some basic hypotheses regarding the variability in body position (Figure 1) of Greenland halibut *Reinhardtius hippoglossoides* during pelagic and demersal phases, respectively. This deepwater flatfish is considered a vigorous swimmer (Smidt 1969; de Groot 1970) and a voracious predator (Woll and Gundersen 2004). Tagging experiments show migrations over thousands of kilometers (Boje 2002), and fast-swimming pelagic fish and cephalopods are

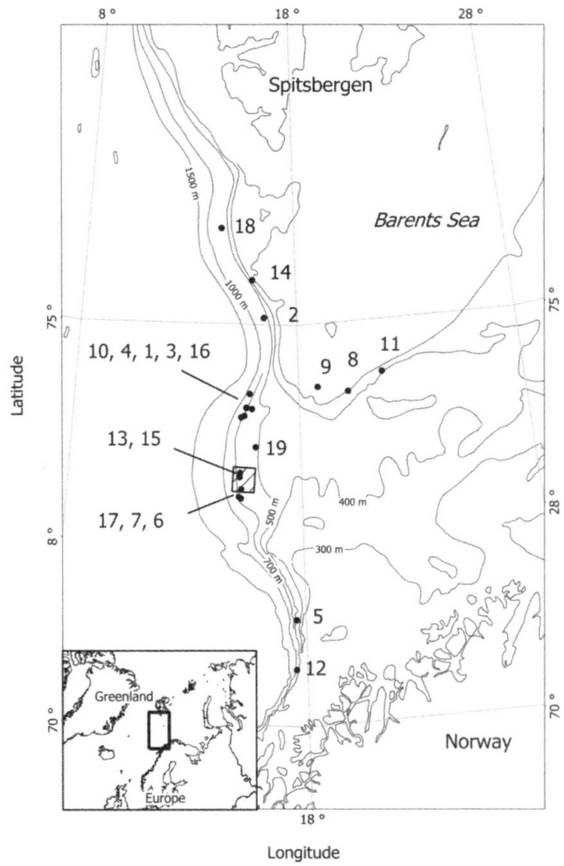

Figure 1. The bathymetry of the study area. Greenland halibut *Reinhardtius hippoglossoides* were caught, tagged and released within the hatched rectangle and recaptured at the positions marked with dots and labeled with tag number.

the most common prey items (Bowering and Lilly 1992; Dawe et al. 1998; Michalsen and Nedreaas 1998; Hovde et al. 2002; Solmundsson 2007). In the Northeast Arctic Atlantic the species is found along the continental slope of the deep Norwegian Sea and the Arctic Ocean from about the shelf break at 500 m and down to the transition between Atlantic water and the Norwegian Sea deep water, at 800–1000 m depths (Godø and Haug 1989; Albert et al. 1998b). It is also found in the deeper trenches that extend into the Barents Sea (Høines and Gundersen 2008), which is a shelf sea with mean depth of 230 m (UNEP/GRID-Arendal 2006). The juveniles are distributed around Svalbard (Haug et al. 1989; Albert et al. 2001a), and spawning occurs in late autumn, along the continental slope between North Norway and Bear Island (Albert et al. 2001b). Length at maturity is estimated to 41 cm for males and 57 cm for females (Morgan et al. 2003), and both juveniles and adults are known to perform extensive pelagic excursions several hundred meters up in the water column (Vollen and Albert 2008).

The first hypothesis (hereafter, "Vertical pelagic orientation") is based on the assumption that Greenland halibut adopts a vertical swimming position, with the ventral side down, during pelagic phases, and behaves more like a roundfish than a flatfish. This has both been discussed in the literature (de Groot 1970; Gibson 2005), and may often be heard from fishermen. This assumption seems to be based mostly on morphological considerations, i.e. the coloration of the blind side and the "un-finished" migration of the left eye over to the right side. The second hypothesis, "Pelagic climb-and-glide" is based on the general swimming behavior of flatfish (Olla et al. 1972). During pelagic excursions they are expected to exhibit gliding behavior at low angle, interrupted by steep ascents (Kawabe et al. 2003, 2004). This behavior can result in significant energy savings (Weihs 1973; Videler and Weihs 1982). The third hypothesis relates to the demersal phase, when they probably mostly swim close to the seabed to increase cost efficiency by taking advantage of the hydrodynamic thrust their tailbeats generate against the seabed (hereafter, "Swimming close to seabed") (Videler 1993; Webb and Gerstner 2000).

If the hypothesis of vertical pelagic orientation is correct, we should expect to find a strong signal in the roll angles (angles in the transversal plane) recorded as fish switch between pelagic and demersal swimming phases, with angles close to 0° in relation to the horizontal plane for fish swimming close to the seabed, and angles close to 90° during pelagic excursions (Figure 2). For fish swimming close to the seabed the expected distribution of pitch angles (angles in the medial plane) will be bell shaped and centered at zero relative to the horizontal plane (Figure 3). In contrast, according to the pelagic climb-and-glide hypothesis, the expected pitch distribution during pelagic phases will be bi-modal, with the lower mode below zero, and the upper mode at relatively steep angles (Figure 3).

Based on field experiments where body angles were recorded and stored in archival tags together with ambient depth data, the present paper evaluates the basic hypotheses outlined above, for selected situations where pelagic or demersal occupancy might be deduced from the depth trajectory alone. The results are then applied for the whole recording periods of all fish recaptured during the experiments.

Materials and Methods

Tagging and recapture

A total of 238 Greenland halibut were tagged with Data Storage Tags (DST) during two cruises in August 2005 and May 2006. The 97 individuals tagged in August 2005

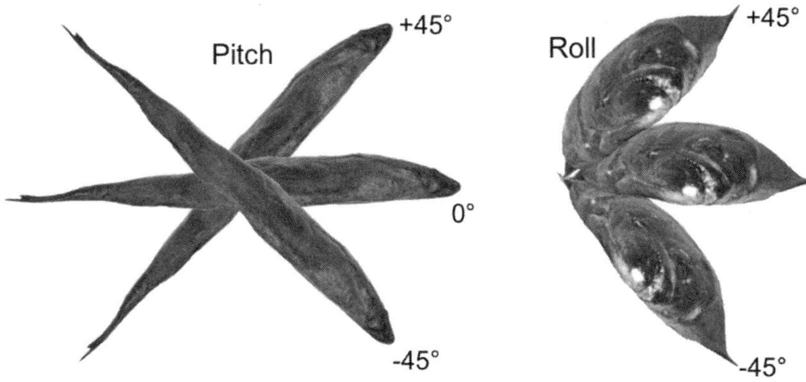

Figure 2. Definition of pitch and roll as measures of body angles (dorsal view left and anterior view right).

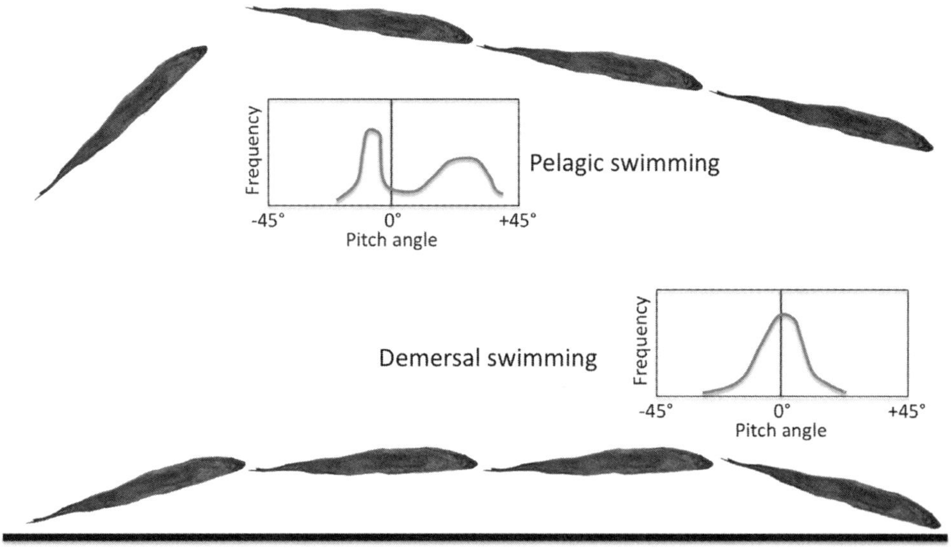

Figure 3. Schematic representation of the two hypotheses "Pelagic climb and glide" (top) and "Swimming close to seabed" (bottom). See text for details.

ranged from 51 to 90 cm total length (mean 66.3 cm, SD 8.0 cm), and the 141 individuals tagged in May 2006 were 42–97 cm (mean 58.0 cm, SD 9.0 cm).

The tags used were DST pitch and roll (Star-Oddi, Reykjavik, Iceland), which record pitch and roll movements of the DST with reference to the earth gravity, as well as ambient pressure (depth) and temperature. The cylindrical tags are 46 mm long with a diameter of 15 mm and weigh 19 g in air and 12 g in water. The resolution (and accuracy) cited by the manufacturer is <1° (±1°) for pitch and roll measurements, and 0.9 m (±12 m) for depth. The range for angle measurements is ± 90° and the tags were calibrated for 0–1,500 m depth. Temperature data were not used. The capacity was 43,000 recordings of all variables, and the tags were programmed to record all variables once every 15 min, corresponding to a maximum recording period of 450 d. This temporal resolution was chosen as a compromise between recording fine scale behavior and seasonal variation in activity. It is a low resolution for behavioral studies, but it was considered sufficient to capture the major shifts in behavior of Greenland halibut, that is known to migrate several hundred meters up in the water column and to perform much of their foraging pelagically (Vollen and Albert 2008). More information on the technical specifications of the tags can be found at www.star-oddi.com.

Greenland halibut were caught with longlines on the continental slope at 450–600 m depth between 72°30–73°00N latitude (Figure 1). Hooks were of type EZ12.0 by O. Mustad & Son Ltd., and bait was a 1/1 mix of Atlantic mackerel *Scomber scombrus* and European flying squid *Todarodes sagittatus*. Soaking time was normally 2–5 h, and hauling was conducted at about half of normal speed to reduce stress to the fish. The fish was lifted aboard by the gangion, which was then cut. The hook was then clipped and gently removed from the fish. Selection of individuals for tagging was based on visual inspections of skin, throat, gills and eyes. Individuals with large wounds, broken jaws, or appearing in general inferior condition were not tagged.

The tag was attached at the eye-side, just below the dorsal fin, with the distance to the head being approximately 1/3 of the fish length. In 2005, tags were attached to the fish using two 0.5 mm stainless steel wires with the tag mounted directly to the fish, whereas in 2006 an attachment cradle was developed in cooperation with the tag manufacturer. The tag was affixed to a plastic plate (i.e., the cradle), which was then attached to the fish with two 0.5 mm stainless steel wires. All parts touching the fish skin were lined with soft silicone pads. For both attachment methods, two syringes (2 mm × 80 mm) were run through the fish from the blind side to the eye-side, and wires were inserted through the holes. The syringes were removed and the wires attached to a plastic plate on the blind side.

To reduce time out of the water, fish were immediately released overboard after tagging; the tag-and-release operation was continuous and in synchrony with the hauling. Handling time was normally 1–2 min.

By February 2008, a total of 19 recaptured tags were returned (Table 1), giving a recapture rate of 8%. All tags had logged data successfully, for periods varying between 8 and 454 d. Eleven of the tags had logged for more than 100 d and the total number of observations was 304,026, representing nearly 80,000 h of recording.

Calibration of angles

The recorded pitch and roll angles were transposed to a scale with zero representing horizontal position, and positive and negative values depicting that the left eye, which is found on the top ridge of the head, was turned upwards or downwards, respectively (Figure

Table 1. Overview of recaptured tags.

Fish no.	Release	Recapture	Recording stop date	Days at large	Days recording at large	No. of recordings at large	Total length at tagging (cm)	Sex
1	16 Aug 2005	09 Sep 2005	09 Oct 2005	54	54	5146	63	F
2	16 Aug 2005	16 Oct 2005	16 Oct 2005	61	61	5803	75	F
3	15 Aug 2005	09 Oct 2005	08 Oct 2005	55	54	5206	66	F
4	15 Aug 2005	21 Oct 2005	21 Oct 2005	67	67	6414	77	F
5	15 Aug 2005	22 Jun 2006	07 Jun 2006	311	296	28351	74	F
6	15 Aug 2005	26 Aug 2005	25 Aug 2005	11	10	960	57	
7	15 Aug 2005	23 Aug 2005	23 Aug 2005	8	8	782	73	F
8	14 Aug 2005	16 Jun 2006	16 Jun 2006	306	306	29327	76	F
9	14 Aug 2005	12 Dec 2005	12 Dec 2005	120	120	11490	63	F
10	15 Aug 2005	05 Oct 2005	05 Oct 2005	51	51	4863	55	F
11	16 Aug 2005	10 Feb 2006	09 Feb 2006	178	177	17028	77	
12	16 May 2006	02 Aug 2007	01 Feb 2007	443	261	25056	67	
13	17 May 2006	29 Nov 2006	29 Nov 2006	196	196	18816	65	
14	16 May 2006	30 Apr 2007	01 Feb 2007	349	261	25056	47	F
15	17 May 2006	23 Aug 2006	23 Aug 2006	98	98	9408	68	F
16	14 Aug 2005	Mar 2007	11 Nov 2006	>454	454	43585	62	
17	16 May 2006	08 Sep 2007	01 Feb 2007	480	261	25056	69	
18	16 May 2006	16 Oct 2007	01 Feb 2007	518	261	25056	60	
19	17 May 2006	31 Oct 2007	01 Feb 2007	532	260	24960	63	F

2). Due to the shape of the fish and slight differences in attachment position, the precise 3D position of the tags varied between individual tagged fish. Therefore, the angles had to be calibrated for each tag individually, so that zero pitch and roll represented a horizontal position of the fish. The calibrated angle A is given by:

$$A = A_i - A'$$

where A_i is the measured angle, and A' is the mean measured angle during periods when the fish was considered to largely be lying on the bottom. For the pitch angle this was achieved by averaging over all sequences of approximately constant depth (range in depth recordings less than 1 m) that lasted for at least 10 h. The roll angle was occasionally observed to drift over periods of several months for unknown reasons. This change could have been related to factors such as bending of the wires in the attachment mechanism, or to changes in body shape due to swelling or muscle contractions, possibly induced by the tag itself. For the roll angles, the standardizing value A' was therefore updated following each sequence of at least 2.5 h (10 consecutive recordings) of constant depth (defined as above). Such sequences were usually found at least once per recorder each day, presumably reflecting that inactive bottom dwelling periods are common for Greenland halibut.

Data analyses

As usual for oceanic tagging studies, there was no independent information about the fish behavior. Inferences about pelagic and demersal distribution had to be based on the recordings themselves. We therefore selected sequences where the depth-recordings alone suggested either pelagic or demersal distribution, and compared the pitch and roll recordings from these sequences with the basic orientation hypotheses described earlier.

Three different behavioral scenarios were readily identifiable within the depth profiles available from recaptured fish. From these, we extracted three "examples" that were used to generate pitch and roll spectra considered characteristic of each behavior, for subsequent comparison to data from other fish and other time periods. The first behavior represents the only data points where we know for sure that the fish were pelagic: the few first recordings after release. During this first pelagic descent the individuals may display a flight reaction and their behavior may not be representative of other pelagic excursions. However, they may still use behavior patterns that are within their natural repertoire during pelagic descents, and readings from this first descent, from each of the 19 recaptured fish, were therefore included as our first example.

The second example of a discretely identifiable swimming behavior was extracted from a 14-d period within one tag record, during which distinct diurnal vertical migrations were apparent (The shelf example; Figure 4). The depth recordings were stable at approximately 400 m throughout the day, highly variable between 350 and 200 m during the six hours associated with night time, before stabilizing again at around 400 m the next day. This repeated pattern might occur from the fish being primarily demersal during day and pelagic at night. The relatively stable daytime depth of 400 m may reflect that the fish was actively seeking a preferred bottom depth during daytime, or that it was located in an area with relatively flat ocean floor. Adult Greenland halibut is not commonly found shallower than 450 m along the coastal banks of North Norway (Albert et al. 1998a), but is common below 350 m in the troughs and channels of the Barents Sea (Bowering and Nedreaas 2000), especially in the gently sloping Bear Island channel extending eastwards from the tagging site (Figure 1). It is therefore considered probable that the trajec-

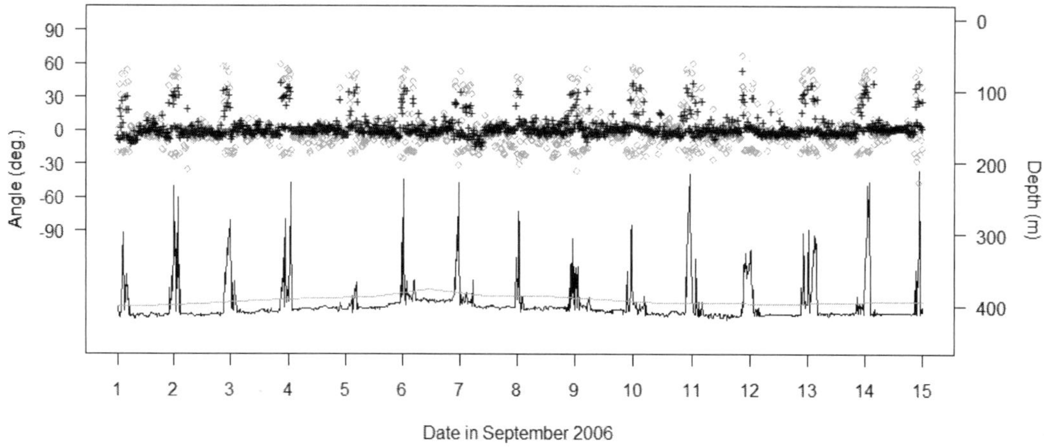

FIGURE 4. The shelf example: a 14 d period of recording of Fish no. 14, indicating diurnal vertical migrations with pelagic excursions up from a stable bottom depth of about 400 m, i.e. probably on gently sloping shelf habitat. Ambient depth indicated by solid black line (right vertical axis), pitch angles by gray circles and roll angles by black crosses (left vertical axis). Solid gray line indicates 15 m above depth at noon each day (used to separate pelagic and along bottom periods).

tory of this example represents a period when the fish was in the Bear Island channel or a similar gently sloping part of the continental shelf. In this case, it was considered that pelagic and demersal phases could be distinguished by ascent to more than 15 m above the depth at noon each day.

The third example consists of a much more irregular depth trajectory, with no protracted periods of stable depth, and with periods of gradual and extensive descents and ascents, as well as distinct periods of high vertical activity (The slope example; Figure 5). It was considered that these recordings were from the slope area since lower depths varies from 600 to 450 m (refer to Figure 1). From this 14-d period, sequences of gradual and possibly along-bottom vertical movements, and of high-frequency changes in depth resembling the pelagic phases in the shelf example, were selected based on visual inspection of the trajectories.

Based on the analyses of these examples, the angular spectrum hypotheses were investigated by comparing the pitch and roll distributions for the different examples, i.e. if they are uni- or bi-modal, and if they are centered at 0 or at 90°. Comparing frequency distributions statistically requires a lot of data points. With our sampling rate (4 readings per hour), a pelagic excursion lasting for six hours generated 24 data points, which may have been close to a minimum requirement. In order to increase the ability to discriminate between the different behavioural phases, we choose to compare the variability in addition to the distribution of the recordings. Comparisons were done using box-plots, graphically displaying the median and the first and third quartiles, as well as range and outliers. Approximate 95% confidence intervals of the medians (McGill et al. 1978) were shown by notches as standardized in the software package "R." Outliers were defined as observations extending beyond 1.5 times the inter quartile range. For each data point we calculated the standard deviation of pitch, roll and differences between succeeding depth recordings (Δ-depth), for that observation together with the two preceding and the two succeeding data points. This mea-

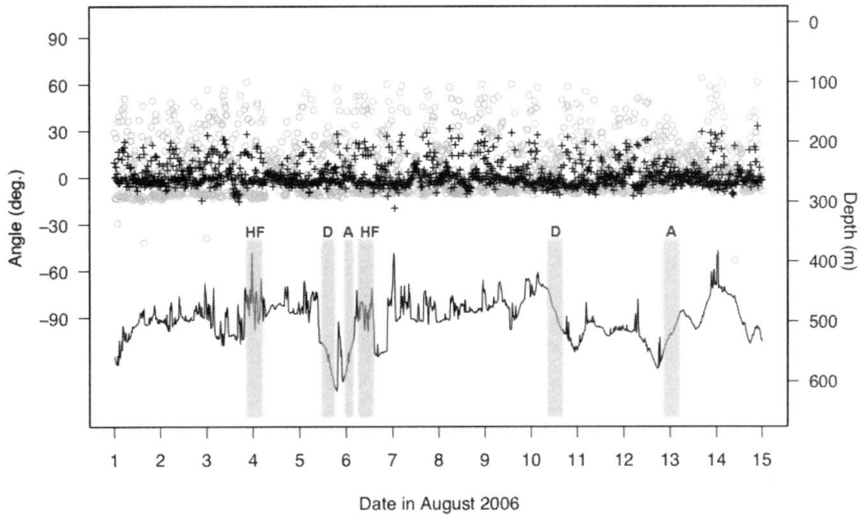

Figure 5. The slope example: a 14 d period of recording of Fish no. 15 showing high vertical activity over a varying bottom depth of 450–600 m, i.e. on the continental slope. Ambient depth, pitch and roll angles indicated as in Figure 4. Gray rectangles denote selected periods of gradual ascents (A) and descents (D) and high frequencies of changes in depth (HF).

sure is referred hereafter as running standard deviation (RnSD). Finally, tentative criteria generated from the example data were used to extract information of pelagic and demersal phases from the total pitch and roll recapture data.

Results

Calibration and saturation

During the horizontal sequences used for calibration, the range of differences between maximum and minimum pitch of individual tags was 35°, but most tags were within a 15° range (Figure 6). Taking the roundness of this species into account (Figure 2), pitch variability may be expected due to only minor variability in the exact positioning of the syringe punctures. The pitch box plots in Figure 6 show that after calibration nearly all pitch recordings were well within the measuring range of the tag, as indicated by the strait lines.

The measured roll angle during protracted periods of stable depth recordings varied extensively, also within the same tag. There was a strong tendency towards negative calibration angles (Figure 6), i.e. with the dorsal fin in a lowered position. Therefore, roll saturation problems may have biased the calibrated values mainly in situations when the fish turned their dorsal side down. The roll box plots in Figure 6 show that for recording periods with calibration roll above –35° nearly all calibrated roll values were well within the measuring range of the tag, as indicated by the straight lines. For periods with lower calibration roll angles, the figure indicates that some recordings that should have had calibrated roll angles between 0° and –50° (i.e., below the lower straight line) may have exceeded the measuring range. According to the technical specifications of the tags, these would have been recorded as positive angles, but with correct absolute values. This may have introduced a double bias in the roll recordings, with more positive values and less negative values. How-

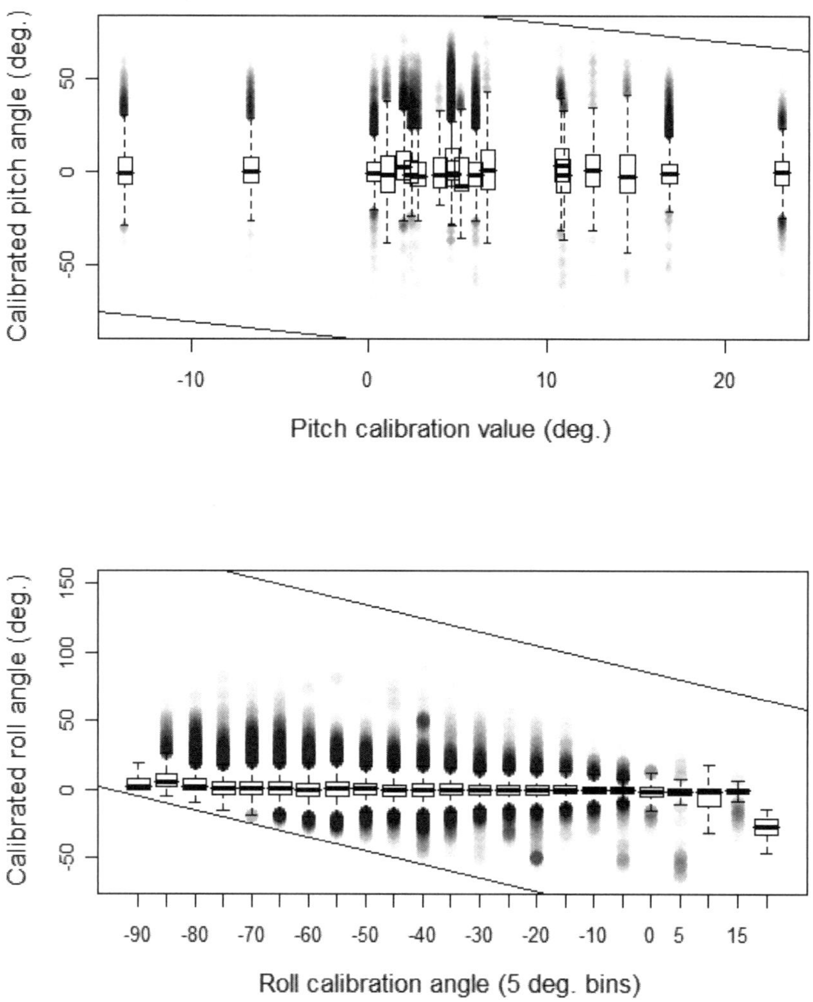

FIGURE 6. Calibrated pitch (upper panel) and roll angles(A) (lower panel) during the whole period at large, plotted against calibration value(A'). The calibration value is the mean measured angle for horizontal fish. There is one pitch calibration value for each tag, but several roll calibration values, as roll calibration value was recalculated during each >2.5 h periods of constant depth. Straight lines show the tag's measuring range. Saturation is indicated where the data are limited by the lines. Transparent dots represent outliers, as defined by >1.5 times inter-quartile range, and darker shades represent more points.

ever, Figure 6 shows that the median roll value and the inter-quartile range were almost identical for all but the most extreme calibration values. It is therefore concluded that actual measuring range of both pitch and roll was sufficient to test the two postulated hypotheses of angular distribution relative to swimming behavior.

Angular spectra of selected examples

Just after the release of tagged individuals, the fish headed steeply downwards. The mean pitch angle (and standard error) for the first three recording intervals (15, 30 and 45 min) after release were –47° (±4), –36° (±5), and –17° (±4), respectively. Since almost all pitch values were negative (Figure 7a), the pitch spectrum was clearly not representative of a voluntary pelagic excursion, which necessarily involves both upward and downward movements. It is still worth noting that there were no tendencies of a vertical swimming position during the descents. The roll angle was centered on zero, with 72% of the recordings within ± 10°.

The pelagic periods of the shelf example were characterized by distinct bimodality in both pitch and roll spectra (Figure 7b). The lower and upper modes in each spectrum corresponded, so that slight negative pitch was associated with zero roll, whereas steep positive pitch was associated with positive roll. Steep positive pitch was strongly positively correlated with positive roll ($p < 0.01$, $r^2 = 0.8$). Thus, the lower modal groups seem to represent gliding phases with little variability in body angles, whereas the upper groups represent climbing phases with increasing twisting as the steepness increases.

The demersal phases of the shelf example were also, at least partly, associated with active swimming. There were often slight and gradual changes in ambient depth over periods of several hours, as well as minor spikes in the depth profiles (Figure 4). Both the pitch and roll spectra were clearly uni-modal with central tendency close to zero (Figure 7c).

Periods of high vertical activity in the slope example were characterized by pitch and roll spectra similar to the pelagic phases in the shelf example (Figure 7 d). Periods of steady ascents or descents had the same roll spectrum as the demersal phases in the shelf example, whereas the pitch spectrum was different from both the pelagic and demersal phases of the shelf example (Figure 7e). The mean pitch was 15°, which was rarely seen in either of the two other situations. Furthermore, there was no steep climbing (i.e., pitch angles >30), but some tendency of gliding behavior (i.e., pitch angles <0).

Low standard deviation in the pitch, roll and Δ-depth (RnSD lower than 15°, 8° and 5 m, respectively) were found in the shelf example in association with the demersal phase. This is shown in Figure 8 by the box plots for those recordings where the recording itself as well as the two preceding and the two succeeding ones were all below the line drawn to define demersal from pelagic phases (0% above the line). In contrast, pelagic behavior, defined as sequences when all of five consecutive recordings were above the line (100% above the line), was associated with significantly higher RnSD in pitch (higher than 20°) and roll (higher than 7°). Change in depth was also significantly higher than during the demersal phase (Figure 8).

During gradual ascents and descents in the slope example, standard deviations in pitch were significantly higher than in the demersal phase (see confidence intervals in Figure 8), but always lower than 20°. Roll was not significantly different from the demersal phase in the shelf example (Figure 8). As in the demersal phase in the shelf example, standard deviations in Δ-depth were smaller than 5 m. Periods of high frequency depth change at the slope were characterized by significantly higher RnSD in pitch (ranging from 13 to 33°), significantly higher RnSD in roll (rang-

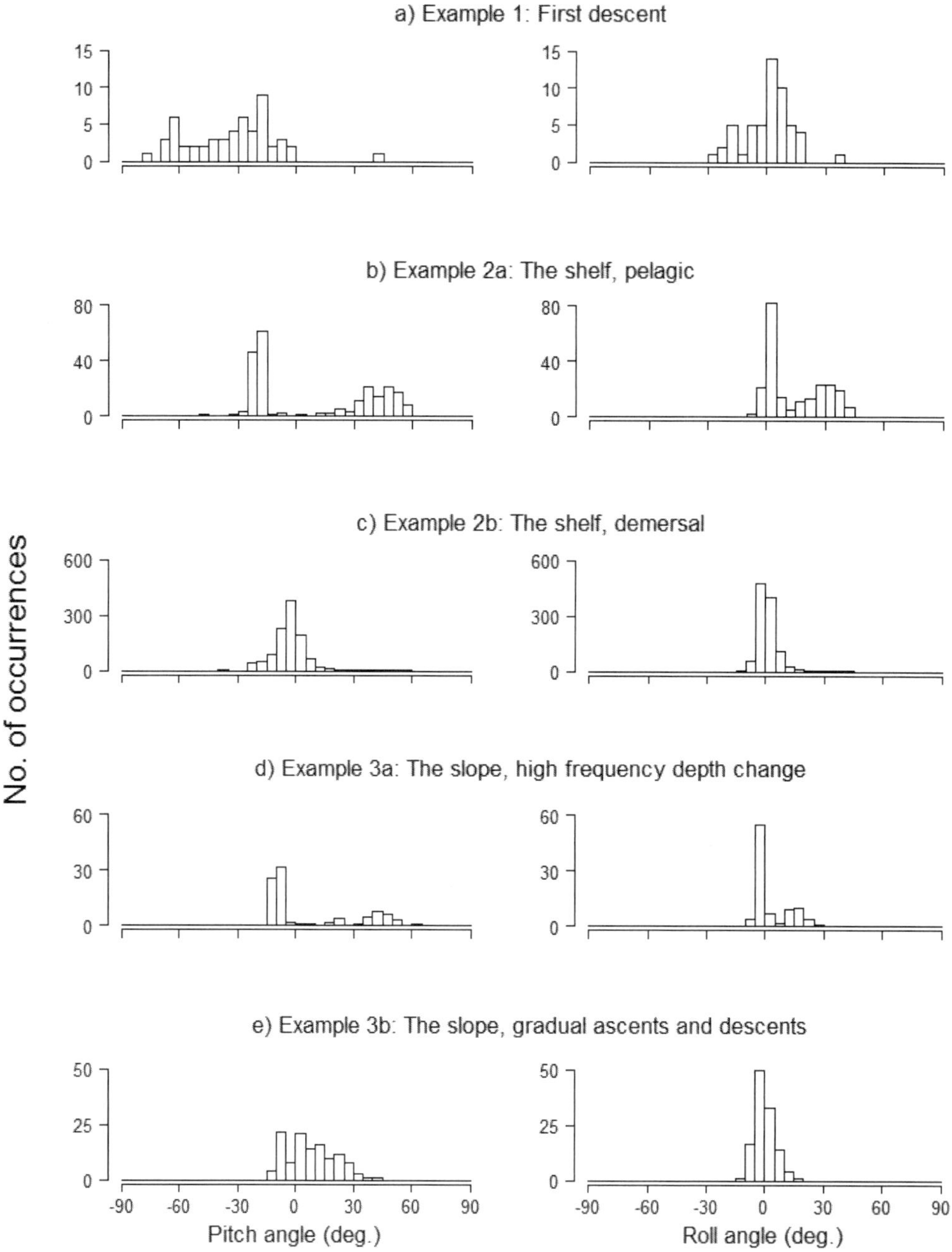

FIGURE 7. Angular spectrum of pitch and roll for selected examples.

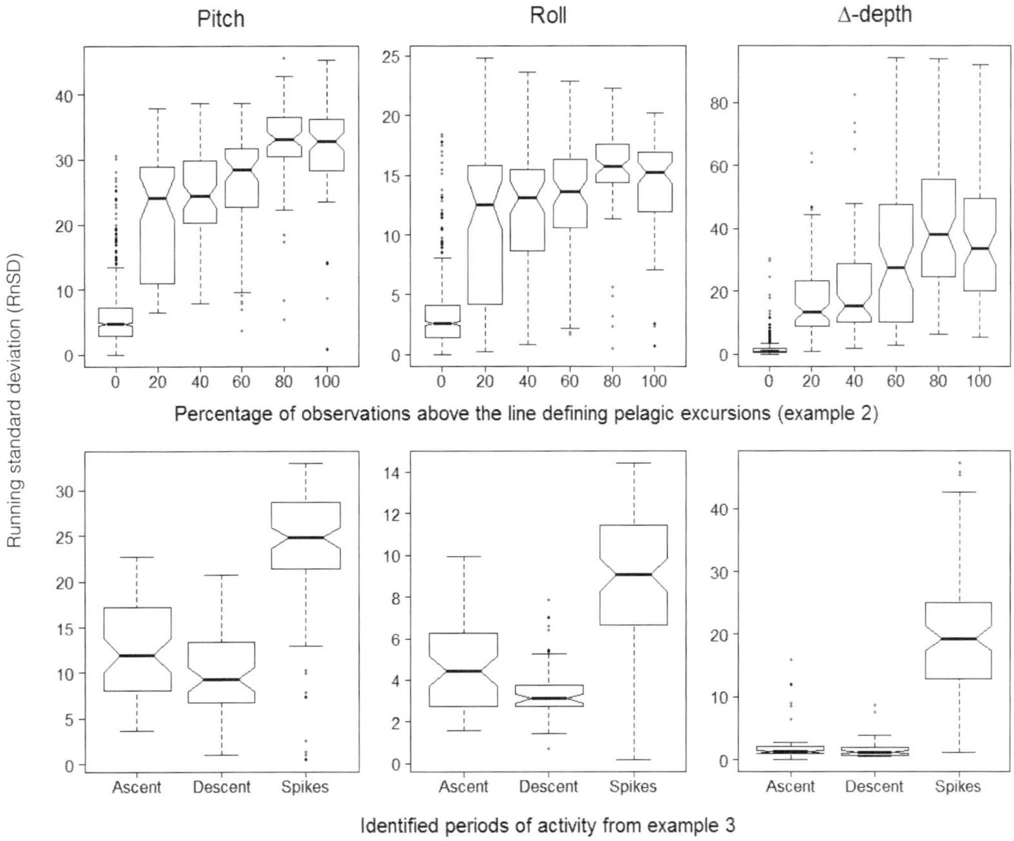

FIGURE 8. Boxplot of the five-point running standard deviation (RnSD) of pitch, roll and Δ-depth for selected situations from the shelf-example (top) and the slope-example (bottom). x-axis in top Figures denote the percentage of observations for which the RnSD were calculated, that were above the line defining pelagic from bottom orientated behavior (see Figure 4). Approximate 95% confidence intervals of median are shown by the notches.

ing all the way from 0 to 14°) and significantly higher Rnsd in Δ-depth (up to 40 m).

Absolute changes in depth during the selected gradual ascents and descents were 10.2 m h^{-1} (median; 1st–3rd quartiles: 4.3–13.7 m) and 7.9 m h^{-1} (median; 1st–3rd quartiles: 3.6–10.8 m), respectively. Assuming a swimming speed of 0.3–1.0 m s^{-1}, corresponding to about 0.5–2.0 body length per second (which is the range of endured swimming speed estimates for European plaice Pleuronectes platessa (Beamish 1978; Gibson 2005), such median changes in depth would be reached by a fish swimming over seabed with bottom slopes of 0.1–0.5°.

Angular spectra from all recordings

Figure 9 shows the pitch, roll and Δ-depth spectra for increasing levels of the RnSD in pitch. The upper and lower panels represent 28 and 6%, respectively, of all the 304,000 recorded intervals, and closely resemble the demersal and the pelagic phases of the shelf example, respectively (see Figure 4). This is true for shape, range, and position of the modes. The upper intermediate distributions resemble gradual ascents and descents from the slope example, whereas the lower intermediate distributions are similar to the periods of high vertical activity in the shelf example. All the

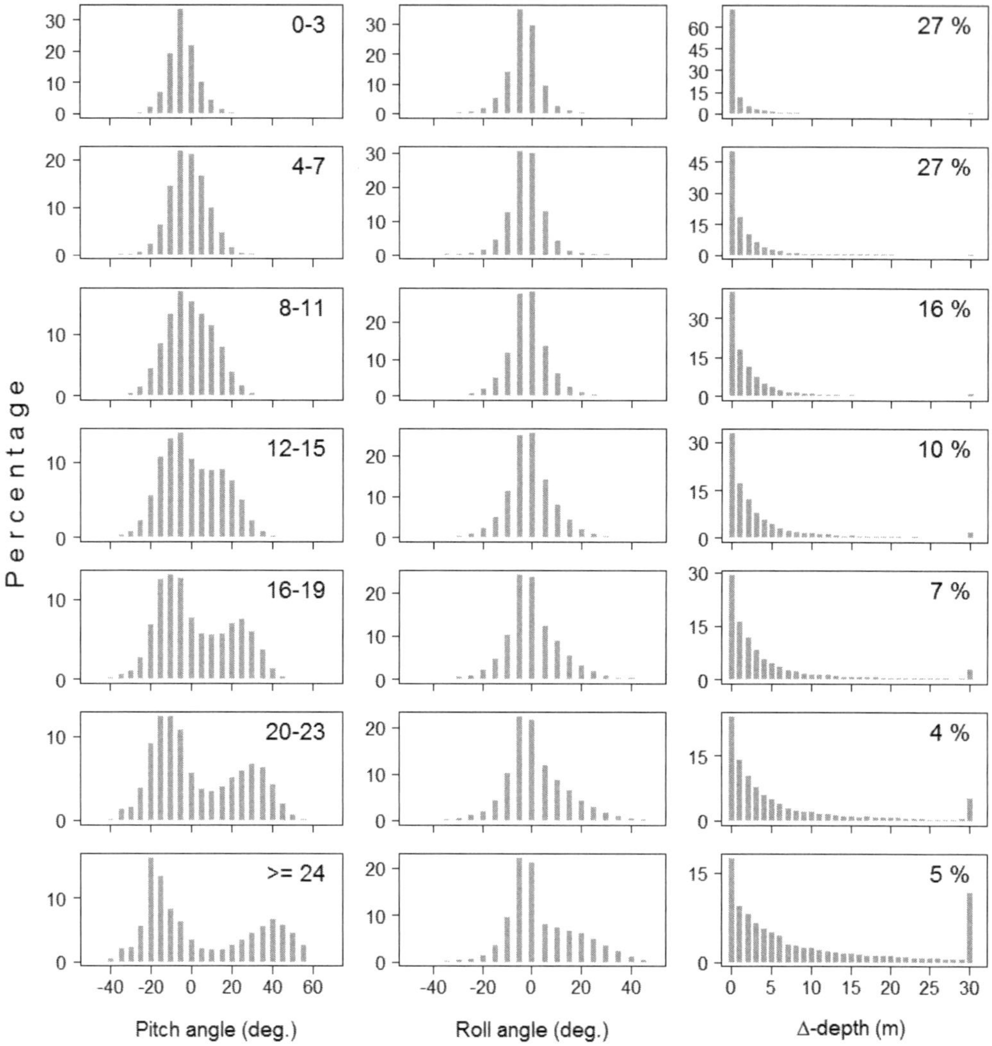

Figure 9. Pitch, roll and Δ-depth spectra for observations with increasing five-points running standard deviation (RnSD) for pitch. RnSD-intervals are given in the pitch angle panels and percentage of all observations in the Δ-depth panels. The last class in the Δ-depth plot is 30+.

three lower distributions (i.e., RnSD(pitch) ≥ 16) are distinctly bimodal, consistent with the pitch hypothesis of pelagic swimming behavior. As RnSD (pitch) increases from the upper to the lower panels, the lower mode in the pitch spectrum decreases towards steeper descents at the same time as the upper mode increases towards steeper ascents.

None of the pitch distributions in the left column of Figure 9 was associated with a vertical swimming position, as shown by the roll distributions in the middle column. Even with the extreme bi-modal pitch distribution, the roll was centered on zero, with a tail extending up to 45° but with 95% of observations below 30°. The analysis of all recordings indicates that only 0.008% of the roll angle values are between 80 and 100°. Roll angle values higher than 60° represent 0.04% of the total recordings. Thus, the "Vertical pelagic orientation" hypothesis implicating 90° roll angle during pelagic swimming is not supported by any of the recorded sequences.

For periods with steady directional ascents lasting 10 h or more, the mean pitch was 10.7° (SE = 1.73) and roll 1.7° (0.65). For equally long periods with steady directional descents, the angles were 5.3° (1.49) and –1.85° (0.56), respectively. Half the difference in pitch (2.7°) is a measure of mean slope of steady ascents and descents during those periods. This compares with the typical bathymetric slope of 1–2° in the region were the fish were tagged and recaptured—and with the steeper areas outside the Lofoten archipelago—with slopes between 2 and 10° (Ottersen and Auran 2007).

The distribution of Δ-depth for the lowest level of RnSD in pitch (Figure 9, upper panel) is dominated by records with no changes in depth between consecutive observations (Δ-depth <1 m), reinforcing the interpretation of this pitch spectrum as representing demersal behavior. An increasing frequency of large Δ-depths values is found for increasing levels of RnSD in pitch, indicating an increasing presence of pelagic behavior. However, the presence of records with no changes in Δ-depth even at the highest levels of RnSD in pitch should be noted. Thus, high variability in pitch angles can also occur with no changes in depth.

Based on the examples and interpretations described in the previous paragraphs, individual trajectories were divided into periods of primarily pelagic and demersal occupancy by using RnSD (pitch)>16 as a proxy for pelagic swimming. Figure 10 shows how these pelagic periods were distributed in time and depth for three example tags. It appears that gradual depth changes that the fish exhibited while occupying slope habitat were largely associated with uni-modal, demersal type pitch spectra, whereas large vertical excursions from a stable bottom depth were characterized by bi-modal, pelagic type pitch spectra. Large vertical activity while in slope habitat may be either demersal or pelagic and occurred generally within the same depth range, i.e. not extending as high up in the water column as the vertical excursions during periods with a stable maximum depth. Figure 10 shows two examples of excursions from 500 to 700 m on the slope, down to below 1000 m depths. They were both part of pelagic periods, and one of them shows a demersal type of return to the previous depths.

The percentage of time with pelagic type of pitch spectra varied between individuals and over time, with a summer peak in mean pelagic activity from July to October and relatively little variation in May and November–December (Figure 11). During the summer peak, Greenland halibut were primarily pelagic for about 25% of the time.

Discussion

Three hypotheses about how body angles relate to swimming behavior were put forward to distinguish pelagic swimming from swimming along the bottom. In the "Vertical pelagic orientation" hypothesis, fish swimming along the seabed would show roll angle values around 0° while roll angles close to 90° would be observed for fish swimming in a vertical position during the pelagic phase. For less than 0.01% of the total recordings examined was a roll angle value close to 90° observed. In fact, very few recordings (i.e., 0.04%) showed roll angle values of more than 60°. We therefore conclude that pelagic swimming in a vertical position (de Groot 1970) is a rare or even absent swimming behavior in Greenland halibut.

The "Swimming close to seabed" hypothesis predicted a uni-modal pitch angle distribution during demersal phases, corresponding to the preferred and energy saving locomotion of flatfish (Kawabe et al. 2003, 2004). The "Pelagic climb-and-glide" hy-

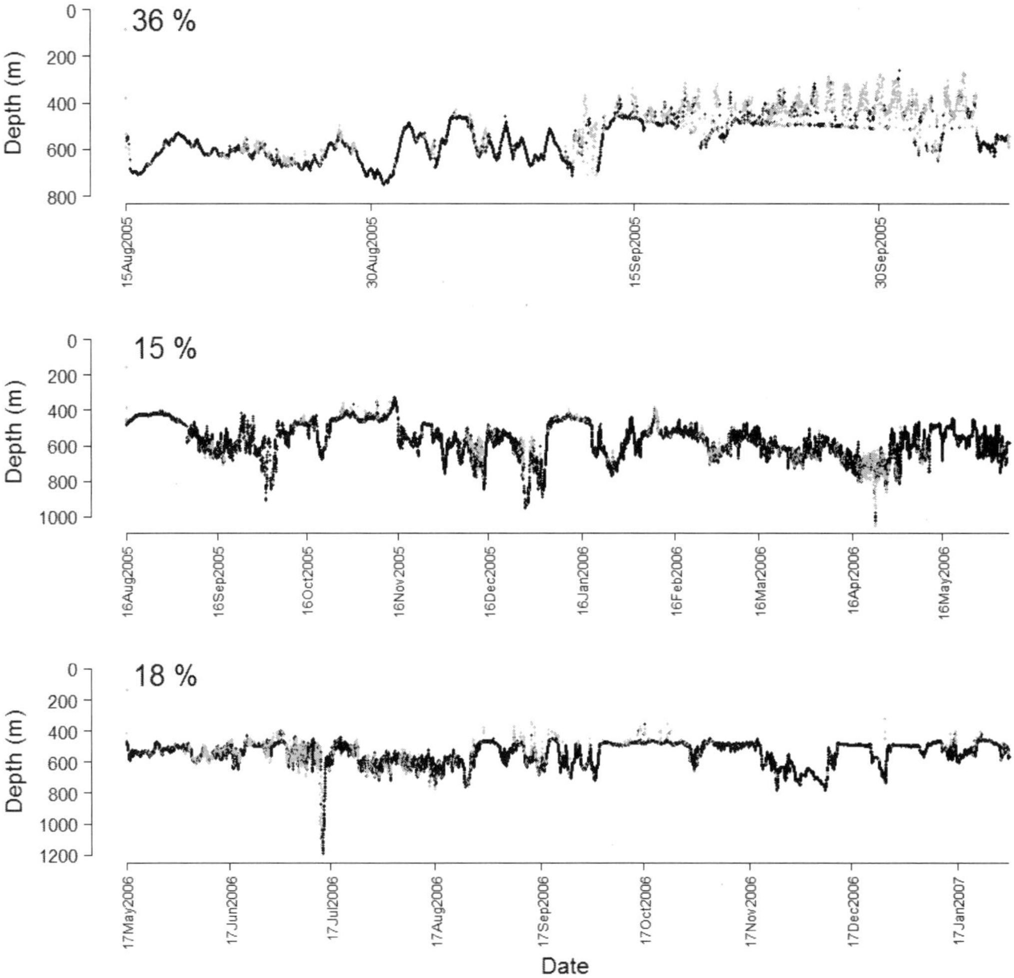

FIGURE 10. Depth trajectories for three recaptures (Tag 3, 5 and 12, respectively) indicating pelagic (gray) and demersal (black) activities, where pelagic activities were defined as five-points running standard deviation (RnSD) of pitch >16. Percentage values given on each panel denote the proportion of putative pelagic observations that were observed for each fish over its entire period at liberty. Scale of both horizontal and vertical axes differ between the three panels.

pothesis predicted a bi-modal pitch angle distribution for pelagic swimming, corresponding to alternating steep ascents and gentle gliding behavior (Kawabe et al. 2003, 2004). Typical pitch-spectra identified for the assumed demersal phase were in accordance with expected distribution of pitch angles consisting of a uni-modal distribution of angles values centered at 0°. Periods of deduced pelagic excursions showed pitch-spectra with a bi-modal distribution of angle values. One modal group consisting of steep positive pitch angles, expected during vertical ascents, and another modal group of lower and less variable negative pitch angles, consistent with a steady gliding behavior.

The recordings made on Greenland halibut during a period with stable daytime depth and high night time vertical activity, were completely in accordance with the "Swim-

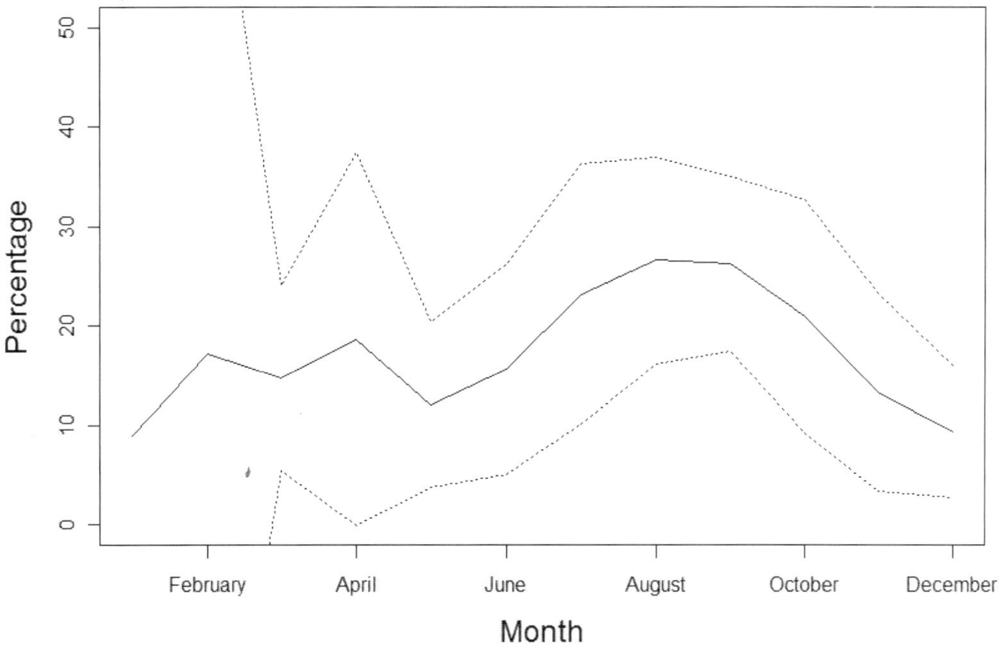

Figure 11. Mean (±2SE) percentage of all observations per tag with five-points running standard deviation (RnSD) of pitch >16, i.e. indicating pelagic swimming. Number of observations each month varies from 9,000–48,000.

ming close to seabed" and the "Pelagic climb-and-glide" hypotheses respectively. The results indicate that RnSD in pitch might be used to delineate pelagic and demersal periods from the archival tag trajectories. Clear diurnal vertical migrations, with relatively shallow excursions lasting for several hours, allowed a precise definition of pitch and roll spectra for both presumed pelagic and presumed demersal swimming. During these diurnal migrations, the assumed pelagic activity was closely associated with changes in depth. However, for fish assumed to be located on the continental slope (1–7° angle), the analysis of changes in depth alone was not sufficient to identify and separate pelagic from demersal activities. Swimming off the slope may result in significant changes in altitude without any important change in depth position. Likewise, demersal swimming along the slope can result in significant changes in depth. For example, a fish swimming up or down a 2.7° slope with a speed of 0.5 m s^{-1} will experience a depth change of 21 m in 15 min. These numbers agree with depth changes and angles seen during steady directional ascents and descents. Nevertheless, for selected data showing gradual or steep ascents and descents, patterns of pitch and roll spectra corresponding to pelagic and demersal activities similar to those observed for fish on the shelf were identified.

However, a flatfish may have several behavior modes associated with the same pitch pattern. Actively foraging periods close to the bottom, with swift take-offs from the sea floor to attack prey from below, followed by gentle landing back on the bottom, may also give a bi-modal spectrum. Such behavior will probably last for shorter periods of time than those extending far from the seabed, and will be interrupted by periods of bottom dwelling. This may give pitch spectra that are combinations of the uni and bi-modal types, yielding a broader uni-modal distribution, as seen in the intermediate levels of RnSD (pitch).

Tags and calibration functioned reasonably well, with observed pitch and roll spectra corresponding to expected patterns. Since the pitch angle values during horizontal sequences for individual tags were fairly stable over time, a single calibration was sufficient for each tag. However, for some fish, a drift in roll angle values was observed with time, and therefore calibration had to be conducted repeatedly. The reason for this drift is not clear but the examination of some of the fish recaptured indicated that some displacement of the tag might have occurred over time with a possible influence on roll values. Controlled experiments with tagged fish for long time periods are necessary to validate this possibility. It is worth mentioning that the calibrated zero may only represent the horizontal position of the fish when it is lying on the bottom. This may not be equal to the horizontal position in water, as both roll and pitch may be influenced by tension or curvature in the body. It seems reasonable to assume that the lying roll angle is turned in the direction of the blind side (negative roll), thus introducing a positive bias in standardized roll angles when the fish is off the bottom. The details of this could only be determined in controlled experiments.

Given the time interval between recordings (i.e., 15 min), the frequency distribution of all recordings may include several activity modes. This could be especially true for pelagic activities where a wider range in recorded pitch and roll angles should be expected during longer vertical movements. As pelagic activity is probably mainly associated with feeding, the behavior associated with searching, chasing, and capturing preys could generate a larger variation in the recorded pitch and roll angles. However, even though the spectra associated with pelagic and demersal behavior types are likely to be a conglomerate of patterns originating from multiple activities, differences in spectral shape between each behavior type may still be seen as signatures for each type. Field studies with increased sampling rate of the DST, as well as controlled experiments using both DST-tags and video recordings could be used to validate pitch and roll distribution of angles for specific swimming gaits (Winger et al. 2004).

The use of RnSD in pitch as a proxy for identifying swimming activity patterns seemed to give a reasonable match to pitch spectra assumed to represent demersal and pelagic activities. Lowest and highest RnSD in pitch showed a correspondence with putative demersal and pelagic activities, respectively. Increasing pitch-variability was also related to increased changes in depth. However, even at the highest levels of RnSD, records with no changes in depth were observed. This means that high variability in pitch angles can also occur with no changes in depth, which probably occur primarily close to the bottom, not representing true pelagic behavior. The tracking of fish at lower sampling intervals together with identified pitch spectrum signatures of activities at finer resolution from laboratory experiments would be needed to better distinguish pelagic excursions from e.g. actively foraging behavior close to the bottom.

Here, pelagic activity was tentatively defined by RnSD in pitch values greater than 16, suggesting that Greenland halibut were using the pelagic zone 10–25% of the time. Monthly variations were observed with higher vertical activity occurring between July and October, supporting previous findings (Vollen and Albert 2008). Roughly 20% of the records we defined as pelagic activity were associated with very low vertical movements (Δ-depth lower than 1 m). This may indicate, as already mentioned, that some bottom-related activity (e.g., foraging behavior close to the seafloor) was included, leading to an overestimation of pelagic excursions. On the other hand, the RnSD limit chosen excluded periods with tendencies of bi-modal pitch distributions that probably partly represent pelagic behavior (i.e., RnSD of 12–16 representing 10% of all recordings). Percentage of time associated with pelagic activities could have been underestimated for some of the individual fish recaptured within a limited time period. For seven fish, the number of days of recordings varied between 8 and 67 d. To our knowledge, no systematic study has been conducted to estimate the time needed for deep-sea fish released with DST-tags to return to a normal behavior. However, fish recaptured after more than 120 d should have returned to a nearly normal behavior as the survival of the fish would certainly have been impeded by the absence of feeding activity for such prolonged period. Long term controlled experiments would be needed to ascertain these assumptions.

With more tag recoveries, seasonal, geographic (slope-shelf), ontogenetic, and sexual effects on pelagic distribution could be studied. Since pelagic activity is probably mainly related to feeding, better knowledge of seasonal variations in feeding intensity, prey selection, and distribution of prey in the water column, would also be necessary to understand the pelagic behavior of Greenland halibut.

Acknowledgments

The paper has greatly benefited from constructive comments and suggestions from three anonymous referees.

References

Albert, O. T., J. E. Eliassen, Å. S. Høines, 1998a. Flatfishes of Norwegian coasts and fjords. Journal of Sea Research, 40:153–171.

Albert, O. T., E. M. Nilssen, A. Stene, A. C. Gundersen and K. H. Nedreaas, 1998. Spawning of the Barents Sea/Norwegian Sea Greenland Halibut (*Reinhardtius hippoglossoides*). ICES CM Doc/O:22, 19pp.

Albert, O. T., Nilssen, E. M., Nedreaas, K. H., and Gundersen, A. C., 2001a. Distribution and abundance of juvenile Northeast Arctic Greenland Halibut (*Reinhardtius hippoglossoides*) in relation to survey coverage and the physical environment. ICES Journal of Marine Science 58:1053–1062.

Albert, O. T., E. M.Nilssen, A. Stene, A. C. Gundersen and K. H. Nedreaas, 2001b. Maturity classes and spawning behaviour of Greenland halibut (*Reinhardtius hippoglossoides*). Fisheries Research 51: 217–228.

Beamish, F. W. H. 1978. Swimming capacity. Pages101–189 *in* W. S. Hoar and D. J. Randall, editors, Fish Physiology, Vol. VII, Locomotion. Academic Press, New York.

Boje, J., 2002. Intermingling and seasonal migrations of Greenland halibut (*Reinhardtius hippoglossoides*) populations determined from tagging studies. Fishery Bulletin 100:414–422

Bowering, W. R., and G. R. Lilly. 1992. Greenland Halibut (*Reinhardtius hippoglossoides*) off southern Labrador and northeastern Newfoundland (Northwest Atlantic) feed primarily on capelin (*Mallotus villosus*). Netherlands Journal of Sea Research 29(1–3):211–222.

Bowering, W. R. and K. Nedreaas, 2000. A comparison of Greenland halibut (*Reinhardtius hippoglossoides* (Walbaum)) fisheries and distribution in the Northwest and Northeast Atlantic. Sarsia 85:61–76.

Dawe, E. G., W. R. Bowering, and J. B. Joy. 1998. Predominance of Arctic squid (*Gonatus* spp.) in the diet of Greenland halibut (*Reinhardtius hippoglossoides*) on the deep slope of the North-east Newfoundland continental shelf. Fisheries Research 36:267–273.

de Groot, S. J. 1970. Some notes on an ambivalent behaviour of Greenland halibut *Reinhardtius hippoglossoides* (Walb.). Pisces: Pleuronectiformes. Journal of Fish Biology 2(3):275–279.

Gibson, R. N., 2005. The behavior of flatfishes. Pages 213–239 *in* R. N. Gibson, editor. Flatfishes: biology and exploitation. Blackwell, Oxford, UK.

Godø, O. R. and T. Haug. 1989. A Review of the Natural History, Fisheries and Management of Greenland halibut (*Reinhardtius hippoglossoides*) in the Eastern Norwegian and Barents Seas. Journal du Conseille International pour l'Exploration de la Mer, 46:62–75.

Haug, T., H. Bjørke, and I.-B. Falk-Petersen. 1989. The distribution, size composition, and feeding of larval Greenland halibut (*Reinhardtius hippoglossoides* Walbaum) in the eastern Norwegian and Barents Seas. Rapports et Procès-Verbaux des Réunions du Conseil International pour l'Exploration de la Mer, 191: 226–232.

Hobson, V. J., D. Righton, J. D. Metcalfe, and G. C. Hays. 2007. Vertical movements of North Sea cod. Marine Ecology Progress Series 347:101–110.

Hovde, S. C., O. T. Albert, and E. M. Nilssen. 2002. Spatial, seasonal and ontogenetic variation in diet of Northeast Arctic Greenland halibut (*Reinhardtius hippoglossoides*). ICES Journal of Marine Science 59(2):421–437.

Høines, Å. S. and Gundersen, A. C., 2008. Rebuilding the Stock of Northeast Arctic Greenland Halibut (*Reinhardtius hippoglossoides*). Journal of Northwest Atlantic Fishery Science 41:107–117.

Hunter, E., J. D. Metcalfe, B. H. Holford, and G. P. Arnold. 2004a. Geolocation of free-ranging fish on the European continental shelf as determined from environmental variables II. Reconstruction of plaice ground tracks. Marine Biology 144(4):787–798.

Hunter, E., J. D. Metcalfe, C. M. O'Brien, G. P. Arnold, and J. D. Reynolds. 2004b. Vertical activity patterns of free-swimming adult plaice in the southern North Sea. Marine Ecology Progress Series 279:261–273.

Hunter, E., J. D. Metcalfe, and J. D. Reynolds. 2003. Migration route and spawning area fidelity by North Sea plaice. Proceedings of the Royal Society of London, Series B: Biological Sciences 270(1529):2097–2103.

Kawabe, R., Y. Naito, K. Sato, K. Miyashita, and N. Yamashita. 2004. Direct measurement of the swimming speed, tailbeat, and body angle of Japanese flounder (*Paralichthys olivaceus*). ICES Journal of Marine Science 61(7):1080–1087.

Kawabe, R., K. Nashimoto, T. Hiraishi, Y. Naito, and K. Sato. 2003. A new device for monitoring the activity of freely swimming flatfish, Japanese flounder *Paralichthys olivaceus*. Fisheries Science 69(1):3–10.

Metcalfe, J. D., and G. P. Arnold, 1997. Tracking fish with electronic tags. Nature 387: 665–666.

Michalsen, K., and K. H. Nedreaas. 1998. Food and feeding of Greenland halibut (*Reinhardtius hippoglossoides*, Walbaum) in the Barents Sea and East Greenland waters. Sarsia 83(5):401–407.

Morgan, M. J., W. R. Bowering, A. C. Gundersen, Å. S. Høines, B. Morin, O. Smirnov, and E. Hjörleifsson. 2003. A comparison of the maturation of Greenland halibut (*Reinhardtius hippoglossoides*) from populations throughout the North Atlantic. Journal of Northwest Atlantic Fishery Science 31:99–112

McGill, R., J. W. Tukey, and W. A. Larsen, 1978. Variations of box plots. The American Statistician 32(1):2–16.

Neat, F. C., and coauthors. 2006. Residency and depth movements of a coastal group of Atlantic cod (*Gadus morhua* L.). Marine Biology 148(3):643–654.

Nichol, D. G., T. Honkalehto, and G. G. Thompson. 2007. Proximity of Pacific cod to the sea floor: using archival tags to estimate fish availability to research bottom trawls. Fisheries Research 86(2–3):129–135.

Olla, B. L., C. E. Samet, and A. L. Studholm. 1972. Activity and feeding behavior of summer flounder (*Paralichthys dentatus*) under controlled laboratory conditions. Fishery Bulletin 70(4):1127–1136.

Ottersen, G. and J. A. Auran, 2007. Helhetlig forvaltningsplan for Norskehavet: Arealrapport med miljø- og naturressursbeskrivelse. Fisken og Havet, no.6, 2007. Institute of Marine Research, Norway (in Norwegian).

Smidt, E. L. B., 1969. The Greenland halibut *Reinhardtius hippoglossoides* (Walb.), Biology and exploitation in Greenland waters. Meddelelser fra Danmarks Fiskeri og Havundersøgelser 6(4):79–148.

Solmundsson, J. 2007. Trophic ecology of Greenland halibut (*Reinhardtius hippoglossoides*) on the Icelandic continental shelf and slope. Marine Biology Research 3:231–242.

Solmundsson, J., H. Karlsson, and J. Palsson. 2003. Sexual differences in spawning behaviour and catchability of plaice (*Pleuronectes platessa*) west of Iceland. Fisheries Research 61(1–3):57–71.

Teo, S. L. H., A. M. Boustany, B. A. Block. 2007. Oceanographic preferences of Atlantic bluefin tuna, *Thunnus thynnus*, on their Gulf of Mexico breeding grounds. Marine Biology 152: 1105–1119.core (*Thunnus alalunga*) in the northwestern North Pacific. Deep-Sea Research 53:419–431.

UNEP/GRID-Arendal, 2006. Barents Region, topography and bathymetry. UNEP/GRID-Arendal Maps and Graphics Library. July 2006. Available: http://maps.grida.no/go/graphic/barents_region_topography_and_bathymetry3 (September 2009).

Videler, J. 1993. Fish Swimming. Chapman & Hall, London.

Videler, J. J., and D. Weihs. 1982. Energetic advantages of burst-and-coast swimming of fish at high speeds. Journal of Experimental Biology 97:169–178.

Vollen, T., and O. T. Albert. 2008. Pelagic behaviour of adult Greenland halibut (*Reinhardtius hippoglossoides*). Fishery Bulletin 106:457–470.

Webb, P. W. and C. L. Gerstner. 2000. Fish swimming behaviour: predictions from physical principles. Pages 59–77 *in* P. Domenici and R.W. Blake, editors. Biomechanics in Animal Behaviour). Bios Scientific Publishers, Oxford, UK.

Weihs, D. 1973. Optimal fish cruising speed. Nature 245(5419):48–50.

West, G. J., and J. D. Stevens. 2001. Archival tagging of school shark, *Galeorhinus galeus*, in Australia: initial results. Environmental Biology of Fishes 60(1–3):283–298.

Winger, P. D., S. J. Walsh, P. G. He, and J. A. Brown. 2004. Simulating trawl herding in flatfish: the role of fish length in behaviour and swimming characteristics. ICES Journal of Marine Science 61(7):1179–1185.

Woll, A. K., and A. C. Gundersen. 2004. Diet composition and intra-specific competition of young Greenland halibut around southern Greenland. Journal of Sea Research 51: 243–249.

American Fisheries Society Symposium 76:529–553, 2012

Through the Looking Glass: Reflections on the Success of Large-Scale Archival Tagging Programs

David Righton* and Julian Metcalfe
Centre for Environment, Fisheries and Aquaculture Science, Lowestoft Laboratory Pakefield Road, Lowestoft, Suffolk, NR33 0HT, UK

Abstract.—Large-scale fish tagging programs are becoming more popular as fishery managers realize the importance of including spatial structure in assessment and management models. Two recent EU-funded projects on plaice and cod have shown how information from electronic tags can be used to gain new insights and add value to historic tagging data. Highlights have been the demonstration of unexpected population sub-structuring in plaice, and the realization that cod behavior is very variable in response to regional environments. Success does not come without planning and management of staff, of data, and of expectations. We share our experiences from the last 10 years of electronic tagging to provide an up-to-date analysis of what makes a good tagging program, and how to get the most from it.

Introduction

Mark–recapture studies

The earliest account of the tagging of fish is usually attributed to Izaak Walton in *The Complete Angler* (first published in 1653). This recounts the tying of ribands [sic] to the tails of young Atlantic salmon to demonstrate that they returned from sea to the same part of their natal river. The tagging of fish on a large scale started at the end of the 19th Century. As a method for studying movement, tagging (more accurately referred to as 'mark–recapture') relies on the tagged fish being re-caught some time after it was tagged and released and that the time and location of recapture, as well as release, is accurately reported. Subsequent analysis of the distribution of recoveries of tagged fish then provides a method for identifying their movements.

*Corresponding author: david.righton@cefas.co.uk

Large-scale tagging studies did not really start until the end of the 19th century (Harden Jones 1968; Arnold and Dewar 2001). There are many methods for marking or tagging fish (see Parker et al. 1990; Jennings et al. 2001). Branding or fin clipping is a quick and simple way to mark large numbers of fish, and chemical tags such as tetracycline (an antibiotic which is deposited specifically in calcified areas and fluoresces under ultraviolet light), can easily be applied to large numbers of fish and remain as a permanent mark. Alternatively, various types of tag can be attached to, or placed into, the fish (Figure 1a). External tags include Petersen discs used to tag plaice (e.g., Houghton and Harding 1976) and other flatfishes, while internal tags include the tiny coded wires that are used for the mass marking of young salmon (Elrod and Schneider 1986).

Mark–recapture methods tell us where individual fish are at two times in their life (i.e., when caught and tagged, and when recap-

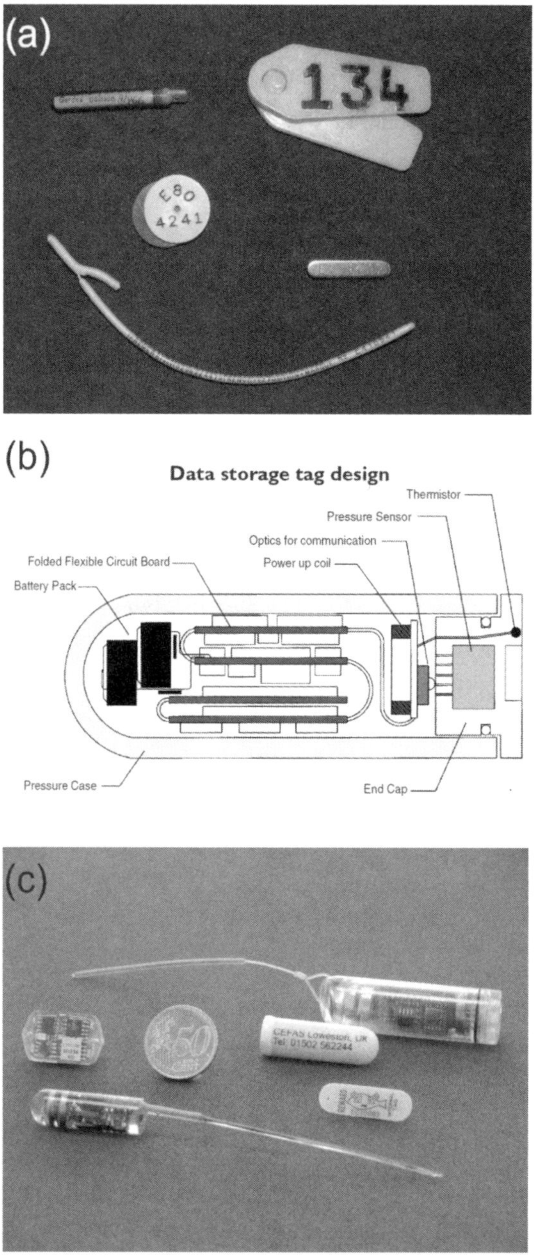

Figure 1. Examples of (a) conventional tags, (b) data storage tag design and (c) the variety of data storage tag casing designs now available on the commercial market.

tured). If tagging and recapture are separated by a suitable amount of time (months or even years), it can provide information on stock identity, movements, migration (both rates and routes), abundance, growth, and mortality (Jones 1976). Some littoral species, like blennies (family Blenniidae), make seasonal inshore and offshore movements that extend no more than a few kilometers. For such species, large-scale mark–recapture programs are unlikely to be cost-effective ways of studying their movements. However, species like herring (*Clupea harengus*), mackerel (*Scomber scombrus*), Atlantic cod (*Gadus morhua*), European plaice (*Pleuronectes platessa*), Atlantic salmon (genus *Salmo*), Pacific salmon (genus *Oncorhynchus*) and eels (*Anguilla* species), various species of tuna, billfishes and large sharks make more extensive movements over several hundreds or thousands of kilometers (Metcalfe et al. 2002). For these species, mark–recapture studies can provide very valuable information about population structure and dynamics.

Although mark–recapture studies can be very useful for describing gross patterns of population movement, the method has its limitations. First, while mark–recapture provides information about when and where the fish was first caught and tagged, and then when and where it was subsequently recaptured, it provides no information about where the fish may have been in between or what it has done. For instance, if a fish is caught, tagged and released on its winter spawning ground, then re-caught in the same location a year later, we cannot know whether it moved away and returned or remained the whole year in the same area. The re-capture of conspecifics that were tagged and released at the same time but subsequently re-caught on a distant feeding ground in summer would indicate some pattern of migration at a population level, but we could still never be sure whether or not the first fish had ever moved (Righton et al. 2007). Moreover, the results of mark–recapture experiments, by their very nature, rely on the re-capture of the fish, usually by commercial fishers, so results are inevitably confounded by the integration of fish behavior and fishing activity (Rijnsdorp and Pastoors 1995; Bolle et al. 2005). This problem is further compounded by inaccuracies in the reported capture location of tagged fish, or biases in reporting rates (McGarvey, 2009).

Although the analysis of mark–recapture data can accommodate spatial variations in fishing effort (Rijnsdorp and Pastoors 1995; Wright et al. 2006a), this is often not known accurately, and movements of fish into unfished areas, or changes in fish behavior that alter their catchability, cannot be accounted for easily. Nor can mark–recapture experiments provide information about how fish migrate (Metcalfe and Arnold 1997). For a more quantitative assessment of these factors, a more detailed understanding of the movements and behavior of a fish in both space and time is required.

Electronic tagging

The advent of microelectronics (more specifically, the development of the solid-state transistor that replaced thermionic valves) in the 1950s paved the way for electronic devices that could be made small enough to attach to, or implant into, fish (see review by Arnold and Dewar 2001). Early electronic fish tags transmitted either radio (for use in freshwater) or acoustic (for use in the sea) signals that allowed individual fish to be tracked, usually from a boat or research vessel. Such tags can also be used in association with static listening stations to record the arrival and departure of fish at particular locations (e.g., Righton et al. 2001). Ship-borne tracking experiments can provide detailed information about the spatial (vertical and horizontal) movements of individual fish and, if environmental data (temperature, water currents, etc.) can be gathered simultaneously by the tracking vessel, make it possible to relate fish behavior to the local environment (Buckley and Arnold 2001). However, active tracking at sea is limited for logistic reasons; usually only one fish can be followed at a time, tag life is limited (usually no more than a week or two) and a ship's time is expensive.

The further development of integrated circuit (silicon chip) technology allowed, by the early 1990s, the development of "data storage" or "archival" tags that intermittently record and store information from on-board sensors that measure environmental variables such as pressure (to give depth), temperature and ambient daylight. Early tags of this type were quite large, (20–40 g), could store comparatively limited amounts of data (30–50k of data points) and were expensive (~£700–800), but technology has advanced rapidly and tags that can store in excess of a million data points, weigh 1–2 g (in water), and cost £200–300 are now available (Metcalfe et al. 2008b).

First results from electronic tagging studies provided exciting insights into the behavior of individual fish, revealing unprecedented levels of detail from fish behaving in their natural environment over extended periods of time. Understandably, early reports described movements over periods of weeks and months (Metcalfe and Arnold 1997; Block et al. 1998), but results spanning multi-year timescales are now appearing in the literature (Hunter et al. 2003b, 2005). Although tag returns for some species and areas may reach as high as 50% (Schaefer et al. 2007), tag returns for most programs are in general no more than 20–30% even for heavily exploited species. In consequence, many hundreds of tags need to be deployed before it starts to become possible to draw conclusions at the scale of fish populations. It is only after many years of study, therefore, that this technology is beginning to deliver meaningful information that can be applied at the level of fish stocks (Hunter et al. 2004b, 2006; Block et al. 2005; Metcalfe 2006; Righton et al. 2007).

Scientists at the Centre for Environment, Fisheries and Aquaculture Science have been at the forefront of electronic tagging studies in the north-east Atlantic since the late 1960s. The studies that have been conducted have shown how electronic tags can provide surprising and important insights into fish behavior. Highlights have been the demonstration of unexpected population sub-structuring in plaice, and the realization that cod behavior is very variable in response to regional environments. In the remainder of this paper, we share our experiences from the last 10 years of electronic tagging to provide an up-to-date analysis of what we believe makes a good tagging program, how to get the most from it, and provide examples of how the data have been integrated into fisheries advice.

Methods

Planning a tagging program

Simply wanting to know where fish move to, or what they do, is not in itself a good enough reason to embark upon a large-scale mark–recapture program. Tailoring a tagging program too closely to existing management regimes and assumptions, however, would be a missed opportunity to collect other useful biological and ecological information. The objectives of a tagging project should therefore be a careful compromise between biologically-driven research and data collection that will deliver information relevant to fisheries management (Metcalfe et al. 2008a). Tagging programs are also multidisciplinary activities, and researchers engaged in them need to be adept in a number of different fields, from publicity to data management.

The information from tagging programmes that can be applied to fisheries management will vary depending on regional legislation and practice, but one core requirement will be knowledge of the areas occupied by the target stock and the seasonal movements of individuals between them (Hilborn 1990, generally; Hunter et al. 2004b, on European plaice; Bolle et al. 2005, on European plaice; Block et al. 2005, on bluefin tuna; McGarvey 2009, generally). There are many objectives that can be nested within this core requirement depending on the specific nature of regional management measures. For instance the locations of key habitats such as spawning areas and feeding grounds (Schopka et al. 2010, on cod; Campana et al. 2010, on porbeagle sharks), the migration routes between them (Robichaud and Rose 2004, on cod; Weng et al. 2005, on salmon sharks; Righton et al. 2008, on cod), and patterns of vertical distribution (Palsson and Thorsteinsson 2003, on cod; Sippel et al. 2007, on striped marlin; Lawson et al. 2010, on tuna). This kind of information may help to define areas that should be closed seasonally, or fishing gears that should be subject to limitations or exemption (Hunter et al. 2006, on thornback rays, Goodyear et al. 2008, on blue marlin).

Investigating the scope for tagging

Having identified the objectives of the tagging program, it is necessary to understand as much as possible about the target stock, in addition to its environment and fishery, before any further plans are made. Key to the success of any tagging program is a good recovery rate of the tag (essential for studies using archival tags), the tagged fish and associated recapture data (Jones 1976; Jennings et al. 2001). For example, it is important to have an approximate idea about the size the "stock," the extent of its geographical distribution, and what tag recovery rates are likely to be. These factors are important in estimating how many fish need to be tagged, where and when, and in what experimental design, to achieve a statistically robust result (Pine et al. 2003). In marine fisheries, the area of encounter is potentially vast but can be reduced significantly with backup information from commercial or research catch data or from earlier tagging studies, and also by talking to those that fish the stock. For electronic tagging programs, pretagging surveys with conventional tags should be carried out to provide a rough estimate of where the electronic tags will be recovered and what the target fisheries are likely to be. It is also important to understand who is likely to catch the fish that will carry the tags; can their help be used in the tagging programme (e.g., the Gulf of Maine cod tagging program; Tallack 2009), or is experimental or charter-boat fishing necessary to get sufficient fish tagged in the appropriate location(s)?

Decide upon your tag type

Data storage tags record and store environmental and behavioral data and, because there is no need for human observers to follow the fish, they make it possible to monitor the behavior and movements of many fish simultaneously over entire migrations (Metcalfe and Arnold 1997, 1998). A variety of such devices (Figure 1) are now being used to study the movements of species as diverse as plaice (Metcalfe and Arnold 1997; Hunter et al. 2004b), cod (Righton et al. 2001; Righton et al. 2007), salmon (Walker et al. 2000), tuna (Gunn 1994; Gunn et al. 1994; Block et al. 1998, Block et al. 2005), thornback rays *Raja clavata* (Hunter et al. 2005) and others.

The data from such tags not only provide fine-scale information about behavior (e.g., vertical movements derived from pressure readings), but can also be used to determine geolocation on many occasions while the fish is at liberty. On the European continental shelf, tidal information (times of high- and low-water and tidal range) derived from pressure measurements can be used to locate fish whenever they remain stationary on the seabed for a full tidal cycle or more (Figure 2; Metcalfe and Arnold 1997, 1998; Hunter et al. 2003a). In the open ocean, records of ambient daylight can be used to derive latitude (from day length) and longitude (from the time of local noon; Wilson et al. 1992; Hill 1994; Gunn et al. 1994; Hill and Braun 2001; Metcalfe 2001). This behavioral and movement information can then be integrated with environmental data that were either recorded by the tag at the same time (e.g., water temperature), or that were collected independently (e.g., satellite data on sea surface temperature or biological productivity) for the same geographical area (Sims et al. 2003).

More recently, tags have been developed that can directly monitor more specific aspects of behavior such as feeding. For example, temperature sensing data storage tags that monitor the difference between visceral and environmental temperatures have been used to monitor feeding events and estimate food intake in southern bluefin tuna *Thunnus thynnus* (Gunn and Block 2001) and tags equipped with movement sensors have been used to detect jaw movements (including feeding events) in penguins (Wilson et al. 2002) and cod (Metcalfe et al. 2009).

Despite such technical advances, the use of data storage tags with many species remains limited because the prospect of the fish being caught and the tags returned is very low. To avoid the need to rely on a commercial fishery, and increase the probability of data recovery, a major area of development has been the "pop-up" tag. These tags are attached externally and have a release mechanism which causes the tag to detach

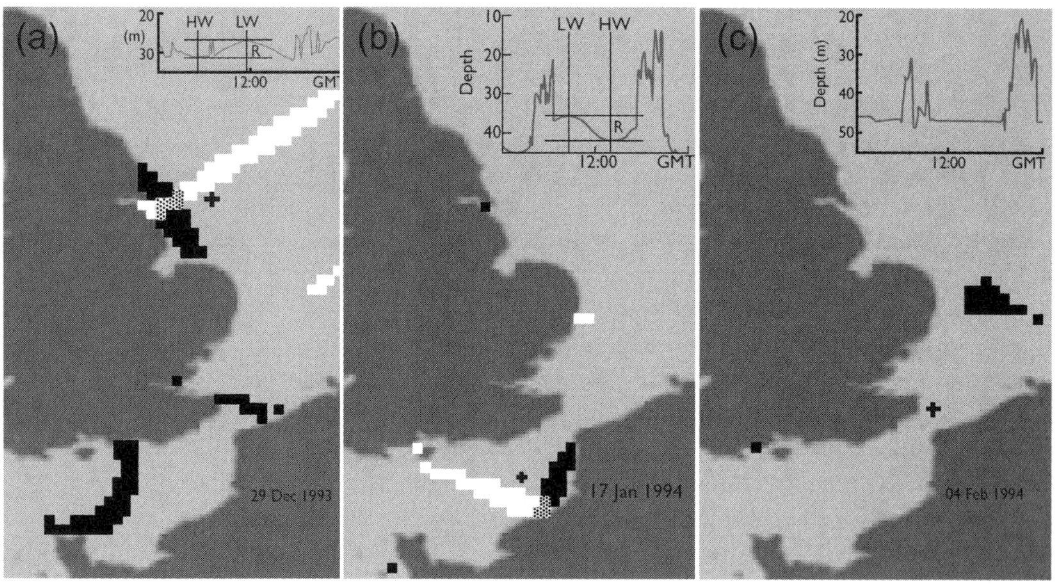

FIGURE 2. Use of tidal data to determine of location a plaice tagged with a data storage tag. The Proudman Oceanographic Laboratory's numerical storm surge model was used to identify areas (in black) with similar tidal range and similar times (in white) of high or low water recorded by the tag (insets at top right of each panel). Areas where *both* tidal range *and* times of high or low water coincide are hatched. Fish locations (+) were derived from the track of the fish reconstructed using a tidal stream simulation model (Arnold and Holford, 1995) assuming the fish migrated using selective tidal stream transport and swan down-tide at 0.6 body lengths s^{-1} while being carried over the ground by the tide. Figure redrawn from Metcalfe and Arnold 1997.

from the fish at a predetermined time and "pop-up" to the sea surface where the data can be recovered by airborne radio or satellite (Nelson 1978; Hunter et al. 1986). Such devices are now commercially available, and are being deployed on large pelagic species such as tuna (Block et al. 1998; Lutcavage et al. 1999, Block et al. 2005), basking sharks *Cetorhinus maximus* (Sims et al. 2003) and larger eel species (Jellyman and Tsukamoto 2005; on the long-finned eel). Although data transmission capabilities are currently very limited, further developments in this field give the prospect of much improved data recovery rates in the future, while further miniaturisation will allow the technology to be applied to smaller species.

Implementing a tagging program

Fish capture, handling, and attachment methods.—It is important to choose capture and tagging methods that will result in the collection of reliable and useful data. The legislation around animal welfare and animal experimentation, and the increasing sophistication of small electronic devices, has rightly changed the face of tag attachment methods (Thorsteinsson, 2002; Metcalfe 2009). The methods used to capture fish are diverse and the method chosen for use in tagging experiments must minimize the stress and damage that any captured fish will experience. Similarly, once caught, fish must be handled gently. They should be tagged and returned to the water as quickly as possible and not dropped on the deck or allowed to strike the side of the boat or the bulkhead. When picked up they should be held horizontally (i.e., not by the tail or gill covers) and the gills should not be touched with the fingers. Only fish in good condition should be tagged and released. This is not only important from a fish welfare point of view, but also because electronic tags are expensive and the long-term survival of the tagged fish is critically important (although see Campana et al. 2009 for an experiment where bycatch mortality of blue sharks was assessed using pop-up satellite tags). However, in field experiments, the ideal conditions for handling fish cannot always be met because setting up facilities for anesthesia and recovery may be difficult either because spatial restrictions or poor weather at sea prevent it. The experimenter must then evaluate the relative difficulties of applying anesthesia against possible trauma and damage caused by handling unanaesthetised fish, although legal considerations may be paramount. When tags can be attached rapidly and nonintrusively, anesthesia has often been replaced by simpler methods of keeping the fish quiet during tagging such as blindfolding. Placing the fish ventral surface up in a tagging cradle often calms them down and in some species induces tonic immobility (Holland et al. 1999, on tiger sharks). For a full review of tagging methods and marks, readers are directed to the report from Thorsteinsson (2002).

Programming tags to sample data.—Where electronic tags are to be used, thought must be given to how the sensors, battery and memory life can be used to best effect. The potential for data collection is not always fully exploited in many archival tagging programs because some tagged fish are caught soon after release and much of the tag's data recording power and memory therefore remains unused. For example, Godø and Michalsen (2000) programmed tags deployed on NE Arctic cod *Gadus morhua* to record pressure and temperature data in a cycle of every two hours for six days, and every 24 h for one day. Righton et al. (2001) and Righton and Metcalfe (2002) programmed tags on North Sea cod to record data every tenminutes for the duration of a expected life of the tag. While, these tagging studies to date have largely achieved the goals set at their inception, relatively few *a priori* or *postposteriori* analyses have been undertaken to determine the most useful data-logging regimes or the limitations of the logging regime used. Instead of using a regular data sampling regime that maximizes the potential deployment time of the tag, a sampling regime that samples more frequently at the beginning of the deployment than it does at the end should be considered (Figure 3). Alternatively, tags that offer time extension or telescoping logs could be used. Fortunately, as tag and battery technology improves, clever or sophisticated recording regimes need not be such a consideration.

Clearly, tags that are programmed to record data only infrequently will not provide as much information as those that record data more frequently. Measuring environmental variables requires data to be recorded frequently enough to minimize sampling error, but not so frequently that tag memory is filled too rapidly. Measurements of rapidly changing variables such as movement rate or feeding behavior requires frequent sampling (Metcalfe 2009), whereas infrequent sampling can be used to measure environmental preferences such as maximum depth or water temperature (Godø and Michalsen 2000; Palsson and Thorsteinsson 2003). Some analytical methods, such as activity or periodicity analysis, are often best conducted with data recorded at high frequency because the statistical power of such analyses is compromised by low frequency data (Hunter et al. 2004a). Examples of analyses requiring a rapid sampling rate include estimation of tilt angle (Heffernan et al. 2004), accurate estimation of the depth of the thermocline and estimation of longitude and latitude from light intensity data (Hill and Braun 2001).

Press, publicity, and reward schemes.— Tagging experiments are usually costly exercises requiring vessel time, experienced staff and, if using electronic tags, expensive

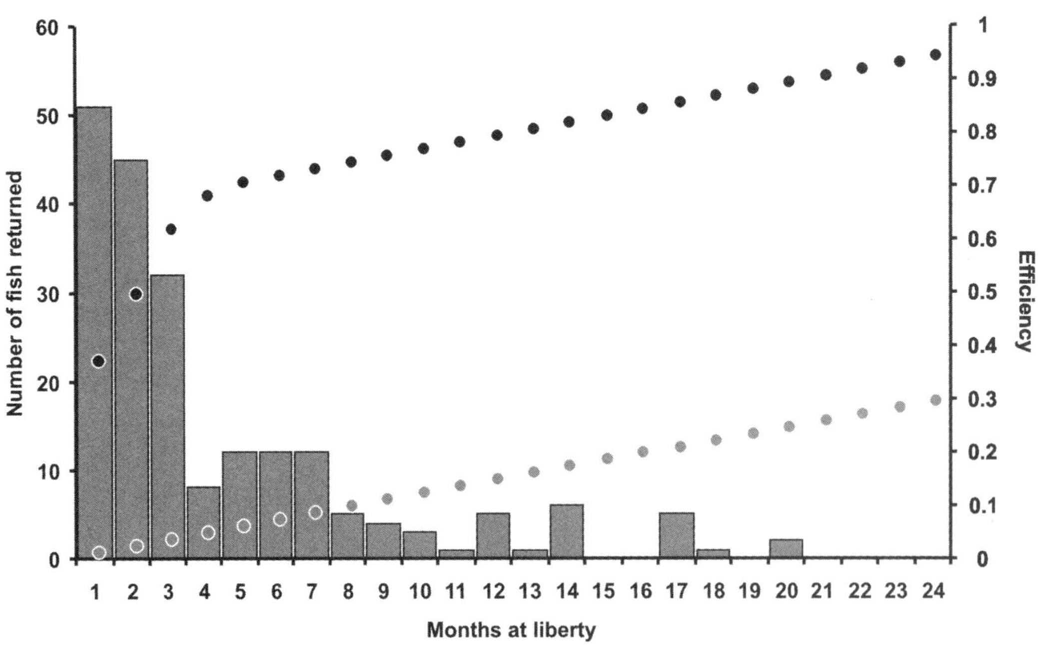

Figure 3. Comparison of the efficiency of data collection by data storage tags. The bar chart shows the number of months at liberty after tagging and release of cod released in UK waters. The symbols show the proportion of tag memory used if the frequency of data collection is constant (gray) or rapid at the start then declining over time (black).

devices. Consequently, tag data are valuable and the data recorded by even a single archival tag can provide significant new insights into fish behavior (Arnold and Dewar 2001). It is therefore paramount the appropriate resources are deployed to encourage fishers to return tags together with accurate tag recapture details and, when appropriate, the fish carcass. In some cases, particularly where tagging has been opportunistic, programs can fail to achieve their full potential because tags returns are poor as a result of insufficient publicity and rewards.

The number of tags recovered will improve considerably with good publicity and reward systems. Initially, the objectives, tag type, secondary tag type (where used) and the rewards (if offered) should be clearly advertised. Prospective individuals who are likely to recover tags or be aware of recovered tags (fishermen, fish processors, sport anglers etc) should be informed by press, posters or presentations that tags of different types may be present in the fish they handle. It is important to emphasize the scientific value of the tagging program, and the value of the data recovered from electronic tags (where used) as well as the overall benefits of the data for protecting and possibly enhancing stock assessment and management. In particular, there should therefore be a good incentive to return tags, particularly if tag recovery is dependent on commercial fishermen or processors.

The value of continual and re-iterative publicity programs is shown by an analysis of tag returns to the Cefas laboratory, UK. When mark–recapture tagging programs were in their relative infancy in the 1960s, the return of plastic marker tags from fish tagged in UK waters was relatively high, at around 30% (Figure 4a). This return rate persisted until the 1980s, but has since fallen to below 10%, probably due to declining fishing opportunities and also a reluctance or apathy on the part of the returnee. The pattern is consistent across a range of tagged species (Figure 4b). In contrast, the return rates of electronic tags remain high, at around 20%, or more than twice the return rate of marker tags. This could be, in part, related to the visibility of the tag, but is also likely to reflect the greater publicity efforts and higher rewards that are offered for electronic tags.

As technology advances, electronic tags are becoming smaller and internal placement rather that external attachment is increasingly possible. While internal placement (usually in the peritoneal activity) is better for fish welfare and may improve tag retention (but see Moore et al. 1990 and references cited therein) internal placement, particularly of small tags, may lead to reduced recovery as a result of the tags not being detected so easily by fishers or fish processors (Righton et al. 2006). If internal placement is to be used, appropriate external tags that indicates the presence of an internal tag, together with suitable publicity, is paramount. Even despite the extra publicity, tag return rates may inevitably be lower. In cod tagging programs with internal tags, the average return rate over 7 years between 2002 and 2009 has ranged between 9% and 23% (923 tagged fish released, 169 tags returned: 18%), compared to a return rate of 33% for externally attached tags (60 tags returned of 166 tags released over three years).

Results and Discussion

Large-scale fish tagging programs have a significant role to play in assisting the development of fisheries management measures and the development of the next generation of stock assessment and management models. We've identified three main avenues for exploiting tagging data in fisheries management: stock structure and distributions, changes in catchability in space and time, and the effects of the environment/climate on fish biology/behavior.

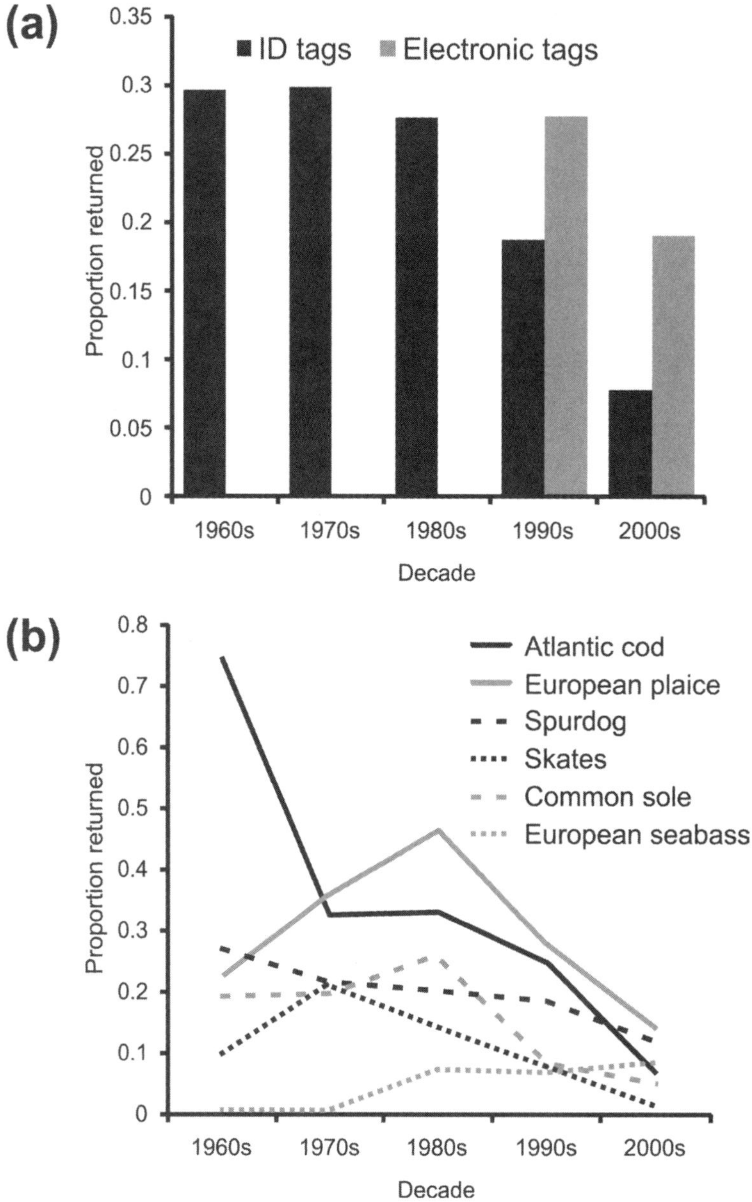

Figure 4. Patterns in the return rate of fish tags. (a) a comparison of the return of identification tags against electronic tags; (b) the general decline in return rates of identification tags from fish tagging experiments since the 1960s.

Stock structure and distribution—population movements

The first step of any fish-tagging program is to assess where individuals have moved. Jones (1976) provides a summary of the statistics that can be used to describe movement and dispersion using simple mark–recapture data. However, to avoid difficulties with interpretation of the results, it is crucial to select data appropriately before applying the statistics. Patterns of space use that differ between seasons or sub-stocks may then emerge, for example in cod in the North Sea (Figure 5; Righton et al. 2007). Similarly, the plaice sub-populations in the North Sea have been identified from previous mark–recapture data (de Veen 1978; Cushing 1990), with results indicating that there are discrete plaice sub-populations that aggregate during the winter spawning then disperse during the summer over distinct but overlapping feeding grounds.

The advent of electronic tags has permitted more detailed analysis of the movements of individuals because the data that are collected permit the geolocation of individuals at frequent intervals (Metcalfe and Arnold

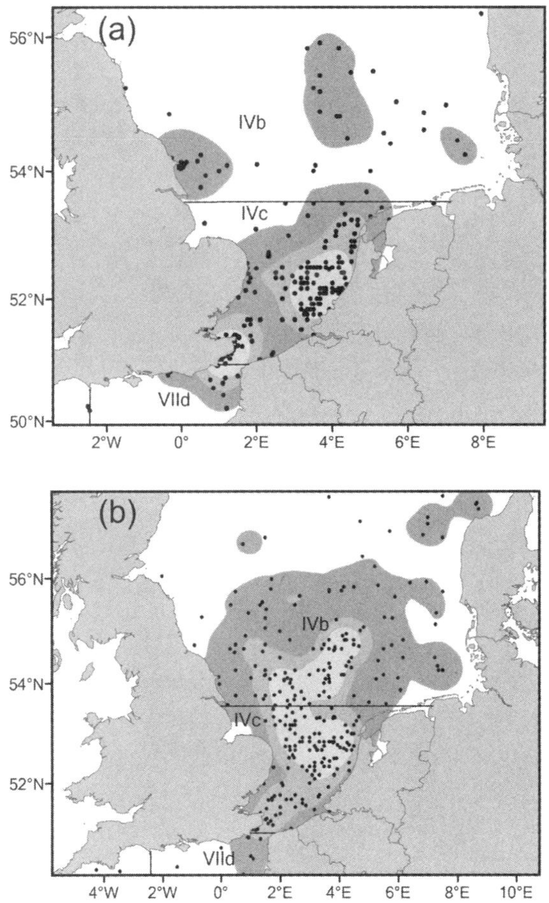

Figure 5. Recapture positions of cod released in ICES area IVc. Solid symbols show exact recapture locations, while shading shows the probability density surfaces for 50% (lightest gray), 75% (mid gray) and 95% (dark gray) of the recaptures. Data shown are for (a) adults recaptured during the spawning season; (b) adults recaptured during the feeding season. Figure reproduced from Righton et al. 2007.

1997; Hunter et al. 2003a; 2003b). In a sense, such geolocations can be viewed as multiple recapture positions, and used to plot the change in spatial distribution of populations over time (Figure 6). Since the early 1990s extensive studies of the movements of plaice equipped with electronic data storage tags have significantly advanced our understanding of the distributions and movements of plaice in the North Sea (Metcalfe and Arnold 1997, 1998; Metcalfe et al. 2002; Hunter et al. 2003a, 2003b; Metcalfe and Pawson 2004; Hunter et al. 2004b; Metcalfe 2006). In contrast to earlier analyses of mark–recapture data, detailed analysis of the data from these electronic tags indicates that the adult plaice population in the central and southern North Sea forms three geographically discrete feeding aggregations during the summer that disperse over the southern North Sea and east-

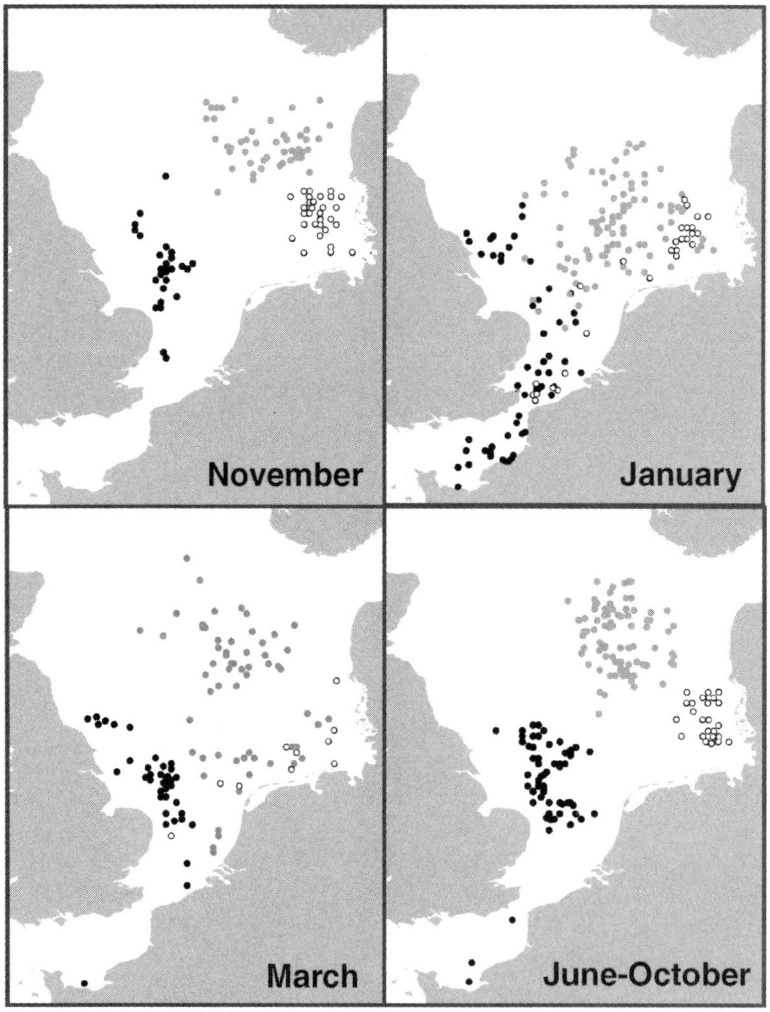

FIGURE 6. Monthly composite plots of geolocations for 144 electronically tagged plaice. During June to October, the horizontal movements of plaice were minimal, therefore these positions were plotted together. Three geographically distinct feeding aggregations were identified on the Indefatigable Banks to the west (black) on the Amrum Ground to the east (hollow) and on Ekofisk Field to the North (gray), which dispersed onto mixed spawning areas during January and February. Figure reproduced from Hunter et al. 2004a.

ern English Channel to spawn in the winter (Hunter et al. 2004b). Conversely, evidence from electronic tags attached to cod suggest the reverse: that populations converge on spawning grounds in the southern North Sea and eastern English Channel in spring, and then disperse to feeding grounds during summer (Righton et al. 2007).

Stock structure and distribution—individual movements

The detailed movements of individuals can be just as revealing and provide insights into migratory mechanisms as population movements. For example, plaice in the central North Sea exhibit clear repeat migrations between feeding and spawning grounds that are remarkably consistent in direction and timing (Hunter et al. 2004b). Detailed results like this can help when interpreting data from mark–recapture studies. For example, the distance from (spring) release positions of cod in the North Sea increases to a peak after six months, before decreasing again (Figure 7). This pattern is repeated in the following 12 months, and is indicative of a significant proportion of the population undertaking repeat migrations (Metcalfe et al. 2005). This is corroborated by findings from electronic tagging studies (Figure 8; Righton and Mills 2008) that show how cod that migrate away from a tagging position in spring subsequently return to the area at spawning time. More recent results show that cod, at least those found in the eastern North Sea,

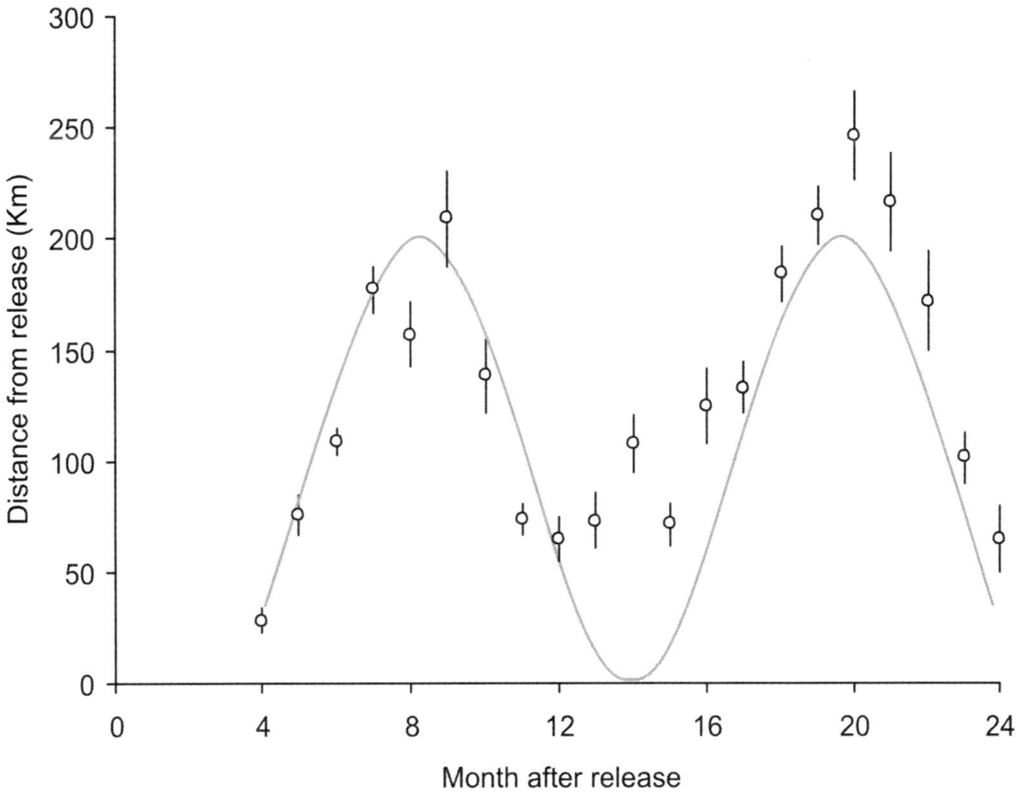

Figure 7. Mean distance between release and recapture positions of cod tagged with conventional tags and released in ICES area IVc in the spawning season. Lines of best fit were calculated by co-varying the parameters of sinusoidal wave (frequency, amplitude and offset) and using a least-squares minimisation routine (as for Bolle et al. 2005). Figure reproduced from Righton et al. 2007.

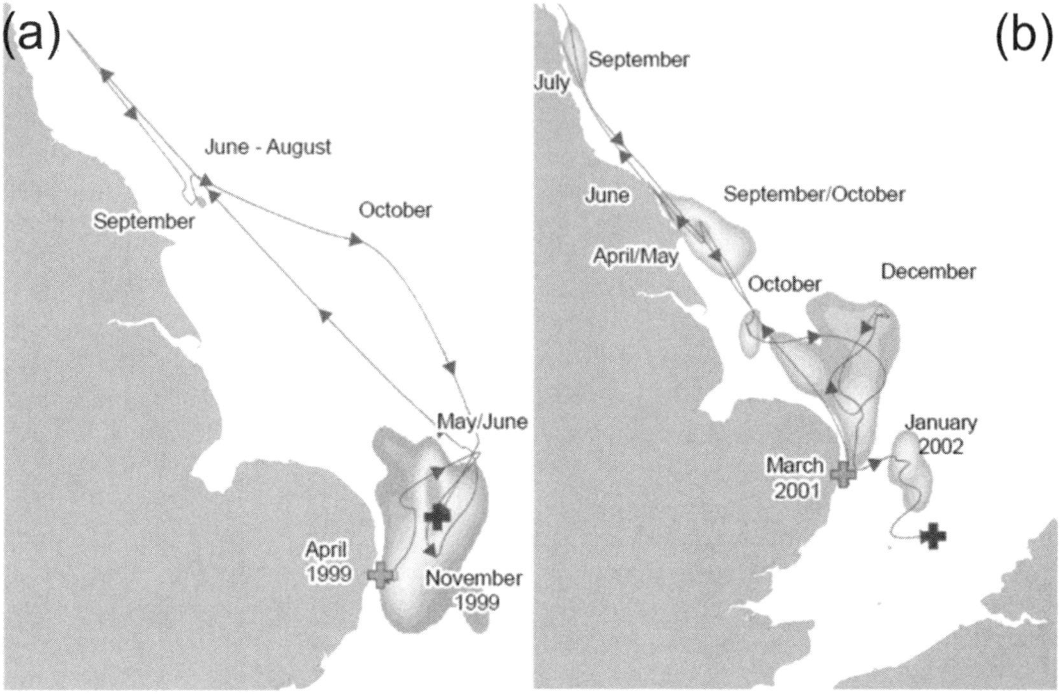

FIGURE 8. A detailed reconstruction of the migration of two cod released in the southern North Sea. (a) The migration path of cod 1432 released in March 1999 was reconstructed using a data simulation method (Righton and Mills, 2008), summarised using a bezier curve and overlaid on a likelihood surface only including 'core' areas; (b) The migration path of cod 2255B, released in March 2001, reconstructed as for cod 1432. Figures reproduced from Righton and Mills 2008.

are capable of natal homing (Svedang et al. 2007), suggesting that mechanisms that define and maintain populations are complex. Results such as these can have far-reaching consequences for the applicability of metapopulation theory to fishery science (McQuinn 1997; Smedbol et al. 2002) and on our perceptions in fisheries management (Smedbol and Stephenson 2001). Importantly, if fish populations act as behavioral entities, degradation of population structures has to be considered as partly irreversible, at least on an ecological time perspective. Further integration of results from tagging studies with new molecular and chemical approaches will help to further clarify the role of behavior in determining stock structure, especially concerning degree of reproductive isolation between spatially closely located spawning aggregations (c.f. Wright et al. 2006b).

Catchability: the vertical dimension

Electronic tagging studies also reveal significant detail of the behavior and activity of individuals. The use of selective tidal stream transport by plaice to aid their pre- and post spawning migrations in the southern North Sea was revealed by extensive acoustic tracking (Greer-Walker et al. 1978; Metcalfe et al. 1992) and data storage tagging studies (Metcalfe and Arnold 1997, 1998). The results can be used to inform more analytical approaches, such as periodogram analysis (Figure 9; Hunter et al. 2004a). Such analyses can reveal the extent to which vertical movements are related to environmental and biological cues (Metcalfe et al. 2006), and give insights into the availability of fish to fishing activity, and how this varies during the year. Clear seasonal changes in vertical movements occur, with

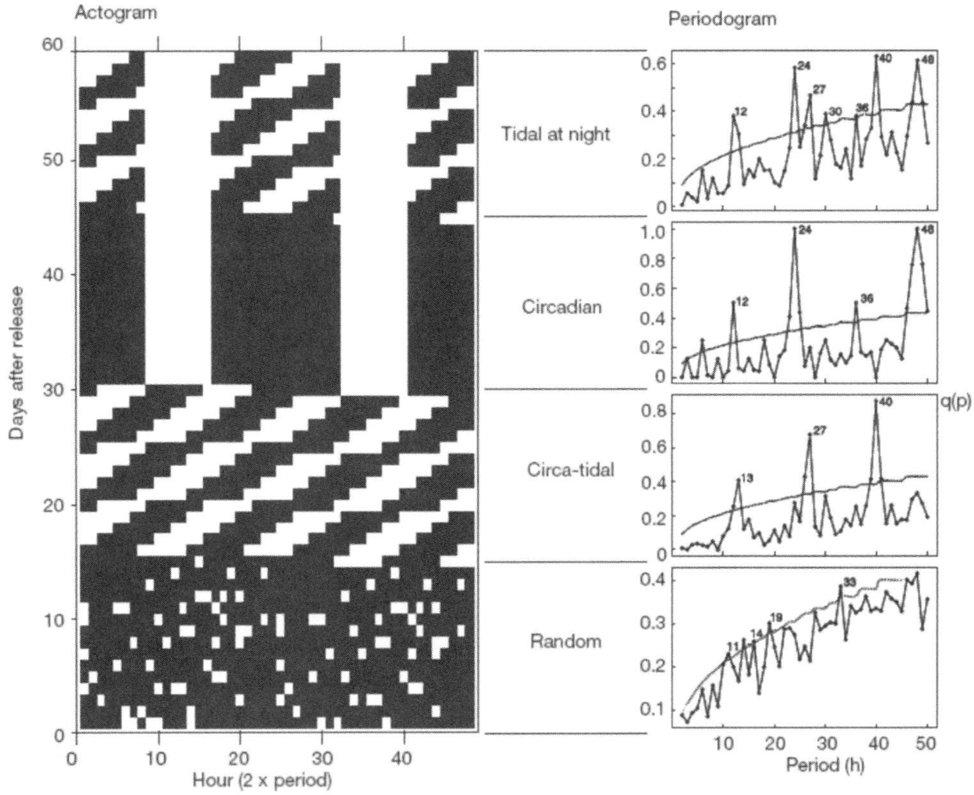

FIGURE 9. Double-plot actogram (left) and periodogram analysis (right) of simulated data containing stereotypical patterns of vertical activity illustrating how these analyses can be used to identify when an individual fish is exhibiting different types of behavior. In the actogram, activity is represented in black, in the periodogram, the dotted line represents ~0.01 significance level and the numbers above the peaks denote significant periodicities (in hours). Figure reproduced from Hunter et al. 2004a.

plaice becoming less vulnerable to trawl gear during their migration to spawning grounds because they spend more time in the upper reaches of the water column to take benefit from the directional flow of rapidly moving tidal currents (Figure 10; Hunter et al. 2004a). This analysis of the DST plaice data not only confirmed the original ship-based observations of plaice migration using selective tidal stream transport, but extended them by revealing further migration-linked changes in patterns of vertical swimming activity. As a result, the new information provided one of few examples where a fish stock's annual behavior patterns have been recorded across a large part of its geographical range. In terms of fisheries management, a better understanding of both when and where plaice are likely to be stationary on the seabed has helped to improve estimates of the likelihood of capture by commercial fishing vessels and scientific surveys of fish abundance can be tuned to allow for behavior.

Turning again to Atlantic cod, it has been shown that, for this species, simple indices of vertical activity can also be used to compare the behavior of individuals in different areas. For example, Righton et al. (2001) used a depth change threshold to classify activity of cod in the North and Irish Seas (Figure 11) and showed that seasonal patterns of activity differed significantly between these two

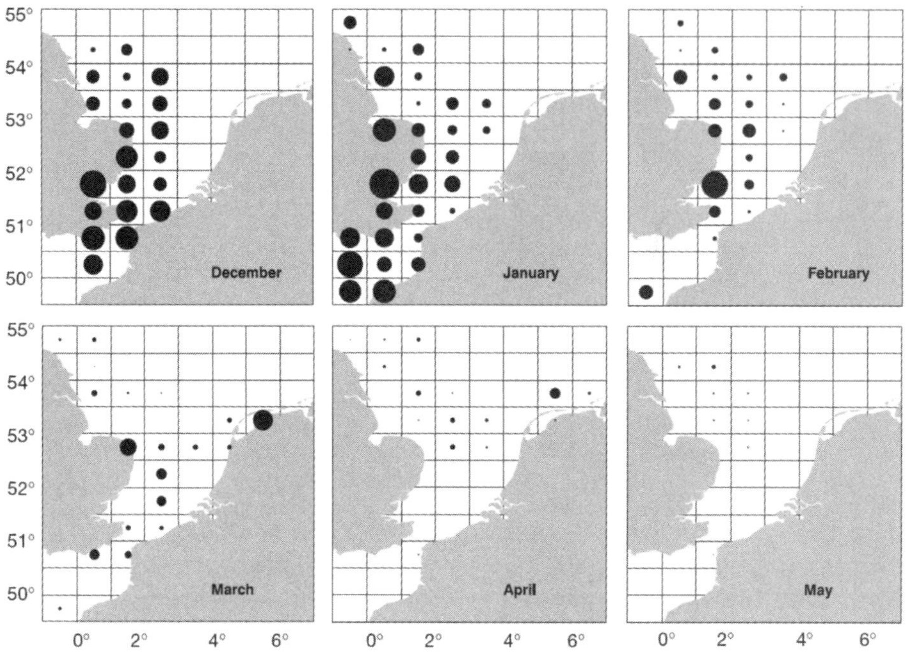

Figure 10. Monthly plots illustrating the proportion of time plaice spent swimming in mid-water per ICES rectangle December to May. Size of the circles is proportional to the average number of hours spent in mid-water per ICES rectangle. Largest circles equal 15 hours per day. Figure reproduced from Hunter et al., 2004b.

habitats. More recently, Hobson et al. (2007) showed how the vertical movement patterns of cod in the North Sea can be characterized using relatively simple algorithms, and showed significant seasonal changes in vertical movement patterns that affect vulnerability to capture (Figure 12). Similar analyses show that the catchability of cod in the North Sea and surrounding areas changes in a predictable way, and that long-term patterns in commercial landings reflect these changes in behavior (Righton et al. 2009).

Thermal tolerance: observing habitat occupation

Previous studies addressing thermal influences on fish distribution (Heessen and Daan 1994; Blanchard et al. 2005; Perry et al. 2005) have been derived from surveys that cut across large (1000's km) geographic areas and long (decadal) timescales. Such studies are insensitive to short-term, individual and regional variation in distribution. This can be a problem when addressing distribution patterns of species like cod that have spatially-structured stocks (Hutchinson et al. 2001; Metcalfe 2006) and show high individual variability in migratory behavior (Righton et al. 2001; Palsson and Thorsteinsson 2003; Neat et al. 2005; Neat and Righton 2007). Direct, individual-based evidence for thermal responses in fish in the field is largely lacking, so the potential offered by electronic tagging programmes is considerable. We used temperature data collected by cod in the North Sea to assess the evidence of climate-linked changes in migration, and the extent to which cod occupy the thermal habitat potentially available to them. Somewhat surprisingly, the majority of temperature experiences of cod

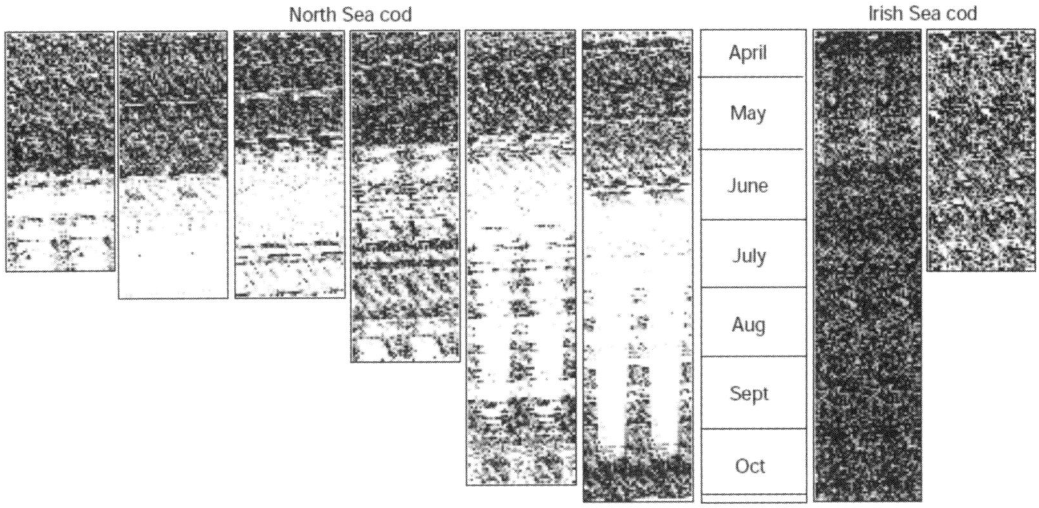

FIGURE 11. Activity patterns of cod at liberty in the North Sea (6 left-hand panels) and Irish Sea (2 right-hand panels) between April and November 1999. Each panel shows a different individual. Active and inactive states of each individual were determined from the archival tag depth (measured every 10 min) record. An individual was judged as 'inactive' on the seabed when the tag recorded only the smooth changes in pressure resulting from the rise and fall of the tide. Individuals were classed as 'active' when vertical movements were more rapid or irregular than could be accounted for by tide alone. For each hour of the day, summed hourly activity held a value between zero (white = inactive) and six (black = most active). Figure reproduced from Righton et al. 2001.

in the North Sea closely followed the upper seasonal trend of sea bottom temperature (Figure 13). In the northern North Sea, cod inhabited the warmest fraction of the area, particularly during the warmer months of the year (Figure 13a). No cod from the northern North Sea moved to cooler waters and the variance between individuals was very low. On occasion, individuals showed abrupt movements into cold fronts and prolonged occupancy of cooler areas (Figure 14a). In the southern North Sea, most individuals released in the Southern Bight (Figure 14b) migrated short distances and experienced temperatures well in excess of 14°C during late summer and autumn with some experiencing temperatures up to 19°C (Figure 14b). Particularly interesting is that the cod in this study do not seem to show any movement away from temperatures that, in a laboratory, would be close to lethal. In fact cod seem to bear these high temperatures for substantial periods of time, suggesting that they are quite capable of tolerating the seasonal change in temperatures that exist in the southern North Sea. The significance of the thermal experience of cod extends beyond growth, and into reproductive biology (Kjesbu et al. 2009).

Conclusions

We have used the example of plaice and cod in European waters to show how mark–recapture experiments using conventional tags, and tagging studies using various types of electronic tag, can provide information about the behavior and geographical movements of exploited fish populations that is relevant to assessment and management. As we enter the 21st century, we are faced with the potential for unparalleled changes in the marine environment as a result of climate change. World fisheries also continue to be

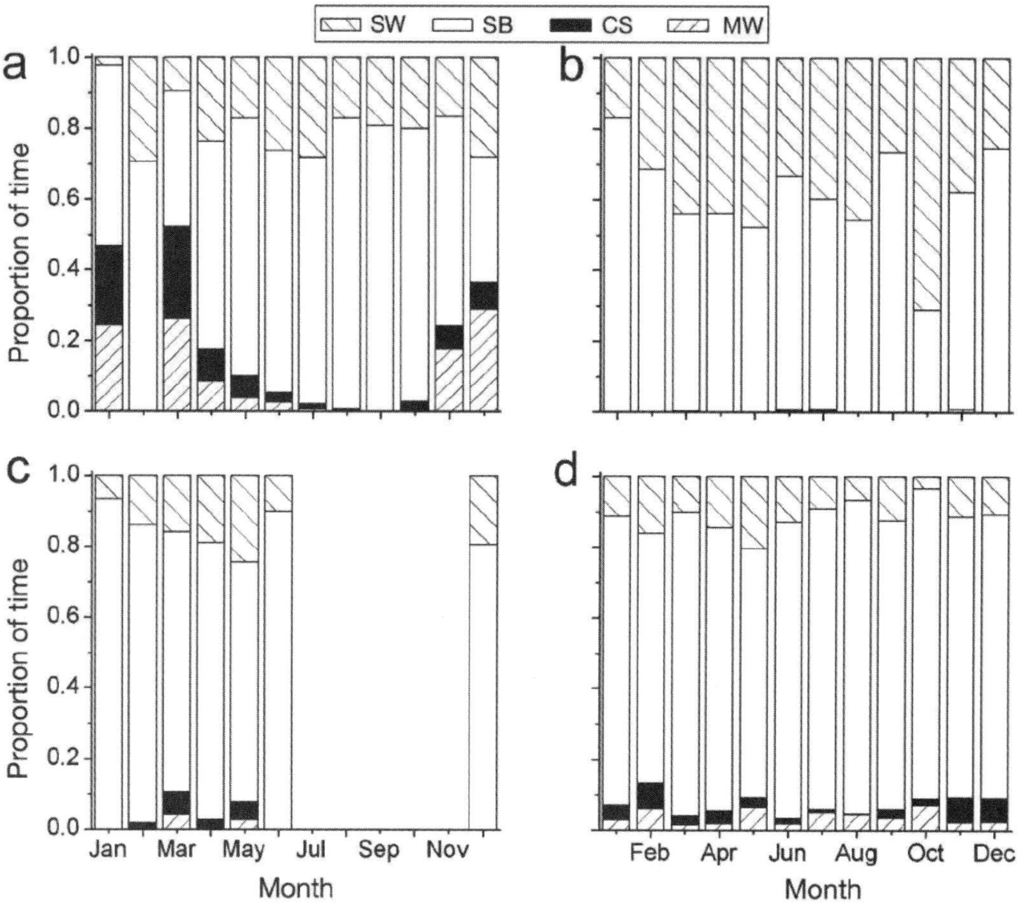

FIGURE 12. The proportion of time classified to different behaviors for fish tagged in (a) southern North Sea, (b) Jutland Bank, (c) Flamborough, (d) Channel. Where midwater (MW) is gray hashed, continuous search (CS) is black, sit and wait (SW) is gray and seabed (SB) is white. Figure reproduced from Hobson et al. 2007.

heavily exploited and drastic reductions in catches, or even closures of entire fisheries (e.g., Newfoundland cod in the 1990s), are considered necessary to conserve stocks. Fortunately, management agencies are becoming increasingly aware of the need for rational management that takes more account of fundamental biology. This includes the need for a better understanding of migration and the spatial dynamics of fish populations and assemblages, not just because it is interesting, but also because it is fundamental to many of the basic elements that underpin the management and conservation of marine fisheries. This applies not only to traditionally exploited species like cod and tuna, but also to newly developing commercial fisheries, like those for deep-water species such as orange roughy *Hoplostethus atlanticus* and roundnosed grenadier *Coryphaenoides rupestris*. For most of these species we know little of their migratory behavior, or of the environmental factors that affect it.

A key challenge for us now is to ensure that this new knowledge and understanding is used to develop and improve future assessment and management methodologies, and that the outputs are taken through into man-

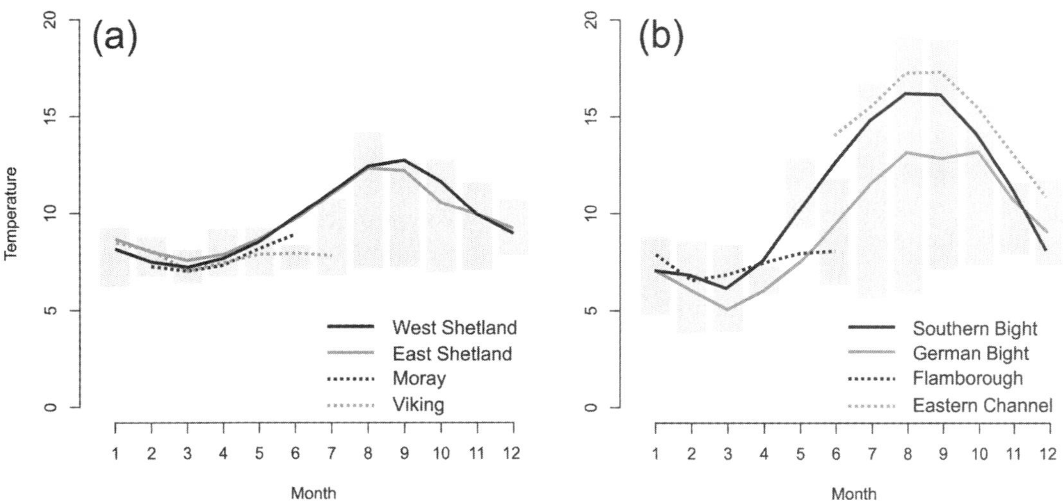

Figure 13. (a) Thermal experience of cod in the northern North Sea. Average monthly temperature experience of cod from each release site are shown by different lines. The range of CTD data in each month is represented by filled gray bars. (b) The same plot for the southern North Sea. Figure modified from Neat and Righton (2007).

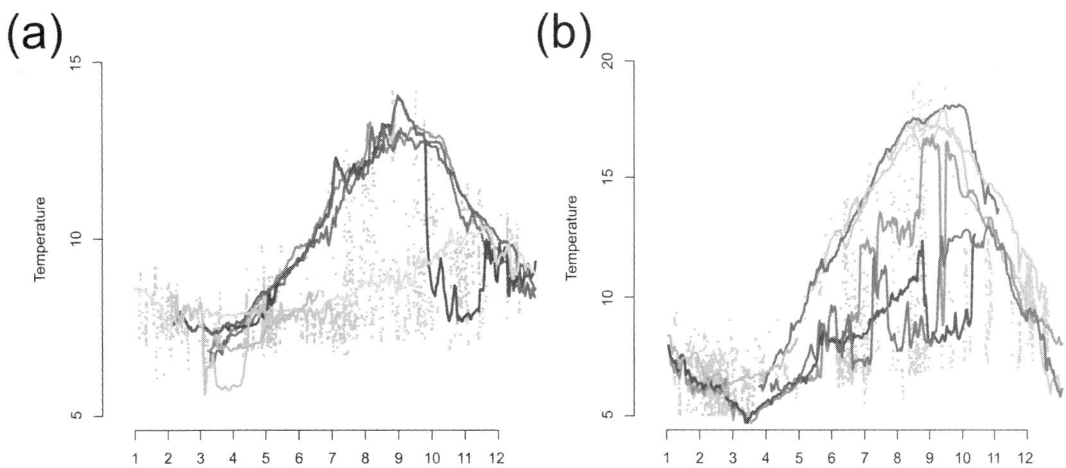

Figure 14. Individual cases illustrating variation in the mean daily thermal experience of cod in (a) the northern North Sea (b) the southern North Sea. Each solid line provides a history of the thermal experience of an individual cod recorded as a daily mean temperature. CTD data (daily resolution) from each sea region are represented by filled gray dots. Figure modified from Neat and Righton (2007).

agement advice (Schnute and Richards 2001). Improved uptake of new biological knowledge will hopefully lead to improved support for fish tagging studies (Metcalfe et al. 2002). However, this is of little value without the necessary tools, so continuing to improve methodologies will be important too. As technology develops, smaller, cheaper, more sophisticated electronic tags will become available. When combined with other techniques used to study population movement, such as genetics and otolith microchemistry, and advanced remote sensing methods that provide detailed environmental information over ocean scales, electronic tags can only improve our understanding of how and why fish migrate, where they go and, ultimately, what environmental factors determine their behavior, movements and distribution.

Acknowledgments

Defra and the European Commission have provided funding for a number of projects from which the authors and colleagues have generated results and publications. Colleagues at the Lowestoft Laboratory have been inspirational, and the collective citations within this paper show how this paper has built on a team effort over the last 25 years. We are also indebted to the fishermen and women who return electronic and identification tags from fish, without which we would not be able to understand many of the ecological and population phenomena described in this paper.

References

Arnold, G. P. and Dewar, H. 2001. Electronic tags in marine fisheries research: a 30 year perspective. Pages 7–64 *in* J. R. Sibert and J. L. Nielsen, editors. Electronic Tagging and Tracking in Marine Fisheries. Kluwer Academic, Dordrecht, Netherlands.

Arnold, G. P. and B. H. Holford. 1995. A computer simulation model for predicting rates and scales of movement of demersal fish on the European continental shelf. ICES Journal of Marine Science. 52:981–990.

Blanchard, J. L., C. Mills, S. Jennings, C. J. Fox, B. D. Rackham, P. D. Eastwood, and C. M. O'Brien. 2005. Distribution-abundance relationships for North Sea Atlantic cod (*Gadus morhua*): observation versus theory. Canadian Journal of Fisheries and Aquatic Science 62:2001–2009.

Block, B. A., S. L. H. Teo, A. Walli, A. Boustany, M. J. W. Stokesbury, C. J. Farwell, K. C. Weng, H. Dewar, and T. D. Williams 2005. Electronic tagging and population structure of Atlantic bluefin tuna. Nature 434:1121–1127.

Block, B. A., H. Dewar, T. Williams,E. D. Prince, C. Farwell, and D. Fudge. 1998. Archival tagging of Atlantic bluefin tuna (*Thunnus thynnus thynnus*). Marine Technology Society Journal. 32:37–46.

Bolle, L. J., E. Hunter, A. D. Rijnsdorp, M. A. Pastoors, J. D. Metcalfe, and J. D. Reynolds. 2005. Do tagging experiments tell the truth? Using electronic tags to evaluate conventional tagging data. ICES Journal of Marine Science 62:236–246.

Buckley, A. A. and G. P. Arnold. 2001. Orientation and swimming speed of plaice migrating by selective tidal stream transport. Pages 263–277 *in* J. R. Sibert and J. L. Nielsen, editors. Electronic Tagging and Tracking in Marine Fisheries. Kluwer Academic, Dordrecht, Netherlands.

Campana, S. E, W. Joyce, and M. Fowler. 2010. Subtropical pupping ground for a cold-water shark. Canadian Journal of Fisheries and Aquatic Science 67:769–773.

Campana, S. E., W. Joyce, M. J. Manning. 2009. Bycatch and discard mortality in commercially caught blue sharks *Prionace glauca* assessed using archival satellite pop-up tags. Marine Ecology Progress Series 387:241–253.

Cushing, D. H. 1990. Hydrographic containment of a spawning group of plaice in the Southern Bight of the North Sea. Marine Ecology Progress Series 58:287–297.

de Veen, J. F. 1978. On selective tidal transport in the migration of North Sea plaice (*Pleuronectes platessa*) and other flatfish species. Netherlands Journal of Sea Reaseach. 12:115–147.

Elrod, J. H., and C. P. Schneider. 1986. Evaluation of coded wire tags for marking lake trout. North American Journal of Fisheries Management 6:264–271.

Godø, O. R. and K. Michalsen. 2000. Migratory behaviour of north-east Arctic cod, studied by use of data storage tags. Fisheries Research 48:127–140.

Goodyear, C. P., J. Luo, E. D. Prince, J. P. Hoolihan, D. Snodgrass, E. S. Orbesen, and J. E. Serafy. 2008. Vertical habitat use of Atlantic blue marlin *Makaira nigricans*: interaction with pelagic longline gear. Marine Ecology Progress Series 365:233–245.

Greer–Walker, M., F. R. Harden Jones, and G. P. Arnold. 1978. The movements of plaice (*Pleuronectes platessa*) tracked in the open sea. Journal du Conseil. Conseil International pour l'Exploration de la Mer 38:58–86.

Gunn, J. 1994. Smart archival tag comes up trumps for tuna. Australian Fisheries. 53:10–11.

Gunn, J. and Block B. 2001. Advances in acoustic, archival, and satellite tagging of tunas. Pages 167–219 *in* B. A. Block and E. D. Stevens, editors. Tuna: physiology, ecology and evolution. Academic Press, London.

Gunn, J. Polacheck, T. Davis, T. Sherlock, M. and Bethlehem, A. 1994. The development and use of archival tags for studying the migration, behaviour and physiology of southern bluefin tuna, with an assessment of the potential for transfer of the technology to groundfish research. Proceedings of the ICES Mini Symposium on Fish Migration 21:23.

Harden Jones, F. R. 1968. Fish Migration. Edward Arnold, London.

Heessen, H. and N. Daan. 1994. Cod distribution and temperature in the North Sea. ICES Marine Science Symposia. 198:244–253.

Heffernan O., D. Righton, and K. Michalsen. 2004. Use of data storage tags to quantify vertical movements of cod: effects on acoustic measures. ICES Journal of Marine Science 61(7):1062–1070.

Hilborn, R. 1990. Determination of fish movement patterns from tag recoveries using maximum likelihood estimators. Canadian Journal of Fisheries and Aquatic Sciences 47:635–643.

Hill, R. D. 1994. Theory of geolocation by light levels. Pages 227–236 *in* B. J. Le Boeuf and R. M. Laws. Elephant seals: population ecology, behavior, and physiology. University of California, Berkeley, California.

Hill, R. D., and M. J. Braun. 2001. Geolocation by light level—The next step: latitude. Pages 315–330 *in* J. R. Sibert and J. L. Nielsen. Electronic tagging and tracking in marine fisheries reviews: methods and technologies in fish biology and fisheries, Vol 1. Kluwer Academic Press, Dordrecht, Netherlands.

Hobson, V., D. Righton, J. Metcalfe, and G. Hays. 2007. Vertical movements of North Sea cod. Marine Ecology Progress Series 347:101–110.

Holland, K. N., B. M. Wetherbee, C. G. Lowe, C. G. Meyer. 1999. Movements of tiger sharks (*Galeocerdo cuvier*) in coastal Hawaiian waters. Marine Biology 134:665–673.

Houghton, R. G., and D. Harding. 1976. The plaice of the English Channel: spawning and migration. Journal du Conseil International pour l'Exploration de la Mer 36:229–239.

Hunter, E., J. N. Aldridge, J. D. Metcalfe, and G. P. Arnold. 2003a. Geolocation of free-ranging fish on the European continental shelf as determined from environmental variables. I. Tidal location method. Marine Biology 142:601–609.

Hunter, E., J. D. Metcalfe, and J. D. Reynolds. 2003b. Migration route and spawning site fidelity by North Sea plaice. Proceedings of the Royal Society of London, Series B: Biological Sciences 270:2097–2103.

Hunter, E., A. A. Buckley, C. Stewart, and J. D. Metcalfe. 2005. Migratory Behaviour of the thornback ray, *Raja clavata* L. in the southern North Sea. Journal of the Marine Biological Association of the United Kingdom 85:1095–1105.

Hunter, E., J. D. Metcalfe, C. M. O'Brien, G. P. Arnold, and J. D. Reynolds. 2004a. Vertical activity patterns of free-swimming adult plaice in the southern North Sea. Marine Ecology Progress Series 279:261–273.

Hunter, E., J. D. Metcalfe, J. D. Reynolds, and G. P. Arnold. 2004b. Impacts of migratory behaviour on population structure of North Sea plaice. Journal of Animal Ecology 74: 377–385.

Hunter, E., F. Berry, A. A. Buckley, C. Stewart, and J. D. Metcalfe. 2006. Seasonal migration of thornback rays and implications for closure management. Journal of Applied Ecology 43:710–720.

Hunter, J. R., A. W. Argue, W. H. Bayliff, A. E. Dizon, A. Fonteneau, D. Goodman, and G. R. Seckel. 1986. The dynamics of tuna movements: an evaluation of past and future research. FAO Fisheries Technical Paper, 277:1–78.

Hutchinson W. F., G. R. Carvalho, and S. I. Rogers. 2001. Marked genetic structuring in localized spawning populations of cod *Gadus morhua* in the North Sea and adjoining waters as revealed by microsatellites. Marine Ecology Progress Series 223:251–260.

Jellyman, D., and K. Tsukamoto. 2005. Swimming depths of offshore migrating longfin eels *Anguilla dieffenbachii*. Marine Ecology Progress Series. 286:261–267.

Jennings, S., M. Kaiser, and J. Reynolds, J. 2001. Marine fisheries ecology. Blackwell, Oxford, UK.

Jones, R. 1976. The use of marking data in fish population analysis. FAO Fisheries Technical Paper. 153: 42pp.

Kjesbu, O. S., D. Righton, M. Kruger Johnsen, A. Thorsen, K. Michalsen, M. Fonn, and P. R. Witthames. 2010. Thermal dynamics of ovarian maturation in Atlantic cod (*Gadus morhua*). Canadian Journal of Fisheries and Aquatic Sciences 67:605–625.

Lawson G. L., M. R. Castleton, and B. A. Block. 2010. Movements and diving behavior of Atlantic bluefin tuna (*Thunnus thynnus*) in relation to water column structure in the Northwestern Atlantic. Marine Ecology Progress Series 400:245–265.

Lutcavage, M. E., R. W. Brill, G. B. Skomal, B. C. Chase, and P. W. Howey. 1999. Results of pop-up satellite tagging of spawning size-class fish in the Gulf of Maine: do North Atlantic bluefin tuna spawn in the mid-Atlantic? Canadian Journal of Fisheries and Aquatic Science 56(2):173–177.

McGarvey, R. 2009. Methods of estimating mortality and movement rates from single-tag recovery data that are unbiased by tag nonreporting. Reviews in Fisheries Science 17:291–304.

McQuinn, I. H. 1997. Metapopulations and Atlantic herring. Reviews in Fish Biology and Fisheries 7:297–329.

Metcalfe, J. D. 2001. Summary report of a workshop on daylight measurements for geolocation in animal telemetry. Pages 331–342 *in* J. R. Sibert and J. L. Nielsen, editors. Electronic tagging and tracking in marine fisheries reviews: methods and technologies in fish biology and fisheries, Vol 1. Kluwer Academic Press, Dordrecht, Netherlands.

Metcalfe, J. D. 2006. Fish population structuring in the North Sea: understanding processes and mechanisms from studies of the movements of adults. Journal of Fish Biology 69 (Supplement C):48–65.

Metcalfe, J. 2009. Fish welfare in wild capture fisheries. Journal of Fish Biology 75:2855–2861.

Metcalfe, J. D., and G. P. Arnold. 1997. Tracking fish with electronic tags. Nature 387:665–666.

Metcalfe, J. D., and G. P. Arnold. 1998. Tracking migrating fish with electronic tags. Pages 199–206 in EEZ technology, a review of advanced technologies for the management of EEZs world-wide (2nd edition). ICG Publishing, London.

Metcalfe, J. D., and M. G. Pawson. 2004. Measuring fish behaviour: the relevance to the managed exploitation of shared stocks. Pages 303–315 *in* A. I. L Payne, C. M. O'Brien, and S. I. Rogers, editors. Blackwell Publishing, Oxford, UK.

Metcalfe, J. D., G. P. Arnold, and P. W. McDowall. 2002. Migration. Pages 175–199 *in* P. J. B. Hart and J. D. Reynolds. Handbook of fish biology and fisheries. Volume 1. Fish Biology. Blackwell Publishing, Oxford, UK.

Metcalfe, J. D. C. J. Fox, D. A. Righton, P. J. Wright, and J. Casey. 2005. A review of the biological evidence for cod stock sub-structure in the North Sea. WP05 STECF Subgroup SGRST. Evaluation of the Cod Recovery Plan Joint Research Centre, 13–17 June 2005, Ispra, Italy.

Metcalfe, J. D., M. Fulcher, and T. J. Storeton-West. 1992. Progress and developments in teletmetry for monitoring the migratory behaviour of plaice in the North Sea. Pages 359–366 in I. G. Priede and S. M. Swift, editors. Wildlife telemetry: remote monitoring and tracking of animals. Ellis Horwood, Chichester, UK.

Metcalfe, J. D., M. C. Fulcher, S. R. Clarke, M. J. Challiss, and S. Hetherington. 2009. An archival tag for monitoring key behaviours (Feeding and Spawning) in fish. *In* J. L. Nielsen, H. Arrizabalaga, N. Fragoso, A. Hobday, M. Lutcavage, J. Sibert, editors. Tagging and tracking of marine animals with electronic devices. Reviews: Methods and Technologies in Fish Biology and Fisheries 9.

Metcalfe, J. D., E. Hunter, and A. A. Buckley. 2006. The migratory behaviour of North Sea plaice: currents, clocks and clues. Marine and Freshwater Behaviour and Physiology 39(1):25–36.

Metcalfe, J. D., D. Righton, P. Eastwood, E. Hunter. 2008a. Migration and habitat choice in marine fishes. *In* C. Magnhagen, V. A. Braithwaite, E. Forsgren, and B. G. Kapoor, editors. Fish behaviour. Science Publishers, New Hampshire.

Metcalfe, J. D., D. Righton, E. Hunter, S. Neville, and D. K. Mills. 2008b. New technologies for the advancement of fisheries science. In: Advances in Fisheries Science: 50 years on from Beverton and Holt. Eds: A. Payne, J. Cotter and E. Potter. Wiley-Blackwell, Oxford, UK.

Moore, A., I. C. Russell, and E. C. E. Potter. 1990. The effects of intraperitoneally implanted dummy acoustic transmitters on the behaviour and physiology of juvenile Atlantic salmon *Salmo salar* L. Journal of Fish Biology, 37:713–721.

Neat, F., and D. Righton. 2007. Warm water occupancy by North Sea cod. Proceedings of the Royal Society of London B: Biological Sciences 274: 789–798.

Neat, F. C., P. J. Wright, A. F. Zuur, I. M. Gibb, F. M. Gibb, D. A. Righton, and R. J. Turner. 2005. Residency and depth movements of a coastal group of Atlantic cod (*Gadus morhua* L.) from the Northern North Sea. Marine Biology 148(3): 643–654

Nelson, D. R. 1978. Telemetering techniques for the study of free-ranging sharks. Pages 419–482 *in* E. S. Hodgson, R. F. Mathewson, editors. Sensory biology of sharks, skates, and rays. Office of Naval Research, Department of the Navy, Government Printing Office, Arlington, Virginia.

Palsson, O.K., and V. Thorsteinsson. 2003. Migration patterns, ambient temperature and growth of Icelandic cod (*Gadus morhua*): evidence from storage tag data. Canadian Journal of Fisheries and Aquatic Science 60:1409–1423

Parker, N. C., A. E. Giorgi, R. C. Heidinger, D. B. Jester, E. D. Prince, and G. A. Winans, editors. 1990. Fish-marking techniques. American Fisheries Society, Symposium 7, Bethesda, Maryland.

Perry, A. L., P. J. Lowe, J. R. Ellis, and J. D. Reynolds. 2005. Climate change and distribution shifts in marine fishes. Science 308:1912–1915.

Pine, W. E., K. H. Pollock, J. E. Hightower, T. J. Kwak, and J. A. Rice. 2003. A review of tagging methods for estimating fish population size and components of mortality. Fisheries 28:10–23.

Righton, D., O. S. Kjesbu, and J. Metcalfe. 2006. A field and experimental evaluation of the effect of data storage tags on the growth of cod (*Gadus morhua* Linné). Journal of Fish Biology 68:385–400.

Righton, D. R. and J. D. Metcalfe. 2002. Multi-torsking: simultaneous measurements of cod behaviour show differences between North Sea and Irish Sea stocks. Hydrobiologica 483:193–200.

Righton, D. R., J. D. Metcalfe, and P. Connolly. 2001. Electronic tags reveal behavioural differences between North Sea and Irish Sea cod. Nature 411:156.

Righton, D. and C. Mills. 2008. A geolocation technique that combines depth, temperature and a swimming speed rule to reconstruct the migrations of fish tagged with electronic tags: an example using cod in the North Sea. Marine Biology 153:507–521.

Righton, D., F. Neat, H. Svedang, V. Bendall, P. Wright, and J. Metcalfe. 2008. Spatial dynamics of Atlantic cod (*Gadus morhua*) in the North Sea: results from a large-scale electronic tagging programme. ICES CM, P:04.

Righton, D., Quayle, V. A., Hetherington, S., and Burt, G. 2007. Movements and distribution of cod (*Gadus morhua* L.) in the southern North Sea and English Channel: results from conventional and electronic tagging experiments. Journal of Marine Biological Association of the United Kingdom 87:599–613.

Righton, D., B. Townhill, and J. van der Kooij. 2009. Catch me if you can: archival tagging studies can help assess changes in the accessibility of Atlantic cod (Gadus morhua) to trawl gears. ICES CM 2009/ J:08.

Rijnsdorp, A. D., and M. A. Pastoors. 1995. Modeling the spatial dynamics of fisheries of North Sea plaice (*Pleuronectes platessa* L.) based on tagging data. ICES Journal of Marine Science 52:963–980.

Robichaud, D. and G. Rose. 2004. Migratory behaviour and range in Atlantic cod: inference from a century of tagging. Fish and Fisheries 5:185–214.

Schaefer, K. M., D. W. Fuller, B. A. Block. 2007. Movements, behavior, and habitat utilization of yellowfin tuna (*Thunnus albacares*) in the northeastern Pacific Ocean, ascertained through archival tag data. Marine Biology 152:503–525.

Schnute, J. T., and L. J. Richards. 2001. Use and abuse of fisheries models. Canadian Journal of Fisheries and Aquatic Science. 58:10–17.

Schopka, S. A., J. Solmundsson, S. A. Ragnarsson, and V. Thorsteinsson. 2010. Using tagging experiments to evaluate the potential of closed areas in protecting migratory Atlantic cod (*Gadus morhua*). ICES Journal of Marine Science 67:1024–1035.

Sims, D. W., E. J. Southall, A. J. Richardson, P. C. Reid, and J. D. Metcalfe. 2003. Seasonal movements and behaviour of basking sharks from archival tagging: no evidence of winter hibernation. Marine Ecology Progress Series 248:187–196.

Sippel, T. J., P. S. Davie, J. C. Holdsworth, and B. A. Block. 2007. Striped marlin (*Tetrapturus audax*) movements and habitat utilization during a summer and autumn in the Southwest Pacific Ocean. Fisheries Oceanography 16:459–472.

Smedbol, R. K. and R. Stephenson. 2001. The importance of managing within-species diversity in cod and herring fisheries of the northwestern Atlantic. Journal of Fish Biology 59 (Supplement A):109–128.

Smedbol, R. K., A. McPherson, M. M. Hansen, and E. Kenchington. 2002. Myths and moderation in marine metapopulations? Fish and Fisheries 3:20–35.

Svedang, H., P. Jonsson, and D. Righton. 2007. Observations on migratory behaviour of Atlantic cod (*Gadus morhua* L) support natal homing as the prime stock separating mechanism. Marine Ecology Progress Series 345:1–12

Tallack, S. M. L. 2009. Regional growth estimates of Atlantic cod, *Gadus morhua*: Applications of the maximum likelihood GROTAG model to tagging data in the Gulf of Maine (USA/Canada) region. Fisheries Research 99:137–150.

Thorsteinsson, V. 2002. Tagging methods for stock assessment and research in fisheries. Report of Concerted Action FAIR CT.96.1394 (CATAG). Marine Research Institute Technical Report, No. 79, Reykjavik, Iceland. Available: hppt://www.hafro.is/catag/ (October 2010).

Walker, R. V., K. W. Myers, N. D. Davis, K. Y. Aydin, F. D. Friedland, H. R. Carlson, G. W. Boehlert, S. Urawa, Y. Ueno, and G. Anma. 2000. Diurnal variation in the thermal environment experienced by salmonids in the north Pacific as indicated by data storage tags. Fisheries Oceanography 9:171–186.

Walton, I. and C. Cotton. 1898. The complete angler: or the contemplative man's recreation. Brown, Little, Boston.

Weng, K. C., P. C. Castilho, J. M. Morrissette, A. Landiera, D. B. Holts, R. J. Schallert, K. J. Goldman, and B. A. Block. 2005. Satellite tagging and cardiac physiology reveal niche expansion in salmon sharks. Science 310:104–106.

Wilson R. P., A. Steinfurth, Y. Ropert-Coudert, A. Kato, and M. Kurita. 2002. Lip-reading in remote subjects: an attempt to quantify and separate ingestion, breathing and vocalisation in free-living animals using penguins as a model. Marine Biology 140:17–27.

Wilson, R. P., J.- J. Ducamp, W. G. Rees, B. M. Culik, and K. Niekamp. 1992. Estimation of location: global coverage using light intensity. Pages 131–134 in I. G. Priede and S. M. Swift, editors. Wildlife Telemetry. Ellis Horwood, New York.

Wright, P. J., E. Galley, I. M. Gibb, F. C. Neat. 2006a. Fidelity of adult cod to spawning grounds in Scottish waters. Fisheries Research 77:148–158.

Wright, P. J., F. C. Neat, F. M. Gibb, I. M. Gibb, and H. Thordarson. 2006b. Evidence for metapopulation structuring in cod from the west of Scotland and North Sea. Journal of Fish Biology 69 (supplement C):181–199.

Panel Discussion Summary

Report on the Panel Discussion on Reconsideration of Stock Assessment Models Using Electronic Tag Data

Andrew C. Seitz

Panel Discussion on the Future Use of Electronic Tag Data in Fisheries Research and Management

The panel consisted of Drs. Roger Hill of Wildlife Computers, Barbara Block of Stanford University, and Julian Melcalfe and David Righton of the Centre for Environment, Fisheries & Aquaculture Science. The purpose of the discussion was to bring together experts from tag manufacturing companies, government agencies, and academia to discuss how to integrate electronic tag data into assessment and management processes. The following summary of the panel discussion is not intended to be an objective and exhaustive literature review, but rather it is a synopsis of the discussion among the expert panel and the participants in the audience.

Before the discussion, the panel agreed to avoid archival tagging subjects euphemistically referred to as "old chestnuts," which are subjects or ideas that have been discussed excessively. These ideas were identified as building smaller and cheaper tags, improved light-based geolocation accuracy, development of additional sensors and memory, and improved tag and data recovery. Several of these "old chestnuts," such as additional sensors and increased memory storage, have become reality.

Rather, the panel suggested focusing the discussion on "new chestnuts," or ideas that are just coming to the forefront of fish and fisheries research, which in this case is integrating our improved understanding of fish behavior, physiology and ecology from archival tagging into fisheries assessment and management processes. Frequently, these new ideas are old issues that must be re-examined because of recent findings from electronic tagging experiments and the development of these new ideas is occurring at an increasingly rapid pace. To keep the discussion focused on this "new chestnut," the following question was posed: "how have we used archival tagging data to challenge current stock assessment models?" This information is essential as modelers develop "next-generation" models that incorporate habitat parameters and population structure. Examples of challenges to stock assessment models were provided from several fish taxa and they generally fall under the three broad categories of dispersal patterns, behavior and habitat use.

Fish dispersal patterns are typically described using electronic tag release and recovery locations, as well as geolocation estimates produced when the fishes are at-liberty. Several electronic tagging studies have revealed previously unknown dispersal patterns that challenge stock assessment assumptions. For example, the total area of ocean commonly occupied by Atlantic bluefin tuna *Thunnus thynnus* and the rate of mixing between the east and west Atlantic Ocean are higher than assumed in the current stock assessment model used by the International Commission for Conservation of Atlantic Tunas (ICCAT) (Block et al. 2001, Block et al. 2005). Although this high level of dispersal in Atlantic bluefin tuna has long been suspected, managers would not consider it prior to the evidence of it provided by electronic tagging data. In another example, electronic tags have been used to document movement by Pacific halibut *Hippoglossus stenolepis* among regulatory areas during the commercial fishing season (Loher and Seitz 2006). During this dispersal, interception of fish that predominantly reside in one regulatory area may occur, a process that would effectively transfer exploitable biomass among areas and even between the United States and Canada. Movement information collected by electronic tags may be used to account for interception of Pacific halibut moving among areas. In the cases of Pacific halibut and Atlantic bluefin tuna, the dispersal information from electronic tagging is especially effective in defining population units appropriate for management when combined with other information such as population genetics (Boustany et al. 2008, Nielsen et al. 2009) and otolith microchemistry (Rooker et al. 2008).

Electronic tags provide a valuable tool for inferring behavioral patterns of fishes, especially those that occupy inhospitable and remote environments. For example, depth readings and body angles may be combined to determine pelagic swimming behavior by Greenland halibut *Reinhardtius hippoglossoides* (Albert et al.). Depth data from electronic tags has also been used to identify spawning behavior in Pacific halibut, Atlantic bluefin tuna and Atlantic cod *Gadus morhua*. For example, Atlantic bluefin tuna perform a distinct diel oscillatory diving behavior that is seen only during its purported spawning season (Teo et al. 2007). Not only is the behavior information gleaned through electronic tagging biologically interesting, it has been used to validate stock assessment surveys. For Greenland halibut, pronounced pelagic behavior and variation in vertical distribution can have important implications for stock indices of abundance if benthic fishing gear is the only component of the assessments. For Atlantic bluefin tuna, the proportion of time spent in surface waters defines their availability to aerial population censuses.

Geolocation estimates, depth, and temperature data collected by electronic tags are often used to characterize fish habitat. For example, spawning habitat has been described for Atlantic bluefin tuna by correlating spawning behavior with temperature and geolocations. Atlantic bluefin tuna spawn from February to June in the central Gulf of Mexico in relatively warm water (~26°C) at a depth of approximately 200 m (Teo et al. 2007). Depth and temperature specific spawning habitat has also been described for lake whitefish *Coregonus clupeaformis* and lake trout *Salvelinus namaycush* in the Laurentian Great Lakes (R. Bergstedt, USGS Great Lakes Science Center, personal communication). Defining in situ fish habitat is critical for several reasons. First, many new assessment models have an arbitrary habitat parameter and improved definitions of habitat are needed for these models. Second, the description of "optimal" fish habitat as gleaned from archival tagging experiments frequently does not match that defined in laboratory experiments and therefore some assumptions used in models may need to be re-visited. Finally, the habitat information derived from electronic tags can be used in models to predict timing and location of behaviors such as spawning. In the face of global climate change, these predictions are especially germane as questions arise about water temperatures on spawning grounds exceeding the thermal tolerance of several important fish species, such as sockeye salmon *Oncorhynchus nerka* in the Fraser River, bigeye tuna *Thunnus obesus* in the western Pacific Ocean, and Atlantic bluefin tuna in the Gulf of Mexico.

Our understanding of the biology of many fish species has been greatly enhanced through combining electronic tagging-derived observations of dispersion, behavior, and habitat use with more "traditional" fisheries data. For example, it appears that up to 25% of sexually mature Pacific halibut skip spawning each year (Loher and Seitz 2008) and that median age-at-maturity of Atlantic bluefin in the Gulf of Mexico is twelve years, rather than eight as originally assumed in ICCAT assessment models (Teo et al. 2007). Other examples of archival tag data refining our understanding of fish biology and ecology provided in the panel discussion included understanding growth patterns of yellowfin tuna *Thunnus albacares* in the eastern Pacific Ocean and delineating age-specific foraging and spawning areas of swordfish *Xiphias gladius* off Australia and New Zealand. The knowledge gained by combining electronic tagging data with traditional fisheries data can be incorporated into management. For example, time/area closures may be used to reduce bycatch of unwanted species and to avoid capturing spawning fish. The knowledge gained by combining electronic tagging data with traditional fisheries data also can be incorporated into stock assessment models. For example, dispersal patterns may be used to parameterize residency of fish in a region and skipped spawning can be used to redefine annual spawning stock biomass.

The validity of extrapolating results from a limited sample size of electronic tags to a population of fish may be tenuous, and a main consideration of future tagging experiments should be how to make individual tag deployments more representative of a population. However, even if sample sizes in tagging experiments are too small or biased to generate accurate model parameters, tagging results can be used to test conceptual assumptions and results from lab experiments to improve assessment and management models.

Although the field of electronic tagging is relatively young, electronic tags are generating large amounts of information. These data are not only providing insight on the direct topic of investigation the data have also lead to many new serendipitous findings. In just the past decade, fisheries scientists and fish biologists have passed the stage of speculating how these data *might* be useful, and are now focusing on describing how they *are* useful. Several examples were discussed that demonstrate how archival tag data may be used to challenge several biological assumptions and fisheries management paradigms.

It is critical to disseminate the information gained through archival tagging and to provide advice and input to modelers as they develop the next generation of models. As such, it was recommended that a formal review be compiled and written that details the types of data gathered from archival tagging experiments and provides examples of how they may be used to improve next-generation stock assessment models. Equally important is for scientists to alert tag manufacturers of their needs for new sensors and measurements, because frequently the engineers designing the tags are not aware of a developing need for new information.

Panel Participant Contact Details

Dr. Roger Hill
Wildlife Computers
8345 154th Avenue NE
Redmond, WA 98052, USA
tags@wildlifecomputers.com

Dr. Barbara Block
Hopkins Marine Station of Stanford University
Oceanview Blvd.
Pacific Grove, CA, 93950, USA
bblock@stanford.edu

Drs. Julian Metcalfe and David Righton
The Centre for Environment, Fisheries & Aquaculture Science
Lowestoft Laboratory, Pakefield Road, Lowestoft, Suffolk, NR33 0HT, UK
julian.metcalfe@cefas.co.uk
david.righton@cefas.co.uk

References

Arnold, G. P., and H. Dewar. 2001. Electronic tags in marine fisheries research: a 30-year perspective. Pages 7–64 *in* J. R. Sibert and J. L. Nielsen, editors. Electronic tagging and tracking of marine fisheries. Kluwer Academic Publishers, Dordrecht, The Netherlands.

Block, B. A., H. Dewar, S. B. Blackwell, T. D. Williams, E. D. Prince, C. J. Farwell, A. Boustany, S. L. H. Teo, A. Seitz, A. Walli and D. Fudge. 2001. Migratory movements, depth preferences, and thermal biology of Atlantic bluefin tuna. Science 293:1310–1314.

Block, B. A., S. L. H. Teo, A. Walli, A. Boustany, M. J. W. Stokesbury, C. Farwell, T. Williams, K. Weng, and H. Dewar. 2005. Electronic tagging and population structure of Atlantic bluefin tuna. Nature 434:1121–1127.

Boustany, A. M., C. A. Reeb, and B. A. Block. 2008. Mitochondrial DNA and electronic tracking reveal population structure of Atlantic bluefin tuna (*Thunnus thynnus*). Marine Biology 156:13–24.

Loher, T. and A. C. Seitz. 2008. Characterization of active spawning season and depth for eastern Pacific halibut (*Hippoglossus stenolepis*), and evidence of probably skipped-spawning. Journal of Northwest Atlantic Fishery Science 41:23–36.

Loher, T. L. and A. C. Seitz. 2006. Seasonal migration and environmental conditions experienced by Pacific halibut in the Gulf of Alaska, elucidated from pop-up archival transmitting (PAT) tags. Marine Ecology Progress Series 317:259–271.

Nielsen, J. L., S. L. Graziano, and A. C. Seitz. 2009. Fine-scale population genetic structure in Alaskan Pacific halibut (*Hippoglossus stenolepis*) reflects differences in ocean condition and dispersal processes. Conservation Genetics DOI 10.1007/s10592-009-9943-8.

Rooker, J. R., D. H. Secor, G. DeMetrio, R. Schloesser, B. A. Block, and J. D. Neilson. 2008. Natal homing and connectivity in Atlantic bluefin tuna populations. Science 322:742–744.

Teo, S. L. H., A. Boustany, H. Dewar, M. J. W. Stokesbury, K. C. Weng, S. Beemer, A. C. Seitz, C. J. Farwell, E. D. Prince and B. A. Block. 2007. Movement patterns, diving behavior and thermal biology of Atlantic bluefin tuna (*Thunnus thynnus*) on their breeding grounds in the Gulf of Mexico. Marine Biology 151(1):1–18.